HANDBOOK OF DIFFERENTIAL EQUATIONS

EVOLUTIONARY EQUATIONS

VOLUME I

HANDBOOK OF DIFFERENTIAL EQUATIONS

EVOLUTIONARY EQUATIONS

Volume I

Edited by

C.M. DAFERMOS
Division of Applied Mathematics
Brown University
Providence, USA

E. FEIREISL
Mathematical Institute AS CR
Praha, Czech Republic

ELSEVIER
NORTH
HOLLAND

2004

Amsterdam • Boston • Heidelberg • London • New York • Oxford •
Paris • San Diego • San Francisco • Singapore • Sydney • Tokyo

ELSEVIER B.V.
Sara Burgerhartstraat 25
P.O. Box 211, 1000 AE
Amsterdam, The Netherlands

ELSEVIER Inc.
525 B Street, Suite 1900
San Diego, CA 92101-4495
USA

ELSEVIER Ltd
The Boulevard, Langford Lane
Kidlington, Oxford OX5 1GB
UK

ELSEVIER Ltd
84 Theobalds Road
London WC1X 8RR
UK

First edition 2004

Library of Congress Cataloging in Publication Data
A catalog record is available from the Library of Congress.

British Library Cataloguing in Publication Data
A catalogue record is available from the British Library.

ISBN: 0 444 51131 8
Set ISBN: 0 444 51744 8

♾ The paper used in this publication meets the requirements of ANSI/NISO Z39.48-1992 (Permanence of Paper).

Transferred to digital print 2008

Printed and bound in Great Britain by CPI Antony Rowe, Eastbourne

Preface

The aim of this Handbook is to acquaint the reader with the current status of the theory of evolutionary partial differential equations, and with some of its applications. This is not an easy task: Unlike other mathematical theories which exhibit a tree-like structure, with clearly distinguishable trunk, main and secondary branches, the theory of partial differential equations has the appearance of a bush with very complex structure. Its roots as well as its flowers are often related to the physical world, and it is fertilized by ideas and techniques borrowed from virtually every other area of mathematics.

Evolutionary partial differential equations made their first appearance in the 18th century, in the endeavor to understand the motion of fluids and other continuous media. It is remarkable that this research program is still ongoing and many fundamental questions remain unanswered. Beyond this area, however, evolutionary partial differential equations have become ubiquitous, as they seem to govern the dynamics of any physical, chemical biological, ecological or economic system whose state is described by spatially dependent variables.

The active research effort over the span of two centuries, combined with the wide variety of physical phenomena that had to be explained, has resulted in an enormous body of literature. Any attempt to produce a comprehensive survey would be futile. The aim here is to collect review articles, written by leading experts, which will highlight the present and expected future directions of development of the field. The emphasis will be on nonlinear equations, which pose the most challenging problems today. The various articles will offer the reader the opportunity to compare and contrast the behavior of hyperbolic and parabolic equations. Hyperbolic equations are typically associated with media exhibiting "elastic" response, and encompass the notoriously difficult class of "hyperbolic conservation laws". Parabolic equations are associated with various diffusive mechanisms. An extremely important and challenging example is the Navier–Stokes equation.

Volume I of this Handbook will focus on the abstract theory of evolutionary equations, addressing questions of existence, uniqueness and other general qualitative properties of solutions. Future volumes will consider more concrete problems relating to specific applications. Our hope is that the handbook will provide a panorama of this amazingly complex and rapidly developing branch of mathematics.

Constantin Dafermos
Eduard Feireisl

List of Contributors

Arendt, W., *Universität Ulm, Ulm, Germany* (Ch. 1)
Bressan, A., *Penn State University, PA, USA* (Ch. 2)
DiBenedetto, E., *Vanderbilt University, Nashville, TN, USA* (Ch. 3)
Hsiao, L., *Academy of Mathematics and Systems Science, Academia Sinica, Beijing, China* (Ch. 4)
Jiang, S., *Institute of Applied Physics and Computational Mathematics, Beijing, China* (Ch. 4)
Lunardi, A., *Università di Parma, Parma, Italy* (Ch. 5)
Perthame, B., *École Normale Supérieure, Paris, France* (Ch. 6)
Serre, D., *ENS Lyon, UMPA, Lyon, France* (Ch. 7)
Urbano, J.M., *Universidade de Coimbra, Coimbra, Portugal* (Ch. 3)
Vespri, V., *Università di Firenze, Firenze, Italy* (Ch. 3)

Contents

Preface v
List of Contributors vii

1. Semigroups and Evolution Equations: Functional Calculus, Regularity and Kernel Estimates 1
   W. Arendt
2. Front Tracking Method for Systems of Conservation Laws 87
   A. Bressan
3. Current Issues on Singular and Degenerate Evolution Equations 169
   E. DiBenedetto, J.M. Urbano and V. Vespri
4. Nonlinear Hyperbolic–Parabolic Coupled Systems 287
   L. Hsiao and S. Jiang
5. Nonlinear Parabolic Equations and Systems 385
   A. Lunardi
6. Kinetic Formulations of Parabolic and Hyperbolic PDEs: From Theory to Numerics 437
   B. Perthame
7. L^1-stability of Nonlinear Waves in Scalar Conservation Laws 473
   D. Serre

Author Index 555

Subject Index 565

CHAPTER 1

Semigroups and Evolution Equations: Functional Calculus, Regularity and Kernel Estimates

Wolfgang Arendt

Abteilung Angewandte Analysis, Universität Ulm, 89069 Ulm, Germany,
E-mail: arendt@mathematik.uni-ulm.de

Contents

Introduction . . . 3
1. Semigroups . . . 4
  1.1. The algebraic approach . . . 4
  1.2. The Cauchy problem . . . 5
  1.3. Semigroups and Laplace transforms . . . 5
  1.4. More general C_0-semigroups . . . 6
  1.5. The inhomogeneous Cauchy problem . . . 6
2. Holomorphic semigroups . . . 7
  2.1. Characterization of bounded holomorphic semigroups . . . 7
  2.2. Characterization of holomorphic semigroups . . . 7
  2.3. Boundary groups . . . 7
  2.4. The Gaussian semigroup . . . 8
  2.5. The Dirichlet Laplacian . . . 8
  2.6. The Neumann Laplacian on $C(\overline{\Omega})$ . . . 10
  2.7. Wentzell boundary conditions . . . 11
  2.8. Dynamic boundary conditions . . . 11
3. Asymptotics . . . 12
  3.1. Exponential stability . . . 12
  3.2. Ergodic semigroups . . . 14
  3.3. Convergence and asymptotically almost periodicity . . . 15
  3.4. Positive semigroups . . . 16
  3.5. Positive irreducible semigroup . . . 17
4. Functional calculus . . . 18
  4.1. Sectorial operators . . . 18
  4.2. The sum of commuting operators . . . 19
  4.3. The elementary functional calculus . . . 20
  4.4. Fractional powers and *BIP* . . . 23
  4.5. Bounded H^∞-calculus for sectorial operators . . . 28

HANDBOOK OF DIFFERENTIAL EQUATIONS
Evolutionary Equations, volume 1
Edited by C.M. Dafermos and E. Feireisl

4.6. Perturbation . . . 33
4.7. Groups and positive contraction semigroups . . . 33
5. Form methods and functional calculus . . . 35
5.1. Bounded H^∞-calculus on Hilbert space . . . 35
5.2. m-accretive operators on Hilbert space . . . 36
5.3. Form methods . . . 38
5.4. Form sums and Trotter's product formula . . . 41
5.5. The square root property . . . 44
5.6. Groups and cosine functions . . . 45
6. Fourier multipliers and maximal regularity . . . 49
6.1. Vector-valued Fourier series and periodic multipliers . . . 49
6.2. Maximal regularity via periodic multipliers . . . 53
7. Gaussian estimates and ultracontractivity . . . 59
7.1. The Beurling–Deny criteria . . . 59
7.2. Extrapolating semigroups . . . 60
7.3. Ultracontractivity, kernels and Sobolev embedding . . . 64
7.4. Gaussian estimates . . . 69
8. Elliptic operators . . . 72
8.1. Boundary conditions . . . 73
8.2. Positivity and irreducibility . . . 73
8.3. Submarkov property: Dirichlet boundary conditions . . . 74
8.4. Quasicontractivity in L^p . . . 74
8.5. Gaussian estimates: real coefficients . . . 74
8.6. Complex second-order coefficients . . . 75
8.7. Further comments on Gaussian estimates . . . 75
8.8. The square root property . . . 76
8.9. The hyperbolic equation . . . 76
8.10. Nondivergence form . . . 77
8.11. Elliptic operators with Banach space-valued coefficients . . . 77
Acknowledgments . . . 78
References . . . 78
Monographs . . . 78
Research Articles . . . 79

Abstract

This is a survey on recent developments of the theory of one-parameter semigroups and evolution equations with special emphasis on functional calculus and kernel estimates. Also other topics as asymptotic behavior for large time and holomorphic semigroups are discussed. As main application we consider elliptic operators with various boundary conditions.

Introduction

The theory of one-parameter semigroups provides a framework and tools to solve evolutionary problems. It is impossible to give an account of this rich and most active field. In this chapter we rather try to present a survey on a particular subject, namely functional calculus, maximal regularity and kernel estimates which, in our eyes, has seen a most spectacular development, and which, so far, is not presented in book form. We comment on these three subjects:

1. *Functional calculus* (Section 4). If A is a self-adjoint operator, one can define $f(A)$ for all bounded complex-valued measurable functions defined on the spectrum of A. It was McIntosh who initiated and developed a theory of functional calculus for a less restricted large class of operators, namely *sectorial operators*; i.e., operators whose spectrum is included in a sector and whose resolvent satisfies a certain estimate. Negative generators of bounded holomorphic semigroups are sectorial operators and are our main subject of investigation. And indeed, for these operators $f(A)$ can be defined for a large class of holomorphic functions defined on a sector containing the spectrum. Taking $f(z) = \mathrm{e}^{-tz}$ leads to the semigroup e^{-tA}, the function $f(z) = z^{\alpha}$ to the fractional power of such an operator A. One important reason to study functional calculus is the Dore–Venni theorem. In its hypotheses functional calculus plays a role; the conclusion is the invertibility of the sum of two operators A and B. Thus, the Dore–Venni theorem asserts that the equation

$$Ax + Bx = y$$

has a unique solution $x \in D(A) \cap D(B)$. To say that the solution is at the same time in both domains can be rephrased by saying that the solution has "*maximal regularity*", a crucial property in many circumstances.

2. *Form methods* (Section 5). On Hilbert space the functional calculus behaves particularly well as we show in Sections 4 and 5. Most interesting is the close connection with form methods. Basically, the following is true: an operator is associated with a form if and only if it has a bounded H^{∞}-calculus. Form methods, based on the fundamental Lax–Milgram lemma, allow a most efficient treatment of elliptic and parabolic problems as we show later.

3. *Maximal regularity and Fourier transform* (Section 6). The following particular problem of maximal regularity is important for solving nonlinear equations: The generator A of a semigroup T is said to have property (*MR*) if $T * f \in W^{1,2}((0,1); X)$ for all $f \in L^2((0,1); X)$. On Hilbert spaces every generator of a holomorphic semigroup has (*MR*); but a striking result of Kalton–Lancien asserts that this fact characterizes Hilbert spaces (among a large class of Banach spaces). On the other hand, in recent years it has been understood which role "*unconditional properties*" play for operator-valued Fourier transform and Cauchy problems. So one may characterize property (*MR*) by R-boundedness, a property defining "unconditional boundedness" of sets of operators.

4. *Kernel estimates* (Section 7). Gaussian estimates for the kernels of parabolic equations have been investigated for many years. It is most interesting in its own right that the solutions of a parabolic equation with measurable coefficients are very close to the Gaussian semigroup. But Gaussian estimates have also striking consequences for the underlying semigroup. For example, we show that they do imply boundedness of the H^∞-calculus.

5. *Elliptic operators* (Section 8). The theory presented here can be applied to elliptic operators with measurable coefficients to which Section 8 is devoted. We will explain *Kato's square root problem*, the most difficult question of coincidence of form domain and the domain of the square root, which has been solved recently by Auscher, Hofmann, Lacey, McIntosh and Tchamitchian.

We start the chapter by putting together some basic properties of semigroups which are particularly useful in the sequel. Special attention is given to holomorphic semigroups (Section 2) and to the theory of asymptotic behavior (Section 3). As prototype example in this account serve the Laplacian with Dirichlet and Neumann boundary conditions: On spaces of continuous functions this operator will be considered in Section 2, later its L^p-properties are established. Concerning the results on asymptotic behavior we concentrate on those which can be applied to parabolic equations in Section 8.

Most of the results are presented without proof, referring to the literature. Frequently, only particular cases which are easy to formulate are presented; and in a few cases we give proofs. Some of them are new, not very well known or particularly elegant. The article presents a special choice, guided by personal taste, even in this narrow subject. We hope that the numerous references allow reader to go beyond that choice and that the list of monographs at the end helps them to view the subject in a broader context.

1. Semigroups

In this introductory section we present semigroups from three different points of view. We mention few properties but refer to the various text books concerning the theory.

1.1. *The algebraic approach*

Let X be a complex Banach space, let $C(\mathbb{R}_+, X)$ be the space of all continuous functions defined on $\mathbb{R}_+ := [0, \infty)$ with values in X. *A C_0-semigroup* is a mapping $T : \mathbb{R}_+ \to \mathcal{L}(X)$ such that

(a) $T(\cdot)x \in C(\mathbb{R}_+, X)$ for all $x \in X$;

(b) $T(0) = I$;

(c) $T(s+t) = T(s)T(t)$, $s, t \in \mathbb{R}_+$.

Given a C_0-semigroup T on X, one defines the *generator* A of T as an unbounded operator on X by

$$Ax = \lim_{t \downarrow 0} \frac{T(t)x - x}{t}$$

with domain $D(A) := \{x \in X\colon \lim_{t\downarrow 0} \frac{T(t)x-x}{t} \text{ exists}\}$. Then $D(A)$ is dense in X and A is closed and linear. In other words, A is the derivative of T in 0 (in the strong sense) and for this reason one also calls A the *infinitesimal generator* of T.

The second approach involves the Cauchy problem.

1.2. *The Cauchy problem*

Let A be a closed linear operator on X. Let $J \subset \mathbb{R}$ be an interval. A *mild solution* of the differential equation

$$\dot{u}(t) = Au(t), \quad t \in J, \tag{1.1}$$

is a function $u(t) \in C(J, X)$ such that $\int_s^t u(r)\,\mathrm{d}r \in D(A)$ for all $s, t \in J$ and $A\int_s^t u(r)\,\mathrm{d}r = u(t) - u(s)$. A *classical solution* is a function $u \in C^1(J, X)$ such that $u(t) \in D(A)$ and $\dot{u}(t) = Au(t)$ for all $t \in J$. Since A is closed, a mild solution u is a classical solution if and only if $u \in C^1(J, X)$.

THEOREM [[ABHN01], 3.1.12]. *Let $J = [0, \infty)$. The following assertions are equivalent*:
(i) *A generates a C_0-semigroup T*;
(ii) *for all $x \in X$, there exists a unique mild solution u of* (1.1) *satisfying $u(0) = x$.*
In that case $u(t) = T(t)x$, $t \geqslant 0$.

The theorem implies in particular that the generator A determines uniquely the semigroup. Here is the third approach.

1.3. *Semigroups and Laplace transforms*

Let T be a C_0-semigroup with generator A. Then the *growth bound*

$$\omega(T) := \inf\{\omega \in \mathbb{R}\colon \exists M \text{ such that } \|T(t)\| \leqslant M\mathrm{e}^{\omega t} \text{ for all } t \geqslant 0\}$$

satisfies $-\infty \leqslant \omega(T) < \infty$, and if $\lambda \in \mathbb{C}$, $\operatorname{Re}\lambda > \omega(T)$, then λ is in the *resolvent set* $\rho(A)$ of A and

$$R(\lambda, A)x = \int_0^\infty \mathrm{e}^{-\lambda t} T(t)x\,\mathrm{d}t := \lim_{\tau\to\infty} \int_0^\tau \mathrm{e}^{-\lambda t} T(t)x\,\mathrm{d}t$$

for all $x \in X$, where $R(\lambda, A) = (\lambda - A)^{-1}$.

THEOREM [[ABHN01], Theorem 3.1.7, p. 113]. *Let $T : \mathbb{R}_+ \to \mathcal{L}(X)$ be strongly continuous. Let A be an operator on X, $\lambda_0 \in \mathbb{R}$ such that $(\lambda_0, \infty) \subset \rho(A)$ and*

$$R(\lambda, A)x = \int_0^\infty \mathrm{e}^{-\lambda t} T(t)x\,\mathrm{d}t, \quad \lambda > \lambda_0,$$

for all $x \in X$. Then T is a C_0-semigroup and A its generator.

Thus generators of C_0-semigroups are precisely those operators whose resolvent is a Laplace transform. Laplace transform techniques play an important role in semigroup theory (see [[ABHN01]] for a systematic theory).

1.4. *More general C_0-semigroups*

In order to talk about Dirichlet boundary conditions we need more general semigroups (cf. Section 2.5). A strongly continuous function $T : (0, \infty) \to \mathcal{L}(X)$ is called a (*nondegenerate locally bounded*) *semigroup* if

(a) $T(t)T(s) = T(t+s)$, $t, s > 0$;
(b) $\sup_{0<t\leqslant 1} \|T(t)\| < \infty$;
(c) $T(t)x = 0$ for all $t > 0$ implies $x = 0$.

As a consequence $\omega(T) < \infty$ and there exists a unique operator A such that $(\omega(T), \infty) \subset \rho(A)$ and $R(\lambda, A)x = \int_0^\infty e^{-\lambda t} T(t)x \, dt$ for all $x \in X$ and $\lambda > \omega(T)$. We call A the *generator* of T (see [[ABHN01], 3.2]). Then T is a C_0-semigroup if and only if $\overline{D(A)} = X$. If X is reflexive, then this is automatically true.

1.5. *The inhomogeneous Cauchy problem*

If A generates a C_0-semigroup T, then also the inhomogeneous Cauchy problem

$$\begin{cases} \dot{u}(t) = Au(t) + f(t), & t \in [0, \tau], \\ u(0) = x \end{cases} \tag{1.2}$$

is well posed. More precisely, let $x \in X$, $f \in L^1((0, \tau); X)$. A *mild solution* of (1.2) is a continuous function $u : [0, \tau] \to X$ such that $\int_0^t u(s) \, ds \in D(A)$ and

$$u(t) - x = A \int_0^t u(s) \, ds + \int_0^t f(s) \, ds$$

for all $t \in [0, \tau]$. Define $T * f$ by $T * f(t) = \int_0^t T(t-s) f(s) \, ds$.

PROPOSITION. *The function u given by $u(t) = T(t)x + T * f(t)$ is the unique mild solution of* (1.2).

1.5.1. *Classical solutions.* Thus, in the particular case where $x = 0$, the mild solution of (1.2) is $u = T * f$. It is very rare that this u is a classical solution for all f. Let T be a C_0-semigroup on X with generator A.

THEOREM (Baillon; see [EG92]). *Assume that X is reflexive or $X = L^1$ (or more generally that $c_0 \not\subset X$). If for all $f \in C([0, \tau]; X)$ one has $T * f \in C^1([0, \tau], X)$, then A is bounded.*

In Section 5 we will devote much attention to the question when

$$T * f \in W^{1,p}\big((0,\tau); X\big) \quad \text{for all } f \in L^p\big((0,\tau); X\big).$$

2. Holomorphic semigroups

A semigroup $T : (0,\infty) \to X$ (in the sense of Section 1.4) is called *holomorphic* if there exists $\theta \in (0, \pi/2]$ such that T has a holomorphic extension $T : \Sigma_\theta \to \mathcal{L}(X)$ which is bounded on $\{z \in \Sigma_\theta \colon |z| \leqslant 1\}$. In that case, this holomorphic extension is unique and satisfies $T(z_1 + z_2) = T(z_1)T(z_2)$ for all $z_1, z_2 \in \Sigma_\theta$. If T is a C_0-semigroup, then

$$\lim_{\substack{z\to 0\\ z\in\Sigma_\theta}} T(z)x = x$$

for all $x \in X$, and we call T a *holomorphic C_0-semigroup*. If the extension T is bounded on Σ_θ, we call T a *bounded holomorphic semigroup*. Thus, for this property, it does not suffice that T is bounded on $[0,\infty)$. Take for example, $T(t) = \mathrm{e}^{\mathrm{i}t}$ on $X = \mathbb{C}$.

2.1. *Characterization of bounded holomorphic semigroups*

Let A be an operator on X. The following assertions are equivalent:

(i) *A generates a bounded holomorphic semigroup T*;

(ii) *one has $\lambda \in \rho(A)$ whenever* $\operatorname{Re}\lambda > 0$ *and* $\sup_{\operatorname{Re}\lambda>0} \|\lambda R(\lambda, A)\| < \infty$;

(iii) *there exists $\alpha \in (0, \pi/2)$ such that $\mathrm{e}^{\pm\mathrm{i}\alpha} A$ generates a bounded semigroup.*

In that case $(T(\mathrm{e}^{\pm\mathrm{i}\alpha} t))_{t\geqslant 0}$ is the semigroup generated by $\mathrm{e}^{\pm\mathrm{i}\alpha} A$. Moreover, T is a C_0-semigroup if and only if $D(A)$ is dense. This is automatic whenever X is reflexive.

2.2. *Characterization of holomorphic semigroups*

An operator A generates a holomorphic semigroup T if and only if there exists ω such that $A - \omega$ generates a bounded holomorphic semigroup S. In that case $T(t) = \mathrm{e}^{\omega t} S(t)$, $t \geqslant 0$.

2.3. *Boundary groups*

Let A be the generator of a C_0-semigroup T having a holomorphic extension to the half-plane $\mathbb{C}_+ = \{\lambda \in \mathbb{C}\colon \operatorname{Re}\lambda > 0\}$. Then $\mathrm{i}A$ generates a C_0-group U if and only if $\sup_{\operatorname{Re} z>0, |z|\leqslant 1} \|T(z)\| < \infty$.

In that case we call U the *boundary group* of T and write $U(s) =: T(\mathrm{i}s)$, $s \in \mathbb{R}$.

2.4. *The Gaussian semigroup*

Consider the *Gaussian semigroup* G defined on $L^1 + L^\infty := L^1(\mathbb{R}^n) + L^\infty(\mathbb{R}^n)$ by

$$\big(G(z)f\big)(x) = (4\pi z)^{-n/2} \int_{\mathbb{R}^n} f(y) \mathrm{e}^{-(x-y)^2/4z} \, \mathrm{d}y \tag{2.1}$$

for all $x \in \mathbb{R}^n$, $f \in L^1 + L^\infty$, $\operatorname{Re} z > 0$. Then G is a holomorphic semigroup which is bounded on Σ_θ for each $0 < \theta < \pi/2$. The generator $\Delta_{1+\infty}$ of G is given by

$$D(\Delta_{1+\infty}) = \big\{f \in L^1 + L^\infty \colon \Delta f \in L^1 + L^\infty\big\},$$
$$\Delta_{1+\infty} f = \Delta f \quad \text{in } \mathcal{D}\big(\mathbb{R}^n\big)'.$$

Let E be one of the spaces $L^p(\mathbb{R}^n)$, $1 \leqslant p \leqslant \infty$, $C^{\mathrm{b}}(\mathbb{R}^n) = \{f \in L^\infty(\mathbb{R}^n)\colon f \text{ is continuous}\}$;

$$BUC\big(\mathbb{R}^n\big) := \big\{f \in C^{\mathrm{b}}\big(\mathbb{R}^n\big)\colon f \text{ is uniformly continuous}\big\},$$
$$C_0\big(\mathbb{R}^n\big) := \Big\{f \in C^{\mathrm{b}}\big(\mathbb{R}^n\big)\colon \lim_{|x|\to\infty} \big|f(x)\big| = 0\Big\},$$

which are all subspaces of $L^1 + L^\infty$. Then the restriction $G(t)|_E$ defines a holomorphic semigroup G_E on E which is bounded on Σ_θ for each $\theta \in (0, \pi/2)$. Its generator is Δ_E given by

$$D(\Delta_E) = \{f \in E\colon \Delta f \in E\},$$
$$\Delta_E f = \Delta f \quad \text{in } \mathcal{D}\big(\mathbb{R}^n\big)'.$$

On $E = L^p(\mathbb{R}^n)$, $1 \leqslant p < \infty$, $BUC(\mathbb{R}^n)$ and $C_0(\mathbb{R}^n)$ the semigroup G_E is a C_0-semigroup.

2.5. *The Dirichlet Laplacian*

In this section we introduce the Laplacian with Dirichlet boundary conditions on the space $C(\overline{\Omega})$. It is a most basic example and we show in an elementary way that it generates a holomorphic semigroup.

Let $\Omega \subset \mathbb{R}^n$ be open and bounded. We assume that Ω is *Dirichlet regular*; i.e., for all $\varphi \in C(\partial\Omega)$ there exists a solution of

$$D(\varphi) \begin{cases} h \in C\big(\overline{\Omega}\big), & h|_{\partial\Omega} = \varphi, \\ \Delta h = 0 & \text{in } \mathcal{D}(\Omega)'. \end{cases}$$

Such a solution is unique and automatically in $C^\infty(\Omega)$. If Ω has Lipschitz boundary, then Ω is Dirichlet regular, but much milder geometric assumptions suffice (see, e.g., [[DL88],

Chapter II]). We consider the operator A defined on $C(\overline{\Omega})$ by

$$D(A) = \{u \in C_0(\Omega)\colon\ \Delta u \in C(\overline{\Omega})\},$$
$$Au = \Delta u \quad \text{in } \mathcal{D}(\Omega)',$$

where $C_0(\Omega) := \{u \in C(\overline{\Omega})\colon\ u_{|\partial\Omega} = 0\}$ and $\mathcal{D}(\Omega)'$ denotes the space of all distributions. We call A the *Dirichlet Laplacian* on $C(\overline{\Omega})$.

THEOREM. *The operator A generates a bounded holomorphic semigroup T on $C(\overline{\Omega})$.*

For the proof we need the following form of the maximum principle.

MAXIMUM PRINCIPLE. *Let $v \in C(\overline{\Omega})$ such that $\lambda v - \Delta v = 0$ in $\mathcal{D}(\Omega)'$, where* $\operatorname{Re}\lambda > 0$. *Then*

$$\sup_{x\in\partial\Omega} |v(x)| = \sup_{x\in\overline{\Omega}} |v(x)|.$$

PROOF. Suppose that $\|v\|_{L^\infty(\Omega)} > \sup_{x\in\partial\Omega} |v(x)|$. Let $K := \{x \in \Omega\colon\ |v(x)| = \|v\|_\infty\}$ and $v_\varepsilon = \rho_\varepsilon * v$, where ρ_ε is a mollifier. Then $v_\varepsilon \to v$, $\Delta v_\varepsilon = \rho_\varepsilon * \Delta v \to \Delta v$ uniformly on compact subsets of Ω as $\varepsilon \downarrow 0$. Let $\Omega_1 \subset \Omega$ be relatively compact such that $K \subset \Omega_1$ and $\overline{\Omega}_1 \subset \Omega$. Then, for small $\varepsilon > 0$, there exists $x_\varepsilon \in \Omega_1$ such that $|v_\varepsilon(x_\varepsilon)| = \sup_{x\in\Omega_1} |v_\varepsilon(x)|$. Then

$$\operatorname{Re}\big[\Delta v_\varepsilon(x_\varepsilon)\overline{v_\varepsilon(x_\varepsilon)}\big] \leqslant 0. \tag{2.2}$$

In fact, consider the function $f(y) = \operatorname{Re} v_\varepsilon(y)\overline{v_\varepsilon(x_\varepsilon)}$. Then f has a local maximum in x_ε. Hence $\Delta f(x_\varepsilon) \leqslant 0$. Let $x_{\varepsilon_n} \to x_0$. Since $v_{\varepsilon_n} \to v$ uniformly on Ω_1, it follows that $x_0 \in K$. From (2.2) we deduce that $\operatorname{Re}[\overline{v(x_0)}\Delta v(x_0)] \leqslant 0$. Hence,

$$\begin{aligned}\operatorname{Re}\lambda|v(x_0)|^2 &\leqslant \operatorname{Re}\lambda|v(x_0)|^2 - \operatorname{Re}\big[\overline{v(x_0)}\Delta v(x_0)\big]\\ &= \operatorname{Re}\big[\overline{v(x_0)}\big(\lambda v(x_0) - \Delta v(x_0)\big)\big] = 0.\end{aligned}$$

Since $x_0 \in K$, it follows that $v = 0$, contradicting the assumption. □

PROOF OF THE THEOREM. (a) Similarly as the Maximum principle above, one shows that A is dissipative.

(b) We show that $0 \in \rho(A)$. Let $f \in C(\overline{\Omega})$. Denote by $\tilde{f}$ the extension of f to $\mathbb{R}^n$ by 0 and let $v = E * \tilde{f}$, where E is the Newtonian potential. Then $v \in C(\mathbb{R}^n)$ and $\Delta v = f$ in $\mathcal{D}(\Omega)'$. Let $\varphi = v|_{\partial\Omega}$ and consider the solution h of the Dirichlet problem $D(\varphi)$. Then $u = v - h \in D(A)$ and $Au = f$. We have shown that A is surjective. Since the solution of $D(0)$ is unique, the operator A is injective. Since A is closed, it follows that $0 \in \rho(A)$.

(c) It follows from (a) and (b) that A is m-dissipative. In particular, $\lambda \in \rho(A)$ whenever $\operatorname{Re}\lambda > 0$.

(d) Denote by Δ_0 the Laplacian on $C_0(\mathbb{R}^n)$ which generates a bounded holomorphic semigroup by 2.4. Thus there exists $M \geqslant 0$ such that

$$\left\|\lambda R(\lambda, \Delta_0)\right\| \leqslant M \qquad \text{if } \operatorname{Re}\lambda > 0.$$

We show that $\|\lambda R(\lambda, A)\| \leqslant 2M$ if $\operatorname{Re}\lambda > 0$ which proves the Theorem. In fact, let $\operatorname{Re}\lambda > 0$, $f \in C(\overline{\Omega})$, $g = R(\lambda, A)f$. Let $\tilde{g} = R(\lambda, \Delta_\infty)\tilde{f}$. Then $v = g - \tilde{g} \in C(\overline{\Omega})$. Moreover, $v|_{\partial\Omega} = -\tilde{g}|_{\partial\Omega}$ and $\lambda v - \Delta v = 0$ in $\mathcal{D}(\Omega)'$. By the Maximum principle, one has

$$\begin{aligned} \sup_{x\in\overline{\Omega}}\left|v(x)\right| &= \sup_{x\in\partial\Omega}\left|v(x)\right| = \sup_{x\in\partial\Omega}\left|\tilde{g}(x)\right| \\ &\leqslant \frac{M}{|\lambda|}\|\tilde{f}\|_{L^\infty(\mathbb{R}^n)} = \frac{M}{|\lambda|}\|f\|_{L^\infty(\Omega)}. \end{aligned}$$

Consequently,

$$\begin{aligned} \|g\|_{L^\infty(\Omega)} &\leqslant \|v\|_{L^\infty(\Omega)} + \left\|\tilde{g}\right\|_{L^\infty(\Omega)} \\ &\leqslant 2\frac{M}{|\lambda|}\|f\|_{L^\infty(\Omega)}. \end{aligned}$$

□

FURTHER PROPERTIES. *One has $\omega(T) < \infty$ and $T(t)$ is positive and compact for all $t > 0$. The restriction T_0 of T to $C_0(\Omega)$ is a C_0-semigroup.*

REFERENCE. The elegant elementary argument above is due to Lumer and Paquet [LP79], where it is given for more general elliptic operators. See also [[ABHN01], Chapter 6] for a different presentation, and [LS99] for more general results.

2.6. *The Neumann Laplacian on $C(\overline{\Omega})$*

Let $\Omega \subset \mathbb{R}^n$ be open. There is a natural realization Δ_Ω^N of the Laplacian on $L^2(\Omega)$ with Neumann boundary conditions. The operator Δ_Ω^N is self-adjoint and generates a bounded, holomorphic, positive C_0-semigroup $(e^{t\Delta_\Omega^N})_{t\geqslant 0}$ on $L^2(\Omega)$. We refer to Section 5.3.3 for the precise definition.

THEOREM [FT95]. *Let Ω be bounded with Lipschitz-boundary. Then $C(\overline{\Omega})$ is invariant under the semigroup $(e^{t\Delta_\Omega^N})_{t\geqslant 0}$ and the restriction is a bounded holomorphic C_0-semigroup T on $C(\overline{\Omega})$.*

The main point in the proof is to show invariance of $C(\overline{\Omega})$ by the semigroup or the resolvent. Holomorphy can be shown with the help of Gaussian estimates (Section 7.4.3). We also mention that the semigroup T induced on $C(\overline{\Omega})$ is *compact*, i.e., each $T(t)$ is compact for $t > 0$. This follows from ultracontractivity (Sections 7.3.3 and 7.3.7).

Not for each open, bounded set Ω the space $C(\overline{\Omega})$ is invariant: the Theorem is false on the domain

$$\Omega = \{(x, y) \in \mathbb{R}^2 \colon |x| < 1, |y| < 1\} \setminus [0, 1) \times \{0\};$$

see [Bie03].

The Theorem also holds for Robin boundary conditions, we refer to [War03a].

2.7. *Wentzell boundary conditions*

As a further example we mention a very different kind of boundary condition. Let $m \in C[0, 1]$ be strictly positive. Let

$$(Lu)(x) = m(x) \cdot x(1 - x)u''(x), \qquad x \in (0, 1),$$

for $u \in C^2(0, 1)$. Define the operator A on $C[0, 1]$ by

$$D(A) = \Big\{u \in C[0, 1] \cap C^2(0, 1) \colon \lim_{\substack{x \to 0 \\ x \to 1}} (Lu)(x) = 0\Big\},$$
$$Au = Lu.$$

Then A generates a holomorphic C_0-semigroup T of angle $\pi/2$. Moreover, T is positive and contractive.

We refer to Campiti and Metafune [CM98] for this and more general degenerate elliptic operators with *Wentzell boundary conditions*.

2.8. *Dynamic boundary conditions*

Let $\Omega \subset \mathbb{R}^n$ be a bounded, open set of class C^2. We will introduce a realization of the Laplacian in $C(\overline{\Omega})$ with Wentzell–Robin boundary conditions. By $C_\nu^1(\overline{\Omega})$ we denote the space of all functions $f \in C(\overline{\Omega})$ for which the outer normal derivative

$$\frac{\partial f}{\partial \nu}(z) = -\lim_{t \downarrow 0} \frac{f(z - t \cdot \nu(z)) - f(z)}{t}$$

exists uniformly for $z \in \partial\Omega$; see [[DL88], Vol. 1, Sect. II.1.3b]. Let $\beta, \gamma \in C(\partial\Omega)$ and suppose that $\beta(z) > 0$ for all $z \in \partial\Omega$. Define the operator A on $C(\overline{\Omega})$ by

$$D(A) := \Big\{f \in C_\nu^1(\overline{\Omega}) \colon \Delta f \in C(\overline{\Omega}), \Delta f + \beta \frac{\partial f}{\partial \nu} + \gamma f = 0 \text{ on } \partial\Omega\Big\},$$
$$Af := \Delta f.$$

THEOREM. *The operator A generates a positive, compact and holomorphic C_0-semigroup on $C(\overline{\Omega})$.*

Favini, Goldstein, Goldstein and Romanelli [FGGR02] were the first to prove that A is a generator with the help of dissipativity. An approach by form methods was then given in [AMPR03]. Warma [War03b] proved analyticity in the case where Ω is an interval. It was Engel [Eng04] who succeeded to prove that the semigroup is holomorphic in the general case.

The boundary conditions incorporated into the domain of A express in fact *dynamic boundary conditions* for the evolution equation. To see this, denote by T the semigroup generated by A. Let $f \in C(\overline{\Omega})$ and let $u(t) = T(t)f$. Then $u \in C^1((0,\infty), C(\overline{\Omega}))$ and $\dot{u}(t) = \Delta u(t)$, $t > 0$, on Ω. Moreover, $\Delta u(t) \in C(\overline{\Omega})$ and $\Delta u(t) = -\beta\frac{\partial u}{\partial \nu}(t) - \gamma u(t)$ on $\partial\Omega$. Hence,

$$\dot{u}(t) = -\beta\frac{\partial u}{\partial \nu}(t) - \gamma u(t) \quad \text{on } \partial\Omega, t > 0.$$

3. Asymptotics

Most important and interesting is the study of the asymptotic behavior of a semigroup $T(t)$ for $t \to \infty$. The philosophy is, as for other questions, that one knows better the generator and its resolvent than the semigroup. Thus the challenge is to deduce the asymptotic behavior from spectral properties of the generator. Here we will describe some principal results with emphasis on those which can be applied to parabolic equations in Section 8. We refer to [[ABHN01]] and [[Nee96]] for a systematic theory of the asymptotic behavior of semigroups and to [[EN00]] for other kinds of examples.

3.1. *Exponential stability*

Let T be a C_0-semigroup with generator A. By

$$s(A) = \sup\{\operatorname{Re}\lambda\colon \lambda \in \sigma(A)\}$$

we denote the *spectral bound* of A. We say that T is *exponentially stable* if $\omega(T) < 0$; i.e., if there exist $\varepsilon > 0$, $M \geqslant 0$ such that

$$\|T(t)\| \leqslant M\mathrm{e}^{-\varepsilon t}, \quad t \geqslant 0.$$

It is easy to see that T is exponentially stable if and only if

$$\lim_{t\to\infty} \|T(t)\| = 0.$$

FUNDAMENTAL QUESTION. *Does $s(A) < 0$ imply that T is exponentially stable*? In general this is not true. The realization of the Cauchy problem

$$\begin{cases} \frac{\partial u}{\partial t}(t,s) = s\frac{\partial u}{\partial s}(t,s), & t > 0, s > 1, \\ u(0,s) = u_0(s), & s > 1, \end{cases}$$

in the Sobolev space $W^{1,2}(1,\infty)$ leads to a C_0-semigroup T whose generator A has spectral bound $s(A) < -\frac{1}{2}$ but T is unbounded [[ABHN01], p. 350]. An example of a hyperbolic equation is given by Renardy [Ren94].

However, additional hypotheses are known, which lead to a positive answer. Here we consider three important cases corresponding to a regularity assumption, semigroups on Hilbert space and a positivity assumption.

3.1.1. *Eventually norm continuous semigroups.* A C_0-semigroup T is called *eventually norm-continuous* if $\lim_{t\downarrow 0} \|T(t_0+t) - T(t_0)\| = 0$ for some $t_0 > 0$, and T is called *norm-continuous* if this holds for all $t_0 > 0$. Of course, each holomorphic semigroup has this property.

THEOREM. *Let A be the generator of an eventually norm-continuous semigroup T. If $s(A) < 0$, then T is exponentially stable.*

For further extensions we refer to [[ABHN01], Chapter 5], [Bla01], [BBN01].

We mention some perturbation results: If A generates an eventually norm continuous C_0-semigroup T on X and $B \in \mathcal{L}(X)$ is compact, then $A + B$ also generates an eventually norm-continuous C_0-semigroup. The compactness assumption cannot be omitted, in general. It can be omitted if T is norm-continuous on $(0,\infty)$. Eventually norm-continuous semigroups appear in models for cell growth. We refer to [[Nag86], A-II.1.30 and C-IV.2.15].

3.1.2. *The Gearhart–Prüss theorem.* *Let H be a Hilbert space. Assume that $s(A) < 0$ and $\sup_{\operatorname{Re}\lambda>0} \|R(\lambda, A)\| < \infty$. Then T is exponentially stable.*

There are several proofs of this result which all depend on the fact that the vector-valued Fourier transform is an isomorphism for Hilbert spaces. Prüss' proof [Prü84] uses Fourier-series. The above result is not true on L^p-spaces for $1 \leqslant p \leqslant \infty$, $p \neq 2$; see [ArBu02], Example 3.7.

3.1.3. *Positive semigroups on L^p-spaces.* In the next result we consider a C_0-semigroup T on $L^p(\Omega, \Sigma, \mu)$, where (Ω, Σ, μ) is a measure space and $1 \leqslant p < \infty$. The semigroup is called *positive* if $T(t)f \geqslant 0$ for each $0 \leqslant f \in L^p(\Omega, \Sigma, \mu)$. Of course, here $f \geqslant 0$ means that $f(x) \geqslant 0$ μ-a.e. For positive semigroups, there is an easy criterion for negative spectral bound: One has

$$s(A) < 0 \text{ if and only if } A \text{ is invertible and } A^{-1} \leqslant 0.$$

THEOREM (Weis [Wei95]). *Let A be the generator of a positive C_0-semigroup T on $L^p(\Omega)$, $1 \leqslant p < \infty$. If $s(A) < 0$, then T is exponentially stable.*

For a proof and further references we refer to [[ABHN01], 5.3.6] and [[Nee96]]. A similar result is true on $C_0(\Omega)$, where Ω is a locally compact space [[ABHN01], 5.3.8], but false on a space $L^p \cap L^q$ [[ABHN01], 5.1.11].

3.2. *Ergodic semigroups*

Let T be a bounded C_0-semigroup on a Banach space X. Denote by A the generator of T and by A^* the adjoint of A. We say that T is *ergodic* if

$$Px = \lim_{t\to\infty} \frac{1}{t} \int_0^t T(s)x \, \mathrm{d}s$$

exists for all $x \in X$.

ERGODIC THEOREM. *The following assertions are equivalent:*

(i) *T is ergodic;*
(ii) $X = \ker A \oplus \overline{R(A)}$;
(iii) $\ker A$ *separates* $\ker A^*$.

In that case P is the projection onto $\ker A$ *along* $\overline{R(A)}$.

Here we denote by $R(A) = \{Ax\colon\ x \in D(A)\}$ the range of A. One has always $\ker A \cap \overline{R(A)} = \{0\}$. To say that $\ker A$ separates $\ker A^*$ means that for all $x^* \in \ker A^*$, $x^* \neq 0$, there exists $x \in \ker A$ such that $\langle x^*, x\rangle \neq 0$. Note that $\ker A^*$ always separates $\ker A$ (by the Hahn–Banach theorem). Thus on reflexive spaces, ergodicity is automatic.

THEOREM. *Every bounded C_0-semigroup on a reflexive space is ergodic.*

It is interesting to know whether reflexivity is the best possible hypothesis on the Banach space in order to guarantee automatic ergodicity. And indeed it is, under some additional hypothesis. In fact, on a general Banach space no method is known to construct nontrivial C_0-semigroups. For this reason one has to suppose some geometric property. We will assume that X has a Schauder basis. See Section 4.5.2 for the precise definition and also for a method to construct diagonal semigroups under this hypothesis. The following theorem is a semigroup version of a result on power bounded operators by Fonf–Lin–Wojtaszczyk [FLW01] which was recently given by Mugnolo [Mug02].

THEOREM. *Let X be a Banach space with a Schauder basis. Assume that each bounded C_0-semigroup on X is ergodic. Then X is reflexive.*

We conclude by an example. Denote by G_p the Gaussian semigroup, on $L^p(\mathbb{R}^n)$, $1 \leqslant p \leqslant \infty$, and by Δ_p its generator (see Section 2.4). Then G_1 is not ergodic since

$\ker \Delta_1 = 0$ and $\ker \Delta_\infty = \mathbb{R} \cdot 1$. The semigroup G_p is ergodic for $1 < p < \infty$. But one can even show that $\lim_{t\to\infty} G_p(t) = 0$ strongly in $L^p(\mathbb{R}^n)$, $1 < p < \infty$. Strong convergence, and not merely convergence in mean, is the subject of the next section and the result for the Gaussian semigroup follows from Section 3.3.2.

3.3. *Convergence and asymptotically almost periodicity*

In the preceding section we described when a semigroup converges in mean. Now we investigate a stronger property, namely strong convergence (see Theorem 2 of Section 3.3.2). More generally, we consider semigroups which can be decomposed into a semigroup converging strongly to zero and an almost periodic group.

Let T be a bounded C_0-semigroup on X with generator A. We say that T is *asymptotically almost periodic* if $X = X_0 \oplus X_{\mathrm{ap}}$, where

$$X_0 := \Bigl\{x \in X\colon \lim_{t\downarrow 0} T(t)x = 0\Bigr\}$$

$$X_{\mathrm{ap}} := \overline{\operatorname{span}}\bigl\{x \in X\colon \exists \eta \in \mathbb{R},\, T(t)x = \mathrm{e}^{\mathrm{i}\eta t}x\bigr\}.$$

Note that X_0 and X_{ap} are invariant under the semigroup. Moreover, there exists a bounded C_0-semigroup U on X_{ap} such that $T(t)|_{X_{\mathrm{ap}}} = U(t)$, $t \geqslant 0$. Let $\sigma_p(A) := \{\lambda \in \mathbb{C}\colon \exists x \in D(A), x \neq 0, Ax = \lambda x\}$ be the *point spectrum* of A. If $\sigma_p(A) \cap \mathrm{i}\mathbb{R} \subset \{0\}$, then $X_{\mathrm{ap}} = \ker A$. In that case T is asymptotically almost periodic if and only if $Px := \lim_{t\to\infty} T(t)x$ converges for all $x \in X$ (and not just only in mean as considered in the previous section).

3.3.1. *Compact resolvent.* Let A be an operator with nonempty resolvent set $\rho(A)$. We say that A has *compact resolvent* if $R(\lambda, A)$ is compact for all (equivalently one) $\lambda \in \rho(A)$. This is equivalent to saying that the injection $D(A) \hookrightarrow X$ is compact where $D(A)$ carries the graph norm. It implies that $\sigma(A) = \sigma_p(A)$ is a sequence converging to ∞ (unless $\dim X < \infty$).

THEOREM [[ABHN01], p. 361]. *Let T be a bounded C_0-semigroup whose generator has compact resolvent. Then T is asymptotically almost periodic.*

This result can be generalized to the case where $\sigma(A) \cap \mathrm{i}\mathbb{R}$ is countable with more involved proofs, though

3.3.2. *Countable spectrum.* Let T be a bounded C_0-semigroup with generator A.

THEOREM 1. *Assume that X is reflexive and $\sigma(A) \cap \mathrm{i}\mathbb{R}$ is countable. Then T is asymptotically almost periodic.*

If X is not reflexive, one needs an ergodicity hypothesis.

THEOREM 2. *Assume that*

(a) $\sigma(A) \cap \mathrm{i}\mathbb{R}$ *is countable*;

(b) $\sigma_p(A^*) \cap i\mathbb{R} \subset \{0\}$;

(c) *T is ergodic.*

Then $Px = \lim_{t\to\infty} T(t)x$ *converges for all* $x \in X$.

REFERENCES: [[ABHN01], Chapter 5], [Vu97]. So far, there does not exist a (spectral) characterization of strong convergence. But we refer to Chill and Tomilov [CT03,CT04] for very interesting (not purely spectral) conditions.

3.4. *Positive semigroups*

The asymptotic behavior of positive semigroups is most interesting and has applications to many areas, for example population dynamics and transport theory (see [[EN00]], [[Mok97]], [[Nag86]]). Here we think more of applications to parabolic equations and establish a result on convergence to a rank-1-projection which will be applied to a second-order parabolic equation (see Section 3.5.1 and (8.5)). We also present some other results which are obtained by putting together the results on countable spectrum of Section 3.3 and Perron–Frobenius theory for positive semigroups. Throughout this section we assume that X is a space of the following two kinds:

(a) $X = L^p(\Omega)$, $1 \leqslant p < \infty$, where (Ω, Σ, μ) is a σ-finite measure space, or

(b) $X = C_0(K)$, where K is locally compact.

Here we let $C_0(K) := \{f : K \to \mathbb{C}$ continuous: for each $\varepsilon > 0$ there exists a compact set $K_\varepsilon \subset K$ such that $|f(x)| \leqslant \varepsilon$ for $x \in K \setminus K_\varepsilon\}$. Of course, if K is compact, then $C_0(K) = C(K)$.

By X_+ we denote the cone of all functions f in X which are positive almost everywhere if $X = L^p$ and everywhere if $X = C_0(K)$. Let T be a positive C_0-semigroup on X; i.e., T satisfies $T(t)X_+ \subset X_+$ for all $t > 0$. Denote by A the generator of T. At first we recall that

$$s(A) = \omega(T) \tag{3.1}$$

and $s(A) \in \sigma(A)$ if $\sigma(A) \neq \emptyset$, two important special properties of positive C_0-semigroups on X (see Section 1.3 for the definition of $\omega(T)$, and [[ABHN01], Theorems 5.3.6 and 5.3.1] for the proofs).

THEOREM. *Let T be a bounded, ergodic, eventually norm-continuous, positive C_0-semigroup on X. Then*

$$Pf = \lim_{t\to\infty} T(t)f$$

exists for all $f \in X$.

As a consequence P is a positive projection, the projection onto $\ker A$ along $\overline{R(A)}$. Recall that each holomorphic semigroup is eventually norm-continuous and ergodicity is automatic if $X = L^p$, $1 < p < \infty$.

It is a remarkable result of Perron–Frobenius theory that $\sigma(A) \cap i\mathbb{R} \subset \{0\}$ whenever T is positive, bounded and eventually norm-continuous, [[Nag86], C-III.2.10, p. 202 and A-II.1.20, p. 38]. Thus the theorem follows from Theorem 2 in Section 3.3.

3.5. *Positive irreducible semigroup*

We assume again that $X = L^p(\Omega)$, $1 \leqslant p < \infty$, or $X = C_0(K)$ as in Section 3.4. An element $f \in L^p(\Omega)$ is called *strictly positive* if $f(x) > 0$ a.e. An element $f \in C_0(K)$ is called *strictly positive* if $f(x) > 0$ for all $x \in K$. Finally, a functional $\varphi \in C_0(K)^*$ is called *strictly positive* if

$$\langle \varphi, f \rangle > 0 \quad \text{for all } f \in C_0(K)_+ \setminus \{0\}.$$

Let $u \in X$ and $\varphi \in X^*$ be strictly positive such that $\langle \varphi, u \rangle = 1$. Then $Pf = \langle \varphi, f \rangle u$ defines a projection on X. We call P a *strictly positive rank-1-projection*.

DEFINITION [[Nag86], p. 306]. A positive C_0-semigroup is *irreducible* if for some (equivalently all) $\lambda > s(A)$ the function $R(\lambda, A)f$ is strictly positive for all $f \in X_+ \setminus \{0\}$.

Irreducibility has many remarkable consequences (see [[Nag86], p. 306]). For example, if $X = C_0(K)$, it implies that $\sigma(A) \neq \emptyset$. This also remains true on $X = L^p(\Omega)$, $1 \leqslant p < \infty$, if in addition we assume that $T(t_0)$ is compact for some $t_0 > 0$. However, this case is more difficult and depends on a deep result of de Pagter [[Nag86], C-III-3.7].

3.5.1. *Convergence to a rank-1-projection.* In the following theorem we assume that $T(t_0)$ is compact for some $t_0 > 0$. This implies that $T(t)$ is compact for all $t > t_0$ and that T is norm-continuous on $[t_0, \infty)$. For example, if T is holomorphic and A has compact resolvent, then $T(t)$ is compact for all $t > 0$.

THEOREM. *Let T be a positive, irreducible C_0-semigroup on X. Assume that $T(t_0)$ is compact for some $t_0 > 0$. Then $s(A) > -\infty$ and there exists a strictly positive rank-1-projection P such that*

$$\left\| e^{-s(A)t} T(t) - P \right\| \leqslant M e^{-\varepsilon t}, \quad t \geqslant 0,$$

for some $M \geqslant 0$, $\varepsilon > 0$.

Thus the rescaled semigroup converges exponentially to a rank-1-projection. Such rescaled convergence is sometimes called *balanced exponential growth* and plays an important role for models describing cell growth (see [Web87]). Here the theorem will be applied to parabolic equations (see Section 8.5).

PROOF OF THEOREM. Since $\sigma(A) \neq \emptyset$ we can assume that $s(A) > -\infty$. Since $T(t_0)$ is compact, T is quasicompact [[Nag86], B-IV.2.8, p. 214]. Now the result follows from [[Nag86], C-IV.2.1, p. 343, and C-III.3.5(d), p. 310]. □

3.5.2. *Convergence to a periodic group.* The following result is a combination of Perron–Frobenius theory for positive semigroups and the results on countable spectrum of Section 3.3. Let T be a bounded C_0-semigroup with generator A. Assume that A has compact resolvent. Then we know from Section 3.3 that $X = X_0 \oplus X_{ap}$, where X_0 and X_{ap} are invariant under T. Moreover, there exists a C_0-group U on X_{ap} such that $U(t) = T(t)|_{X_{ap}}$ for $t \geqslant 0$.

THEOREM. *In addition to the assumptions made above, if T is positive and irreducible, then U is periodic.*

PROOF. If $s(A) < 0$, then $X_{ap} = \{0\}$. If $\sigma(A) \cap i\mathbb{R} = \{0\}$, then $X_{ap} = \ker A$; i.e., $U(t) \equiv I$. If $\sigma(A) \cap i\mathbb{R} \neq \{0\}$, it follows from [[Nag86], C-III.3.8, p. 313] that $\sigma(A) \cap i\mathbb{R} = i\frac{2\pi}{\tau}\mathbb{Z}$ for some $\tau > 0$. It follows from the definition of X_{ap} that $T(t+\tau)x = T(t)x$, $t \geqslant 0$, for all $x \in X_{ap}$. □

4. Functional calculus

In this section we consider sectorial operators. These are unbounded operators whose spectra lie in a sector. Generators of bounded C_0-semigroups are of this type. If A is such an operator, we will define closed operators $f(A)$ for a large class of holomorphic functions defined on a sector. Of particular importance are fractional powers A^α. We will introduce the class *BIP* which is important to establish theorems of maximal regularity but also for results on interpolation. Throughout this chapter X is a complex Banach space.

4.1. *Sectorial operators*

Given an angle $0 < \varphi < \pi$, we consider the open sector

$$\Sigma_\varphi := \{re^{i\alpha}: r > 0, |\alpha| < \varphi\}.$$

Let X be a Banach space. An operator A is called *φ-sectorial* if

$$\sigma(A) \subset \overline{\Sigma_\varphi} \tag{4.1}$$

and

$$\sup_{\lambda \in \mathbb{C} \setminus \Sigma_\varphi} \|\lambda R(\lambda, A)\| < \infty. \tag{4.2}$$

An operator A on X is called *sectorial* if there exists $0 < \varphi < \pi$ such that A is φ-sectorial. In that case we define the *sectoriality angle* $\varphi_{sec}(A)$ of A by

$$\varphi_{sec}(A) := \inf\{\varphi \in (0, \pi): \text{ (4.1) and (4.2) are valid}\}. \tag{4.3}$$

Note that $\varphi_{sec}(A) \in [0, \pi)$.

4.1.1. *Simple criterion.* If $(-\infty, 0) \subset \rho(A)$ and

$$\|\lambda R(\lambda, A)\| \leqslant c \text{ for all } \lambda < 0 \text{ and some } c \geqslant 0, \text{ then } A \text{ is sectorial.}$$

This follows from the power series expansion of the resolvent.

4.1.2. *Bounded holomorphic semigroups.* An operator A is sectorial with $\varphi_{\mathrm{sec}}(A) < \frac{\pi}{2}$ if and only if $-A$ generates a bounded holomorphic semigroup T. It is a C_0-semigroup if and only if $D(A)$ is dense. This is automatically the case if X is reflexive.

4.1.3. *Bounded semigroups.* If $-A$ generates a bounded C_0-semigroup, then A is sectorial and $\varphi_{\mathrm{sec}}(A) \leqslant \frac{\pi}{2}$.

4.1.4. *Injective operators.* Let A be a sectorial operator with range $R(A) := \{Ax\colon x \in D(A)\}$. If A is injective, then A^{-1} with domain $D(A^{-1}) = R(A)$ is sectorial and $\varphi_{\mathrm{sec}}(A^{-1}) = \varphi_{\mathrm{sec}}(A)$.

4.1.5. *Reflexive spaces.* If A is an injective sectorial operator on a reflexive space, then $D(A) \cap R(A)$ is dense.

4.1.6. *Warning.* The notion of *sectorial* operators is not universal in the literature. In particular, it does not coincide with Kato's definition in [[Kat66]]. In [PS90], [DHP01] and [[Prü93]] the additional assumption that A is injective, densely defined with dense image is incorporated into the definition of sectorial. This implies in particular that $D(A) \cap R(A)$ is dense.

4.2. *The sum of commuting operators*

Many interesting problems can be formulated in the following form. Let A and B be operators on a Banach space: Under which conditions is the problem

$$Ax + Bx = y \tag{P}$$

well posed? This means that for all $y \in X$ there exists a unique $x \in D(A) \cap D(B)$ solving (P). In this section we will establish a spectral theoretical approach which leads to a "weak solution". We define the operator $A + B$ on the domain $D(A + B) := D(A) \cap D(B)$ by $(A + B)x = Ax + Bx$.

Assume that $\rho(A) \cap \rho(B) \neq \emptyset$. We say that A and B *commute* if $R(\lambda, A)R(\lambda, B) = R(\lambda, B)R(\lambda, A)$ for all (equivalently one) $\lambda \in \rho(A) \cap \rho(B)$. This is equivalent to saying that for some (equivalently all) $\lambda \in \rho(A)$ one has $R(\lambda, A)D(B) \subset D(B)$ and $BR(\lambda, A)x = R(\lambda, A)Bx$ for all $x \in D(B)$. For example, if A and B generate C_0-semigroups S and T, then A and B commute if and only if $S(t)T(t) = T(t)S(t)$ for all $t \geqslant 0$. In that case $U(t) = T(t)S(t)$ is a C_0-semigroup, the operator $A + B$ is closable and $\overline{A + B}$ is the generator of U.

THEOREM. *Let A, B be two commuting sectorial operators such that*

$$\varphi_{\mathrm{sec}}(A) + \varphi_{\mathrm{sec}}(B) < \pi.$$

Then $A + B$ has a unique extension $(A+B)^{\sim}$ such that $(\omega, \infty) \subset \rho((A+B)^{\sim})$ for some $\omega \in \mathbb{R}$ and such that $(A+B)^{\sim}$ commutes with A. Moreover,

$$\sigma\big((A+B)^{\sim}\big) \subset \sigma(A) + \sigma(B). \tag{4.4}$$

If $D(A)$ is dense in X, then $A + B$ is closable and $(A+B)^{\sim}$ is the closure of $A + B$.

Note that by (4.4), $(A+B)^{\sim}$ is invertible if $0 \in \rho(A)$ or $0 \in \rho(B)$. This result is due to Da Prato and Grisvard [DPG75], Théorème 3.7, under the assumption that $D(A)$ is dense. The spectral inclusion (4.2.1) was proved in [[Prü93], Theorem 8.5] and in [ARS94], where also nondensely defined operators are considered.

If A or B is invertible, then by the Theorem, for each $y \in X$, there exists a unique $x \in D((A+B)^{\sim})$ such that

$$(A+B)^{\sim} x = y.$$

One aim of the functional calculus to be developed here is to establish conditions under which $x \in D(A) \cap D(B)$, which is expressed by saying that the solution has *maximal regularity*.

4.3. *The elementary functional calculus*

Let A be a sectorial operator and let $\varphi_{\mathrm{sec}}(A) < \varphi < \pi$.

4.3.1. *Invertible operators.* By $H_1^{\infty}(\Sigma_{\varphi})$ we denote the space of all bounded holomorphic functions $f : \Sigma_{\varphi} \to \mathbb{C}$ such that

$$\big|f(z)\big| \leqslant c|z|^{-\gamma}, \quad |z| \geqslant 1, \tag{4.5}$$

where $c \geqslant 0$, $\gamma > 0$ depend on f. Then $H_1^{\infty}(\Sigma_{\varphi})$ is an algebra for pointwise operations. Let $\varphi_{\mathrm{sec}}(A) < \varphi_1 < \varphi$. Consider the path

$$\Gamma(r) = \begin{cases} -re^{-\mathrm{i}\varphi_1} & \text{if } r < 0, \\ re^{\mathrm{i}\varphi_1} & \text{if } r \geqslant 0. \end{cases}$$

For $f \in H_1^{\infty}(\Sigma_{\varphi})$ we define $f(A) \in \mathcal{L}(X)$ by

$$f(A) = \frac{1}{2\pi \mathrm{i}} \int_{\Gamma} f(\lambda) R(\lambda, A) \, \mathrm{d}\lambda. \tag{4.6}$$

This definition does not depend on the choice of φ_1. The mapping $f \mapsto f(A)$ is an algebra homomorphism from $H_1^\infty(\Sigma_\varphi)$ into $\mathcal{L}(X)$ such that

$$f_\lambda(A) = R(\lambda, A) \quad \text{for } \lambda \in \mathbb{C} \setminus \Sigma_\varphi,$$

where $f_\lambda(z) = \frac{1}{\lambda - z}$. Now let $x \in D(A)$. Then

$$f(A)x = (I + A)g(A)x, \tag{4.7}$$

where $g(\lambda) = \frac{f(\lambda)}{1+\lambda}$. This follows from Cauchy's theorem since $R(\lambda, A) - \frac{1}{1+\lambda} R(\lambda, A)(I + A) = \frac{1}{1+\lambda} I$. For $f \in H^\infty(\Sigma_\varphi)$ we define $f(A)$ by

$$f(A) = (I + A)g(A) \tag{4.8}$$

with $g(\lambda) = \frac{f(\lambda)}{1+\lambda}$. Since $g \in H_1^\infty(\Sigma_\varphi)$, one has $g(A) \in \mathcal{L}(X)$ and so $f(A)$ is a closed operator with domain

$$D\big(f(A)\big) := \big\{x \in X\colon\ g(A)x \in D(A)\big\}.$$

This definition is consistent with (4.6) if $f \in H_1^\infty(\Sigma_\varphi)$. Next we relax the hypothesis of invertibility.

4.3.2. *Injective operators.* Let A be a sectorial operator and $\varphi > \varphi_{\mathrm{sec}}(A)$. Denote by $H_0^\infty(\Sigma_\varphi)$ the algebra of all bounded holomorphic functions $f : \Sigma_\varphi \to \mathbb{C}$ satisfying an estimate

$$\begin{aligned} \big|f(z)\big| &\leqslant c|z|^{-\gamma}, \quad |z| > 1, \\ \big|f(z)\big| &\leqslant c|z|^{\gamma}, \quad |z| \leqslant 1, \end{aligned}$$

where $c \geqslant 0$ and $\gamma > 0$ depend on f. For $f \in H_0^\infty(\Sigma_\varphi)$ we define $f(A) \in \mathcal{L}(X)$ by (4.6). Now assume that A is injective. The operator $B = A(I + A)^{-2}$ is bounded and injective with range $D(A) \cap R(A)$. Hence, $A + 2I + A^{-1} = B^{-1}$ with domain $D(A) \cap R(A)$ is a closed, injective operator. Now let $f \in H^\infty(\Sigma_\varphi)$. Let $g(\lambda) = \frac{\lambda}{(1+\lambda)^2} f(\lambda)$. Then $g \in H_0^\infty(\Sigma_\varphi)$. Hence, $g(A) \in \mathcal{L}(X)$. We define $f(A)$ by

$$\begin{aligned} f(A) &= \big(A + 2I + A^{-1}\big)g(A) \\ &= (I + A)^2 A^{-1} g(A). \end{aligned} \tag{4.9}$$

Then $f(A)$ is a closed operator. A similar argument as in Section 4.3.1 shows that this new definition is consistent with the previous one (4.6) if $f \in H_0^\infty(\Sigma_\varphi)$. Moreover, if A is invertible, then the definition is consistent with Section 4.3.1.

In this way we defined a closed operator $f(A)$ for each $f \in H^\infty(\Sigma_\varphi)$. We may extend the definition to an even larger class of holomorphic functions. Let $\Psi(\lambda) = \frac{\lambda}{(1+\lambda)^2}$ and let

$L(A) = (I + A)^2 A^{-1}$. Denote by $B(\Sigma_\varphi)$ the space of all holomorphic functions f on Σ_φ such that $\Psi^n f \in H_0^\infty(\Sigma_\varphi)$ for some $n \in \mathbb{N}$. Then we define the closed operator $f(A)$ by

$$f(A) = L(A)^n \big(\Psi^n f\big)(A). \tag{4.10}$$

This definition is consistent with the previous ones.

We call the functional calculus defined in Sections 4.3.1 or 4.3.2 the *elementary functional calculus*. The following properties justify this name.

4.3.3. *Properties of the functional calculus.* Let A be an injective sectorial operator. For $\lambda \in \mathbb{C} \setminus \Sigma_{\varphi_{\mathrm{sec}}(A)}$ one has

$$R(\lambda, A) = f_\lambda(A),$$

where $f_\lambda(z) = \frac{1}{z-\lambda}$. The set $H(A) := \{f \in B(\Sigma_\varphi)\colon\ f(A) \in \mathcal{L}(X)\}$ is a subalgebra of $B(\Sigma_\varphi)$ and

$$f \mapsto f(A)$$

is an algebra homomorphism [McI86,LeM98b].

4.3.4. *Strip-type operators.* Instead of sectors we consider here a horizontal strip

$$St_\omega := \big\{z \in \mathbb{C}\colon\ |\operatorname{Im} z| < \omega\big\}$$

where $\omega > 0$. We say that an operator B on X is of *strip type*, if there exists a strip St_ω such that

(a) $\sigma(B) \subset St_\omega$ and
(b) $\sup_{|\operatorname{Im}\lambda| \geqslant \omega} \|R(\lambda, B)\| < \infty$.

We denote by

$$\omega_{\mathrm{st}}(B) = \inf\big\{\omega > 0\colon\ \text{(a) and (b) hold}\big\}$$

the strip-type of B. Two kinds of examples are important: the case where $\mathrm{i}B$ generates a C_0-group and $B = \log A$ to which the two following subsections are devoted.

4.3.5. *Groups.* Let B be an operator such that $\mathrm{i}B$ generates a C_0-group U. Denote by

$$\omega_U := \inf\big\{\omega \in \mathbb{R}\colon\ \exists M \geqslant 0 \text{ such that } \big\|U(t)\big\| \leqslant M\mathrm{e}^{\omega|t|} \text{ for all } t \in \mathbb{R}\big\}$$

the *group type* of U. Since the resolvent of $\pm\mathrm{i}B$ is the Laplace transform of the semigroup $(U(\pm t))_{t \geqslant 0}$, it follows that B is a strip type operator and

$$\omega_{\mathrm{st}}(B) \leqslant \omega_U. \tag{4.11}$$

In general, it may happen that

$$\omega_{\mathrm{st}}(B) < \omega_U$$

(see [Wol81]). But on Hilbert space, strip type and group type coincide by the Gearhart–Prüss theorem; i.e.,

$$\omega_{\mathrm{st}}(B) = \omega_U \tag{4.12}$$

whenever $\mathrm{i}B$ generates a C_0-group U on a Hilbert space. Next we give another important example of a strip-type operator.

4.3.6. *The logarithm.* Let A be an injective sectorial operator. Let $\varphi > \varphi_{\mathrm{sec}}(A)$. Let $g(\lambda) = \frac{\lambda}{(1+\lambda)^2} \log \lambda$. Then $g \in H_0^\infty(\Sigma_\varphi)$. We define

$$\log A = (I + A)^2 A^{-1} g(A).$$

Thus $\log A$ is a closed operator by the same argument as in Section 4.3.2. We note some of the properties of the logarithm. Recall that A^{-1} is sectorial. One has

$$\log A^{-1} = -\log A. \tag{4.13}$$

THEOREM (Haase [Haa03a], Nollau [Nol69]). *Let A be a sectorial operator. Then* $\log A$ *is a strip-type operator and*

$$\omega_{\mathrm{st}}(\log A) = \varphi_{\mathrm{sec}}(A).$$

Now it may happen that $\mathrm{i} \log A$ generates a C_0-group. Then this group is given by the imaginary powers of A, which we consider in Section 4.4.

Further references: [Oka99,Oka00a,Oka00b].

4.4. *Fractional powers and BIP*

In the section we introduce fractional powers A^α of a sectorial operator A. We first consider the case where A is invertible which is simpler to present.

4.4.1. *Fractional powers of invertible operators.* Let A be an invertible sectorial operator. Let $\alpha \in \mathbb{C}$ such that $\operatorname{Re} \alpha > 0$. Then the function $f_\alpha(z) = z^{-\alpha}$ is in $H_1^\infty(\Sigma_\varphi)$ for all $0 < \varphi < \pi$. Thus we may define

$$A^{-\alpha} = f_\alpha(A) \in \mathcal{L}(X), \quad \operatorname{Re} \alpha > 0, \tag{4.14}$$

by (4.6). It follows that

$$A^{-\alpha} A^{-\beta} = A^{-(\alpha+\beta)}, \quad \operatorname{Re} \alpha, \operatorname{Re} \beta > 0. \tag{4.15}$$

In fact, $(A^{-\alpha})_{\mathrm{Re}\,\alpha>0}$ is a bounded holomorphic semigroup. Its generator is $-\log A$ (which was defined in Section 4.3.6). The operator A is densely defined if and only if $\log A$ is so (i.e., if and only if $(A^{-\alpha})_{\mathrm{Re}\,\alpha>0}$ is a C_0-semigroup).

A particular case of Section 4.4.1 occurs when $-A$ generates an exponentially stable C_0-semigroup T (i.e., $\|T(t)\| \leqslant M\mathrm{e}^{-\varepsilon t}$ for all $t \geqslant 0$ and some $M \geqslant 0$, $\varepsilon > 0$). Then A is invertible and $A^{-\alpha}$ can be expressed in terms of the semigroup instead of the resolvent as above, namely,

$$A^{-\alpha} = \frac{1}{\Gamma(\alpha)} \int_0^\infty t^{\alpha-1} T(t)\,\mathrm{d}t \tag{4.16}$$

for $\mathrm{Re}\,\alpha > 0$.

Next we consider a more general class of sectorial operators.

4.4.2. *Fractional powers of injective sectorial operators.* Let A be an injective sectorial operator and let $\pi > \varphi > \varphi_{\mathrm{sec}}(A)$. Let $\alpha \in \mathbb{C}$, $f_\alpha(z) = z^\alpha$, $z \in \Sigma_\varphi$. Then $f_\alpha \in B(\Sigma_\varphi)$. Thus

$$A^\alpha = f_\alpha(A), \quad \alpha \in \mathbb{C},$$

is defined according to Section 4.3.2. It is a closed injective operator and the following properties are valid:

$$A^{-\alpha} = \left(A^\alpha\right)^{-1} = \left(A^\alpha\right)^{-1}, \quad \alpha \in \mathbb{C}; \tag{4.17}$$

$$A^\alpha A^\beta \subset A^{\alpha+\beta}, \quad \alpha, \beta \in \mathbb{C}; \tag{4.18}$$

$$A^\alpha A^\beta = A^{\alpha+\beta}, \quad \mathrm{Re}\,\alpha, \mathrm{Re}\,\beta > 0. \tag{4.19}$$

If $\mathrm{Re}\,\alpha > 0$, then

$$D\big((A+\omega)^\alpha\big) = D\big(A^\alpha\big) \quad \text{for all } \omega \geqslant 0. \tag{4.20}$$

Moreover, for $0 < \alpha < \frac{\pi}{\varphi_{\mathrm{sec}}(A)}$, the operator A^α is sectorial and

$$\varphi_{\mathrm{sec}}\big(A^\alpha\big) = \alpha\varphi_{\mathrm{sec}}(A) \quad \text{and} \quad \log\big(A^\alpha\big) = \alpha \log A; \tag{4.21}$$

moreover,

$$\big(A^\alpha\big)^\beta = A^{\alpha\beta}, \quad \mathrm{Re}\,\beta > 0. \tag{4.22}$$

Formula (4.21) is interesting: it shows in particular that $-A^\alpha$ generates a bounded holomorphic semigroup if $0 < \alpha$ is small enough.

There is an enormous amount of literature on fractional powers and we refer in particular to the recent monograph [[MS01]].

Next we consider imaginary powers $A^{\mathrm{i}s}$ of an injective sectorial operator. They play an important role for regularity theory and also for interpolation theory.

4.4.3. *Characterization of BIP.* Let A be a sectorial injective operator on X. Then by Section 4.4.2 we can define the closed operators A^{is}, $s \in \mathbb{R}$, and also the closed operator $\log A$ (by Section 4.3.6).

THEOREM. *The following assertions are equivalent*:
(i) *$D(A) \cap R(A)$ is dense and $A^{is} \in \mathcal{L}(X)$ for all $s \in \mathbb{R}$.*
(ii) *The operator* $\mathrm{i}\log A$ *generates a C_0-group U.*
In that case $U(s) = A^{is}$, $s \in \mathbb{R}$.

DEFINITION. We say that A has *bounded imaginary powers* and write $A \in BIP$, if A is injective, sectorial and if the two equivalent conditions of Theorem are satisfied.

Note that this includes the hypothesis that $D(A) \cap R(A)$ be dense in X. Recall however, that $D(A) \cap R(A)$ is automatically dense in X if X is reflexive and A is a sectorial, injective operator. It is obvious that

$$A \in BIP \quad \text{if and only if} \quad A^{-1} \in BIP, \tag{4.23}$$

References: [Haa03a], [[Prü93]].

4.4.4. *BIP for invertible operators.* Property *BIP* can be formulated in a different way if the operator is invertible. Let A be sectorial of dense domain. Assume that $0 \in \rho(A)$. Then, for $\operatorname{Re}\alpha > 0$, the operator $A^{-\alpha} \in \mathcal{L}(X)$ was defined in Section 4.4.1, and we had seen that $(A^{-\alpha})_{\operatorname{Re}\alpha>0}$ is a holomorphic C_0-semigroup which is bounded on Σ_φ for each $0 < \varphi < \frac{\pi}{2}$. The generator of this semigroup is $-\log A$. Now it follows from Section 2.3 that $A \in BIP$ if and only if the holomorphic C_0-semigroup $(A^{-\alpha})_{\operatorname{Re}\alpha>0}$ has a trace. We formulate this as a theorem.

THEOREM. *Let A be a sectorial densely defined operator with $0 \in \rho(A)$. The following assertion are equivalent*:
(i) *$A \in BIP$*;
(ii) *there exists $c > 0$ such that $\|A^{-\alpha}\| \leqslant c$ whenever* $\operatorname{Re}\alpha > 0$, $|\alpha| \leqslant 1$.

4.4.5. *The BIP-type.* Let $A \in BIP$. Then we denote by

$$\varphi_{bip}(A) = \inf\left\{\omega \in \mathbb{R}\colon \exists M \geqslant 0 \text{ such that } \left\|A^{is}\right\| \leqslant M e^{\omega|s|} \text{ for all } s \in \mathbb{R}\right\}$$

the group-type of the C_0-group $(A^{is})_{s\in\mathbb{R}}$. We call $\varphi_{bip}(A)$ the *BIP-type* of A. One always has

$$\varphi_{\mathrm{sec}}(A) \leqslant \varphi_{bip}(A). \tag{4.24}$$

Recall from (4.10) that $\varphi_{\mathrm{sec}}(A) = \omega_{\mathrm{st}}(\log A)$. If X is a Hilbert space, then $\omega_{\mathrm{st}}(\log A) = \varphi_{bip}(A)$ (by (4.12)). Thus we obtain

$$\varphi_{bip}(A) = \varphi_{\mathrm{sec}}(A) \quad \text{on Hilbert space.} \tag{4.25}$$

This is McIntosh's result with a new proof due to Haase. The identity (4.25) is no longer true on Banach spaces. Recently, an example is given by Haase [Haa03a,Haa03b] where

$$\varphi_{\mathrm{sec}}(A) < \pi < \varphi_{bip}(A) \quad \text{on a UMD-space } X. \tag{4.26}$$

In fact, $X = L^p(\mathbb{R}, \omega_1\,\mathrm{d}x) \cap L^q(\mathbb{R}, \omega_2\,\mathrm{d}x)$ for $1 < p < 2 < q < \infty$ and ω_1, ω_2 strictly positive measurable functions. Another example of different sectorial and *BIP*-angle is given independently by Kalton [Kal03], where the operator has a bounded H^∞-calculus, in addition.

4.4.6. *An inverse theorem for BIP operators.* If $A \in BIP$, then $(A^{\mathrm{i}s})_{s\in\mathbb{R}}$ is a C_0-group. *Which groups occur in this manner*? There is a very satisfying answer if the underlying Banach space has some geometric properties. We refer to Section 6.1.3 for the definition of UMD-spaces. Here we just recall that each space L^p, $1 < p < \infty$, is a UMD-space. Let $(U(s))_{s\in\mathbb{R}}$ be a C_0-group with generator $\mathrm{i}B$. Then B is a strip-type operator. Recall the definition of $\omega_{\mathrm{st}}(B)$ from Section 4.3.4.

THEOREM (Monniaux). *Let* $\mathrm{i}B$ *be the generator of a* C_0*-group* U *on a UMD-space* X. *Assume that*

$$\omega_{\mathrm{st}}(B) < \pi.$$

Then there exists a unique sectorial operator A *such that* $A^{\mathrm{i}s} = U(s)$ *for all* $s \in \mathbb{R}$. *We call* A *the analytic generator of* U.

This theorem is due to Monniaux [Mon99] for the case where $\omega_U < \pi$ (the group-type of U). It was Haase [Haa03b] who observed that the weaker assumption $\omega_{\mathrm{st}}(B) < \pi$ suffices. Of course, in the situation of the theorem one has $B = \log A$. One may ask more generally which strip type operators B are of the form $\log A$ for some sectorial operator A. This seems to be unknown. It is known though, that the hypothesis on the Banach space cannot be omitted in the above theorem (see [Mon99] for an example). More precisely, on each Banach space X, to each C_0-group U one can associate the analytic generator A. If A is sectorial, then $A^{\mathrm{i}s} = U(s)$, $s \in \mathbb{R}$. But if the space X is not a UMD-space, then it may happen that $\rho(A) = \emptyset$. In particular, A is not sectorial. The notion of analytic generators is due to Cioranescu and Zsidó [CZ76] before the notion and properties of UMD-spaces were known. Their aim was to treat Tomita–Takesaki theory for von Neumann algebras.

4.4.7. *Bounded analytic generators.* We now describe when the C_0-group $(A^{\mathrm{i}s})_{s\in\mathbb{R}}$ is the boundary group of a holomorphic semigroup (see Section 2.3 for this notion).

PROPOSITION [Mon99]. *Let* U *be a* C_0*-group of type* $< \pi$ *on a UMD-space* X. *The following assertions are equivalent*:

(i) *The analytic generator* A *of* U *is bounded*;

(ii) *there exists a holomorphic* C_0*-semigroup* $(T(z))_{\mathrm{Re}\,z>0}$ *such that* $U(t)x = \lim_{r\downarrow 0} T(\mathrm{i}t + r)x$ *for all* $x \in X$.

In that case one has $A = T(1)$.

4.4.8. *The Dore–Venni theorem.* Next we establish the famous Dore–Venni theorem [DoVe87] on maximal regularity. It gives an answer to the question of closedness of $A + B$ which arose in Section 4.2. We refer to Section 6.1.3 for the definition and properties of UMD-spaces.

The following version of the Dore–Venni theorem is due to Prüss and Sohr [PS90].

THEOREM. *Let X be a UMD-space. Let $A, B \in BIP$ such that $\varphi_{bip}(A) + \varphi_{bip}(B) < \pi$. Assume that A and B commute. Then $A + B$ is closed. Moreover, $A + B \in BIP$ and*

$$\varphi_{bip}(A + B) \leqslant \max\{\varphi_{bip}(A), \varphi_{bip}(B)\}.$$

If one of the operators is invertible, it is possible to deduce this result from Monniaux' inverse theorem (Section 4.4.6). We sketch the proof.

PROOF [MON99]. Assume that $0 \in \rho(B)$. Then by Section 4.4.7 the group $(B^{-is})_{s\in\mathbb{R}}$ is the boundary of a holomorphic C_0-semigroup T and $B^{-1} = T(1)$. Let $W(t) = A^{it}B^{-it}$. By Monniaux' inverse theorem there exists a sectorial operator C such that $C^{it} = A^{it}B^{-it}$. It is not difficult to show that $C = AB^{-1}$ with domain $D(C) = \{x \in X\colon\ B^{-1}x \in D(A)\}$. In particular, $(AB^{-1} + I)$ is invertible. Let $y \in X$. We have to find $x \in D(A) + D(B)$ such that $Ax + Bx = y$; i.e., $(AB^{-1} + I)Bx = y$. Thus $x = B^{-1}(AB^{-1} + I)^{-1}y$ is a solution. □

4.4.9. *A noncommutative Dore–Venni theorem.* The assumption that A and B commute can be relaxed by imposing a growth condition on the commutator of the resolvent.

THEOREM (Monniaux and Prüss [MP97]). *Assume that X is a UMD-space. Let $A, B \in BIP$ with $\varphi_{bip}(A) + \varphi_{bip}(B) < \pi$. Assume that $0 \in \rho(A)$. Let $\psi_A > \varphi_{bip}(A)$, $\psi_B > \varphi_{bip}(B)$ be angles such that $\psi_A + \psi_B < \pi$. Assume that there exist constants $0 \leqslant \alpha < \beta < 1$ and $c \geqslant 0$ such that*

$$\left\|AR(\lambda, A)\left[A^{-1}R(\mu, B) - R(\mu, B)A^{-1}\right]\right\| \leqslant \frac{c}{(1 + |\lambda|^{1-\alpha})|\mu|^{1+\beta}}$$

for all $\lambda \in \Sigma_{\pi-\psi_A}$ and all $\mu \in \Sigma_{\pi-\psi_B}$. Then there exists $\omega \in \mathbb{R}$ such that $A + B + \omega$ is closed and sectorial.

This result can be applied to Volterra equations [MP97], to nonautonomous evolution equations [HM00a,HM00b] and also to prove maximal regularity of the Ornstein–Uhlenbeck operator [MPRS00].

4.4.10. *Interpolation spaces and BIP.* Let A be an injective sectorial operator. Then $D(A^\alpha)$ is a Banach space for the norm

$$\|x\|_{D(A^\alpha)} := \|x\| + \left\|A^\alpha x\right\|,$$

$0 < \alpha < 1$.

PROPOSITION ([[Tri95], p. 103], [Yag84], [[MS01], Theorem 11.6.1]). *Let $A \in BIP$. Then the complex interpolation space $[X, D(A)]_\alpha$ is isomorphic to $D(A^\alpha)$ for $0 < \alpha < 1$.*

Now assume in addition that A is invertible. Then Dore [Dor99a] (see also [Dor99b] for the noninvertible case) showed that the part of A in the real interpolation space $(X, D(A))_{\alpha,p}$ has a bounded H^∞-calculus whenever $0 < \alpha < 1$, $1 < p < \infty$. (See Section 4.5 for the definition of H^∞-calculus.) If X is a Hilbert space, then $(X, D(A))_{\alpha,2} = [X, D(A)]_\alpha$. Thus, if X is a Hilbert space, the part of A in $[X, D(A)]_\alpha$ has a bounded H^∞-calculus. Observe that the part of A in $D(A^\alpha)$ is similar to A. Thus, if $D(A^\alpha) = [X, D(A)]_\alpha$, then A has a bounded H^∞-calculus.

We have proved the converse of the above proposition and may formulate the following characterization on Hilbert spaces.

THEOREM. *Let A be a sectorial, invertible operator on a Hilbert space and let $0 < \alpha < 1$. The following assertions are equivalent:*

(i) $A \in BIP$;
(ii) $D(A^\alpha) = [X, D(A)]_\alpha$;
(iii) *the H^∞-calculus is bounded.*

For further results on Hilbert spaces we refer to Section 5 and to the paper by Auscher, McIntosh and Nahmrod [AMN97].

This interesting relation between *BIP* and interpolation spaces seemed to be the motivation of much research on operators with bounded imaginary powers in the seventies (see, e.g., Seeley's paper [See67]). Long time after this, the Dore–Venni theorem lead to a different direction, namely regularity theory.

4.5. *Bounded H^∞-calculus for sectorial operators*

Let A be an injective sectorial operator on a Banach space X and let $\pi \geqslant \varphi > \varphi_{\mathrm{sec}}(A)$. Then for $f \in H^\infty(\Sigma_\varphi)$ the closed operator $f(A)$ was defined in Section 4.3.

DEFINITION. We say that *A has a bounded $H^\infty(\Sigma_\varphi)$-calculus* if

$$f(A) \in \mathcal{L}(X) \quad \text{for all } f \in H^\infty(\Sigma_\varphi). \tag{4.27}$$

In that case, the mapping

$$f \mapsto f(A)$$

is a continuous algebra homomorphism from $H^\infty(\Sigma_\varphi)$ into $\mathcal{L}(X)$.

We note the following: Let $\varphi_{\mathrm{sec}}(A) < \varphi_1 < \varphi_2 \leqslant \pi$. If A has a bounded $H^\infty(\Sigma_{\varphi_1})$-calculus then also the $H^\infty(\Sigma_{\varphi_2})$-calculus is bounded. Thus, the weakest property in this context is to have a bounded $H^\infty(\Sigma_\pi)$-calculus for the cut plane $\Sigma_\pi = \mathbb{C} \setminus \{x \in \mathbb{R}\colon x < 0\}$.

DEFINITION. We write $A \in H^\infty$ to say that A is injective, sectorial and has a bounded $H^\infty(\Sigma_\pi)$-calculus. We let

$$\varphi_{H^\infty}(A) := \inf\{\sigma \in (\varphi_{\text{sec}}(A), \pi]:\ A \text{ has a bounded } H^\infty(\Sigma_\varphi)\text{-calculus}\}.$$

4.5.1. *Characterization via H_0^∞.* Let A be an injective sectorial operator with dense domain and dense range, and let $\varphi > \varphi_{\text{sec}}(A)$. Then A has a bounded $H^\infty(\Sigma_\varphi)$-calculus if and only if there exists $c > 0$ such that

$$\|f(A)\|_{\mathcal{L}(X)} \leqslant c\|f\|_{H^\infty(\Sigma_\varphi)} \tag{4.28}$$

for all $f \in H_0^\infty(\Sigma_\varphi)$. Note that $H_0^\infty(\Sigma_\varphi)$ is not norm-dense in $H^\infty(\Sigma_\varphi)$. Still, there is a canonical way to extend the calculus.

4.5.2. *Unbounded H^∞-calculus.* It is not an easy matter to decide whether a given operator has a bounded H^∞-calculus. Later we will give several criteria and examples. Here we show a way to construct counterexamples. Let X be a Banach space. A sequence $(e_n)_{n\in\mathbb{N}}$ is called a *Schauder basis* of X if, for each $x \in X$, there exists a unique sequence $(x_n)_{n\in\mathbb{N}} \subset \mathbb{C}$ such that

$$\sum_{n=1}^\infty x_n e_n = x. \tag{4.29}$$

The Schauder basis is called *unconditional* if for each $x = \sum_{n=1}^\infty x_n e_n \in X$ the sequence $\sum_{n=1}^\infty c_n x_n e_n$ converges in X for each $c = (c_n)_{n\in\mathbb{N}} \in \ell^\infty$. Otherwise we call $(e_n)_{n\in\mathbb{N}}$ a *conditional* Schauder basis. If X has an unconditional Schauder basis it also has a conditional Schauder basis [[Sin70], Chapter II, Theorem. 23.2]. In particular, each separable Hilbert space has a conditional Schauder basis. Thus the following example is in particular valid in a separable Hilbert space.

EXAMPLE (Unbounded $H^\infty(\Sigma_\pi)$-calculus). Let X be a Banach space with a conditional Schauder basis $(e_n)_{n\in\mathbb{N}}$. Define the operator A on X by

$$Ax = \sum_{n=1}^\infty 2^n x_n e_n \tag{4.30}$$

with domain the set of all $x = \sum_{n=1}^\infty x_n e_n \in X$ such that (4.30) converges. Then A is sectorial with $\varphi_{\text{sec}}(A) = 0$. However, A does not have a bounded $H^\infty(\Sigma_\pi)$-calculus.

For the proof it is helpful to consider more general diagonal operators. Denote by BV the space of all scalar sequences $\alpha = (\alpha_n)_{n\in\mathbb{N}}$ such that

$$\|\alpha\|_{BV} = |\alpha_1| + \sum_{n=1}^\infty |\alpha_{n+1} - \alpha_n| < \infty.$$

Then BV is a Banach space (isomorphic to ℓ^1). If $\sum_{n=1}^{\infty} y_n$ converges in X, then $\sum_{n=1}^{\infty} \alpha_n y_n$ converges for each $\alpha \in BV$. This follows from Abel's partial summation rule

$$\sum_{n=1}^{m} \alpha_n y_n = \sum_{n=1}^{m} \alpha_n (s_{n+1} - s_n) = \sum_{n=2}^{m} s_n(\alpha_{n-1} - \alpha_n) + s_{m+1}\alpha_m - s_1\alpha_1,$$

where $s_1 = 0$, $s_n = \sum_{k=1}^{n-1} y_k$ for $n \geqslant 2$. Thus, if $\alpha \in BV$, then

$$D_\alpha x = \sum_{n=1}^{\infty} \alpha_n x_n e_n \tag{4.31}$$

for $x = \sum_{n=1}^{\infty} x_n e_n$ defines an operator $D_\alpha \in \mathcal{L}(X)$. Moreover,

$$\|D_\alpha\| \leqslant c\|\alpha\|_{BV}, \quad \alpha \in BV, \tag{4.32}$$

by the closed graph theorem (or a direct estimate). Next we consider unbounded diagonal operators. Let $(\alpha_n)_{n\in\mathbb{N}} \subset \mathbb{C}$. Then

$$D_\alpha x = \sum_{n=1}^{\infty} \alpha_n x_n e_n \tag{4.33}$$

with $D(D_\alpha) := \{x = \sum_{n=1}^{\infty} x_n e_n\colon$ (4.33) converges$\}$ defines a closed operator on X. This is obvious since the coordinate functionals

$$\sum_{n=1}^{\infty} x_n e_n \mapsto x_m$$

are continuous.

PROPOSITION. *Let* $0 < \alpha_1 < \alpha_n < \alpha_{n+1}$ *such that* $\lim_{n\to\infty} \alpha_n = \infty$. *Then* D_α *is a sectorial operator of type* $\varphi_{\mathrm{sec}}(D_\alpha) = 0$. *Moreover,* $0 \in \rho(D_\alpha)$ *and* D_α *has compact resolvent.*

PROOF. Let $0 < \varphi < \pi$. Let $\pi \geqslant |\theta| > \varphi$, $\lambda = re^{\mathrm{i}\theta}$, $r > 0$. Consider the sequence $\beta_n = \frac{1}{re^{\mathrm{i}\theta} - \alpha_n}$. Then

$$\beta_n - \beta_{n+1} = \frac{1}{r}\left\{\frac{1}{\mathrm{e}^{\mathrm{i}\theta} - \alpha_n/r} - \frac{1}{\mathrm{e}^{\mathrm{i}\theta} - \alpha_{n+1}/r}\right\},$$

$$\begin{aligned}\sum_{n=1}^{\infty} |\beta_n - \beta_{n+1}| &\leqslant \frac{1}{r}\sum_{n=1}^{\infty}\int_{\alpha_n/r}^{\alpha_{n+1}/r} \frac{1}{|\mathrm{e}^{\mathrm{i}\theta} - x|^2}\,\mathrm{d}x \\ &\leqslant \frac{1}{r}\int_0^{\infty} \frac{1}{|\mathrm{e}^{\mathrm{i}\theta} - x|^2}\,\mathrm{d}x = \frac{c_\varphi}{r},\end{aligned}$$

where c_φ does not depend on θ or r. Thus $\|D_\beta\| \leqslant \frac{c}{r}$. It is easy to see that $D_\beta = (\lambda - D_\alpha)^{-1}$. This shows that D_α is sectorial of type 0. We omit the proof of compactness of D_β (see [LeM00]). □

Now let A be the operator of the example; i.e., $A = D_\alpha$ with $\alpha_n = 2^n$. In order to show that A has no bounded $H^\infty(\Sigma_\pi)$-calculus we use the following (see [[Gar81], Theorem 1.1. on pp. 287 and 284 ff]).

INTERPOLATION LEMMA. *Let $b = (b_n)_{n\in\mathbb{Z}} \in \ell^\infty(\mathbb{Z})$. Then there exists $f \in H^\infty(\Sigma_\pi)$ such that*

$$f(2^n) = b_n \quad \textit{for all } n \in \mathbb{Z}.$$

Let $f \in H^\infty(\Sigma_\pi)$. It is easy to see from the definition that $e_n \in D(f(A))$ and

$$f(A)e_n = f(2^n)e_n.$$

Now assume that $f(A) \in \mathcal{L}(X)$. Then it follows that

$$\begin{aligned}\lim_{m\to\infty} \sum_{n=1}^{m} f(2^n)x_n e_n &= \lim_{m\to\infty} f(A)\sum_{n=1}^{m} x_n e_n \\ &= f(A)x\end{aligned}$$

converges for all $x \in X$. Thus, by the Interpolation Lemma, $\sum_{n=1}^{\infty} b_n x_n e_n$ converges for all $b \in \ell^\infty$ and all $x \in X$. This contradicts the fact that the basis is conditional.

A first example of this kind had been given by Baillon and Clément [BC91]. Further developments are contained in [Ven93], [LeM00] and [Lan98].

4.5.3. *BIP versus bounded H^∞-calculus.* It is clear from the definition that $A \in H^\infty$ implies $A \in BIP$ whenever A is a sectorial, injective operator. It is remarkable that in that case

$$\varphi_{bip}(A) = \varphi_{H^\infty}(A) \tag{4.34}$$

(see [CDMY96]). Conversely, if X is a Hilbert space, then *BIP* implies boundedness of the H^∞-calculus by Section 4.4.10. This is not true on L^p-spaces for $p \neq 2$ as the following example shows.

EXAMPLE [Lan98]. Let $1 < p < \infty$, $p \neq 2$. There exists a sectorial injective operator $A \in BIP$ on L^p which does not have a bounded $H^\infty(\Sigma_\pi)$-calculus. We consider the space $L^p_{2\pi}$ of all scalar-valued 2π-periodic measurable functions on $\mathbb{R}$ such that

$\|f\|_p := \frac{1}{2\pi}(\int_0^{2\pi} |f(t)|^p \,\mathrm{d}t)^{1/p} < \infty$. By e_n we denote the nth trigonometric polynomial $e_n(t) = \mathrm{e}^{\mathrm{i}nt}$, $n \in \mathbb{Z}$. For $f \in L^p_{2\pi}$, we denote by

$$\hat{f}(n) := \frac{1}{2\pi}\int_0^{2\pi} f(t)\mathrm{e}^{\mathrm{i}nt}\,\mathrm{d}t$$

the nth Fourier coefficient. Then $\{e_n: n \in \mathbb{Z}\}$ is a Schauder basis of $L^p_{2\pi}$; i.e.,

$$f = \lim_{n\to\infty}\sum_{n=-m}^{m}\hat{f}(n)e_n =: \sum_{n=-\infty}^{+\infty}\hat{f}(n)e_n$$

for all $f \in L^p$. Moreover, the *Riesz-projection*

$$R: f \mapsto \sum_{n=0}^{\infty}\hat{f}(n)e_n$$

is bounded on $L^p_{2\pi}$. Consider the operator A on $L^p_{2\pi}$ given by

$$Af = \sum_{n=-\infty}^{+\infty} 2^n\hat{f}(n)e_n \tag{4.35}$$

with domain $D(A) = \{f \in L^p_{2\pi}\colon (4.35) \text{ converges}\}$. Since the Riesz-projection is bounded we can write A as the direct sum of $A_1 \oplus A_2$, where

$$A_1 f = \sum_{n=0}^{\infty} 2^n\hat{f}(n)e_n,$$

$$A_2 f = \sum_{n=1}^{\infty} 2^{-n}\hat{f}(-n)e_{-n}$$

with appropriate domains. The first is injective and sectorial by Section 4.5.2 the second is the inverse of an operator of the Proposition in Section 4.5.2. Thus A is sectorial and injective. By the same argument as in Section 4.5.2 one sees that A has no bounded $H^\infty(\Sigma_\pi)$-calculus if $p \neq 2$. Now for $a \in \mathbb{R}$ consider the shift operator L_a given by $(L_a u)(t) = u(t+a)$ for $u \in L^p_{2\pi}$. Then $L_a \in \mathcal{L}(L^p_{2\pi})$ and $\|L_a\| = 1$. Moreover, $(L_a u)^\wedge(n) = \mathrm{e}^{\mathrm{i}na}\hat{u}(n)$. Let $a = s\log 2$. Then it follows that

$$A^{\mathrm{i}s} = L_a.$$

Thus $A \in BIP$ and $\|A^{\mathrm{i}s}\|_{\mathcal{L}(L^p_{2\pi})} = 1$, $1 < p < \infty$.

4.6. *Perturbation*

Let A be a sectorial operator on a Banach space X. We consider perturbations $A+B$ where $B: D(A) \to X$ is linear. We say that B is *A-bounded* if B is continuous with respect to the graph norm on $D(A)$; i.e., if there exist constants $a \geqslant 0$, $b \geqslant 0$ such that

$$\|Bx\| \leqslant a\|Ax\| + b\|x\|, \quad x \in D(A). \tag{4.36}$$

The infimum over all $a > 0$ such that there exists $b > 0$ such that (4.36) holds is called the *A-bound* of B. A *small perturbation* of A is a linear mapping $B: D(A) \to X$ with A-bound 0. This is equivalent to $\lim_{\lambda\to\infty} \|BR(\lambda, A)\| = 0$ as is easy to see. Here we consider the following two stronger conditions.

$$\|BR(\lambda, A)\| \leqslant \frac{c}{\lambda^\beta}, \quad \lambda \geqslant 1; \tag{4.37}$$

$$\|R(\lambda, A)Bx\| \leqslant \frac{c}{\lambda^\beta}\|x\|, \quad \lambda \geqslant 1, x \in D(A). \tag{4.38}$$

THEOREM. *Assume that $B: D(A) \to X$ is linear such that* (4.37) *or* (4.38) *is satisfied for some $\beta > 0$, $c \geqslant 0$. Assume that A and $A + B$ are invertible and sectorial. Then the following holds*:

(a) $A \in H^\infty \Rightarrow A + B \in H^\infty$;
(b) $A \in BIP \Rightarrow A + B \in BIP$.

We refer to [AHS94] or [ABH01] for the proof. Given a sectorial operator A there exists a constant $\varepsilon_A > 0$ such that $A + B + \omega$ is sectorial for some $\omega \in \mathbb{R}$ whenever B has an A-bound smaller than ε_A. However, the properties H^∞ and *BIP* are not preserved.

COUNTEREXAMPLE [MY90]. There exists an invertible operator $A \in H^\infty$ on a Hilbert space X such that, for each $\varepsilon > 0$, there exists an operator $B_\varepsilon : D(A) \to X$ with A-bound less than $\varepsilon > 0$ such that $A + B_\varepsilon + \omega \notin BIP$ for any $\omega \in \mathbb{R}$.

In view of the counterexample one would not expect that the properties H^∞ or *BIP* are preserved under small perturbations. This seems to be unknown, though.

4.7. *Groups and positive contraction semigroups*

In this section we give two classes of examples of operators with bounded H^∞-calculus.

4.7.1. *Groups.* Generators of bounded C_0-groups and their squares yield an interesting class of examples.

THEOREM [HP98]. *Let B be the generator of a bounded C_0-group on a UMD-space X. Then*

(a) *B has a bounded* $H^\infty(\Sigma_\varphi)$*-calculus for each* $\varphi > \pi/2$;
(b) $-B^2$ *has a bounded* $H^\infty(\Sigma_\varphi)$*-calculus for each* $\varphi > 0$.

We give an example.

4.7.2. *The abstract Laplace operator.* Let U_j, $j = 1, \dots, n$, be n commuting C_0-groups on a UMD-space X with generators B_j, $j = 1, \dots, n$. By Section 4.7.1 the operators $-B_j^2$ have an $H^\infty(\Sigma_\varphi)$-calculus for all $\varphi > 0$, $j = 1, \dots, n$. It follows from the Dore–Venni theorem (Section 4.4.8) that the operator $A := B_1^2 + \cdots + B_n^2$ is closed with domain

$$D(A) = \bigcap_{j=1}^{n} D(B_j^2).$$

Moreover, $-A$ has a bounded $H^\infty(\Sigma_\varphi)$-calculus for all $\varphi > 0$. We may consider A as an abstract Laplace operator. Indeed, a concrete example is given by $(U_j(t)f)(x) = f(x + te_j)$ on $L^p(\mathbb{R}^n)$, $1 < p < \infty$, where e_j denotes the jth unit vector in $\mathbb{R}^n$. Thus $Af = \Delta f$ with $D(A) = W^{2,p}(\mathbb{R}^n)$. For a more general noncommutative setting on Lie groups see ter Elst and Robinson [tER96a].

4.7.3. *Positive contraction semigroups on* L^p. Next we consider positive semigroups. The following result is due to Duong [Duo89] after a more special result was proved by Cowling [Cow83]. Vector-valued versions are given in [HP98].

THEOREM. *Let* $-A$ *be the generator of a positive, contractive* C_0*-semigroup on* $L^p(\Omega)$, $1 < p < \infty$, *where* (Ω, Σ, μ) *is a measure space. Then A has a bounded* $H^\infty(\Sigma_\varphi)$*-calculus for each* $\varphi > \pi/2$.

This result can be extended to semigroups on L^p which are dominated by a positive contraction semigroup. The proof can be reduced to the group case (Section 4.7.1) by an interesting dilation theorem due to Fendler extending the Akcoglu dilation theorem for single operators to semigroups. We give more details.

4.7.4. *r-contractive semigroups on* L^p. Let S be an operator on $L^p(\Omega)$, $1 \leqslant p \leqslant \infty$. We say that S is *regular*, if S is a linear combination of positive operators; or equivalently, if S is dominated by a positive operator T on $L^p(\Omega)$, i.e.,

$$|Sf| \leqslant T|f|, \quad f \in L^p(\Omega),$$

where $|f|(x) = |f(x)|$, $x \in \Omega$. In that case there exists a smallest positive operator $|S|$ which dominates S. We call $\|S\|_r := \||S|\|$ the r-norm of S. If $p = 1$ or ∞, then every bounded operator S on $L^p(\Omega)$ is regular and $\|S\| = \|S\|_r$, but if $1 < p < \infty$ this is false (if $L^p(\Omega)$ has infinite dimension). We refer to [[Sch74], Chapter IV].

THEOREM. *Let* $1 < p < \infty$. *Let* T *be a* C_0*-semigroup on* $L^p(\Omega)$ *such that* $\|T(t)\|_r \leqslant 1$ *for all* $t \geqslant 0$. *Denote by* $-A$ *the generator of* T. *Then A has a bounded* $H^\infty(\Sigma_\varphi)$*-calculus for each* $\varphi > \pi/2$.

PROOF [LeM98b]. By Fendler's dilation theorem [Fen97] there exist a space $L^p(\widehat{\Omega})$, contractions $j: L^p(\Omega) \to L^p(\widehat{\Omega})$, $P: L^p(\widehat{\Omega}) \to L^p(\Omega)$ and a contraction C_0-group U on $L^p(\widehat{\Omega})$ such that $T(t) = PU(t)j$ for all $t \geqslant 0$. Let $-\widehat{A}$ be the generator of U. By Section 4.7.1 the $H^\infty(\Sigma_\varphi)$-calculus for $\widehat{A}$ is bounded ($\varphi > \pi/2$). It is obvious that this carries over to the $H^\infty(\Sigma_\varphi)$-calculus of A, cf. Section 5.2.2. □

4.7.5. *Holomorphic positive contraction semigroups.* The angle obtained in Section 4.7.3 can be improved if the semigroup is bounded holomorphic. In fact, the following result due to Kalton and Weis [KW01], Corollary 5.2, is proved with the help of R-boundedness techniques which will be presented in Section 6.

THEOREM. *Let $1 < p < \infty$. Let $-A$ be the generator of a bounded holomorphic C_0-semigroup T on $L^p(\Omega)$. Assume that*

$$T(t) \geqslant 0, \quad \|T(t)\|_{\mathcal{L}(L^p(\Omega))} \leqslant 1 \quad \textit{for all } t \geqslant 0.$$

Then A has a bounded H^∞-calculus and $\varphi_{H^\infty}(A) < \pi/2$.

5. Form methods and functional calculus

On Hilbert space generators of contraction C_0-semigroups are characterized by m-accretivity. This most convenient criterion also implies boundedness of the H^∞-calculus, as an easy consequence of the Spectral Theorem after a dilation to a unitary group. If the semigroup is holomorphic, then conversely, a bounded H^∞-calculus also implies m-accretivity after rescaling and changing the scalar product. In addition, the generator is associated with a closed form. The interesting interplay of forms, m-accretivity and H^∞-calculus on Hilbert space, this is the subject of the present section.

5.1. *Bounded H^∞-calculus on Hilbert space*

Things are much simpler on Hilbert space than on general Banach spaces. The *BIP* property is equivalent to bounded H^∞-calculus and all angles can be chosen optimal as was shown in Section 4.4.10, (4.26) and (4.34). We reformulate this more formally.

THEOREM (McIntosh). *Let A be an injective, sectorial operator on a Hilbert space. The following are equivalent*:

(i) $A \in BIP$;
(ii) *for all $\varphi > \varphi_{\sec}(A)$, the $H^\infty(\Sigma_\varphi)$-calculus is bounded*;
(iii) *the $H^\infty(\Sigma_\pi)$-calculus is bounded.*

In that case $\varphi_{bip}(A) = \varphi_{\sec}(A)$.

This optimal situation in Hilbert space contrasts the L^p-case, $p \neq 2$, where we had seen that property *BIP* does not imply boundedness of the $H^\infty(\Sigma_\pi)$-calculus (see Section 4.5.3). Also had we seen that it can happen that $\varphi_{\sec}(A) < \pi < \varphi_{bip}(A)$ on L^p, $p \neq 2$.

Also the Dore–Venni theorem needs fewer hypotheses on Hilbert space.

THEOREM (Dore–Venni on Hilbert space [DoVe87]). *Let A and B be two commuting injective, sectorial operators on a Hilbert space such that $\varphi_{\mathrm{sec}}(A) + \varphi_{\mathrm{sec}}(B) < \pi$. Assume that $A \in BIP$. Then $A + B$ is closed.*

We mention that this result does not hold on L^p for $p \neq 2$ ([Lan98], Theorem 2.3).

5.2. *m-accretive operators on Hilbert space*

The simplest unbounded operators are multiplication operators. Let (Ω, Σ, μ) be a measure space and let $m : \Omega \to \mathbb{R}$ be a measurable function. Define A_m on $L^2(\Omega)$ by $A_m u = mu$ with domain

$$D(A_m) = \{u \in L^2(\Omega) \colon m \cdot u \in L^2(\Omega)\}.$$

For the operator A_m one has the best possible spectral calculus. For each $f \in L^\infty(\mathbb{R})$, we may define $f(A) = A_{f \circ m} \in \mathcal{L}(L^2(\Omega))$. It is not difficult to see that this definition is consistent with our previous one whenever $f \in H^\infty(\Sigma_\varphi)$ for some $\varphi > \pi/2$.

5.2.1. *The Spectral theorem.* The spectral theorem says that each self-adjoint operator is unitarily equivalent to a multiplication operator.

SPECTRAL THEOREM. *Let A be a self-adjoint operator on H. Then there exists a measure space (Ω, Σ, μ), a measurable function $m : \Omega \to \mathbb{R}$ and a unitary operator $U : H \to L^2(\Omega)$ such that*

(a) $UD(A) = D(A_m)$;

(b) $A_m Ux = UAx$, $x \in D(A)$.

Thus, multiplication operators and self-adjoint operators are just the same thing.

Now recall that an operator A generates a *unitary group* (i.e., a C_0-group of unitary operators) if and only if $\mathrm{i}A$ is self-adjoint. Thus $\mathrm{i}A$ is equivalent to a multiplication operator. This shows in particular that each generator A of a unitary group on H has a bounded $H^\infty(\Sigma_\varphi)$-calculus for all $\varphi > \pi/2$. In fact, it is easy to show that for $f \in H^\infty(\Sigma_\varphi)$ one has

$$f(A) = U^{-1} A_{f \circ m} U.$$

5.2.2. *Bounded H^∞-calculus for m-accretive operators.* Recall that an operator A on a Hilbert space H is called *m-accretive* if

(a) $\mathrm{Re}(Ax|x) \geqslant 0$ for all $x \in D(A)$;

(b) $I + A$ is surjective.

The Lumer–Phillips theorem asserts that an operator A is m-accretive if and only if $-A$ generates a contractive C_0-semigroup. In particular, if A is m-accretive, then A is sectorial and $\varphi_{\mathrm{sec}}(A) \leqslant \frac{\pi}{2}$.

THEOREM. *Let A be an m-accretive operator on H. Then A has a bounded $H^\infty(\Sigma_\varphi)$-calculus for each $\varphi > \varphi_{\mathrm{sec}}(A)$.*

PROOF. Denote by T the semigroup generated by $-A$. By the dilation theorem [[Dav80], p. 157] there exist a Hilbert space $\widehat{H}$ containing H as a closed subspace and a unitary group U on $\widehat{H}$ such that

$$P \circ U(t) \circ i = T(t), \quad t \geqslant 0,$$

where $i: H \to \widehat{H}$ is the injection of H into $\widehat{H}$ and $P: \widehat{H} \to H$ is the orthogonal projection. Denote by $\widehat{A}$ the generator of U. Then $\mathrm{i}\widehat{A}$ is selfadjoint. Thus $\widehat{A}$ has a bounded $H^\infty(\Sigma_\varphi)$-calculus for any $\varphi > \pi/2$. It is easy to see from the definitions that

$$f(A) = P \circ f(\widehat{A}) \circ i.$$

Thus $f(A) \in \mathcal{L}(H)$ for all $f \in H^\infty(\Sigma_\varphi)$ (and even $\|f(A)\| \leqslant \|f(\widehat{A})\| \leqslant \|f\|_{L^\infty(\Sigma_{\pi/2})}$). Thus A has a bounded $H^\infty(\Sigma_\varphi)$-calculus for all $\varphi > \frac{\pi}{2}$. Now by McIntosh's result in Section 5.1 the $H^\infty(\Sigma_\varphi)$-calculus is also bounded for $\varphi > \varphi_{\mathrm{sec}}(A)$. □

5.2.3. *Equivalence of bounded H^∞-calculus and m-accretivity.* Let $(\cdot|\cdot)_1$ be an equivalent scalar product on H, i.e., there exist $\alpha > 0$, $\beta > 0$ such that

$$\alpha\|u\|_H^2 \leqslant (u|u)_1 \leqslant \beta\|u\|_H^2$$

for all $u \in H$. Of course, having a bounded H^∞-calculus is independent of the equivalent norm on H we choose. Thus, if A is a sectorial operator on H which is m-accretive with respect to some equivalent scalar product, then $A \in H^\infty$. This describes already the class of all operators A with bounded H^∞-calculus.

THEOREM (Le Merdy [LeM98a]). *Let A be an injective, sectorial operator on H such that $\varphi_{\mathrm{sec}}(A) < \pi/2$. The following are equivalent*:

(i) $A \in H^\infty$,

(ii) *there exists an equivalent scalar product $(\cdot|\cdot)_1$ on H such that* $\mathrm{Re}(Au|u)_1 \geqslant 0$ *for all $u \in D(A)$*;

(iii) *let $0 \leqslant \psi < \varphi_{\mathrm{sec}}(A)$; then there exists an equivalent scalar product $(\cdot|\cdot)_1$ on H such that $(Au|u) \in \Sigma_\psi$ for all $u \in D(A)$.*

Denote by T the C_0-semigroup generated by $-A$. Then (ii) means that

$$\|T(t)u\|_1 \leqslant \|u\|_1 := \sqrt{(u|u)_1}$$

for all $u \in H$, $t \geqslant 0$, and (iii) means that

$$\|T(z)u\|_1 \leqslant \|u\|_1$$

for all $u \in H$ and $z \in \Sigma_{\pi/2-\varphi}$.

The proof of Le Merdy [LeM98a] uses the theory of completely bounded operators and a deep theorem of Paulsen. A more direct proof, based directly on "*quadratic estimates*", is given by Haase [Haa02].

The theorem gives a very satisfying characterization of the boundedness of the H^∞-calculus; see also Section 5.3 for a description by forms. However, it is restricted to sectorial operators of sectorial type smaller than $\pi/2$:

5.2.4. *Counterexample.* There exists an invertible, sectorial operator A on a Hilbert space H such that $-A$ generates a bounded C_0-semigroup T on H. Moreover, A has a bounded $H^\infty(\Sigma_\varphi)$-calculus for any $\varphi > \pi/2$. But for all $\omega > 0$, $A + \omega$ is not accretive with respect to any equivalent scalar product.

The counterexample is due to Le Merdy [LeM98a] (based on a famous example of Pisier solving the Halmos problem) with an additional argument by Haase [Haa02] (taking care of the case where $\omega > 0$).

5.3. *Form methods*

A most efficient way to define generators of holomorphic semigroups on a Hilbert space H is the *form method* (sometimes it is also called "*variational method*").

Let H be a Hilbert space. If V is another Hilbert space we write $V \hookrightarrow H$ if V is a subspace of H such that the embedding is continuous. We write $V \hookrightarrow_d H$ if in addition V is dense in H.

DEFINITION. A *closed form* on H is a sesquilinear form $a : V \times V \to \mathbb{C}$ which is *continuous*, i.e.,

$$|a(u, v)| \leqslant M\|u\|_V \|v\|_V, \quad u, v \in V, \tag{5.1}$$

for some $M \geqslant 0$ and *elliptic*, i.e.,

$$\operatorname{Re} a(u, u) + \omega\|u\|_H^2 \geqslant \alpha\|u\|_V^2, \quad u \in V, \tag{5.2}$$

for some $\omega \in \mathbb{R}$, $\alpha > 0$. The space V is called the *domain* of a.

Properly speaking, a closed form is a couple (a, V) with the properties stipulated in the above definition. Note that by (5.2),

$$\|u\|_a := \left(\operatorname{Re} a(u, u) + \omega\|u\|_H^2\right)^{1/2}$$

defines an equivalent norm on V. Sometimes we will write $D(a)$ instead of V and will use the norm $\|\cdot\|_a$ on $D(a)$ for which $D(a)$ is complete by hypothesis.

REMARK. Let $a: D(a) \times D(a) \to \mathbb{C}$ be sesquilinear, where $D(a)$ is a subspace of H. Assume that a is *bounded below*, i.e.,

$$\|u\|_a^2 := \operatorname{Re} a(u,u) + \omega\|u\|_H^2 \geqslant \|u\|_H^2$$

for all $u \in D(a)$ and some $\omega \in \mathbb{R}$. Then $\|\cdot\|_a$ defines a norm on $D(a)$. The form is continuous with respect to this norm if and only if $a(u,u) \in \omega_1 + \Sigma_\theta$ for some $\omega_1 \in \mathbb{R}$ and some $\theta \in (0, \pi/2]$, i.e., if and only if the form a is *sectorial* in the terminology of Kato [[Kat66]].

5.3.1. *The operator associated with a closed form.* Let a be a closed form on H with dense domain V. Then we associate an operator A with a, given by

$$D(A) = \{u \in V \colon \exists v \in H \text{ such that } a(u,\varphi) = (v|\varphi)_H \text{ for all } \varphi \in V\},$$
$$Au = v.$$

Note that v is well defined since V is dense in H. We call A *the operator associated with* a. It is not difficult to see that $A + \omega$ is sectorial with angle $\varphi_{\text{sec}}(A+\omega) < \pi/2$. Moreover, $A + \omega$ is m-accretive (where ω is the constant occurring in (5.2)). Thus we obtain the following.

THEOREM. *Let A be an operator associated with a closed form. Then there exists $\omega \in \mathbb{R}$ such that $A + \omega \in H^\infty$.*

5.3.2. *Example*: *self-adjoint operators.* Let a be a densely defined closed form on H with domain V. Assume that a is symmetric; i.e., $a(u,v) = \overline{a(v,u)}$ for all $u, v \in V$. Let A be the operator associated with a. Then A is *bounded below* (i.e., $(Au|u) \geqslant -\omega\|u\|_H^2$ for all $u \in D(A)$ and some $\omega \in \mathbb{R}$) and self-adjoint. Conversely, let A be a self-adjoint operator which is bounded below. Then A is associated with a densely defined symmetric closed form.

We give a classical concrete example right now. Later we will study elliptic operators in more detail.

5.3.3. *Example* (*The Laplacian with Dirichlet and with Neumann boundary conditions*). Let $\Omega \subset \mathbb{R}^n$ be open and $H = L^2(\Omega)$. Let $W^{1,2}(\Omega) := \{u \in L^2(\Omega)\colon D_j u \in L^2(\Omega), j = 1, \dots, n\}$ be the first Sobolev space and let $W_0^{1,2}(\Omega)$ be the closure of the space $\mathcal{D}(\Omega)$ of all test functions in $W^{1,2}(\Omega)$. Define $a: W^{1,2}(\Omega) \times W^{1,2}(\Omega) \to \mathbb{C}$ by

$$a(u,v) = \int_\Omega \nabla u \overline{\nabla v}\, dx.$$

(a) Let $V = W_0^{1,2}(\Omega)$. Then a is a closed symmetric form. Denote by $-\Delta_\Omega^D$ the operator associated with a. Then it is easy to see that $D(\Delta_\Omega^D) = \{u \in W_0^{1,2}(\Omega)\colon \Delta u \in L^2(\Omega)\}$, $\Delta_\Omega^D u = \Delta u$, in $\mathcal{D}(\Omega)'$. We call Δ_Ω^D the *Dirichlet Laplacian* on $L^2(\Omega)$.

(b) Let $V = W^{1,2}(\Omega)$. Then a is closed and symmetric. Denote by $-\Delta_\Omega^N$ the operator associated with a. Then Δ_Ω^N is a self-adjoint operator, called the *Neumann Laplacian*. This can be justified if Ω is bounded with Lipschitz boundary. Then $D(\Delta_\Omega^N) = \{u \in W^{1,2}(\Omega)\colon\ \Delta u \in L^2(\Omega);\ \frac{\partial u}{\partial \nu} = 0 \text{ weakly}\}$, where the weak sense has to be explained. We omit the details. And indeed, for many purposes it is not important to know in which sense the Neumann boundary condition $\frac{\partial u}{\partial \nu} = 0$ is realized. The interesting point is that the Neumann–Laplacian can be defined for arbitrary open sets without using the normal derivative.

5.3.4. *Characterization of operators defined by forms.* Let A be an operator on H. We say that A *is induced by a form*, if there exists a densely defined closed form a on H such that A is associated with a. It is easy to describe such operators by an accretivity-condition.

THEOREM [[Kat66]]. *Let A be an operator on H. The following are equivalent:*

(i) *A is induced by a form;*

(ii) *there exist $\omega \in \mathbb{R}$ and $0 < \theta < \frac{\pi}{2}$ such that*

(a) $(\omega + A)D(A) = H$;

(b) $e^{\pm i\theta}(\omega + A)$ *is accretive.*

(iii) *$-A$ generates a holomorphic C_0-semigroup T of angle $\theta \in (0, \frac{\pi}{2})$ such that for some $\omega \in \mathbb{R}$,*

$$\|T(z)\| \leqslant e^{\omega|z|}, \quad z \in \Sigma_\theta;$$

(iv) *the numerical range $W(A) := \{(Ax|x)\colon\ x \in D(A), \|x\| = 1\}$ is contained in $\Sigma_\theta + \omega$ for some $\theta \in (0, \pi/2)$, $\omega \in \mathbb{R}$ and $(-\infty, \omega) \cap \rho(A) \neq \emptyset$.*

Property (ii) is sometimes formulated by saying that A is *strongly quasi-m-accretive* (and *strongly m-accretive* if $\omega = 0$).

5.3.5. *Changing scalar products.* Let $a\colon V \times V \to \mathbb{C}$ be a closed densely defined form on H. The definition of the associated operator A depends on the scalar product on H. Let $(\cdot|\cdot)_1$ be another equivalent scalar product on H. Then there exists a self-adjoint operator $Q \in \mathcal{L}(H)$ such that

$$(Qu|v)_1 = (u|v) \quad \text{for all } u, v \in H,$$

and Q is *strictly form-positive*. By this we mean that $(Qu|u) \geqslant \varepsilon\|u\|_H^2$ for all $u \in H$ and some $\varepsilon > 0$. Let $u \in D(A)$. Then $u \in V$ and $a(u, \varphi) = (Au|\varphi) = (QAu|\varphi)_1$ for all $\varphi \in V$. This shows that the operator A_1 associated with a on $(H, (\cdot|\cdot)_1)$ is given by $A_1 = QA$. With the help of Le Merdy's theorem in Section 5.2.3 one obtains the following characterization of the class H^∞.

THEOREM [ABH01]. *Let A be an operator on H. The following are equivalent:*

(i) *$A + \omega \in H^\infty$ for some $\omega \geqslant 0$;*

(ii) *there exists an equivalent scalar product* $(\cdot|\cdot)_1$ *on* H *such that* A *is induced by a form on* $(H, (\cdot|\cdot)_1)$;

(iii) *there exists a strictly form-positive self-adjoint operator* $Q \in \mathcal{L}(H)$ *such that* QA *is induced by a form.*

We remark that the rescaling $A + \omega$ is needed to obtain a sectorial operator. From Section 4.6 we know that for an invertible sectorial operator A and $\omega > 0$ one has

$$A \in H^\infty \quad \text{if and only if} \quad A + \omega \in H^\infty.$$

5.3.6. *More on equivalent scalar products.* In the Section 5.3.5 we characterized those operators A which come from a form on $(H, (\cdot|\cdot)_1)$ for some equivalent scalar product. What happens if we replace "*some*" by "*all*"?

THEOREM (Matolcsi [Mat03]). *Let A be an operator on H. The following are equivalent*:

(i) *A is bounded*;

(ii) *for each equivalent scalar product* $(\cdot|\cdot)_1$ *on* H, *the operator A is induced by a form on* $(H, (\cdot|\cdot)_1)$.

5.4. *Form sums and Trotter's product formula*

There is a natural way to define the sum of closed forms leading to a closed form again. Here we want to allow also nondense form domains. Let H be a Hilbert space.

5.4.1. *Closed forms with nondense domain.* We define closed forms on H as before but omit the assumption that the form domain be dense in H. Thus, a *closed form* a on H is a sesquilinear form $a : D(a) \times D(a) \to \mathbb{C}$, where $D(a)$ is a subspace of H (the *form domain*) such that, for some $\omega \in \mathbb{R}$, $\alpha \in [0, \pi/2)$:

(a) $a(u,u) + (\omega - 1)\|u\|_H^2 \in \Sigma_\alpha$, $u \in D(a)$;

(b) $(D(a), \|\cdot\|_a)$ is complete, where $\|u\|_a^2 = \operatorname{Re} a(u,u) + \omega\|u\|_H^2$.

Then one can show that a is a continuous sesquilinear form on $V = (D(a), \|\cdot\|_a)$. Thus (5.1) and (5.2) are satisfied.

Denote by A the operator on $\overline{D(a)}$ (the closure of $D(a)$ in H) associated with a. Then $-A$ generates a holomorphic C_0-semigroup $(e^{-tA})_{t\geqslant 0}$ on $D(a)$. We extend this semigroup by 0 to H defining

$$e^{-ta}x := \begin{cases} e^{-tA}x & \text{if } x \in \overline{D(a)}, \\ 0 & \text{if } x \in D(a)^\perp, \end{cases} \tag{5.3}$$

for $t > 0$ and letting e^{-0a} the orthogonal projection onto $\overline{D(a)}$. Then $t \mapsto e^{-ta} : [0,\infty) \to \mathcal{L}(H)$ is strongly continuous and satisfies the semigroup property

$$e^{-ta}e^{-sa} = e^{-(t+s)a}, \quad t, s \geqslant 0.$$

We call $(e^{-ta})_{t\geqslant 0}$ *the semigroup on H associated with the closed form a*. Before giving examples we consider form sums.

5.4.2. *Form sums.* Now let a and b be two closed forms on H. Then $D(a) \cap D(b)$ is a Hilbert space for the norm $(\|u\|_a^2 + \|u\|_b^2)^{1/2}$. Hence, $a + b$ with domain $D(a + b) = D(a) \cap D(b)$ is a closed form again.

THEOREM. *Let a and b be two closed forms on H. Then Trotter's formula*

$$e^{-t(a+b)}x = \lim_{n\to\infty} \left(e^{-(t/n)a}e^{-(t/n)b}\right)^n x \tag{5.4}$$

holds for all $x \in H$, $t \geqslant 0$.

This result is due to Kato [Kat78] for the case of symmetric forms and it was generalized to closed forms by Simon (see [Kat78], Theorem and Addendum).

It is interesting that this result can be applied in particular when $e^{-tb} \equiv P$, where P is an orthogonal projection. This is done in the following section.

5.4.3. *Induced semigroups.* Let a be a closed form on H. Let H_1 be a closed subspace of H. Even if H_1 is not invariant under $(e^{-ta})_{t\geqslant 0}$ there is a natural induced semigroup on H_1. In fact, let $D(b) = H_1$ and $b \equiv 0$. Then $e^{-tb} \equiv P$, the orthogonal projection onto H_1. Thus by Trotter's formula (5.4),

$$e^{-ta_1}x = \lim_{n\to\infty} \left(e^{-ta/n}P\right)^n x \tag{5.5}$$

converges for all $x \in H$, where $D(a_1) = D(a) \cap H_1$ and $a_1(u, v) = a(u, v)$ for all $u, v \in D(a_1)$. Then $(e^{-ta_1})_{t\geqslant 0}$ is a C_0-semigroup on H_1 if and only if $D(a) \cap H_1$ is dense in H_1.

5.4.4. *Nonconvergence of Trotter's formula.* Before giving a concrete example we want to point out that this way to induce a semigroup on closed subspaces via (5.5) does not work for arbitrary contraction semigroups.

THEOREM (Matolcsi [Mat03]). *Let $-A$ be the generator of a contraction C_0-semigroup T on H. The following assertion are equivalent:*

(i) *A is induced by a form;*

(ii) *$\lim_{n\to\infty}(T(t/n)P)^n x$ converges for each $t > 0$, $x \in H$, and each orthogonal projection P.*

5.4.5. *From the Gaussian semigroup to the Dirichlet Laplacian.* Denote by G the Gaussian semigroup on $L^2(\mathbb{R}^n)$, i.e.,

$$G(t)f(x) = (4\pi t)^{-n/2} \int_{\mathbb{R}^n} e^{-(x-y)^2/4t} f(y)\,dy$$

for all $t > 0$, $f \in L^2(\mathbb{R}^n)$, $x \in \mathbb{R}^n$. Then G is associated with the form $a(u,v) = \int_{\mathbb{R}^n} \nabla u \overline{\nabla v}\,dx$ with domain $D(a) = W^{1,2}(\mathbb{R}^n)$. Now let $\Omega \subset \mathbb{R}^n$ be an open subset of $\mathbb{R}^n$. We identify $L^2(\Omega)$ with the space $\{f \in L^2(\mathbb{R}^n)\colon f(x) = 0 \text{ a.e. on } \Omega^c\}$. Then the orthogonal projection P onto $L^2(\Omega)$ is given by $Pf = 1_\Omega f$ for all $f \in L^2(\mathbb{R}^n)$. Now let $W_0^{1,2}(\overline{\Omega}) = W^{1,2}(\mathbb{R}^n) \cap L^2(\Omega) = \{u \in W^{1,2}(\mathbb{R}^n)\colon u(x) = 0 \text{ a.e. on } \Omega^c\}$. Consider the form b on $W_0^{1,2}(\overline{\Omega})$ given by $b(u,v) = \int_\Omega \nabla u \nabla v$. Then, by (5.5), we have

$$\mathrm{e}^{-tb} f = \lim_{n\to\infty} \bigl(G(t/n)1_\Omega\bigr)^n f \tag{5.6}$$

for all $f \in L^2(\mathbb{R}^n)$. Now we consider the more familiar space $W_0^{1,2}(\Omega)$. For $u \in W_0^{1,2}(\Omega)$ let $\tilde{u}(x) = u(x)$ for $x \in \Omega$ and $\tilde{u}(x) = 0$ for $x \in \Omega^c$. Then $\tilde{u} \in W_0^{1,2}(\overline{\Omega})$ and $D_j\tilde{u} = \widetilde{D_j u}$. Thus we may identify $W_0^{1,2}(\Omega)$ with a subspace of $W_0^{1,2}(\overline{\Omega})$. One says that Ω is *stable*, if $W_0^{1,2}(\Omega) = W_0^{1,2}(\overline{\Omega})$. For example, if Ω has Lipschitz boundary, then Ω is stable. Thus Ω is stable if and only if $(\mathrm{e}^{-tb})_{t\geqslant 0}$ is the semigroup generated by the Dirichlet Laplacian with the canonical extension to $L^2(\mathbb{R}^n)$, i.e.,

$$\bigl(\mathrm{e}^{-tb} f\bigr)(x) = \begin{cases} \bigl(\mathrm{e}^{t\Delta_\Omega^D} f|_\Omega\bigr)(x), & x \in \Omega, \\ 0, & x \notin \Omega. \end{cases}$$

Thus, if Ω is stable, then the semigroup generated by the Dirichlet Laplacian is obtained from the Gaussian semigroup via Trotter's formula (5.6).

Stable open sets can be characterized as follows. Let Ω be an open, bounded set in $\mathbb{R}^n$. We assume that the boundary $\partial\Omega$ of Ω is a null-set (for the n-dimensional Lebesgue measure), and that $\mathring{\overline{\Omega}} = \Omega$. By

$$\operatorname{cap}(A) = \inf\bigl\{\|u\|_{H^1}^2\colon u \in H^1\bigl(\mathbb{R}^n\bigr), u \geqslant 1, \text{ a.e. in a neighborhood of } A\bigr\}$$

we denote the *capacity* of a subset A of $\mathbb{R}^n$.

THEOREM. *The following assertions are equivalent.*

(i) $W_0^{1,2}(\Omega) = W_0^{1,2}(\overline{\Omega})$;

(ii) $\operatorname{cap}(G \setminus \overline{\Omega}) = \operatorname{cap}(G \setminus \Omega)$ *for every open set* $G \subset \mathbb{R}^n$;

(iii) *for each function* $u \in C(\overline{\Omega})$, *which is harmonic on* Ω, *there exist* u_n *harmonic on an open neighborhood of* $\overline{\Omega}$ *which converge to* u *uniformly on* $\overline{\Omega}$ *as* $n \to \infty$.

It is interesting that these properties do not imply Dirichlet regularity. In fact, if Ω is the Lebesgue cusp, then $W_0^{1,2}(\Omega) = W_0^{1,2}(\overline{\Omega})$. References: [Hed93] and [Kel66].

5.5. *The square root property*

Let H be a Hilbert space. During this section we consider a continuous sesquilinear form $a : V \times V \to \mathbb{C}$ which is coercive with respect to H, i.e.,

$$\operatorname{Re} a(u,u) \geqslant \nu \|u\|_V^2, \quad u \in V,$$

where $V \hookrightarrow_d H$ and $\nu > 0$. This means that the ellipticity condition (5.2) is satisfied for $\omega = 0$. We consider the operator A on H which is associated with a. Thus A is a sectorial operator and $0 \in \rho(A)$.

5.5.1. *The square root property.* We say that the operator A (or the form a) has the *square root property* if $V = D(A^{1/2})$. It is easy to see that this is the case if a is symmetric (see the Example below). But an example of McIntosh [McI82] shows that the square root property does not hold for all closed, coercive forms. Here is an easy characterization of the square root property.

PROPOSITION. *The following are equivalent*:
(i) $D(A^{1/2}) = V$,
(ii) $D(A^{*1/2}) = V$,
(iii) $D(A^{1/2}) = D(A^{*1/2})$.

In that case the three norms $\|\cdot\|_V$, $\|A^{1/2}x\|$ and $\|A^{*1/2}x\|$ are equivalent. This is an immediate consequence of the Closed Graph theorem since all three spaces are continuously embedded into H. Of course one may also ask whether (iii) is valid for other powers α than $\alpha = \frac{1}{2}$. Surprisingly, by a result of Kato [[Kat66]] one always has

$$D(A^\alpha) = D(A^{*\alpha})$$

for all $0 < \alpha < \frac{1}{2}$ for each m-accretive operator. Thus $\alpha = \frac{1}{2}$ is the critical power for which things may go wrong.

5.5.2. *Induced operators on V and V'.* Since $V \hookrightarrow_d H$, we may identify H with a subspace of V'. Indeed, given $x \in H$, we define $j_x \in V'$ by $j_x(y) = (x \mid y)_H$. Then $j : H \to V'$, $x \mapsto j_x$ is linear, continuous and has dense image. We will identify x and j_x so that

$$V \underset{d}{\hookrightarrow} H \underset{d}{\hookrightarrow} V'.$$

By the Lax–Milgram theorem, there is an isomorphism $A_{V'} : V \to V'$ given by $\langle A_{V'}u, v\rangle = a(u,v)$, $u, v \in V$. Here V' denotes the space of all continuous, anti-linear forms on V (i.e., $V' = \{\bar{\varphi}\colon \varphi \in V^*\}$, V^* the dual space of V). The operator $-A_{V'}$ generates a bounded holomorphic C_0-semigroup $T_{V'}$ on V' (see [[Tan79], Section 3.5]). Moreover, $T_{V'}(t)|_H = T(t)$, the semigroup generated by $-A$ on H and, clearly, A is the part of $A_{V'}$ in H. Finally, the part of $A_{V'}$ in V, i.e., the operator A_V given by $A_V u = Au$ with domain $D(A_V) = \{u \in D(A)\colon Au \in V\}$ is similar to $A_{V'}$ and $-A_V$ generates a holomorphic

C_0-semigroup T_V which is the restriction of T to V. Thus, A_V and $A_{V'}$ are always similar; but A_V and A are not, unless A has the square root property:

THEOREM. *The following are equivalent*:
(i) $D(A^{1/2}) = V$;
(ii) $A_V \in H^\infty$;
(iii) $A_{V'} \in H^\infty$;
(iv) A_V *and* A *are similar.*

Thus, one can describe the square root property by the boundedness of the H^∞-calculus of A_V (or $A_{V'}$).

Before giving a proof of the Theorem we determine the diverse notions in the case of self-adjoint operators.

EXAMPLE (Self-adjoint operators). Assume that the form a defined at the beginning of the section is symmetric. Then the associated operator A is self-adjoint. Thus, by the spectral theorem, we may assume that $H = L^2(\Omega, \Sigma, \mu)$ and $Au = m \cdot u$, $D(A) = \{u \in L^2(\Omega)$: $mu \in L^2(\Omega)\}$, where $m : \Omega \to [\nu, \infty)$ is measurable. Then $V = L^2(\Omega, m\,\mathrm{d}\mu)$ and $a(u, v) = \int_\omega mu\bar{v}\,\mathrm{d}\mu$. Moreover, $V' = L^2(\Omega, 1/m\,\mathrm{d}\mu)$ and $\langle u, v\rangle = \int_\Omega u\bar{v}\,\mathrm{d}\mu$ for all $u \in V'$, $v \in V$. The operator $A_{V'}$ is given by $D(A_{V'}) = V = L^2(\Omega, m\,\mathrm{d}\mu)$, $A_{V'}u = mu$, and A_V is given by $A_V u = mu$ with domain $D(A_V) = \{u \in V\colon mu \in V\} = L^2(\Omega, m^3\,\mathrm{d}\mu)$. Now, $A^{-\alpha}$ is a bounded operator on H given by $A^{-\alpha}u = m^{-\alpha}u$. Thus $D(A^\alpha) = \{m^{-\alpha}u\colon u \in L^2(\Omega)\} = L^2(\Omega, m^{2\alpha}\,\mathrm{d}\mu)$. It is clear that $D(A^\alpha) = [H, D(A)]_\alpha$ for all $0 < \alpha < 1$.

REMARK (Definition of the complex interpolation space). Let $V \hookrightarrow_d H$. It is possible to define the complex interpolation spaces with help of the spectral theorem. Indeed, $a(u, v) = (u|v)_V$ defines a continuous, coercive form on V. Then we can assume that $V = L^2(\Omega, m\,\mathrm{d}\mu)$ and $H = L^2(\Omega, \mathrm{d}\mu)$. Then $[H, V]_\alpha = L^2(\Omega, m^\alpha\,\mathrm{d}\mu)$ for $0 < \alpha < 1$, where $[H, V]_\alpha$ denotes the complex interpolation space.

PROOF OF THE THEOREM. The operator $A^{-1/2} : H \to D(A^{1/2})$ is an isomorphism. If $V = D(A^{1/2})$, then $A_V = A^{-1/2}AA^{1/2}$, i.e., A and A_V are similar. This shows that (i) $\Rightarrow$ (iv). Since $A \in H^\infty$, (iv) implies (ii). And (ii) implies (iii) since A_V and $A_{V'}$ are always similar. Finally we show that (iii) implies (i). From the previous remark it is clear that $H = [V', V]_{1/2}$. Now if $A_{V'} \in H^\infty$, then, by Section 4.4.10, $D(A_{V'}^{1/2}) = [V', D(A_{V'})]_{1/2} = [V', V]_{1/2} = H$, i.e., $H = A_{V'}^{-1/2}(V')$. This implies that

$$D(A^{1/2}) = A^{-1/2}H = A_{V'}^{-1}(V') = V. \qquad \square$$

5.6. *Groups and cosine functions*

In the preceding sections we have characterized those sectorial operators A on a Hilbert space with sectorial angle $\varphi_{\sec}(A) < \pi/2$ which have a bounded H^∞-calculus. This is equivalent to saying that A is defined by a closed form on $(H, (\cdot|\cdot)_1)$ for some equivalent scalar product $(\cdot|\cdot)_1$. Now we will see that squares of group generators are always of this form.

5.6.1. *Groups on Hilbert spaces.* Let U be a bounded C_0-group on a Hilbert space H. Then, by a classical result of Sz.-Nagy, there exists an equivalent scalar product $(\cdot|\cdot)_1$ on H such that the group is unitary in $(H(\cdot|\cdot)_1)$. Concerning arbitrary C_0-groups the following holds (see, e.g., [Haa03a]).

PROPOSITION. *Let B be the generator of a C_0-group on H. Then there exists a bounded operator C on H such that $B + C$ generates a bounded C_0-group.*

Thus, on Hilbert space, up to equivalent scalar product, each generator of a C_0-group B is of the form

$$B = \mathrm{i}B_0 + C,$$

where B_0 is self-adjoint and C bounded.

5.6.2. *Squares of generators of C_0-groups.* Let B be the generator of a C_0-group. Then B^2 generates a holomorphic C_0-semigroup of angle $\pi/2$. Thus, for each $0 < \theta < \pi/2$, there exists $\omega \in \mathbb{R}$ such that $-B^2 + \omega I$ is sectorial and $\varphi_{\mathrm{sec}}(-B^2 + \omega I) < \theta$.

THEOREM (Haase [Haa03a]). *Let $A = -B^2$, where B generates a C_0-group on a Hilbert space. Then there exists $\omega \in \mathbb{R}$ such that $A + \omega \in H^\infty$.*

Operators of the form $B^2 - \omega$ with B a group generator are the same as generators of cosine functions. This holds in Hilbert spaces, and more generally on UMD-spaces. We describe these facts, starting on arbitrary Banach spaces first and concluding with the square root property.

5.6.3. *Cosine functions.* Let X be a Banach space. A *cosine function* on X is a strongly continuous function $C : \mathbb{R} \to \mathcal{L}(X)$ satisfying

$$2C(t)C(s) = C(t+s) + C(t-s), \quad s, t \in \mathbb{R}, \qquad C(0) = I.$$

In that case there exist $M \geqslant 1$, $\omega \in \mathbb{R}$ such that

$$\|C(t)\| \leqslant M\mathrm{e}^{\omega t}, \quad t \geqslant 0.$$

Moreover, there is a unique operator A such that $(\omega, \infty) \subset \rho(A)$ and

$$\lambda R(\lambda^2, A)x = \int_0^\infty \mathrm{e}^{-\lambda t} C(t)x \,\mathrm{d}t$$

for all $x \in X$, $\lambda > \omega$. We call A the *generator* of C. The cosine function C generated by A governs the second-order Cauchy problem defined by A

$$\begin{cases} \ddot{u}(t) = Au(t), & t \in \mathbb{R}, \\ u(0) = x, & \dot{u}(0) = y. \end{cases} \tag{5.7}$$

We will make this more precise. Introducing the function $v = \dot{u}$, problem (5.7) can be reformulated by

$$\begin{pmatrix} u \\ v \end{pmatrix}' = \begin{pmatrix} 0 & I \\ A & 0 \end{pmatrix} \begin{pmatrix} u \\ v \end{pmatrix}, \qquad u(0) = x, \quad v(0) = y.$$

However, the operator $\left(\begin{smallmatrix} 0 & I \\ A & 0 \end{smallmatrix}\right)$ with domain $D(A) \times X$ generates a C_0-semigroup on $X \times X$ if and only if A is bounded (see [[ABHN01], Section 3.14.9]). Thus, another space than $X \times X$ has to be considered.

THEOREM [[ABHN01], Section 3.14.11]. *Let A be an operator on a Banach space X. The following assertions are equivalent*:

(i) *A generates a cosine function*;

(ii) *there exists a Banach space W such that $D(A) \subset W \hookrightarrow X$, and the operator*

$$\mathcal{A} = \begin{pmatrix} 0 & I \\ A & 0 \end{pmatrix}$$

with domain $D(A) \times W$ generates a C_0-semigroup on $W \times X$.

In that case the Banach space W is uniquely determined by property (ii). We call $W \times X$ the *phase space* of problem (5.7). The phase space is important to produce classical solutions:

PROPOSITION [[ABHN01], Section 3.14.12]. *Let A be the generator of a cosine function, with phase space $W \times X$. Then, for each $x \in D(A)$, $y \in W$, there exists a unique solution $u \in C^2(\mathbb{R}, X) \cap C(\mathbb{R}, D(A))$ of* (5.7).

So far we have described well-posedness of the second-order Cauchy problem (5.7) in arbitrary Banach spaces. Now we come back to the initial subject of this section.

5.6.4. *Squares of group generators and cosine functions.* On UMD-spaces one has the following characterization.

THEOREM (Fattorini [[ABHN01], Section 3.16.8]). *Let A be an operator on a UMD-space X. The following assertions are equivalent*:

(i) *A generates a cosine function*;

(ii) *there exist a generator B of a C_0-group on X and $\omega \geqslant 0$ such that*

$$A = B^2 + \omega I.$$

In that case, the phase space is $D(B) \times X$, where $D(B)$ carries the graph norm.

This theorem explains why L^p-spaces for $p \neq 2$ are not so good as functional framework for hyperbolic problems: There are not many groups operating on L^p for $p \neq 2$. For example, the Laplacian Δ generates a cosine function on $L^p(\mathbb{R}^n)$ if and only if $p = 2$ or $n = 1$. On the other hand, on Hilbert spaces there are many C_0-groups since there are many self-adjoint operators. This explains from an abstract point of view why many examples of well-posed second-order problems exist on Hilbert space (see Section 8.9). In view of the results mentioned so far, they all are related to self-adjoint operators.

5.6.5. *Forms and cosine functions.* Let H be a Hilbert space. If $-A$ generates a cosine function, then it follows from Sections 5.6.4 and 5.6.2 that $A + \omega \in H^\infty$ for ω large enough. Thus, after changing the scalar product on H, the operator A is associated with a closed form. Now let us assume that A is associated with a coercive, continuous form $a : V \times V \to \mathbb{C}$, where $V \hookrightarrow_d H$. Assume in addition that $-A$ generates a cosine function. Then by Fattorini's theorem [[ABHN01], Section 3.16.7] $B = \mathrm{i}A^{1/2}$ generates a C_0-group. Thus, the phase space is $D(A^{1/2}) \times H$. Now it becomes apparent why the square root property is interesting in this context. The given known space is V (for example in applications, V may be a Sobolev space). One would like to know whether $V \times H$ is the phase space of the second-order Cauchy problem.

5.6.6. *Numerical range in a parabola.* If A is an operator on a Hilbert space, then A comes from a form if and only if the numerical range of A

$$W(A) = \{(Ax|x)_H\colon\ x \in D(A),\ \|x\| = 1\}$$

is contained in a sector $\Sigma_\theta + \omega$ for some $\omega \in \mathbb{R}$, $\theta \in [0, \pi/2)$ and $(-\infty, \omega) \cap \rho(A) \neq \emptyset$ (see Section 5.3.4). Now assume that $-A$ is the generator of a cosine function. Then the spectrum $\sigma(A)$ is contained in a parabola

$$P_\omega = \{\xi + \mathrm{i}\eta \in \mathbb{C}\colon\ \xi \leqslant \omega^2 - \eta^2/4\omega^2\}$$

for some $\omega \in \mathbb{R}$ [[ABHN01], Section 3.14.18]. Recall that $\sigma(A) \subset \overline{W(A)}$ by [[Kat66], Section V.3.2].

THEOREM (McIntosh [McI82], Theorems A and C). *Let a be a closed densely defined form on H with associated operator A. Assume that the numerical range $W(A)$ of A is contained in a parabola P_ω for some $\omega \in \mathbb{R}$. Then A has the square root property.*

Generators of cosine functions can be characterized by a real condition on the resolvent [[ABHN01], Section 3.15.3] which is however of little practical use. Much more interesting is the following most remarkable new criterion:

THEOREM (Crouzeix [Cro03]). *Let $a: V \times V \to \mathbb{C}$ be a closed densely defined form on a Hilbert space H such that*

$$a(u,u) \in P_\omega, \quad u \in V, \|u\|_H = 1,$$

for some $\omega \in \mathbb{R}$. Then the associated operator generates a cosine function.

Haase [Haa03a], Corollary 5.18, proved that also the converse implication holds: If A generates a cosine function on a Hilbert space, then there exist an equivalent scalar product $(\cdot|\cdot)_1$ on H, a closed form $a: V \times V \to \mathbb{C}$ on H with $V \hookrightarrow_d H$ such that $-A$ is associated with the form a on $(H, (\cdot|\cdot)_1)$ and

$$a(u,u) \in P_\omega, \quad u \in V, \|u\|_H = 1,$$

for some $\omega \in \mathbb{R}$. Thus, in view of Section 5.3.4 we can formulate the following beautiful characterization.

COROLLARY. *Let A be an operator on a Hilbert space H. The following assertions are equivalent*:

(i) *A generates a cosine function*;

(ii) *there exists an equivalent scalar product $(\cdot|\cdot)_1$ on H and $\omega \in \mathbb{R}$ such that*

$$-(Au|u)_1 \in P_\omega, \quad u \in D(A), \|u\|_H = 1,$$

and $\rho(-A) \cap P_\omega \neq \emptyset$.

6. Fourier multipliers and maximal regularity

The Fourier transform is a classical tool to treat differential equations. In order to apply it to partial differential equations, vector-valued functions have to be considered. Problems of regularity, but also of asymptotic behavior, can frequently be reformulated as the question whether a certain operator is an L^p-Fourier multiplier. Michlin's theorem gives a most convenient criterion for this. It was Weis [Wei00a] (after previous work by Clément et al. [CPSW00]) who discovered the right formulation of Michlin's theorem in the vector-valued case.

We describe the situation in the periodic case which is technically easier, and ideas become more transparent in this case. Still, the periodic case leads to the main application, namely characterization of maximal regularity for the nonhomogeneous Cauchy problem.

6.1. *Vector-valued Fourier series and periodic multipliers*

It was the study of the heat equation which lead Fourier to the introduction of one of the most fundamental concepts of Analysis. In the same spirit, vector-valued Fourier series help us to understand arbitrary abstract evolution equations.

Let X be a Banach space. For $1 \leqslant p \leqslant \infty$ denote by $L^p_{2\pi} := L^p_{2\pi}(X)$ the space of all equivalence classes of 2π-periodic measurable functions $f : \mathbb{R} \to X$ such that

$$\|f\|_{L^p_{2\pi}} := \left(\int_0^{2\pi} \|f(t)\|^p \,\mathrm{d}t \right)^{1/p} < \infty,$$

if $1 \leqslant p < \infty$, and such that

$$\|f\|_{L^\infty_{2\pi}} = \operatorname*{ess\,sup}_{t \in \mathbb{R}} \|f(t)\| < \infty,$$

where functions are identified if they coincide a.e. Then $L^p_{2\pi}$ is a Banach space and $L^p_{2\pi} \hookrightarrow L^1_{2\pi}$, $1 \leqslant p \leqslant \infty$. For $f \in L^1_{2\pi}$, denote by

$$\hat{f}(k) = \frac{1}{2\pi} \int_0^{2\pi} \mathrm{e}^{-\mathrm{i}kt} f(t) \,\mathrm{d}t$$

the kth *Fourier coefficient* of f, where $k \in \mathbb{Z}$. The Fourier coefficients determine the function f; i.e.,

$$\hat{f}(k) = 0 \quad \text{for all } k \in \mathbb{Z} \quad \text{if and only if} \quad f(t) = 0 \text{ a.e.} \tag{6.1}$$

DEFINITION. Let $1 \leqslant p \leqslant \infty$. A sequence $(M_k)_{k \in \mathbb{Z}} \subset \mathcal{L}(X)$ is an $L^p_{2\pi}$-*multiplier* if, for each $f \in L^p_{2\pi}(X)$, there exists a function $g \in L^p_{2\pi}(X)$ such that

$$M_k \hat{f}(k) = \hat{g}(k), \quad k \in \mathbb{Z}.$$

In that case, by the closed graph theorem, the mapping $M := (f \mapsto g)$ defines a bounded linear operator from $L^p_{2\pi}(X)$ into $L^p_{2\pi}(X)$, which we call the *operator associated with* $(M_k)_{k \in \mathbb{Z}}$.

Let $e_k(t) = \mathrm{e}^{\mathrm{i}kt}$, $t \in \mathbb{R}$, $k \in \mathbb{Z}$, and for $x \in X$, denote by $e_k \otimes x$ the function $t \mapsto \mathrm{e}^{\mathrm{i}kt} x$. Linear combinations of such functions are called *trigonometric polynomials*. By Fejér's theorem [[ABHN01], Theorem 4.2.19], the space of all trigonometric polynomials is dense in $L^p_{2\pi}(X)$. Note that the associated operator M acts on trigonometric polynomials by

$$M \sum_{k=-m}^{m} e_k \otimes x_k = \sum_{k=-m}^{m} e_k \otimes M_k x_k. \tag{6.2}$$

6.1.1. *Multipliers on Hilbert spaces.* If H is a Hilbert space, then the Fourier transform $f \mapsto \hat{f}$ is an isometric isomorphism of $L^2_{2\pi}(H)$ onto $\ell^2(H)$ (the space of all sequences $(x_k)_{k \in \mathbb{Z}}$ in $\ell^2(H)$ such that $\sum_{k \in \mathbb{Z}} \|x_k\|^2 < \infty$). This follows easily from the scalar case by considering an orthonormal basis of H. As a consequence, if H is a Hilbert space, then a sequence $(M_k)_{k \in \mathbb{Z}} \subset \mathcal{L}(H)$ is an $L^2_{2\pi}$-multiplier if and only if it is bounded. If $1 < p < \infty$ and $p \neq 2$, then multipliers can no longer be characterized in a satisfying way, even in the scalar case. An important subject of harmonic analysis is to find sufficient conditions. In the scalar case the following is a special case of Marcinkiewicz' multiplier theorem. We state it on Hilbert space (where it is a special case of the Theorem in Section 6.1.5).

THEOREM. *Let H be a Hilbert space and let $(M_k)_{k\in\mathbb{Z}} \subset \mathcal{L}(H)$ be a bounded sequence satisfying*

$$\sup_{k\in\mathbb{Z}} \left\| k(M_{k+1} - M_k) \right\| < \infty. \tag{M1}$$

Then $(M_k)_{k\in\mathbb{Z}}$ is an $L^p_{2\pi}$-multiplier for $1 < p < \infty$.

6.1.2. *Characterization of Hilbert spaces.* Even if $p = 2$, the Theorem in Section 6.1.1 does no longer hold for Banach spaces other than Hilbert space.

THEOREM. *Let X be a Banach space and let $1 < p < \infty$. Assume that each bounded sequence $(M_k)_{k\in\mathbb{Z}} \subset \mathcal{L}(X)$ satisfying $(M1)$ is an $L^p_{2\pi}$-multiplier. Then X is isomorphic to Hilbert space.*

An explicit proof of this theorem is given in [ArBu03b], Proposition 1.17. It is based on a deep characterization of Hilbert spaces due to Kwapien. It was Pisier (unpublished) who had discovered that certain classical operator-valued multiplier theorems hold merely on Hilbert spaces.

6.1.3. *UMD-spaces and the Riesz-projection.* In order to obtain an operator-valued multiplier result one has to replace boundedness in operator norm in $(M1)$ by a stronger assumption (R-boundedness), and a restricted class of Banach spaces has to be considered.

DEFINITION. Let X be a Banach space. For $k \in \mathbb{Z}$, let

$$M_k = \begin{cases} I & \text{if } k \geqslant 0, \\ 0 & \text{if } k < 0. \end{cases} \tag{6.3}$$

We say that X is a *UMD-space* if the sequence $(M_k)_{k\in\mathbb{Z}}$ is an $L^p_{2\pi}(X)$-multiplier for all (equivalently one) $p \in (1, \infty)$. The associated operator R is called the *Riesz-projection*. Note that the sequence (6.3) satisfies condition $(M1)$ of Section 6.1.1.

The letters UMD stay for "*unconditional martingale differences*" and refer to an equivalent property introduced and studied by Burkholder [Bur83]. Here we will not be confronted with martingales and use the term UMD just for saying that the Riesz-projection is bounded.

EXAMPLES. (a) The space $L^q(Y)$ is a UMD-space if $1 < q < \infty$ for any σ-finite measure space (Y, Σ, μ).

(b) Every closed subspace of a UMD-space is a UMD-space.

(c) Each UMD-space is reflexive (even superreflexive).

(d) It follows from (a)–(c) that, for any nonempty open subset Ω of $\mathbb{R}^n$, the Sobolev space $W^{k,p}(\Omega)$ is a UMD-space if and only if $1 < p < \infty$, where $k \in \mathbb{N}_0$.

(e) If $\Omega \subset \mathbb{R}^n$ is a nonempty open bounded set, then $C(\overline{\Omega})$ is not a UMD-space.

6.1.4. *R-boundedness.* In order to formulate an operator-valued version of Marcinkiewicz's theorem we need a new notion of boundedness for sets of operators. Recall that a series $\sum_{k=1}^{\infty} x_k$ in a Banach space is called *unconditionally convergent* if the series $\sum_{k=1}^{\infty} \varepsilon_k x_k$ converges for any choice of signs $(\varepsilon_k)_{k\in\mathbb{N}} \in \{-1,1\}^{\mathbb{N}}$. This implies convergences of $\sum_{k=1}^{\infty} \alpha_k x_k$ for each $\alpha = (\alpha_k)_{k\in\mathbb{N}} \in \ell^\infty$. But in general, this does not imply that $\sum_{k=1}^{\infty} \|x_k\| < \infty$, unless X is finite dimensional. We introduce the means over all signs

$$\left\|(x_1,\ldots,x_n)\right\|_R := \frac{1}{2^n} \sum_{\varepsilon\in\{-1,1\}^n} \left\| \sum_{j=1}^n \varepsilon_j x_j \right\|. \tag{6.4}$$

DEFINITION. Let X, Y be Banach spaces. A subset $\mathcal{T}$ of $\mathcal{L}(X,Y)$ is called *R-bounded* if there exists a constant $c \geqslant 0$ such that

$$\left\|(T_1x_1,\ldots,T_nx_n)\right\|_R \leqslant c\left\|(x_1,\ldots,x_n)\right\|_R \tag{6.5}$$

for all $T_1,\ldots,T_n \in \mathcal{T}$, $x_1,\ldots,x_n \in X$, $n \in \mathbb{N}$. The least constant c such that (6.5) is satisfied is called the *R-bound* of $\mathcal{T}$ and is denoted by $R(\mathcal{T})$.

The notion of R-boundedness was introduced by Berkson and Gillespie [BG94], where the "R" stands for 'Riesz'. Since the mean

$$\left\|(x_1,\ldots,x_n)\right\|_R$$

can be expressed in terms of Rademacher functions some pronounce it "Rademacher bounded". Finally, the Rademacher functions may be replaced by other independent random variables which leads some to use "randomized boundedness". In any case, it is some sort of unconditional boundedness.

R-boundedness clearly implies boundedness. But if $X = Y$, the notion of R-boundedness is strictly stronger than boundedness unless the underlying space is a Hilbert space [ArBu02], Proposition 1.17.

6.1.5. *The operator-valued Marcinkiewicz theorem.* On an arbitrary Banach space, in general it is a difficult task to verify R-boundedness (we will come back to this point later). Nevertheless, for multipliers it is the right notion as the following two results show. First of all, we state that it is a necessary condition.

PROPOSITION ([ArBu02], [CP01]). *Let $(M_k)_{k\in\mathbb{Z}} \subset \mathcal{L}(X)$ be an $L^p_{2\pi}$-multiplier for some $1 < p < \infty$. Then the set $\{M_k\colon k \in \mathbb{Z}\}$ is R-bounded.*

If in Marcinkiewicz's theorem we replace boundedness by R-boundedness, the theorem remains true on UMD-spaces (see [ArBu02], Theorem 1.3).

THEOREM (The operator-valued Marcinkiewicz theorem). *Let X be a UMD-space. Let $(M_k)_{k\in\mathbb{Z}} \subset \mathcal{L}(X)$ be R-bounded. Assume that*

$$\{k(M_{k+1} - M_k)\colon\ k \in \mathbb{Z}\}$$

is R-bounded. Then $(M_k)_{k\in\mathbb{Z}}$ is an $L^p_{2\pi}$-multiplier whenever $1 < p < \infty$.

We remark that a set of scalar operators (i.e., operators of the form $c \cdot I$, where $c \in \mathbb{C}$) is bounded if and only if it is R-bounded. In that case, the Theorem had been proved by Clément, de Pagter, Sukochev and Witvliet [CPSW00]. It was Weis [Wei00a] who proved the first operator-valued multiplier theorem on UMD-spaces, namely the operator-valued version of Michlin's theorem on the real line.

6.1.6. *The variational Marcinkiewicz' condition.* Finally we state the original, more general version of Marcinkiewicz's theorem. It holds in Hilbert spaces as was shown by Schwartz [Schw61].

THEOREM. *Let X be a Hilbert space and let $(M_k)_{k\in\mathbb{Z}} \subset \mathcal{L}(X)$ be a bounded sequence satisfying*

$$\sup_{n\in\mathbb{N}} \sum_{2^{n-1}\leqslant|k|<2^n} \|M_{k+1} - M_k\| < \infty. \tag{MV}$$

Then $(M_k)_{k\in\mathbb{Z}}$ is an $L^p_{2\pi}$-multiplier for $1 < p < \infty$.

We call (*MV*) the *variational Marcinkiewicz condition*. There is a version on UMD-spaces due to Strkalj and Weis [SW00]. But the "R-version" of being of uniform bounded variation on dyadic intervals is more complicated to formulate.

Finally we mention a remarkable difference between the variational Marcinkiewicz condition and the stronger condition $(M1)$ of Section 6.1.1. In the scalar case, if $(M1)$ is satisfied the associated operator is of weak type $(1, 1)$. This does not hold, if merely the more general condition (MV) is satisfied (see Fournier's example [[EG77], Section 7.5]).

6.2. *Maximal regularity via periodic multipliers*

Let A be a closed operator and let $\tau > 0$. For $f \in L^1((0, \tau); X)$ and $x \in X$ we consider the problem

$$P_x(f)\begin{cases} u'(t) = Au(t) + f(t), & t \in [0, \tau],\\ u(0) = x. \end{cases}$$

6.2.1. *Mild and strong L^p-solutions.* Recall the notion of mild solution of $P_x(f)$ given in Section 1.5. We now define a stronger notion of solution. Let $1 \leqslant p < \infty$. By $W^{1,p}((0,\tau);X)$ we denote the space of all $u \in C([0,\tau];X)$ such that there exists $u' \in L^p((0,\tau);X)$ such that

$$u(t) = u(0) + \int_0^t u'(s)\,\mathrm{d}s, \quad t \in [0,\tau]. \tag{6.6}$$

Then $u'(t)$ is the derivative of u a.e. Let $f \in L^p((0,\tau);X)$. A *strong L^p-solution* u of $P_x(f)$ is a function $u \in W^{1,p}((0,\tau);X) \cap L^p((0,\tau);D(A))$ such that $u(0) = x$ and $u'(t) = Au(t) + f(t)$ a.e. on $(0,\tau)$. Each strong solution is also a mild solution. Conversely, since A is closed, a mild solution u of $P_x(f)$ is a strong L^p-solution if and only if $u \in W^{1,p}((0,\tau);X)$. Now assume that A is the generator of a C_0-semigroup T. Then, by Section 1.5, $u = T(\cdot)x + T * f$ is the unique mild solution of $P_x(f)$. Thus, we are lead to investigate under which conditions $T(\cdot)x \in W^{1,p}((0,\tau);X)$. We start to consider the case $f = 0$.

6.2.2. *An interpolation space and strong L^p-solutions of the homogeneous problem.* Let X, Y be Banach spaces such that $Y \hookrightarrow X$. For $1 < p < \infty$ one may define the real interpolation space

$$(X,Y)_{1/p^*,p} = \bigl\{u(0)\colon\ u \in W^{1,p}\bigl((0,\tau);X\bigr) \cap L^p\bigl((0,\tau);Y\bigr)\bigr\},$$

where $1/p^* + 1/p = 1$. In particular, if A generates a holomorphic C_0-semigroup T on X, then

$$\bigl(X, D(A)\bigr)_{1/p^*,p} = \bigl\{x \in X\colon\ AT(\cdot)x \in L^p\bigl((0,\tau);X\bigr)\bigr\}. \tag{6.7}$$

Thus, the mild solution $u = T(\cdot)x$ of $P_x(0)$ is in $W^{1,p}((0,\tau);X))$ if and only if $x \in (D(X),A)_{1/p^*,p}$.

6.2.3. *Strong solutions of periodic problems.* Let A be a closed operator on a Banach space X. For $f \in L^p((0,2\pi);X)$, we consider the periodic problem

$$P_{\mathrm{per}}(f)\begin{cases} u'(t) = Au(t) + f(t), & t \in [0,2\pi], \\ u(0) = u(2\pi). \end{cases}$$

A *strong L^p-solution* of $P_{\mathrm{per}}(f)$ is a function $u \in W^{1,p}((0,\tau);X) \cap L^p((0,\tau);D(A))$ such that (P_{per}) is satisfied t-a.e. We may identify $L^p((0,2\pi);X)$ and $L^p_{2\pi}(\mathbb{R},X)$. Then it is not difficult to see the following.

PROPOSITION. *Let $1 < p < \infty$. The following assertions are equivalent:*

(i) *For each $f \in L^p((0,2\pi);X)$, there exists a unique strong L^p-solution of $P_{\mathrm{per}}(f)$;*

(ii) *$\mathrm{i}\mathbb{Z} \subset \rho(A)$ and $(kR(\mathrm{i}k,A))_{k\in\mathbb{Z}}$ is an L^p-multiplier.*

Let us consider a Hilbert space H. Then we obtain the following theorem.

THEOREM. *Let $1 < p < \infty$. Let A be a closed operator on a Hilbert space H. The following are equivalent*:

(i) $i\mathbb{Z} \subset \rho(A)$ *and* $\sup_{k\in\mathbb{Z}} \|kR(ik, A)\| < \infty$;

(ii) *for all $f \in L^p$, there exists a unique strong L^p-solution of $P_{\text{per}}(f)$.*

For the implication (i) $\Rightarrow$ (ii) one considers $M_k = kR(ik, A)$. The resolvent identity implies that $(k(M_{k+1} - M_k))_{k\in\mathbb{Z}}$ is bounded. Thus assertion (ii) follows from Section 6.1.1. Similarly, one obtains from Section 6.1.5 the following characterization on UMD-spaces.

THEOREM [ArBu02]. *Let A be a closed operator on a UMD-space X and let $1 < p < \infty$. The following are equivalent*:

(i) $i\mathbb{Z} \subset \rho(A)$ *and* $\{kR(ik, A)\colon k \in \mathbb{Z}\}$ *is R-bounded*;

(ii) *for all $f \in L^p((0, 2\pi); X)$, there exists a unique strong L^p-solution of (P_{per}).*

Two things are remarkable:

1. Condition (ii), i.e., well-posedness of the periodic problem in the sense of strong L^p-solutions, is p-independent for $1 < p < \infty$.
2. Whereas no characterization of L^p-multipliers seems possible in general (if $1 < p < \infty$, $p \neq 2$), in the context of resolvents, it is.

Finally, we remark that A satisfies the equivalent conditions of the theorem if and only if $-A$ does so. In particular, A need not be the generator of a C_0-semigroup.

6.2.4. *Maximal regularity on Hilbert space.* Let X be a Banach space. Assume that A generates a holomorphic semigroup T. Then, for $f \in L^1((0, \tau); X)$, $x \in X$,

$$u = T(\cdot)x + T * f$$

is the unique mild solution of $P_x(f)$. We want to investigate when the solution is strong. If $f \equiv 0$, this is done in Section 6.2.2. Thus we may consider $x = 0$, i.e., we want to investigate $P_0(f)$.

THEOREM. *Let A be the generator of a holomorphic C_0-semigroup T on a Hilbert space H. Then*

$$T * f \in W^{1,p}\big((0, \tau); X\big) \quad \textit{for all } f \in L^p\big((0, \tau); X\big), \quad 1 < p < \infty. \tag{6.8}$$

PROOF. Let $\tau = 2\pi$. Replacing A by $A - \omega$ we may assume that $\lambda \in \rho(A)$ whenever $\operatorname{Re}\lambda \geqslant 0$ and $\|\lambda R(\lambda, A)\| \leqslant M$. Let $f \in L^p((0, \tau); H)$. Then by the first theorem of Section 6.2.3 there exists a unique strong L^p-solution v of $P_{\text{per}}(f)$. Then $v(0) \in (H, D(A))_{1/p^*,p}$ by Section 6.2.2. Let $u(t) = v - T(t)v(0)$. Then u is a strong L^p-solution of $P_0(f)$. $\square$

Property (6.8) is frequently called *maximal regularity*, or (*MR*) for short. Thus, on Hilbert space, every generator of a holomorphic C_0-semigroup does enjoy this property (*MR*). It is also known that (*MR*) implies holomorphy of the semigroup (see Section 6.2.6).

6.2.5. *Characterization of Hilbert space.* For a long time it has been an open problem whether the preceding theorem holds more generally on UMD-spaces. However, the answer is negative.

THEOREM (Kalton and Lancien [KL00]). *Let X be a Banach space possessing an unconditional basis. Let $1 < p < \infty$. If X is not isomorphic to a Hilbert space, then there exist a holomorphic C_0-semigroup T and $f \in L^p((0,\tau); X)$ such that $T * f \notin W^{1,p}((0,\tau); X)$.*

The semigroup T is not constructed directly, the existence of such semigroup T is proved implicitly using several deep results of geometry of Banach spaces. However one can say that the generator A of T is a block-diagonal operator with respect to some conditional basis in X.

It is an open problem whether the Theorem does also hold in Banach spaces with conditional bases. Also, so far no explicit counterexample, and in particular, no model is known for which (*MR*) fails.

6.2.6. *Maximal regularity on UMD-spaces.* On UMD-spaces, with the same proof as in Section 6.2.4 based on the periodic operator-valued multiplier theorem of Section 6.1.5 one obtains the following characterization.

THEOREM (Weis). *Assume that X is a UMD-space. Let A be a closed operator, $\tau > 0$, $1 < p < \infty$. The following are equivalent:*

(i) *for all $f \in L^p((0,\tau); X)$, there exists a unique $u \in W^{1,p}((0,\tau); X) \cap L^p((0,\tau); D(A))$ such that*

$$\begin{cases} u'(t) = Au(t) + f(t) & a.e., \\ u(0) = 0; \end{cases}$$

(ii) *there exist $\omega \in \mathbb{R}$ such that $\lambda \in \rho(A)$ whenever $\operatorname{Re}\lambda > \omega$ and the set $\{\lambda R(\lambda, A): \operatorname{Re}\lambda > \omega\}$ is R-bounded.*

It was Weis [Wei00a] who proved that (ii) implies (i), Clément and Prüss [CP01] showed necessity of the condition. Weis gave a different proof (than we indicated here), based on his operator-valued Michlin's theorem on the real line.

DEFINITION. We say that an operator A satisfies *condition* (*MR*) (for maximal regularity) if condition (i) is satisfied.

It can be seen from the Theorem that property (*MR*) is independent of $p \in (1, \infty)$ and of $\tau > 0$. Moreover, (*MR*) implies that A generates a holomorphic semigroup T. Thus, the solution u of (i) is given by $u = T * f$. If A satisfies (*MR*) then the existence of strong L^p-solutions on the entire half-line $(0, \infty)$ is merely a question of the asymptotic behavior of T as the following result shows. See Section 1.3 for the definition of $\omega(T)$.

THEOREM. *Let A be the generator of a holomorphic C_0-semigroup T on a Banach space X. The following assertions are equivalent:*
(i) *A satisfies (MR) and $\omega(T) < 0$;*
(ii) *$T * f \in W^{1,p}((0,\infty), X) \cap L^p((0,\infty); D(A))$ for all $f \in L^p((0,\infty); X)$.*

PROOF. (ii) $\Rightarrow$ (i) follows from Datko's theorem [[ABHN01], p. 336]. For (ii) $\Rightarrow$ (i), see [Dor00], Theorem 5.2. □

Finally, we mention that the restriction to $p \in (1, \infty)$ is essential. If A generates a C_0-semigroup T on a reflexive space X such that $T * f \in W^{1,1}((0,1); X)$ for all $f \in L^1((0,1); X)$, then A is bounded (see [Gue95]).

6.2.7. *Perturbation of* (*MR*). In general it is not at all easy to prove R-boundedness. However, it is not difficult to see that condition (ii) of the preceding theorem is stable under small perturbation (see [KuWe01], [ArBu02]).

THEOREM. *Let A be an operator satisfying (MR) on a UMD-space X. Then there exists $\varepsilon > 0$ such that, for each linear $B : D(A) \to X$,*

$$\|Bx\| \leqslant \varepsilon \|Ax\| + b\|x\|, \quad x \in D(A),$$

for some $b \geqslant 0$, also $A + B$ satisfies (MR).

6.2.8. *BIP and maximal regularity.* An important criterion for maximal regularity is property *BIP*. The following theorem is due to Dore and Venni [DoVe87].

THEOREM. *Let $A \in BIP$ on a UMD-space X such that $\varphi_{bip}(A) < \pi/2$. Then $-A$ has property (MR).*

We sketch two different proofs.

FIRST PROOF. One can show directly that *BIP* implies that for $\varphi_{bip} < \varphi < \pi/2$ one has $\sigma(A) \subset \Sigma_\varphi$ and that the set $\{\lambda R(\lambda, A) \colon \lambda \in \mathbb{C} \setminus \Sigma_\varphi\}$ is R-bounded, [DHP01], Theorem 4.5. Thus the first theorem of Section 6.2.6 can be applied. □

SECOND PROOF [DOVE87]. The space $Y = L^p((0,\tau); X)$ has the UMD-property and the operator $\mathcal{A}$ on $L^p((0,\tau); X)$ given by $(\mathcal{A}u)(t) = Au(t)$ with domain $D(\mathcal{A}) = L^p((0,\tau); D(A))$ obviously inherits *BIP* from A with angle $\varphi_{bip}(\mathcal{A}) \leqslant \varphi_{bip}(A)$. On the other hand, the operator $\mathcal{B}$ on $L^p((0,\tau); X)$ given by $D(\mathcal{B}) = W_0^{1,p}((0,\tau); X)$, $\mathcal{B}u = u'$, has *BIP* with angle $\varphi_{bip}(\mathcal{B}) = \pi/2$ (see [HP97] or [DoVe87]). Assuming that $0 \in \rho(A)$, it follows from the Dore–Venni theorem (Section 4.4.8) that $\mathcal{A} + \mathcal{B}$ is closed. The Theorem of Section 4.2 shows that $\mathcal{A} + \mathcal{B}$ is invertible. This is exactly property (*MR*). □

6.2.9. *Maximal regularity for positive contraction semigroups on $L^p(\Omega)$.* The characterization of maximal regularity by R-sectoriality 6.2.11 (and Section 4.7.5) has the following interesting consequence [Wei00b].

THEOREM (Lamberton and Weis). *Let $1 < p < \infty$. Let $-A$ be the generator of a positive contractive C_0-semigroup T on $L^p(\Omega)$. If T is holomorphic, then A has (MR).*

This result had been first proved by Lamberton [Lam87] in the case where T is also contractive for the L^∞- and L^1-norm.

6.2.10. *Maximal regularity and quasi-linear problems.* The property of maximal regularity is most important in order to solve quasi-linear problems. Here we give a result on local existence.

Let X be a UMD-space and let D be a Banach space such that $D \hookrightarrow_d X$. Let $1 < p < \infty$ and $Y = (X, D)_{1/p^*,p}$, where $1/p + 1/p^* = 1$. Let $A : Y \to \mathcal{L}(D, X)$ be Lipschitz continuous on each bounded subset of Y. Let $u_0 \in Y$ and assume that the operator $-A(u_0)$ on X with domain $D(A(u_0)) = D$ has property (*MR*). Then the following result holds.

THEOREM (Clément and Li [CL94]). *For each $f \in L^p_{\text{loc}}((0, \infty); X)$, there exist $\tau > 0$ and a solution $u \in W^{1,p}((0, \tau); X) \cap L^p((0, \tau); D)$ of*

$$\begin{cases} u'(t) + A\big(u(t)\big)u(t) = f(t), & t \in (0, \tau), \\ u(0) = u_0. \end{cases} \tag{P}$$

Observe that $W^{1,p}((0, \tau); X) \cap L^p((0, \tau); D) \subset C([0, \tau]; Y)$ so that the condition on the initial value of u makes sense.

SKETCH OF THE PROOF OF THE THEOREM. We show how maximal regularity is used to give a fixed point argument. Rewrite the problem as

$$u'(t) + A(u_0)u(t) = f(t) + \big(A(u_0) - A(u)\big)u.$$

Let $MR(\tau) := L^p((0, \tau); D) \cap W^{1,p}((0, \tau); X)$. Then $S_1 v = f + (A(u_0) - A(v))v$ defines a mapping $S_1 : MR(\tau) \to L^p((0, \tau); X)$. Consider $S_2 : L^p((0, \tau); X) \to MR(\tau)$ defined by $S_2 g = u$, where $u \in MR(\tau)$ is the solution of $\dot{u} + A(u_0)u = g$, $u(0) = u_0$. Then $S = S_2 \circ S_1$ is a mapping from $MR(\tau)$ into $MR(\tau)$. One can show that S is a strict contraction if $\tau > 0$ is small enough. Thus the Banach fixed point theorem shows that S has a fixed point which is a solution of (P). □

6.2.11. *R-sectorial operators.* As we mentioned before, on Hilbert space (and only on Hilbert space), boundedness and R-boundedness are the same. On the other hand, it turns out that many results known for Hilbert spaces can be carried over to L^p-spaces, $1 < p < \infty$, if boundedness is replaced by R-boundedness. The operator-valued Marcinkiewicz of Section 6.1.5 and the Michlin multiplier theorem [Wei00a] are of this kind. We mention some further results. We say that an operator A is *R-sectorial* if there exists $0 < \varphi < \pi$ such that $\sigma(A) \subset \Sigma_\varphi$ and such that the set $\{\lambda R(\lambda, A)\colon \lambda \in \mathbb{C} \setminus \Sigma_\varphi\}$ is R-bounded. In that case let $\varphi_{R\,\mathrm{sec}}(A)$ be the infimum of all angles with these properties.

On a UMD-space an operator A has property (MR) if and only if $A + \omega$ is R-sectorial for some $\omega \in \mathbb{R}$.

In analogy to Section 5.1 one has the following results. Assume that $X = L^p(\Omega)$, $1 < p < \infty$.

If A has a bounded H^∞-calculus, then A is R-sectorial and $\varphi_{H^\infty}(A) = \varphi_{R\,\mathrm{sec}}(A)$. Moreover, if B is a second R-sectorial operator commuting with A such that $\varphi_{H^\infty}(A) + \varphi_{R\,\mathrm{sec}}(B) < \pi$, then $A + B$ is closed.

For this and many other interesting properties we refer to [Wei00a], [Wei00b] and [KW01].

7. Gaussian estimates and ultracontractivity

Frequently it is easy to define a semigroup on $L^2(\Omega)$ with the help of form methods for example. In this section we discuss under which conditions such a semigroup can be extended to $L^p(\Omega)$. An important item is to investigate whether semigroup properties (as holomorphy, maximal regularity, H^∞-calculus) extrapolate to L^p. Integral representations by Gaussian kernels play a big role.

7.1. *The Beurling–Deny criteria*

For simplicity we consider real spaces in this section. Let H be a real Hilbert space. Let $V \hookrightarrow_d H$ be another Hilbert space and $a : V \times V \to \mathbb{R}$ a continuous bilinear form which is *H-elliptic*, i.e.,

$$a(u, u) + \omega\|u\|_H^2 \geqslant \alpha\|u\|_V^2, \quad u \in V,$$

for some $\omega \in \mathbb{R}$, $\alpha > 0$. Recall that this is the same as saying that a is a closed form with form domain $D(a) = V$. Denote by A the operator associated with a (see Section 5.3.1). Then $-A$ generates a C_0-semigroup T on H which has a holomorphic extension to $H_\mathbb{C}$, the complexification of H. In the following we assume that $H = L^2(\Omega)$ for some measure space (Ω, Σ, μ) so that $H_\mathbb{C} = L^2(\Omega, \mathbb{C})$. We let $V_+ = V \cap L^2(\Omega)_+$.

7.1.1. Theorem (*First Beurling–Deny criterion*). *The semigroup T is positive if and only if $u \in V$ implies $u^+ \in V$ and*

$$a(u^+, u^-) \geqslant 0.$$

Here, for $u \in L^2(\Omega)$, we let $u^+(x) = \max\{u(x), 0\}$, $u^- = (-u)^+$. The next condition characterizes L^∞-contractivity. An operator S on $L^2(\Omega)$ is called *submarkovian* if $S \geqslant 0$ and $\|Sf\|_\infty \leqslant \|f\|_\infty$ for all $f \in L^2(\Omega) \cap L^\infty(\Omega)$. A semigroup T on $L^2(\Omega)$ is called *submarkovian* if each $T(t)$ is submarkovian.

7.1.2. *Theorem (Second Beurling–Deny criterion). Assume that the semigroup T is positive. Then T is submarkovian if and only if $u \wedge 1 \in V$ for each $u \in V_+$ and*

$$a\big(u \wedge 1, (u-1)^+\big) \geqslant 0.$$

Here we denote by 1 the function identically equal to 1. Note that $u \wedge 1 + (u-1)^+ = u$. We refer to [Ouh92], [[Ouh04]] for a proof of these criteria. Note that T^* is associated with $a^* : V \times V \to \mathbb{R}$ given by $a^*(u, v) = a(v, u)$. If T and T^* are submarkovian, then

$$\big\|T(t)f\big\|_{L^p} \leqslant \|f\|_{L^p} \tag{7.1}$$

for all $f \in L^p(\Omega) \cap L^2(\Omega)$, $t \geqslant 0$ and $p = 1, \infty$ at first; but by interpolation also for all $p \in (1, \infty)$. The form a is called a *Dirichlet form* if the two criteria of Beurling–Deny are satisfied. If in addition the form a is *symmetric* (i.e., $a = a^*$), then the semigroup T consists of self-adjoint operators satisfying (7.1). Each C_0-semigroup of symmetric submarkovian operators is associated with a symmetric Dirichlet form. The monographs [[FOT94]], [[MR86]], [[BH91]] are devoted to the theory of Dirichlet forms and their important and interesting interplay with stochastic processes. Here we are merely interested in analytical properties.

7.1.3. *Domination* [Ouh96]. *Let a and b be two closed forms on $L^2(\Omega)$ with associated semigroups S and T, respectively. Assume that S and T are positive. The following assertions are equivalent*:

(i) $S(t) \leqslant T(t)$, $t \geqslant 0$;

(ii) (a) $D(a) \subset D(b)$ *and for* $u \in D(b)$, *if* $0 \leqslant u \leqslant v \in D(a)$, *then* $u \in D(a)$ (ideal property), *and*

(b) $a(u, v) \leqslant b(u, v)$ *for all* $0 \leqslant u, v \in D(a)$ (monotonicity).

We mention two prototype examples.

7.1.4. *Examples.* Let $\Omega \subset \mathbb{R}^n$ be an arbitrary open set. The Dirichlet Laplacian Δ_Ω^D and the Neumann Laplacian Δ_Ω^N generate symmetric submarkovian semigroups satisfying $0 \leqslant e^{t\Delta_\Omega^D} \leqslant e^{t\Delta_\Omega^N}$, $t \geqslant 0$.

7.2. *Extrapolating semigroups*

Now we consider complex spaces. Let (Ω, Σ, μ) be a measure space and T a C_0-semigroup on $L^2(\Omega)$. We assume throughout this section that

$$\big\|T(t)\big\|_{\mathcal{L}(L^p)} \leqslant M, \quad 0 < t \leqslant 1, \tag{7.2}$$

for $p = 1, \infty$ and hence for all $p \in [1, \infty]$ by interpolation. Here, given an operator $S \in \mathcal{L}(L^2(\Omega))$ and $1 \leqslant p, q \leqslant \infty$, we let

$$\|S\|_{\mathcal{L}(L^p, L^q)} = \sup\{\|Sf\|_{L^q}\colon f \in L^p \cap L^2, \|f\|_{L^p} \leqslant 1\}. \tag{7.3}$$

Note that T and T^* are submarkovian if and only if $M = 1$. It follows from (7.2) that for suitable constants M_1, ω_1,

$$\|T(t)\|_{\mathcal{L}(L^p)} \leqslant M_1 e^{\omega_1 t}, \quad t \geqslant 0, \tag{7.4}$$

for all $p \in [1, \infty]$. From this we deduce that, for $1 \leqslant p < \infty$, there exist bounded operators $T_p(t) \in \mathcal{L}(L^p(\Omega))$ which are *consistent*, i.e.,

$$T_p(t)f = T_q(t)f, \quad f \in L^p \cap L^q, t > 0, \tag{7.5}$$

and such that $T_2(t) = T(t)$, $t > 0$. It is clear that $T_p(t+s) = T_p(t)T_p(s)$ for $t, s \geqslant 0$, $1 \leqslant p < \infty$. If $1 < p < 2$, we deduce from the interpolation inequality

$$\|T_p(t)f - f\|_{L^p} \leqslant \|T_1(t)f - f\|_{L^1}^{\theta} \|T_2(t)f - f\|_{L^2}^{1-\theta},$$

where $\frac{1}{p} = \frac{\theta}{1} + \frac{1-\theta}{2}$, so that $\lim_{t\downarrow 0} T_p(t)f = f$ in L^p for $f \in L^2 \cap L^1$. Thus T_p is a C_0-semigroup for $1 < p \leqslant 2$, and also for $2 < p < \infty$ by a similar argument. It is clear that $T_1 : (0, \infty) \to \mathcal{L}(L^1(\Omega))$ is strongly measurable, and hence strongly continuous [[Dav80], p. 18]. However, it seems to be unknown whether T_1 is a C_0-semigroup, in general. It is if one of the following conditions is satisfied.

7.2.1. *Conditions for T_1 being a C_0-semigroup.* *Assume that one of the following conditions is satisfied*:

(a) $M = 1$;

(b) *Ω has finite measure*;

(c) $T(t) \geqslant 0$, $t > 0$;

(d) *there exist an open set $\Omega' \supset \Omega$ and a C_0-semigroup S on $L^1(\Omega')$ such that*

$$|T(t)f| \leqslant S(t)|f| \quad \text{on } \Omega \text{ for all } f \in L^1(\Omega). \tag{7.6}$$

Then T_1 is a C_0-semigroup.

See [Voi92] for the cases (a)–(c) and [AtE97] for (d).

We assume throughout that one of the four conditions (a)–(d) is satisfied. Then T_1 is a C_0-semigroup. Observe that also T^* satisfies (7.2) and one of these conditions. This allows us to define the extension of T to $L^\infty(\Omega)$. In fact, $(T^*)_1$ is a C_0-semigroup. Now define $T_\infty(t) = (T^*)_1(t)^*$. Then T_∞ is a weak*-continuous semigroup whose generator we define by $A_\infty = (A_1^*)^*$. The consistency property (7.5) remains valid for $p = 1$ and $1 \leqslant q \leqslant \infty$. If Ω has finite measure, then of course, $T_\infty(t) = T(t)|_{L^\infty(\Omega)}$.

We call T_p the *extrapolation semigroup* of T on L^p, $1 \leqslant p \leqslant \infty$.

Now we want to investigate how properties of the semigroup $T = T_2$ are inherited by the extrapolation semigroups T_p. We denote by A_p the generator of T_p. Note that $A_\infty = (A_1^*)^*$.

7.2.2. *The Heritage list*

Property of T_2	Inherited by
(a) bounded generator	T_p for $1 < p < \infty$
(b) holomorphy	T_p for $1 < p < \infty$, not by T_1
(c) norm continuous on (t_0, ∞)	T_p for $1 < p < \infty$, not by T_1
(d) A_2 has compact resolvent	A_p for $1 < p < \infty$, not by A_1
(e) spectrum	$\sigma(A_p) \neq \sigma(A_q)$, $p \neq q$ in general, but $\sigma(A_2) = \sigma(A_p)$ for $1 < p < \infty$ if A_2 has compact resolvent
(f) positivity	T_p for $1 \leqslant p \leqslant \infty$
(g) positivity and irreducibility	T_p for $1 \leqslant p < \infty$
(h) (*MR*) for A_2	unknown
(i) H^∞-calculus for A_2	unknown
(j) *BIP*	unknown

7.2.3. *Proofs and comments.* We comment on the diverse properties. The first three (a)–(c) concern regularity of the semigroup.

(a) Let $1 < p < 2$, $\frac{1}{p} = \frac{\theta}{1} + \frac{1-\theta}{2}$. Then by the Riesz–Thorin theorem $\|T_p(t) - I\| \leqslant \|T_1(t) - I\|^\theta \|T_2(t) - I\|^{1-\theta} \to 0$ as $t \downarrow 0$. This implies that A_p is bounded.

(b) This is a consequence of Stein's interpolation theorem (cf. [[Dav90], Theorem 1.4.2]). But a similar proof as for (a) can be given for contraction semigroups using the following result (see [[Paz83], p. 68]).

THEOREM (Kato–Neuberger–Pazy). *Let $1 < p < \infty$ and let T be a C_0-semigroup on $L^p(\Omega)$ such that $\|T(t)\| \leqslant 1$, $t \geqslant 0$. Then T is holomorphic if and only if $\overline{\lim}_{t\downarrow 0}\|T(t) - I\| < 2$.*

Now the argument of (a) also gives a proof of (b).

(c) The argument of (a) also gives a proof of (c).

(d) This is [[Dav90], Theorem 1.6.1 and Corollary 1.6.2]. Properties (f) and (g) are trivial. Here a semigroup is called *positive*, if it leaves the real space invariant and its restriction to the real space is positive.

(i) Example 4.5.3 provides a counterexample where the semigroups extrapolate to L^p merely for $1 < p < \infty$, the operator A_2 has H^∞-calculus, but A_p has not H^∞-calculus for $p \neq 2$. We do not know such an example where the semigroups extrapolate to L^p for $1 \leqslant p \leqslant \infty$.

Next we give some interesting examples showing, in particular, that the positive results in (b), (c) or (d) cannot be extended to $p = 1$.

7.2.4. *The harmonic oscillator* [[Dav90], Section 4.3]. Consider the operator given by

$$A_p u = \frac{1}{2}\left(-u'' + x^2 u - u\right)$$

on $L^p(\mathbb{R}, \mathrm{e}^{-x^2}\,\mathrm{d}x)$ with maximal domain. Then A_p generates a C_0-semigroup T_p on $L^p(\mathbb{R}, \mathrm{e}^{-x^2}\,\mathrm{d}x)$, $1 \leqslant p < \infty$. This family is consistent. Moreover, T_p is holomorphic for $1 < p < \infty$, but T_1 is not norm continuous on $[t_0, \infty)$ for any $t_0 > 0$. A_p has a compact resolvent for $1 < p < \infty$. One has $\sigma(A_p) = \mathbb{N}_0$ for $1 < p < \infty$ but $\sigma(A_1) = \{\lambda \in \mathbb{C}\colon \operatorname{Re}\lambda \geqslant 0\}$. Thus A_1 has not a compact resolvent.

7.2.5. *Black–Scholes equation.* For $1 \leqslant p < \infty$, let the operator B_p on $L^p(0, \infty)$ be given by

$$(B_p)(x) = x^2 u''(x) + 2xu'(x),$$
$$D(B_p) = \left\{u \in L^p(0, \infty)\colon xu' \in L^p(0, \infty), x^2 u'' \in L^p(0, \infty)\right\}.$$

Then B_2 is associated with the symmetric Dirichlet form a given by

$$a(u, v) = \int_0^\infty u'v'x^2\,\mathrm{d}x,$$
$$D(a) = \left\{u \in L^2(0, \infty)\colon xu' \in L^2(0, \infty)\right\}$$

and thus generates a symmetric Markovian semigroup T_2. The extrapolation semigroup T_p has B_p as generator. The spectrum of B_p is the parabola $\sigma(B_p) = \{(1/p - 1/2 + \mathrm{i}s)^2 - 1/4\colon s \in \mathbb{R}\}$, $1 \leqslant p < \infty$. In particular,

$$\sigma(B_p) \cap \sigma(B_q) = \emptyset \quad \text{if } 1 \leqslant p < q \leqslant 2.$$

Note that T_p governs the Black–Scholes partial differential equation

$$u_t = x^2 u_{xx} + 2xu_x.$$

Reference: [Are94], Section 3, Example 3.

7.2.6. *Symmetric submarkovian semigroups, optimal angles and the Neumann Laplacian on horn domains.* Let A_2 be the generator of a symmetric submarkovian semigroup T_2 on $L^2(\Omega)$ and denote by T_p the C_0-semigroup on $L^p(\Omega)$ with generator A_p extrapolating T_2, $1 \leqslant p < \infty$. Then Liskevich and Perelmuter [LP95] showed that the angle obtained by the Stein interpolation theorem can be improved. In fact, T_p has a holomorphic, contractive extension to the sector Σ_{θ_p}, where $0 \leqslant \theta_p < \frac{\pi}{2}$ such that $\cos\theta_p = |1 - 2/p|$. In particular, the spectrum of A_p is contained in the sector

$$S_p := \left\{r\mathrm{e}^{\mathrm{i}\theta}\colon r \geqslant 0, \frac{\pi}{2} + \theta_p \leqslant |\theta| \leqslant 2\pi\right\}, \quad 1 \leqslant p \leqslant \infty.$$

It was Voigt [Voi96] who first showed that this sector is optimal. In fact, modifying Example 7.2.5 he constructed a degenerate elliptic operator A_p on $L^p(0,\infty)^2$ such that the spectrum of A_p is equal to S_p.

Another interesting example in this context is the symmetric Ornstein–Uhlenbeck operator A_p on $L^p(\mathbb{R}^n,\mu)$, where μ is the invariant measure. It generates a symmetric submarkovian semigroup T_p on $L^p(\mathbb{R}^n,\mu)$. Metafune, Pallara and Priola [MPP02] computed explicitly the spectrum and showed that it is independent of $p \in [1,\infty)$. In particular, the spectrum of A_p is real for all $p \in [1,\infty)$. However, Chill, Fašangová, Metafune and Pallara [CFMP03] showed that Σ_{θ_p} is the sector of holomorphy of T_p, i.e., the worst case is realized for the symmetric Ornstein–Uhlenstein semigroup.

Kunstmann [Kun02] constructed an open unbounded domain Ω in $\mathbb{R}^n$ (which is composed by infinitely many "horn domains") such that the Neumann Laplacian has bad interpolation properties. In particular, for a domain Ω composed by infinitely many horn domains the spectrum of the Neumann Laplacian $\Delta^N_{\Omega,p}$ in $L^p(\Omega)$ depends on p. Moreover, the semigroup $(e^{t\Delta^N_{\Omega,1}})_{t\geqslant 0}$ is not holomorphic in $L^1(\Omega)$. In fact, it is not even eventually norm continuous since the spectrum of $\Delta^N_{\Omega,1}$ is the left-half plane.

Concerning the functional calculus, one can say the following: given a symmetric submarkovian semigroup T_2 with extension semigroups T_p on $L^p(\Omega)$, by Section 4.7.5 the negative generator A_p of T_p has a bounded H^∞-calculus and

$$\varphi_{R\sec}(A_p) = \varphi_{H^\infty}(A_p), \qquad 1 < p < \infty.$$

Cowling [Cow83] showed that $\varphi_{H^\infty}(A_p) \leqslant \varphi_s(A_p)$, where $\varphi_s(A_p) = \pi/2 - \pi|1/2 - 1/p|$ is the angle obtained by the Stein interpolation theorem. However Kunstmann and Strkalj [KS03] showed that $\varphi_{H^\infty}(A_p) < \varphi_s(A_p)$ for $1 < p < \infty$, $p \neq 2$. Still, it seems to be open whether always $\varphi_{H^\infty}(A_p) = \theta_p$.

7.3. *Ultracontractivity, kernels and Sobolev embedding*

In this section we consider a semigroup T on $L^2(\Omega)$ which is regularizing in the L^p-sense: We will ask that $T(t)L^2(\Omega) \subset L^q(\Omega)$ for some $q > 2$. This property implies in particular that $T(t)$ is an integral operator. Throughout this section (Ω, Σ, μ) is a measure space.

7.3.1. *The Dunford–Pettis criterion.* Let $K \in L^\infty(\Omega \times \Omega)$. Then

$$(S_K f)(x) = \int_\Omega K(x,y) f(y)\,\mathrm{d}y \tag{7.7}$$

defines a bounded operator $S_K \in \mathcal{L}(L^1(\Omega), L^\infty(\Omega))$.

THEOREM (see, e.g., [AB94]). *The mapping $K \mapsto S_K$ is an isometric isomorphism from $L^\infty(\Omega\times\Omega)$ onto $\mathcal{L}(L^1(\Omega), L^\infty(\Omega))$. Moreover, $K(x,y) \geqslant 0$ a.e. if and only if $S_K \geqslant 0$.*

7.3.2. *Ultracontractive semigroups.* Let T be a C_0-semigroup on $L^2(\Omega)$. We assume that (7.2) is satisfied and denote by T_p the extrapolation semigroup of T on $L^p(\Omega)$. Thus T_p is a C_0-semigroup for $1 < p < \infty$, we suppose that this be true also for $p = 1$ (cf. Section 7.2.1). Denote by A_p the generator of T_p. It follows from (7.2) that

$$\|T_p(t)\| \leqslant M_1 e^{\omega t}, \quad t \geqslant 0, \tag{7.8}$$

for some $M_1 \geqslant 1$, $\omega \in \mathbb{R}$ and all $p \in [1, \infty]$. Thus $(\omega - A_p)$ is sectorial for all $1 \leqslant p < \infty$ by Section 4.1.3 and the fractional powers $(\omega - A_p)^\alpha$ are defined for all $\alpha \geqslant 0$. Note that

$$D\big((\omega - A_p)^\alpha\big) = D\big((\omega_1 - A_p)^\alpha\big) = (\omega_1 - A_p)^{-\alpha} L^p(\Omega)$$

for all $\omega_1 > \omega$.

THEOREM. *Let $n > 0$ be a real number. Consider the following conditions.*

(i) *There exist $c > 0$, $1 \leqslant p < q \leqslant \infty$ such that*

$$\|T(t)\|_{\mathcal{L}(L^p, L^q)} \leqslant c t^{-\frac{n}{2}|\frac{1}{p} - \frac{1}{q}|} \quad \textit{for all } 0 < t \leqslant 1.$$

(ii) *There exists a constant $c > 0$ such that for all $1 \leqslant p < q \leqslant \infty$,*

$$\|T(t)\|_{\mathcal{L}(L^p, L^q)} \leqslant c t^{-\frac{n}{2}|\frac{1}{p} - \frac{1}{q}|}, \quad 0 < t \leqslant 1.$$

(iii) *There exist $1 < p < \infty$ and $0 < \alpha < \frac{n}{2p}$ such that $D((\omega - A_p)^\alpha) \subset L^q$, where q is defined by $\alpha = \frac{n}{2}(\frac{1}{p} - \frac{1}{q})$.*

(iv) *For all $0 < \alpha < \frac{n}{2p}$ and all $1 < p < \infty$ one has $D((\omega - A_p)^\alpha) \subset L^q$, where q is defined by $\alpha = \frac{n}{2}(\frac{1}{p} - \frac{1}{q})$.*

Then (i) $\Leftrightarrow$ (ii) $\Rightarrow$ (iv) $\Rightarrow$ (iii). *If T is holomorphic then all four assertions are equivalent.*

If the generator A of T is associated with a closed form with form domain $V \hookrightarrow L^2(\Omega)$ such that $V \cap L^1(\Omega)$ is dense in $L^1(\Omega)$ and if $n > 2$, then these four conditions are also equivalent to

(v) $V \hookrightarrow L^{2n/(n-2)}(\Omega)$.

We call a semigroup *ultracontractive* if the equivalent conditions (i) and (ii) are satisfied. The number $\dim(T) = \inf\{n > 0\colon \text{(i) is valid}\}$ is called the *semigroup dimension*. Note that n is not entire, in general. It was Varopoulos who startet a systematic investigation of the dimension of a semigroup, mainly in the framework of symmetric submarkovian semigroups. We refer to [[VSC93]], [[Sal02]], [[Dav90]] for the historical references and further results.

PROOF OF THE THEOREM. (i) $\Leftrightarrow$ (ii): Assertion (ii) for $p = 1, q = \infty$ becomes

$$\text{(ii)}' \quad \|T(t)\|_{\mathcal{L}(L^1, L^\infty)} \leqslant \text{const} \cdot t^{-n/2}, \quad 0 < t \leqslant 1.$$

The Riesz–Thorin theorem allows one to go from (ii)$'$ to (ii). The following trick due to Coulhon [Cou90] allows one to show that (i) implies (ii)$'$. It is done in two steps. We assume (i).

(a) We show that $\|T(t)\|_{\mathcal{L}(L^1,L^q)} \leqslant \text{const}\cdot t^{-\frac{n}{2}(1-\frac{1}{q})}$, $0 \leqslant t \leqslant 1$. Let $\alpha = \frac{n}{2}(\frac{1}{p}-\frac{1}{q})$, $\beta = \frac{n}{2}(1-\frac{1}{q})$. Choose $0<\theta<1$ such that $\frac{1}{p} = \frac{\theta}{1} + \frac{1-\theta}{q}$. Then $\frac{1}{p}-\frac{1}{q} = \theta(1-\frac{1}{q})$, i.e., $\alpha = \theta\beta$. By the hypothesis (i), $\|T(t)\|_{\mathcal{L}(L^p,L^q)} \leqslant \text{const}\cdot t^{-\alpha}$ and we want to show that

$$\|T(t)\|_{\mathcal{L}(L^1,L^q)} \leqslant \text{const}\cdot t^{-\beta}.$$

Let

$$f \in L^1 \cap L^\infty, \quad \|f\|_{L^1} \leqslant 1,$$
$$c_f := \sup_{0<t\leqslant 1} t^\beta \|T(t)f\|_{L^q}.$$

Then

$$\begin{aligned}\|T(t)f\|_{L^q} &= \|T(t/2)T(t/2)f\|_{L^q}\\ &\leqslant \text{const}\cdot t^{-\alpha}\|T(t/2)f\|_{L^p}\\ &\leqslant \text{const}\cdot t^{-\alpha}\|T(t/2)f\|_{L^1}^{\theta}\|T(t/2)f\|_{L^q}^{1-\theta}\\ &\leqslant \text{const}\cdot t^{-\alpha}t^{-\beta(1-\theta)}(c_f)^{1-\theta}\\ &\leqslant \text{const}\cdot t^{-\beta}(c_f)^{1-\theta}.\end{aligned}$$

Hence, $c_f = \sup t^\beta\|T(t)f\|_{L^q} \leqslant \text{const}\cdot(c_f)^{1-\theta}$. Thus $c_f \leqslant \text{const}^{1/\theta}$, where const does not depend of f. This proves the claim.

(b) It follows from (a) by duality that

$$\begin{aligned}\|T(t)\|_{\mathcal{L}(L^{q'},L^\infty)} &\le \text{const}\cdot t^{-\frac{n}{2}(1-\frac{1}{q})}\\ &= \text{const}\cdot t^{-\frac{n}{2}(\frac{1}{q'}-\frac{1}{\infty})}.\end{aligned}$$

Applying (a) to this gives (ii)$'$.

(i) $\Rightarrow$ (iv). The proof of [[Dav90], Theorem 2.4.2] based on the Marcinkiewicz interpolation theorem carries over to this case.

(iii) $\Rightarrow$ (i). Assume that T is holomorphic. Replacing A by $A-\omega$ we can assume that T is exponentially stable. Let $q = \frac{np}{n-2\alpha p}$. Since $\|t^\alpha A_p^\alpha T_p(t)f\| \leqslant \text{const}\cdot\|f\|_p$, $0<t\leqslant 1$, and $D(A_p^\alpha) \hookrightarrow L^q$ by hypothesis, we have

$$\|T_p(t)f\|_q \le \text{const}\cdot\|T_p(t)f\|_{D(A_p^\alpha)}$$

$$= \text{const} \cdot \left\| A_p^\alpha T_p(t) f \right\|_{L^p}$$
$$\leq \text{const} \cdot t^{-\alpha} \| f \|_{L^p}.$$

Since $\alpha = \frac{n}{2}(\frac{1}{p} - \frac{1}{q})$, the claim follows.

Now assume that A is associated with a closed form.

(v) $\Rightarrow$ (i). Since $\|T(t)\|_{\mathcal{L}(L^2,V)} \leqslant \text{const} \cdot t^{-1/2}$, $0 < t \leqslant 1$ (see [[Tan79]]), the argument is as in the previous implication for $p = 2$.

(i) $\Rightarrow$ (v). As we know from Section 5.5.1 it may happen that $V \neq D(A^{1/2})$. Thus we cannot use the previous implication (i) $\Rightarrow$ (iv). However, for forms, ultracontractivity can be characterized by Nash's inequality

$$\|f\|_2^{2+4/n} \leqslant \text{const} \cdot \|f\|_V^2 \|f\|_1^{4/n}, \quad f \in V \cap L^1(\Omega),$$

by the proof of [[Dav90], Theorem 2.4.6]. Thus, assuming (i), Nash's inequality holds. Consider the symmetric form $b(u, v) = \frac{1}{2}(a(u, v) + \overline{a(v, u)})$ with domain V. Denote by B the operator associated with b and by S the semigroup generated by $-B$. It follows from the characterization via Nash's inequality we just mentioned that S satisfies (i). Hence, $D(B^{1/2}) \subset L^{2n/n-2}$ by the implication (i) $\Rightarrow$ (iv). But $V = D(B^{1/2})$ since B is self-adjoint. □

Now we show some further properties of ultracontractive semigroups and give examples.

7.3.3. *Kernels and compactness.* Assume that T is an ultracontractive semigroup on $L^2(\Omega)$. Then

$$\left\|T(t)\right\|_{\mathcal{L}(L^1,L^\infty)} \leqslant M_2 t^{-n/2} e^{\omega_2 t} \quad \text{for all } t > 0 \tag{7.9}$$

and for some $M_2 \geqslant 1$, $\omega_2 \in \mathbb{R}$. Thus, by the Dunford–Pettis criterion there exists a kernel $K_t \in L^\infty(\Omega \times \Omega)$ such that

$$\big(T_p(t) f\big)(x) = \int_\Omega K_t(x, y) f(y)\, dy, \quad x\text{-a.e.} \tag{7.10}$$

for all $f \in L^p(\Omega) \cap L^1(\Omega)$. Moreover,

$$\left|K_t(x, y)\right| \leqslant M_2 e^{\omega_2 t} \cdot t^{-n/2}, \quad t > 0. \tag{7.11}$$

Now assume that Ω has finite measure. Then $T(t)$ is a Hilbert–Schmidt operator and hence compact. This property extends to the entire L^p scale, $1 \leqslant p \leqslant \infty$, in virtue of ultracontractivity. In fact, given $t > 0$, $1 \leqslant p \leqslant \infty$, we can factorize $T_p(t)$ as follows:

$$L^p \overset{T_p(t/3)}{\longrightarrow} L^\infty \hookrightarrow L^2 \overset{T_2(t/3)}{\longrightarrow} L^2 \overset{T_2(t/3)}{\longrightarrow} L^\infty \hookrightarrow L^p.$$

Thus $T_p(t)$ is compact. We have shown the following.

PROPOSITION. *Let T be an ultracontractive semigroup on $L^2(\Omega)$ with extrapolating semigroups T_p, $1 \leqslant p \leqslant \infty$. If Ω has finite measure, then $T_p(t)$ is a compact operator on $L^p(\Omega)$ for all $t > 0$, $1 \leqslant p \leqslant \infty$.*

7.3.4. *The Gaussian semigroup and Sobolev embedding.* As an illustrating example we consider the Gaussian semigroup G_p on $L^p(\mathbb{R}^n)$ given by the kernel

$$K_t^G(x, y) = (4\pi t)^{-n/2} \mathrm{e}^{-|x-y|^2/4t}. \tag{7.12}$$

Thus (i) is satisfied for $p = 1, q = \infty$, and in this case the semigroup dimension n is the dimension of the space $\mathbb{R}^n$. Denote the generator of G_p by Δ_p. Then $D(\Delta_p) = W^{2,p}(\mathbb{R}^n)$, $1 < p < \infty$.

PROOF. The operator $G_p(t)$ is a Fourier multiplier with symbol $\mathrm{e}^{-\xi^2 t}$. Thus $R(1, \Delta_p)$ has symbol $(1 + \xi^2)^{-1}$. Since for

$$f \in \varphi(\mathbb{R}^n)', \quad \mathcal{F}(D_j f) = \mathrm{i}\xi_j \mathcal{F} f,$$

we have to show that $\xi_i \xi_j (1 + \xi^2)^{-1}$ is an L^p-Fourier multiplier (in order to deduce that $D(\Delta_p) = R(1, \Delta_p) L^p(\mathbb{R}^n) \subset W^{2,p}(\mathbb{R}^n)$). This follows from Michlin's theorem. □

Similarly, one sees with the help of Michlin's theorem that

$$D(\Delta_p^m) = W^{2m,p}(\mathbb{R}^n), \quad 1 < p < \infty, m \in \mathbb{N}.$$

Thus, the conclusion (iv) of the Theorem in Section 7.3.2 is the usual Sobolev embedding theorem. We remark that such results are of particular interest if they are applied to the Laplace–Beltrami operator on a Riemannian manifold (see [[VSC93]]).

7.3.5. *The Dirichlet Laplacian.* Let $\Omega \subset \mathbb{R}^n$ be an arbitrary open set and consider the Dirichlet Laplacian Δ_Ω^D on $L^2(\Omega)$. Then $(\mathrm{e}^{t\Delta_\Omega^D})$ is self-adjoint, submarkovian and ultracontractive since the form domain $W_0^{1,2}(\Omega)$ is in L^q, where $q = \frac{2n}{n-2}$ if $n > 2$ and arbitrary $q < \infty$ if $n \leqslant 2$. Denote by $\Delta_{\Omega,p}^D$ the generator of the extrapolation semigroup, $1 \leqslant p \leqslant \infty$. Then we deduce from the Theorem in Section 7.3.2 that

$$D\big((-\Delta_{\Omega,p}^D)^\alpha\big) \subset L^{np/(n-2\alpha p)} \tag{7.13}$$

whenever $0 < \alpha < \frac{n}{2p}$. On the other hand, if Ω is irregular, then it can happen that $D(\Delta_{\Omega,p}^D) \not\subset W^{1,p}(\Omega)$ for $p > 2$, see [[Gri92]]. Thus Sobolev embedding results cannot be used here to prove (7.13).

7.3.6. *The extension property.* An open set $\Omega \subset \mathbb{R}^n$ has the *extension property* if the restriction mapping $R: W^{1,2}(\mathbb{R}^n) \to W^{1,2}(\Omega)$ is surjective. Then $R_0: (\ker R)^\perp \to W^{1,2}(\Omega)$ is an isomorphism and $E := R_0^{-1}$ is a bounded operator from $W^{1,2}(\Omega)$ into $W^{1,2}(\mathbb{R}^n)$ such that $(Eu)|_\Omega = u$ for all $u \in W^{1,2}(\Omega)$. We call E an *extension operator*. If Ω is bounded and has Lipschitz boundary, then Ω has the extension property. If Ω has the extension property then

$$W^{1,2}(\Omega) \subset L^q(\Omega) \tag{7.14}$$

for $q = \frac{2n}{n-2}$ if $n > 2$ and for all $q < \infty$ if $n \leqslant 2$. This follows from the case $\Omega = \mathbb{R}^n$. Property (7.14) implies ultracontractivity in many interesting examples. Here we consider the Neumann Laplacian as prototype:

7.3.7. *The Neumann Laplacian.* Let $\Omega \subset \mathbb{R}^N$ be open. The Neumann Laplacian Δ_Ω^N generates a symmetric submarkovian semigroup on $L^2(\Omega)$. If Ω has the extension property then $(e^{t\Delta_\Omega^N})_{t\geqslant 0}$ is ultracontractive. However, this is not true without this hypothesis. For example, in dimension 1, we may consider $\Omega = (0,1) \setminus \{1/n : n \in \mathbb{N}\}$. Then 0 is an eigenvalue of infinite multiplicity of Δ_Ω^N, hence, the resolvent is not compact. In particular, the semigroup is not ultracontractive.

7.3.8. *Asymptotic behavior of $T_p(t)$ as $t \to \infty$. Let T be an ultracontractive, positive, irreducible C_0-semigroup on $L^2(\Omega,\mu)$, where Ω has finite measure. Then there exist $u, \varphi \in L^\infty(\Omega)$ such that $u(x) > 0$, $\varphi(x) > 0$ a.e., $\int_\Omega u(x)\varphi(x)\,d\mu = 1$ and $\omega \in \mathbb{R}$, $\varepsilon > 0$, $M \geqslant 0$ such that*

$$\left\| e^{-\omega t} T_p(t) - P \right\|_{\mathcal{L}(L^p(\Omega))} \leqslant M e^{-\varepsilon t}, \quad t \geqslant 0,$$

for $1 \leqslant p \leqslant \infty$, where $Pf = \int_\Omega f(x)\varphi(x)\,d\mu(x) \cdot u$, $f \in L^1(\Omega)$.

PROOF. By Section 7.3.3 the operators $T_p(t)$ are compact for $1 \leqslant p \leqslant \infty$, $t > 0$. It follows that A_p has compact resolvent. Thus $\sigma(A_p)$ is independent of $p \in [1,\infty]$ (see [[Dav90], p. 36]). In particular, $\omega := s(A_p)$ is independent of $p \in [1,\infty]$. Now the claim follows from Section 3.5.1 for $1 \leqslant p < \infty$. For $p = \infty$ one may use a duality argument. □

A particular example is $T(t) = e^{t\Delta_\Omega^n}$ where Ω is a connected, bounded open set in $\mathbb{R}^n$ with Lipschitz boundary. In this case $Pf = \frac{1}{|\Omega|}\int_\Omega f\,dx \cdot 1_\Omega$ for all $f \in L^p(\Omega)$.

7.4. *Gaussian estimates*

Let $\Omega \subset \mathbb{R}^n$ be an open set and let T be a C_0-semigroup on $L^2(\Omega)$ (the L^2-space of complex-valued functions). We identify $L^2(\Omega)$ with a subspace of $L^2(\mathbb{R}^n)$ extending functions on Ω by 0 on $\mathbb{R}^N \setminus \Omega$. Consider the Gaussian semigroup G on $L^2(\mathbb{R}^n)$.

DEFINITION. The semigroup T admits a *Gaussian estimate* if there exist constants $c > 0$, $b > 0$ such that

$$\left|T(t)f\right| \leqslant cG(bt)|f|, \quad 0 < t \leqslant 1, \tag{7.15}$$

for all $f \in L^2(\Omega)$.

Note that (7.15) is an inequality between measurable functions in the almost everywhere sense. It is not difficult to show that it implies an estimate of the form

$$|T(t)f| \leqslant M\mathrm{e}^{\omega t} G(bt)|f| \tag{7.16}$$

for all $t > 0$, $f \in L^2(\Omega)$. By the Dunford–Pettis criterion this is equivalent to saying that $T(t) = S_{K_t}$, where $K_t \in L^\infty(\Omega \times \Omega)$, is a kernel satisfying

$$|K_t(x, y)| \leqslant \text{const} \cdot \mathrm{e}^{\omega t} t^{-n/2} \mathrm{e}^{-|x-y|^2/4bt}, \quad (x, y)\text{-a.e.} \tag{7.17}$$

for all $t > 0$. If T admits a Gaussian estimate then by Section 7.2 (and in particular, Section 7.2.1(d)) there exist consistent extrapolation semigroups T_p on $L^p(\Omega)$, $1 \leqslant p \leqslant \infty$, such that $T_2 = T$. For $1 \leqslant p < \infty$, T_p is a C_0-semigroup, T_∞ is w^*-continuous. By A_p we denote the generator of T_p. These notations will be used in the following without further notice. Our aim is to establish consequences of Gaussian estimates for T_p which allow us in particular to replace some of the negative answers in the heritage list of Section 7.2.2 by positive assertions. Later we will see that a large class of elliptic operators generate semigroups which satisfy Gaussian estimates. But before that we give two prototype examples.

7.4.1. *Examples.* (a) *Let $\Omega \subset \mathbb{R}^n$ be an arbitrary open set. The semigroup generated by the Dirichlet Laplacian Δ_Ω^D on $L^2(\Omega)$ satisfies*

$$|\mathrm{e}^{t\Delta_\Omega^D} f| \leqslant G(t)|f|, \quad t > 0, \tag{7.18}$$

for all $f \in L^2(\Omega)$.

(b) *Assume that Ω has the extension property. Then the semigroup $(\mathrm{e}^{t\Delta_\Omega^N})_{t\geqslant 0}$ generated by the Neumann Δ_Ω^N Laplacian admits a Gaussian estimate.*

PROOF. (a) Let $0 \leqslant f \in L^2(\mathbb{R}^n)$. It suffices to show that $u := R(\lambda, \Delta_\Omega^D)f \leqslant R(\lambda, \Delta)f =: v$ on Ω. By the definition via forms we have

$$\lambda \int_\Omega u\varphi + \int_\Omega \nabla u \nabla \varphi = \int_\Omega f\varphi, \varphi \in W_0^{1,2}(\Omega),$$

$$\lambda \int_{\mathbb{R}^n} v\varphi + \int_{\mathbb{R}^n} \nabla v \nabla \varphi = \int_{\mathbb{R}^n} f\varphi, \varphi \in W^{1,2}(\mathbb{R}^n).$$

Hence, $\lambda \int_\Omega (u - v)\varphi + \int_{\mathbb{R}^n} \nabla(u - v)\nabla\varphi = 0$ for all $\varphi \in W_0^{1,2}(\Omega)$. Taking $\varphi = (u - v)^+$ one obtains that $(u - v)^+ = 0$.

(b) In this generality, this is due to Ouhabaz [Ouh03]. For a stronger version of the extension property the result is proved in [[Dav90], p. 90]. □

7.4.2. *Remark* (*Proper domination*). There is a significant difference between the two examples: For the Dirichlet Laplacian we have *proper domination*; i.e., the constant c in (7.15) is equal to 1. But this is the only realization of the Laplacian with this property:

Let $\Omega \subset \mathbb{R}^n$ be open. We assume that Ω is stable (*see Section* 5.4.5). *Let S be a positive, symmetric C_0-semigroup on $L^2(\Omega)$ such that $S(t) \leqslant G_2(t)$, $0 < t \leqslant 1$. Assume that the generator A of S satisfies $\mathcal{D}(\Omega) \subset D(A)$ and $Au = \Delta u$ for all $u \in \mathcal{D}(\Omega)$. Then $A = \Delta_\Omega^D$.*

PROOF. (a) We consider $L^2(\Omega)$ as a subspace of $L^2(\mathbb{R}^n)$, extending functions by 0. Then $S(t) = (1_\Omega S(t/n))^n \leqslant (1_\Omega G_2(t/n))^n$ for all $n \in \mathbb{N}$. Since $\lim_{n\to\infty}(1_\Omega G(t/n))^n = e^{t\Delta_\Omega^D}$ strongly by Section 5.4.5, we conclude that $S(t) \leqslant e^{t\Delta_\Omega^D}$, $0 < t < \infty$.

(b) Denote by $a: V \times V \to \mathbb{R}$ the symmetric closed form associated with $-A$. It follows from (a) and Section 7.1.3 that $V \subset H_0^1(\Omega)$. For $u, v \in \mathcal{D}(\Omega)$, we have $a(u, v) = -(Au|v)_{L^2} = \int \nabla u \nabla v \, dx$. Note that $\|u\|_{L^2}^2 + a(u, u)$ defines an equivalent norm on V. Thus $\|\cdot\|_V$ and $\|\cdot\|_{H^1}$ are equivalent. Since $\mathcal{D}(\Omega)$ is dense in $H_0^1(\Omega)$, it follows that $H_0^1(\Omega) \subset V$. Thus $V = H_0^1(\Omega)$ and $a(u, v) = \int_\Omega \nabla u \nabla v \, dx$ for all $u, v \in H_0^1(\Omega)$. □

The fact that the constant c in (7.15) might be larger than 1 makes it difficult to prove Gaussian estimates, and no such simple criteria as in Section 7.1.3 are available. We comment on techniques to prove Gaussian estimates in Section 8.7. Now we establish heritage properties made possible by Gaussian estimates.

7.4.3. *Gaussian estimates and extrapolation of holomorphy. Let T be a C_0-semigroup on $L^2(\Omega)$ which admits a Gaussian estimate. Assume that T is holomorphic of angle $\theta \in (0, \pi/2]$. Then also the extrapolation semigroups T_p are holomorphic of the same angle θ for each $p \in [1, \infty]$.*

This result due to Ouhabaz [Ouh95] (see [AtE97], Theorem 5.4, in the nonsymmetric case) contrasts the examples in Sections 7.2.4 and 7.2.6, where, in absence of Gaussian estimates, the angle depends on p and where T_1 is not holomorphic.

7.4.4. *Gaussian estimates and maximal regularity. Let T be a C_0-semigroup on $L^2(\Omega)$ which has Gaussian estimates. If T is holomorphic, then T_p has property* (*MR*) *for $1 < p < \infty$.*

This result is most important for applications to nonlinear equations. We refer to Section 6.2.6 for the definition of (*MR*) and to [HP97], [CP01], [CD00], [Wei00b] and [ArBu03a], Corollary 4.5, for proofs of this result.

Recall that holomorphy is a necessary condition for (*MR*). It may happen that a semigroup T on $L^2(\Omega)$ has Gaussian estimates without being holomorphic. In fact, Voigt showed that the operator $\Delta + ix$ on $L^2(\mathbb{R})$ with suitable domain generates a C_0-semigroup S such that $|S(t)f| \leqslant G(t)|f|$, but S is not holomorphic, cf. [LM97], p. 303. However, so far no example of a nonholomorphic positive semigroup with Gaussian estimates is known.

7.4.5. *Gaussian estimates and H^∞-calculus. Let $-A$ be the generator of a holomorphic C_0-semigroup on $L^2(\Omega)$ admitting a Gaussian estimate. Assume that $(A + \omega)$ has* *a bounded H^∞-calculus for some $\omega \in \mathbb{R}$. Then also $(A_p+\omega)$ has a bounded H^∞-calculus for some $\omega \in \mathbb{R}$, for all $1 < p < \infty$.*

This result is due to Duong and Robinson [DR96]. The restriction on Ω (doubling property) made there was removed later by Duong and McIntosh [DM99]. We do not try to optimize the rescaling constant ω in our formulation, but rather refer to [DR96] and [AtE97], pp. 118 and 122, for this.

We recall that on Hilbert space it suffices that the semigroup T be quasicontractive to insure $(A+\omega)$ to have a bounded H^∞-calculus for some ω.

7.4.6. *Gaussian estimates and p-independence of the spectrum.* *Let T be a C_0-semigroup on $L^p(\Omega)$ admitting a Gaussian estimate. Denote by T_p the extrapolation C_0-semigroup on $L^p(\Omega)$ and by A_p its generator. Then $\sigma(A_p)=\sigma(A_2)$, $1 \leqslant p \leqslant \infty$.*

In [Are94] it was proved that the connected component of $\rho(A_p)$ is p-independent, the general result was obtained by Kunstmann [Kun99]. An example of special interest are Schrödinger operators for which the result had been obtained before by Hempel and Voigt [HV86].

8. Elliptic operators

In this section we apply the results obtained before to semigroups generated by elliptic operators of second order. If the coefficients and the domain in $\mathbb{R}^n$ are smooth, then classical estimates of Agmon–Douglis–Nirenberg give precise information on the domain. Here we consider merely measurable coefficients. Then the domain can no longer be determined. Under some mild conditions, however, Gaussian estimates can still be proved and lead to a variety of semigroup and spectral properties. At the end of this chapter we mention also some results for higher order operators and systems.

Let $\Omega \subset \mathbb{R}^n$ be an open set. Let $a_{ij}, b_i, c_i, a_0 \in L^\infty(\Omega)$, $i, j = 1, \dots, n$, be complex-valued coefficients. We assume the ellipticity condition

$$\operatorname{Re} \sum_{i,j=1}^{n} a_{ij}(x)\xi_i\bar{\xi}_j \geqslant \alpha|\xi|^2 \quad \text{for all } \xi \in \mathbb{C}^n,\ x\text{-a.e.},$$

where $\alpha > 0$. Then we consider the elliptic operator

$$L : W^{1,2}_{\mathrm{loc}}(\Omega) \to \mathcal{D}(\Omega)'$$

given by

$$Lu = -\sum_{i,j=1}^{n} D_i(a_{ij}D_j u) + \sum_{i=1}^{n}\bigl(b_i D_i u - D_i(c_i u)\bigr) + a_0 u.$$

With the help of forms we will define various realizations of L in $L^2(\Omega)$ corresponding to diverse boundary conditions.

Let V be a closed subspace of $W^{1,2}(\Omega)$ containing $W_0^{1,2}(\Omega)$. We define the form $a_V : V \times V \to \mathbb{C}$ by

$$a_V(u, v) = \int_\Omega \left[\sum_{i,j=1}^{n} a_{ij}(x) D_i u \overline{D_j v} + \sum_{i=1}^{n} \left(b_i D_i u \bar{v} + c_i u \overline{D_i v} \right) + a_0 u \bar{v} \right] \mathrm{d}x.$$

Then a_V is continuous and $L^2(\Omega)$-elliptic. Denote by A_V the operator on $L^2(\Omega)$ associated with a_V. Then $-A_V$ generates a holomorphic C_0-semigroup T_V on $L^2(\Omega)$, and we will investigate various properties of T_V. At first we describe the operator A_V more precisely. It follows from the definition of the associated operator that

$$A_V u = Lu$$

for all $u \in D(A_V)$. We will consider in particular three different boundary conditions which we describe in the following section.

8.1. *Boundary conditions*

(a) Let $V = W_0^{1,2}(\Omega)$. Then we call A_V the elliptic operator with *Dirichlet boundary conditions*. In that case one has $D(A_V) = \{u \in W_0^{1,2}(\Omega) \colon Lu \in L^2(\Omega)\}$.

(b) Let $V = W^{1,2}(\Omega)$. Then we call A_V the elliptic operator with *Neumann boundary conditions*. If coefficients and domain are smooth, then for $u \in C^2(\overline{\Omega})$ one has $u \in D(A_V)$ if and only if

$$\sum_{j=1}^{n} \left(\sum_{i=1}^{n} a_{ij} D_i u + c_i u \right) \nu_j = 0 \quad \text{on } \partial\Omega,$$

where $\nu = (\nu_1, \ldots, \nu_n)$ is the outer normal. This is a consequence of Green's formula.

(c) *Mixed boundary conditions* are realized by taking

$$V = \left\{ u|_\Omega \colon u \in \mathcal{D}\left(\mathbb{R}^n \setminus \Gamma\right) \right\}^{-W^{1,2}(\Omega)}, \quad \text{where } \Gamma \text{ is a closed subset of } \partial\Omega.$$

8.2. *Positivity and irreducibility*

By Section 7.1.1 the semigroup T_V generated by $-A_V$ is positive if and only if all coefficients are real valued and

$$u \in V \quad \text{implies} \quad (\operatorname{Re} u)^+ \in V. \tag{8.1}$$

This is, in particular, the case for all three boundary conditions considered in Section 8.1. Moreover, if Ω is connected, then the semigroup is irreducible.

8.3. *Submarkov property: Dirichlet boundary conditions*

Let $V = W_0^{1,2}(\Omega)$. Assume that all coefficients are real. Then T_V is submarkovian if and only if

$$\sum_{j=1}^{n} D_j c_j \leqslant a_0 \quad \text{in } \mathcal{D}(\Omega)'. \tag{8.2}$$

See [ABBO00], Théorème 2.1.

8.4. *Quasicontractivity in* L^p

Assume that all coefficients are real valued. Assume furthermore that (8.1) and

$$u \in V_+ \Rightarrow 1 \wedge u \in V \tag{8.3}$$

hold. Then for each $p \in (1, \infty)$ there exists $\omega_p \in \mathbb{R}$ such that

$$\|T_V(t)\|_{\mathcal{L}(L^p)} \leqslant \mathrm{e}^{\omega_p t}, \quad t \geqslant 0. \tag{8.4}$$

This assertion is false though for $p = 1$ or $p = \infty$. However, if $b_j,\ c_j \in W^{1,\infty}(\Omega)$, $j = 1, \dots, n$, then there exists $\omega \in \mathbb{R}$ such that

$$\|T_V(t)\|_{\mathcal{L}(L^p)} \leqslant \mathrm{e}^{\omega t}, \quad t \geqslant 0, \tag{8.5}$$

for all $p \in [1, \infty]$.

References: [Ouh92], [ABBO00], [Dan00], [[Ouh04]].

8.5. *Gaussian estimates: real coefficients*

Assume that all coefficients are real valued. Let V be one of the three spaces considered in Section 8.1. In the case of Neumann or mixed boundary conditions assume that Ω has the extension property (in the weak sense of Section 7.3.6). Then T_V admits Gaussian estimates; see [AtE97], [Dan00], [Ouh03]. In particular, we have the following consequences: Denote by T_p the extrapolation semigroup of T_V in $L^p(\Omega)$ and by $-A_p$ its generator. Then

(a) T_p is a holomorphic C_0-semigroup, $1 \leqslant p < \infty$.
(b) The operator A_p satisfies (*MR*), $1 < p < \infty$.
(c) The operator $(A_p + \omega)$ has a bounded H^∞-calculus for ω large enough and $\varphi_{H^\infty}(A_p + \omega) < \pi/2$, where $1 < p < \infty$.
(d) The spectrum $\sigma(A_p)$ is independent on $p \in [1, \infty]$.

(e) Asymptotic behavior: Assume that Ω is connected and bounded. Then there exist u, $\varphi \in L^\infty(\Omega)$ such that

$$u(x) > 0, \varphi(x) > 0 \text{ a.e.}, \quad \int_\Omega u(x)\varphi(x)\,\mathrm{d}x = 1,$$

and such that

$$\left\| \mathrm{e}^{-\omega t} T_p(t) - P \right\|_{\mathcal{L}(L^p(\Omega))} \leqslant M\mathrm{e}^{-\varepsilon t}, \quad t \geqslant 0,$$

for all $1 \leqslant p < \infty$ and some $\varepsilon > 0$, $M \geqslant 0$, where $\omega = s(A_p)$ and $Pf = \int_\Omega f(x)\varphi(x)\,\mathrm{d}x \cdot u$.

This follows from Sections 7.4.3 for (a), 7.4.4 for (b), 7.4.5 for (c), 7.4.6 for (d) and 7.3.8 for (e) observing that Gaussian estimates imply ultracontractivity. We remark that assertion (c) can already be deduced from (8.4) by Section 4.7.5.

8.6. *Complex second-order coefficients*

If the coefficients a_{ij} are complex-valued, the situation is more complicated. For $\Omega = \mathbb{R}^n$, there are always Gaussian estimates if the a_{ij} are uniformly continuous [Aus96] or if $n \leqslant 2$; but otherwise there are counterexamples (see Auscher, Coulhon and Tchamitchian [ACT96] and Davies [Dav97]). If Ω is a Lipschitz domain and $V = W_0^{1,2}(\Omega)$ or $W^{1,2}(\Omega)$, then the existence of Gaussian estimates depends on the Lipschitz constant even for constant complex a_{ij}: For small Lipschitz constant Gaussian estimates are valid in that case [AT01a], but a counterexample based on [MNP85] is given for large Lipschitz constant. However, if the imaginary parts of the a_{ij} are symmetric then Gaussian estimates are valid:

THEOREM ([Ouh03], Theorem 5.5). *Assume that the imaginary parts of the coefficients satisfy*

$$\operatorname{Im} a_{ij} \in W^{1,\infty}, \quad \operatorname{Im}(a_{ij} + a_{ji}) = 0.$$

Let $V = W_0^{1,2}(\Omega)$ or $W^{1,2}(\Omega)$ assuming the extension property in the latter case. Then T_V admits Gaussian estimates.

8.7. *Further comments on Gaussian estimates*

Gaussian estimates were first proved by Aronson [Aro67] for real nonsymmetric elliptic operators on $\mathbb{R}^n$ with measurable coefficients. He used Moser's parabolic Harnack inequality [Mos64]. New impetus to the subject came from Davies [Dav87] who introduced a perturbation method ("Davies' trick") which provides an efficient tool to prove Gaussian estimates via ultracontractivity. One of Davies' motivations was to find optimal constants in the estimates, and the results are presented for symmetric operators in his book [[Dav90]].

Here we do not explain how the Gaussian estimates are proved but refer to the corresponding articles mentioned above and to [[Fri64]], [AER94], [AMT98], [DHZ94]. Some of the semigroup properties put together here carry over to estimates valid for higher-order equations, see, for example, [Are97], [Hie96], [HP97], [tER98].

8.8. *The square root property*

Assume that $\Omega \subset \mathbb{R}^n$ is open, bounded with Lipschitz boundary. Consider the operator A_V with

$$V = W_0^{1,2}(\Omega) \quad \text{or} \quad V = W^{1,2}(\Omega),$$

i.e., Dirichlet or Neumann boundary conditions are imposed. Let $\omega \in \mathbb{R}$ be large so that $A_V + \omega$ is sectorial. Then

$$D\big((A_V + \omega)^{1/2}\big) = V.$$

This is an extremely deep result. For $\mathbb{R}^n$, it is the famous *Kato's conjecture* which was solved recently by Auscher, Hofmann, Lacey, McIntosh and Tchamitchian [AHLMT02]. The result for the boundary conditions mentioned here is proved in [AT01a]. We refer to these articles as well as [[AT98]] for the sophisticated techniques from harmonic analysis leading to this result and also for various connections of these results with other areas. We mention that Kato's square root problem has also been solved for higher order elliptic operators in $\mathbb{R}^n$ [AHMT01]. Finally, we mention an elegant short proof of the square root property by ter Elst and Robinson [tER96b] (after previous work by McIntosh [McI85]) in the special case where the second-order coefficients are Hölder continuous.

8.9. *The hyperbolic equation*

We keep the notations and assumptions made in the beginning of this section, but we assume in addition that

$$a_{ij} = \overline{a_{ji}}, \quad i, j = 1, \ldots, n.$$

Then the following holds.

THEOREM. *The operator $-A_V$ generates a cosine function on $L^2(\Omega)$ with phase space $V \times L^2(\Omega)$.*

The proof [[ABHN01], Theorem 7.1] given for $V = W_0^{1,2}(\Omega)$ carries over to arbitrary closed subspaces V of $W^{1,2}(\Omega)$ containing $W_0^{1,2}(\Omega)$. The assertion of the Theorem implies in particular the square root property $D((A_V + \omega)^{1/2}) = V$ (for ω large). However, here A_V is a certain perturbation of a self-adjoint operator [[ABHN01], Corollary 3.14.12] for which the square root property is easy in comparison with the general result mentioned in Section 8.8.

8.10. *Nondivergence form*

Let $a_{ij} \in C(\mathbb{R}^n)$ such that

$$\operatorname{Re} \sum a_{ij}(x)\xi_i\xi_j \geqslant \alpha|\xi|^2$$

for all $x \in \mathbb{R}^n$, $\xi \in \mathbb{R}^n$ and some $\alpha > 0$. Consider the operator A on $L^2(\mathbb{R}^n)$ given by

$$Af = \sum_{i,j=1}^{n} a_{ij} D_i D_j u$$

with domain $D(A) = H^2(\mathbb{R}^n)$. If the coefficients a_{ij} are Hölder continuous, then A generates a holomorphic C_0-semigroup on $L^2(\mathbb{R}^n)$ admitting Gaussian estimates [[Fri64], Theorem 9.4.2]. However, there are no Gaussian estimates in general, if the coefficients are merely uniformly continuous and bounded, even if they are real [Bau84]. However, by other methods (viz. the $T1$ Theorem) Duong and Simonett [DS97] show that the L^p-realization of the above operator is sectorial and has a bounded H^∞-calculus. Further generalizations to *VMO*-coefficients are given by Duong and Yan [DY01] by wavelet methods and by Heck and Hieber [HH03] via weighted estimates.

8.11. *Elliptic operators with Banach space-valued coefficients*

Elliptic operators with coefficients in a Banach space and the associated parabolic equations were introduced and investigated by Amann [Ama01]. Further regularity results were obtained by Denk, Hieber and Prüss [DHP01] where a comprehensive presentation of the subject is given. A basic result is the following.

THEOREM ([DHP01], Section 5.5). *Let X be a UMD-space, $n, m \in \mathbb{N}$, $1 < p < \infty$, $a_\alpha \in \mathcal{L}(X)$ for multiindices α of order $|\alpha| = m$. Assume the ellipticity condition*

$$\sigma\big(\mathcal{A}(\xi)\big) \in \Sigma_\theta$$

for all $\xi \in \mathbb{R}^n$ such that $|\xi| = 1$, and some $\theta \in [0, \pi/2)$, where we set

$$\mathcal{A}(\xi) = \sum_{|\alpha|=m} a_\alpha \xi^\alpha, \quad \xi \in \mathbb{R}^n.$$

Consider the operator A on $L^p(\mathbb{R}^n, X)$ with domain $D(A) = W^{m,p}(\mathbb{R}^n, X)$ given by

$$Au = \sum_{|\alpha|=m} a_\alpha D^\alpha u.$$

Then A has a bounded $H^\infty(\Sigma_\theta)$-calculus. In particular, A is R-sectorial.

This result can be extended to elliptic operators on $L^p(\Omega, X)$ with diverse boundary conditions. It is remarkable that it is possible to characterize precisely those boundary conditions for which maximal regularity (*MR*) is valid; viz. by the Lopatinskii–Shapiro condition; see [DDHPV04].

Acknowledgments

We are most grateful to many colleagues who helped us with their advise and knowledge; among them we would like to mention. P. Auscher, S. Bu, R. Chill, M. Duelli, T. ter Elst, E. Fašangová, M. Haase, M. Hieber, P. Kunstmann, G. Lancien, A. McIntosh, D. Mugnolo, R. Nagel and J. Prüss.

References

Monographs

[[ABHN01]] W. Arendt, C.J.K. Batty, M. Hieber and F. Neubrander, *Vector-Valued Laplace Transforms and Cauchy Problems*, Monographs in Mathematics, Birkhäuser, Basel (2001).

[[AT98]] P. Auscher and Ph. Tchamitchian, *Square Root Problem for Divergence Operators and Related Topics*, Astérisque, Vol. 249, SMF, Paris (1998).

[[BH91]] N. Bouleau and F. Hirsch, *Dirichlet Forms and Analysis on Wiener Space*, de Gruyter, Berlin (1991).

[[DL88]] R. Dautray and J.L. Lions, *Mathematical Analysis and Numerical Methods for Science and Technology*, Vols 1–3, Springer, Berlin (1988).

[[Dav80]] E.B. Davies, *One-Parameter Semigroups*, Academic Press, London (1980).

[[Dav90]] E.B. Davies, *Heat Kernels and Spectral Theory*, Cambridge University Press, Cambridge (1990).

[[DER03]] N. Dungey, A.F.M. ter Elst and D.W. Robinson, *Analysis on Lie Groups with Polynomial Growth*, Birkhäuser, Basel (2003).

[[EG77]] R.E. Edwards and G.I. Gaudry, *Littlewood-Paley and Multiplier Theory*, Springer, Berlin (1977).

[[EN00]] K.J. Engel and R. Nagel, *One-Parameter Semigroups for Linear Evolution Equations*, Springer, Berlin (2000).

[[FOT94]] M. Fukushima, Y. Oshima and T. Takeda, *Dirichlet Forms and Symmetric Markov Processes*, North-Holland, Amsterdam (1994).

[[Fri64]] A. Friedman, *Partial Differential Equations of Parabolic Type*, Prentice-Hall, Englewood Cliffs, NJ (1964).

[[Gar81]] J.B. Garnett, *Bounded Analytic Functions*, Academic Press, New York (1981).

[[Gri92]] P. Grisvard, *Singularities in Boundary Value Problems*, Springer, Berlin (1992).

[[Kat66]] T. Kato, *Perturbation Theory of Linear Operators*, Springer, Berlin (1996).

[[LT96]] J. Lindenstrauss and L. Tzafriri, *Classical Banach Spaces II*, Springer, Berlin (1996).

[[Lun95]] A. Lunardi, *Analytic Semigroups and Optimal Regularity in Parabolic Problems*, Birkhäuser, Basel (1995).

[[MR86]] Z.M. Ma and M. Röckner, *Introduction to the Theory on Non-Symmetric Dirichlet Forms*, Springer, Berlin (1986).

[[MS01]] C. Martinez and M. Sanz, *The Theory of Fractional Powers of Operators*, North-Holland, Amsterdam (2001).

[[Mok97]] M. Mokhtar-Kharroubi, *Mathematical Topics in Neutron Transport Theory: New Aspects*, World Scientific, Singapore (1997).

[[Nag86]] R. Nagel, ed., *One-Parameter Semigroups of Positive Operators*, Lecture Notes in Math., Vol. 1184, Springer, Berlin (1986).

[[Nee96]] J.M.A.M. van Neerven, *The Asymptotic Behaviour of Semigroups of Linear Operators*, Birkhäuser, Basel (1996).

[[Ouh04]] E. Ouhabaz, *Analysis of Heat Equations on Domains*, London Math. Soc. Monographs, Princeton University Press, Princeton (2004), to appear.

[[Paz83]] A. Pazy, *Semigroups of Linear Operators and Applications to Partial Differential Equations*, Springer, Berlin (1983).

[[Prü93]] J. Prüss, *Evolutionary Integral Equations and Applications*, Birkhäuser, Basel (1993).

[[Sal02]] L. Saloff-Coste, *Aspects of Sobolev-Type Inequalities*, Cambridge Univ. Press (2002).

[[Sch74]] H.H. Schaefer, *Banach Lattices and Positive Operators*, Springer, Berlin (1974).

[[Sin70]] I. Singer, *Bases in Banach Spaces I*, Springer, Berlin (1970).

[[Tan79]] H. Tanabe, *Equations of Evolution*, Pitman Monographs Surveys Pure Appl. Math., Vol. 6, Pitman, London (1979).

[[Tri95]] H. Triebel, *Interpolation Theory, Function Spaces, Differential Operators*, Barth-Verlag, Heidelberg (1995).

[[VSC93]] N. Varopoulos, L. Saloff-Coste and T. Coulhon, *Geometry and Analysis on Groups*, Cambridge Univ. Press (1993).

Research Articles

[ADM96] D. Albrecht, X. Duong and A. McIntosh, *Operator theory and harmonic analysis*, Instructional Workshop on Analysis and Geometry, Part. III (Canberra 1995), ANU, Canberra (1996), 77–136.

[Ama97] H. Amann, *Operator-valued Fourier multipliers, vector-valued Besov spaces, and applications*, Math. Nachr. **186** (1997), 5–56.

[Ama01] H. Amann, *Elliptic operators with infinite-dimensional state spaces*, J. Evol. Equ. **2** (2001), 143–188.

[AHS94] H. Amann, M. Hieber and G. Simonett, *Bounded H^∞-calculus for elliptic operators*, Differential Integral Equations **7** (1994), 613–653.

[Are97] W. Arendt, *Semigroup properties by Gaussian estimates*, Nonlinear Evolution Equations and Applications, RIMS, Kyoto, Kokyuroku **1009** (1997), 162–180.

[Are94] W. Arendt, *Gaussian estimates and interpolation of the spectrum in L^p*, Differential Integral Equations **7** (1994), 1153–1168.

[ABB03] W. Arendt, C. Batty and S. Bu, *Fourier multipliers for Hölder continuous functions and maximal regularity*, Studia Math. **160** (2004), 23–51.

[ArBu02] W. Arendt and S. Bu, *The operator-valued Marcinkiewicz multiplier theorem and maximal regularity*, Math. Z. **240** (2002), 311–343.

[ArBu03a] W. Arendt and S. Bu, *Tools for maximal regularity*, Math. Proc. Cambridge Philos. Soc. **134** (2003), 317–336.

[ArBu03b] W. Arendt and S. Bu, *Operator-valued multiplier theorems characterizing Hilbert space*, Austr. J. Math., to appear.

[ABH01] W. Arendt, S. Bu and M. Haase, *Functional calculus, variational methods and Liapunov's theorem*, Arch. Math. **77** (2001), 65–75.

[AB94] W. Arendt and A.V. Bukhvalov, *Integral representations of resolvents and semigroups*, Forum Math. **6** (1994), 111–135.

[AMPR03] W. Arendt, G. Metafune, D. Pallara and S. Romanelli, *The Laplacian with Wentzell–Robin boundary conditions on spaces of continuous functions*, Semigroup Forum **67** (2003), 247–261.

[ARS94] W. Arendt, F. Räbiger and A.R. Sourour, *Spectral properties of the operator equation $AX + XB = Y$*, Quart. J. Math. Oxford **45** (1994), 133–149.

[AtE97] W. Arendt and A.F.M. ter Elst, *Gaussian estimates for second order elliptic operators with boundary conditions*, J. Operator Theory **38** (1997), 87–130.

[Aro67] D.G. Aronson, *Bounds for the fundamental solution of a parabolic equation*, Bull. Amer. Math. Soc. (N.S.) **73** (1967), 890–896.

[Aus96] P. Auscher, *Regularity theorems and heat kernel for elliptic operators*, J. London Math. Soc. **54** (1996), 284–296.

[ABBO00] P. Auscher, L. Barthélemy, Ph. Bénilan and E. Ouhabaz, *Absence de la L^{∞}-contractivité pour les semi-groupes associés aux opérateurs elliptiques complexes sous forme divergence*, Potential Anal. **12** (2000), 169–189.

[ACT96] P. Auscher, Th. Coulhon and Ph. Tchamitchian, *Absence du principe du maximum pour certaines équations paraboliques complexes*, Coll. Math. **171** (1996), 87–95.

[AER94] P. Auscher, A.F.M. ter Elst and D.W. Robinson, *On positive Rockland operators*, Coll. Math. **67** (1994), 197–216.

[AHLMT02] P. Auscher, S. Hofmann, M. Lacey, A. McIntosh and Ph. Tchamitchian, *The solution of the Kato square root problems for second order elliptic operators on R^n*, Ann. Math. **156** (2002), 633–654.

[AHMT01] P. Auscher, S. Hofmann, A. McIntosh and Ph. Tchamitchian, *The Kato square root problem for higher order elliptic operators and systems in $\mathbb{R}^n$*, J. Evol. Equ. **1** (2001), 361–385.

[AMN97] P. Auscher, A. McIntosh and A. Nahmrod, *Holomorphic functional calculi of operators, quadratic estimates and interpolation*, Indiana Univ. Math. J. **46** (1997), 375–403.

[AMT98] P. Auscher, A. McIntosh and Ph. Tchamitchian, *Heat kernels of second order complex elliptic operators and their applications*, J. Funct. Anal. **152** (1998), 22–73.

[AT01a] P. Auscher and Ph. Tchamitchian, *Gaussian estimates for second order elliptic divergence operators on Lipschitz and C^1-domains*, Evolution Equations and Their Applications in Physical and Life Sciences (Bad Herrenalb 1998), G. Lumer and L. Weis, eds, Lecture Notes in Pure and Applied Mathematics, Vol. 215, Dekker, New York (2001), 15–52.

[AT01b] P. Auscher and Ph. Tchamitchian, *Square roots of elliptic second order divergence operators on strongly Lipschitz domains: L^2-theory*, J. Anal. Math. **90** (2003), 1–12.

[BC91] J.B. Baillon and Ph. Clément, *Examples of unbounded imaginary powers of operators*, J. Funct. Anal. **100** (1991), 419–434.

[Bau84] P. Baumann, *Equivalence of the Green's function for diffusion operators in $\mathbb{R}^n$: A counterexample*, Proc. Amer. Math. Soc. **91** (1984), 64–68.

[BG94] E. Berkson and T.A. Gillespie, *Spectral decompositions and harmonic analysis on UMD-spaces*, Studia Math. **112** (1994), 13–49.

[Bie03] M. Biegert, *The Neumann Laplacian on spaces of continuous functions*, Note Mat. **22** (2003).

[Bla01] M. Blake, *A spectral bound for asymptotically norm-continuous semigroups*, J. Operator Theory **45** (2001), 111–130.

[BBN01] M. Blake, S. Brendle and R. Nagel, *On the structure of the critical spectrum of strongly continuous semigroups*, Evolution Equations and Their Applications in Physical and Life Sciences (Bad Herrenalb, 1998), G. Lumer and L. Weis, eds, Lecture Notes in Pure and Applied Mathematics, Vol. 215, Dekker, New York (2001), 55–65.

[Bou86] J. Bourgain, *Vector-valued singular integrals and the H^1–BMO duality*, Probability Theory and Harmonic Analysis, D. Burkholder, ed., Dekker, New York (1986), 1–19.

[BGD97] S. Bu and S. Guerre-Delabrière, *Imaginary powers and interpolation spaces of unbounded operators*, Rend. Istit. Mat. Univ. Trieste, Suppl. **XXVIII** (1997), 91–100.

[Bur83] D. Burkholder, *A geometric condition that implies the existence of certain singular integrals of Banach-space-valued functions.* Proc. Conf. Harmonic Analysis in Honor of Antoni Zygmund, Chicago 1981, Wadsworth, Belmont, CA (1983), 270–286.

[CM98] M. Campiti and G. Metafune, *Ventcel's boundary conditions and analytic semigroups*, Arch. Math. **70** (1998), 327–424.

[CFMP03] R. Chill, E. Fašangová, G. Metafune and D. Pallara, *The sector of analyticity of the Ornstein–Uhlenbeck semigroup on L_p-spaces with respect to invariant measure*, Preprint, Lecce (2003).

[CT03] R. Chill and Y. Tomilov, *Stability of C_0-semigroups and geometry of Banach spaces*, Math. Proc. Cambridge Philos. Soc. **135** (2003), 493–511.

[CT04] R. Chill and Y. Tomilov, *Analytic continuation and stability of C_0-semigroups*, J. Anal. Math. (2004), to appear.

[CZ76] I. Cioranescu and L. Zsidó, *Analytic generators for one-parameter groups*, Tohoku Math. J. **28** (1976), 327–362.

[CL94] Ph. Clément and S. Li, *Abstract parabolic quasilinear equations and applications to a groundwater flow problem*, Adv. Math. Sci. Appl. **3** (1994), 17–32.

[CPSW00] Ph. Clément, B. de Pagter, F.A. Sukochev and M. Witvliet, *Schauder decomposition and multiplier theorems*, Studia Math. **138** (2000), 135–163.

[CP01] Ph. Clément and J. Prüss, *An operator-valued transference principle and maximal regularity on vector-valued L_p-spaces*, Evolution Equations and Their Applications in Physical and Life Sciences, G. Lumer and L. Weis, eds, Lecture Notes in Pure Appl. Math., Vol. 215, Dekker, New York (2001), 67–87.

[Cou90] T. Coulhon, *Dimension à l'infini d'un semi-groupe analytique*, Bull. Sci. Math. **114** (1990), 485–500.

[CD00] T. Coulhon and X.T. Duong, *Maximal regularity and kernel bounds: Observations on a theorem of Hieber and Prüss*, Adv. Differential Equations **5** (2000), 343–368.

[Cow83] M. Cowling, *Harmonic analysis on semigroups*, Ann. Math. **117** (1983), 267–283.

[CDMY96] M. Cowling, I. Doust, A. McIntosh and A. Yagi, *Banach space operators with a bounded H^∞ functional calculus*, J. Austral. Math. Soc. Ser. A **60** (1996), 51–89.

[Cro03] M. Crouzeix, *Operators with numerical range in a parabola*, Arch. Math., to appear.

[Dan00] D. Daners, *Heat kernel estimates for operators with boundary conditions*, Math. Nachr. **217** (2000), 13–41.

[DPG75] G. Da Prato and P. Grisvard, *Sommes d'opérateurs linéaires et équations différentielles opérationnelles*, J. Math. Pures Appl. **54** (1975), 305–387.

[Dav87] E.B. Davies, *Explicit constants for Gaussian upper bounds on heat kernels*, Amer. J. Math. **109** (1987), 319–333.

[Dav97] E.B. Davies, *Limits on L^p-regularity of self-adjoint elliptic operators*, J. Differential Equations **135** (1997), 83–102.

[DDHPV04] R. Denk, G. Dore, M. Hieber, J. Prüss and A. Venni, *New thoughts on old results of R.T. Seeley*, Math. Ann. **328** (2004), 545–583.

[DHP01] R. Denk, M. Hieber and J. Prüss, *$\mathcal{R}$-Boundedness, Fourier Multipliers and Problems of Elliptic and Parabolic Type*, Mem. Amer. Math. Soc., Vol. 788, AMS, Providence, RI (2003).

[Dor93] G. Dore, *L^p-regularity for abstract differential equations*, Functional Analysis and Related Topics, H. Komatsu, ed., Lecture Notes in Math., Vol. 1540, Springer-Verlag, Berlin (1993), 25–38.

[Dor99a] G. Dore, *H^∞ functional calculus in real interpolation spaces*, Studia Math. **137** (1999), 161–167.

[Dor00] G. Dore, *Maximal regularity in L^p-spaces for an abstract Cauchy problem*, Adv. Differential Equations **5** (2000), 293–322.

[Dor99b] G. Dore, *H^∞ functional calculus in real interpolation spaces, II*, Studia Math. **145** (2001), 75–83.

[DoVe87] G. Dore and A. Venni, *On closedness of the sum of two closed operators*, Math. Z. **196** (1987), 189–201.

[Duo89] X.T. Duong, *H_∞-functional calculus for second order elliptic partial differential operators on L^p-spaces*, Miniconference on Operators in Analysis, I. Doust, B. Jefferies, C. Li and A. McIntosh, eds, ANU, Canberra (1989), 91–102.

[DM99] X.T. Duong and A. McIntosh, *Singular integral operators with nonsmooth kernels on irregular domains*, Rev. Mat. Iberoamericana **15** (1999), 233–265.

[DR96] X.T. Duong and D.W. Robinson, *Semigroup kernels, Poissons bounds, and holomorphic functional calculus*, J. Funct. Anal. **142** (1996), 89–128.

[DS97] X.T. Duong and G. Simonett, *H^∞-calculus for elliptic operators with nonsmooth coefficients*, Differential Integral Equations **10** (1997), 201–217.

[DY01] X.T. Duong and L.X. Yan, *Bounded holomnorphic functional calculus for non-divergence form differential operators*, Differential Integral Equations **15** (2002), 709–730.

[DHZ94] J. Dziubański, W. Hebisch and J. Zinkiewitz, *Note on semigroups generated by positive Rockland operators on graded homogeneous groups*, Studia Math. **110** (1994), 115–126.

[EG92] B. Eberhardt and G. Greiner, *Baillon's theorem on maximal regularity*, Acta Appl. Math. **27** (1992), 47–54.

[tER96a] A.F.M. ter Elst and D.W. Robinson, *Elliptic operators on Lie groups*, Acta Appl. Math. **44** (1996), 133–150.

[tER96b] A.F.M. ter Elst and D.W. Robinson, *On Kato's square root problem*, Hokkaido Math. J. **26** (1997), 365–376.

[tER98] A.F.M. ter Elst and D.W. Robinson, *Weighted subcoercive operators on Lie groups*, J. Funct. Anal. **157** (1998), 88–163.

[Eng04] K.-J. Engel, *The Laplacian on $C(\overline{\Omega})$ with generalized Wentzell boundary conditions*, Arch. Math. **81** (2003), 761–779.

[FGGR02] A. Favini, G.R. Goldstein, J.A. Goldstein and S. Romanelli, *The heat equation with generalized Wentzell boundary condition*, J. Evol. Equ. **2** (2002), 1–19.

[Fen97] G. Fendler, *Dilations of one-parameter semigroups of positive contractions on L_p-spaces*, Canad. J. Math. **49** (1997), 736–748.

[FLW01] V.P. Fonf, M. Lin and P. Wojtaszcyk, *Ergodic characterizations of reflexivity of Banach spaces*, J. Funct. Anal. **187** (2001), 146–162.

[FT95] M. Fukushima and M. Tomisaki, *Reflecting diffusions on Lipschitz domains with cusps: Analytic construction and Skorokod representation*, Potential Anal. **4** (1995), 377–408.

[Gri67] P. Grisvard, *Caractérisation de quelques espaces d'interpolation*, Arch. Ration. Mech. Anal. **25** (1967), 40–63.

[Gue95] S. Guerre-Delabrière, *L_p-regularity of Cauchy problems and the geometry of Banach spaces*, Illinois J. Math. **39** (1995), 556–566.

[Haa03a] M. Haase, *The functional calculus for sectorial operators and similarity methods*, Ph.D. Thesis, Ulm (2003).

[Haa02] M. Haase, *A characterization of group generators on Hilbert spaces and H^∞-calculus*, Semigroup Forum **66** (2003), 288–304.

[Haa03b] M. Haase, *Spectral properties of operator logarithms*, Math. Z. **245** (2003), 761–779.

[Haa03c] M. Haase, *A decomposition theorem for generators of strongly continuous groups on Hilbert space*, J. Operator Theory, to appear.

[HH03] H. Heck and M. Hieber, *Maximal L^p-regularity for elliptic operators with VMO-coefficients*, J. Evol. Equ. **3** (2003), 33–361.

[Hed93] L.I. Hedberg, *Approximation by harmonic functions, and stability of the Dirichlet problem*, Expo. Math. **11** (1993), 193–259.

[HV86] P. Hempel and J. Voigt, *The spectrum of a Schrödinger operator in $L^p(\mathbb{R}^\nu)$ is p-independent*, Comm. Math. Phys. **104** (1986), 243–250.

[Hie96] M. Hieber, *Gaussian estimates and holomorphy of semigroups on L^p-spaces*, J. London Math. Soc. **54** (1996), 148–160.

[HM00a] M. Hieber and S. Monniaux, *Pseudodifferential operators and maximal regularity results for non-autonomous parabolic equations*, Proc. Amer. Math. Soc. **128** (2000), 1047–1058.

[HM00b] M. Hieber and S. Monniaux, *Heat kernels and maximal L_p–L_q-estimates, the non-autonomous case*, J. Fourier Anal. Appl. **6** (2000), 467–481.

[HP97] M. Hieber and J. Prüss, *Heat kernels and maximal L_p–L_q estimates for parabolic evolution equations*, Comm. Partial Differential Equations **22** (1997), 1647–1669.

[HP98] M. Hieber and J. Prüss, *Functional calculi for linear operators in vector-valued L^p-spaces via transference principle*, Adv. Differential Equations **3** (1998), 847–872.

[Kal03] N.J. Kalton, *A remark on sectorial operators with an H^∞-calculus*, Preprint, Columbia, MO (2003).

[KL00] N.J. Kalton and G. Lancien, *A solution to the problem of the L^p-maximal regularity*, Math. Z. **235** (2000), 559–568.

[KW01] N.J. Kalton and L. Weis, *The H^∞-calculus and sums of closed operators*, Math. Ann. **321** (2001), 319–345.

[Kat78] T. Kato, *Trotter's product for an arbitrary pair of self-adjoint contraction semigroups*, Topics in Functional Analysis, I. Gohberg and M. Kac, eds, Academic Press, New York (1978), 185–195.

[Kel66] M.V. Keldyš, *On the solvability and stability of the Dirichlet problem*, Uspekhi Mat. Nauk. **8** (1941), 171–231; English transl.: Amer. Math. Soc. Transl. Ser. 2 **51** (1966), 1–73.

[Kun99] P.C. Kunstmann, *Heat kernel estimates and L^p-spectral independence of elliptic operators*, Bull. London Math. Soc. **31** (1999), 345–353.

[Kun02] P.C. Kunstmann, *L^p-spectral properties of the Neumann Laplacian on horns, comets and stars*, Math. Z. **242** (2002), 183–201.

[KS03] P.C. Kunstmann and Z. Strkalj, *H^∞-calculus for submarkovian semigroups*, Proc. Amer. Math. Soc. **131** (2002), 2081–2088.

[KuWe01] P.C. Kunstmann and L. Weis, *Perturbation Theorems for maximal L_p-regularity*, Ann. Sc. Norm. Sup. Pisa Cl. Sci. **XXX** (2001), 415–435.

[Kwa72] S. Kwapien, *Isomorphic characterizations of inner product spaces by orthogonal series with vector valued coefficients*, Studia Math. **44** (1972), 583–595.

[Lam87] D. Lamberton, *Equations d'évolution linéaires associées à des semigroupes de contractions dans les espaces L^p*, J. Funct. Anal. **72** (1987), 252–262.

[Lan98] G. Lancien, *Counterexamples concerning sectorial operators*, Arch. Math. **71** (1998), 388–398.

[LeM98a] C. Le Merdy, *The similarity problem for bounded analytic semigroups on Hilbert space*, Semigroup Forum **56** (1998), 205–224.

[LeM98b] C. Le Merdy, *H^∞-functional calculus and applications to maximal regularity*, Publications Mathématiques de Besançon Fascicule **16** (1998), 41–78.

[LeM00] C. Le Merdy, *A bounded compact semigroup on Hilbert space not similar to a contraction one*, Progress in Nonlinear Differential Equations and Their Applications, Vol. 42, Birkhäuser, Basel (2000), 213–216.

[LM97] V. Liskevich and A. Manavi, *Dominated semigroups with singular complex potentials*, J. Funct. Anal. **151** (1997), 281–305.

[LP95] V.A. Liskevich and M.A. Perelmuter, *Analyticity of submarkovian semigroups*, Proc. Amer. Math. Soc. **123** (1995), 1097–1104.

[LP79] G. Lumer and L. Paquet, *Semi-groupes holomorphes, produit tensoriel de semigroupes et équations d'évolution*, Sém. du Potentiel **4** (Paris 1977/78), Lecture Notes in Math., Vol. 713, Springer, Berlin (1979), 156–177.

[LS99] G. Lumer and R. Schnaubelt, *Local operator methods and time dependent parabolic equations on non-cylindrial domains*, Evolution Equations, Feshbach Resonances, Singular Hodge Theory, Wiley–VCH, Berlin (1999), 58–130.

[Mat03] M. Matolcsi, *On the relation of closed forms and Trotter's product formula*, Preprint, Budapest (2003).

[MNP85] V.G. Maz'ja, S.A. Nazarov and B.A. Plamensevskii, *Absence of De Giorgi type theorems for strongly elliptic equations with complex coefficients*, J. Soviet. Math. **28** (1985), 726–739.

[McI82] A. McIntosh, *On representing closed accretive sesquilinear forms as $(A^{1/2}u, A^{1/2}v)$*, Collège de France Seminar, Vol. III, H. Brézis and J.L. Lions, eds, Pitman Res. Notes Math. Ser., Vol. 70, Longman, Harlow (1982), 252–267.

[McI85] A. McIntosh, *Square roots of elliptic operators*, J. Funct. Anal. **61** (1985), 307–327.

[McI86] A. McIntosh, *Operators which have an H_∞-functional calculus*, Miniconference on Operator Theory and Partial Differential Equations (North Ryde 1986), ANU, Canbera (1986), 210–231.

[McI90] A. McIntosh, *The square root problem for elliptic operators: A survey*, Functional Analytical Methods for Partial Differential Equations (Tokyo 1989), Lectures Notes in Math., Vol. 1450, Springer-Verlag, Berlin (1990), 122–140.

[MHL02] A. McIntosh, S. Hoffmann and M. Lacey, *The solution of the Kato square root problem for divergence form elliptic operators with Gaussian heat kernel bounds*, Ann. of Math. **156** (2002), 623–631.

[MY90] A. McIntosch and A. Yagi, *Operators of type ω without a bounded H_∞-functional calculus*, Miniconference on Operators in Analysis (Sydney 1989), ANU, Canberra (1990), 159–172.

[MPP02] G. Metafune, D. Pallara and E. Priola, *Spectrum of Ornstein–Uhlenbeck operators in L^p spaces with invariant measures*, J. Funct. Anal. **196** (2002), 40–60.

[MPRS00] G. Metafune, J. Prüss, A. Rhandi and R. Schnaubelt, *L^p-regularity for elliptic operators with unbounded coefficients*, Preprint, Halle (2002).

[Mie87] A. Mielke, *Über maximale L^p-Regularität für Differentialgleichungen in Banach- und Hilberträumen*, Math. Ann. **277** (1987), 121–133.

[Mon99] S. Monniaux, *A new approach to the Dore–Venni theorem*, Math. Nachr. **204** (1999), 163–183.

[MP97] S. Monniaux and J. Prüss, *A theorem of the Dore–Venni type for non-commuting operators*, Trans. Amer. Math. Soc. **349** (1997), 4787–4814.

[Mos64] J. Moser, *A Harnack inequality for parabolic differential equations*, Comm. Pure Appl. Math. **17** (1964), 101–134.

[Mug02] D. Mugnolo, *A semigroup analogue of the Fonf–Lin–Wojtaszcyk ergodic characterization of reflexive Banach spaces with basis*, Tübinger Berichte zur Funktionalanalysis, Heft 11 (2001/2002), 219–226.

[Nol69] V. Nollau, *Über den Logarithmus abgeschlossener Operatoren in Banachschen Räumen*, Acta. Sci. Math. (Szeged) **30** (1969), 161–174.

[Oka99] N. Okazawa, *Logarithmic approach to bounded imaginary powers*, Publ. Math. UFR Sci. Tech. Besançon **16** (1999).

[Oka00a] N. Okazawa, *Logarithm Characterization of Bounded Imaginary Powers*, Progress in Nonlinear Differential Equations and Their Applications, Vol. 42, Birkhäuser, Basel (2000).

[Oka00b] N. Okazawa, *Logarithms and imaginary powers of closed linear operators*, Integral Equations Operator Theory **38** (2000), 458–500.

[Ouh92] E.M. Ouhabaz, *L^∞-contractivity of semigroups generated by sectorial forms*, Proc. London Math. Soc. **46** (1992), 529–542.

[Ouh95] E.M. Ouhabaz, *Gaussian estimates and holomorphy of semigroups*, Proc. Amer. Math. Soc. **123** (1995), 1465–1474.

[Ouh96] E.M. Ouhabaz, *Invariance of closed convex sets and domination criteria for semigroups*, Potential Anal. **5** (1996), 611–625.

[Ouh03] E.M. Ouhabaz, *Gaussian upper bounds for heat kernels of second order elliptic operators with complex coefficients on arbitrary domains*, J. Operator Theory (2004), to appear.

[Prü84] J. Prüss, *On the spectrum of C_0-semigroups*, Trans. Amer. Math. Soc. **284** (1984), 847–857.

[PS90] J. Prüss and H. Sohr, *On operators with bounded imaginary powers in Banach spaces*, Math. Z. **203** (1990), 429–452.

[PS93] J. Prüss and H. Sohr, *Imaginary powers of elliptic second order differential operators in L^p-spaces*, Hiroshima Math. J. **23** (1993), 161–192.

[Ren94] M. Renardy, *On the linear stability of hyperbolic PDE's and viscoelastic flows*, Z. Angew. Math. Phys. **45** (1994), 854–865.

[Schw61] J. Schwartz, *A remark on inequalities of Calderon–Zygmund type for vector-valued functions*, Comm. Pure Appl. Math. **14** (1961), 785–799.

[See67] R. Seely, *Complex powers of an elliptic operator*, Singular Integrals, Proc. Sympos. Pure Math., Vol. 10, Amer. Math. Soc., Providence, RI (1967), 288–307.

[SW00] Z. Štrkalj and L. Weis, *On operator-valued Fourier multiplier theorems*, Preprint (2000).

[Ven93] A. Venni, *A counterexample concerning imaginary powers of linear operators*, Functional Analysis and Related Topics, Lecture Notes in Math., Vol. 1540, Springer, Berlin (1993), 361–387.

[Voi92] J. Voigt, *One-parameter semigroups acting simultaneously on different L_p-spaces*, Bull. Soc. Roy. Sci. Liège **61** (1992), 465–470.

[Voi96] J. Voigt, *The sector of holomorphy for symmetric submarkovian semigroups*, Functional Analysis Proc. Trier 1994, S. Dierolf et al., eds, de Gruyter, Berlin (1996), 449–453.

[Vu97] Q.P. Vũ, *Almost periodic and strongly stable semigroups of operators*, Linear Operators, Banach Center Publ., Vol. 38, Polish Acad. Sci., Warsaw (1997), 401–426.

[War03a] M. Warma, *The Laplacian with general Robin boundary condition*, Ph.D. Thesis, Ulm (2002).

[War03b] M. Warma, *Wentzell–Robin boundary conditions on $C[0, 1]$*, Semigroup Forum **66** (2003), 162–170.

[Web87] G.F. Webb, *An operator theoretic formulation of asynchronous exponential growth*, Trans. Amer. Math. Soc. **303** (1987), 751–763.

[Wei95] L. Weis, *The stability of positive semigroups on L_p-spaces*, Proc. Amer. Math. Soc. **123** (1995), 3089–3094.

[Wei00b] L. Weis, *A new approach to maximal L_p-regularity*, Proc. 6th International Conference on Evolution Equations, G. Lumer and L. Weis, eds, Dekker, New York (2000), 195–214.

[Wei00a] L. Weis, *Operator-valued Fourier multiplier theorems and maximal L_p-regularity*, Math. Ann. **319** (2001), 735–758.

[Wol81] M. Wolff, *A remark on the spectral bound of the generator of semigroups of positive operators with applications to stability theory*, Functional Analysis and Approximations (Oberwolfach 1980), Internat. Ser. Numer. Math., Vol. 60, Birkhäuser, Basel (1981), 39–50.

[Yag84] A. Yagi, *Coïncidence entre des espaces d'interpolation et des domaines de puissances fractionaires d'opérateurs*, C. R. Acad. Sci. Paris **299** (1984), 173–176.

[Zim89] F. Zimmermann, *On vector-valued Fourier multiplier theorems*, Studia Math. (1989), 201–222.

CHAPTER 2

Front Tracking Method for Systems of Conservation Laws

Alberto Bressan

Department of Mathematics, Penn State University, University Park, PA 16802, USA

Contents

1. Introduction . . . 89
2. Weak solutions . . . 94
   2.1. Basic definitions . . . 94
   2.2. Rankine–Hugoniot conditions . . . 95
   2.3. Admissibility conditions . . . 99
3. The Riemann problem . . . 101
   3.1. Shock and rarefaction waves . . . 103
   3.2. Three special cases . . . 104
   3.3. General solution of the Riemann problem . . . 106
   3.4. Some estimates . . . 108
4. Front tracking approximations . . . 111
   4.1. Global existence of entropy weak solutions . . . 111
   4.2. A naive front tracking algorithm . . . 113
   4.3. Bounds on the total variation . . . 115
   4.4. Convergence to a limit solution . . . 117
   4.4. A modified front tracking algorithm . . . 120
   Simplified Riemann Solver . . . 122
   Notes . . . 130
5. Uniqueness and stability . . . 132
   5.1. A Lyapunov functional . . . 132
   5.2. Proofs of the main estimates . . . 141
   5.3. A semigroup of solutions . . . 150
   5.4. Notes . . . 152
6. Further properties of solutions . . . 152
   6.1. Viscosity solutions and characterization of semigroup trajectories . . . 152
   6.2. Error estimates . . . 154
   6.3. Uniqueness of entropy weak solutions . . . 156
   6.4. Qualitative structure of solutions . . . 158
   6.5. Notes . . . 159

HANDBOOK OF DIFFERENTIAL EQUATIONS
Evolutionary Equations, volume 1
Edited by C.M. Dafermos and E. Feireisl

Appendix 159
A.1. A compactness theorem 160
A.2. An elementary error estimate 160
A.3. Taylor estimates 162
References 166

1. Introduction

A scalar conservation law in one space dimension is a first-order partial differential equation of the form

$$u_t + f(u)_x = 0. \tag{1.1}$$

Here u is called the *conserved quantity* while f is the *flux*. Equations of this type often describe transport phenomena. Integrating (1.1) over a given interval $[a, b]$ one obtains

$$\begin{aligned}\frac{\mathrm{d}}{\mathrm{d}t}\int_a^b u(t,x)\,\mathrm{d}x &= \int_a^b u_t(t,x)\,\mathrm{d}x \\ &= -\int_a^b f\big(u(t,x)\big)_x\,\mathrm{d}x \\ &= f\big(u(t,a)\big) - f\big(u(t,b)\big) \\ &= [\text{inflow at } a] - [\text{outflow at } b].\end{aligned}$$

In other words, the quantity u is neither created nor destroyed: the total amount of u contained inside any given interval $[a, b]$ can change only due to the flow of u across boundary points (see Figure 1). By the knowledge of the flux function $f(u)$, one should thus be able to determine the evolution of u in future times.

Throughout the following, we shall be concerned with an $n \times n$ system of conservation laws

$$\begin{cases} \frac{\partial}{\partial t}u_1 + \frac{\partial}{\partial x} f_1(u_1, \ldots, u_n) = 0, \\ \quad\vdots \\ \frac{\partial}{\partial t}u_n + \frac{\partial}{\partial x} f_n(u_1, \ldots, u_n) = 0. \end{cases}$$

For simplicity, this will still be written in the form (1.1), but keeping in mind that now $u = (u_1, \ldots, u_n)$ is a vector in $\mathbb{R}^n$ and that $f = (f_1, \ldots, f_n)$ is a map from $\mathbb{R}^n$ into itself. The components $u_1, \ldots, u_n$ are now called the *conserved quantities*, while $f_1, \ldots, f_n$ are the *fluxes*.

Systems of this form are often used in order to express the fundamental balance laws of continuum physics, assuming that small viscosity or dissipation effects are neglected. A primary example is provided by the Euler equations describing the evolution of a compressible, nonviscous fluid:

$$\begin{cases} \rho_t + (\rho v)_x = 0 & \text{(conservation of mass)}, \\ (\rho v)_t + \big(\rho v^2 + p\big)_x = 0 & \text{(conservation of momentum)}, \\ (\rho E)_t + (\rho E v + p v)_x = 0 & \text{(conservation of energy)}. \end{cases}$$

Here ρ is the mass density, v is the velocity while $\rho E = \rho e + \rho v^2/2$ is the energy density, obtained as the sum of an internal energy ρe and the kinetic energy $\rho v^2/2$. The system is

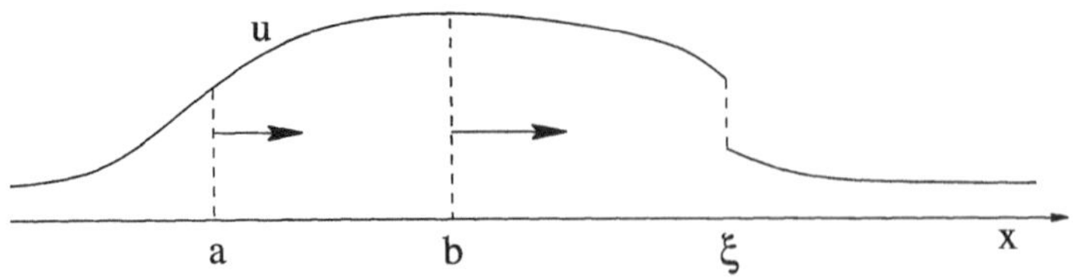

Fig. 1. Solution to a conservation law.

closed by a *constitutive relation* of the form $p = p(\rho, e)$, giving the pressure as a function of the density and of the internal energy. The particular choice of p depends on the gas under consideration.

Using the chain rule, the system (1.1) can be written in the *quasilinear* form

$$u_t + A(u)u_x = 0, \tag{1.2}$$

where $A(u) \doteq Df(u)$ is the Jacobian matrix of first-order partial derivatives of f. For smooth solutions, the two equations (1.1) and (1.2) are entirely equivalent. However, if u has a jump at a point ξ (see Figure 1), the left-hand side of (1.2) will contain a product of the discontinuous function $A(u)$ with the distributional derivative u_x, which in this case contains a Dirac mass at the point ξ. In general, such a product is not well defined. Hence, (1.2) is meaningful only within a class of continuous functions. On the other hand, the system (1.1) is in divergence form and can be always interpreted in distributional sense. More precisely, a locally integrable function $u = u(t, x)$ is a *weak solution* of (1.1) provided that

$$\iint \{u\phi_t + f(u)\phi_x\}\,\mathrm{d}x\,\mathrm{d}t = 0 \tag{1.3}$$

for every differentiable function with compact support $\phi \in \mathcal{C}_c^1$.

We say that the above system is *strictly hyperbolic* if every matrix $A(u)$ has n real, distinct eigenvalues, say $\lambda_1(u) < \cdots < \lambda_n(u)$. In this case, by a standard result of linear algebra one can find dual bases of left and right eigenvectors of $A(u)$, denoted by $l_1(u), \ldots, l_n(u)$ and $r_1(u), \ldots, r_n(u)$, normalized according to

$$|r_i(u)| = 1, \qquad l_i(u) \cdot r_j(u) = \begin{cases} 1 & \text{if } i = j, \\ 0 & \text{if } i \neq j. \end{cases} \tag{1.4}$$

To appreciate the consequences of the nonlinearity of the equations, consider first the case of a linear system with constant coefficients:

$$u_t + Au_x = 0. \tag{1.5}$$

As before, we call $\lambda_1 < \cdots < \lambda_n$ the eigenvalues of the matrix A and let $\{l_1, \ldots, l_n\}$, $\{r_1, \ldots, r_n\}$ be dual bases of left and right eigenvectors, as in (1.4). One can then write the general solution of (1.5) as a superposition of independent linear waves:

$$u(t,x) = \sum_i \phi_i(x - \lambda_i t) r_i. \tag{1.6}$$

The functions ϕ_i can here be determined by decomposing the initial data $u(0,\cdot)$ with respect to the given basis of eigenvectors:

$$u(0,y) = \sum_i \phi_i(y) r_i, \qquad \phi_i(y) \doteq l_i \cdot u(0,y).$$

Notice that the solution (1.6) is completely decoupled along the eigenspaces of A. Each component travels with constant speed, given by the corresponding eigenvalue of A.

In the nonlinear case (1.2) where the matrix A is not constant but depends on the state u, new features will appear.

(i) The eigenvalues λ_i now depend on u. Since these eigenvalues determine the speeds of propagation of the wave fronts, as a result the shape of the various components in the solution will vary in time. In particular, rarefaction waves will decay, and compression waves will become steeper, possibly leading to shock formation in finite time (see Figure 2).

(ii) The eigenvectors r_i also depend on u. These vectors determine the basis along which the solution is decomposed. As a result, the various components are no longer independent, and nontrivial interactions between waves of different families will occur. In particular, the strength of the interacting waves may change and new waves can be created, as the result of an interaction.

Of all these features, the most important for the mathematical theory is the possible loss of regularity in the solutions, related to the formation of shocks [J]. This happens also because in our model equations we neglect small second-order terms that could provide a smoothing effect. The strong nonlinearity of the equations, together with this lack of regularity in the solutions, is responsible for many of the difficulties encountered in a rigorous mathematical analysis. Indeed, most of the well established, powerful techniques of functional analysis do not apply in this context. The theory of hyperbolic conservation laws has thus progressed mainly by developing ad hoc methods.

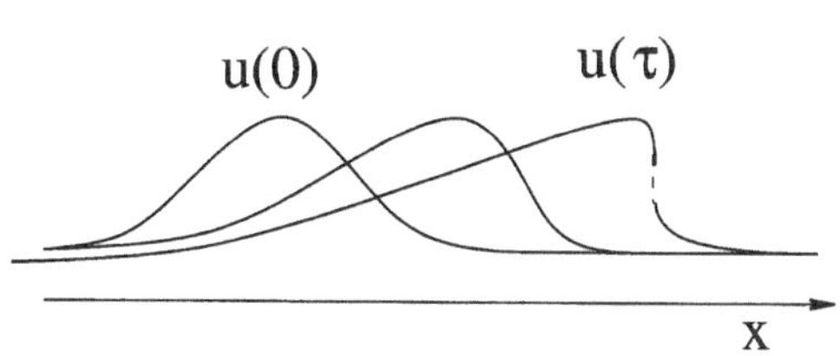

Fig. 2. Shock formation in a nonlinear wave.

Toward the construction as well as the analysis of solutions, a basic building block is the so called *Riemann problem*, where the initial data is piecewise constant:

$$u(0,x) = \begin{cases} u^- & \text{if } x < 0, \\ u^+ & \text{if } x > 0. \end{cases} \tag{1.8}$$

This particular initial value problem was first introduced in 1860 in a fundamental paper of Riemann [Rm], who studied its solution in the context of isentropic gas dynamics. The case of more general $n \times n$ systems was considered by Lax [Lx] nearly a century later. The solution is assumed to be self-similar, having the form

$$u(t,x) = U(x/t).$$

The central position taken by the Riemann problem is related to a symmetry of the equations (1.1). If $u = u(t,x)$ is any solution, then, for each $\theta > 0$, the rescaled function

$$u^\theta(t,x) \doteq u(\theta t, \theta x)$$

provides another solution. The special solutions which are invariant under the above stretching of coordinates are precisely the ones which solve a Riemann problem.

For the Cauchy problem with general initial data

$$u(0,x) \doteq \bar{u}(x), \tag{1.7}$$

the first global existence theorem was proved in a fundamental paper by Glimm [G]. Approximate solutions are constructed by solving a family of Riemann problems, one for each node of a fixed grid in the t–x plane. The convergence of a sequence of approximations is guaranteed by a compactness argument, based on a priori bounds on the total variation. A key idea in Glimm's approach, which remained fundamental in nearly all subsequent developments, is to control the total variation by means of a quadratic interaction functional. Roughly speaking, this functional measures the amount of possible future interactions among waves in a given solution. It is monotonically decreasing in time, provided that the total variation is small enough. For this reason Glimm's result, as well as most of the present theory for $n \times n$ conservation laws, is restricted to the case of small BV data.

An alternative method for constructing approximate solutions to the general Cauchy problem is by *front tracking*. This method was first proposed by Dafermos [D1] to study scalar conservation laws, then adapted by DiPerna to the case of 2×2 systems. More recently, it has been refined and extended to $n \times n$ systems in [B2,Ri,BaJ,AM2].

The underlying idea is quite simple. We seek approximate solutions which are piecewise constant, having jumps along a finite number of straight lines in the t–x plane. For this purpose, at time $t = 0$ we start with an initial data which is piecewise constant. At each point of jump, we construct a piecewise constant approximate solution of the corresponding Riemann problem. Piecing together these local solutions, we obtain a solution $u = u(t,x)$ which is well defined until the first time t_1, where two lines of discontinuity interact. In this case, the solution can be further prolonged in time by solving the new Riemann problem determined by the interaction. This yields an approximation valid up to the next time $t_2 > t_1$ where two more fronts interact. Once again we solve the corresponding Riemann problem, thus extending the solution further in time, etc.

This approach shares many features in common with the Glimm scheme. The Riemann solutions remain the basic building blocks, and the control of the total variation relies on exactly the same wave interaction functional introduced in [G]. However, there are some distinct advantages:

- In the Glimm scheme, at each time step $t_j \doteq j\Delta t$ one has to replace the constructed solution $u(t_j-,\cdot)$ with a piecewise constant function $u(t_j,\cdot)$, having jumps exactly at the nodes of a fixed grid. This is achieved by a random sampling procedure. As a consequence, the approximate solution $t \mapsto u(t,\cdot)$ is not continuous as a map with values in $\mathbf{L}^1_{\mathrm{loc}}$. Moreover, the sum of all the errors

$$\sum_{j\geqslant 1} \big\| u(t_j) - u(t_j-) \big\|_{\mathbf{L}^1}$$

  remains uniformly large as the mesh size Δt tends to zero. The convergence to an exact solution stems from the fact that the differences $u(t_j) - u(t_j-)$ are somehow "random": in the long run they tend to cancel each other out, according to the law of large numbers. However, the analysis of the convergence rate is not an easy task.

  The front tracking algorithm, on the other hand, constructs approximate solutions $u(t,\cdot)$ whose jumps move with constant speed. Therefore, the map $t \mapsto u(t,\cdot)$ is now Lipschitz continuous with respect to the $\mathbf{L}^1$ distance. This makes it much easier to derive error estimates and properties of the limit solution.
- For the Glimm scheme, the dependence of the approximate solution on the initial data is not continuous, and extremely difficult to study. On the other hand, for a solution u constructed by front tracking there is a natural class of perturbations which can be analyzed in detail. Namely, one can slightly shift the locations of the jumps in the initial data $u(0,\cdot)$ and estimate at a later time the corresponding shifts of the wave fronts in $u(t,\cdot)$. This approach yields a basic understanding of the Lipschitz continuous dependence of solutions on their initial data, in the $\mathbf{L}^1$ norm. In the 2×2 case, a rigorous proof along these lines was worked out in [BC1].
- A very effective tool in the analysis of smooth solutions to hyperbolic systems is the method of characteristics. For a fixed $i \in \{1,\dots,n\}$, one looks at the curves $x = x_i(t)$ in the t–x plane which satisfy the ordinary differential equation (ODE)

$$\dot{x}(t) = \lambda_i\big(u\big(t, x(t)\big)\big).$$

  This method works perfectly well also for solutions constructed by front tracking. In this case, the analysis is particularly simple because the characteristic curves are polygonals.
- For a front tracking solution, to each front one can assign a *generation order*, counting the number of interactions needed to produce it. Fronts originating at time $t = 0$ are the "primal ancestors", and have order 1. When these interact, to the new fronts emerging from the interaction we assign order 2, etc. This provides an additional, useful tool for the analysis of solutions.

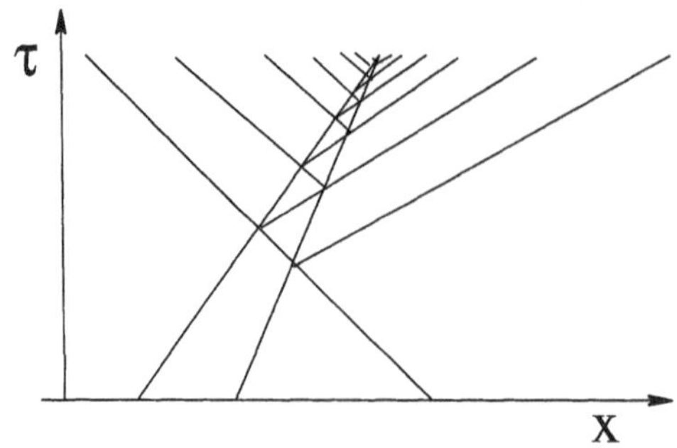

Fig. 3. Infinitely many wave fronts.

- Front tracking approximations have an elementary structure, being piecewise constant in the t–x plane, with exactly 2 incoming wave fronts at each interaction point. Hence, all of the relevant estimates can be obtained by the careful analysis of one single building block: the Riemann problem generated by the interaction of two wave fronts. By proving suitable estimates on front tracking approximations and passing to the limit, relying on a general uniqueness theorem one can now obtain detailed results on the existence, stability, local and global structure, and asymptotic behavior of weak solutions.

On the negative side, one must mention one technical problem that has slowed down the development of the front tracking method for several years. If the algorithm is implemented in a straightforward way, it may well produce a solution with infinitely many wave fronts within finite time (see Figure 3). When this happens, the whole construction breaks down. This unpleasant situation does not occur for scalar conservation laws, and can be easily avoided in the 2×2 case. But for general $n \times n$ systems it represents a real difficulty. To prevent the number of wave fronts from blowing up, some new technical provision must be inserted. As shown in [B2,BaJ], this can be achieved by occasionally solving a Riemann problem in a simplified way, which is not very accurate but involves a minimum number of outgoing fronts. In the end, this modification of the algorithm is a purely technical device, which does not affect in any way the results which can be obtained by the method. Yet, it lengthens a bit every proof, requiring one more case to consider before the punch line. For this reason, we feel it somewhat detracts from the elegance of the whole approach.

2. Weak solutions

2.1. *Basic definitions*

A basic feature of nonlinear hyperbolic systems is the possible loss of regularity: solutions which are initially smooth may become discontinuous within finite time. In order to construct solutions globally in time, we are thus forced to work in a space of discontinuous functions, and interpret the conservation equations in a distributional sense.

DEFINITION 2.1. Let $f:\mathbb{R}^n \mapsto \mathbb{R}^n$ be a smooth vector field. A measurable function $u = u(t,x)$ from an open set $\Omega \subseteq \mathbb{R} \times \mathbb{R}$ into $\mathbb{R}^n$ is a *distributional solution* of the system of conservation laws

$$u_t + f(u)_x = 0 \tag{2.1}$$

if, for every C^1 function $\phi : \Omega \mapsto \mathbb{R}^n$ with compact support, one has

$$\iint_\Omega \{u\phi_t + f(u)\phi_x\} \,\mathrm{d}x \,\mathrm{d}t = 0. \tag{2.2}$$

Observe that no continuity assumption is made on u. We only require u and $f(u)$ to be locally integrable in Ω. Notice also that weak solutions are defined up to $\mathbf{L}^1$ equivalence. A solution is not affected by changing its values on a set of measure zero in the t–x plane.

An easy consequence of the above definition is the closure of the set of solutions with respect to convergence in $\mathbf{L}^1_{\mathrm{loc}}$.

LEMMA 2.1. *Let $(u_m)_{m\geqslant 1}$ be a uniformly bounded sequence of distributional solutions of* (2.1). *If $u_m \to u$ in $\mathbf{L}^1_{\mathrm{loc}}$, then the limit function u is itself a distributional solution.*

PROOF. Indeed, the assumption of uniform boundedness implies $f(u_m) \to f(u)$ in $\mathbf{L}^1_{\mathrm{loc}}$. For every $\phi \in C^1_c$, we now have

$$\iint_\Omega \{u\phi_t + f(u)\phi_x\} \,\mathrm{d}x \,\mathrm{d}t = \lim_{m\to\infty} \iint_\Omega \{u_m\phi_t + f(u_m)\phi_x\} \,\mathrm{d}x \,\mathrm{d}t = 0. \qquad \square$$

In the following, we shall be mainly interested in solutions defined on a strip $[0,T] \times \mathbb{R}$, with an assigned initial condition

$$u(0,x) = \bar{u}(x). \tag{2.3}$$

In this case, it is convenient to require some additional regularity with respect to time.

DEFINITION 2.2. A function $u:[0,T] \times \mathbb{R} \mapsto \mathbb{R}^n$ is a *weak solution* of the Cauchy problem (2.1), (2.3) if u is continuous as a function from $[0,T]$ into $\mathbf{L}^1_{\mathrm{loc}}$, the initial condition (2.3) holds and the restriction of u to the open strip $]0,T[\times \mathbb{R}$ is a distributional solution of (2.1).

2.2. *Rankine–Hugoniot conditions*

Let u be a weak solution of (2.1). If u is continuously differentiable restricted to an open domain Ω', then at every point $(t,x) \in \Omega'$, the function u satisfies the quasilinear system

$$u_t + A(u)u_x = 0, \tag{2.4}$$

with $A(u) \doteq Df(u)$. Indeed, from (2.2) an integration by parts yields

$$\iint [u_t + A(u)u_x]\phi \, dx \, dt = 0.$$

Since this holds for every $\phi \in \mathcal{C}_c^1(\Omega')$, the identity (2.4) follows.

Next, we look at a discontinuous solution and derive some conditions which must be satisfied at points of jump. Consider first the simple case of a piecewise constant function, say

$$U(t,x) = \begin{cases} u^+ & \text{if } x > \lambda t, \\ u^- & \text{if } x < \lambda t, \end{cases} \tag{2.5}$$

for some $u^-, u^+ \in \mathbb{R}^n$, $\lambda \in \mathbb{R}$.

LEMMA 2.2. *If the function U in* (2.5) *is a weak solution of the system of conservation laws* (2.1), *then*

$$\lambda(u^+ - u^-) = f(u^+) - f(u^-). \tag{2.6}$$

PROOF. Let $\phi = \phi(t,x)$ be any continuously differentiable function with compact support. Let Ω be a ball containing the support of ϕ and consider the two domains

$$\Omega^+ \doteq \Omega \cap \{x > \lambda t\}, \qquad \Omega^- \doteq \Omega \cap \{x < \lambda t\},$$

as in Figure 4. Introducing the vector field $\mathbf{v} \doteq (U\phi, f(U)\phi)$, the identity (2.2) can be rewritten as

$$\iint_{\Omega^+ \cup \Omega^-} \operatorname{div} \mathbf{v} \, dx \, dt = 0. \tag{2.7}$$

We now apply the divergence theorem separately on the two domains Ω^+, Ω^-. Call $\mathbf{n}^+, \mathbf{n}^-$ the outer unit normals to Ω^+, Ω^-, respectively. Observe that $\phi = 0$ on the boundary $\partial\Omega$. Denoting by ds the differential of the arc-length, along the line $x = \lambda t$ we have

$$\mathbf{n}^+ ds = (\lambda, -1)\, dt, \qquad \mathbf{n}^- ds = (-\lambda, 1)\, dt,$$

$$\begin{aligned} &\iint_{\Omega^+ \cup \Omega^-} \operatorname{div} \mathbf{v} \, dx \, dt \\ &\quad = \int_{\partial\Omega^+} \mathbf{n}^+ \cdot \mathbf{v} \, ds + \int_{\partial\Omega^-} \mathbf{n}^- \cdot \mathbf{v} \, ds \\ &\quad = \int [\lambda u^+ - f(u^+)]\phi(t, \lambda t)\, dt + \int [-\lambda u^- + f(u^-)]\phi(t, \lambda t)\, dt. \end{aligned}$$

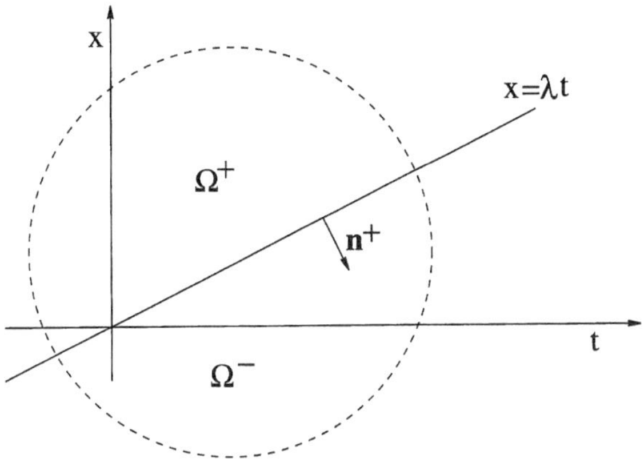

Fig. 4. A single shock solution.

By (2.7), the identity

$$\int \left[\lambda\left(u^+ - u^-\right) - f\left(u^+\right) + f\left(u^-\right)\right]\phi(t, \lambda t)\,\mathrm{d}t = 0$$

must hold for every function $\phi \in \mathcal{C}_c^1$. This implies (2.6). □

The vector equations (2.6) are the famous *Rankine–Hugoniot conditions*. They form a set of n scalar equations relating the right and left states $u^+, u^- \in \mathbb{R}^n$ and the speed λ of the discontinuity. An alternative way of writing these conditions is as follows. Denote by $A(u) = Df(u)$ the $n \times n$ Jacobian matrix of f at u. For any u, v, define the averaged matrix

$$A(u, v) \doteq \int_0^1 A\big(\theta u + (1-\theta)v\big)\,\mathrm{d}\theta \tag{2.8}$$

and call $\lambda_i(u, v)$, $i = 1, \ldots, n$, its eigenvalues. We can then write (2.6) in the equivalent form

$$\begin{aligned}\lambda\left(u^+ - u^-\right) = f\left(u^+\right) - f\left(u^-\right) &= \int_0^1 Df\left(\theta u^+ + (1-\theta)u^-\right)\cdot\left(u^+ - u^-\right)\mathrm{d}\theta \\ &= A\left(u^+, u^-\right)\cdot\left(u^+ - u^-\right).\end{aligned} \tag{2.9}$$

In other words, the Rankine–Hugoniot conditions hold iff the jump $u^+ - u^-$ is an eigenvector of the averaged matrix $A(u^+, u^-)$ and the speed λ coincides with the corresponding eigenvalue.

We now consider a more general solution $u = u(t, x)$ of (2.1) and show that the Rankine–Hugoniot equations are still satisfied at every point (τ, ξ) where u has an approximate jump, in the following sense.

DEFINITION 2.3. We say that a function $u = u(t,x)$ has an *approximate jump discontinuity* at the point (τ,ξ) if there exist vectors $u^+ \neq u^-$ and a speed λ such that, defining U as in (2.5), there holds

$$\lim_{r\to 0+} \frac{1}{r^2}\int_{-r}^{r}\int_{-r}^{r} \big|u(\tau+t,\xi+x) - U(t,x)\big|\,\mathrm{d}x\,\mathrm{d}t = 0. \tag{2.10}$$

Moreover, we say that u is *approximately continuous* at the point (τ,ξ) if the above relations hold with $u^+ = u^-$ (and λ arbitrary).

Observe that the above definitions depend only on the $\mathbf{L}^1$ equivalence class of u. Indeed, the limit in (2.10) is unaffected if the values of u are changed on a set $\mathcal{N}\subset\mathbb{R}^2$ of Lebesgue measure zero.

EXAMPLE 2.1. Let $f^-, f^+ : \mathbb{R}^2 \mapsto \mathbb{R}^n$ be any two continuous functions and let $x = \gamma(t)$ be a smooth curve. Define the function

$$u(t,x) \doteq \begin{cases} f^-(t,x) & \text{if } x < \gamma(t),\\ f^+(t,x) & \text{if } x > \gamma(t). \end{cases}$$

At a point (τ,ξ), with $\xi = \gamma(\tau)$, call $u^- \doteq f^-(\tau,\xi)$, $u^+ \doteq f^+(\tau,\xi)$. If $u^+ = u^-$, then u is continuous at (τ,ξ), hence also approximately continuous. On the other hand, if $u^+ \neq u^-$, then u has an approximate jump at (τ,ξ). Indeed, the limit (2.10) holds with U as in (2.5) and $\lambda = \dot\gamma(\tau)$.

We now prove the Rankine–Hugoniot conditions in the more general case of a point of approximate jump.

THEOREM 2.1. *Let u be a bounded distributional solution of* (2.1) *having an approximate jump at a point (τ,ξ). In other words, assume that* (2.10) *holds, for some states u^-, u^+ and a speed λ, with U as in* (2.5). *Then the Rankine–Hugoniot equations* (2.6) *hold.*

PROOF. For any given $\theta > 0$, one easily checks that the rescaled function

$$u^\theta(t,x) \doteq u(\tau+\theta t,\ \xi+\theta x)$$

is also a solution to the system of conservation laws. We claim that, as $\theta\to 0$, the convergence $u^\theta \to U$ holds in $\mathbf{L}^1_{\mathrm{loc}}(\mathbb{R}^2;\mathbb{R}^n)$. Indeed, for any $R > 0$, one has

$$\begin{aligned} &\lim_{\theta\to 0}\int_{-R}^{R}\int_{-R}^{R} \big|u^\theta(t,x) - U(t,x)\big|\,\mathrm{d}x\,\mathrm{d}t \\ &\quad = \lim_{\theta\to 0}\frac{1}{\theta^2}\int_{-\theta R}^{\theta R}\int_{-\theta R}^{\theta R} \big|u(\tau+t,\xi+x) - U(t,x)\big|\,\mathrm{d}x\,\mathrm{d}t = 0 \end{aligned}$$

because of (2.10). Lemma 2.1 now implies that U itself is a distributional solution of (2.1), hence, by Lemma 2.2, the Rankine–Hugoniot equations (2.6) hold. □

2.3. *Admissibility conditions*

To motivate the following discussion, we first observe that the concept of weak solution is usually not stringent enough to achieve uniqueness for a Cauchy problem. In some cases, infinitely many weak solutions can be found.

EXAMPLE 2.2. For Burgers' equation

$$u_t + (u^2/2)_x = 0, \tag{2.11}$$

consider the Cauchy problem with initial data

$$u(0,x) = \begin{cases} 1 & \text{if } x \geqslant 0, \\ 0 & \text{if } x < 0. \end{cases}$$

For every $\alpha \in]0,1[$, a weak solution is

$$u_\alpha(t,x) = \begin{cases} 0 & \text{if } x < \alpha t/2, \\ \alpha & \text{if } \alpha t/2 \leqslant x < (1+\alpha)t/2, \\ 1 & \text{if } x \geqslant (1+\alpha)t/2. \end{cases}$$

Indeed, the piecewise constant function u_α trivially satisfies the equation outside the jumps. Moreover, the Rankine–Hugoniot conditions hold along the two lines of discontinuity $\{x = \alpha t/2\}$ and $\{x = (1+\alpha)t/2\}$.

From the previous example it is clear that, in order to achieve the uniqueness and continuous dependence on the initial data, the notion of weak solution must be supplemented with further "admissibility conditions" along shocks, possibly motivated by physical considerations. A standard way to impose additional conditions is through the use of convex entropies.

DEFINITION 2.4. A continuously differentiable function $\eta : \mathbb{R}^n \mapsto \mathbb{R}$ is called an *entropy* for the system of conservation laws (2.1), with *entropy flux* $q : \mathbb{R}^n \mapsto \mathbb{R}$, if for all $u \in \mathbb{R}^n$ there holds

$$D\eta(u) \cdot Df(u) = Dq(u). \tag{2.12}$$

An immediate consequence of (2.12) is that, if $u = u(t,x)$ is a C^1 solution of (2.1), then

$$\eta(u)_t + q(u)_x = 0. \tag{2.13}$$

Indeed,

$$D\eta(u)u_t + Dq(u)u_x = D\eta(u)\big(-Df(u)u_x\big) + Dq(u)u_x = 0.$$

In other words, for a smooth solution u, not only the quantities $u_1, \ldots, u_n$ are conserved but the additional conservation law (2.13) holds as well. However, one should be aware that, when u is discontinuous, the quantity $\eta(u)$ may not be conserved.

A standard admissibility condition for weak solutions of (1.1) can now be formulated as follows.

DEFINITION 2.5. Let η be a convex entropy for the system (2.1), with entropy flux q. A weak solution u is *entropy-admissible* if

$$\eta(u)_t + q(u)_x \leqslant 0 \tag{2.14}$$

in distribution sense, i.e.,

$$\iint \big\{\eta(u)\varphi_t + q(u)\varphi_x\big\}\, dx\, dt \geqslant 0 \tag{2.15}$$

for every function $\varphi \geqslant 0$ continuously differentiable with compact support.

For a bounded solution, we now show that the entropy admissibility condition (2.14) poses additional restrictions every point of approximate jump.

THEOREM 2.2. *Let $u = u(t,x)$ be a bounded, entropy admissible solution of the system of conservation laws* (2.1). *Let η be a convex entropy with flux q. Assume that u has an approximate jump at the point (τ, ξ), so that* (2.10) *holds. Then the left and right states u^-, u^+ and the speed λ satisfy*

$$\lambda\big[\eta(u^+) - \eta(u^-)\big] \geqslant q(u^+) - q(u^-). \tag{2.16}$$

PROOF. By the same rescaling argument used in the proof of Theorem 2.3, we conclude that the function U in (2.5) is itself an entropy admissible solution. Given any nonnegative function $\varphi \in \mathcal{C}_c^1$, using the divergence theorem we find

$$\begin{aligned} 0 &\leqslant \iint \eta(U)\varphi_t + q(U)\varphi_x\, dx\, dt \\ &= \int \big\{\lambda\big[\eta(u^+) - \eta(u^-)\big] - q(u^+) + q(u^-)\big\}\varphi(t, \lambda t)\, dt. \end{aligned}$$

Therefore (2.16) must hold. □

Of course, the entropy admissibility condition can be useful only if some nontrivial convex entropy for the system (2.1) is known. Observe that (2.12) represents a first-order system of n partial differential equations for the two scalar variables η, q. For $n \geqslant 3$, this system is overdetermined. In general, one should thus expect to find solutions only in the case $n \leqslant 2$. In mathematical physics, however, one encounters several special cases of systems where a nontrivial convex entropy exists.

An alternative admissibility condition, due to Lax [Lx], is particularly useful because it can be applied to any system and has a simple geometrical interpretation. We recall that a function u has an approximate jump at a point (τ, ξ) if (2.10) holds, for some U defined as in (2.5). In this case, by Theorem 2.3 the right and left states u^-, u^+ and the speed λ of the jump satisfy the Rankine–Hugoniot equations. In particular, λ must be an eigenvalue of the averaged matrix $A(u^-, u^+)$ defined at (2.8), i.e., $\lambda = \lambda_i(u^-, u^+)$ for some $i \in \{1, \ldots, n\}$.

DEFINITION 2.6. A solution $u = u(t, x)$ of (2.1) satisfies the *Lax admissibility condition* if, at each point (τ, ξ) of approximate jump, the left and right states u^-, u^+ and the speed $\lambda = \lambda_i(u^-, u^+)$ of the jump satisfy

$$\lambda_i(u^-) \geqslant \lambda \geqslant \lambda_i(u^+). \tag{2.17}$$

To appreciate the geometric meaning of this condition, consider a piecewise smooth solution, having a discontinuity along the line $x = \gamma(t)$, where the solution jumps from a left state u^- to a right state u^+. According to (2.9), this discontinuity must travel with a speed $\lambda \doteq \dot{\gamma} = \lambda_i(u^-, u^+)$ equal to an eigenvalue of the averaged matrix $A(u^-, u^+)$. If we now look at the *i-characteristics*, i.e., at the solutions of the ODE

$$\dot{x} = \lambda_i\big(u(t, x)\big),$$

we see that the Lax condition requires that these lines run into the shock, from both sides.

3. The Riemann problem

In this section we construct the solution to the *Riemann problem*, consisting of the system of conservation laws

$$u_t + f(u)_x = 0 \tag{3.1}$$

with the simple, piecewise constant initial data

$$u(0, x) = \bar{u}(x) \doteq \begin{cases} u^- & \text{if } x < 0, \\ u^+ & \text{if } x > 0. \end{cases} \tag{3.2}$$

This will provide the basic building block toward the solution of the Cauchy problem with more general initial data. Throughout our analysis, we shall adopt the following standard assumption, introduced by Lax [Lx].

(H) For each $i = 1, \ldots, n$, the ith field is either *genuinely nonlinear*, so that $D\lambda_i(u) \cdot r_i(u) > 0$ for all u, or *linearly degenerate*, with $D\lambda_i(u) \cdot r_i(u) = 0$ for all u.

Notice that, in the genuinely nonlinear case, the ith eigenvalue λ_i is strictly increasing along each integral curve of the corresponding field of eigenvectors r_i. In the linearly degenerate case, on the other hand, the eigenvalue λ_i is constant along each such curve. With the above assumption we are ruling out the possibility that, along some integral curve of an eigenvector r_i, the corresponding eigenvalue λ_i may partly increase and partly decrease, having several local maxima and minima.

EXAMPLE 3.1. Denote by ρ the density of a gas, by $v = \rho^{-1}$ its specific volume and by u its velocity. A simple model for isentropic gas dynamics is then provided by the system

$$\begin{cases} v_t - u_x = 0, \\ u_t + p(v)_x = 0. \end{cases}$$

Here $p = p(v)$ is a function which determines the pressure in terms of the specific volume. An appropriate choice is $p(v) = kv^{-\gamma}$, with $1 \leqslant \gamma \leqslant 3$. In the region where $v > 0$, the Jacobian matrix of the system is

$$A \doteq Df = \begin{pmatrix} 0 & -1 \\ p'(v) & 0 \end{pmatrix}.$$

The eigenvalues and eigenvectors are found to be

$$\begin{aligned} \lambda_1 &= -\sqrt{-p'(v)}, & \lambda_2 &= \sqrt{-p'(v)}, \\ r_1 &= \bigl(1, \sqrt{-p'(v)}\bigr)^T, & r_2 &= \bigl(-1, \sqrt{-p'(v)}\bigr)^T. \end{aligned}$$

It is now clear that the system is strictly hyperbolic provided that $p'(v) < 0$ for all $v > 0$. Moreover, observing that

$$D\lambda_1 \cdot r_1 = \frac{p''(v)}{2\sqrt{-p'(v)}} = D\lambda_2 \cdot r_2,$$

we conclude that both characteristic fields are genuinely nonlinear if $p''(v) > 0$ for all $v > 0$.

As we shall see in the sequel, if the assumption (H) holds, then the solution of the Riemann problem has a simple structure consisting of the superposition of n elementary waves: shocks, rarefactions or contact discontinuities. This considerably simplifies all further analysis. On the other hand, for strictly hyperbolic systems that do not satisfy the condition (H), basic existence and stability results can still be obtained but at the price of heavier technicalities. See [L4] for an approach based on the Glimm scheme, or the recent paper [BiB] for the vanishing viscosity approach.

3.1. *Shock and rarefaction waves*

Fix a state $u_0 \in \mathbb{R}^n$ and an index $i \in \{1, \ldots, n\}$. As before, let $r_i(u)$ be an i-eigenvector of the Jacobian matrix $A(u) = Df(u)$. The integral curve of the vector field r_i through the point u_0 is called the *i-rarefaction curve* through u_0. It is obtained by solving the Cauchy problem in state space:

$$\frac{\mathrm{d}u}{\mathrm{d}\sigma} = r_i(u), \qquad u(0) = u_0. \tag{3.3}$$

We shall denote this curve as

$$\sigma \mapsto R_i(\sigma)(u_0). \tag{3.4}$$

Clearly, the parametrization depends on the choice of the eigenvectors r_i. In particular, if we impose the normalization $|r_i(u)| \equiv 1$, then the rarefaction curve (3.4) will be parametrized by arc-length.

Next, for a fixed $u_0 \in \mathbb{R}^n$ and $i \in \{1, \ldots, n\}$, we consider the curve of states u which can be connected to the right of u_0 by an i-shock, satisfying the Rankine–Hugoniot equations

$$\lambda(u - u_0) = f(u) - f(u_0). \tag{3.5}$$

As in (2.9), we can write these equations in the form

$$A(u, u_0)(u - u_0) = \lambda(u - u_0), \tag{3.6}$$

showing that $u - u_0$ must be a right i-eigenvector of the averaged matrix

$$A(u, u_0) \doteq \int_0^1 A\big(su + (1-s)u_0\big)\,\mathrm{d}s.$$

By a theorem of linear algebra, this holds if and only if $u - u_0$ is orthogonal to every left j-eigenvector of $A(u, u_0)$, with $j \neq i$. The Rankine–Hugoniot conditions can thus be written in the equivalent form

$$l_j(u, u_0) \cdot (u - u_0) = 0 \quad \text{for all } j \neq i, \tag{3.7}$$

together with $\lambda = \lambda_i(u, u_0)$. Notice that (3.7) is a system of $n-1$ scalar equations in n variables (the n components of the vector u). Linearizing at the point $u = u_0$ one obtains the linear system

$$l_j(u_0) \cdot (w - u_0) = 0, \quad j \neq i,$$

whose solutions are all the points $w = u_0 + c r_i(u_0)$, $c \in \mathbb{R}$. Since the vectors $l_j(u_0)$ are linearly independent, we can apply the implicit function theorem and conclude that the set

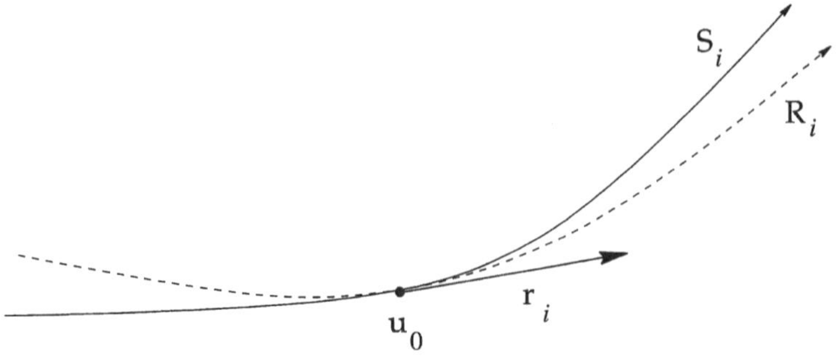

Fig. 5. Shock and rarefaction curves.

of solutions of (3.7) is a smooth curve, tangent to the vector r_i at the point u_0. This will be called the *i-shock curve* through the point u_0 and denoted as

$$\sigma \mapsto S_i(\sigma)(u_0). \tag{3.8}$$

Using a suitable parametrization (say, by arc-length), one can show that the two curves R_i, S_i have a second-order contact at the point u_0 (see Figure 5). More precisely, the following estimates hold.

$$\begin{cases} R_i(\sigma)(u_0) = u_0 + \sigma r_i(u_0) + \mathcal{O}(1) \cdot \sigma^2, \\ S_i(\sigma)(u_0) = u_0 + \sigma r_i(u_0) + \mathcal{O}(1) \cdot \sigma^2, \end{cases} \tag{3.9}$$

$$\left| R_i(\sigma)(u_0) - S_i(\sigma)(u_0) \right| = \mathcal{O}(1) \cdot \sigma^3, \tag{3.10}$$

$$\lambda_i\big(S_i(\sigma)(u_0),\ u_0\big) = \lambda_i(u_0) + \frac{\sigma}{2}\big(D\lambda_i(u_0)\big) \cdot r_i(u_0) + \mathcal{O}(1) \cdot \sigma^2. \tag{3.11}$$

Here and throughout the following, the Landau symbol $\mathcal{O}(1)$ denotes a quantity whose absolute value satisfies a uniform bound, depending only on the system (3.1).

3.2. *Three special cases*

Before constructing the solution of the Riemann problem (3.1) and (3.2) with arbitrary data u^-, u^+, we first study three special cases.

Case 1 – *Centered rarefaction waves.* Let the ith field be genuinely nonlinear, and assume that u^+ lies on the positive i-rarefaction curve through u^-, i.e., $u^+ = R_i(\sigma)(u^-)$ for some $\sigma > 0$. For each $s \in [0, \sigma]$, define

$$\lambda_i(s) = \lambda_i\big(R_i(s)(u^-)\big).$$

Observe that, by genuine nonlinearity, the map $s \mapsto \lambda_i(s)$ is strictly increasing. Hence, for every $\lambda \in [\lambda_i(u^-), \lambda_i(u^+)]$, there is a unique value $s \in [0, \sigma]$ such that $\lambda = \lambda_i(s)$. For $t \geqslant 0$, we claim that the function

$$u(t,x) = \begin{cases} u^- & \text{if } x/t < \lambda_i(u^-), \\ R_i(s)(u^-) & \text{if } x/t = \lambda_i(s) \in [\lambda_i(u^-), \lambda_i(u^+)], \\ u^+ & \text{if } x > t\lambda_i(u^+), \end{cases} \tag{3.12}$$

is a piecewise smooth solution of the Riemann problem, continuous for $t > 0$. Indeed, from the definition it follows

$$\lim_{t \to 0+} \| u(t, \cdot) - \bar{u} \|_{\mathbf{L}^1} = 0.$$

Moreover, (3.1) is trivially satisfied in the sectors where $x < t\lambda_i(u^-)$ or $x > t\lambda_i(u^+)$, because here $u_t = u_x = 0$. Next, assume $x = t\lambda_i(s)$ for some $s \in]0, \sigma[$. Since u is constant along each ray through the origin $\{x/t = c\}$, we have

$$u_t(t,x) + \frac{x}{t} u_x(t,x) = 0. \tag{3.13}$$

We now observe that the definition (3.12) implies $x/t = \lambda_i(u(t,x))$. By construction, the vector u_x has the same direction as $r_i(u)$, hence it is an eigenvector of the Jacobian matrix $A(u) \doteq Df(u)$ with eigenvalue $\lambda_i(u)$. On the sector $\{\lambda_i(u^-) < x/t < \lambda_i(u^+)\}$ we thus have

$$u_t + A(u)u_x = u_t + \lambda_i(u)u_x = 0,$$

proving our claim. Notice that the assumption $\sigma > 0$ is essential for the validity of this construction. In the opposite case $\sigma < 0$, the definition (3.12) would yield a triple-valued function in the region where $x/t \in [\lambda_i(u^+), \lambda_i(u^-)]$.

Case 2 – Shocks. Assume again that the ith family is genuinely nonlinear and that the state u^+ is connected to the right of u^- by an i-shock, i.e., $u^+ = S_i(\sigma)(u^-)$. Then, calling $\lambda \doteq \lambda_i(u^+, u^-)$ the Rankine–Hugoniot speed of the shock, the function

$$u(t,x) = \begin{cases} u^- & \text{if } x < \lambda t, \\ u^+ & \text{if } x > \lambda t, \end{cases} \tag{3.14}$$

provides a piecewise constant solution to the Riemann problem. Observe that, if $\sigma < 0$, then this solution is entropy admissible in the sense of Lax. Indeed, since the speed is monotonically increasing along the shock curve, recalling (3.11) we have

$$\lambda_i(u^+) < \lambda_i(u^-, u^+) < \lambda_i(u^-). \tag{3.15}$$

If η is a strictly convex entropy with flux q, one can also prove that the entropy admissibility conditions (2.16) hold, for all $\sigma \leqslant 0$ small. In the case $\sigma > 0$, however, one has $\lambda_i(u^-) < \lambda_i(u^+)$ and the admissibility conditions (2.16) and (2.17) are both violated.

Case 3 – *Contact discontinuities.* Assume that the ith field is linearly degenerate and that the state u^+ lies on the ith rarefaction curve through u^-, i.e., $u^+ = R_i(\sigma)(u^-)$ for some σ. By assumption, the ith characteristic speed λ_i is constant along this curve. Choosing $\lambda = \lambda(u^-)$, the piecewise constant function (3.13) then provides a solution to our Riemann problem. Indeed, the Rankine–Hugoniot conditions hold at the point of jump:

$$\begin{aligned} f(u^+) - f(u^-) &= \int_0^\sigma Df\big(R_i(s)(u^-)\big)r_i\big(R_i(s)(u^-)\big)\,ds \\ &= \lambda_i(u^-)\cdot\big[R_i(\sigma)(u^-) - u^-\big]. \end{aligned} \tag{3.16}$$

In this case, the Lax entropy condition holds regardless of the sign of σ. Indeed,

$$\lambda_i(u^+) = \lambda_i(u^-, u^+) = \lambda_i(u^-). \tag{3.17}$$

Moreover, if η is a strictly convex entropy with flux q, one can show that the entropy admissibility conditions (2.16) hold, for all values of σ.

Observe that, according to (3.16), for linearly degenerate fields the shock and rarefaction curves actually coincide, i.e., $S_i(\sigma)(u_0) = R_i(\sigma)(u_0)$ for all σ.

The above results can be summarized as follows. For a fixed left state u^- and $i \in \{1, \dots, n\}$ define the mixed curve

$$\Psi_i(\sigma)(u^-) = \begin{cases} R_i(\sigma)(u^-) & \text{if } \sigma \geqslant 0, \\ S_i(\sigma)(u^-) & \text{if } \sigma < 0. \end{cases} \tag{3.18}$$

In the special case where $u^+ = \Psi_i(\sigma)(u^-)$ for some σ, the Riemann problem can then be solved by an elementary wave: a rarefaction, a shock or a contact discontinuity.

3.3. *General solution of the Riemann problem*

Relying on the previous analysis, the solution of the general Riemann problem (3.1) and (3.2) can be obtained by finding intermediate states $\omega_0 = u^-, \omega_1, \dots, \omega_n = u^+$ such that each pair of adjacent states ω_{i-1}, ω_i can be connected by an elementary wave, i.e.,

$$\omega_i = \Psi_i(\sigma_i)(\omega_{i-1}), \quad i = 1, \dots, n. \tag{3.19}$$

This can be done whenever u^+ is sufficiently close to u^-. Indeed, for $|u^+ - u^-|$ small, the implicit function theorem provides the existence of unique wave strengths $\sigma_1, \dots, \sigma_n$, such that

$$u^+ = \Psi_n(\sigma_n) \circ \cdots \circ \Psi_1(\sigma_1)(u^-). \tag{3.20}$$

In turn, these determine the intermediate states ω_i in (3.19). The complete solution is now obtained by piecing together the solutions of the n Riemann problems

$$u_t + f(u)_x = 0, \qquad u(0,x) = \begin{cases} \omega_{i-1} & \text{if } x < 0, \\ \omega_i & \text{if } x > 0, \end{cases} \tag{3.21}$$

on different sectors of the t–x plane. By construction, each of these problems has an entropy-admissible solution consisting of a simple wave of the ith characteristic family. More precisely:

Case 1. The ith characteristic field is genuinely nonlinear and $\sigma_i > 0$. Then the solution of (3.21) consists of a centered rarefaction wave. The ith characteristic speeds range over the interval $[\lambda_i^-, \lambda_i^+]$, defined as

$$\lambda_i^- \doteq \lambda_i(\omega_{i-1}), \qquad \lambda_i^+ \doteq \lambda_i(\omega_i).$$

Case 2. Either the ith characteristic field is genuinely nonlinear and $\sigma_i \leqslant 0$, or else the ith characteristic field is linearly degenerate (with σ_i arbitrary). Then the solution of (3.21) consists of an admissible shock or a contact discontinuity, traveling with Rankine–Hugoniot speed

$$\lambda_i^- \doteq \lambda_i^+ \doteq \lambda_i(\omega_{i-1}, \omega_i).$$

The solution to the original problem (3.1) and (3.2) can now be constructed (see Figure 6) by piecing together the solutions of the n Riemann problems (3.21), $i = 1, \dots, n$. Indeed, for $\sigma_1, \dots, \sigma_n$ sufficiently small, the speeds λ_i^-, λ_i^+ introduced above remain close to the corresponding eigenvalues $\lambda_i(u^-)$ of the matrix $A(u^-)$. By strict hyperbolicity and continuity, we can thus assume that the intervals $[\lambda_i^-, \lambda_i^+]$ are disjoint, i.e.,

$$\lambda_1^- \leqslant \lambda_1^+ < \lambda_2^- \leqslant \lambda_2^+ < \cdots < \lambda_n^- \leqslant \lambda_n^+.$$

Therefore, a piecewise smooth solution $u : [0, \infty) \times \mathbb{R} \mapsto \mathbb{R}^n$ is well defined by the assignment

$$u(t,x) = \begin{cases} u^- = \omega_0 & \text{if } \frac{x}{t} \in \,]-\infty, \lambda_1^-[, \\ R_i(s)(\omega_{i-1}) & \text{if } \frac{x}{t} = \lambda_i\big(R_i(s)(\omega_{i-1})\big) \in [\lambda_i^-, \lambda_i^+[, \\ \omega_i & \text{if } \frac{x}{t} \in [\lambda_i^+, \lambda_{i+1}^-[, \\ u^+ = \omega_n & \text{if } \frac{x}{t} \in [\lambda_n^+, \infty[. \end{cases} \tag{3.22}$$

Observe that this solution is self-similar, having the form $u(t,x) = \psi(x/t)$, with $\psi : \mathbb{R} \mapsto \mathbb{R}^n$ possibly discontinuous.

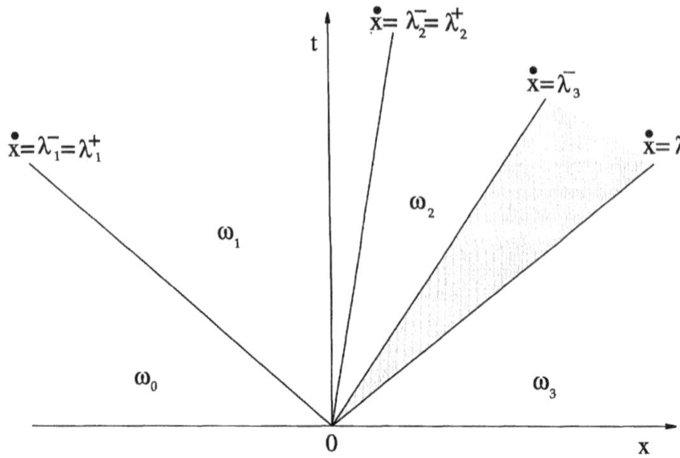

Fig. 6. Solution of a Riemann problem.

3.4. *Some estimates*

We conclude this section by proving two types of estimates, which will play a key role in the analysis of front tracking approximations.

Fix a left state u^- and consider a right state $u^+ \doteq R_k(\sigma)(u^-)$ on the k-rarefaction curve. In general, the function

$$u(t,x) \doteq \begin{cases} u^- & \text{if } x < \lambda_k(u^-), \\ u^+ & \text{if } x > \lambda_k(u^-) \end{cases} \tag{3.23}$$

is not an exact solution of the system (3.1). However, we now show that the error in the Rankine–Hugoniot equations and the possible increase in a convex entropy are indeed very small.

LEMMA 3.1 (Error estimates). *For $\sigma > 0$ small, one has the estimate*

$$\lambda_k(u^-)\big[R_k(\sigma)(u^-) - u^-\big] - \big[f\big(R_k(\sigma)(u^-)\big) - f(u^-)\big] = \mathcal{O}(1)\cdot\sigma^2. \tag{3.24}$$

Moreover, if η is a convex entropy with entropy flux q, then

$$\lambda_k(u^-)\big[\eta\big(R_k(\sigma)(u^-)\big) - \eta(u^-)\big] - \big[q\big(R_k(\sigma)(u^-)\big) - q(u^-)\big] = \mathcal{O}(1)\cdot\sigma^2. \tag{3.25}$$

Here $\mathcal{O}(1)$ denotes a quantity which remains uniformly bounded as u^- ranges on bounded sets.

PROOF. Call $E(\sigma)$ the left-hand side of (3.24). Clearly $E(0) = 0$. Differentiating with respect to σ at the point $\sigma = 0$ and recalling that $\mathrm{d}R_k/\mathrm{d}\sigma = r_k$, we find

$$\frac{\mathrm{d}E}{\mathrm{d}\sigma}\bigg|_{\sigma=0} = \lambda_k(u^-)r_k(u^-) - Df(u^-)r_k(u^-) = 0.$$

Since E varies smoothly with u^- and σ, the estimate (3.24) easily follows by Taylor's formula.

The second estimate is proved similarly. Call $E'(\sigma)$ the left-hand side of (3.25) and observe that $E'(0) = 0$. Differentiating with respect to σ we now obtain

$$\frac{\mathrm{d}E'}{\mathrm{d}\sigma}\bigg|_{\sigma=0} = \lambda_k(u^-)D\eta(u^-)r_k(u^-) - Dq(u^-)r_k(u^-) = 0,$$

because $D\eta\lambda_k r_k = D\eta Df r_k = Dq r_k$. Taylor's formula now yields (3.25). □

Next, consider a left state u^l, a middle state u^m and a right state u^r (see Figure 7). Assume that the pair (u^l, u^m) is connected by a j-wave of strength σ', while the pair (u^l, u^m) is connected by an i-wave of strength σ'', with $i < j$. We are interested in the strength of the waves $(\sigma_1, \ldots, \sigma_n)$ in the solution of the Riemann problem where $u^- = u^l$ and $u^+ = u^r$. Roughly speaking, these are the waves determined by the interaction of the σ' and σ''. The next lemma shows that $\sigma_i \approx \sigma''$, $\sigma_j \approx \sigma'$ while $\sigma_k \approx 0$ for $k \neq i, j$.

A different type of interaction is considered in Figure 8. Here the pair (u^l, u^m) is connected by an i-wave of strength σ', while the pair (u^l, u^m) is connected by a second i-wave, say of strength σ''. In this case, the strengths $(\sigma_1, \ldots, \sigma_n)$ of the outgoing waves satisfy $\sigma_i \approx \sigma' + \sigma''$ while $\sigma_k \approx 0$ for $k \neq i$.

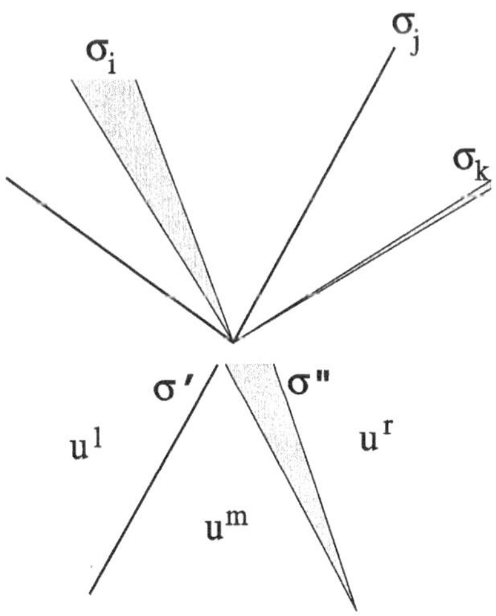

Fig. 7. Interaction of waves of different families.

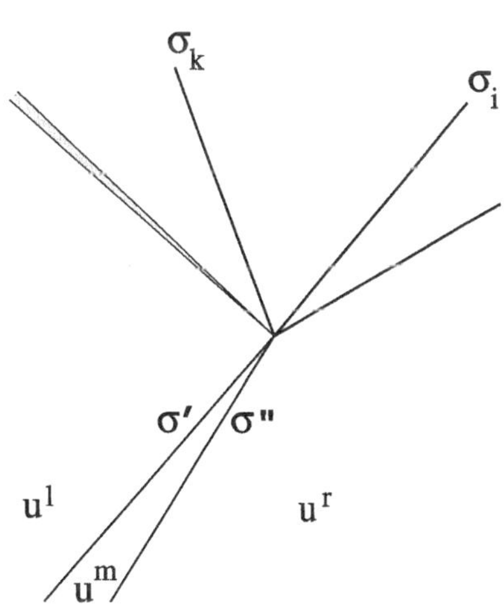

Fig. 8. Interaction of waves of the same family.

LEMMA 3.2 (Interaction estimates).

(i) *Consider the Riemann problem* (3.1) *and* (3.2). *Recalling* (3.18), *assume that the right state is given by*

$$u^+ = \Psi_i(\sigma'') \circ \Psi_j(\sigma')(u^-). \tag{3.26}$$

Let the solution consist of waves of size $(\sigma_1, \ldots, \sigma_n)$, *as in* (3.20). *Then*

$$|\sigma_i - \sigma''| + |\sigma_j - \sigma'| \sum_{k \neq i,j} |\sigma_k| = \mathcal{O}(1) \cdot |\sigma'\sigma''|. \tag{3.27}$$

(ii) *Next, assume that the right state is given by*

$$u^+ = \Psi_i(\sigma_i') \circ \Psi_i(\sigma_i'')(u^-). \tag{3.28}$$

Then the waves $(\sigma_1, \ldots, \sigma_n)$ *in the solution of the Riemann problem can be estimated by*

$$|\sigma_i - \sigma' - \sigma''| + \sum_{k \neq i} |\sigma_k| = \mathcal{O}(1) \cdot |\sigma'\sigma''|(|\sigma'| + |\sigma''|). \tag{3.29}$$

Here $\mathcal{O}(1)$ *denotes a quantity which remains uniformly bounded as* u^- *ranges n bounded sets.*

PROOF. For $u^-, u^+ \in \mathbb{R}^n$, $k = 1, \ldots, n$, call $E_k(u^-, u^+) \doteq \sigma_k$ the size of the kth wave in the solution of the corresponding Riemann problem. By our earlier analysis, the functions E_k are $\mathcal{C}^2$ with Lipschitz continuous second derivatives.

For a given left state u^-, we now define the composed functions Φ_k, $k = 1, \ldots, n$, by setting

$$\begin{aligned}
\Phi_i(\sigma', \sigma'') &\doteq \sigma_i - \sigma'' = E_i(u^-, \Psi_j(\sigma') \circ \Psi_i(\sigma'')(u^-)) - \sigma'', \\
\Phi_j(\sigma', \sigma'') &\doteq \sigma_j - \sigma' = E_j(u^-, \Psi_j(\sigma') \circ \Psi_i(\sigma'')(u^-)) - \sigma_j', \\
\Phi_k(\sigma', \sigma'') &\doteq \sigma_k = E_k(u^-, \Psi_j(\sigma') \circ \Psi_i(\sigma'')(u^-)) \quad \text{if } k \neq i, j.
\end{aligned}$$

The Φ_k are $\mathcal{C}^2$ functions of σ', σ'' with Lipschitz continuous second derivatives, depending continuously also on the left state u^-. Observing that

$$\Phi_k(\sigma', 0) = \Phi_k(0, \sigma'') = 0 \quad \text{for all } \sigma', \sigma'', \ k = 1, \ldots, n,$$

by Taylor's formula (see Lemma A.2 in the Appendix) we conclude that

$$\Phi_k(\sigma', \sigma'') = \mathcal{O}(1) \cdot |\sigma'\sigma''|$$

for all $k = 1, \ldots, n$. This establishes (3.27).

To prove (3.29), we consider the functions

$$\Phi_i(\sigma', \sigma'') \doteq \sigma_i - \sigma' - \sigma'' = E_i(u^-, \Psi_i(\sigma'') \circ \Psi_i(\sigma')(u^-)) - \sigma' - \sigma'',$$
$$\Phi_k(\sigma', \sigma'') \doteq \sigma_k = E_k(u^-, \Psi_i(\sigma'') \circ \Psi_i(\sigma')(u^-)) \quad \text{if } k \neq i.$$

In the case where $\sigma', \sigma'' \geqslant 0$, recalling (3.18) we have

$$u^+ = R_i(\sigma'') \circ R_i(\sigma')(u^-) = R_i(\sigma'' + \sigma')(u^-).$$

Hence, the Riemann problem is solved exactly by an i-rarefaction of strength $\sigma' + \sigma''$. Therefore

$$\Phi_k(\sigma', \sigma'') = 0 \quad \text{whenever } \sigma', \sigma'' \geqslant 0, \tag{3.30}$$

for all $k = 1, \ldots, n$. As before, the Φ_k are $\mathcal{C}^2$ functions of σ', σ'' with Lipschitz continuous second derivatives, depending continuously also on the left state u^-. We observe that

$$\Phi_k(\sigma', 0) = \Phi_k(0, \sigma'') = 0. \tag{3.31}$$

Moreover, by (3.30) the continuity of the second derivatives implies

$$\frac{\partial^2 \Phi_k}{\partial \sigma' \partial \sigma''}(0, 0) = 0. \tag{3.32}$$

By (3.31) and (3.32), using the estimate (A.16) in the Appendix we now conclude

$$\Phi_k(\sigma, \sigma') = \mathcal{O}(1) \cdot |\sigma'\sigma''|(|\sigma'| + |\sigma''|)$$

for all $k = 1, \ldots, n$. This establishes (ii). □

4. Front tracking approximations

4.1. *Global existence of entropy weak solutions*

In this section we describe in detail the construction of front tracking approximations and give the main application of this technique, proving the global existence of weak solutions. Consider the Cauchy problem

$$u_t + f(u)_x = 0, \tag{4.1}$$
$$u(0, x) = \bar{u}(x), \tag{4.2}$$

where the flux function f is smooth, defined on a neighborhood of the origin. We always assume that the system is strictly hyperbolic, and the assumption (H) holds, so that each characteristic field is either genuinely nonlinear or linearly degenerate. The following basic theorem provides the global existence of an entropy weak solution, for all initial data with suitably small total variation.

THEOREM 4.1. *There exists a constant* $\delta_0 > 0$ *such that, for every initial condition* $\bar{u} \in \mathbf{L}^\infty(\mathbb{R};\mathbb{R}^n)$ *with*

$$\text{Tot.Var}(\bar{u}) \leqslant \delta_0, \qquad \|\bar{u}\|_{\mathbf{L}^\infty} \leqslant \delta_0, \tag{4.3}$$

the Cauchy problem (4.1) *and* (4.2) *has a weak solution* $u = u(t,x)$ *defined for all* $t \geqslant 0$. *If* η *is a convex entropy with entropy flux* q, *one can construct a solution which also satisfies*

$$\eta(u)_t + q(u)_x \leqslant 0. \tag{4.4}$$

The above theorem was first proved in the fundamental paper of Glimm [G], where approximate solutions are constructed by piecing together Riemann solutions on the nodes of a fixed grid. Here we shall describe an alternative construction, based on wave-front tracking.

Roughly speaking, a front tracking ε-approximate solution (see Figure 9) is a piecewise constant function $u = u(t,x)$ whose jumps are located along finitely many segments $x = x_\alpha(t)$ in the t–x plane. At any given time $t > 0$, all these jumps should approximately satisfy the Rankine–Hugoniot conditions, namely

$$\sum_\alpha \left| \dot{x}_\alpha \left[u(t, x_\alpha+) - u(t, x_\alpha-) \right] - \left[f\left(u(t, x_\alpha+)\right) - f\left(u(t, x_\alpha-)\right) \right] \right| = \mathcal{O}(1) \cdot \varepsilon.$$

Moreover, if η is a convex entropy with flux q, at each time $t > 0$ in view of (2.16) we also

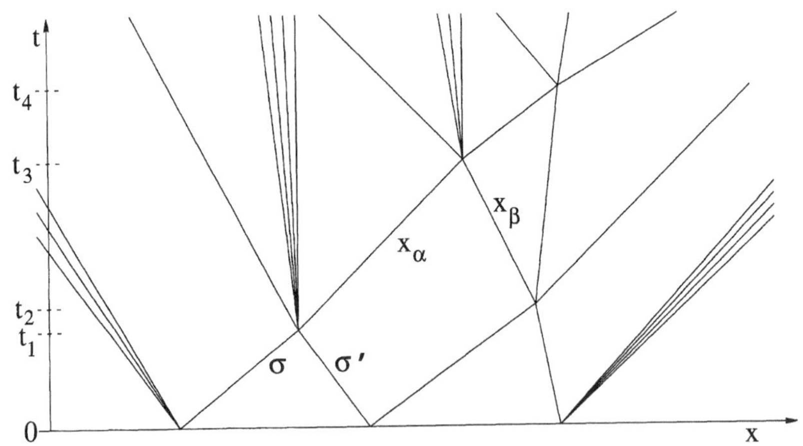

Fig. 9. Front tracking approximation.

require the approximate admissibility conditions

$$\sum_\alpha \{[q(u(t,x_\alpha+)) - q(u(t,x_\alpha-))] - \dot{x}_\alpha[\eta(u(t,x_\alpha+)) - \eta(u(t,x_\alpha-))]\}$$
$$\leqslant \mathcal{O}(1)\cdot\varepsilon.$$

For didactical purposes, the proof of Theorem 4.1 will be given in two steps. In the first part of the proof we describe a *naive front tracking algorithm* and derive uniform bounds on the total variation of the approximate solutions. Relying on Helly's compactness theorem, we also prove that a suitable subsequence of front tracking approximations converges to an entropy weak solution. This first step thus contains all the "heart of the matter". It would provide a complete proof, except for one gap: we are not considering here the possibility that infinitely many wave-fronts appear within finite time (see Figure 3). If this happens, the naive front tracking algorithm breaks down, and solutions cannot be further prolonged in time.

The only purpose of the second part of the proof is to fix this technical problem. We show that a suitable *modified front tracking algorithm* will always produce approximate solutions having a finite number of wave-fronts for all times $t \geqslant 0$. This will achieve a completely rigorous proof of Theorem 4.1.

4.2. *A naive front tracking algorithm*

Let the initial condition $\bar{u}$ be given and fix $\varepsilon > 0$. We now describe an algorithm which produces a piecewise constant approximate solution to the Cauchy problem (4.1) and (4.2). The construction (see Figure 9) starts at time $t = 0$ by taking a piecewise constant approximation $u(0,\cdot)$ of $\bar{u}$, such that

$$\text{Tot.Var}\{u(0,\cdot)\} \leqslant \text{Tot.Var}\{\bar{u}\}, \qquad \int_{-1/\varepsilon}^{1/\varepsilon} |u(0,x) - \bar{u}(x)|\,dx \leqslant \varepsilon. \tag{4.5}$$

Let $x_1 < \cdots < x_N$ be the points where $u(0,\cdot)$ is discontinuous. For each $\alpha = 1,\ldots,N$, the Riemann problem generated by the jump $(u(0,x_\alpha-), u(0,x_\alpha+))$ is approximately solved on a forward neighborhood of $(0,x_\alpha)$ in the t–x plane by a piecewise constant function, according to the following procedure.

Accurate Riemann Solver. Consider the general Riemann problem at a point $(\bar{t},\bar{x})$,

$$v_t + f(v)_x = 0, \qquad v(\bar{t},x) = \begin{cases} u^- & \text{if } x < \bar{x}, \\ u^+ & \text{if } x > \bar{x}. \end{cases} \tag{4.6}$$

Recalling (3.18), let $\omega_0,\ldots,\omega_n$ be the intermediate states and $\sigma_1,\ldots,\sigma_n$ be the strengths of the waves in the solution, so that

$$\omega_0 = u^-, \qquad \omega_n = u^+, \qquad \omega_i = \Psi_i(\sigma_i)(\omega_{i-1}), \quad i = 1,\ldots,n. \tag{4.7}$$

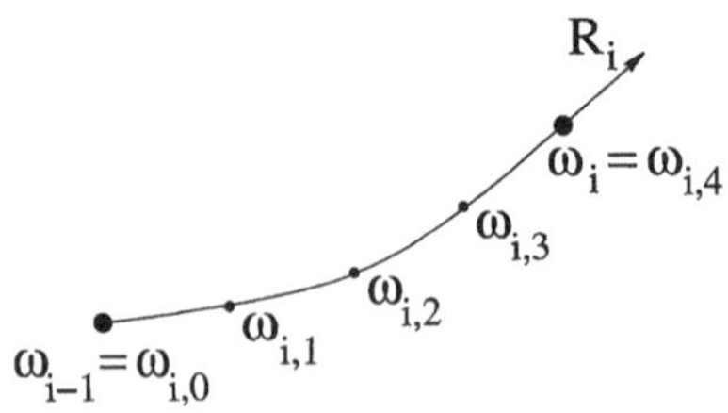

Fig. 10. Partitioning a rarefront wave.

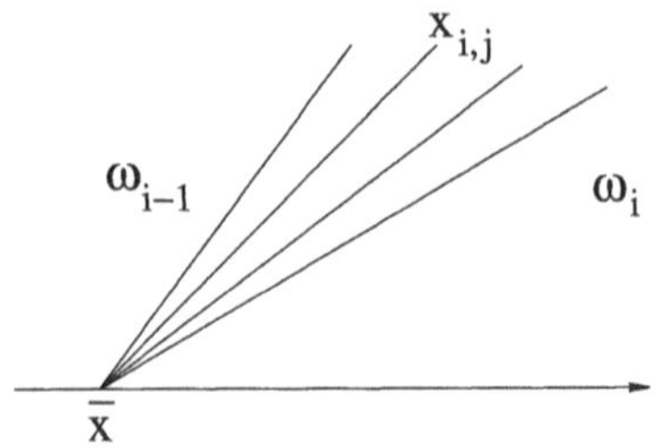

Fig. 11. See Figure 10.

If all jumps (ω_{i-1}, ω_i) were shocks or contact discontinuities, then this solution would be already piecewise constant. In general, the exact solution of (4.6) is not piecewise constant, because of the presence of centered rarefaction waves. These will be approximated by piecewise constant rarefaction fans, inserting additional states $\omega_{i,j}$ as follows (see Figure 10).

If the ith characteristic field is genuinely nonlinear and $\sigma_i > 0$, we divide the centered i-rarefaction into a number p_i of smaller i-waves, each with strength σ_i/p_i. Since we want $\sigma/p_i < \varepsilon$, we choose the integer

$$p_i \doteq 1 + [\![\sigma_i/\varepsilon]\!], \tag{4.8}$$

where $[\![s]\!]$ denotes the largest integer less or equal to s. For $j = 1, \ldots, p_i$, we now define the intermediate states and wave-fronts (see Figure 11)

$$\omega_{i,j} = \Psi_i(j\sigma_i/p_i)(\omega_{i-1}), \qquad x_{i,j}(t) = \bar{x} + (t - \bar{t})\lambda_i(\omega_{i,j-1}). \tag{4.9}$$

For notational convenience, if the ith characteristic field is genuinely nonlinear and $\sigma_i \leqslant 0$, or if the ith characteristic field is linearly degenerate (with σ_i arbitrary), we define $p_i \doteq 1$ and set

$$\omega_{i,1} = \omega_i, \qquad x_{i,1}(t) = \bar{x} + (t - \bar{t})\lambda_i(\omega_{i-1}, \omega_i). \tag{4.10}$$

Here $\lambda_i(\omega_{i-1}, \omega_i)$ is the Rankine–Hugoniot speed of the jump connecting ω_{i-1} with ω_i, so that

$$\lambda_i(\omega_{i-1}, \omega_i) \cdot (\omega_i - \omega_{i-1}) = f(\omega_i) - f(\omega_{i-1}). \tag{4.11}$$

As soon as the intermediate states $\omega_{i,j}$ and the locations $x_{i,j}(t)$ of the jumps have been determined by (4.9) or (4.10), we can define a piecewise constant approximate solution to the Riemann problem (4.6) by setting (see Figure 12)

$$v(t,x) = \begin{cases} u^- & \text{if } x < x_{1,1}(t), \\ u^+ & \text{if } x > x_{n,p_n}(t), \\ \omega_i \ (= \omega_{i,p_i}) & \text{if } x_{i,p_i}(t) < x < x_{i+1,1}(t), \\ \omega_{i,j} & \text{if } x_{i,j}(t) < x < x_{i,j+1}(t), \quad j = 1, \ldots, p_i - 1. \end{cases} \tag{4.12}$$

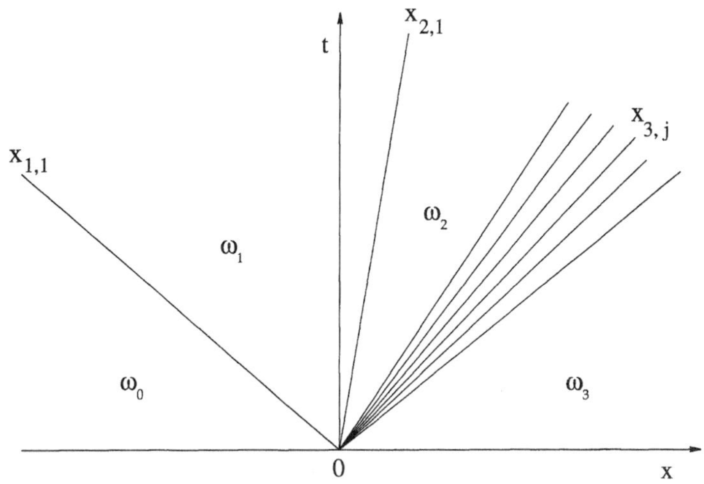

Fig. 12. Approximate solution to a Riemann problem.

We observe that the difference between this function v and the exact solution is only due to the fact that every centered i-rarefaction wave is here divided into equal parts and replaced by a rarefaction fan containing p_i wave-fronts. Because of (4.8), the strength of each one of these fronts is less than ε.

We now resume the construction of a front tracking solution to the original Cauchy problem (4.1) and (4.2). Having solved all the Riemann problems at time $t = 0$, the approximate solution u can be prolonged until a first time t_1 is reached, when two wave-fronts interact (see Figure 9). Since $u(t_1, \cdot)$ is still a piecewise constant function, the corresponding Riemann problems can again be approximately solved within the class of piecewise constant functions. The solution u is then continued up to a time t_2 where a second interaction takes place, etc. We remark that, by an arbitrary small change in the speed of one of the wave fronts, it is not restrictive to assume that at most two incoming fronts collide, at each given time $t > 0$. This will considerably simplify all subsequent analysis, since we do not need to consider the case where three or more incoming fronts meet together.

4.3. *Bounds on the total variation*

In this section we derive bounds on the total variation of a front tracking approximation $u(t, \cdot)$, uniformly valid for all $t \geqslant 0$. These estimates will be obtained from Lemma 3.2, using an interaction functional.

We begin by introducing some notation. At a fixed time t let x_α, $\alpha = 1, \ldots, N$, be the locations of the fronts in $u(t, \cdot)$. Moreover, let $|\sigma_\alpha|$ be the strength of the wave-front at x_α, say of the family $k_\alpha \in \{1, \ldots, n\}$. Consider the two functionals

$$V(t) \doteq V\big(u(t)\big) \doteq \sum_\alpha |\sigma_\alpha|, \tag{4.13}$$

measuring the *total strength of waves* in $u(t,\cdot)$, and

$$Q(t) \doteq Q\big(u(t)\big) \doteq \sum_{(\alpha,\beta)\in\mathcal{A}} |\sigma_\alpha \sigma_\beta|, \tag{4.14}$$

measuring the *wave interaction potential*. In (4.14), the summation ranges over the set $\mathcal{A}$ of all couples of approaching wave-fronts.

DEFINITION 4.1. Two fronts, located at points $x_\alpha < x_\beta$ and belonging to the characteristic families $k_\alpha, k_\beta \in \{1,\dots,n\}$, respectively, are *approaching* if $k_\alpha > k_\beta$ or else if $k_\alpha = k_\beta$ and at least one of the wave-fronts is a shock of a genuinely nonlinear family.

Roughly speaking, two fronts are approaching if the one behind has the larger speed (and hence it can collide with the other, at a future time).

Now consider the approximate solution $u = u(t,x)$ constructed by the front tracking algorithm. It is clear that the quantities $V(u(t))$, $Q(u(t))$ remain constant except at times where an interaction occurs. At a time τ where two fronts of strength $|\sigma'|$, $|\sigma''|$ collide, the interaction estimates (3.27) or (3.29) yield

$$\Delta V(\tau) \doteq V(\tau+) - V(\tau-) = \mathcal{O}(1)\cdot \big|\sigma'\sigma''\big|, \tag{4.15}$$

$$\Delta Q(\tau) \doteq Q(\tau+) - Q(\tau-) = -\big|\sigma'\sigma''\big| + \mathcal{O}(1)\cdot \big|\sigma'\sigma''\big|\cdot V(\tau-). \tag{4.16}$$

Indeed (see Figure 13), after time τ the two colliding fronts σ', σ'' are no longer approaching. Hence the product $|\sigma'\sigma''|$ is no longer counted within the summation (4.14). On the other hand, the new waves emerging from the interaction (having strength $\mathcal{O}(1)\cdot|\sigma'\sigma''|$) can approach all the other fronts not involved in the interaction (which have total strength $\leqslant V(\tau-)$).

If V remains sufficiently small, so that $\mathcal{O}(1)\cdot V(\tau-)\leqslant 1/2$, from (4.16) it follows

$$Q(\tau+) - Q(\tau-) \leqslant -\frac{|\sigma'\sigma''|}{2}. \tag{4.17}$$

By (4.15) and (4.17) we can thus choose a constant C_0 large enough so that the quantity

$$\Upsilon(t) \doteq V(t) + C_0 Q(t)$$

decreases at every interaction time, provided that V remains sufficiently small.

We now observe that the total strength of waves is an equivalent way of measuring the total variation. Indeed, for some constant κ one has

$$\text{Tot.Var}\big\{u(t)\big\} \leqslant V\big(u(t)\big) \leqslant \kappa\cdot \text{Tot.Var}\big\{u(t)\big\}. \tag{4.18}$$

Moreover, the definitions (4.13) and (4.14) trivially imply $Q \leqslant V^2$. If the total variation of the initial data $u(0,\cdot)$ is sufficiently small, the previous estimates show that the quantity $V + C_0 Q$ is nonincreasing in time. Therefore

$$\text{Tot.Var}\big\{u(t)\big\} \leqslant V\big(u(t)\big) \leqslant V\big(u(0)\big) + C_0 Q\big(u(0)\big). \tag{4.19}$$

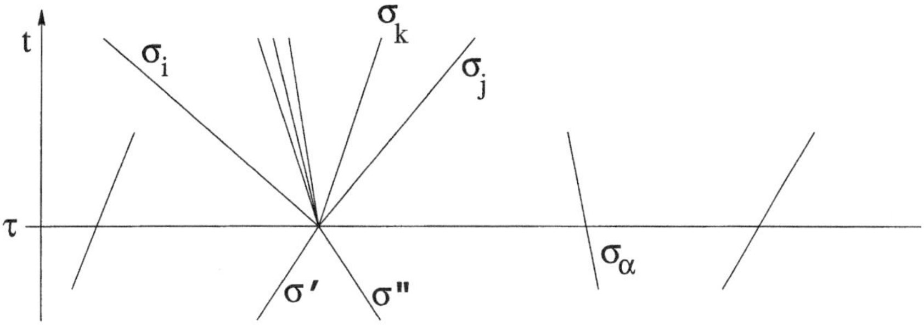

Fig. 13. Interaction of two wave-fronts.

This provides a uniform bound on the total variation of $u(t, \cdot)$ valid for all times $t \geqslant 0$.

An important consequence of the bound (4.19) is that, at every time τ where two fronts interact, the corresponding Riemann problem can always be solved. Indeed, the left and right states differ by the quantity

$$\left|u^+ - u^-\right| \leqslant \text{Tot.Var}\left\{u(\tau)\right\},$$

which remains small.

Another consequence of the bound on the total variation is the continuity of $t \mapsto u(t, \cdot)$ as a function with values in $\mathbf{L}^1_{\text{loc}}$. More precisely, there exists a Lipschitz constant L' such that

$$\int_{-\infty}^{\infty} \left|u(t, x) - u\left(t', x\right)\right| \mathrm{d}x \leqslant L'\left|t - t'\right| \quad \text{for all } t, t' \geqslant 0. \tag{4.20}$$

Indeed, if no interaction occurs inside the interval $[t, t']$, the left-hand side of (4.20) can be estimated simply as

$$\begin{aligned} &\left\|u(t) - u\left(t'\right)\right\|_{\mathbf{L}^1} \\ &\quad \leqslant \left|t - t'\right| \sum_{\alpha} |\sigma_\alpha||\dot{x}_\alpha| \\ &\quad \leqslant \left|t - t'\right| \cdot \text{[total strength of all wave fronts]} \cdot \text{[maximum speed]} \\ &\quad \leqslant L' \left|t - t'\right|, \end{aligned} \tag{4.21}$$

for some uniform constant L'. The case where one or more interactions take place within $[t, t']$ is handled in the same way, observing that the map $t \mapsto u(t, \cdot)$ is continuous across interaction times.

4.4. *Convergence to a limit solution*

Fix any sequence $\varepsilon_\nu \to 0+$. For every $\nu \geqslant 1$ we apply the front tracking algorithm and construct an ε_ν-approximate solution u_ν of the Cauchy problem (4.1) and (4.2). By the previous analysis, the total variation of $u_\nu(t,\cdot)$ remains bounded, uniformly for all $t \geqslant 0$ and $\nu \geqslant 1$. Moreover, by (4.20) the maps $t \mapsto u_\nu(t,\cdot)$ are uniformly Lipschitz continuous with respect to the $\mathbf{L}^1$ distance. We can thus apply Helly's compactness theorem (see Theorem A.1 in the Appendix) and extract a subsequence which converges to some limit function u in $\mathbf{L}^1_{\text{loc}}$, also satisfying (4.20).

Since $u_\nu(0) \to \bar{u}$ in $\mathbf{L}^1_{\text{loc}}$, the initial condition (4.2) is clearly attained. To prove that u is a weak solution of the Cauchy problem, it remains to show that, for every $\phi \in \mathcal{C}^1_c$ with compact support contained in the open half plane where $t > 0$, one has

$$\int_0^\infty \int_{-\infty}^\infty \phi_t(t,x)u(t,x) + \phi_x(t,x) f\big(u(t,x)\big)\,\mathrm{d}x\,\mathrm{d}t = 0. \tag{4.22}$$

Since the u_ν are uniformly bounded and f is uniformly continuous on bounded sets, it suffices to prove that

$$\lim_{\nu\to 0}\left[\int_0^\infty \int_{-\infty}^\infty \big\{\phi_t(t,x)u_\nu(t,x) + \phi_x(t,x)\big(u_\nu(t,x)\big)\big\}\,\mathrm{d}x\,\mathrm{d}t\right] = 0. \tag{4.23}$$

Choose $T > 0$ such that $\phi(t,x) = 0$ whenever $t \notin [0,T]$. For a fixed ν, at any time t call $x_1(t) < \cdots < x_N(t)$ the points where $u_\nu(t,\cdot)$ has a jump, and set

$$\begin{aligned}
&\Delta u_\nu(t,x_\alpha) \doteq u_\nu(t,x_\alpha+) - u_\nu(t,x_\alpha-),\\
&\Delta f\big(u_\nu(t,x_\alpha)\big) \doteq f\big(u_\nu(t,x_\alpha+)\big) - f\big(u_\nu(t,x_\alpha-)\big).
\end{aligned}$$

Observe that the polygonal lines $x = x_\alpha(t)$ subdivide the strip $[0,T]\times\mathbb{R}$ into finitely many regions Γ_j where u_ν is constant (see Figure 14). Introducing the vector

$$\Phi \doteq \big(\phi\cdot u_\nu, \phi\cdot f(u_\nu)\big),$$

by the divergence theorem the double integral in (4.23) can be written as

$$\sum_j \iint_{\Gamma_j} \operatorname{div}\Phi(t,x)\,\mathrm{d}t\,\mathrm{d}x = \sum_j \int_{\partial\Gamma_j} \Phi\cdot\mathbf{n}\,\mathrm{d}\sigma. \tag{4.24}$$

Here $\partial\Gamma_j$ is the oriented boundary of Γ_j, while $\mathbf{n}$ denotes an outer normal. Observe that $\mathbf{n}\,\mathrm{d}\sigma = \pm(\dot{x}_\alpha, -1)\,\mathrm{d}t$ along each polygonal line $x = x_\alpha(t)$, while $\phi(t,x) = 0$ along the lines $t = 0$, $t = T$. By (4.24) the expression within square brackets in (4.23) is computed by

$$\int_0^T \sum_\alpha \big[\dot{x}_\alpha(t)\cdot\Delta u_\nu(t,x_\alpha) - \Delta f\big(u_\nu(t,x_\alpha)\big)\big]\phi\big(t,x_\alpha(t)\big)\,\mathrm{d}t. \tag{4.25}$$

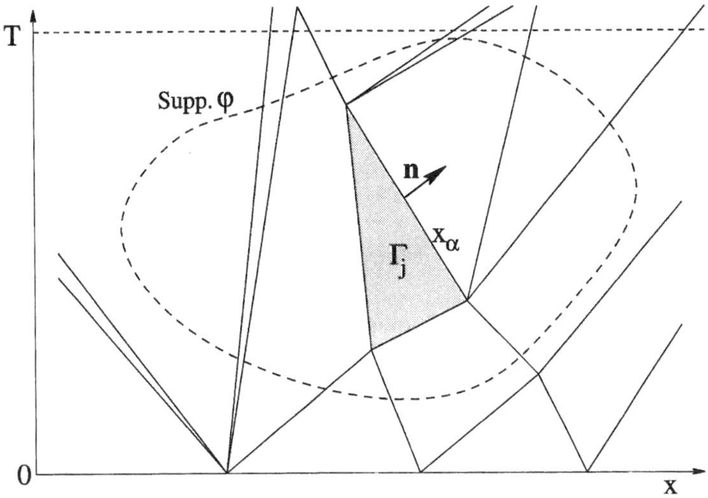

Fig. 14. Applying the divergence theorem.

Here, for each $t \in [0, T]$, the sum ranges over all fronts of $u_\nu(t, \cdot)$. To estimate the above integral, let $|\sigma_\alpha|$ be the strength of the wave-front at x_α. If this wave is a shock or contact discontinuity, by construction the Rankine–Hugoniot equations are satisfied exactly, i.e.,

$$\dot{x}_\alpha(t) \cdot \Delta u_\nu(t, x_\alpha) - \Delta f\big(u_\nu(t, x_\alpha)\big) = 0. \tag{4.26}$$

On the other hand, if the wave at x_α is a rarefaction front, its strength will satisfy $\sigma_\alpha \in\,]0, \varepsilon_\nu[$. Therefore, the error estimate (3.24) yields

$$\big|\dot{x}_\alpha(t) \cdot \Delta u_\nu(t, x_\alpha) - \Delta f\big(u_\nu(t, x_\alpha)\big)\big| = \mathcal{O}(1) \cdot |\sigma_\alpha|^2 \leqslant C \cdot \varepsilon_\nu |\sigma_\alpha| \tag{4.27}$$

for some constant C. Summing over all wave-fronts and recalling that the total strength of waves in $u_\nu(t, \cdot)$ satisfies a uniform bound independent of t, ν, we obtain

$$\begin{aligned} &\limsup_{\nu \to \infty} \left| \sum_\alpha \big[\dot{x}_\alpha(t) \cdot \Delta u_\nu(t, x_\alpha) - \Delta f\big(u_\nu(t, x_\alpha)\big)\big] \phi\big(t, x_\alpha(t)\big) \right| \\ &\quad \leqslant \Big(\max_{t,x} |\phi(t,x)|\Big) \cdot \limsup_{\nu \to \infty} \left\{ \sum_{\alpha \in \mathcal{R}} C\varepsilon_\nu |\sigma_\alpha| \right\} = 0. \end{aligned} \tag{4.28}$$

The limit (4.23) is now an easy consequence of (4.28). This shows that u is a weak solution to the Cauchy problem.

To conclude the proof, assume that a convex entropy η is given, with entropy flux q. To prove the admissibility condition (4.4) we need to show that

$$\liminf_{\nu \to \infty} \int_0^\infty \int_{-\infty}^\infty \big\{\eta(u_\nu)\varphi_t + q(u_\nu)\varphi_x\big\}\, \mathrm{d}x\, \mathrm{d}t \geqslant 0 \tag{4.29}$$

for every nonnegative $\varphi \in \mathcal{C}_c^1$ with compact support contained in the half plane where $t > 0$. Choose $T > 0$ so that φ vanishes outside the strip $[0, T] \times \mathbb{R}$. Using again the divergence theorem, for every ν, the double integral in (4.29) can be computed as

$$\int_0^T \sum_\alpha \left[\dot{x}_\alpha(t) \cdot \Delta\eta\big(u_\nu(t, x_\alpha)\big) - \Delta q\big(u_\nu(t, x_\alpha)\big)\right]\varphi(t, x_\alpha)\, dt, \tag{4.30}$$

where, for each $t \in [0, T]$, the sum ranges over all jumps of $u_\nu(t, \cdot)$. We use here the notation

$$\Delta\eta \doteq \eta\big(u_\nu(t, x_\alpha+)\big) - \eta\big(u_\nu(t, x_\alpha-)\big),$$
$$\Delta q \doteq q\big(u_\nu(t, x_\alpha+)\big) - q\big(u_\nu(t, x_\alpha-)\big).$$

To estimate the integral (4.30), let $|\sigma_\alpha|$ be the strength of the wave-front at x_α. If this wave is a shock or contact discontinuity, by construction it satisfies the entropy condition

$$\dot{x}_\alpha(t) \cdot \Delta\eta\big(u_\nu(t, x_\alpha)\big) - \Delta q\big(u_\nu(t, x_\alpha)\big) \geqslant 0. \tag{4.31}$$

On the other hand, if the wave at x_α is a rarefaction front, its strength will satisfy $\sigma_\alpha \in\ ]0, \varepsilon_\nu[$. Therefore, the estimate (3.25) yields

$$\dot{x}_\alpha(t) \cdot \Delta\eta\big(u_\nu(t, x_\alpha)\big) - \Delta q\big(u_\nu(t, x_\alpha)\big) = \mathcal{O}(1) \cdot \sigma_\alpha^2 \geqslant -C \cdot \varepsilon_\nu |\sigma_\alpha| \tag{4.32}$$

for some constant C. Summing over all wave-fronts and recalling that the total strength of waves in $u_\nu(t, \cdot)$ satisfies a uniform bound independent of t, ν, we obtain

$$\begin{aligned} &\liminf_{\nu \to \infty} \sum_\alpha \left[\dot{x}_\alpha(t) \cdot \Delta\eta\big(u_\nu(t, x_\alpha)\big) - \Delta q\big(u_\nu(t, x_\alpha)\big)\right]\varphi(t, x_\alpha)\, dt \\ &\quad \geqslant \Big(\max_{t,x} \varphi(t, x)\Big) \liminf_{\nu \to \infty} \Big\{ -\sum_{\alpha \in \mathcal{R}} C\varepsilon_\nu |\sigma_\alpha| \Big\} \\ &\quad \geqslant 0. \end{aligned} \tag{4.33}$$

The relation (4.29) is now an easy consequence of (4.33).

4.4. *A modified front tracking algorithm*

For general $n \times n$ systems, the analysis given in the previous sections does not provide a complete proof of Theorem 4.1, because the naive front tracking algorithm can generate in finite time an infinite number of wave-fronts, in which case the whole construction breaks down.

We thus need to modify the algorithm, to ensure that the total number of fronts remains uniformly bounded. Following [B2], we shall use two different procedures for

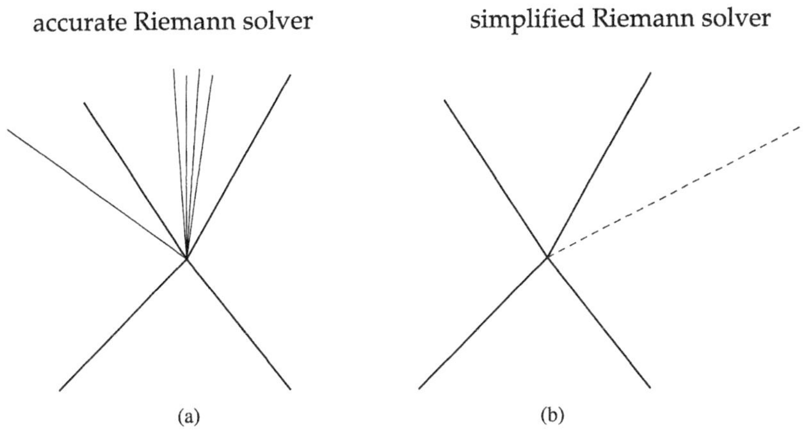

Fig. 15. Different Riemann solvers.

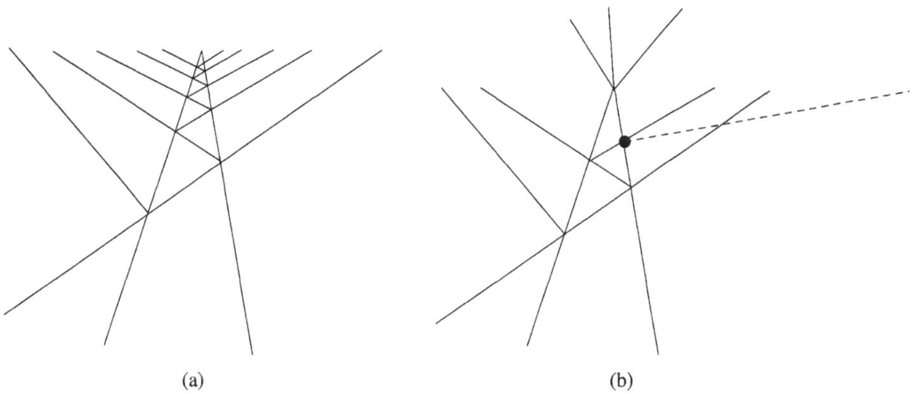

Fig. 16. Reducing the number of wave-fronts.

solving a Riemann problem within the class of piecewise constant functions: the *Accurate Riemann Solver* (see Figure 15(a)), which introduces several new wave-fronts, and a *Simplified Riemann Solver* (see Figure 15(b)), which involves a minimum number of outgoing fronts. In this second case, all new waves are lumped together in a single *nonphysical* front, traveling with a fixed speed $\hat{\lambda}$ strictly larger than all characteristic speeds. The main feature of this algorithm is illustrated in Figures 16(a) and (b). If all Riemann problems are solved accurately, the number of wave-fronts becomes infinite (see Figure 16(a)). On the other hand, if at a certain point we use the Simplified Riemann Solver, the total number of fronts remains bounded for all times (see Figure 16(b)).

The Accurate Riemann Solver was described at (4.12). Next, we introduce a simplified way of solving a Riemann problem, with $\leqslant 3$ outgoing fronts. Throughout the following, $\hat{\lambda}$ will denote a fixed constant, strictly larger than all characteristic speeds $\lambda_j(u)$.

Simplified Riemann Solver

Consider again the Riemann problem (4.6) at a point $(\bar{t}, \bar{x})$, say generated by the interaction of two incoming fronts, of strength σ, σ'. We distinguish two cases.

Case 1. Let $j, j' \in \{1, \ldots, n\}$ be the families of the two incoming wave-fronts, with $j \geqslant j'$.

Assume that the left, middle and right states u_l, u_m, u_r before the interaction are related by

$$u_m = \Psi_j(\sigma)(u_l), \qquad u_r = \Psi_{j'}(\sigma')(u_m). \tag{4.34}$$

Define the auxiliary right state

$$\tilde{u}_r = \begin{cases} \Psi_j(\sigma) \circ \Psi_{j'}(\sigma')(u_l) & \text{if } j > j', \\ \Psi_j(\sigma + \sigma')(u_r) & \text{if } j = j'. \end{cases} \tag{4.35}$$

Let $\tilde{v} = \tilde{v}(t, x)$ be the piecewise constant solution of the Riemann problem with data $u_l, \tilde{u}_r$, constructed as in (4.12). Because of (4.35), the piecewise constant function $\tilde{v}$ contains exactly two wave-fronts of sizes σ', σ if $j > j'$, or a single wave-front of size $\sigma + \sigma'$ if $j = j'$. It is important to observe that $\tilde{u}_r \neq u_r$, in general. We let the jump $(\tilde{u}_r, u_r)$ travel with the fixed speed $\hat{\lambda}$, strictly bigger than all characteristic speeds. In a forward neighborhood of the point $(\bar{t}, \bar{x})$, we thus define an approximate solution v as follows (see Figures 17(a) and (b))

$$v(t, x) = \begin{cases} \tilde{v}(t, x) & \text{if } x - \bar{x} < (t - \bar{t})\hat{\lambda}, \\ u_r & \text{if } x - \bar{x} > (t - \bar{t})\hat{\lambda}. \end{cases} \tag{4.36}$$

Notice that this Simplified Riemann Solver introduces a new *nonphysical* wave-front, joining the states $\tilde{u}_r, u_r$ and traveling with constant speed $\hat{\lambda}$. In turn, this front may interact with other wave-fronts. One more case of interaction thus needs to be considered.

Case 2. A nonphysical front hits from the left a wave front of the $i$the family (see Figure 17(c)), for some $i \in \{1, \ldots, n\}$.

Let u_l, u_m, u_r be the left, middle and right state before the interaction. If

$$u_r = \Psi_i(\sigma)(u_m), \tag{4.37}$$

define the auxiliary right state

$$\tilde{u}_r = \Psi_i(\sigma)(u_l). \tag{4.38}$$

Call $\tilde{v}$ the solution to the Riemann problem with data $u_l, \tilde{u}_r$, constructed as in (4.12). Because of (4.38), $\tilde{v}$ will contain a single i-wave with size σ. Since $\tilde{u}_r \neq u_r$ in general, we

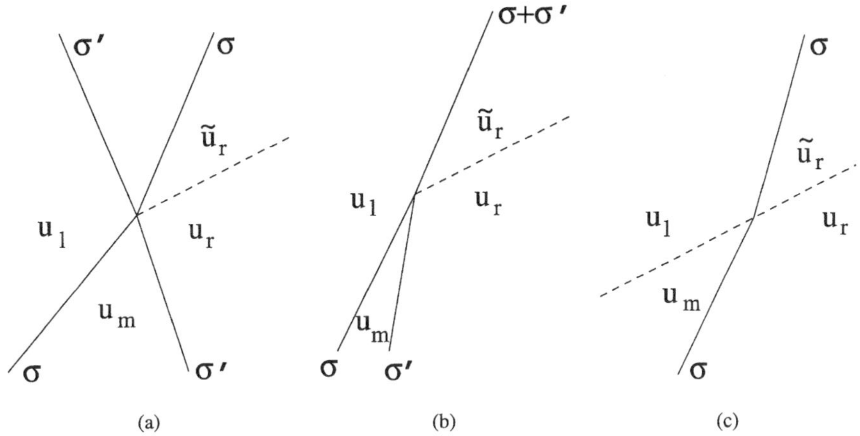

Fig. 17. Interactions involving two nonphysical wave-fronts.

let the jump $(\tilde{u}_r, u_r)$ travel with the fixed speed $\hat{\lambda}$. In a forward neighborhood of the point $(\bar{t}, \bar{x})$, the approximate solution u is then defined again according to (4.36).

By construction, all nonphysical fronts travel with the same speed $\hat{\lambda}$, hence they never interact with each other. The above cases therefore cover all possible interactions between two wave-fronts.

To complete the description of the algorithm, it remains to specify which Riemann Solver is used at any given interaction. For this purpose, to each wave-front we attach a *generation order*, counting how many interactions were needed to produce it.

DEFINITION 4.2. The *generation order* of a front is inductively defined as follows (see Figure 18).

- All fronts generated by the Riemann problems at the initial time $t = 0$ are the "primary ancestors" and have generation order $k = 1$.
- Let two incoming fronts interact, say of the families $i, i' \in \{1, \dots, n\}$, with generation orders k, k'. The orders of the outgoing fronts are then defined as follows.

Case 1: $i \neq i'$. Then

- the outgoing i-wave and i'-wave have the same orders k, k' as the incoming ones,
- the outgoing fronts of every other family $j \neq i, i'$ have order $\max\{k, k'\} + 1$.

Case 2: $i = i'$. Then

- the outgoing front of the ith family has order $\min\{k, k'\}$;
- the outgoing fronts of every family $j \neq i$ have order $\max\{k, k'\} + 1$.

In the above, we tacitly assumed that at most two incoming wave-fronts interact at any given time. This can always be achieved by an arbitrarily small change in the speed of the fronts. We shall also adopt the following important provision:

(P) In every Riemann Solver, rarefaction waves of the same family as one of the incoming fronts are never partitioned.

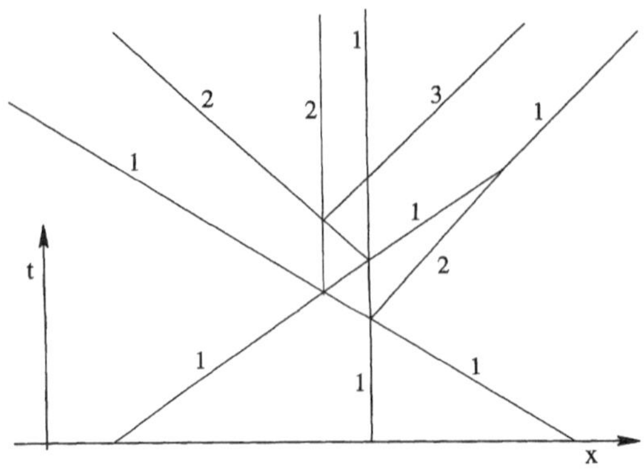

Fig. 18. Generation orders of various wave-fronts.

In other words, if one of the incoming fronts belongs to the ith family, we set $p_i = 1$ regardless of the strength of the ith outgoing front. This guarantees that every wave-front can be uniquely continued forward in time, unless it gets completely canceled by interacting with other fronts of the same family and opposite sign.

To complete the description of the algorithm, given an integer $\nu \geqslant 1$, at each point of interaction we choose:

- The Accurate Riemann Solver, if the two incoming fronts both have generation order less than ν.
- The Simplified Riemann Solver, if one of the incoming fronts has generation order greater than or equal to ν, or if it is a nonphysical front.

In the remainder of this section we will show that, with the above modifications, the front tracking algorithm yields a completely rigorous proof of Theorem 4.1. The proof is given in several steps.

(1) *Bounds on the number of wave-fronts.* Let N_0 be the number of jumps in the initial condition $u(0,\cdot)$. At every point where the Accurate Riemann Solver is used, we clearly have an upper bound on the number of outgoing fronts, say less than or equal to M. By the provision (P), after its birth no front is ever partitioned any further. Therefore,

$$[\text{total number of fronts of first generation}] \leqslant M N_0.$$

Fronts of second generation can appear only because of an interaction among two fronts of the first generation. Hence,

$$[\text{number of fronts of second generation}] \leqslant M(M N_0)^2.$$

By induction, we find

$$\begin{aligned}&[\#\text{ of fronts of generation } k+1]\\&\quad \leqslant M[\#\text{ of fronts of generation } k]\cdot[\#\text{ of fronts of generation } \leqslant k].\end{aligned}$$

Hence, for each $k = 1, \dots, \nu + 1$, there can be at most finitely many fronts of generation k. We now observe that all physical fronts must have generation $\leqslant \nu$, while the fronts of generation $\nu + 1$ are precisely the nonphysical ones. Moreover, when a nonphysical front interacts with a physical one, no additional front is generated (see Figure 17(c)). We thus conclude that the total number of all wave-fronts is bounded.

(2) *Bounds on the total variation.* The uniform estimates on the total variation are obtained as in Section 4.2, with only minor changes that we describe here. The functional (4.13), measuring the *total strength of waves*, must now include also the nonphysical fronts. The strength of such fronts is simply defined as the size of the jump, i.e.,

$$|\sigma_\alpha| \doteq \left|u(x_\alpha+) - u(x_\alpha-)\right|.$$

The *wave interaction potential* (4.14) now takes into account also the nonphysical fronts. For this purpose, our Definition 4.1 of *approaching waves* is supplemented by

- A nonphysical front located at a point x_α approaches all other physical wave-fronts (shocks, rarefactions or contact discontinuities) located at points $x_\beta > x_\alpha$.

Observe that all nonphysical fronts travel with exactly the same speed $\hat{\lambda}$. Hence (by definition), they do not approach each other.

To establish the inequalities (4.15)–(4.19) in this more general case, the interaction estimates given in Lemma 3.2 must be supplemented with two additional estimates, holding when the Simplified Riemann Solver is used.

LEMMA 4.1.

(i) *Let σ, σ' be the sizes of two incoming fronts, say of the families j, j'. Let u_l, u_m and u_r be the left, middle and right states before interaction, so that* (4.34) *holds. Introducing the auxiliary right state $\tilde{u}_r$ as in* (4.35), *one has (see Figures* 17(a) *and* (b))

$$\left|\tilde{u}_r - u_r\right| = \mathcal{O}(1) \cdot \left|\sigma\sigma'\right|. \tag{4.39}$$

(ii) *Let a nonphysical front connecting the states u_l, u_m interact with an i-wave of size σ, connecting u_m, u_r, so that* (4.37) *holds. Defining the auxiliary right state $\tilde{u}_r$ as in* (4.38), *one has (see Figure* 17(c))

$$\left|u_r - u_r\right| - \left|u_m - u_l\right| = \mathcal{O}(1) \cdot |\sigma|\left|u_m - u_l\right|. \tag{4.40}$$

PROOF. To establish the first estimate, in the case where $j > j'$ we consider the map

$$\begin{aligned}(\sigma, \sigma') \mapsto \Phi(\sigma, \sigma') &\doteq \tilde{u}_r - u_r \\ &= \Psi_{j'}(\sigma') \circ \Psi_j(\sigma)(u_l) - \Psi_j(\sigma) \circ \Psi_{j'}(\sigma')(u_l). \end{aligned} \tag{4.41}$$

This map is $\mathcal{C}^2$ with Lipschitz continuous second derivatives and satisfies

$$\Phi(\sigma, 0) = \Phi(0, \sigma') = 0 \quad \text{for all } \sigma, \sigma'. \tag{4.42}$$

The estimate (4.39) is thus a consequence of Lemma A.2 in the Appendix.

In the case where $j = j'$, we apply the same arguments to the map

$$(\sigma, \sigma') \mapsto \Phi(\sigma, \sigma') \doteq \tilde{u}_r - u_r = \Psi_j(\sigma + \sigma')(u_l) - \Psi_j(\sigma') \circ \Psi_j(\sigma)(u_l). \tag{4.43}$$

To establish the second estimate, for a given left state u_l, consider the map $\Phi : \mathbb{R}^n \times \mathbb{R} \mapsto \mathbb{R}^n$,

$$(v, \sigma) \mapsto \Phi(v, \sigma) \doteq \big(\Psi_i(\sigma)(u_l) + v\big) - \Psi_i(\sigma)(u_l + v).$$

Observe that Φ is well defined in a neighborhood of the origin, and twice continuously differentiable with Lipschitz continuous second derivatives. Moreover,

$$\Phi(v, 0) = \Phi(0, \sigma) = 0 \quad \text{for all } v, \sigma. \tag{4.44}$$

Applying once again Lemma A.2, we conclude

$$\Phi(v, \sigma) = \mathcal{O}(1) \cdot |v||\sigma|. \tag{4.45}$$

Choosing $v = u_m - u_l$, we have

$$\Phi(v, \sigma) = \big(\tilde{u}_r + (u_m - u_l)\big) - u_r = \big(\tilde{u}_r - u_r\big) - (u_l - u_m)$$

and (4.40) follows immediately.

We observe that, in the above proofs, the various functions Φ together with their first two derivatives depend continuously on the left state u_l. By Remark 2 in the Appendix, the quantities $\mathcal{O}(1)$ thus remain uniformly bounded as u_l ranges over a compact set. □

Using the above estimates at every point where the Simplified Riemann Solver is used, we can repeat exactly the same arguments in Section 4.2 and deduce the uniform bound (4.19) on the total variation. In turn, this again implies the estimate (4.21) on the uniform Lipschitz continuity with respect to time.

(3) *Strength of each rarefaction front is small.* Whenever a new rarefaction front is introduced, its strength is always less than ε. Because of the provision (P), this front is never again partitioned, even if, as a result of subsequent interactions, its size becomes greater than ε. The following analysis will show that, if the total strength of waves remains small, the size of each rarefaction front never grows bigger than 2ε.

By construction, at a time t_0 where a new rarefaction front is introduced by the Accurate Riemann Solver, its size is $\sigma_\alpha(t_0) \in \,]0, \varepsilon]$. We now examine how its strength can change at times $t > t_0$. Clearly, our front will never interact with rarefaction fronts of the same family. When it hits a shock of the same family, its size will decrease due to a cancellation. When it crosses a nonphysical wave, by construction its size remains unchanged. On the other hand, by subsequent interactions with fronts of other families, our rarefaction front may increase its initial strength. To keep track of its size $\sigma_\alpha(t)$ at any time $t > t_0$, consider the quantity

$$V_\alpha(t) = \sum_{\beta \in \mathcal{A}(\alpha)} |\sigma_\beta|,$$

where the summation is restricted to the set $\mathcal{A}(\alpha)$ of all wave-fronts which are approaching σ_α. Consider any interaction time $\tau > t_0$.

Case 1: *The interaction does not involve the front σ_α.* Using the interaction estimates in Lemma 3.2 or in Lemma 4.1, we obtain

$$\Delta\sigma_\alpha(\tau) = 0, \qquad \Delta V_\alpha(\tau) + C_0 \Delta Q(\tau) \leqslant 0.$$

Case 2: *At time τ, the rarefaction front σ_α interacts with another front of strength $|\sigma_\beta|$.* In this case, the interaction estimates imply that for some constant C' there holds

$$\Delta V_\alpha(\tau) = -|\sigma_\beta|, \qquad \Delta Q(\tau) < 0, \qquad \Delta\sigma_\alpha(\tau) \leqslant C' \big|\sigma_\alpha(\tau-)\big| |\sigma_\beta|.$$

By the previous analysis, the map

$$t \mapsto \sigma_\alpha(t) \exp\big\{C'\big[V_\alpha(t) + C_0 Q(t)\big]\big\}$$

is nonincreasing in time. As a consequence, for all $t > t_0$, we have

$$\begin{aligned} \sigma_\alpha(t) &\leqslant \sigma_\alpha(t_0) \exp\big\{C'\big[V_\alpha(t_0) + C_0 Q(t_0)\big]\big\} \\ &\leqslant \varepsilon \exp\big\{C'\big[V(0) + C_0 Q(0)\big]\big\} \\ &\leqslant 2\varepsilon, \end{aligned}$$

provided that the total strength of waves remains sufficiently small.

(4) *Total strength of all nonphysical fronts is small.* Let u be a front tracking approximation constructed by using the Simplified Riemann Solver whenever one of the incoming fronts has generation greater than or equal to ν. Observing that an interaction involving an incoming nonphysical front does not produce any new front, is becomes clear that:

- Physical fronts are precisely those of generation less than or equal to ν.
- Nonphysical fronts have generation order $\nu + 1$.
- No front of generation order greater than or equal to $n + 2$ is ever created.

In the following, we show that the total strength of fronts of generation order $\geqslant k$ satisfies a uniform bound which approaches zero as $k \to \infty$. For $k \geqslant 1$, call

$$V_k(t) \doteq \sum_{\text{order}(\alpha) \geqslant k} |\sigma_\alpha|$$

the sum of the strengths of all waves of order greater than or equal to k in $u(t,\cdot)$. To obtain a priori bounds on V_k we also define

$$Q_k(t) \doteq \sum_{\max\{\mathrm{order}(\alpha),\mathrm{order}(\beta)\}\geqslant k} |\sigma_\alpha \sigma_\beta|,$$

where the sum extends over all couples of approaching waves in $u(t,\cdot)$, say of order k_α, k_β, with $\max\{k_\alpha, k_\beta\} \geqslant k$. Moreover, call I_k the set of times where two waves of order k_α, k_β interact, with $\max\{k_\alpha, k_\beta\} = k$. Repeating the arguments used to derive (4.15) and (4.16), but now keeping track of the order of the wave-fronts involved in the interactions, we obtain

$$\begin{aligned}
&\Delta V_k(t) = 0, && t \in I_1 \cup \cdots \cup I_{k-2},\\
&\Delta V_k(t) + C_0 \Delta Q_{k-1}(t) \leqslant 0, && t \in I_{k-1} \cup I_k \cup I_{k+1} \cup \cdots,\\
&\Delta Q_k(t) + C_0 \Delta Q(t) V_k(t-) \leqslant 0, && t \in I_1 \cup \cdots \cup I_{k-2},\\
&\Delta Q_k(t) + C_0 \Delta Q_{k-1}(t) V(t-) \leqslant 0, && t \in I_{k-1},\\
&\Delta Q_k(t) \leqslant 0, && t \in I_k \cup I_{k+1} \cup \cdots.
\end{aligned} \tag{4.46}$$

Observe that $(4.46)_1$ says that the total strength of waves of order greater than or equal to k is unaffected by interactions involving wave-fronts of order less than or equal to $k-2$. When one of the incoming waves has order greater than or equal to $k-1$, according to $(4.46)_2$ the growth of V_k is counterbalanced by the decrease of the interaction potential Q_{k-1}. The remaining estimates $(4.46)_{3\text{–}5}$ provide bounds on the increase of the potentials Q_k.

In the following, we write $[s]_+ = \max\{s, 0\}$ and $[s]_- = \max\{-s, 0\}$ for the positive and negative parts of a real number s. Observing that $V_1 = V$, $Q_1 = Q$ and $V_k(0+) = Q_k(0+) = 0$ if $k \geqslant 2$, from (4.46) we deduce the two estimates

$$V_k(t) \leqslant C_0 \sum_{0<\tau\leqslant t} \big[\Delta Q_{k-1}(\tau)\big]_-,$$

$$\begin{aligned}
Q_k(t) &\leqslant \sum_{0<\tau\leqslant t} \big[\Delta Q_k(\tau)\big]_+\\
&\leqslant C_0 \sum_{0<\tau\leqslant t} \big[\Delta Q(\tau)\big]_- \sup_t V_k(t) + C_0 \sum_{0<\tau\leqslant t} \big[\Delta Q_{k-1}(\tau)\big]_- \sup_t V(t),
\end{aligned}$$

both valid for every $t > 0$ and $k \geqslant 2$. Moreover,

$$0 \leqslant Q_k(t) = \sum_{0<\tau\leqslant t} \big\{\big[\Delta Q_k(\tau)\big]_+ - \big[\Delta Q_k(\tau)\big]_-\big\}.$$

Since the quantity $\Upsilon(t) \doteq V(t) + C_0 Q(t)$ is nonincreasing, one has

$$V_k(t) \leqslant V(t) \leqslant \Upsilon(t) \leqslant \Upsilon(0),$$
$$\sum_{0<\tau<\infty} [\Delta Q(\tau)]_- \leqslant Q(0) \leqslant \Upsilon(0).$$

Defining $\widetilde{Q}_1 \doteq Q_1(0)$ and, for $k \geqslant 2$,

$$\widetilde{Q}_k \doteq \sum_{t>0} [\Delta Q_k(t)]_+, \qquad \widetilde{V}_k \doteq \sup_{t>0} V_k(t),$$

from the above estimates we deduce the sequence of inequalities (valid for $k \geqslant 2$)

$$\begin{cases} \widetilde{V}_k \leqslant C_0 \widetilde{Q}_{k-1}, \\ \widetilde{Q}_k \leqslant C_0 \Upsilon(0) \widetilde{V}_k + C_0 \widetilde{Q}_{k-1} \Upsilon(0) \leqslant (C_0^2 + C_0) \Upsilon(0) \widetilde{Q}_{k-1}. \end{cases} \tag{4.47}$$

If Tot.Var$(u(0+, \cdot))$ is sufficiently small, $\Upsilon(0) \doteq V(0) + C_0 Q(0)$ will satisfy

$$\gamma \doteq (C_0^2 + C_0) \Upsilon(0) < 1.$$

By induction on k, for every $t > 0$ and $k \geqslant 2$ the two previous estimates (4.47) yield

$$\begin{aligned} &Q_k(t) \leqslant \widetilde{Q}_k \leqslant \widetilde{Q}_1 \gamma^{k-1} \leqslant V^2(0) \gamma^{k-1}, \\ &V_k(t) \leqslant \widetilde{V}_k \leqslant C_0 \widetilde{Q}_{k-1} \leqslant C_0 V^2(0) \gamma^{k-2}. \end{aligned} \tag{4.48}$$

From (4.48) we deduce

$$\begin{aligned} &[\text{total strength of nonphysical fronts}] \\ &\quad = [\text{total strength fronts of order } \nu + 1] \leqslant C_0 V^2(0) \gamma^{\nu-1}. \end{aligned} \tag{4.49}$$

Since $\gamma < 1$, the right-hand side of (4.49) approaches zero as $\nu \to \infty$.

(5) *Convergence to an entropy weak solution.* The remainder of the proof again follows closely the arguments given in Section 4.4. Fix any sequence $\varepsilon_\nu \to 0+$. For every $\nu \geqslant 1$, we apply the modified front tracking algorithm and construct an approximate solution u_ν, dividing all new rarefactions into fronts of size $< \varepsilon_\nu$ and using the Simplified Riemann Solver whenever one of the incoming fronts has generation order $\geqslant \nu$. By the previous analysis, the total variation of $u_\nu(t, \cdot)$ remains bounded, uniformly for all $t \geqslant 0$ and $\nu \geqslant 1$. Moreover, by (4.21) the maps $t \mapsto u_\nu(t, \cdot)$ are uniformly Lipschitz continuous with respect to the $\mathbf{L}^1$ distance. We can thus apply Theorem A.1 in the Appendix and extract a subsequence which converges to some limit function u in $\mathbf{L}^1_{\text{loc}}$, satisfying (4.20).

To show that u is actually an entropy weak solution, we first need to supplement Lemma 3.2 with new error estimates concerning nonphysical fronts. Consider any such front, say located at x_α, connecting the left and right states u^-, u^+, with strength $|\sigma_\alpha| = |u^+ - u^-|$. The Lipschitz continuity of the flux function f now trivially implies

$$|E_\alpha| \doteq \left|\left[f(u^+) - f(u^-)\right] - \dot{x}_\alpha\left[u^+ - u^-\right]\right| \leqslant C|\sigma_\alpha| \tag{4.50}$$

for some constant C. Moreover, if η is a convex entropy with flux q, again by Lipschitz continuity we obtain

$$|E'_\alpha| \doteq \left|\left[q(u^+) - q(u^-)\right] - \dot{x}_\alpha\left[\eta(u^+) - \eta(u^-)\right]\right| \leqslant C|\sigma_\alpha|. \tag{4.51}$$

To prove (4.22), or equivalently (4.23), we use again the divergence theorem. We estimate the summation in (4.25), considering physical (shocks or rarefactions) and non-physical fronts separately:

$$\begin{aligned} &\limsup_{\nu\to\infty}\left|\sum_{\alpha\in\mathcal{S}\cup\mathcal{R}\cup\mathcal{NP}}\left[\dot{x}_\alpha(t)\cdot\Delta u_\nu(t,x_\alpha) - \Delta f\big(u_\nu(t,x_\alpha)\big)\right]\phi\big(t,x_\alpha(t)\big)\right| \\ &\quad\leqslant \Big(\max_{t,x}\left|\phi(t,x)\right|\Big)\limsup_{\nu\to\infty}\left\{\sum_{\alpha\in\mathcal{R}} C\varepsilon_\nu|\sigma_\alpha| + \sum_{\alpha\in\mathcal{NP}} C|\sigma_\alpha|\right\} \\ &\quad= 0. \end{aligned} \tag{4.52}$$

Indeed, the total strength of waves in u_ν remains uniformly bounded, while the total strength of nonphysical fronts approaches zero as $\nu \to \infty$. The limit (4.23) now follows from (4.52). Therefore, u is a weak solution to the Cauchy problem.

Finally, consider a convex entropy η with flux q. Given a nonnegative test function $\varphi \in \mathcal{C}^1_c$, an entirely similar argument now yields

$$\begin{aligned} &\liminf_{\nu\to\infty}\sum_{\alpha\in\mathcal{S}\cup\mathcal{R}\cup\mathcal{NP}}\left[\dot{x}_\alpha(t)\cdot\Delta\eta\big(u_\nu(t,x_\alpha)\big) - \Delta q\big(u_\nu(t,x_\alpha)\big)\right]\varphi(t,x_\alpha)\,dt \\ &\quad\geqslant \Big(\max_{t,x}\varphi(t,x)\Big)\liminf_{\nu\to\infty}\left\{-\sum_{\alpha\in\mathcal{R}} C\varepsilon_\nu|\sigma_\alpha| - \sum_{\alpha\in\mathcal{NP}} C|\sigma_\alpha|\right\} \\ &\quad\geqslant 0. \end{aligned} \tag{4.53}$$

The relation (4.29) is now an easy consequence of (4.53).

Notes

The first proof of global existence for weak solutions with small total variation appeared in the fundamental paper of Glimm [G]. The striking new ideas introduced in this work have been deeply influential, providing the foundation for much of the subsequent literature on the subject. The original proof in [G] was based on the construction of approximate solutions generated by Riemann problems along the nodes of a fixed grid, with a restarting procedure defined through a random sampling. The convergence of a deterministic version of this algorithm was later proved by Liu [L1]. An excellent presentation of the Glimm scheme can be found in the book of Smoller [Sm].

The front tracking method for scalar conservation laws was introduced by Dafermos [D1]. The technique was extended to 2×2 systems by DiPerna [DP1]. In this case, a naive form of the algorithm already suffices, since it produces only finitely many wave-fronts. For general $n \times n$ systems, some technical adjustment in order to prevent the blow up of the number of fronts becomes necessary. The use of a Simplified Riemann Solver, and the definition of generation number of a wave-front were introduced in [B2]. Several variants of this technique are found in the literature.

In [Sc2], Schochet chooses the speed of nonphysical fronts to be $\hat{\lambda} = +\infty$. In other words, when the Simplified Riemann Solver is used at a point P (see Figure 19), the nonphysical front is immediately sent to $+\infty$, by solving a sequence of Riemann problems at all points $P_1, P_2, \ldots$ to the right of P. This has the advantage of eliminating nonphysical fronts altogether. The price to pay is that now the solutions have no longer finite propagation speed, while the map $t \mapsto u(t, \cdot)$ is no longer continuous with values in $\mathbf{L}^1_{\mathrm{loc}}$.

In [BaJ], Baiti and Jenssen show that, instead counting the generation number, the choice between Accurate and Simplified Riemann Solver can be based solely on the strengths σ, σ' of the two incoming waves. Given a small threshold value $\rho > 0$, one can use the Accurate Solver if $|\sigma\sigma'| \geqslant \rho$ and the Simplified Solver if $|\sigma\sigma'| < \rho$.

Yet another possibility, considered in [Ri,HR], is to define a Simplified Riemann Solver by attaching all new waves to one of the two main fronts involved in the interaction (see Figure 20). This method appears to be numerically efficient, because it completely avoids the introduction of nonphysical wave-fronts, and all the computations connected with them. A rigorous analysis of this algorithm requires interaction estimates slightly more general than Lemma 3.2. Indeed, one must keep in mind that each front can now contain waves of several different families.

Early applications of the front tracking algorithm to special systems are found in the papers of Alber [Al], Lin [Ln] and Wendroff [W].

The above global existence result has been generalized by Liu [L4] to general $n \times n$ systems, without the hypothesis (H) of genuine nonlinearity or linear degeneracy, using the Glimm scheme. A front tracking algorithm for these more general systems was recently developed in [AM2].

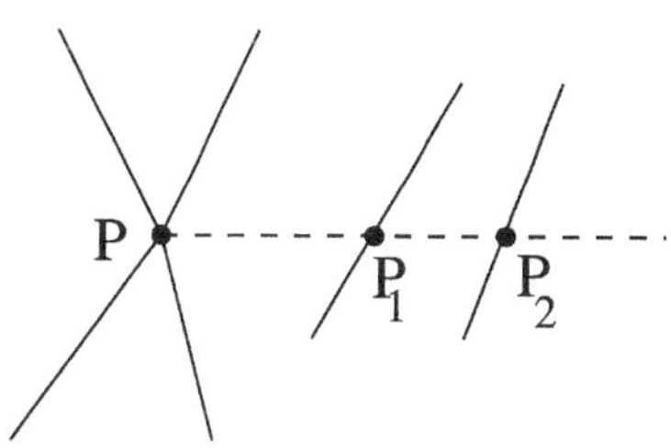

Fig. 19. Alternative front tracking algorithms.

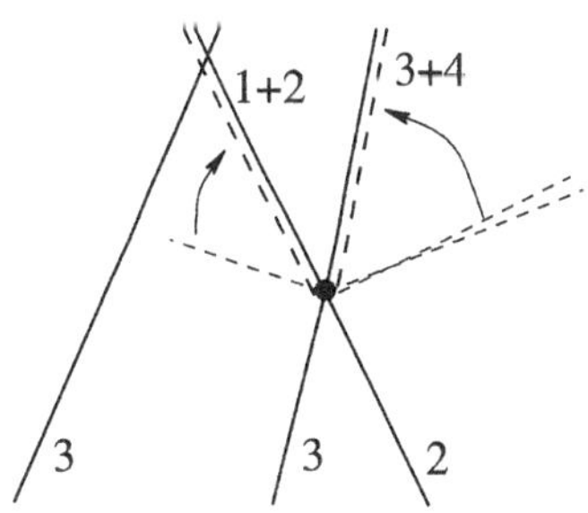

Fig. 20. See Figure 19.

The smallness assumption on the total variation is essential. For large initial data, a local existence theorem was proved by Schochet [Sc1]. A counterexample to global existence, showing that the $\mathbf{L}^\infty$ norm of solutions can blow up in finite time, was constructed by Jenssen [Je].

5. Uniqueness and stability

In the previous section, the existence of a global weak solution was proved by a compactness argument, which did not provide any information on the uniqueness of the limit. Our next goal is to show that front tracking approximations converge to a unique limit solution, continuously depending on the initial data.

5.1. *A Lyapunov functional*

In order to estimate the distance between two front tracking solutions, we introduce a functional $\Phi = \Phi(u, v)$, uniformly equivalent to the $\mathbf{L}^1$ distance, which is "almost decreasing" along pairs of solutions. Throughout the following, it will be convenient to use an alternative parametrization of shock and rarefaction curves, in terms of the characteristic speeds.

- If the ith field is genuinely nonlinear, the i-shock and i-rarefaction curve through a point u_0 are parametrized so that

$$\begin{aligned} &\lambda_i\big(S_i(\sigma)(u_0)\big) - \lambda_i(u_0) = \sigma, \\ &\lambda_i\big(R_i(\sigma)(u_0)\big) - \lambda_i(u_0) = \sigma. \end{aligned} \tag{5.1}$$

- For a linearly degenerate field, the shock and rarefaction curves are always parametrized by arc-length.

Thanks to this choice of the parametrization, we now have the identity

$$S_i(-\sigma) \circ S_i(\sigma)(u_0) = u_0, \tag{5.2}$$

which could have failed otherwise. We also observe that the second-order tangency condition (3.10) between shock and rarefaction curves still holds, while (3.11) now takes the form

$$\lambda_i\big(S_i(\sigma)(u_0),\ u_0\big) = \lambda_i(u_0) + \frac{\sigma}{2} + \mathcal{O}(1) \cdot \sigma^2.$$

We now begin the construction of the functional Φ. Given two piecewise constant functions u, v, consider the scalar functions q_i defined implicitly by (see Figure 21)

$$v(x) = S_n\big(q_n(x)\big) \circ \cdots \circ S_1\big(q_1(x)\big)\big(u(x)\big). \tag{5.3}$$

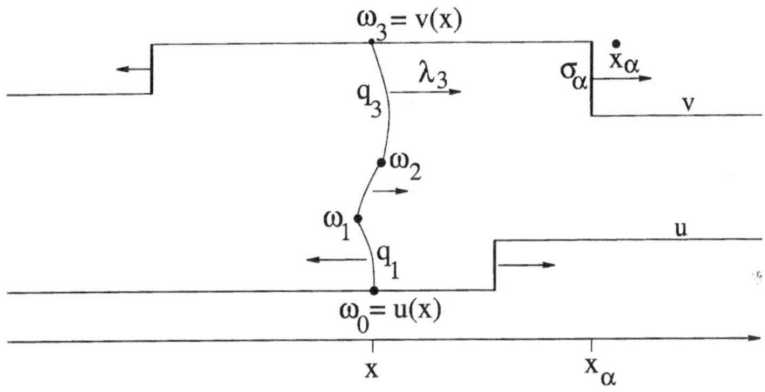

Fig. 21. Jump decomposition.

Intuitively, $q_i(x)$ can be regarded as the strength of the ith component of the jump $(u(x), v(x))$, measured along shock curves. On a compact neighborhood of the origin we clearly have

$$\frac{1}{C_1}\big|v(x)-u(x)\big| \leqslant \sum_{i=1}^{n}\big|q_i(x)\big| \leqslant C_1\big|v(x)-u(x)\big| \tag{5.4}$$

for some constant C_1. Hence, the integral

$$\int \sum_{i=1}^{n}\big|q_i(x)\big|\,\mathrm{d}x \tag{5.5}$$

provides an equivalent way to measure the $\mathbf{L}^1$ distance between u and v. Unfortunately, this integral may well increase in time, along a pair of front tracking solutions. To achieve our goal, suitable weights W_i must be inserted. We thus consider the functional

$$\Phi(u,v) \doteq \sum_{i=1}^{n}\int_{-\infty}^{\infty}\big|q_i(x)\big|W_i(x)\,\mathrm{d}x, \tag{5.6}$$

where the weights W_i are defined by setting

$$\begin{aligned} W_i(x) &\doteq 1+\kappa_1\cdot\big[\text{total strength of waves in } u \text{ and}\\ &\qquad\qquad \text{in } v \text{ which approach the } i\text{-wave } q_i(x)\big]\\ &\quad + \kappa_1\kappa_2\cdot[\text{wave interaction potentials of } u \text{ and of } v]\\ &\doteq 1+\kappa_1 A_i(x)+\kappa_1\kappa_2\big[Q(u)+Q(v)\big]. \end{aligned} \tag{5.7}$$

We recall that Q is the interaction potential introduced at (4.14). The total strength of wave-fronts approaching $q_i(x)$ is defined as follows. If the ith field is linearly degenerate, we simply take

$$A_i(x) \doteq \left[\sum_{x_\alpha<x,\, i<k_\alpha\leqslant n} + \sum_{x_\alpha>x,\, 1\leqslant k_\alpha<i}\right]|\sigma_\alpha|. \tag{5.8}$$

The summations here extend to waves both of u and of v. Notice that here we are considering fronts located behind x with a faster speed, and fronts located ahead of x with slower speed.

On the other hand, if the ith field is genuinely nonlinear, our definition of A_i will contain an additional term, accounting for waves in u and in v of the same ith family:

$$\begin{aligned} A_i(x) \doteq{}& \left[\sum_{\substack{\alpha\in\mathcal{J}(u)\cup\mathcal{J}(v)\\ x_\alpha<x,\, i<k_\alpha\leqslant n}} + \sum_{\substack{\alpha\in\mathcal{J}(u)\cup\mathcal{J}(v)\\ x_\alpha>x,\, 1\leqslant k_\alpha<i}}\right]|\sigma_\alpha| \\ &+ \begin{cases} \left[\sum\limits_{\substack{k_\alpha=i\\ \alpha\in\mathcal{J}(u),\, x_\alpha<x}} + \sum\limits_{\substack{k_\alpha=i\\ \alpha\in\mathcal{J}(v),\, x_\alpha>x}}\right]|\sigma_\alpha| & \text{if } q_i(x)<0,\\ \left[\sum\limits_{\substack{k_\alpha=i\\ \alpha\in\mathcal{J}(v),\, x_\alpha<x}} + \sum\limits_{\substack{k_\alpha=i\\ \alpha\in\mathcal{J}(u),\, x_\alpha>x}}\right]|\sigma_\alpha| & \text{if } q_i(x)>0. \end{cases} \end{aligned} \tag{5.9}$$

Here and in the sequel, $\mathcal{J}(u)$ and $\mathcal{J}(v)$ denote the sets of all jumps in u and in v, while $\mathcal{J} \doteq \mathcal{J}(u)\cup\mathcal{J}(v)$. Notice that the strengths of nonphysical waves do enter in the definition of Q. Indeed, a nonphysical front located at x_α approaches all shock and rarefaction fronts located at points $x_\beta > x_\alpha$. On the other hand, nonphysical fronts play no role in the definition of A_i.

The values of the large constants κ_1, κ_2 in (5.7) will be specified later. Observe that, as soon as these constants have been assigned, we can then impose a suitably small bound on the total variation of u, v so that

$$1 \leqslant W_i(x) \leqslant 2 \quad \text{for all } i, x. \tag{5.10}$$

From (5.4), (5.6) and (5.10) it thus follows

$$\frac{1}{C_1}\|v-u\|_{\mathbf{L}^1} \leqslant \Phi(u,v) \leqslant 2C_1\|v-u\|_{\mathbf{L}^1}. \tag{5.11}$$

In the following, by an *ε-approximate front tracking solution* of the system (4.1) we mean a piecewise constant function $u = u(t,x)$ constructed as in the previous section, such that:

- Every shock front satisfies the Rankine–Hugoniot equations exactly, and is entropy admissible.

- Every rarefaction front has strength $\sigma_\alpha \in]0, \varepsilon]$ and travels with speed $\dot{x}_\alpha = \lambda_i(u(x_\alpha-))$, i.e., with the characteristic speed of its left state.
- All nonphysical front travel with the fixed speed $\hat{\lambda}$. The total strength of all nonphysical fronts in $u(t, \cdot)$ is always lees than or equal to ε.

The existence of ε-approximate solutions was discussed in the previous section. One should mention that, to prevent three or more fronts interacting at a same point, we may need to slightly change the speed of one of the incoming fronts. To avoid pointless technicalities, in all future estimates we neglect this additional error term, since it clearly can be made arbitrarily small. Recalling the definition $\Upsilon(u) \doteq V(u) + C_0 Q(u)$, the basic $\mathbf{L}^1$ stability estimate for front tracking approximations can now be stated as follows.

THEOREM 5.1. *For suitable constants* $C_2, \kappa_1, \kappa_2, \delta_0 > 0$ *the following holds. Let* u, v *be* ε*-approximate front tracking solutions of* (4.1) *having small total variation, so that*

$$\Upsilon(u(t)) < \delta_0, \qquad \Upsilon(v(t)) < \delta_0 \quad \textit{for all } t \geqslant 0. \tag{5.12}$$

Then the functional Φ *in* (5.6)–(5.9) *satisfies*

$$\Phi(u(t), v(t)) - \Phi(u(s), v(s)) \leqslant C_2\varepsilon(t-s) \quad \textit{for all } 0 \leqslant s < t. \tag{5.13}$$

Toward a proof, the key point is to understand how the functional Φ evolves in time. Outside interaction times, the change in Φ is only due to the fact that the locations x_α of the jumps in u and v move with constant speeds $\dot{x}_\alpha$. Therefore

$$\begin{aligned}
&\frac{\mathrm{d}}{\mathrm{d}t}\Phi(u(t), v(t)) \\
&\quad = \sum_{\alpha\in\mathcal{J}} \sum_{i=1}^{n} \{|q_i(x_\alpha-)|W_i(x_\alpha-) - |q_i(x_\alpha+)|W_i(x_\alpha+)\} \cdot \dot{x}_\alpha.
\end{aligned} \tag{5.14}$$

Recalling the decomposition (5.3), at each point x we define the intermediate states $\omega_0(x) = u(x), \omega_1(x), \ldots, \omega_n(x) = v(x)$ by setting (see Figure 21)

$$\omega_i(x) \doteq S_i(q_i(x)) \circ \cdots \circ S_1(q_1(x))(u(x)). \tag{5.15}$$

Moreover, we call

$$\lambda_i(x) \doteq \lambda_i(\omega_{i-1}(x), \omega_i(x)),$$

the speed of the i-shock connecting $\omega_{i-1}(x)$ with $\omega_i(x)$. For convenience, we shall use the notations $W_i^{\alpha+} \doteq W_i(x_\alpha+)$, $q_i^{\alpha-} \doteq q_i(x_\alpha-)$, etc. Since all our functions are constant on each open interval $]x_{\alpha-1}, x_\alpha[$ for $x_{\alpha-1} < x < x_\alpha$, we clearly have

$$\begin{aligned}
|q_i(x)|\lambda_i(x)W_i(x) &= |q_i^{(\alpha-1)+}|\lambda_i^{(\alpha-1)+}W_i^{(\alpha-1)+} \\
&= |q_i^{\alpha-}|\lambda_i^{\alpha-}W_i^{\alpha-}.
\end{aligned} \tag{5.16}$$

The assumption that $u(t,\cdot)$ and $v(t,\cdot)$ are integrable and piecewise constant implies $q_i(t,x)\equiv 0$ for x outside a bounded interval. We can add and subtract the terms (5.16) in the summation (5.14), thus obtaining

$$\frac{\mathrm{d}}{\mathrm{d}t}\Phi\big(u(t),v(t)\big)$$

$$=\sum_{\alpha\in\mathcal{J}}\sum_{i=1}^{n}\big\{|q_i^{\alpha+}|W_i^{\alpha+}(\lambda_i^{\alpha+}-\dot{x}_\alpha)-|q_i^{\alpha-}|W_i^{\alpha-}(\lambda_i^{\alpha-}-\dot{x}_\alpha)\big\}. \tag{5.17}$$

We regard the quantity $|q_i(x)|\lambda_i(x)$ as the flux of the ith component of $|v-u|$ at x.

In connection with (5.17), for each jump point $\alpha\in\mathcal{J}$ and every $i=1,\dots,n$, define

$$E_{\alpha,i}\doteq|q_i^{\alpha+}|W_i^{\alpha+}(\lambda_i^{\alpha+}-\dot{x}_\alpha)-|q_i^{\alpha-}|W_i^{\alpha-}(\lambda_i^{\alpha-}-\dot{x}_\alpha). \tag{5.18}$$

Our main goal will be to establish the bounds

$$\sum_{i=1}^{n}E_{\alpha,i}=\mathcal{O}(1)\cdot|\sigma_\alpha|\quad\text{if }\alpha\in\mathcal{NP}, \tag{5.19}$$

$$\sum_{i=1}^{n}E_{\alpha,i}\leqslant\mathcal{O}(1)\cdot\varepsilon|\sigma_\alpha|\quad\text{if }\alpha\in\mathcal{R}\cup\mathcal{S}. \tag{5.20}$$

As usual, by the Landau symbol $\mathcal{O}(1)$ we denote a quantity whose absolute value satisfies a uniform bound, depending only on the system (4.1). In particular, this bound should not depend on ε or on the functions u, v.

From (5.19) and (5.20), recalling that the total strength of nonphysical fronts is less than or equal to ε and the total strength of shock and rarefaction fronts satisfies the uniform bound (5.12), we obtain

$$\frac{\mathrm{d}}{\mathrm{d}t}\Phi\big(u(t),\ v(t)\big)\leqslant\mathcal{O}(1)\cdot\varepsilon. \tag{5.21}$$

Moreover, at each time τ where two fronts in u or two fronts in v interact, by (4.17) the corresponding functional $Q(u)$ or $Q(v)$ is strictly decreasing. Therefore, if the constant κ_2 in (5.7) is chosen large enough, all weight functions $W_i(x)$ will decrease at each time τ where two fronts of u or two fronts of v interact. Observing that the maps $t\mapsto q_i(t,\cdot)$ are continuous with values in $\mathbf{L}^1$ also at interaction times, we conclude

$$\Phi\big(u(\tau+),\ v(\tau+)\big)\leqslant\Phi\big(u(\tau-),\ v(\tau-)\big)$$

at every time τ where an interaction occurs. Integrating (5.21) over the interval $[s,t]$ we therefore obtain

$$\Phi\big(u(t),\ v(t)\big)\leqslant\Phi\big(u(s),\ v(s)\big)+\mathcal{O}(1)\cdot\varepsilon(t-s), \tag{5.22}$$

proving the theorem. All the remaining work is thus aimed at establishing the key estimates (5.19) and (5.20). In the remainder of this section we illustrate some features of the functional Φ, pointing out the leading terms that contribute to its decrease in time.

Given two piecewise constant functions u, v with compact support, for $i = 1, \ldots, n$, we can define the scalar components u_i, v_i by induction on the jump points of u, v. We start by setting $u_i(-\infty) = v_i(-\infty) = 0$. If $x_\alpha \in \mathcal{J}(u)$ is a jump point of u, then we let v_i be constant across x_α and set

$$u_i(x_\alpha+) \doteq u_i(x_\alpha-) - \big[q_i(x_\alpha+) - q_i(x_\alpha-)\big].$$

On the other hand, if $x_\alpha \in \mathcal{J}(v)$ is a jump point of v, then we let u_i be constant across x_α and set

$$v_i(x_\alpha+) \doteq v_i(x_\alpha-) + \big[q_i(x_\alpha+) - q_i(x_\alpha-)\big].$$

These definitions trivially imply

$$q_i(x) = v_i(x) - u_i(x) \quad \text{for all } x \in \mathbb{R},\ i = 1, \ldots, n.$$

Observe that, according to the definition (5.9), the i-waves in u and v which approach $q_i(x)$ are those located within the thick portions of the graphs of u_i, v_i in Figure 22. Vice versa, for a given i-wave σ_α located at x_α, the regions where the jumps $q_i(x)$ approach σ_α are represented by the shaded areas in Figure 23.

Inserting the expressions for the weights (5.8) or (5.9) inside (5.7) and interchanging the summation with the integral, the functional Φ in (5.6) can be written in the equivalent form

$$\begin{aligned}\Phi(u, v) = {}& \big[1 + \kappa_1\kappa_2 Q(u) + \kappa_1\kappa_2 Q(v)\big] \cdot \sum_{i=1}^{n} \int_{-\infty}^{\infty} \big|q_i(x)\big|\,\mathrm{d}x \\ & + \kappa_1 \cdot \sum_{\sigma_\alpha \in \mathcal{J}(u)\cup\mathcal{J}(v)} |\sigma_\alpha| \cdot \sum_i \int_{q_i(x)\ \text{approaches}\ \sigma_\alpha} \big|q_i(x)\big|\,\mathrm{d}x. \end{aligned} \tag{5.23}$$

To understand why the expression (5.23) should decrease in time, assume first that v has a wave-front at x_α with strength σ_α, say in the kth family. To fix the ideas, assume that $u_k(x_\alpha) < v_k(x_\alpha+) < v_k(x_\alpha-)$. In connection with this front (see Figure 24), for every $i < k$, the functional $\Phi(u, v)$ contains a term of the form

$$\Phi_{\alpha,i} \doteq \kappa_1|\sigma_\alpha|[\text{area of the region between the graphs of } u_i \text{ and } v_i, \text{ to the right of } x_\alpha].$$

By strict hyperbolicity, the ith and kth characteristic speeds are strictly separated, say $\lambda_k - \lambda_i \geqslant c > 0$. We thus expect

$$\frac{\mathrm{d}\Phi_{\alpha,i}}{\mathrm{d}t} \approx -\kappa_1|\sigma_\alpha|\big|q_i^\alpha\big|\big(\dot{x}_\alpha - \lambda_i^\alpha\big) \leqslant -c\kappa_1|\sigma_\alpha|\big|q_i^{\alpha+}\big| \tag{5.24}$$

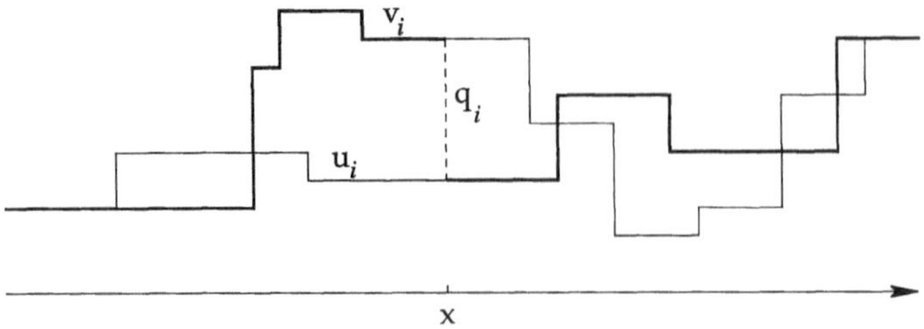

Fig. 22. Fronts approaching a jump q_i.

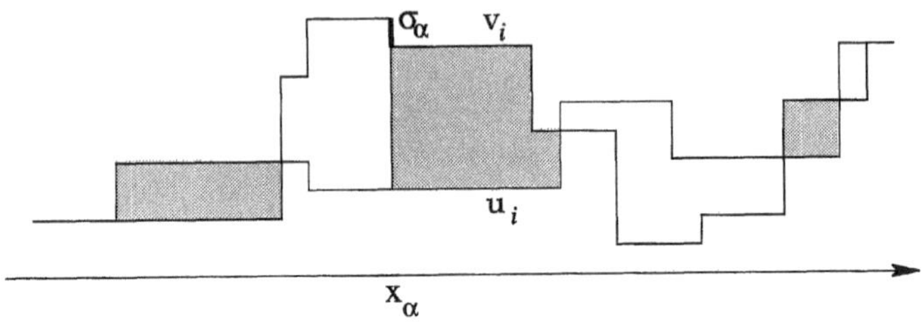

Fig. 23. Area approaching a front σ_α.

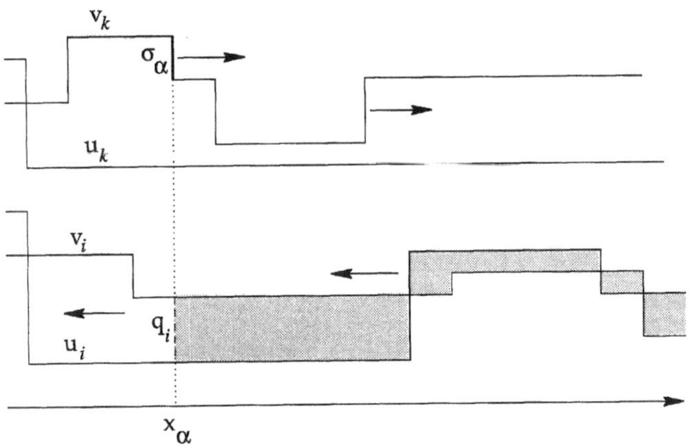

Fig. 24. Decrease of the functional due to strict hyperbolicity.

up to higher-order terms. Here $\lambda_i^\alpha \doteq \lambda_i(\omega_{i-1}(x_\alpha),\ \omega_i(x_\alpha))$ is the speed of the i-shock q_i^α. In addition, for every $i > k$, the functional $\Phi(u, v)$ contains a term of the form

$$\Phi_{\alpha,i} \doteq \kappa_1 |\sigma_\alpha| [\text{area of the region between the graphs of } u_i \text{ and } v_i \text{, to the left of } x_\alpha].$$

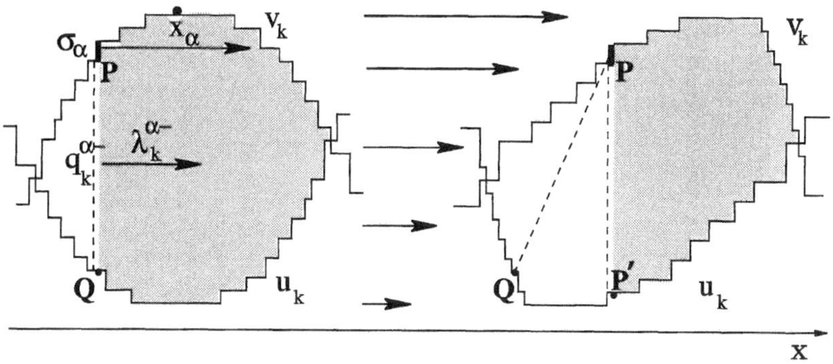

Fig. 25. Decrease of the functional due to genuine nonlinearity.

Entirely similar estimates can be proved also for these terms. In our future estimates (5.50) and (5.54), the leading negative terms on the right-hand sides correspond to (5.24) and are motivated by Figure 24.

Next, if the kth field is genuinely nonlinear, according to (5.23) the functional $\Phi(u, v)$ also contains a term of the form (see Figure 25)

$$\Phi_{\alpha,k} \doteq \kappa_1 |\sigma_\alpha| [\text{area of the region between the graphs of } u_k \text{ and } v_k, \text{ to the right of } x_\alpha].$$

Notice that, because of genuine nonlinearity, the points on the graphs of u_k and v_k move with different speeds. As a consequence, the shape of these graphs changes in time. In particular, while the area enclosed by the two graphs may remain constant, the portion of this area located to the right of a given front σ_α will decrease. In Figure 25, the two points P, Q lie initially on the same vertical line. At a later time this is no longer true. The area of the shaded region, enclosed by the graphs of u_k and v_k and by a vertical line through P, has decreased by an amount roughly equal to the area of the triangular region with vertices P, P', Q.

Recalling the parametrization (5.1), the difference between the speed $\dot{x}_\alpha$ of the jump σ_α at P and the average speed $\lambda_k^{\alpha-}$ of points between P and Q is estimated by

$$\dot{x}_\alpha - \lambda_k^{\alpha-} \approx \frac{1}{2} |q_k^{\alpha-} + \sigma_\alpha|.$$

Hence, up to higher-order terms, the time derivative of $\Phi_{\alpha,k}$ satisfies

$$\frac{d\Phi_{\alpha,k}}{dt} \approx -\kappa_1 |\sigma_\alpha| |q_k^\alpha| (\dot{x}_\alpha - \lambda_k^{\alpha-}) \approx -\kappa_1 |\sigma_\alpha| |q_k^{\alpha-}| \cdot \frac{1}{2} |q_k^{\alpha-} + \sigma_\alpha|. \tag{5.25}$$

In our future estimates (5.62), the leading negative term on the right-hand sides corresponds to (5.25) and is motivated by Figure 25.

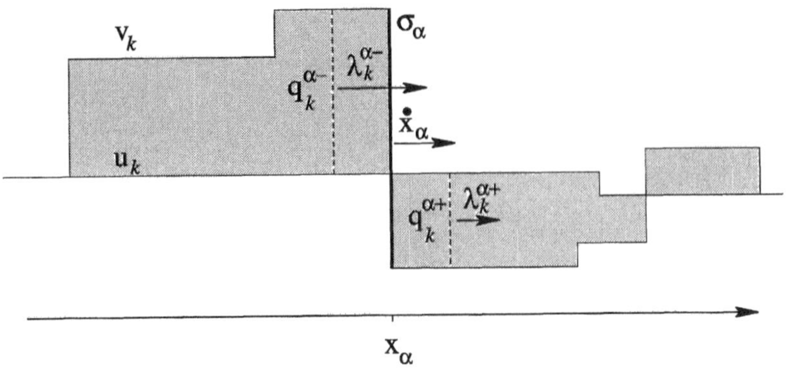

Fig. 26. Decrease of the functional due to a shock.

Yet another kind of estimate is needed in the case where the jump in v_k crosses the graph of u_k, say $v_k(x_\alpha+) < u_k(x_\alpha) < v_k(x_\alpha-)$. This happens when v has a shock of the kth family, say of strength σ_α, with

$$q_k^{\alpha-} > 0, \qquad q_k^{\alpha+} < 0, \qquad \sigma_\alpha \approx q_k^{\alpha+} - q_k^{\alpha-} < 0 \tag{5.26}$$

as in Figure 26. In this case, the estimates (5.24) remain valid. In connection with the kth field, the functional Φ contains a term of the form

$$\Phi_k^* \doteq 1 \cdot [\text{area of the region between the graphs of } u_k \text{ and } v_k],$$

where the above area includes points both on the right and on the left of x_α.

For convenience, call $\bar{\lambda}_k \doteq \lambda_k(v(x_\alpha+))$. Since $q_k^{\alpha+} \approx q_k^{\alpha-} + \sigma_\alpha$, the genuine nonlinearity of the kth characteristic field and the parametrization (5.1) yield

$$\lambda_k^{\alpha-} \approx \bar{\lambda}_k + |q_k^{\alpha+}| + \frac{1}{2}|q_k^{\alpha-}|, \qquad \lambda_k^{\alpha+} \approx \bar{\lambda}_k + \frac{1}{2}|q_k^{\alpha+}|,$$

$$\dot{x}_\alpha \approx \bar{\lambda}_k + \frac{1}{2}\big(|q_k^{\alpha+}| + |q_k^{\alpha-}|\big).$$

Neglecting higher-order terms, the shaded area in Figure 26 thus decreases at the rate

$$\begin{aligned} \frac{d\Phi_k^*}{dt} &\approx |q_k^{\alpha-}|(\lambda_k^{\alpha-} - \dot{x}_\alpha) + |q_k^{\alpha+}|(\dot{x}_\alpha - \lambda_k^{\alpha+}) \\ &\approx |q_k^{\alpha-}||q_k^{\alpha+}| \approx |q_k^{\alpha-}||\sigma_\alpha + q_k^{\alpha-}|. \end{aligned} \tag{5.27}$$

In our future estimate (5.66), the leading negative term on the right-hand side corresponds to (5.27) and is motivated by Figure 26. Notice that in this last case the decrease of the functional Φ is simply due to the decrease of the $\mathbf{L}^1$ distance between the kth components of u and v. In this occasion, the weights W_k play no role.

5.2. *Proofs of the main estimates*

In this section we establish the key estimates (5.19) and (5.20), thus completing the proof of Theorem 5.1. We shall always assume that x_α is a point where v has a jump in the k_αth family, of size σ_α. The case of a jump in u is entirely similar. In the following, since all computations refer to a fixed jump $\alpha \in \mathcal{J}(v)$, we shall often drop the superscript "α" and simply write $W_i^+ \doteq W_i^{\alpha+}$, $q_{k_\alpha}^- \doteq q_{k_\alpha}^{\alpha-}$, $v^- \doteq v(x_\alpha-)$, $v^+ \doteq v(x_\alpha+)$, etc. The proof will be given in several steps.

(1) *Nonphysical fronts.* If $\alpha \in \mathcal{NP}$ is a nonphysical front, the jump travels with the fixed speed $\dot{x}_\alpha = \hat{\lambda}$ and $\sigma_\alpha \doteq |v^+ - v^-|$. Hence, for $i = 1, \ldots, n$, one has

$$\begin{aligned} q_i^+ - q_i^- &= \mathcal{O}(1) \cdot \sigma_\alpha, \\ \lambda_i^+ - \lambda_i^- &= \mathcal{O}(1) \cdot \sigma_\alpha. \end{aligned} \tag{5.28}$$

Moreover, from the definitions (5.7)–(5.9) and the bound (5.10) it follows

$$\left|W_i^+ - W_i^-\right| \leqslant 1 \quad \text{for all } i. \tag{5.29}$$

If q_i^- and q_i^+ have the same sign, then

$$W_i^+ = W_i^-.$$

If q_i^- and q_i^+ have opposite signs, then (5.28) implies

$$|q_i^-| + |q_i^+| = \mathcal{O}(1) \cdot \sigma_\alpha.$$

Writing

$$\begin{aligned} E_{\alpha,i} = {} & W_i^+ \left(|q_i^+| - |q_i^-|\right)\left(\lambda_i^+ - \dot{x}_\alpha\right) + \left(W_i^+ - W_i^-\right)|q_i^-|\left(\lambda_i^+ - \dot{x}_\alpha\right) \\ & + W_i^- |q_i^-|\left(\lambda_i^+ - \lambda_i^-\right), \end{aligned}$$

the estimate (5.19) is now clear.

(2) *Reduction to the shock case.* We now work toward the estimate (5.20). In order to reduce the number of cases that must be considered, we now show that rarefaction fronts can be replaced by shock fronts of the same size. More precisely, in the case of a physical front $\alpha \in \mathcal{R} \cup \mathcal{S}$ of strength σ_α, consider the auxiliary state and speed

$$v^\oplus \doteq S_{k_\alpha}(\sigma_\alpha)(v^-), \qquad \dot{x}_\alpha^* \doteq \lambda_{k_\alpha}\left(v^-, v^\oplus\right). \tag{5.30}$$

Define the components $q_i^\oplus$ in terms of the implicit relation

$$v^\oplus = S_n\left(q_n^\oplus\right) \circ \cdots \circ S_1\left(q_1^\oplus\right)\left(u(x_\alpha)\right). \tag{5.31}$$

Similarly, define the intermediate states $\omega_0^\oplus = u(x_\alpha)$, $\omega_1^\oplus, \dots, \omega_n^\oplus = v^\oplus$ and the shock speeds $\lambda_i^\oplus$ by setting

$$\begin{aligned}
&\omega_i^\oplus \doteq S_i\big(q_i^\oplus\big) \circ S_{i-1}\big(q_{i-1}^\oplus\big) \circ \cdots \circ S_1\big(q_1^\oplus\big)\big(u(x_\alpha)\big),\\
&\lambda_i^\oplus \doteq \lambda_i\big(\omega_{i-1}^\oplus, \omega_i^\oplus\big).
\end{aligned} \tag{5.32}$$

We now consider two cases.

Case 1: *The jump at* x_α *is a shock or a contact discontinuity.* In this case $v^\oplus = v^+$ and hence we trivially have

$$\dot{x}_\alpha^* = \dot{x}_\alpha, \qquad q_i^\oplus = q_i^+, \qquad \omega_i^\oplus = \omega_i^+, \qquad \lambda_i^\oplus = \lambda_i^+ \tag{5.33}$$

for all $i = 1, \dots, n$.

Case 2: *The jump at* x_α *is a rarefaction, hence* $0 < \sigma_\alpha \leqslant \varepsilon$. In this case, since shock and rarefaction curves have a tangency of second order, we have $v^\oplus - v^+ = \mathcal{O}(1) \cdot |\sigma_\alpha|^3$ and hence

$$\begin{aligned}
&q_i^\oplus - q_i^+ = \mathcal{O}(1) \cdot |\sigma_\alpha|^3,\\
&\omega_i^\oplus - \omega_i^+ = \mathcal{O}(1) \cdot |\sigma_\alpha|^3,\\
&\lambda_i^\oplus - \lambda_i^+ = \mathcal{O}(1) \cdot |\sigma_\alpha|^3
\end{aligned} \tag{5.34}$$

for all i. Moreover,

$$\big|\dot{x}_\alpha^* - \dot{x}_\alpha\big| = \big|\lambda_k\big(v^\oplus, v^-\big) - \lambda_k(v^-)\big| = \frac{1}{2}\sigma_\alpha + \mathcal{O}(1) \cdot \sigma_\alpha^2 < \varepsilon. \tag{5.35}$$

We now compute

$$\begin{aligned}
E_{\alpha,i} &= W_i^+|q_i^+|\big(\lambda_i^+ - \dot{x}_\alpha\big) - W_i^-|q_i^-|\big(\lambda_i^- - \dot{x}_\alpha\big)\\
&= W_i^+|q_i^+|\big(\lambda_i^+ - \dot{x}_\alpha^*\big) - W_i^-|q_i^-|\big(\lambda_i^- - \dot{x}_\alpha^*\big)\\
&\quad + \big(\dot{x}_\alpha^\oplus - \dot{x}_\alpha\big)\big\{W_i^+|q_i^+| - W_i^-|q_i^-|\big\}\\
&= \big\{W_i^+|q_i^\oplus|\big(\lambda_i^\oplus - \dot{x}_\alpha^*\big) - W_i^-|q_i^-|\big(\lambda_i^- - \dot{x}_\alpha^*\big)\big\}\\
&\quad + \big\{W_i^+|q_i^\oplus|\big(\lambda_i^+ - \lambda_i^\oplus\big) + W_i^+\big(|q_i^+| - |q_i^\oplus|\big)\big(\lambda_i^+ - \dot{x}_\alpha^*\big)\big\}\\
&\quad + \big(\dot{x}_\alpha^* - \dot{x}_\alpha\big)\big\{W_i^+\big(|q_i^+| - |q_i^-|\big) + \big(W_i^+ - W_i^-\big)|q_i^-|\big\}\\
&\doteq E'_{\alpha,i} + E''_{\alpha,i} + E'''_{\alpha,i}.
\end{aligned} \tag{5.36}$$

In view of the estimates (5.33) and (5.34), for each $i = 1, \ldots, n$, we have

$$E''_{\alpha,i} = \begin{cases} 0 & \text{if } \sigma_\alpha < 0, \\ \mathcal{O}(1) \cdot |\sigma_\alpha|^3 & \text{if } \sigma_\alpha \in [0, \varepsilon]. \end{cases} \tag{5.37}$$

Toward a bound for $E'''_{\alpha,i}$, observe that the definitions (5.7)–(5.9) of the weights imply

$$W_i^+ - W_i^- = \mathcal{O}(1) \cdot \kappa_1 \sigma_\alpha \quad \text{if } q_i^+ q_i^- > 0.$$

Moreover, we have the estimates

$$q_i^+ - q_i^- = \mathcal{O}(1) \cdot |\sigma_\alpha|, \qquad q_i^- = \mathcal{O}(1) \cdot |\sigma_\alpha| \quad \text{if } q_i^+ q_i^- \leqslant 0.$$

By (5.10), the bound (5.29) always holds. Using (5.35) together with the above estimates, we deduce

$$E'''_{\alpha,i} = \mathcal{O}(1) \cdot \varepsilon |\sigma_\alpha|. \tag{5.38}$$

Thanks to (5.37) and (5.38), the proof of (5.20) is reduced to showing that

$$\sum_{i=1}^{n} E'_{\alpha,i} \leqslant \mathcal{O}(1) \cdot \varepsilon |\sigma_\alpha|, \tag{5.39}$$

with $E'_{\alpha,i}$ defined at (5.36). In our future estimates, we can thus replace v^+ with $v^\oplus$ and $\dot{x}_\alpha$ with $\dot{x}^*_\alpha$. In essence, this reduces the problem to the case (5.32) where the right state $v^\oplus$ is connected to the left state v^- by a k_α-shock, traveling with the exact Rankine–Hugoniot speed. From now on, we will work toward a proof of (5.39) in the case $\alpha \in \mathcal{R} \cup \mathcal{S}$.

(3) *A technical tool.* We describe here a simple technique for deriving a number of a priori bounds. First of all, observe that, as soon as the state $u^- = u(x_\alpha -)$ is assigned, all quantities $v^-, v^+, \lambda_i^-, q_i^\oplus, \lambda_i^\oplus, \omega_i^-, \omega_i^\oplus, \ldots$ can then be recovered as functions of $q_1^-, \ldots, q_n^-$ and σ_α. Indeed, v^- is obtained from (5.3), while $v^\oplus$ and $\dot{x}^*_\alpha$ are obtained by (5.30). In turn, the components $q_i^\oplus$ are implicitly determined by (5.31), while (5.14) and (5.32) define the intermediate states ω_i^- and $\omega_i^\oplus$, respectively. Introducing the variables

$$\tilde{q} \doteq (q_1^-, \ldots, q_{k_\alpha - 1}^-, q_{k_\alpha + 1}^-, \ldots, q_n^-) \in \mathbb{R}^{n-1}, \qquad q^* \doteq q_{k_\alpha}^-, \qquad \sigma \doteq \sigma_\alpha,$$

one can thus apply Lemma A.3 or Lemma A.4 to functions of the form

$$\Psi = \Psi(q_1^-, \ldots, q_n^-, \sigma_\alpha) = \Psi(\tilde{q}, q^*, \sigma)$$

and derive useful a priori estimates.

For example, if the k_αth field is linearly degenerate, then each of the functions

$$\Psi_i \doteq q_i^\oplus - q_i^-, \quad i \neq k_\alpha,$$
$$\Psi_{k_\alpha} \doteq q_{k_\alpha}^\oplus - q_{k_\alpha}^- - \sigma_\alpha,$$

satisfies the assumptions (A.18) and (A.20) of Lemma A.3. Therefore, (A.21) yields the estimate

$$\left|q_{k_\alpha}^\oplus - q_{k_\alpha}^- - \sigma_\alpha\right| + \sum_{i \neq k_\alpha} \left|q_i^\oplus - q_i^-\right| = \mathcal{O}(1) \cdot \left(\sum_{i \neq k_\alpha} |q_i^-|\right) |\sigma_\alpha|. \tag{5.40}$$

On the other hand, if k_αth field is genuinely nonlinear, then the above functions Ψ_i, Ψ_{k_α} satisfy only the assumptions (A.18) of Lemma A.3. In this case, (A.19) yields the weaker estimate

$$\begin{aligned} &\left|q_{k_\alpha}^\oplus - q_{k_\alpha}^- - \sigma_\alpha\right| + \sum_{i \neq k_\alpha} \left|q_i^\oplus - q_i^-\right| \\ &\quad = \mathcal{O}(1) \cdot \left(|q_{k_\alpha}^-||q_{k_\alpha}^- + \sigma_\alpha| + \sum_{i \neq k_\alpha} |q_i^-|\right) |\sigma_\alpha|. \end{aligned} \tag{5.41}$$

As a further example, observe that the two functions

$$\Psi_\alpha'(q_1^-, \ldots, q_n^-, \sigma_\alpha) \doteq \dot{x}_\alpha^* - \lambda_{k_\alpha}^- \quad \text{and} \quad \Psi_\alpha''(q_1^-, \ldots, q_n^-, \sigma_\alpha) \doteq \dot{x}_\alpha^* - \lambda_{k_\alpha}^\oplus$$

satisfy the assumptions (A.25) and (A.27), respectively. Using Lemma A.4, from (A.26) and (A.28) one obtains

$$\dot{x}_\alpha^* - \lambda_{k_\alpha}^- = \frac{q_{k_\alpha}^- + \sigma_\alpha}{2} + \mathcal{O}(1) \cdot \left(|q_{k_\alpha}^- + \sigma_\alpha| \left(|q_{k_\alpha}^-| + |\sigma_\alpha|\right) + \sum_{i \neq k_\alpha} |q_i^-|\right), \tag{5.42}$$

$$\dot{x}_\alpha^* - \lambda_{k_\alpha}^\oplus = \frac{q_{k_\alpha}^-}{2} + \mathcal{O}(1) \cdot \left(|q_{k_\alpha}^-| \left(|q_{k_\alpha}^-| + |\sigma_\alpha|\right) + \sum_{i \neq k_\alpha} |q_i^-|\right). \tag{5.43}$$

(4) *Linearly degenerate fields.* Here and in the next step, our goal is to estimate the terms

$$E_{\alpha,i}' \doteq W_i^+ |q_i^\oplus| \left(\lambda_i^\oplus - \dot{x}_\alpha^*\right) - W_i^- |q_i^-| \left(\lambda_i^- - \dot{x}_\alpha^*\right)$$

introduced at (5.36). We start with the case where the k_αth field is linearly degenerate, so that the identities in (5.33) hold. For notational convenience, consider the two sets of indices

$$\begin{aligned} \mathcal{I} &\doteq \left\{ i \in \{1, \ldots, n\},\ i \neq k_\alpha,\ q_i^+ q_i^- > 0 \right\}, \\ \mathcal{I}' &\doteq \left\{ i \in \{1, \ldots, n\},\ i \neq k_\alpha,\ q_i^+ q_i^- \leqslant 0 \right\}. \end{aligned} \tag{5.44}$$

Our definition of weights implies $W^+_{k_\alpha} = W^-_{k_\alpha}$ while

$$W_i^+ = W_i^- - \kappa_1 |\sigma_\alpha| \operatorname{sgn}(i - k_\alpha) \quad \text{if } i \in \mathcal{I}. \tag{5.45}$$

On the other hand, if $i \in \mathcal{I}'$, by (5.40) there exists a constant C such that

$$|q_i^\oplus| + |q_i^-| \leqslant C \cdot \Big(\sum_{i \neq k_\alpha} |q_i^-|\Big) |\sigma_\alpha|. \tag{5.46}$$

Hence,

$$E'_{\alpha,i} = \mathcal{O}(1) \cdot \Big(\sum_{i \neq k_\alpha} |q_i^-|\Big) |\sigma_\alpha|. \tag{5.47}$$

A bound on E'_{α,k_α} will be obtained using the estimates

$$\lambda^-_{k_\alpha} - \dot{x}^*_\alpha = \lambda_{k_\alpha}(\omega^-_{k_\alpha}) - \lambda_{k_\alpha}(\omega^-_n) = \mathcal{O}(1) \cdot \sum_{i > k_\alpha} |q_i^-|,$$

$$q^\oplus_{k_\alpha} - q^-_{k_\alpha} = \mathcal{O}(1) \cdot |\sigma_\alpha|.$$

Observing that the quantity

$$\Psi_{k_\alpha}(q_1^-, \dots, q_n^-, \sigma_\alpha) \doteq \lambda^-_{k_\alpha} - \lambda^\oplus_{k_\alpha} = \lambda_{k_\alpha}(\omega^-_{k_\alpha}) - \lambda_{k_\alpha}(\omega^\oplus_{k_\alpha})$$

satisfies both assumptions (A.18) and (A.20), using Lemma A.3 and the above estimates, we obtain

$$\begin{aligned} E'_{\alpha,k_\alpha} &= W^+_{k_\alpha} \cdot \big\{ |q^\oplus_{k_\alpha}| (\lambda^\oplus_{k_\alpha} - \dot{x}^*_\alpha) - |q^-_{k_\alpha}| (\lambda^-_{k_\alpha} - \dot{x}^*_\alpha) \big\} \\ &\leqslant W^+_{k_\alpha} \cdot \big\{ |q^\oplus_{k_\alpha}| |\lambda^-_{k_\alpha} - \lambda^\oplus_{k_\alpha}| + \big| |q^\oplus_{k_\alpha}| - |q^-_{k_\alpha}| \big| |\lambda^-_{k_\alpha} - \dot{x}^\oplus_\alpha| \big\} \\ &= \mathcal{O}(1) \cdot |\sigma_\alpha| \sum_{i \neq k_\alpha} |q_i^-|. \end{aligned} \tag{5.48}$$

For $\iota \in \mathcal{I}$, observing that the quantity

$$\Psi_i(q_1^-, \dots, q_n^-, \sigma_\alpha) \doteq q_i^\oplus (\lambda_i^\oplus - \dot{x}^*_\alpha) - q_i^- (\lambda_i^- - \dot{x}^\oplus_\alpha) \tag{5.49}$$

satisfies both assumptions (A.18) and (A.20), using Lemma A.3 we obtain

$$\begin{aligned} E'_{\alpha,i} &= -\kappa_1 |\sigma_\alpha| |q_i^-| |\lambda_i^- - \dot{x}^*_\alpha| + W_i^+ \big\{ |q_i^\oplus| (\lambda_i^\oplus - \dot{x}^*_\alpha) - |q_i^-| (\lambda_i^- - \dot{x}^*_\alpha) \big\} \\ &\leqslant -c\kappa_1 |\sigma_\alpha| |q_i^-| + \mathcal{O}(1) \cdot \Big(\sum_{i \neq k_\alpha} |q_i^-|\Big) |\sigma_\alpha|. \end{aligned} \tag{5.50}$$

Since the total variation of u, v is small, we can assume that the constant C in (5.46) satisfies $nC|\sigma_\alpha| < 1/2$. This implies

$$\sum_{i\in\mathcal{I}'}|q_i^-| \leqslant nC\Bigl(\sum_{i\neq k_\alpha}|q_i^-|\Bigr)|\sigma_\alpha| < \frac{1}{2}\sum_{i\in\mathcal{I}\cup\mathcal{I}'}|q_i^-|,$$
$$\sum_{i\in\mathcal{I}}|q_i^-| > \frac{1}{2}\cdot\sum_{i\neq k_\alpha}|q_i^-|. \tag{5.51}$$

Summing together the inequalities (5.47), (5.48) and (5.50) and using (5.51), we obtain

$$\sum_{i=1}^{n}E'_{\alpha,i} \leqslant -c\kappa_1|\sigma_\alpha|\cdot\sum_{i\in\mathcal{I}}|q_i^-| + \mathcal{O}(1)\cdot\Bigl(\sum_{i\neq k_\alpha}|q_i^-|\Bigr)|\sigma_\alpha|$$
$$= -c\kappa_1|\sigma_\alpha|\cdot\sum_{i\in\mathcal{I}}|q_i^-| + \mathcal{O}(1)\cdot\Bigl(\sum_{i\in\mathcal{I}}|q_i^-|\Bigr)|\sigma_\alpha| \leqslant 0, \tag{5.52}$$

provided that the constant κ_1 is chosen large enough.

(5) *Genuinely nonlinear fields.* We now assume that the k_αth field is genuinely nonlinear. Define the two sets of indices

$$\mathcal{I} \doteq \bigl\{i \in \{1,\dots,n\},\ i \neq k_\alpha,\ q_i^-, q_i^+, q_i^\oplus \text{ all have the same sign}\bigr\},$$
$$\mathcal{I}' \doteq \bigl\{i \in \{1,\dots,n\},\ i \neq k_\alpha,\ i \notin \mathcal{I}\bigr\}. \tag{5.53}$$

For $i \in \mathcal{I}$, the weights $W_i^\pm$ still satisfy (5.45). Furthermore, the quantity Ψ_i in (5.49) now satisfies only the assumptions (A.18) of Lemma A.3, hence it can estimated in terms of (A.19). Repeating the argument at (5.50) we deduce

$$E'_{\alpha,i} = -\kappa_1|\sigma_\alpha||q_i^-||\lambda_i^- - \dot{x}_\alpha^*| + W_i^+\bigl\{|q_i^\oplus|(\lambda_i^\oplus - \dot{x}_\alpha^*) - |q_i^-|(\lambda_i^- - \dot{x}_\alpha^*)\bigr\}$$
$$\leqslant -c\kappa_1|\sigma_\alpha||q_i^-| + \mathcal{O}(1)\cdot\Bigl(|q_{k_\alpha}^-||q_{k_\alpha}^- + \sigma_\alpha| + \sum_{i\neq k_\alpha}|q_i^-|\Bigr)|\sigma_\alpha|. \tag{5.54}$$

On the other hand, if $i \in \mathcal{I}'$, by (5.33), (5.34) and (5.41) we deduce

$$|q_i^\oplus| + |q_i^-| \leqslant 2\bigl(|q_i^- - q_i^\oplus| + |q_i^+ - q_i^\oplus|\bigr)$$
$$\leqslant C\cdot\Bigl(\varepsilon|\sigma_\alpha| + |q_{k_\alpha}^-||q_{k_\alpha}^- + \sigma_\alpha| + \sum_{i\neq k_\alpha}|q_i^-|\Bigr)|\sigma_\alpha| \tag{5.55}$$

for some constant C. Therefore,

$$E'_{\alpha,i} = \mathcal{O}(1)\cdot\Bigl(\varepsilon + |q_{k_\alpha}^-||q_{k_\alpha}^- + \sigma_\alpha| + \sum_{i\neq k_\alpha}|q_i^-|\Bigr)|\sigma_\alpha|. \tag{5.56}$$

The term E'_{α,k_α} will be estimated separately in three different cases.

Case 1: $|\sigma_\alpha| \leqslant \varepsilon$, $|q^-_{k_\alpha}| \leqslant 2|\sigma_\alpha|$. In this case, using the bounds

$$q^{\oplus}_{k_\alpha} - q^-_{k_\alpha} = \mathcal{O}(1) \cdot |\sigma_\alpha|,$$
$$\lambda^{\oplus}_{k_\alpha} - \lambda^-_{k_\alpha} = \mathcal{O}(1) \cdot |\sigma_\alpha|,$$
$$\lambda^{\oplus}_{k_\alpha} - \dot{x}^*_\alpha = \mathcal{O}(1) \cdot \left(|\sigma_\alpha| + \sum_{i=1}^{n} |q^-_i|\right) = \mathcal{O}(1) \cdot \left(\varepsilon + \sum_{i \neq k_\alpha} |q^-_i|\right),$$

we can write

$$\begin{aligned} E'_{\alpha,k_\alpha} &\leqslant W^+_{k_\alpha} |q^{\oplus}_{k_\alpha} - q^-_{k_\alpha}| |\lambda^{\oplus}_{k_\alpha} - \dot{x}^*_\alpha| \\ &\quad + |W^+_{k_\alpha} - W^-_{k_\alpha}| |q^-_{k_\alpha}| |\lambda^{\oplus}_{k_\alpha} - \dot{x}^*_\alpha| + W^-_{k_\alpha} |q^-_{k_\alpha}| |\lambda^{\oplus}_{k_\alpha} - \lambda^-_{k_\alpha}| \\ &= \mathcal{O}(1) \cdot \left(\varepsilon + \sum_{i \neq k_\alpha} |q^-_i|\right) |\sigma_\alpha|. \end{aligned} \tag{5.57}$$

By (5.55), when $i \in \mathcal{I}'$ it follows

$$|q^-_i| \leqslant C \cdot \left(\varepsilon|\sigma_\alpha| + \sum_{i \neq k_\alpha} |q^-_i|\right) |\sigma_\alpha|$$

for some constant C. Since the total variation is small, we can again assume $nC|\sigma_\alpha| < 1/2$. This yields

$$\begin{aligned} &\sum_{i \in \mathcal{I}'} |q^-_i| \leqslant \frac{1}{2}\left(\varepsilon|\sigma_\alpha| + \sum_{i \in \mathcal{I} \cup \mathcal{I}'} |q^-_i|\right), \\ &\sum_{i \neq k_\alpha} |q^-_i| \leqslant \varepsilon|\sigma_\alpha| + 2\sum_{i \in \mathcal{I}} |q^-_i|. \end{aligned} \tag{5.58}$$

From (5.54), (5.56) and (5.57), using (5.58) we conclude

$$\begin{aligned} \sum_{i=1}^{n} E'_{\alpha,i} &\leqslant -c\kappa_1 |\sigma_\alpha| \sum_{i \in \mathcal{I}} |q^-_i| + \mathcal{O}(1) \cdot |\sigma_\alpha| \left(\varepsilon + \sum_{i \in \mathcal{I}} |q^-_i|\right) \\ &\leqslant \mathcal{O}(1) \cdot \varepsilon |\sigma_\alpha|, \end{aligned} \tag{5.59}$$

provided that the constant κ_1 is large enough.

Case 2: $q_{k_\alpha}^-, q_{k_\alpha}^+, q_{k_\alpha}^\oplus$ all have the same sign. In this case we have

$$W_{k_\alpha}^+ = W_{k_\alpha}^- + \kappa_1 |\sigma_\alpha| \operatorname{sgn}(q_{k_\alpha}^-).$$

Since the quantity

$$\Psi_{k_\alpha}(q_1^-, \dots, q_n^-, \sigma_\alpha) \doteq q_{k_\alpha}^\oplus (\lambda_{k_\alpha}^\oplus - \dot{x}_\alpha^*) - q_{k_\alpha}^- (\lambda_{k_\alpha}^- - \dot{x}_\alpha^\oplus)$$

satisfies the assumptions (A.18), by Lemma A.3 it can be estimated as

$$\begin{aligned} & q_{k_\alpha}^\oplus (\lambda_{k_\alpha}^\oplus - \dot{x}_\alpha^*) - q_{k_\alpha}^- (\lambda_{k_\alpha}^- - \dot{x}_\alpha^*) \\ &\quad = \mathcal{O}(1) \cdot \Big(|q_{k_\alpha}^-| |q_{k_\alpha}^- + \sigma_\alpha| + \sum_{i \neq k_\alpha} |q_i^-| \Big) |\sigma_\alpha|. \end{aligned} \tag{5.60}$$

From the estimate (5.42) we deduce

$$\begin{aligned} & \kappa_1 |\sigma_\alpha| \operatorname{sgn}(q_{k_\alpha}^-) |q_{k_\alpha}^-| (\lambda_{k_\alpha}^- - \dot{x}_\alpha^*) \\ &\quad \leqslant -\frac{1}{2} \kappa_1 |\sigma_\alpha| |q_{k_\alpha}^-| |q_{k_\alpha}^- + \sigma_\alpha| \\ &\qquad + \mathcal{O}(1) \cdot \Big(|q_{k_\alpha}^-| |q_{k_\alpha}^- + \sigma_\alpha| + \sum_{i \neq k_\alpha} |q_i^-| \Big) |\sigma_\alpha|. \end{aligned} \tag{5.61}$$

Using (5.60) and (5.61) we obtain

$$\begin{aligned} E'_{\alpha, k_\alpha} &= (W_{k_\alpha}^+ - W_{k_\alpha}^-) |q_{k_\alpha}^-| (\lambda_{k_\alpha}^- - \dot{x}_\alpha^*) \\ &\quad + W_{k_\alpha}^+ \cdot \big\{ |q_{k_\alpha}^\oplus| (\lambda_{k_\alpha}^\oplus - \dot{x}_\alpha^*) - |q_{k_\alpha}^-| (\lambda_{k_\alpha}^- - \dot{x}_\alpha^*) \big\} \\ &\leqslant -\frac{\kappa_1}{2} |\sigma_\alpha| |q_{k_\alpha}^-| |q_{k_\alpha}^- + \sigma_\alpha| \\ &\quad + \mathcal{O}(1) \cdot \Big(|q_{k_\alpha}^-| |q_{k_\alpha}^- + \sigma_\alpha| + \sum_{i \neq k_\alpha} |q_i^-| \Big) |\sigma_\alpha|. \end{aligned} \tag{5.62}$$

Since the total variation is small, in (5.55) we can assume $nC|\sigma_\alpha| < 1/2$. Hence,

$$\begin{aligned} & \sum_{i \in \mathcal{I}'} |q_i^-| \leqslant \frac{1}{2} \Big(\varepsilon |\sigma_\alpha| + |q_{k_\alpha}^-| |q_{k_\alpha}^- + \sigma_\alpha| + \sum_{i \in \mathcal{I} \cup \mathcal{I}'} |q_i^-| \Big), \\ & \sum_{i \neq k_\alpha} |q_i^-| \leqslant \varepsilon |\sigma_\alpha| + |q_{k_\alpha}^-| |q_{k_\alpha}^- + \sigma_\alpha| + 2 \sum_{i \in \mathcal{I}} |q_i^-|. \end{aligned} \tag{5.63}$$

From (5.54), (5.56) and (5.62), using (5.63) we conclude

$$\sum_{i=1}^{n} E'_{\alpha,i} \leqslant -c\kappa_1|\sigma_\alpha|\sum_{i\in\mathcal{I}}|q_i^-| - \frac{\kappa_1}{2}|\sigma_\alpha||q_{k_\alpha}^-||q_{k_\alpha}^- + \sigma_\alpha|$$
$$+ \mathcal{O}(1)\cdot|\sigma_\alpha|\Big(\varepsilon + |q_{k_\alpha}^-||q_{k_\alpha}^- + \sigma_\alpha| + \sum_{i\in\mathcal{I}}|q_i^-|\Big)$$
$$\leqslant \mathcal{O}(1)\cdot\varepsilon|\sigma_\alpha|, \tag{5.64}$$

provided that the constant κ_1 is large enough.

Case 3: $q_{k_\alpha}^+ < 0 < q_{k_\alpha}^-$. Notice that in this case $\sigma_\alpha < 0$, hence, the front is a shock or contact discontinuity and the identities (5.33) hold. By (5.41) we can also assume

$$\frac{1}{2}|\sigma_\alpha| < |q_{k_\alpha}^-| + |q_{k_\alpha}^+| < 2|\sigma_\alpha|. \tag{5.65}$$

Recalling that $1 \leqslant W_{k_\alpha}^\pm \leqslant 2$, using (5.42), (5.43), (5.65) and the fact that $q_{k_\alpha}^\oplus = q_{k_\alpha}^+ < 0$, one obtains

$$E'_{\alpha,k_\alpha} = W_{k_\alpha}^+|q_{k_\alpha}^+|(\lambda_{k_\alpha}^+ - \dot{x}_\alpha) - W_{k_\alpha}^-|q_{k_\alpha}^-|(\lambda_{k_\alpha}^- - \dot{x}_\alpha)$$
$$= W_{k_\alpha}^+|q_{k_\alpha}^+|\Big(-\frac{q_{k_\alpha}^-}{2} + \mathcal{O}(1)\cdot|q_{k_\alpha}^-|\big(|q_{k_\alpha}^-| + |\sigma_\alpha|\big) + \mathcal{O}(1)\cdot\sum_{i\neq k_\alpha}|q_i^-|\Big)$$
$$- W_{k_\alpha}^-|q_{k_\alpha}^-|\Big(-\frac{q_{k_\alpha}^- + \sigma_\alpha}{2} + \mathcal{O}(1)\cdot|q_{k_\alpha}^- + \sigma_\alpha|\big(|q_{k_\alpha}^-| + |\sigma_\alpha|\big)$$
$$+ \mathcal{O}(1)\cdot\sum_{i\neq k_\alpha}|q_i^-|\Big)$$
$$\leqslant -|q_{k_\alpha}^- + \sigma_\alpha||q_{k_\alpha}^-| + \mathcal{O}(1)\cdot|q_{k_\alpha}^-||q_{k_\alpha}^+ - q_{k_\alpha}^- - \sigma_\alpha|$$
$$+ \mathcal{O}(1)\cdot|q_{k_\alpha}^- + \sigma_\alpha||q_{k_\alpha}^-|\big(|q_{k_\alpha}^-| + |\sigma_\alpha|\big)$$
$$+ \mathcal{O}(1)\cdot\big(|q_{k_\alpha}^+| + |q_{k_\alpha}^-|\big)\sum_{i\neq k_\alpha}|q_i^-|$$
$$\leqslant -|q_{k_\alpha}^- + \sigma_\alpha||q_{k_\alpha}^-| + \mathcal{O}(1)\cdot|\sigma_\alpha|\Big(|q_{k_\alpha}^-||q_{k_\alpha}^- + \sigma_\alpha| + \sum_{i\neq k_\alpha}|q_i^-|\Big). \tag{5.66}$$

When $i \in \mathcal{I}$, the estimates (5.54) remain valid. When $i \in \mathcal{I}'$, by (5.33) and (5.41) we deduce

$$|q_i^+| + |q_i^-| = |q_i^+ - q_i^-| \leqslant C\cdot\Big(|q_{k_\alpha}^-||q_{k_\alpha}^- + \sigma_\alpha| + \sum_{i\neq k_\alpha}|q_i^-|\Big)|\sigma_\alpha| \tag{5.67}$$

for some constant C. Therefore,

$$E'_{\alpha,i} = \mathcal{O}(1) \cdot \left(|q_{k_\alpha}^-||q_{k_\alpha}^- + \sigma_\alpha| + \sum_{i \neq k_\alpha} |q_i^-| \right) |\sigma_\alpha|. \tag{5.68}$$

Since the total variation is small, in (5.67) we can assume $nC|\sigma_\alpha| < 1/2$. Hence,

$$\begin{aligned} &\sum_{i \in \mathcal{I}'} |q_i^-| \leqslant \frac{1}{2} \left(|q_{k_\alpha}^-||q_{k_\alpha}^- + \sigma_\alpha| + \sum_{i \in \mathcal{I} \cup \mathcal{I}'} |q_i^-| \right), \\ &\sum_{i \neq k_\alpha} |q_i^-| \leqslant |q_{k_\alpha}^-||q_{k_\alpha}^- + \sigma_\alpha| + 2 \sum_{i \in \mathcal{I}} |q_i^-|. \end{aligned} \tag{5.69}$$

By the smallness of the total variation we can also assume

$$\mathcal{O}(1) \cdot |\sigma_\alpha| \ll 1.$$

Summing together (5.54), (5.66) and (5.68) and using (5.69), we obtain

$$\begin{aligned} \sum_{i=1}^{n} E'_{\alpha,i} &\leqslant -c\kappa_1 |\sigma_\alpha| \sum_{i \in \mathcal{I}} |q_i^-| - |q_{k_\alpha}^-||q_{k_\alpha}^- + \sigma_\alpha| \\ &\quad + \mathcal{O}(1) \cdot |\sigma_\alpha| \left(|q_{k_\alpha}^-||q_{k_\alpha}^- + \sigma_\alpha| + \sum_{i \in \mathcal{I}} |q_i^-| \right) \\ &\leqslant 0, \end{aligned} \tag{5.70}$$

provided that the constant κ_1 is chosen suitably large.

The above three cases cover all possible situations. Indeed, if $q_{k_\alpha}^- \leqslant 0 \leqslant q_{k_\alpha}^+$, then the front is a rarefaction and has strength $\sigma_\alpha \in]0, \varepsilon]$. Hence Case 1 applies.

In all cases, the above analysis shows that the bound (5.39) holds, provided that we first choose the constants κ_1, κ_2 suitably large, then we let the total variation be sufficiently small. This completes the proof of Theorem 5.1.

5.3. *A semigroup of solutions*

Recalling the definitions (4.13) and (4.14), consider a domain of the form

$$\begin{aligned} \mathcal{D} = \mathrm{cl}\big\{ &u \in \mathbf{L}^1(\mathbb{R}; \mathbb{R}^n);\ u \text{ is piecewise constant}, \\ &\Upsilon(u) \doteq V(u) + C_0 \cdot Q(u) < \delta_0 \big\}, \end{aligned} \tag{5.71}$$

where cl denotes closure in $\mathbf{L}^1$. For a suitable choice of the constants C_0 and $\delta_0 > 0$, within the proof of Theorem 4.1 we showed that, for every $\bar{u} \in \mathcal{D}$, one can construct a convergent sequence of front tracking approximations taking values inside $\mathcal{D}$. Relying on Theorem 5.1, we now show that the limit of these approximate solutions is unique and depends Lipschitz continuously on the initial data.

THEOREM 5.2. *Fix $\bar{u} \in \mathcal{D}$ and consider a sequence $\varepsilon_\nu \to 0$. Then every sequence of ε_ν-approximate solutions $u_\nu : [0, \infty[\mapsto \mathcal{D}$ of the Cauchy problem* (4.1) *and* (4.2) *converges to a unique limit solution $u : [0, \infty[\mapsto \mathcal{D}$. The map $(\bar{u}, t) \mapsto u(t, \cdot) \doteq S_t\bar{u}$ is a uniformly Lipschitz semigroup, satisfying*

$$S_0\bar{u} = \bar{u}, \qquad S_s(S_t\bar{u}) = S_{s+t}\bar{u}, \tag{5.72}$$

$$\|S_t\bar{u} - S_s\bar{v}\|_{\mathbf{L}^1} \leqslant L\|\bar{u} - \bar{v}\|_{\mathbf{L}^1} + L'|t - s| \quad \textit{for all } \bar{u}, \bar{v} \in \mathcal{D},\ s, t \geqslant 0. \tag{5.73}$$

PROOF. Let $\bar{u} \in \mathcal{D}$ be given. Consider any sequence $\{u_\nu\}_{\nu \geqslant 1}$, such that each u_ν is a front tracking ε_ν-approximate solution of (4.1), with $\varepsilon_\nu \to 0$ and

$$\|u_\nu(0, \cdot) - \bar{u}\|_{\mathbf{L}^1} \leqslant \varepsilon_\nu, \qquad \Upsilon(u_\nu(t)) < \delta_0 \quad \text{for all } t \geqslant 0,\ \nu \geqslant 1.$$

For every $\mu, \nu \geqslant 1$ and $t \geqslant 0$, by (5.4), (5.10) and (5.13), it now follows

$$\begin{aligned}\|u_\mu(t) - u_\nu(t)\|_{\mathbf{L}^1} &\leqslant C_1 \cdot \Phi(u_\mu(t),\ u_\nu(t)) \\ &\leqslant C_1 \cdot [\Phi(u_\mu(0),\ u_\nu(0)) + C_2 t \max\{\varepsilon_\mu,\ \varepsilon_\nu\}] \\ &\leqslant 2C_1^2\|u_\mu(0) - u_\nu(0)\|_{\mathbf{L}^1} + C_1C_2 t \max\{\varepsilon_\mu,\ \varepsilon_\nu\}.\end{aligned} \tag{5.74}$$

Since the right-hand side of (5.74) approaches zero as $\mu, \nu \to \infty$, the sequence is Cauchy and converges to a unique limit.

The semigroup property (5.72) is an immediate consequence of uniqueness. To prove the continuous dependence on the initial data, let $\bar{u}, \bar{v} \in \mathcal{D}$ be given. For each $\nu \geqslant 1$, let u_ν, v_ν be front tracking ε_ν-approximate solutions of (4.1) with

$$\|u_\nu(0) - \bar{u}\|_{\mathbf{L}^1} < \varepsilon_\nu, \qquad \|v_\nu(0) - \bar{v}\|_{\mathbf{L}^1} < \varepsilon_\nu. \tag{5.75}$$

Using again (5.4) and (5.13), we deduce

$$\begin{aligned}\|u_\nu(t) - v_\nu(t)\|_{\mathbf{L}^1} &\leqslant C_1 \cdot \Phi(u_\nu(t),\ v_\nu(t)) \\ &\leqslant C_1 \cdot [\Phi(u_\nu(0), v_\nu(0)) + C_2 t\varepsilon_\nu] \\ &\leqslant 2C_1^2\|u_\nu(0) - v_\nu(0)\|_{\mathbf{L}^1} + C_1C_2 t\varepsilon_\nu.\end{aligned}$$

Letting $\nu \to \infty$, by (5.75) it follows

$$\|u(t) - v(t)\|_{\mathbf{L}^1} \leqslant 2C_1^2 \cdot \|\bar{u} - \bar{v}\|_{\mathbf{L}^1}. \tag{5.76}$$

This establishes the uniform Lipschitz continuity of the semigroup with respect to the initial data. The uniform Lipschitz continuity with respect to time is an immediate consequence of (4.20). □

5.4. *Notes*

The continuous dependence of solutions on the initial data was first proved in [BC1], in the case of 2×2 systems. A construction valid for $n \times n$ was then given in [BCP]. These earlier works rely on a technique based on linearization + homotopy. As a first step, one estimates the distance between a reference solution u and an infinitesimal perturbation (see [B1,B4]). This is achieved by constructing a Lyapunov functional $\widetilde{\Phi}$ which is nonincreasing along all solutions to a linearized system, describing the evolution of a first-order perturbation (see [B1,B4]). In a second step, to compare two solutions u, v, one constructs a one-parameter family of solutions u^θ connecting u with v. For each time t, the distance $\|u(t) - v(t)\|_{\mathbf{L}^1}$ can then be bounded in terms of the length of the curve $t \mapsto u^\theta(t)$. A drawback of this approach comes from the possible loss of regularity of the solutions u^θ. In order to retain the minimal regularity (piecewise Lipschitz continuity) required for the existence of tangent vectors, in [BC1] and [BCP] various approximation and restarting procedures had to be devised. These yield entirely rigorous proofs, but at the price of heavy technicalities.

An entirely different approach was introduced by Liu and Yang [LY2,LY3], defining a functional $\Phi(u, v)$ which is equivalent to the $\mathbf{L}^1$ distance and decreases along couples of solutions of the hyperbolic system. In their construction, a key role is played by a new entropy functional for genuinely nonlinear scalar fields, introduced in [LY1] and here illustrated in Figure 25. This approach was developed into its final form in [BLY]. For solutions generated by the Glimm scheme, an equivalent version of these estimates can be found in [LY4]. Our proof of Theorem 5.1 follows [BLY] with some refinements suggested by Guerra. For yet another proof of continuous dependence, see [HLF] or [LF]. Further results and references in this direction can be found in the monograph [B5].

All of the above results on continuous dependence rely on the assumption (H) of genuine nonlinearity or linear degeneracy of each characteristic field. If this condition fails, the analysis becomes far more difficult. For 2×2 systems, a proof was worked out by Ancona and Marson [AM1]. More recently, the uniform Lipschitz continuous dependence on initial data has been established in [BiB] for all $n \times n$ strictly hyperbolic systems, by a vanishing viscosity approach.

6. Further properties of solutions

This final section collects various results concerning the uniqueness and qualitative properties of entropy weak solutions. For the proofs we mainly refer to [B5].

6.1. *Viscosity solutions and characterization of semigroup trajectories*

We begin by reviewing some properties that characterize the weak solutions obtained as limits of front tracking approximations. In [B3] the author introduced a definition of "viscosity solution" for a system of conservation laws, based on local integral estimates. This definition was fully validated only recently, when in [BiB] it was proved that these viscosity solutions actually coincide with the limits of vanishing viscosity approximations. These results will now be briefly described.

In connection with the hyperbolic system (4.1), we first introduce some notations. Given a function $u = u(t, x)$ and a point (τ, ξ), we denote by $U^{\sharp}_{(u;\tau,\xi)}$ the solution of the Riemann problem with initial data

$$u(0, x) = \begin{cases} u^- & \text{if } x < 0, \\ u^+ & \text{if } x > 0, \end{cases} \tag{6.1}$$

where

$$u^- = \lim_{x \to \xi-} u(\tau, x), \qquad u^+ = \lim_{x \to \xi+} u(\tau, x). \tag{6.2}$$

Assuming that the two states u^+, u^- are sufficiently close, this solution was constructed at (3.22).

In addition, we define $U^{\flat}_{(u;\tau,\xi)}$ as the solution of a linear hyperbolic Cauchy problem with constant coefficients:

$$w_t + \widehat{A} w_x = 0, \qquad w(0, x) = u(\tau, x), \tag{6.3}$$

with $\widehat{A} \doteq Df(u(\tau, \xi))$. Observe that (6.3) is obtained from the quasilinear system

$$u_t + A(u) u_x = 0, \qquad A(u) \doteq Df(u), \tag{6.4}$$

by "freezing" the coefficients of the matrix $A(u)$ at the point (τ, ξ) and choosing $u(\tau)$ as initial data.

The notion of "viscosity solution" is now defined by locally comparing a function u with the self-similar solution of a Riemann problem and with the solution of a linear hyperbolic system with constant coefficients.

DEFINITION 6.1. A function $u = u(t, x)$ is a *viscosity solution* of the system (4.1) if $t \mapsto u(t, \cdot)$ is continuous as a map with values into $\mathbf{L}^1_{\text{loc}}$, and moreover, the following integral estimates hold.

(i) At every point (τ, ξ), for every $\beta' > 0$, one has

$$\lim_{h \to 0+} \frac{1}{h} \int_{\xi - \beta' h}^{\xi + \beta' h} \left| u(\tau + h,\ x) - U^{\sharp}_{(u;\tau,\xi)}(h,\ x - \xi) \right| \mathrm{d}x = 0. \tag{6.5}$$

(ii) There exist constants $C, \beta > 0$ such that, for every $\tau \geqslant 0$ and $a < \xi < b$, one has

$$\limsup_{h \to 0+} \frac{1}{h} \int_{a + \beta h}^{b - \beta h} \left| u(\tau + h,\ x) - U^{\flat}_{(u;\tau,\xi)}(h, x) \right| \mathrm{d}x$$
$$\leqslant C \cdot \left(\text{Tot.Var}\{u(\tau);\ ]a, b[\ \} \right)^2. \tag{6.6}$$

The next result shows that the viscosity solutions coincide precisely with the limits of front tracking approximations.

THEOREM 6.1 (Characterization of semigroup trajectories). *Let $S:\mathcal{D}\times[0,\infty[\mapsto\mathcal{D}$ be the semigroup whose trajectories are the limits of front tracking approximations for the system* (4.1). *A map $t\mapsto u(t,\cdot)$ taking values inside the domain $\mathcal{D}$ is a viscosity solution of* (4.1) *if and only if it coincides with a semigroup trajectory, i.e.,*

$$u(t)=S_t u(0)\quad \text{for all } t\geqslant 0. \tag{6.7}$$

For a proof see [B5], p. 178.

Together with the hyperbolic system (4.1), consider the viscous approximations

$$u_t^\varepsilon+f(u^\varepsilon)_x=\varepsilon u_{xx}^\varepsilon. \tag{6.8}$$

Letting the viscosity parameter $\varepsilon\to 0+$, it is natural to expect that the viscous solutions u^ε should converge to an entropy solution of the hyperbolic system (4.1). This long standing conjecture was indeed proved in [BiB]. We summarize here the main results in this direction.

THEOREM 6.2 (Convergence of vanishing viscosity approximations). *For every $\varepsilon>0$ and every initial data $u(0)=\bar u$ with small total variation, the viscous solutions u^ε of* (6.8) *are well defined, for all $t\geqslant 0$. Their total variation remains uniformly bounded. Moreover, letting $\varepsilon\to 0$, for each $t\geqslant 0$, one has*

$$u(t)=\lim_{\varepsilon\to 0+}u^\varepsilon(t)=S_t\bar u, \tag{6.9}$$

the convergence taking place in $\mathbf{L}^1_{\mathrm{loc}}$. Indeed, the limit coincides with the unique solution of the Cauchy problem (4.1) *and* (4.2) *obtained as limit of front tracking approximations.*

6.2. *Error estimates*

According to Theorem 5.2, the hyperbolic system (4.1) generates a continuous semigroup $S:\mathcal{D}\times[0,\infty[\mapsto\mathcal{D}$, defined on a domain $\mathcal{D}$ of functions with small total variation. Thanks to the uniform Lipschitz continuity of the semigroup, error estimates for front tracking approximations can be derived.

In the following, by an *ε-approximate front tracking solution* of the Cauchy problem (4.1) and (4.2) we mean an approximation $u=u(t,x)$ constructed as in Section 4, such that: $\|u(0,\cdot)-\bar u\|_{\mathbf{L}^1}\leqslant\varepsilon$, each shock front satisfies the Rankine–Hugoniot and entropy conditions, each rarefaction front has strength less than or equal to ε and the total strength of all nonphysical fronts is less than or equal to ε.

THEOREM 6.3 (Error estimates for front tracking approximations). *Let $S:\mathcal{D}\times[0,\infty[\mapsto\mathcal{D}$ be the semigroup generated by the system* (4.1) *as in* (5.72) *and* (5.73). *For some constant C, every ε-approximate front tracking solution $u:[0,\infty[\mapsto\mathcal{D}$ satisfies*

$$\big\|u(\tau,\cdot)-S_\tau\bar u\big\|_{\mathbf{L}^1}=C\varepsilon(1+\tau),\qquad \tau\geqslant 0. \tag{6.10}$$

PROOF. Using (5.73) and Theorem A.2 in the Appendix, we can write

$$\begin{aligned}
&\left\|u(\tau,\cdot) - S_\tau \bar{u}\right\|_{\mathbf{L}^1} \\
&\quad \leqslant \left\|S_\tau u(0,\cdot) - S_\tau \bar{u}\right\|_{\mathbf{L}^1} + \left\|u(\tau,\cdot) - S_\tau u(0,\cdot)\right\|_{\mathbf{L}^1} \\
&\quad \leqslant L\cdot \left\|u(0,\cdot) - \bar{u}\right\|_{\mathbf{L}^1} + L\cdot \int_0^\tau \left\{ \liminf_{h\to 0+} \frac{\|u(t+h) - S_h u(t)\|_{\mathbf{L}^1}}{h} \right\} \mathrm{d}t.
\end{aligned}$$

To prove the theorem it now suffices to show that, for some constant C' independent of ε, one has

$$\lim_{h\to 0+} \frac{\|u(t+h) - S_h u(t)\|_{\mathbf{L}^1}}{h} \leqslant C'\varepsilon \tag{6.11}$$

at each time $t > 0$ where no wave-front interaction takes place. To fix the ideas, let $u(t,\cdot)$ have jumps at points $x_1 < \cdots < x_N$. Call $\mathcal{S}$ the set of indices $\alpha \in \{1,\dots,N\}$ such that $u(t, x_\alpha-)$ and $u(t, x_\alpha+)$ are connected by a shock or by a contact discontinuity. Moreover, call $\mathcal{R}$ the set of indices α corresponding to a rarefaction wave of a genuinely nonlinear family. Finally, call $\mathcal{NP}$ the set of indices α corresponding to nonphysical fronts.

For each α, call σ_α the strength of the jump at x_α and let $\omega_\alpha(\cdot)$ be the self-similar solution of the Riemann problem with data $u^\pm = u(t,\ x_\alpha\pm)$. We observe that, for $h > 0$ small enough, the semigroup trajectory $h \mapsto S_h u(t)$ is a solution of (4.1), obtained by piecing together the solutions of these Riemann problems (Figure 27). Choosing $\rho > 0$ suitably small, we can write

$$\begin{aligned}
&\lim_{h\to 0+} \frac{\|u(t+h) - S_h u(t)\|_{\mathbf{L}^1}}{h} \\
&\quad = \sum_{\alpha\in\mathcal{R}\cup\mathcal{S}\cup\mathcal{NP}} \left(\lim_{h\to 0+} \frac{1}{h} \int_{x_\alpha-\rho}^{x_\alpha+\rho} \left|u(t+h,x) - \omega_\alpha(h, x - x_\alpha)\right| \mathrm{d}x \right) \\
&\quad = \sum_{\alpha\in\mathcal{S}} 0 + \sum_{\alpha\in\mathcal{R}} \mathcal{O}(1)\cdot|\sigma_\alpha|^2 + \sum_{\alpha\in\mathcal{NP}} \mathcal{O}(1)\cdot|\sigma_\alpha| \\
&\quad = \mathcal{O}(1)\cdot\varepsilon.
\end{aligned}$$

Indeed, along shocks the front tracking approximation is exact, while along rarefactions the error is quadratic with respect to the front strength. The last estimate follows from the definition of ε-approximate solution. This establishes (6.11), completing the proof. □

For approximate solutions constructed by the Glimm scheme, the above approach does not yield useful information, because of the additional errors introduced by the restarting procedures at times $t_k \doteq k\Delta t$. However, relying on a careful analysis of Liu [L1], one can

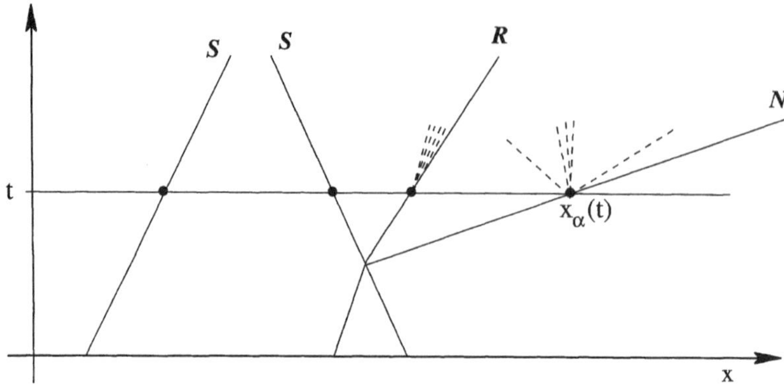

Fig. 27. Local error rate.

construct a front tracking approximate solution having the same initial and terminal values as the Glimm solution. By this technique, in [BM] the authors proved the estimate

$$\lim_{\Delta x \to 0} \frac{\|u^{\text{Glimm}}(\tau, \cdot) - u^{\text{exact}}(\tau, \cdot)\|_{\mathbf{L}^1}}{\sqrt{\Delta x} \cdot |\ln \Delta x|} = 0 \tag{6.12}$$

for every $\tau \geqslant 0$. In other words, letting the mesh sizes $\Delta x, \Delta t \to 0$ while keeping their ratio $\Delta x / \Delta t$ constant, the $\mathbf{L}^1$ norm of the error in the Glimm approximate solution tends to zero at a rate slightly slower than $\sqrt{\Delta x}$.

6.3. *Uniqueness of entropy weak solutions*

For scalar conservation laws, a very general uniqueness and stability result was proved in the fundamental paper of Kruzhkov [K]. In the case of $n \times n$ systems, under some mild regularity assumptions, the uniqueness of entropy weak solutions with small total variation has also been established. For sake of clarity, a complete set of assumptions is listed below.

(A1) (Conservation equations.) The function $u = u(t, x)$ is a weak solution of the Cauchy problem (4.1) and (4.2), taking values within the domain $\mathcal{D}$ of a semigroup S. More precisely, $u : [0, T] \mapsto \mathcal{D}$ is continuous with respect to the $\mathbf{L}^1$ distance. The initial condition (2.2) holds, together with

$$\iint \left(u \varphi_t + f(u) \varphi_x\right) \mathrm{d}x \, \mathrm{d}t = 0 \tag{6.13}$$

for every $\mathcal{C}^1$ function φ with compact support contained inside the open strip $]0, T[\times \mathbb{R}$.

(A2) (Lax entropy condition.) Let u have an approximate jump discontinuity at some point $(\tau, \xi) \in]0, T[\times \mathbb{R}$. More precisely, let there exists states $u^-, u^+ \in \Omega$ and a speed $\lambda \in \mathbb{R}$ such that, calling

$$U(t,x) \doteq \begin{cases} u^- & \text{if } x < \lambda t, \\ u^+ & \text{if } x > \lambda t, \end{cases} \tag{6.14}$$

there holds

$$\lim_{r \to 0+} \frac{1}{r^2} \int_{-r}^{r} \int_{-r}^{r} \left| u(\tau + t,\ \xi + x) - U(t,x) \right| \mathrm{d}x\, \mathrm{d}t = 0. \tag{6.15}$$

Then, for some $i \in \{1, \ldots, n\}$, one has the inequalities

$$\lambda_i(u^-) \geqslant \lambda \geqslant \lambda_i\big(u^+\big). \tag{6.16}$$

(A3) (Tame oscillation condition.) For some constants C, β the following holds. For every point $x \in \mathbb{R}$ and every $t, h > 0$, one has

$$\left| u(t+h, x) - u(t,x) \right| \leqslant C \cdot \text{Tot.Var}\big\{u(t, \cdot); [x - \beta h, x + \beta h]\big\}. \tag{6.17}$$

(A4) (Bounded variation condition.) There exists $\delta > 0$ such that, for every space-like curve $\{t = \tau(x)\}$ with $|\mathrm{d}\tau/\mathrm{d}x| \leqslant \delta$ a.e., the function $x \mapsto u(\tau(x), x)$ has locally bounded variation.

REMARK. The condition (A3) restricts the oscillation of the solution. An equivalent, more intuitive formulation is the following. For some constant $\hat{\lambda}$ larger than all characteristic speeds, given any interval $[a, b]$ and $t \geqslant 0$, the oscillation of u on the triangle (Figure 28) $\Delta \doteq \{(s, y) : s \geqslant t, a + \beta(s - t) < y < b - \beta(s - t)\}$, defined as

$$\text{Osc}\{u; \Delta\} \doteq \sup_{(s,y),(s',y') \in \Delta} \left| u(s,y) - u\big(s', y'\big) \right|,$$

is bounded by a constant multiple of the total variation of $u(t, \cdot)$ on $[a, b]$.

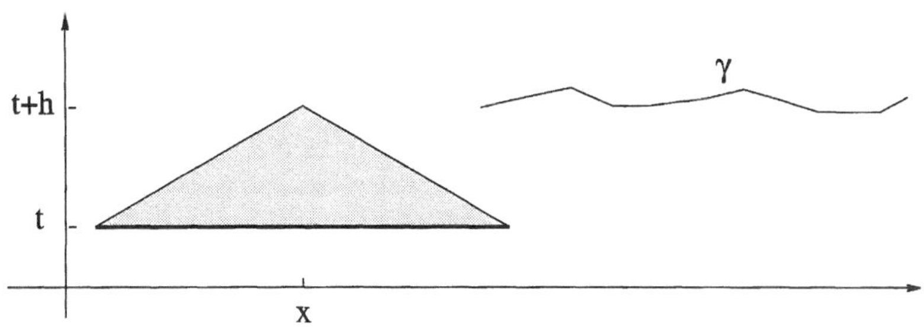

Fig. 28. Geometry of the regularity assumptions.

The assumption (A4) simply requires that, for some fixed $\delta > 0$, the function u has bounded variation along every space-like curve γ as in Figure 28, of the form $\{t = \tau(x);\ x \in [a,b]\}$ with slope $\leqslant \delta$, i.e., with

$$|\tau(x) - \tau(x')| \leqslant \delta |x - x'| \quad \text{for all } x, x' \in [a,b].$$

One can prove that all of the above assumptions are satisfied by weak solutions obtained as limits of front tracking approximations (see [B5], p. 85). The following result shows that the entropy weak solution of the Cauchy problem (4.1) and (4.2) is unique within the class of functions that satisfy either the additional regularity condition (A3) or (A4).

THEOREM 6.4 (Uniqueness of entropy weak solutions). *Let the map $u : [0,T] \mapsto \mathcal{D}$ be continuous (with respect to the $\mathbf{L}^1$ distance), taking values in the domain of the semigroup S generated by the system* (2.1). *If* (A1), (A2) *and* (A3) *hold, then*

$$u(t,\cdot) = S_t \bar{u} \qquad \text{for all } t \in [0,T]. \tag{6.18}$$

In particular, the weak solution that satisfies these conditions is unique. The same conclusion holds if the assumption (A3) *is replaced by* (A4).

The proof relies on the error estimate

$$\|u(\tau) - S_\tau u(0)\|_{\mathbf{L}^1} \leqslant L \int_0^\tau \left\{ \liminf_{h \to 0+} \frac{\|u(t+h) - S_h u(t)\|_{\mathbf{L}^1}}{h} \right\} dt, \tag{6.19}$$

proved in the Appendix. With the above assumptions, one can show that the entropy weak solution u satisfies

$$\lim_{h \to 0+} \frac{\|u(t+h) - S_h u(t)\|_{\mathbf{L}^1}}{h} = 0$$

for almost every $t \in [0,T]$. For details we again refer to [B5], p. 188.

6.4. *Qualitative structure of solutions*

In the following we assume that every characteristic field is genuinely nonlinear. All results refer to solutions of (4.1) with small total variation, obtained as limits of front tracking approximations.

THEOREM 6.5 (Local structure of solutions). *Let $u = u(t,x)$ be a solution of* (1.1). *Fix a point (τ, ξ) and consider the rescaled functions*

$$u^\eta(t,x) \doteq u(\tau + \eta t, \xi + \eta x), \qquad \eta > 0. \tag{6.20}$$

Then, as $\eta \to 0+$, *the functions* u^η *converge in* $\mathbf{L}^1_{\mathrm{loc}}$ *for all* $t \in \mathbb{R}$ *to a self-similar solution* $\tilde{u} : \mathbb{R} \times \mathbb{R} \mapsto \mathbb{R}^n$ *of* (4.1). *On the upper half plane* $\{t \geqslant 0\}$, *the function* $\tilde{u}$ *coincides with the standard solution of the Riemann problem with data* $u^- \doteq u(\tau, \xi-)$ *and* $u^+ \doteq u(\tau, \xi+)$. *On the lower half plane* $\{t < 0\}$, $\tilde{u}$ *contains only negative waves, i.e., shock waves and centered compression waves.*

A proof can be found in [BLF2].

THEOREM 6.6 (Global structure of solutions). *Let* u *be a solution of the Cauchy problem* (4.1) *and* (4.2) *obtained as limit of front tracking approximations. Then there exists a countable set* $\Theta \doteq \{(t_\ell, x_\ell);\ \ell \geqslant 1\}$ *of interaction points and a countable family of Lipschitz continuous curves (shocks or contact discontinuities)* $\Gamma \doteq \{x = y_m(t);\ t \in]a_m, b_m[,\ m \geqslant 1\}$ *such that the following holds. For each* m *and each* $t \in]a_m, b_m[$ *with* $(t, y_m(t)) \notin \Theta$, *there exist the derivative* $\dot{y}_m(t)$ *and the right and left limits*

$$u^- \doteq \lim_{\substack{(s,y)\to(t,y_m(t))\\ y<y_m(s)}} u(s,y), \qquad u^+ \doteq \lim_{\substack{(s,y)\to(t,y_m(t))\\ y>y_m(s)}} u(s,y). \tag{6.21}$$

These limits satisfy the Rankine–Hugoniot equations and the Lax entropy conditions

$$\dot{y}_m\left[u^+ - u^-\right] = \left[f\left(u^+\right) - f\left(u^-\right)\right], \tag{6.22}$$

$$\lambda_i\left(u^+\right) \leqslant \dot{y}_m \leqslant \lambda_i\left(u^-\right) \quad \textit{for some } i \in \{1, \dots, n\}.$$

Moreover, u *is continuous outside the set* $\Theta \cup \Gamma$.

A proof can be found in [B5], p. 219.

6.5. *Notes*

The definition of viscosity solution for hyperbolic systems of conservation laws, based on local integral estimates, was introduced in [B3]. In the same paper the author proved the error estimate (6.10) for front tracking approximations, and the uniqueness of solutions obtained as limits of Glimm approximations. The uniqueness results collected in Theorem 6.4 were obtained in [BLF1,BG] and [BLw]. The qualitative structure of solutions constructed by the Glimm scheme was first studied by DiPerna [DP2] and Liu [L4]. Theorem 6.5 was proved in [BLF2], while Theorem 6.6 is taken from [B5]. A detailed analysis of the large time asymptotic behavior of solutions can be found in the papers of Liu [L2,L3].

Appendix

We collect here some results of basic mathematical analysis, which were used in previous sections.

A.1. *A compactness theorem*

We state below a version of Helly's compactness theorem, which provides the basic tool in the proof of existence of weak solutions.

THEOREM A.1. *Consider a sequence of functions* $u_\nu : [0, \infty[\times \mathbb{R} \mapsto \mathbb{R}^n$ *with the following properties.*

$$\text{Tot.Var}\{u_\nu(t,\cdot)\} \leqslant C, \qquad |u_\nu(t,x)| \leqslant M \quad \textit{for all } t, x, \tag{A.1}$$

$$\int_{-\infty}^{\infty} |u_\nu(t,x) - u_\nu(s,x)|\,\mathrm{d}x \leqslant L|t-s| \quad \textit{for all } t, s \geqslant 0, \tag{A.2}$$

for some constants C, M, L. *Then there exists a subsequence* u_μ *which converges to some function* u *in* $\mathbf{L}^1_{\text{loc}}([0,\infty) \times \mathbb{R}; \mathbb{R}^n)$. *This limit function satisfies*

$$\int_{-\infty}^{\infty} |u(t,x) - u(s,x)|\,\mathrm{d}x \leqslant L|t-s| \quad \textit{for all } t, s \geqslant 0. \tag{A.3}$$

The point values of the limit function u can be uniquely determined by requiring that

$$u(t,x) = u(t,x+) \doteq \lim_{y \to x+} u(t,y) \quad \textit{for all } t, x. \tag{A.4}$$

In this case, one has

$$\text{Tot.Var}\{u(t,\cdot)\} \leqslant C, \qquad |u(t,x)| \leqslant M \quad \textit{for all } t, x. \tag{A.5}$$

A detailed proof can be found in [B5], p. 16.

A.2. *An elementary error estimate*

Let $\mathcal{D}$ be a closed subset of a Banach space E and consider a Lipschitz continuous semigroup $S : \mathcal{D} \times [0, \infty[\mapsto \mathcal{D}$. More precisely, assume that:

(i) $S_0 u = u$, $S_s S_t u = S_{s+t} u$;

(ii) $\|S_t u - S_s v\| \leqslant L \cdot \|u - v\| + L' \cdot |t - s|$.

Given a Lipschitz continuous map $w : [0, T] \mapsto \mathcal{D}$, we wish to estimate the difference between w and the trajectory of the semigroup S starting at $w(0)$.

THEOREM A.2. *Let* $S : \mathcal{D} \times [0, \infty[\mapsto \mathcal{D}$ *be a continuous flow satisfying the properties* (i) *and* (ii). *For every Lipschitz continuous map* $w : [0, T] \mapsto \mathcal{D}$, *one then has the estimate*

$$\|w(T) - S_T w(0)\| \leqslant L \cdot \int_0^T \left\{ \liminf_{h \to 0+} \frac{\|w(t+h) - S_h w(t)\|}{h} \right\} \mathrm{d}t. \tag{A.6}$$

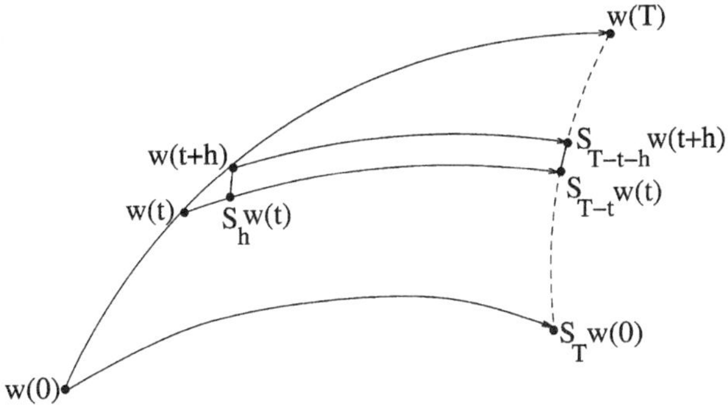

Fig. 29. The basic error estimates.

REMARK 1. The integrand in (A.6) can be regarded as the instantaneous error rate for w at time t. Since the flow is uniformly Lipschitz continuous, during the time interval $[t, T]$ this error is amplified at most by a factor L (see Figure 29).

PROOF OF THEOREM A.2. We first show that the integrand function

$$\phi(t) \doteq \liminf_{h\to 0+} \frac{\|w(t+h) - S_h w(t)\|}{h}$$

is bounded and measurable. Indeed, for every $h > 0$, the function

$$\phi_h(t) \doteq \frac{\|w(t+h) - S_h w(t)\|}{h} \tag{A.7}$$

is continuous. By the continuity of the maps $h \mapsto \phi_h(t)$, we have

$$\phi(t) = \lim_{\varepsilon\to 0+} \inf_{h\in\mathbf{Q}\cap]0,\varepsilon]} \phi_h(t),$$

where the infimum is taken only over rational values of h. Hence the function ϕ is Borel measurable. Its boundedness is a consequence of the uniform Lipschitz continuity of w and of all trajectories of the semigroup S.

Next, define

$$\Phi(t) \doteq \left\| S_{T-t} w(t) - S_T w(0) \right\|, \tag{A.8}$$

$$z(t) \doteq \Phi(t) - L\int_0^t \phi(s)\,\mathrm{d}s. \tag{A.9}$$

We need to show that $z(T) \leqslant 0$. Since z is Lipschitz continuous and $z(0) = 0$, it suffices to show that the time derivative satisfies $\dot z(s) \leqslant 0$ for almost every $s \in [0, T]$. By the theorems of Rademacher and of Lebesgue, there exists a null set $\mathcal{N} \subset [0, T]$ such that, for each

$s \notin \mathcal{N}$, both functions z, Φ are differentiable at time s while ϕ is quasicontinuous at s. For $s \notin \mathcal{N}$ we thus have

$$\dot{z}(s) = \lim_{h\to 0} \frac{z(s+h) - z(s)}{h} = \lim_{h\to 0} \frac{\Phi(s+h) - \Phi(s)}{h} - L \cdot \phi(s). \tag{A.10}$$

By the definition of Φ and the properties of S it follows

$$\begin{aligned}\Phi(s+h) - \Phi(s) &= \|S_{T-s-h}w(s+h) - S_T w(0)\| - \|S_{T-s}w(s) - S_T w(0)\| \\ &\leqslant \|S_{T-s-h}w(s+h) - S_{T-s}w(s)\| \\ &\leqslant \|S_{T-s-h}w(s+h) - S_{T-s-h}S_h w(s)\| \\ &\leqslant L \cdot \|w(s+h) - S_h w(s)\|.\end{aligned}$$

Therefore

$$\lim_{h\to 0+} \frac{\Phi(s+h) - \Phi(s)}{h} \leqslant L \cdot \liminf_{h\to 0+} \frac{\|w(s+h) - S_h w(s)\|}{h} = L \cdot \phi(s).$$

By (A.10) this implies

$$\dot{z}(s) \leqslant 0 \quad \text{for all } s \notin \mathcal{N}.$$

Hence, $z(T) \leqslant 0$, proving the theorem. □

A.3. *Taylor estimates*

This final section contains some useful bounds on the size of a function Ψ of several variables, valid in a neighborhood of the origin. The assumption that Ψ vanishes on certain sets, such as the coordinate axes or the diagonals, implies that some of the Taylor coefficients of Ψ at the origin are equal to zero. Writing a Taylor approximation and using this additional information, one obtains the desired estimates. As usual, the Landau notation $\mathcal{O}(1)$ here indicates a function whose absolute value remains uniformly bounded.

LEMMA A.1. *Let $\Phi : \mathbb{R}^m \mapsto \mathbb{R}^n$ be differentiable with Lipschitz continuous derivative. If*

$$\Phi(0) = 0, \qquad \frac{\partial \Phi}{\partial x}(0) = 0, \tag{A.11}$$

then one has

$$\Phi(x) = \mathcal{O}(1) \cdot |x|^2. \tag{A.12}$$

LEMMA A.2. *Let $\Phi : \mathbb{R}^p \times \mathbb{R}^q \mapsto \mathbb{R}^n$ be twice differentiable, with Lipschitz continuous second derivatives. If*

$$\Phi(x,0) = \Phi(0,y) = 0 \quad \text{for all } x \in \mathbb{R}^p,\ y \in \mathbb{R}^q, \tag{A.13}$$

then

$$\Phi(x,y) = \mathcal{O}(1) \cdot |x||y| \tag{A.14}$$

for all x, y in a neighborhood of the origin. If in addition there holds

$$\frac{\partial^2 \Phi}{\partial x\, \partial y}(0,0) = 0, \tag{A.15}$$

then one has the sharper estimate

$$\Phi(x,y) = \mathcal{O}(1) \cdot |x||y|\big(|x| + |y|\big). \tag{A.16}$$

PROOF. Observe that (A.13) implies

$$\begin{aligned}\Phi(x,y) &= \int_0^1 \left(\frac{\partial \Phi}{\partial y}(x, \theta y)\right) y \, \mathrm{d}\theta \\ &= \int_0^1 \int_0^1 \left(\frac{\partial^2 \Phi}{\partial x\, \partial y}(\theta' x,\ \theta y)\right)(x \otimes y)\, \mathrm{d}\theta' \, \mathrm{d}\theta \\ &= \mathcal{O}(1) \cdot |x||y|. \end{aligned} \tag{A.17}$$

This gives (A.14). If (A.15) also holds, then the Lipschitz continuity of the second derivative implies

$$\left| \frac{\partial^2 \Phi}{\partial x\, \partial y}(x,y) \right| = \mathcal{O}(1) \cdot \big(|x| + |y|\big).$$

Using this bound inside the double integral in (A.17), we obtain (A.16). □

LEMMA A.3. *Let $\Phi = \Phi(\tilde{q}, q^*, \sigma)$ be a $\mathcal{C}^2$ mapping from $\mathbb{R}^{n-1} \times \mathbb{R} \times \mathbb{R}$ into $\mathbb{R}$, with Lipschitz continuous second derivatives. Assume that, for all $\tilde{q} \in \mathbb{R}^{n-1}$, $s, q^*, \sigma \in \mathbb{R}$ there holds*

$$\Phi\big(\tilde{q}, q^*, 0\big) = \Phi(0, s, -s) = \Phi(0,0,\sigma) = 0. \tag{A.18}$$

Then one has the estimate

$$\Phi\big(\tilde{q}, q^*, \sigma\big) = \mathcal{O}(1) \cdot \big(|\tilde{q}||\sigma| + |q^*||\sigma||q^* + \sigma|\big). \tag{A.19}$$

If, in addition, for all q^*, σ there holds

$$\Phi(0, q^*, \sigma) = 0, \tag{A.20}$$

then one has

$$\Phi(\tilde{q}, q^*, \sigma) = \mathcal{O}(1) \cdot |\tilde{q}||\sigma|. \tag{A.21}$$

PROOF. To derive (2.19) we write

$$\Phi(\tilde{q}, q^*, \sigma) = \Phi(0, q^*, \sigma) + [\Phi(\tilde{q}, q^*, \sigma) - \Phi(0, q^*, \sigma)]. \tag{A.22}$$

By (A.18), for all $\tilde{q}, q^*$ one has

$$\frac{\partial \Phi}{\partial \tilde{q}}(\tilde{q}, q^*, 0) = 0.$$

Hence, by the Lipschitz continuity of the first derivative,

$$\frac{\partial \Phi}{\partial \tilde{q}}(\tilde{q}, q^*, \sigma) = \mathcal{O}(1) \cdot |\sigma|.$$

Using this last estimate we obtain

$$\begin{aligned} \Phi(\tilde{q}, q^*, \sigma) - \Phi(0, q^*, \sigma) &= \int_0^1 \left(\frac{\partial \Phi}{\partial \tilde{q}}(\theta\tilde{q}, q^*, \sigma)\right)\tilde{q}\, \mathrm{d}\theta \\ &= \mathcal{O}(1) \cdot |\tilde{q}||\sigma|. \end{aligned} \tag{A.23}$$

To establish (A.19) it thus remains to prove

$$\Phi(0, q^*, \sigma) = \mathcal{O}(1) \cdot |q^*||\sigma||q^* + \sigma|. \tag{A.24}$$

Observe that (A.18) implies

$$\frac{\partial^2 \Phi}{\partial q^* \, \partial \sigma}(0, 0, 0) = 0.$$

In the case where q^* and σ have the same sign, one has $|q^* + \sigma| = |q^*| + |\sigma|$. Therefore (A.24) follows from Lemma A.2, being equivalent to the estimate (A.16).

To see what happens when q^* and σ have opposite signs, to fix the ideas assume $q^* < 0 < \sigma$, $|q^*| < \sigma$, the other cases being entirely similar. We consider two possibilities:

If $|q^*| < \sigma/2$, then

$$|q^* + \sigma| \geqslant \frac{|\sigma|}{2} \geqslant \frac{|q^*| + |\sigma|}{3}$$

and (A.24) again follows from (A.16) in Lemma A.2.

In the remaining case where $\sigma/2 \leqslant |q^*| \leqslant \sigma$, we write

$$\begin{aligned}
\Phi(0,q^*,\sigma) &= \Phi(0,-\sigma,\sigma) + \int_{-\sigma}^{q^*} \frac{\partial \Phi}{\partial q^*}(0,s,\sigma)\,ds \\
&= 0 + \int_{-\sigma}^{q^*}\int_0^{\sigma} \frac{\partial^2 \Phi}{\partial q^*\,\partial\sigma}(0,s,s')\,ds\,ds' \\
&= \int_{-\sigma}^{q^*}\int_0^{\sigma} \mathcal{O}(1)\cdot\big(|s|+|s'|\big)\,ds\,ds' \\
&= \mathcal{O}(1)\cdot\big|\sigma+q^*\big||\sigma|\big(|\sigma|+|\sigma|\big).
\end{aligned}$$

This yields again (A.24), completing the proof of (A.19).

If the additional assumption (A.20) holds, then the estimate (A.21) is an immediate consequence of (A.22) and (A.23). □

LEMMA A.4. *Let $\Phi = \Phi(\tilde q, q^*, \sigma)$ be a C^1 mapping from $\mathbb{R}^{n-1}\times\mathbb{R}\times\mathbb{R}$ into $\mathbb{R}$, with Lipschitz continuous derivatives.*

(a) *If Φ satisfies the assumptions*

$$\frac{\partial \Phi}{\partial q^*}(0,0,0) = \frac{\partial \Phi}{\partial \sigma}(0,0,0) = \frac{1}{2}, \qquad \Phi(0,s,-s) = 0 \quad \textit{for all } s, \tag{A.25}$$

then one has the estimate

$$\Phi(\tilde q, q^*, \sigma) = \frac{q^*+\sigma}{2} + \mathcal{O}(1)\cdot\big(|\tilde q| + |q^*+\sigma|\big(|q^*|+|\sigma|\big)\big). \tag{A.26}$$

(b) *If instead Φ satisfies the assumptions*

$$\frac{\partial \Phi}{\partial q^*}(0,0,0) = \frac{1}{2}, \qquad \Phi(0,0,\sigma) = 0 \quad \textit{for all } \sigma, \tag{A.27}$$

then one has the estimate

$$\Phi(\tilde q, q^*, \sigma) = \frac{q^*}{2} + \mathcal{O}(1)\cdot\big(|\tilde q| + |q^*|\big(|q^*|+|\sigma|\big)\big). \tag{A.28}$$

PROOF. We write Φ in the form

$$\Phi(\tilde q, q^*, \sigma) = \big[\Phi(\tilde q, q^*, \sigma) - \Phi(0, q^*, \sigma)\big] + \Phi(0, q^*, \sigma).$$

The first term on the right-hand side clearly satisfies

$$\Phi(\tilde q, q^*, \sigma) - \Phi(0, q^*, \sigma) = \mathcal{O}(1)\cdot|\tilde q|. \tag{A.29}$$

To estimate the second term, in case (a) we define $s \doteq (q^* - \sigma)/2$ and compute

$$\begin{aligned}\Phi(0, q^*, \sigma) &= \Phi(0, s, -s) + \int_0^{(q^*+\sigma)/2} \left[\frac{\mathrm{d}}{\mathrm{d}\theta}\Phi(0, s+\theta, -s+\theta)\right] \mathrm{d}\theta \\ &= \int_0^{(q^*+\sigma)/2} \left[1 + \mathcal{O}(1) \cdot \left(|q^*| + |\sigma|\right)\right] \mathrm{d}\theta \\ &= \frac{q^* + \sigma}{2} + \mathcal{O}(1) \cdot |q^* + \sigma| \left(|q^*| + |\sigma|\right). \qquad \text{(A.30)}\end{aligned}$$

Combining (A.29) with (A.30) one obtains (A.26).

In case (b), we write

$$\begin{aligned}\Phi(0, q^*, \sigma) &= \int_0^{q^*} \left[\frac{\mathrm{d}}{\mathrm{d}\theta}\Phi(0, \theta, \sigma)\right] \mathrm{d}\theta \\ &= \int_0^{q^*} \left[\frac{1}{2} + \mathcal{O}(1) \cdot \left(|q^*| + |\sigma|\right)\right] \mathrm{d}\theta \\ &= \frac{q^*}{2} + \mathcal{O}(1) \cdot |q^*| \left(|q^*| + |\sigma|\right). \qquad \text{(A.31)}\end{aligned}$$

Combining (A.29) with (A.31) one obtains (A.28). □

REMARK 2. In the above lemmas, assume that the functions Φ and their derivatives depend continuously on an additional parameter ω and satisfy the given assumptions for all values of ω. Then all of the above estimates are still valid, for quantities $\mathcal{O}(1)$ which remain uniformly bounded as ω ranges on compact sets.

References

[Al] H.D. Alber, *A local existence theorem for quasilinear wave equations with initial values of bounded variation*, Lecture Notes in Math., Vol. 1151, Springer-Verlag, New York (1985), 9–24.

[AM1] F. Ancona and A. Marson, *Well posedness for general* 2×2 *systems of conservation laws*, Mem. Amer. Math. Soc. **801** (2004).

[AM2] F. Ancona and A. Marson, *Front tracking solutions for general systems of conservation laws*, to appear.

[BaJ] P. Baiti and H.K. Jenssen, *On the front tracking algorithm*, J. Math. Anal. Appl. **217** (1998), 395–404.

[BiB] S. Bianchini and A. Bressan, *Vanishing viscosity solutions of nonlinear hyperbolic systems*, Ann. of Math., to appear.

[B1] A. Bressan, *Contractive metrics for nonlinear hyperbolic systems*, Indiana Univ. Math. J. **37** (1988), 409–421.

[B2] A. Bressan, *Global solutions of systems of conservation laws by wave-front tracking*, J. Math. Anal. Appl. **170** (1992), 414–432.

[B3] A. Bressan, *The unique limit of the Glimm scheme*, Arch. Ration. Mech. Anal. **130** (1995), 205–230.

[B4] A. Bressan, *A locally contractive metric for systems of conservation laws*, Ann. Scuola Norm. Sup. Pisa Cl. Sci. (4) **22** (1995), 109–135.

[B5] A. Bressan, *Hyperbolic Systems of Conservation Laws. The One Dimensional Cauchy Problem*, Oxford University Press (2000).

[BC1] A. Bressan and R.M. Colombo, *The semigroup generated by* 2×2 *conservation laws*, Arch. Ration. Mech. Anal. **133** (1995), 1–75.
[BC2] A. Bressan and R.M. Colombo, *Unique solutions of* 2×2 *conservation laws with large data*, Indiana Univ. Math. J. **44** (1995), 677–725.
[BCP] A. Bressan, G. Crasta and B. Piccoli, *Well posedness of the Cauchy problem for* $n \times n$ *systems of conservation laws*, Mem. Amer. Math. Soc. **694** (2000).
[BG] A. Bressan and P. Goatin, *Oleinik type estimates and uniqueness for* $n \times n$ *conservation laws*, J. Differential Equations **156** (1999), 26–49.
[BLF1] A. Bressan and P. LeFloch, *Uniqueness of weak solutions to hyperbolic systems of conservation laws*, Arch. Ration. Mech. Anal. **140** (1997), 301–317.
[BLF2] A. Bressan and P. LeFloch, *Structural stability and regularity of entropy solutions to hyperbolic systems of conservation laws*, Indiana Univ. Math. J. **48** (1999), 43–84.
[BLw] A. Bressan and M. Lewicka, *A uniqueness condition for hyperbolic systems of conservation laws*, Discrete Contin. Dyn. Syst. **6** (2000), 673–682.
[BLY] A. Bressan, T.P. Liu and T. Yang, L^1 *stability estimates for* $n \times n$ *conservation laws*, Arch. Ration. Mech. Anal. **149** (1999), 1–22.
[BM] A. Bressan and A. Marson, *Error bounds for a deterministic version of the Glimm scheme*, Arch. Ration. Mech. Anal. **142** (1998), 155–176.
[D1] C. Dafermos, *Polygonal approximations of solutions of the initial value problem for a conservation law*, J. Math. Anal. Appl. **38** (1972), 33–41.
[D2] C. Dafermos, *Hyperbolic Conservation Laws in Continuum Physics*, Springer-Verlag, Berlin (1999).
[DP1] R.J. DiPerna, *Global existence of solutions to nonlinear hyperbolic systems of conservation laws*, J. Differential Equations **20** (1976), 187–212.
[DP2] R.J. DiPerna, *Singularities of solutions of nonlinear hyperbolic systems of conservation laws*, Arch. Ration. Mech. Anal. **60** (1975), 75–100.
[G] J. Glimm, *Solutions in the large for nonlinear hyperbolic systems of equations*, Comm. Pure Appl. Math. **18** (1965), 697–715.
[HR] H. Holden and N.H. Risebro, *Front Tracking for Hyperbolic Systems of Conservation Laws*, Springer-Verlag, New York (2002).
[HLF] J. Hu and P.G. LeFloch, L^1 *continuous dependence for systems of conservation laws*, Arch. Ration. Mech. Anal. **151** (2000), 45–93.
[Je] H.K. Jenssen, *Blowup for systems of conservation laws*, SIAM J. Math. Anal. **31** (2000), 894–908.
[J] F. John, *Formation of singularities in one-dimensional nonlinear wave propagation*, Comm. Pure Appl. Math. **27** (1974), 377–405.
[K] S. Kruzhkov, *First-order quasilinear equations with several space variables*, Mat. Sb. **123** (1970), 228–255; English transl.: Math. USSR-Sb. **10** (1970), 217–273.
[Lx] P.D. Lax, *Hyperbolic systems of conservation laws II*, Comm. Pure Appl. Math. **10** (1957), 537–566.
[LF] P.G. LeFloch, *Hyperbolic Systems of Conservation Laws*, Lecture Notes in Mathematics, ETH Zürich, Birkhäuser, Basel (2002).
[Ln] L. Lin, *On the vacuum state for the equations of isentropic gas dynamics*, J. Math. Anal. Appl. **121** (1987), 406–425.
[L1] T.P. Liu, *The deterministic version of the Glimm scheme*, Comm. Math. Phys. **57** (1977), 135–148.
[L2] T.P. Liu, *Decay to N-waves of solutions of general systems of nonlinear hyperbolic conservation laws*, Comm. Pure Appl. Math. **30** (1977), 585–610.
[L3] T.P. Liu, *Linear and nonlinear large-time behavior of solutions of general systems of hyperbolic conservation laws*, Comm. Pure Appl. Math. **30** (1977), 767–796.
[L4] T.P. Liu, *Admissible solutions of hyperbolic conservation laws*, Mem. Amer. Math. Soc. **240** (1981).
[LY1] T.P. Liu and T. Yang, *A new entropy functional for scalar conservation laws*, Comm. Pure Appl. Math. **52** (1999), 1427–1442.
[LY2] T.P. Liu and T. Yang, L^1 *stability of conservation laws with coinciding Hugoniot and characteristic curves*, Indiana Univ. Math. J. **48** (1999), 237–247.
[LY3] T.P. Liu and T. Yang, L^1 *stability for* 2×2 *systems of hyperbolic conservation laws*, J. Amer. Math. Soc. **12** (1999), 729–774.

[LY4] T.P. Liu and T. Yang, *Well posedness theory for hyperbolic conservation laws*, Comm. Pure Appl. Math. **52** (1999), 1553–1586.

[Rm] B. Riemann, *Über die Fortpflanzung ebener Luftwellen von endlicher Schwingungsweite*, Göttingen Abh. Math. Cl. **8** (1860), 43–65.

[Ri] N.H. Risebro, *A front-tracking alternative to the random choice method*, Proc. Amer. Math. Soc. **117** (1993), 1125–1139.

[Sc1] S. Schochet, *Sufficient conditions for local existence via Glimm's scheme for large BV data*, J. Differential Equations **89** (1991), 317–354.

[Sc2] S. Schochet, *The essence of Glimm's scheme*, Nonlinear Evolutionary Partial Differential Equations, X. Ding and T.P. Liu, eds, Amer. Math. Soc.–International Press (1997), 355–362.

[Se] D. Serre, *Systems of Conservation Laws*, Vols 1, 2, Cambridge University Press (2000).

[Sm] J. Smoller, *Shock Waves and Reaction–Diffusion Equations*, Springer-Verlag, New York (1983).

[V] A.I. Volpert, *The spaces BV and quasilinear equations*, Math. USSR-Sb. **2** (1967), 225–267.

[W] B. Wendroff, *An analysis of front tracking for chromatography*, Acta Appl. Math. **30** (1993), 265–285.

CHAPTER 3

Current Issues on Singular and Degenerate Evolution Equations

Emmanuele DiBenedetto

Department of Mathematics, Vanderbilt University, 1326 Stevenson Center, Nashville, TN 37240-0001, USA

José Miguel Urbano

Departamento de Matemática, Universidade de Coimbra, 3001-454 Coimbra, Portugal

Vincenzo Vespri

Dipartimento di Matematica, Università di Firenze, viale Morgagni 67a, 50134 Firenze, Italy

Contents

1. Introduction 171
   1.1. Historical background on regularity and Harnack estimates 174
   1.2. A new approach to regularity 176
   1.3. Limiting cases and miscellanea remarks 177
   1.4. Singular equations of the Stefan-type 177
   1.5. Outline of these notes 178
2. Regularity of weak solutions 179
   2.1. Weak solutions and local estimates 180
   2.2. The classical approach of De Giorgi 185
   2.3. The degenerate case $p > 2$ 187
   2.4. The singular case $1 < p < 2$ 209
   2.5. The porous medium equation and other generalizations 224
3. Boundedness of weak solutions 225
   3.1. The degenerate case $p > 2$ 226
   3.2. The singular case $1 < p < 2$ 228
4. Intrinsic Harnack estimates 230
   4.1. Harnack estimates: the degenerate case 236
   4.2. Harnack estimates: the singular case 240

HANDBOOK OF DIFFERENTIAL EQUATIONS
Evolutionary Equations, volume 1
Edited by C.M. Dafermos and E. Feireisl

4.3. Elliptic-type Harnack estimates and extinction time . . . 242
4.4. Raleigh quotient and extinction profile . . . 249
5. Stefan-like problems . . . 253
5.1. The continuity of weak solutions . . . 255
5.2. A bridge between singular and degenerate equations . . . 257
5.3. Parabolic equations with one-point singularity . . . 259
5.4. Parabolic equations with multiple singularities . . . 261
Acknowledgments . . . 279
References . . . 279

1. Introduction

Let Ω be an open set in $\mathbb{R}^N$ and consider the quasilinear parabolic partial differential equation of the second order

$$\begin{cases} u \in L^\infty_{\mathrm{loc}}\big(0,T; W^{1,p}_{\mathrm{loc}}(\Omega)\big), & p>1, \\ u_t - \operatorname{div}\mathbf{A}(x,t,u,\nabla u) = B(x,t,u,\nabla u) & \text{weakly in } \Omega_T. \end{cases} \tag{1.1$_p$}$$

Here $T>0$ is given, $\Omega_T = \Omega \times (0,T)$ and ∇ denotes the gradient with respect to the space variables $x=(x_1,\dots,x_N)$. The functions $\mathbf{A}=(A_1,\dots,A_N)$ and B are real valued, measurable with respect to their arguments, and satisfying the structure conditions

$$\begin{cases} C_0|\nabla u|^{p-2}|\nabla u|^2 - C \leqslant \mathbf{A}(x,t,u,\nabla u)\cdot\nabla u, \\ \big|\mathbf{A}(x,t,u,\nabla u)\big| + \big|B(x,t,u,\nabla u)\big| \leqslant C\big(1+|\nabla u|^{p-1}\big), \end{cases} \tag{1.2$_p$}$$

where C_0 and C are given positive constants. The quantity $C_0|\nabla u|^{p-2}$ is the modulus of ellipticity of the equation. If $p>2$ it vanishes whenever $|\nabla u|=0$ and the equation is said to be *degenerate* at those $(x,t)\in\Omega_T$, where this occurs. If $1<p<2$ the modulus of ellipticity becomes infinity whenever $|\nabla u|=0$ and the equation is said to be *singular* at those $(x,t)\in\Omega_T$, where $|\nabla u|=0$.

Along with $(1.1)_p$–$(1.2)_p$, consider also the quasilinear equation,

$$\begin{cases} u \in L^\infty_{\mathrm{loc}}\big(0,T; W^{1,2}_{\mathrm{loc}}(\Omega)\big), \\ u_t - \operatorname{div}\mathbf{A}(x,t,u,\nabla u) = B(x,t,u,\nabla u) \quad \text{weakly in } \Omega_T, \end{cases} \tag{1.1$_m$}$$

with structure conditions,

$$\begin{cases} C_0|u|^{m-1}|\nabla u|^2 - C \leqslant \mathbf{A}(x,t,u,\nabla u)\cdot\nabla u, \\ \big|\mathbf{A}(x,t,u,\nabla u)\big| + \big|B(x,t,u,\nabla u)\big| \leqslant C|u|^{m-1}\big(1+|\nabla u|\big), \end{cases} \tag{1.2$_m$}$$

where m is a given positive number. The prototype example of $(1.1)_m$–$(1.2)_m$ is

$$u_t - \Delta|u|^{m-1}u = 0 \quad \text{for some } m>0, \tag{1.3$_m$}$$

and the prototype example of $(1.1)_p$–$(1.2)_p$ is

$$u_t - \operatorname{div}|\nabla u|^{p-2}\nabla u = 0 \quad \text{for some } p>1. \tag{1.3$_p$}$$

The first is called the *porous medium* equation. If $m>1$ the modulus of ellipticity vanishes for $u=0$ and the equation is degenerate at those points of Ω_T where the solution u vanishes. If $0<m<1$ the modulus of ellipticity is infinity whenever $u=0$ and the equation is *singular* at those points of Ω_T where $u=0$. If $m>1$ the equation is referred to as the *slow diffusion* case of the porous medium equation. The case $0<m<1$ is the *fast diffusion*.

The equation in $(1.3)_p$ is the p-Laplacian equation and its modulus of ellipticity is $|\nabla u|^{p-2}$. If $p>2$ such a modulus vanishes whenever $|\nabla u|=0$ and at such points the equation is degenerate. If $1<p<2$ the modulus of ellipticity becomes infinity whenever $|\nabla u|=0$ and the equation is singular at these points.

If $m=1$ in $(1.3)_m$ or $p=2$ in $(1.3)_p$ one recovers the classical heat equation for which information are essentially encoded in the fundamental solution

$$\Gamma(x,y;t,\tau)=\frac{1}{(t-\tau)^{N/2}}\exp\left\{-\frac{|x-y|^2}{4(t-\tau)}\right\},\quad x,y\in\mathbb{R}^N,t>\tau. \tag{1.4}$$

The porous medium equation in $(1.3)_m$ admits an explicit similarity solution that "resembles" the fundamental solution of the heat equation. Such a solution is called the *Barenblatt* similarity solution and it is given by [17]

$$\Gamma_m(x,y;t,\tau)=\frac{1}{(t-\tau)^{N/\lambda}}\left\{1-\gamma_m\left[\frac{|x-y|}{(t-\tau)^{1/\lambda}}\right]^2\right\}_+^{1/(m-1)},\quad t>\tau, \qquad (1.4)_m$$

where, for a real number α, $\{\alpha\}_+=\max\{\alpha;0\}$, and

$$\lambda=N(m-1)+2,\qquad \gamma_m=\frac{1}{\lambda}\frac{m-1}{2}. \qquad (1.4)'_m$$

An examination of this solution reveals that it is well defined for all positive values of m for which $\lambda>0$. One also verifies that $\Gamma_m\to\Gamma$ as $m\to 1$. In this sense Γ_m is the *fundamental solution* of the porous medium equation.

Also the p-Laplacian equation $(1.3)_p$ admits explicit similarity solutions,

$$\Gamma_p(x,y;t,\tau)=\frac{1}{(t-\tau)^{N/\kappa}}\left\{1-\gamma_p\left[\frac{|x-y|}{(t-\tau)^{1/\kappa}}\right]^{p/(p-1)}\right\}_+^{(p-1)/(p-2)},\quad t>\tau, \qquad (1.4)_p$$

where

$$\kappa=N(p-2)+p,\qquad \gamma_p=\left(\frac{1}{\kappa}\right)^{1/(p-1)}\frac{p-2}{p}. \qquad (1.4)'_p$$

This is well defined for all $p>1$ such that $\kappa>0$ and $\Gamma_p\to\Gamma$ as $p\to 2$. In this sense, Γ_p is the *fundamental solution* of the p-Laplacian equation.

Issues of compact support and regularity. Assume first $m>1$ and $p>2$. The first difference between these *fundamental solutions* and the fundamental solution of the heat equation is in their support with respect to the space variables. For fixed $t>\tau$ and $y\in\mathbb{R}^N$, the functions $x\mapsto\Gamma_m(x),\Gamma_p(x)$ are compactly supported in $\mathbb{R}^N$, whereas $x\mapsto\Gamma(x)$ is positive in the whole $\mathbb{R}^N$. The *moving boundaries* separating the regions where Γ_m and Γ_p are positive from the regions where they vanish are,

$$\begin{aligned} (t-\tau)^{2/\lambda} &= \gamma_m |x-y|^2 && \text{for } \Gamma_m, \\ (t-\tau)^{p/\kappa} &= \gamma_p^{p-1} |x-y|^p && \text{for } \Gamma_p, \end{aligned} \qquad \tau \in \mathbb{R},\ y \in \mathbb{R}^N \text{ fixed.}$$

The second difference is in their degree of regularity. For fixed $t > \tau$ and $y \in \mathbb{R}^N$, the function $x \mapsto \Gamma_m(x)$ is Hölder continuous with Hölder exponent $1/(m-1)$. The function $x \mapsto \Gamma_p(x)$ is differentiable and its partial derivatives are Hölder continuous with Hölder exponent $1/(p-2)$. Such a modest degree of regularity is in contrast to the fundamental solution of the heat equation which is analytic in the space variables.

Let now $0 < m < 1$ by keeping $\lambda > 0$. Then Γ_m is positive and locally analytic in the whole $\mathbb{R}^N \times \{t > \tau\}$. Thus Γ_m seems to share the same properties as Γ.

If $1 < p < 2$ by keeping $\kappa > 0$ then Γ_p is positive in the whole $\mathbb{R}^N \times \{t > \tau\}$ but still it maintains a limited degree of regularity.

Issues of Harnack inequalities. Nonnegative, local solutions of the heat equation in Ω_T satisfy the Harnack inequality. This is a celebrated result of Hadamard [89] and Pini [144] and it takes the following form. Fix $(x_0, t_0) \in \Omega_T$ and for $\rho > 0$ consider the ball $B_\rho(x_0)$ centered at x_0 and with radius ρ and the cylindrical domain

$$Q_\rho(x_0, t_0) = B_\rho(x_0) \times \left(t_0 - \rho^2, t_0 + \rho^2\right). \tag{1.5}$$

These are the *parabolic cylinders* associated with the heat equation and also to $(1.3)_m$ since these equations remain unchanged under a similarity transformation of the space–time variables that keeps constant the ratio $|x|^2/t$.

There exists a constant C depending only upon N and independent of (x_0, t_0) and ρ, such that

$$Cu(x_0, t_0) \geqslant \sup_{B_\rho(x_0)} u\left(x_0, t_0 - \rho^2\right) \quad \text{provided} \quad Q_{2\rho}(x_0, t_0) \subset \Omega_T. \tag{1.6}$$

The proof is based on local representations by means of the heat potentials (1.4). In particular, for fixed $\tau \in \mathbb{R}$ and $y \in \mathbb{R}^N$, the fundamental solution $(x, t) \mapsto \Gamma(x, t)$ satisfies such a Harnack estimate. It is then natural to ask whether the *fundamental solution* $(1.4)_m$ would satisfy a Harnack estimate. Take for example the Γ_m for $\tau = 0$ and y the origin of $\mathbb{R}^N$. Assume first that $m > 1$ and fix (x_0, t_0) on the moving boundary, so that $\Gamma_m(x_0, t_0) = 0$. If ρ and $|x_0|$ are sufficiently large, the ball $B_\rho(x_0)$ intersects the support of Γ_m at $t = t_0 - \rho^2$. Therefore for such choices the left-hand side of (1.6) is zero and the right-hand side would be positive.

If $0 < m < 1$ and $\lambda > 0$, take $x_0 = 0$ and $t_0 > 4\rho^2$. Then a direct calculation shows that (1.6) cannot be verified for a constant C independent of ρ and t_0.

Consider now $(1.3)_p$ and the corresponding Γ_p. The similarity rescaling that keeps $(1.3)_p$ invariant is $|x|^p/t = \text{const}$. Therefore the natural *parabolic cylinders* associated with $(1.3)_p$ are of the type

$$Q_\rho(x_0, t_0) = B_\rho(x_0) \times \bigl(t_0 - \rho^p, t_0 + \rho^p\bigr). \tag{1.5$_p$}$$

Similar arguments show that Γ_p does not satisfy the Harnack inequality for all $p > 1$ such that $\kappa > 0$, even in its own natural parabolic geometry $(1.5)_p$.

Local behavior of solutions. These issues suggest a unifying theory of the local behavior of weak solutions of degenerate or singular parabolic equations. A cornerstone of such a unifying theory would be that weak solutions of $(1.1)_{m,p}$–$(1.2)_{m,p}$ are Hölder continuous.

Another key component would be an understanding of the Harnack estimate in the degenerate or singular setting of $(1.1)_{m,p}$–$(1.2)_{m,p}$. Whether, for example, there is a form of such an estimate that replaces (1.6) and that would reduce to it when either $m \to 1$ or $p \to 2$. The general structure in $(1.1)_{m,p}$–$(1.2)_{m,p}$ is not an artificial requirement. To illustrate this point we return briefly on the issue of regularity of solutions of $(1.3)_p$. We have observed that the space–gradient of the *fundamental solution* Γ_p, for $p > 2$, is locally Hölder continuous in $\mathbb{R}^N \times (\tau, \infty)$. It is then natural to conjecture that the same would be true for local solutions u of $(1.3)_p$. For such solutions, it turns out that $v = |\nabla u|^2$ formally satisfies [55,165]

$$v_t - \bigl(a_{ij} v^{(p-2)/2} v_{x_i}\bigr)_{x_j} \leqslant 0, \quad \text{where } a_{ij} = \delta_{ij} + (p-2)\frac{u_{x_i} u_{x_j}}{|\nabla u|^2}.$$

This is a quasilinear version of $(1.3)_m$ with $m = p/2$. Thus an investigation of the local regularity of solutions of $(1.3)_p$ requires an understanding of degenerate or singular equations with the general quasilinear structure $(1.1)_m$ and $(1.2)_m$.

1.1. *Historical background on regularity and Harnack estimates*

Considerable progress was made in the early 1950s and mid 1960s in the theory of elliptic equations, due to the discoveries of De Giorgi [47] and Moser [132,133]. Consider local weak solutions of

$$(a_{ij} u_{x_i})_{x_j} = 0 \quad \text{weakly in } \Omega, u \in W^{1,2}_{\text{loc}}(\Omega), \tag{1.7}$$

where the coefficients $x \mapsto a_{ij}(x), i, j = 1, 2, \ldots, N$, are only bounded and measurable and satisfying the ellipticity condition

$$a_{ji}\xi_i\xi_j \geqslant C_0|\xi|^2 \quad \text{a.e. in } \Omega\ \forall \xi \in \mathbb{R}^N, \text{for some } C_0 > 0. \tag{1.8}$$

De Giorgi established that local solutions are Hölder continuous and Moser proved that nonnegative solutions satisfy the Harnack inequality. Such inequality can be used, in turn, to prove the Hölder continuity of solutions. Both authors worked with *linear* PDEs. However the linearity has no bearing in the proofs. This permits an extension of these results to elliptic quasilinear equations of the type

$$\begin{aligned}&\operatorname{div}\mathbf{A}(x,u,\nabla u)+B(x,u,\nabla u)=0\\&\quad\text{weakly in }\Omega;\ u\in W^{1,p}_{\mathrm{loc}}(\Omega),\ p>1,\end{aligned}\tag{1.9}$$

with structure conditions

$$\begin{cases}C_0|\nabla u|^p-C\leqslant\mathbf{A}(x,u,\nabla u)\cdot\nabla u,\\ \big|\mathbf{A}(x,u,\nabla u)\big|+\big|B(x,u,\nabla u)\big|\leqslant C\big(1+|\nabla u|^{p-1}\big),\end{cases}\tag{1.10}$$

for a given $C_0>0$ and a given nonnegative constant C. By using the methods of De Giorgi, Ladyzhenskaja and Ural'tzeva [121] established that weak solutions of (1.9)–(1.10) are Hölder continuous, whereas Serrin [157] and Trudinger [163], following the methods of Moser, proved that nonnegative solutions satisfy a Harnack principle. The generalization is twofold, i.e., the principal part $\mathbf{A}(x,u,\nabla u)$ is permitted to have a nonlinear dependence with respect to ∇u, and a *nonlinear growth* with respect to $|\nabla u|$. The latter is of particular interest since the equation in (1.9) might be either degenerate or singular.

A striking result of Moser [134] is that the Harnack estimate (1.6) continues to hold for nonnegative, local, weak solutions of

$$\begin{cases}u\in L^\infty\big(0,T;L^2(\Omega)\big)\cap L^2\big(0,T;W^{1,2}(\Omega)\big),\\ u_t-\big(a_{ij}(x,t)u_{x_i}\big)_{x_j}=0\quad\text{in }\Omega_T,\end{cases}\tag{1.11}$$

where $a_{ij}\in L^\infty(\Omega_T)$ satisfy the analog of the ellipticity condition (1.8). As before, it can be used to prove that weak solutions are locally Hölder continuous in Ω_T. Since the linearity of (1.11) is immaterial to the proof, one might expect, as in the elliptic case, an extension of these results to quasilinear equations of the type

$$u_t-\operatorname{div}\mathbf{A}(x,t,u,\nabla u)=B(x,t,u,\nabla u)\quad\text{in }\Omega_T,\tag{1.12}$$

where the structure condition is as in (1.10). Surprisingly, however, Moser's proof could be extended only for the case $p=2$, i.e., for equations whose principal part has a *linear growth* with respect to $|\nabla u|$. This appears in the work of Aronson and Serrin [16] and Trudinger [164]. The methods of De Giorgi also could not be extended. Ladyzhenskaja et al. [120] proved that solutions of (1.12) are Hölder continuous, provided the principal part has exactly a *linear growth* with respect to $|\nabla u|$. Analogous results were established by Kruzkov [110–112] and by Nash [136] by entirely different methods. Thus it appears that unlike the elliptic case, the degeneracy or singularity of the principal part plays a peculiar role, and for example, for the p-Laplacian equation in $(1.3)_p$ one could not establish whether nonnegative weak solutions satisfy the Harnack estimate or whether a solution is locally Hölder continuous.

In the mid-1980, some progress was made in the theory of degenerate partial differential equations (PDEs) of the type of (1.12), for $p>2$. It was shown that the solutions are locally

Hölder continuous (see [51]). Surprisingly, the same techniques can be suitably modified to establish the *local* Hölder continuity of *any local* solution of quasilinear porous medium-type equations. These modified methods in turn, are crucial in proving that weak solutions of the p-Laplacian equation $(1.3)_p$ are of class $C^{1,\alpha}_{\rm loc}(\Omega_T)$.

Therefore understanding the local structure of the solutions of (1.12) has implications to the theory of equations with degeneracies quite different than (1.12).

In the early 1990s the theory was completed [37] by establishing that solutions of $(1.1)_{m,p}$–$(1.2)_{m,p}$ are Hölder continuous for all $p > 1$ and all $m > 0$. For a complete account see [55].

1.2. *A new approach to regularity*

These results follow, one way or another, from a single unifying idea which we call *intrinsic rescaling*. The diffusion processes in $(1.3)_{m,p}$ evolve in a time scale determined instant by instant by the solution itself, so that, loosely speaking, they can be regarded as the heat equation in their own intrinsic time-configuration. A precise description of this fact as well as its effectiveness is linked to its technical implementations which we will present in Section 2.

The indicated regularity results assume the solutions to be locally or globally bounded. A theory of boundedness of weak solutions of $(1.1)_{m,p}$–$(1.2)_{m,p}$ is quite different from the linear theory and it is presented in Section 3. For example, weak solutions of $(1.1)_p$–$(1.2)_p$ are locally bounded only if $\kappa = N(p-2) + p > 0$ and weak solutions of $(1.1)_m$–$(1.2)_m$ are locally bounded only if $\lambda = N(m-1) + 2 > 0$. It is shown by counterexamples that these conditions are sharp.

The same notion of intrinsic rescaling is at the basis of a new notion of Harnack inequality for nonnegative solutions of $(1.3)_{m,p}$ established in the late 1980s and early 1990s [54,66]. Consider nonnegative weak solutions of $(1.3)_m$ for $m > 1$. The Harnack inequality (1.6) continues to hold for such solutions provided the time is rescaled by the quantity $u^{m-1}(x_0, t_0)$. Similar statements hold for $(1.3)_p$ in their intrinsic parabolic geometry $(1.5)_p$. In Section 4 we present these *intrinsic* versions of the Harnack inequality and trace their connection to the Hölder continuity of solutions.

A major open problem is to establish the Harnack estimate for nonnegative solutions of $(1.1)_{m,p}$ with the full quasilinear structure $(1.2)_{m,p}$. The proofs in [54,66] use in an essential way the structure of $(1.3)_{m,p}$ as well as their corresponding *fundamental solutions* $\Gamma_{m,p}$. The leap forward of Moser's Harnack inequality was in bypassing the classical approaches based on heat potentials, by introducing new harmonic analysis methods and techniques. It is our belief that a proof of the intrinsic Harnack estimate for nonnegative solutions of $(1.1)_{m,p}$–$(1.2)_{m,p}$ that would bypass the *potentials* $\Gamma_{m,p}$, would have the same impact.

The values of $p > 1$ for which nonnegative solutions of $(1.3)_p$ satisfy Harnack's inequality are those for which $\kappa = N(p-2) + p > 0$. Likewise the values of $m > 0$ for which nonnegative solutions of $(1.3)_m$ satisfy Harnack's inequality are those for which $\lambda = N(m-1) + 2 > 0$. These limitations are sharp for a Harnack estimate to hold (Section 4).

1.3. *Limiting cases and miscellanea remarks*

The cases $\kappa, \lambda \leqslant 0$ are not well understood and form the object of current investigations. The case $1 < p \leqslant 2N/(N+1)$ seems to suggest questions similar to those of the limiting Sobolev exponent for elliptic equations (see Brézis [30]) and questions in differential geometry. As $p \searrow 1$, $(1.3)_p$ tends *formally* to a PDE of the type of motion by mean curvature. Investigations in this directions are due to Evans and Spruck [79]. As $m \to 0$ the porous medium equation $(1.3)_m$ for $u \geqslant 0$ tends to the singular equation

$$u_t - \Delta \ln u = 0 \quad \text{weakly in } \Omega_T. \tag{1.13}$$

When $N = 2$ the Cauchy problem for this equation is related to the Ricci flow associated to a complete metric in $\mathbb{R}^2$ [59,60,90,182]. A characterization of the initial data for which (1.13) is solvable has been identified and the theory seems fairly complete [46,59,78]. The case $N \geqslant 3$, however, is still not understood and while solvability has been established for a rather large class of data [59], a precise characterization of such a class still eludes the investigators.

Degenerate and singular elliptic and parabolic equations are one of the branches of modern analysis both in view of the physical significance of the equations at hand [8,10,11, 91,92,113,114,119,126,127,161,183] and the novel analytical techniques that they generate [55].

The class of such equations is large, ranging from flows by mean curvature to Monge–Ampére equations to infinity-Laplacian. These are *implicitly* degenerate or singular equations in that the solution itself determines, implicitly, the regions of degeneracy. Explicitly degenerate equations would be those for which the degeneracy or singularity is a priori prescribed in the coefficients. For example if the modulus of ellipticity C_0 in (1.8) were a nonnegative function of x vanishing at some specified value x_*, such a point would be a point of explicit degeneracy. There is a vast literature on all these aspects of degenerate equations. We have chosen to present a subsection of the theory that has a unifying set of techniques, issues, physical relevance, and future directions.

1.4. *Singular equations of the Stefan-type*

In this framework fall singular parabolic evolution equations where the singularity occurs on the time-part of the operator. These take the form

$$\begin{cases} u \in C_{\mathrm{loc}}\big(0, T; W^{1,2}_{\mathrm{loc}}(\Omega_T)\big), \\ \beta(u)_t - \operatorname{div} \mathbf{A}(x, t, u, \nabla u) \ni B(x, t, u, \nabla u) \quad \text{in } \mathcal{D}'(\Omega_T), \end{cases} \tag{1.14}$$

where $\mathbf{A}$ and B have the same structure conditions as $(1.2)_m$ for $m = 1$ and $\beta(\cdot)$ is a coercive, maximal monotone graph in $\mathbb{R} \times \mathbb{R}$. The prototype example is

$$\beta(u)_t - \Delta u \ni 0 \quad \text{in } \mathcal{D}'(\Omega_T) \tag{1.15}$$

for a $\beta(\cdot)$ given by

$$\beta(s)=\begin{cases} s & \text{for } s<0,\\ [0,1] & \text{for } s=0,\\ 1+s & \text{for } s>0. \end{cases} \tag{1.16}$$

Graphs $\beta(\cdot)$ such as this one, i.e., exhibiting a single jump at the origin, arise from a weak formulation of the classical Stefan problem modeling a solid/liquid phase transition such as water–ice. In the latter case a natural question would be to ask whether the transition of phase occurs with a continuous temperature across water/ice interface. This issue, raised initially by Oleinik in the 1950s and reported in the book [120] is at the origin of the modern and current theory of local regularity and local behavior of solutions of degenerate and/or singular evolution equations. The coercivity of $\beta(\cdot)$ for a solution to be continuous is essential, as pointed out by examples and counterexamples in [63].

It was established in [31,48,149,150,184] that for $\beta(\cdot)$ exhibiting a single jump, the solutions of (1.14) are continuous with a given quantitative modulus of continuity (not Hölder). This raises naturally the question of a graph $\beta(\cdot)$ exhibiting multiple jumps and or singularities of other nature (Section 5). For these rather general graphs, in the mid 1990s it was established in [72] that solutions of (1.14) are continuous provided $N=2$. For dimension $N\geqslant 3$ the same conclusion holds provided the principal part of the differential equation is *exactly* the Laplacian, as in the first of (1.15). Several recent investigations have extended and improved these results for specific graphs [86,87]. It is still an open question however, whether solutions of (1.14) with its full quasilinear structure and for a general coercive maximal monotone graph $\beta(\cdot)$ and for $N\geqslant 3$, are continuous in their domain of definition.

1.5. *Outline of these notes*

The issues touched on here will be expanded in the next sections. We will provide precise statements and self-sufficient structure of proofs.

In Section 2 we deal with the question of the regularity of the weak solutions of singular and degenerate quasilinear parabolic equations, proving their Hölder character. We start with the precise definition of weak solution and the derivation of the building blocks of the theory: the local energy and logarithmic estimates. In Section 2.2 we briefly present the classical approach of De Giorgi to uniformly elliptic equations. We introduce De Giorgi's class and show that functions in De Giorgi's class are Hölder continuous. The two main Sections 2.3 and 2.4 deal, respectively, with the degenerate and the singular case. There we present in full detail the idea of intrinsic scaling and, at least in the degenerate case, prove all the results leading to the Hölder continuity. We have decided to present the theory for the model case of the p-Laplace equation to bring to light what is really essential in the method, leaving aside technical refinements needed to deal with more general equations. We close the section with remarks on the possible generalizations, namely to porous medium type equations.

Section 3 addresses the boundedness of weak solutions. The theory discriminates between the degenerate and the singular case. If $p>2$, a local bound for the solution is implicit in the notion of weak solution. If $1 < p < 2$, local or global solutions need not be bounded in general.

In Section 4, we first give a review about classical results concerning Harnack inequalities. Then we consider the degenerate case and we point out the differences with respect to the nondegenerate one. We sketch a proof of the Harnack inequality both in the degenerate and singular case. We show that for positive solutions of the singular p-Laplace equation an "elliptic" Harnack inequality holds. We also analyze the phenomenon of the extinction of the solution in finite time. Through a suitable use of the Raleigh quotient, we are also able to give sharp estimates on the extinction time and to describe the asymptotic profile of the extinction. In the whole section we point out the major open questions about Harnack inequalities for singular and degenerate parabolic equations.

In Section 5 we give physical motivations concerning Stefan-like equations and show, through the Kruzkov–Sukorjanski transformation, the deep links between degenerate equations and Stefan-like equations. Then, we describe the approaches made by Aronson, Caffarelli, DiBenedetto, Sachs and Ziemer in the 1980s. Thanks to their contributions the case of only one singularity was completely solved. Lastly we analyze the new pioneering approach of [72] where, through a lemma of measure theory, the case of multiple singularities was totally solved in the case $N = 2$. Moreover, we show that this approach also works in the case $N \geqslant 3$ but only under strong assumptions. In this section we also point out the major open questions.

We have chosen not to present existence theorems for boundary value problems associated with these equations. Theorems of this kind are mostly based on Galerkin approximations and appear in the literature in a variety of forms. We refer, for example, to [120] or [123]. Given the a priori estimates presented here these can be obtained alternatively by a limiting process in a family of approximating problems and an application of Minty's lemma.

These notes can be ideally divided in three parts:

1. Hölder continuity and boundedness of solutions (Sections 2 and 3).
2. Harnack-type estimates (Section 4).
3. Stefan-like problems (Section 5).

These parts are technically linked but they are conceptually independent, in the sense that they deal with issues that have developed in independent directions. We have attempted to present them in such a way that they can be approached independently.

2. Regularity of weak solutions

We address the question of the regularity of weak solutions of singular and degenerate parabolic equations by proving that they are Hölder continuous.

We will concentrate on quasilinear parabolic equations, with principal part in divergence form, of the type

$$u_t - \operatorname{div} |\nabla u|^{p-2} \nabla u = 0, \quad p > 1. \tag{2.1}$$

If $p > 2$, the equation is degenerate in the space part, due to the vanishing of its modulus of ellipticity $|\nabla u|^{p-2}$ at points where $|\nabla u| = 0$. The singular case corresponds to $1 < p < 2$: the modulus of ellipticity becomes infinity at points where $|\nabla u| = 0$.

The results in this section extend to a variety of equations and, in particular, to equations with general principal parts satisfying appropriate structure assumptions and with lower-order terms. We have chosen to present the results and the proofs for the particular model case (2.1) to bring to light what we feel are the essential features of the theory. Remarks on generalizations, which in some way or another correspond to more or less sophisticated technical improvements, are made at the end of the section.

Results on the continuity of solutions at a point consist basically in constructing a sequence of nested and shrinking cylinders with vertex at that point, such that the essential oscillation of the function in those cylinders converges to zero when the cylinders shrink to zero. At the basis of the proof is an iteration technique, that is a refinement of the technique by De Giorgi and Moser (cf. [47,120,132]), based on energy (also known as Cacciopolli) and logarithmic estimates for the solution, that we briefly review in Section 2.2. In the degenerate or singular cases these estimates are not homogeneous in the sense that they involve integral norms corresponding to different powers, namely the powers 2 and p. The key idea is then to look at the equation in its own geometry, i.e., in a geometry dictated by its intrinsic structure. This amounts to rescale the standard parabolic cylinders by a factor that depends on the oscillation of the solution. This procedure, which can be called accommodation of the degeneracy, allows one to recover the homogeneity in the energy estimates written over these rescaled cylinders. We can say heuristically that the equation behaves in its own geometry like the heat equation. In the sequel, we first treat the degenerate case in Section 2.3 and then the more involved singular case in Section 2.4. We conclude the section with some remarks on generalizations, namely to porous medium-type equations.

2.1. *Weak solutions and local estimates*

A local weak sub(super)-solution of (2.1) is a measurable function

$$u \in C_{\text{loc}}\big(0, T; L^2_{\text{loc}}(\Omega)\big) \cap L^p_{\text{loc}}\big(0, T; W^{1,p}_{\text{loc}}(\Omega)\big)$$

such that, for every compact $K \subset \Omega$ and for every subinterval $[t_1, t_2]$ of $(0, T]$,

$$\int_K u\varphi \,\mathrm{d}x \bigg|_{t_1}^{t_2} + \int_{t_1}^{t_2} \int_K \big\{-u\varphi_t + |\nabla u|^{p-2}\nabla u \cdot \nabla\varphi\big\} \,\mathrm{d}x\,\mathrm{d}t \leqslant (\geqslant) 0 \tag{2.2}$$

for all $\varphi \in W^{1,2}_{\text{loc}}(0, T; L^2(K)) \cap L^p_{\text{loc}}(0, T; W^{1,p}_0(K))$, $\varphi \geqslant 0$. A function that is both a local subsolution and a local supersolution of (2.1) is a local solution of (2.1).

It would be technically convenient to have at hand a formulation of weak solution involving the time derivative u_t. Unfortunately, solutions of (2.1), whenever they exist, possess a modest degree of time-regularity and in general u_t has a meaning only in the sense of distributions. To overcome this limitation we introduce the Steklov average of a function $v \in L^1(\Omega_T)$, defined for $0 < h < T$ by

$$v_h = \begin{cases} \frac{1}{h}\int_t^{t+h} v(\cdot,\tau)\,d\tau & \text{if } t \in (0, T-h], \\ 0 & \text{if } t \in (T-h, T], \end{cases}$$

and observe that the notion (2.2) of solution is equivalent to:

$$\begin{aligned} &\text{for every compact } K \subset \Omega \text{ and for all } 0 < t < T-h, \\ &\int_{K\times\{t\}} \{(u_h)_t\varphi + (|\nabla u|^{p-2}\nabla u)_h \cdot \nabla\varphi\}\,dx \leqslant (\geqslant)\,0 \end{aligned} \tag{2.3}$$

for all $\varphi \in W_0^{1,p}(K) \cap L^\infty_{\mathrm{loc}}(\Omega)$, $\varphi \geqslant 0$.

We will show that locally bounded solutions of (2.1) are locally Hölder continuous within their domain of definition. No specific boundary or initial values need to be prescribed for u. Although the arguments below are of local nature, to simplify the presentation we assume that u is a.e. defined and bounded in Ω_T and set

$$M \equiv \|u\|_{L^\infty(\Omega_T)}.$$

See Section 3 for results on the boundedness of weak solutions.

2.1.1. *Local energy and logarithmic estimates.* Given a point $x_0 \in \mathbb{R}^N$, denote by $K_\rho(x_0)$ the N-dimensional cube with centre at x_0 and wedge 2ρ:

$$K_\rho(x_0) := \left\{x \in \mathbb{R}^N : \max_{1\leqslant i\leqslant N} |x_i - x_{0i}| < \rho\right\},$$

given a point $(x_0, t_0) \in \mathbb{R}^{N+1}$, the cylinder of radius ρ and height $\tau > 0$ is

$$(x_0, t_0) + Q(\tau, \rho) := K_\rho(x_0) \times (t_0 - \tau, t_0).$$

Consider a cylinder $(x_0, t_0) + Q(\tau, \rho) \subset \Omega_T$ and let $0 \leqslant \zeta \leqslant 1$ be a piecewise smooth cutoff function in $(x_0, t_0) + Q(\tau, \rho)$ such that

$$|\nabla\zeta| < \infty \quad \text{and} \quad \zeta(x,t) = 0, \quad x \notin K_\rho(x_0). \tag{2.4}$$

We start with the energy estimates. Without loss of generality, we will state them for cylinders with "vertex" at the origin $(0, 0)$, the changes being obvious for cylinders with "vertex" at a generic (x_0, t_0).

PROPOSITION 1. *Let u be a local weak solution of* (2.1). *There exists a constant $C \equiv C(p) > 0$ such that, for every cylinder $Q(\tau, \rho) \subset \Omega_T$,*

$$\begin{aligned}
&\sup_{-\tau<t<0} \int_{K_\rho \times \{t\}} (u-k)_\pm^2 \zeta^p \, dx + \int_{-\tau}^0 \int_{K_\rho} \left| \nabla (u-k)_\pm \zeta \right|^p dx\, dt \\
&\quad \leqslant \int_{K_\rho \times \{-\tau\}} (u-k)_\pm^2 \zeta^p \, dx + C \int_{-\tau}^0 \int_{K_\rho} (u-k)_\pm^p |\nabla \zeta|^p \, dx\, dt \\
&\qquad + p \int_{-\tau}^0 \int_{K_\rho} (u-k)_\pm^2 \zeta^{p-1} \partial_t \zeta \, dx\, dt. \qquad (2.5)
\end{aligned}$$

PROOF. Use $\varphi = \pm(u_h - k)_\pm \zeta^p$ as a testing function in (2.3) and perform standard energy estimates (cf. [55], pp. 24–27). □

Given constants a, b, c, with $0 < c < a$, define the nonnegative function

$$\begin{aligned}
\psi^\pm_{\{a,b,c\}}(s) &\equiv \left(\ln \left\{ \frac{a}{(a+c) - (s-b)_\pm} \right\} \right)_+ \\
&= \begin{cases} \ln\left\{ \frac{a}{(a+c) \pm (b-s)} \right\} & \text{if } b \pm c \lesseqgtr s \lesseqgtr b \pm (a+c), \\ 0 & \text{if } s \lesseqgtr b \pm c, \end{cases}
\end{aligned}$$

whose first derivative is

$$\left(\psi^\pm_{\{a,b,c\}}\right)'(s) = \begin{cases} \frac{1}{(b-s) \pm (a+c)} & \text{if } b \pm c \lesseqgtr s \lesseqgtr b \pm (a+c), \\ 0 & \text{if } s \lesseqgtr b \pm c, \end{cases} \gtreqless 0,$$

and second derivative, off $s = b \pm c$, is $(\psi^\pm_{\{a,b,c\}})'' = \{(\psi^\pm_{\{a,b,c\}})'\}^2 \geqslant 0$.

Now, given a bounded function u in a cylinder $(x_0, t_0) + Q(\tau, \rho)$ and a number k, define the constant

$$H^\pm_{u,k} \equiv \operatorname*{ess\,sup}_{(x_0,t_0)+Q(\tau,\rho)} \left| (u-k)_\pm \right|.$$

The following function was introduced in [48] and since then has been used as a recurrent tool in the proof of results concerning the local behavior of solutions of degenerate PDEs:

$$\psi^\pm\left(H^\pm_{u,k}, (u-k)_\pm, c\right) \equiv \psi^\pm_{\{H^\pm_{u,k}, k, c\}}(u), \qquad 0 < c < H^\pm_{u,k}.$$

From now on, when referring to this function we will write it as $\psi^\pm(u)$, omitting the subscripts whose meaning will be clear from the context.

Let $x \mapsto \zeta(x)$ be a time-independent cutoff function in $K_\rho(x_0)$ satisfying (2.4). The logarithmic estimates in cylinders $Q(\tau, \rho)$ with "vertex" at $(0, 0)$, are the following.

PROPOSITION 2. *Let u be a local weak solution of* (2.1), $k \in \mathbb{R}$ *and* $0 < c < H^{\pm}_{u,k}$. *There exists a constant $C > 0$ such that, for every cylinder $Q(\tau, \rho) \subset \Omega_T$,*

$$\sup_{-\tau<t<0} \int_{K_\rho \times \{t\}} \left[\psi^{\pm}(u)\right]^2 \zeta^p \, dx \leqslant \int_{K_\rho \times \{-\tau\}} \left[\psi^{\pm}(u)\right]^2 \zeta^p \, dx + C \int_{-\tau}^{0} \int_{K_\rho} \psi^{\pm}(u) \left|(\psi^{\pm})'(u)\right|^{2-p} |\nabla \zeta|^p \, dx \, dt. \tag{2.6}$$

PROOF. Take $\varphi = 2\psi^{\pm}(u_h)[(\psi^{\pm})'(u_h)]\zeta^p$ as a testing function in (2.3) and integrate in time over $(-\tau, t)$ for $t \in (-\tau, 0)$. Since $\partial_t \zeta \equiv 0$,

$$\begin{aligned} &\int_{-\tau}^{t} \int_{K_\rho} \partial_t u_h \left\{2\psi^{\pm}(u_h)\left[(\psi^{\pm})'(u_h)\right]\zeta^p\right\} dx \, dt \\ &\quad = \int_{-\tau}^{t} \int_{K_\rho} \partial_t \left\{\left[\psi^{\pm}(u_h)\right]^2\right\} \zeta^p \, dx \, dt \\ &\quad = \int_{K_\rho \times \{t\}} \left[\psi^{\pm}(u_h)\right]^2 \zeta^p \, dx - \int_{K_\rho \times \{-\tau\}} \left[\psi^{\pm}(u_h)\right]^2 \zeta^p \, dx. \end{aligned}$$

From this, letting $h \to 0$,

$$\begin{aligned} &\int_{-\tau}^{t} \int_{K_\rho} \partial_t u_h \left\{2\psi^{\pm}(u_h)\left[(\psi^{\pm})'(u_h)\right]\zeta^p\right\} dx \, dt \\ &\quad \to \int_{K_\rho \times \{t\}} \left[\psi^{\pm}(u)\right]^2 \zeta^p \, dx - \int_{K_\rho \times \{-\tau\}} \left[\psi^{\pm}(u)\right]^2 \zeta^p \, dx. \end{aligned}$$

As for the remaining term, we first let $h \to 0$, to obtain

$$\begin{aligned} &\int_{-\tau}^{t} \int_{K_\rho} |\nabla u|^{p-2} \nabla u \cdot \nabla \left\{2\psi^{\pm}(u)\left[(\psi^{\pm})'(u)\right]\zeta^p\right\} dx \, dt \\ &\quad = \int_{-\tau}^{t} \int_{K_\rho} |\nabla u|^p \left\{2\left(1 + \psi^{\pm}(u)\right)\left[(\psi^{\pm})'(u)\right]^2 \zeta^p\right\} dx \, dt \\ &\qquad + p \int_{-\tau}^{t} \int_{K_\rho} |\nabla u|^{p-2} \nabla u \cdot \nabla \zeta \left\{2\psi^{\pm}(u)\left[(\psi^{\pm})'(u)\right]\zeta^{p-1}\right\} dx \, dt \\ &\quad \geqslant \int_{-\tau}^{t} \int_{K_\rho} |\nabla u|^p \left\{2\left(1 + \psi^{\pm}(u) - \psi^{\pm}(u)\right)\left[(\psi^{\pm})'(u)\right]^2 \zeta^p\right\} dx \, dt \end{aligned}$$

$$\begin{aligned}&-2(p-1)^{p-1}\int_{-\tau}^{t}\int_{K_\rho}\psi^{\pm}(u)\bigl|(\psi^{\pm})'(u)\bigr|^{2-p}|\nabla\zeta|^p\,\mathrm{d}x\,\mathrm{d}t\\ &\geqslant -C\int_{-\tau}^{t}\int_{K_\rho}\psi^{\pm}(u)\bigl|(\psi^{\pm})'(u)\bigr|^{2-p}|\nabla\zeta|^p\,\mathrm{d}x\,\mathrm{d}t.\end{aligned}$$

Since $t\in(-\tau,0)$ is arbitrary, we can combine both estimates to obtain (2.6). □

2.1.2. *Some technical tools.* We gather a few technical facts that, although marginal to the theory, are essential in establishing its main results.

Given a continuous function $v:\Omega\to\mathbb{R}$ and two real numbers $k<l$, we define

$$\begin{aligned}&[v>l]\equiv\{x\in\Omega:\ v(x)>l\},\\ &[v<k]\equiv\{x\in\Omega:\ v(x)<k\},\\ &[k<v<l]\equiv\{x\in\Omega:\ k<v(x)<l\}.\end{aligned}\tag{2.7}$$

LEMMA 1 (De Giorgi [47]). *Let $v\in W^{1,1}(B_\rho(x_0))\cap C(B_\rho(x_0))$, with $\rho>0$ and $x_0\in\mathbb{R}^N$ and $k<l\in\mathbb{R}$. There exists a constant C, depending only on N and p (so independent of ρ, x_0, v, k and l), such that*

$$(l-k)\bigl|[v>l]\bigr|\leqslant C\frac{\rho^{N+1}}{|[v<k]|}\int_{[k<v<l]}|\nabla v|\,\mathrm{d}x.$$

REMARK 1. The conclusion of the lemma remains valid, provided Ω is convex, for functions $v\in W^{1,1}(\Omega)\cap C(\Omega)$. We will use it in the case Ω is a cube. In fact, the continuity is not essential for the result to hold. For a function merely in $v\in W^{1,1}(\Omega)$, we define the sets (2.7) through any representative in the equivalence class. It can be shown that the conclusion of the lemma is independent of that choice.

The following lemma concerns the geometric convergence of sequences.

LEMMA 2. *Let $\{X_n\}$, $n=0,1,2,\dots$, be a sequence of positive real numbers satisfying the recurrence relation*

$$X_{n+1}\leqslant Cb^nX_n^{1+\alpha},$$

where $C,b>1$ and $\alpha>0$ are given. If

$$X_0\leqslant C^{-1/\alpha}b^{-1/\alpha^2},$$

then $X_n\to0$ as $n\to\infty$.

Let $V_0^p(\Omega_T)$ denote the space

$$V_0^p(\Omega_T)=L^\infty\bigl(0,T;L^p(\Omega)\bigr)\cap L^p\bigl(0,T;W_0^{1,p}(\Omega)\bigr)$$

endowed with the norm

$$\|u\|^p_{V^p(\Omega_T)} = \operatorname*{ess\,sup}_{0\leqslant t\leqslant T} \|u(\cdot, t)\|^p_{p,\Omega} + \|\nabla u\|^p_{p,\Omega_T},$$

for which the following embedding theorem holds (cf. [55], p. 9):

THEOREM 1. *Let $p > 1$. There exists a constant γ, depending only upon N and p, such that, for every $v \in V^p_0(\Omega_T)$,*

$$\|v\|^p_{p,\Omega_T} \leqslant \gamma \big| |v| > 0 \big|^{p/(N+p)} \|v\|^p_{V^p(\Omega_T)}.$$

With C or C_j we denote constants that depend only on N and p and that might be different in different contexts.

2.2. *The classical approach of De Giorgi*

Results concerning the Hölder continuity of weak solutions u consist essentially in showing that for every point $(x_0, t_0) \in \Omega_T$ we can find a sequence of nested and shrinking cylinders $(x_0, t_0) + Q(\tau_n, \rho_n)$ such that the essential oscillation of u in these cylinders approaches zero as $n \to \infty$ in a way that can be quantified.

The approach to regularity introduced by De Giorgi is based on the following embedding theorem (see [47] for the elliptic case and [120] for the parabolic case):

PROPOSITION 3. *Assume that $u \in L^2_{\text{loc}}(0, T; W^{1,2}_{\text{loc}}(\Omega)) \cap W^{1,2}_{\text{loc}}(0, T; L^2_{\text{loc}}(\Omega))$ is locally bounded and satisfies the Caccioppoli inequalities* (2.5) *with $p = 2$. Then u is locally Hölder continuous, with the modulus of continuity depending only upon the data.*

A solution of a nondegenerate parabolic equation with the full quasilinear structure satisfies these inequalities. One uses the structure of the equation to prove the Caccioppoli estimates for a solution u and this is the only role of the equation. Once such inequalities are derived, the Hölder continuity of u is solely a consequence of (2.5) for $p = 2$.

Alternative approaches are in Kruzkov [110–112] and, through the use of the Harnack inequality, in Moser [132,134], Trudinger [164], Aronson and Serrin [16] and, for equations in nondivergence form, in Krylov and Safonov [117].

Set $Q_r = Q(r^2, r)$, fix a point $(x_0, t_0) \in \Omega_T$, and let ρ_0 be the largest radius so that $(x_0, t_0) + Q_{\rho_0}$ is contained in Ω_T. For a constant $\delta \in (0, 1)$, consider the sequence of decreasing radii,

$$\rho_n = \delta^n \rho_0, \quad n = 0, 1, 2, \ldots, \tag{2.8}$$

and the family of nested shrinking cylinders, with the same "vertex"

$$(x_0, t_0) + Q_{\rho_n}, \quad n = 0, 1, 2, \ldots. \tag{2.9}$$

Set

$$\mu_n^- := \operatorname*{ess\,inf}_{(x_0,t_0)+Q_{\rho_n}} u, \qquad \mu_n^+ := \operatorname*{ess\,sup}_{(x_0,t_0)+Q_{\rho_n}} u,$$

$$\omega_n := \operatorname*{ess\,osc}_{(x_0,t_0)+Q_{\rho_n}} u = \mu_n^+ - \mu_n^-.$$

PROPOSITION 4. *Let u satisfy the Caccioppoli inequalities (2.5) for $p = 2$. Then there exist constants $C > 1$ and $\delta, \eta \in (0, \frac{1}{2})$, that can be determined a priori only in terms of the data, such that, for every $(x_0, t_0) \in \Omega_T$ and every $n \in \mathbb{N}$, at least one of the following two inequalities holds*

$$\operatorname*{ess\,sup}_{(x_0,t_0)+Q_{\rho_{n+1}}} u \leqslant \mu_n^+ - \eta\omega_n, \tag{2.10}$$

$$\operatorname*{ess\,inf}_{(x_0,t_0)+Q_{\rho_{n+1}}} u \geqslant \mu_n^- + \eta\omega_n. \tag{2.11}$$

A nontrivial proof can be found in [120]. This proposition can be interpreted as a weak maximum principle. For example (2.10) asserts that the supremum of u over the cylinder $Q_{\rho_{n+1}}$ is strictly less than the supremum of u over the larger coaxial cylinder Q_{ρ_n}. In other words, the supremum of u over Q_{ρ_n} can only be achieved in the parabolic shell $Q_{\rho_n} \setminus Q_{\rho_{n+1}}$ that can be considered as a sort of parabolic boundary of Q_{ρ_n}.

A consequence of such a weak maximum principle is:

PROPOSITION 5. *Let u be as above. Then there exist constants $C > 1$ and $\delta, \eta \in (0, \frac{1}{2})$, that can be determined a priori only in terms of the data, such that, for every $(x_0, t_0) \in \Omega_T$ and every $n \in \mathbb{N}$,*

$$\omega_{n+1} \leqslant (1 - \eta)\omega_n. \tag{2.12}$$

This in turn implies that u is locally Hölder continuous in Ω_T.

PROOF. Fix $(x_0, t_0) \in \Omega_T$. Assume that (2.10) holds. By subtracting μ_{n+1}^- from the left-hand side and μ_n^- from the right-hand side,

$$\omega_{n+1} = \mu_{n+1}^+ - \mu_{n+1}^- \leqslant \mu_n^+ - \mu_n^- - \eta\omega_n = (1 - \eta)\omega_n.$$

If (2.11) holds one can argue in a similar way. By iteration,

$$\omega_n \leqslant (1 - \eta)^n \omega_0 \quad \forall n \in \mathbb{N}. \tag{2.13}$$

The numbers η and δ are related by $(1 - \eta) = \delta^\alpha$ and $\alpha = \frac{\ln(1-\eta)}{\ln \delta} \in (0, 1)$. Therefore,

$$\omega_n \leqslant \omega_0 \left(\frac{\rho_n}{\rho_0}\right)^\alpha \quad \forall n \in \mathbb{N}. \tag{2.14}$$

Since $(x_0, t_0) \in \Omega_T$ is arbitrary, we conclude that u is locally Hölder continuous in Ω_T with exponent α. □

REMARK 2. The cylinder $(x_0, t_0) + Q_{\rho_0}$ must be contained in Ω_T. Thus from (2.14) it follows that the Hölder continuity can be claimed only within compact subsets of Ω_T and that the Hölder constant $\omega_0 \rho_0^{-\alpha}$ deteriorates as (x_0, t_0) approaches the parabolic boundary of Ω_T.

2.3. *The degenerate case $p > 2$*

We go back to equation (2.1) and focus on the degenerate case $p > 2$. The energy and logarithmic estimates of Section 2.1 are not homogeneous in the space and time parts due to the presence of the power $p \neq 2$. To go about this difficulty we will consider the equation in a geometry dictated by its own structure, which is designed, roughly speaking, to restore the homogeneity of the various parts of the Caccioppoli inequalities (2.5). This means that, instead of the usual cylinders, we have to work in cylinders whose dimensions take the degeneracy of the equation into account, in a process that we call *intrinsic rescaling*. Let us make this idea precise.

2.3.1. *The geometric setting and the alternative.* Consider $R > 0$ such that $Q(R^2, R) \subset \Omega_T$, define

$$\mu_+ := \operatorname*{ess\,sup}_{Q(R^2,R)} u, \qquad \mu_- := \operatorname*{ess\,inf}_{Q(R^2,R)} u, \qquad \omega := \operatorname*{ess\,osc}_{Q(R^2,R)} u = \mu_+ - \mu_-$$

and construct the cylinder

$$Q\big(a_0 R^p, R\big) \equiv K_R(0) \times \big(-a_0 R^p, 0\big) \quad \text{with } a_0 = \left(\frac{\omega}{2^\lambda}\right)^{2-p}, \tag{2.15}$$

where $\lambda > 1$ is to be fixed later depending only on the data (see (2.55)). Note that for $p = 2$, i.e., in the nondegenerate case, these are the standard parabolic cylinders reflecting the natural homogeneity between the space and time variables.

We will assume, without loss of generality, that

$$R < \frac{\omega}{2^\lambda} \tag{2.16}$$

because if this does not hold there is nothing to prove since the oscillation is comparable to the radius.

Now, (2.16) implies the inclusion

$$Q\big(a_0 R^p, R\big) \subset Q\big(R^2, R\big)$$

and the relation

$$\operatorname*{ess\,osc}_{Q(a_0R^p,R)} u \leqslant \omega, \tag{2.17}$$

which will be the starting point of an iteration process that leads to our main results. Note that we had to consider the cylinder $Q(R^2, R)$ and assume (2.16), so that (2.17) would hold for the rescaled cylinder $Q(a_0R^p, R)$. This is in general not true for a given cylinder since its dimensions would have to be intrinsically defined in terms of the essential oscillation of the function within it – the stretching procedure is commonly referred to as an *accommodation of the degeneracy.*

We now consider subcylinders of $Q(a_0R^p, R)$ of the form

$$(0, t^*) + Q(dR^p, R) \quad \text{with } d = \left(\frac{\omega}{2}\right)^{2-p} \tag{2.18}$$

that are contained in $Q(a_0R^p, R)$ if

$$\left(2^{p-2} - 2^{\lambda(p-2)}\right)\frac{R^p}{\omega^{p-2}} < t^* < 0. \tag{2.19}$$

The proof now follows from the analysis of two complementary cases. We briefly describe them in the following way: in the first case we assume that there is a cylinder of the type $(0, t^*) + Q(dR^p, R)$ where u is essentially away from its infimum. We show that going down to a smaller cylinder the oscillation decreases by a small factor that we can exhibit. If that cylinder cannot be found then u is essentially away from its supremum in all cylinders of that type and we can compound this information to reach the same conclusion as in the previous case. We state this in a precise way.

For a constant $\nu_0 \in (0, 1)$, that will be determined depending only on the data, we will assume that either:

The first alternative. There is a cylinder of the type $(0, t^*) + Q(dR^p, R)$ for which

$$\frac{|\{(x,t) \in (0, t^*) + Q(dR^p, R)\colon u(x,t) < \mu_- + \omega/2\}|}{|Q(dR^p, R)|} \leqslant \nu_0 \tag{2.20}$$

or

The second alternative. For every cylinder of the type $(0, t^*) + Q(dR^p, R)$,

$$\frac{|\{(x,t) \in (0, t^*) + Q(dR^p, R)\colon u(x,t) > \mu_+ - \omega/2\}|}{|Q(dR^p, R)|} < 1 - \nu_0. \tag{2.21}$$

2.3.2. *Analysis of the first alternative.*

LEMMA 3. *Assume* (2.20) *holds for some* t^* *as in* (2.19) *and that* (2.16) *is in force. There exists a constant* $\nu_0 \in (0,1)$, *depending only on the data, such that*

$$u(x,t) > \mu^- + \frac{\omega}{4}, \quad \text{a.e. } (x,t) \in (0,t^*) + Q\left(d\left(\frac{R}{2}\right)^p, \frac{R}{2}\right).$$

PROOF. Take the cylinder for which (2.20) holds and assume, by translation, that $t^* = 0$. Let

$$R_n = \frac{R}{2} + \frac{R}{2^{n+1}}, \quad n = 0, 1, \ldots,$$

and construct the family of nested and shrinking cylinders $Q(dR_n^p, R_n)$. Consider piecewise smooth cutoff functions $0 < \zeta_n \leqslant 1$, defined in these cylinders, and satisfying the following set of assumptions

$$\zeta_n = 1 \quad \text{in } Q\left(dR_{n+1}^p, R_{n+1}\right), \qquad \zeta_n = 0 \quad \text{on } \partial_p Q\left(dR_n^p, R_n\right),$$
$$|\nabla \zeta_n| \leqslant \frac{2^{n+1}}{R}, \qquad 0 \leqslant \partial_t \zeta_n \leqslant \frac{2^{p(n+1)}}{dR^p}.$$

Write the energy inequality (2.5) for the functions $(u - k_n)_-$, with

$$k_n = \mu_- + \frac{\omega}{4} + \frac{\omega}{2^{n+2}}, \quad n = 0, 1, \ldots,$$

in the cylinders $Q(dR_n^p, R_n)$ and with $\zeta = \zeta_n$. They read

$$\begin{aligned}
&\sup_{-dR_n^p < t < 0} \int_{K_{R_n} \times \{t\}} (u - k_n)_-^2 \zeta_n^p \, dx + \int_{-dR_n^p}^0 \int_{K_{R_n}} |\nabla (u - k_n)_- \zeta_n|^p \, dx \, dt \\
&\quad \leqslant C \int_{-dR_n^p}^0 \int_{K_{R_n}} (u - k_n)_-^p |\nabla \zeta_n|^p \, dx \, dt \\
&\qquad + p \int_{-dR_n^p}^0 \int_{K_{R_n}} (u - k_n)_-^2 \zeta_n^{p-1} \partial_t \zeta_n \, dx \, dt \\
&\quad \leqslant C \frac{2^{p(n+1)}}{R^p} \left\{ \int_{-dR_n^p}^0 \int_{K_{R_n}} (u - k_n)_-^p \, dx \, dt + \frac{1}{d} \int_{-dR_n^p}^0 \int_{K_{R_n}} (u - k_n)_-^2 \, dx \, dt \right\}.
\end{aligned} \tag{2.22}$$

Next, observing that

$$(u - k_n)_- = (\mu_- - u) + \frac{\omega}{4} + \frac{\omega}{2^{n+2}} \leqslant \frac{\omega}{2}$$

and

$$(u-k_n)_-^2 = (u-k_n)_-^{2-p}(u-k_n)_-^p \geqslant \left(\frac{\omega}{2}\right)^{2-p}(u-k_n)_-^p,$$

we obtain from (2.22)

$$\begin{aligned}
&\left(\frac{\omega}{2}\right)^{2-p} \sup_{-dR_n^p<t<0} \int_{K_{R_n}\times\{t\}} (u-k_n)_-^p \zeta_n^p \,\mathrm{d}x \\
&\quad + \int_{-dR_n^p}^0 \int_{K_{R_n}} \left|\nabla(u-k_n)_-\zeta_n\right|^p \mathrm{d}x\,\mathrm{d}t \\
&\quad \leqslant C\frac{2^{p(n+1)}}{R^p}\left\{\left(\frac{\omega}{2}\right)^p + \frac{1}{d}\left(\frac{\omega}{2}\right)^2\right\} \int_{-dR_n^p}^0 \int_{K_{R_n}} \chi_{\{(u-k_n)_->0\}} \,\mathrm{d}x\,\mathrm{d}t. \qquad (2.23)
\end{aligned}$$

Recall that $d = (\frac{\omega}{2})^{2-p}$ and divide (2.23) by d to get

$$\begin{aligned}
&\sup_{-dR_n^p<t<0} \int_{K_{R_n}\times\{t\}} (u-k_n)_-^p \zeta_n^p \,\mathrm{d}x + \frac{1}{d}\int_{-dR_n^p}^0 \int_{K_{R_n}} \left|\nabla(u-k_n)_-\zeta_n\right|^p \mathrm{d}x\,\mathrm{d}t \\
&\quad \leqslant C\frac{2^{p(n+1)}}{R^p}\left(\frac{\omega}{2}\right)^p \frac{1}{d} \int_{-dR_n^p}^0 \int_{K_{R_n}} \chi_{\{(u-k_n)_->0\}} \,\mathrm{d}x\,\mathrm{d}t. \qquad (2.24)
\end{aligned}$$

Now we perform a change of the time variable in (2.24), putting $\bar{t} = t/d$ and defining

$$\bar{u}(\cdot,\bar{t}) = u(\cdot,t) \quad \text{and} \quad \bar{\zeta}_n(\cdot,\bar{t}) = \zeta_n(\cdot,t),$$

and obtain the simplified inequality

$$\begin{aligned}
&\left\|(\bar{u}-k_n)_-\bar{\zeta}_n\right\|_{V_0^p(Q(R_n^p,R_n))}^p \\
&\quad \leqslant C\frac{2^{pn}}{R^p}\left(\frac{\omega}{2}\right)^p \int_{-R_n^p}^0 \int_{K_{R_n}} \chi_{\{(\bar{u}-k_n)_->0\}} \,\mathrm{d}x\,\mathrm{d}t. \qquad (2.25)
\end{aligned}$$

Define, for each n,

$$A_n = \int_{-R_n^p}^0 \int_{K_{R_n}} \chi_{\{(\bar{u}-k_n)_->0\}} \,\mathrm{d}x\,\mathrm{d}\bar{t}$$

and observe that

$$\begin{aligned}\frac{1}{2^{p(n+2)}}\left(\frac{\omega}{2}\right)^{p} A_{n+1} &\leqslant |k_n - k_{n+1}|^p A_{n+1}\\ &\leqslant \left\|(\bar{u}-k_n)_-\right\|^p_{p,Q(R^p_{n+1},R_{n+1})}\\ &\leqslant \left\|(\bar{u}-k_n)_-\bar{\zeta}_n\right\|^p_{p,Q(R^p_n,R_n)}\\ &\leqslant C\left\|(\bar{u}-k_n)_-\bar{\zeta}_n\right\|^p_{V_0^p(Q(R^p_n,R_n))} A_n^{p/(N+p)}\\ &\leqslant C\frac{2^{pn}}{R^p}\left(\frac{\omega}{2}\right)^p A_n^{1+p/(N+p)}. \end{aligned} \tag{2.26}$$

The first three of these inequalities follow from the definition of A_n and $k_{n+1} < k_n$; the fourth inequality is a consequence of Theorem 1 and the last one follows from (2.25). Next, define the numbers

$$X_n = \frac{A_n}{|Q(R^p_n, R_n)|},$$

divide (2.26) by $|Q(R^p_{n+1}, R_{n+1})|$ and obtain the recursive relation

$$X_{n+1} \leqslant C4^{pn} X_n^{1+p/(N+p)},$$

for a constant C depending only upon N and p. By Lemma 2 on fast geometric convergence, if

$$X_0 \leqslant C^{-(N+p)/p} 4^{-(N+p)^2/p} \equiv \nu_0 \tag{2.27}$$

then

$$X_n \to 0. \tag{2.28}$$

Therefore,

$$\left|\left\{(x,t)\in Q\left(d\left(\frac{R}{2}\right)^p, \frac{R}{2}\right): u(x,t) \leqslant \mu_- + \frac{\omega}{4}\right\}\right| = 0. \qquad \square$$

Our next aim is to show that the conclusion of Lemma 3 holds in a full cylinder $Q(\tau, \rho)$. The idea is to use the fact that at the time level

$$-\hat{t} := t^* - d\left(\frac{R}{2}\right)^p \tag{2.29}$$

the function $u(x, -\hat{t})$ is strictly above the level $\mu^- + \frac{\omega}{4}$ in the cube $K_{R/2}$, and look at this time level as an initial condition to make the conclusion hold up to $t = 0$. As an intermediate step we need the following lemma.

LEMMA 4. *Assume* (2.20) *holds for some* t^* *as in* (2.19) *and that* (2.16) *is in force. Given* $\nu_* \in (0, 1)$, *there exists* $s_* \in \mathbb{N}$, *depending only on the data, such that*

$$\left|\left\{x \in K_{R/4}\colon u(x,t) < \mu_- + \frac{\omega}{2^{s_*}}\right\}\right| \leqslant \nu_* |K_{R/4}| \quad \forall t \in (-\hat{t}, 0).$$

PROOF. We use the logarithmic estimate (2.6) applied to the function $(u - k)_-$ in the cylinder $Q(\hat{t}, \frac{R}{2})$, with the choices

$$k = \mu_- + \frac{\omega}{4} \quad \text{and} \quad c = \frac{\omega}{2^{n+2}},$$

where $n \in \mathbb{N}$ will be chosen later. We have

$$k - u \leqslant H^-_{u,k} = \operatorname*{ess\,sup}_{Q(\hat{t}, \frac{R}{2})} \left|\left(u - \mu_- - \frac{\omega}{4}\right)_-\right| \leqslant \frac{\omega}{4}. \tag{2.30}$$

If $H^-_{u,k} \leqslant \frac{\omega}{8}$ the result is trivial for the choice $s^* = 3$. Assuming $H^-_{u,k} > \frac{\omega}{8}$ the logarithmic function is defined in the whole $Q(\hat{t}, \frac{R}{2})$ and it is given by

$$\Psi^- = \psi^-_{\{H^-_{u,k}, k, \omega/2^{n+2}\}}(u) = \begin{cases} \ln\left\{\frac{H^-_{u,k}}{H^-_{u,k} + u - k + \omega/2^{n+2}}\right\} & \text{if } u < k - \frac{\omega}{2^{n+2}}, \\ 0 & \text{if } u \geqslant k - \frac{\omega}{2^{n+2}}. \end{cases}$$

From (2.30), we estimate

$$\Psi^- \leqslant n \ln 2 \quad \text{since} \quad \frac{H^-_{u,k}}{H^-_{u,k} + u - k + \omega/2^{n+2}} \leqslant \frac{\omega/4}{\omega/2^{n+2}} = 2^n \tag{2.31}$$

and

$$\left|(\psi^-)'(u)\right|^{2-p} = \left(H^-_{u,k} + u - k + c\right)^{p-2} \leqslant \left(\frac{\omega}{2}\right)^{p-2}. \tag{2.32}$$

Now observe that as a consequence of Lemma 3, we have $u(x, -\hat{t}) > k$ in the cube $K_{R/2}$, which implies that

$$\Psi^-(x, -\hat{t}) = 0, \quad x \in K_{R/2}.$$

Choosing a piecewise smooth cutoff function $0 < \zeta(x) \leqslant 1$, defined in $K_{R/2}$ and such that

$$\zeta = 1 \quad \text{in } K_{R/4} \quad \text{and} \quad |\nabla\zeta| \leqslant \frac{8}{R},$$

inequality (2.6) reads

$$\sup_{-\hat{t}<t<0} \int_{K_{R/2}\times\{t\}} \left[\psi^{-}(u)\right]^2 \zeta^p \,\mathrm{d}x \leqslant C \int_{-\hat{t}}^{0} \int_{K_{R/2}} \psi^{-}(u) \left|(\psi^{-})'(u)\right|^{2-p} |\nabla\zeta|^p \,\mathrm{d}x\,\mathrm{d}t. \tag{2.33}$$

The right-hand side is estimated above, using (2.31) and (2.32), by

$$Cn(\ln 2)\left(\frac{\omega}{2}\right)^{p-2}\left(\frac{8}{R}\right)^{p} \hat{t}|K_{R/2}| \leqslant Cn2^{\lambda(p-2)}|K_{R/4}|,$$

since by (2.29),

$$\hat{t} \leqslant a_0 R^p = \left(\frac{\omega}{2^\lambda}\right)^{2-p} R^p.$$

We estimate below the left-hand side of (2.33) by integrating over the smaller set

$$S(t) = \left\{x \in K_{R/4}\colon\ u(x,t) < \mu_- + \frac{\omega}{2^{n+2}}\right\} \subset K_{R/2}$$

and observing that in S, $\zeta = 1$ and

$$\frac{H^{-}_{u,k}}{H^{-}_{u,k} + u - k + \omega/2^{n+2}} \geqslant \frac{\omega/8}{(H^{-}_{u,k} - \omega/4) + \omega/2^{n+2}} \geqslant \frac{\omega/2^3}{\omega/2^{n+2}} = 2^{n-1},$$

since $(H^{-}_{u,k} - \omega/4) \leqslant 0$. Therefore,

$$\left[\psi^{-}(u)\right]^2 \geqslant \left[\ln\left(2^{n-1}\right)\right]^2 = (n-1)^2(\ln 2)^2 \quad \text{on } S(t).$$

Combining these estimates in (2.33) we get

$$\left|\left\{x \in K_{R/4}\colon\ u(x,t) < \mu_- + \frac{\omega}{2^{n+2}}\right\}\right| \leqslant C\frac{n}{(n-1)^2} 2^{\lambda(p-2)}|K_{R/4}|,$$

for all $t \in (-\hat{t}, 0)$ and to prove the lemma we choose

$$s_* = n + 2 \quad \text{with } n > 1 + \frac{2C}{\nu_*} 2^{\lambda(p-2)}. \tag{2.34}$$

□

We now state the main result of this section.

PROPOSITION 6. *Assume* (2.20) *holds for some* t^* *as in* (2.19) *and that* (2.16) *is in force. There exist constants* $\nu_0 \in (0,1)$, $s_1 \in \mathbb{N}$, *depending only on the data, such that*

$$u(x,t) > \mu^- + \frac{\omega}{2^{s_1+1}}, \quad \text{a.e. } (x,t) \in Q\left(\hat{t}, \frac{R}{8}\right).$$

PROOF. Consider the cylinder for which (2.20) holds, let

$$R_n = \frac{R}{8} + \frac{R}{2^{n+3}}, \quad n = 0, 1, \ldots,$$

and construct the family of nested and shrinking cylinders $Q(\hat{t}, R_n)$, where $\hat{t}$ is given by (2.29). Take piecewise smooth cutoff functions $0 < \zeta_n(x) \leqslant 1$, independent of t, defined in K_{R_n} and satisfying

$$\zeta_n = 1 \quad \text{in } K_{R_{n+1}}, \qquad |\nabla \zeta_n| \leqslant \frac{2^{n+4}}{R}.$$

Write the local energy inequalities (2.5) for the functions $(u - k_n)_-$ in the cylinders $Q(\hat{t}, R_n)$, with

$$k_n = \mu_- + \frac{\omega}{2^{s_1+1}} + \frac{\omega}{2^{s_1+1+n}}, \quad n = 0, 1, \ldots,$$

and $\zeta = \zeta_n$. Observing that, due to Lemma 3, we have $u(x, -\hat{t}) > \mu_- + \frac{\omega}{4} \geqslant k_n$ in the cube $K_{R/2} \supset K_{R_n}$, which implies that

$$(u - k_n)_-(x, -\hat{t}) = 0, \quad x \in K_{R_n}, n = 0, 1, \ldots,$$

they read

$$\begin{aligned}
&\sup_{-\hat{t}<t<0} \int_{K_{R_n} \times \{t\}} (u - k_n)_-^2 \zeta_n^p \, dx + \int_{-\hat{t}}^0 \int_{K_{R_n}} |\nabla (u - k_n)_- \zeta_n|^p \, dx \, dt \\
&\quad \leqslant C \int_{-\hat{t}}^0 \int_{K_{R_n}} (u - k_n)_-^p |\nabla \zeta_n|^p \, dx \, dt \\
&\quad \leqslant C \frac{2^{p(n+4)}}{R^p} \int_{-\hat{t}}^0 \int_{K_{R_n}} (u - k_n)_-^p \, dx \, dt.
\end{aligned} \tag{2.35}$$

From (2.29) we estimate $\hat{t} \leqslant a_0 R^p = (\frac{\omega}{2^\lambda})^{2-p} R^p$, where a_0 is defined in (2.15). From this,

$$\begin{aligned}
(u - k_n)_-^2 &\geqslant \left(\frac{\omega}{2^{s_1}}\right)^{2-p} (u - k_n)_-^p \geqslant \left(\frac{2^{s_1}}{2^\lambda}\right)^{p-2} \frac{\hat{t}}{R^p} (u - k_n)_-^p \\
&\geqslant \frac{\hat{t}}{R^p} (u - k_n)_-^p,
\end{aligned}$$

provided $s_1 > \lambda$. Dividing now (2.35), by $\hat{t}/(\frac{R}{2})^p$ gives

$$\sup_{-\hat{t}<t<0} \int_{K_{R_n}\times\{t\}} (u-k_n)_-^p \zeta_n^p \, dx + \frac{R^p}{\hat{t}} \int_{-\hat{t}}^0 \int_{K_{R_n}} \left|\nabla(u-k_n)_-\zeta_n\right|^p dx\, dt$$
$$\leqslant C\frac{2^{pn}}{\hat{t}} \int_{-\hat{t}}^0 \int_{K_{R_n}} (u-k_n)_-^p \, dx\, dt.$$

The change of the time variable $\bar{t} = t(\frac{R}{2})^p/\hat{t}$, along with the new function

$$\bar{u}(\cdot,\bar{t}) = u(\cdot,t),$$

leads to the simplified inequality

$$\left\|(\bar{u}-k_n)_-\zeta_n\right\|^p_{V_0^p(Q((R/2)^p,R_n))}$$
$$\leqslant C\frac{2^{pn}}{(R/2)^p}\left(\frac{\omega}{2^{s_1}}\right)^p \int_{-(R/2)^p}^0 \int_{K_{R_n}} \chi_{\{(\bar{u}-k_n)_->0\}} \, dx\, d\bar{t}.$$

Define, for each n,

$$A_n = \int_{-(R/2)^p}^0 \int_{K_{R_n}} \chi_{\{(\bar{u}-k_n)_->0\}} \, dx\, d\bar{t}.$$

By a reasoning similar to the one leading to (2.26):

$$\frac{1}{2^{p(n+2)}}\left(\frac{\omega}{2^{s_1}}\right)^p A_{n+1} \leqslant |k_n - k_{n+1}|^p A_{n+1}$$
$$\leqslant \left\|(\bar{u}-k_n)_-\right\|^p_{p,Q((R/2)^p,R_{n+1})}$$
$$\leqslant \left\|(\bar{u}-k_n)_-\zeta_n\right\|^p_{p,Q((R/2)^p,R_n)}$$
$$\leqslant C\left\|(\bar{u}-k_n)_-\zeta_n\right\|^p_{V_0^p(Q((R/2)^p,R_n))} A_n^{p/(N+p)}$$
$$\leqslant C\frac{2^{pn}}{(R/2)^p}\left(\frac{\omega}{2^{s_1}}\right)^p A_n^{1+p/(N+p)}.$$

Next, define the numbers

$$X_n = \frac{A_n}{|Q((R/2)^p, R_n)|},$$

divide the previous inequality by $|Q((R/2)^p, R_{n+1})|$ to obtain the recursive relations

$$X_{n+1} \leqslant C4^{pn} X_n^{1+p/(N+p)}.$$

By Lemma 2 on fast geometric convergence, if

$$X_0 \leqslant C^{-(N+p)/p} 4^{-(N+p)^2/p} \equiv \nu_* \in (0,1), \tag{2.36}$$

then

$$X_n \to 0. \tag{2.37}$$

To verify (2.36), apply Lemma 4 with such a ν_* and conclude that there exists $s_* \equiv s_1$, depending only on the data, such that

$$\left|\left\{x \in K_{R/4}\colon u(x,t) < \mu_- + \frac{\omega}{2^{s_1}}\right\}\right| \leqslant \nu_* |K_{R/4}| \quad \forall t \in (-\hat{t}, 0).$$

Since (2.37) implies that $A_n \to 0$, we conclude that

$$\left|\left\{(x,\bar{t}) \in Q\left(\left(\frac{R}{2}\right)^p, \frac{R}{8}\right)\colon \bar{u}(x,\bar{t}) \leqslant \mu_- + \frac{\omega}{2^{s_1+1}}\right\}\right|$$
$$= \left|\left\{(x,t) \in Q\left(\hat{t}, \frac{R}{8}\right)\colon u(x,t) \leqslant \mu_- + \frac{\omega}{2^{s_1+1}}\right\}\right| = 0. \qquad \square$$

COROLLARY 1. *Assume* (2.20) *holds for some* t^* *as in* (2.19) *and that* (2.16) *is in force. There exist constants* $\nu_0, \sigma_0 \in (0,1)$, *depending only on the data, such that*

$$\operatorname*{ess\,osc}_{Q(d(R/8)^p, R/8)} u \leqslant \sigma_0 \omega. \tag{2.38}$$

PROOF. By Proposition 6, there exists $s_1 \in \mathbb{N}$ such that

$$\operatorname*{ess\,inf}_{Q(\hat{t}, R/8)} u \geqslant \mu^- + \frac{\omega}{2^{s_1+1}}.$$

From this,

$$\operatorname*{ess\,osc}_{Q(\hat{t}, R/8)} u \leqslant \left(1 - \frac{1}{2^{s_1+1}}\right)\omega.$$

Since $d(R/8)^p \leqslant \hat{t} = -t^* + d(R/2)^p$, $t^* < 0$, we have

$$Q\left(d\left(\frac{R}{8}\right)^p, \frac{R}{8}\right) \subset Q\left(\hat{t}, \frac{R}{8}\right),$$

and the corollary follows with $\sigma_0 = (1 - 1/2^{s_1+1})$. $\square$

2.3.3. *Analysis of the second alternative.* Assume now that the second alternative (2.21) holds true. We will show that a conclusion similar to (2.38) can be reached. Recall that the constant ν_0 has already been quantitatively determined by (2.27), and it is fixed. We continue to assume that (2.16) is in force.

LEMMA 5. *Assume* (2.21) *holds and let* (2.16) *be in force. Fix a cylinder* $(0, t^*) + Q(dR^p, R) \subset Q(a_0 R^p, R)$ *for which* (2.21) *holds. There exists a time level*

$$t^\circ \in \left[t^* - dR^p, t^* - \frac{\nu_0}{2} dR^p\right]$$

such that

$$\left|\left\{x \in K_R \colon u\big(x, t^\circ\big) > \mu_+ - \frac{\omega}{2}\right\}\right| \leqslant \left(\frac{1-\nu_0}{1-\nu_0/2}\right)|K_R|.$$

PROOF. Suppose not. Then, for all $t \in [t^* - dR^p, t^* - \frac{\nu_0}{2} dR^p]$,

$$\begin{aligned}
&\left|\left\{(x,t) \in \big(0, t^*\big) + Q\big(dR^p, R\big) \colon u(x,t) > \mu_+ - \frac{\omega}{2}\right\}\right| \\
&\quad\geqslant \int_{t^*-dR^p}^{t^*-\frac{\nu_0}{2}dR^p} \left|\left\{x \in K_R \colon u(x,\tau) > \mu_+ - \frac{\omega}{2}\right\}\right| d\tau \\
&\quad> (1-\nu_0)\big|Q\big(dR^p, R\big)\big|,
\end{aligned}$$

which contradicts (2.21). □

The next lemma asserts that the set where $u(\cdot, t)$ is close to its supremum is small, not only at the specific time level t°, but for all time levels near the top of the cylinder $(0, t^*) + Q(dR^p, R)$.

LEMMA 6. *Assume* (2.21) *holds and let* (2.16) *be in force. There exists* $1 < s_2 \in \mathbb{N}$, *depending only on the data, such that*

$$\left|\left\{x \in K_R \colon u(x,t) > \mu_+ - \frac{\omega}{2^{s_2}}\right\}\right| \leqslant \left(1 - \left(\frac{\nu_0}{2}\right)^2\right)|K_R|$$

for all $t \in [t^* - \frac{\nu_0}{2} dR^p, t^*]$.

PROOF. The proof consists in using the logarithmic inequalities (2.6) applied to the function $(u-k)_+$ in the cylinder $K_R \times (t^\circ, t^*)$, with the choices

$$k = \mu_+ - \frac{\omega}{2} \quad \text{and} \quad c = \frac{\omega}{2^{n+1}},$$

where $n \in \mathbb{N}$ will be chosen later. We have

$$u - k \leqslant H^+_{u,k} = \operatorname*{ess\,sup}_{K_R\times(t^\circ,t^*)} \left| \left(u - \mu_+ + \frac{\omega}{2} \right)_+ \right| \leqslant \frac{\omega}{2}. \tag{2.39}$$

If $H^+_{u,k} \leqslant \omega/8$ the result is trivial for the choice of $s_2 = 3$. Assuming $H^+_{u,k} > \omega/8$ the logarithmic function Ψ^+ is defined in the whole $K_R \times (t^\circ, t^*)$, and it is given by

$$\Psi^+ = \psi^+_{\{H^+_{u,k},k,\omega/2^{n+1}\}}(u) = \begin{cases} \ln\left\{\frac{H^+_{u,k}}{H^+_{u,k}-u+k+\omega/2^{n+1}}\right\} & \text{if } u > k + \frac{\omega}{2^{n+1}}, \\ 0 & \text{if } u \leqslant k + \frac{\omega}{2^{n+1}}. \end{cases}$$

From (2.39), estimate

$$\Psi^+ \leqslant n \ln 2 \quad \text{since} \quad \frac{H^+_{u,k}}{H^+_{u,k} - u + k + \omega/2^{n+1}} \leqslant \frac{\omega/2}{\omega/2^{n+1}} = 2^n, \tag{2.40}$$

and

$$\left|(\psi^+)'(u)\right|^{2-p} = \left(H^+_{u,k} - u + k + c\right)^{p-2} \leqslant \left(\frac{\omega}{2}\right)^{p-2}. \tag{2.41}$$

Choosing a piecewise smooth cutoff function $0 < \zeta(x) \leqslant 1$, defined in K_R and such that, for some $\sigma \in (0, 1)$,

$$\zeta = 1 \quad \text{in } K_{(1-\sigma)R} \quad \text{and} \quad |\nabla\zeta| \leqslant (\sigma R)^{-1},$$

inequality (2.6) reads

$$\begin{aligned} &\sup_{t^\circ<t<t^*} \int_{K_R\times\{t\}} \left[\psi^+(u)\right]^2 \zeta^p \, dx \\ &\quad \leqslant \int_{K_R\times\{t^\circ\}} \left[\psi^+(u)\right]^2 \zeta^p \, dx + C \int_{t^\circ}^{t^*} \int_{K_R} \psi^+(u) \left|(\psi^+)'(u)\right|^{2-p} |\nabla\zeta|^p \, dx \, dt. \end{aligned} \tag{2.42}$$

The first integral on the right-hand side can be bounded above using Lemma 5 and taking into account that Ψ^+ vanishes on the set $\{x \in K_R\colon\ u(x, \cdot) < k\}$. This gives

$$\int_{K_R\times\{t^\circ\}} \left[\psi^+(u)\right]^2 dx \leqslant n^2 (\ln 2)^2 \left(\frac{1 - \nu_0}{1 - \nu_0/2} \right) |K_R|.$$

To bound the second integral we use (2.40) and (2.41):

$$\begin{aligned}
&C\int_{t^\circ}^{t^*}\int_{K_R}\psi^+(u)\big|(\psi^+)'(u)\big|^{2-p}|\nabla\zeta|^p\,dx\,dt\\
&\quad\leqslant Cn(\ln 2)\left(\frac{\omega}{2}\right)^{p-2}(\sigma R)^{-p}\big(t^*-t^\circ\big)|K_R|\\
&\quad\leqslant Cn\left(\frac{\omega}{2}\right)^{p-2}\left(\frac{1}{\sigma R}\right)^p dR^p|K_R|\\
&\quad\leqslant Cn\frac{1}{\sigma^p}|K_R|
\end{aligned}$$

since $t^*-t^\circ\leqslant dR^p$. The left-hand side is estimated below by integrating over the smaller set

$$S=\left\{x\in K_{(1-\sigma)R}\colon\ u(x,t)>\mu_+-\frac{\omega}{2^{n+2}}\right\}\subset K_R$$

and observing that in S, $\zeta=1$ and

$$\begin{aligned}
\frac{H_{u,k}^+}{H_{u,k}^+-u+k+\omega/2^{n+1}}&\geqslant\frac{\omega/8}{(H_{u,k}^+-\omega/2)+\omega/2^{n+2}}\\
&\geqslant\frac{\omega/2}{\omega/2^n}=2^{n-1},
\end{aligned}$$

since $(H_{u,k}^+-\omega/2)\leqslant 0$. Therefore,

$$\big[\psi^+(u)\big]^2\geqslant\big[\ln\big(2^{n-1}\big)\big]^2=(n-1)^2(\ln 2)^2.$$

From this

$$\sup_{t^\circ<t<t^*}\int_{K_R\times\{t\}}\big[\psi^+(u)\big]^2\zeta^p\,dx\geqslant(n-1)^2(\ln 2)^2|S|.$$

Combining these three estimates, we arrive at

$$\begin{aligned}
|S|&\leqslant\left(\frac{n}{n-1}\right)^2\left(\frac{1-\nu_0}{1-\nu_0/2}\right)|K_R|+C\frac{n}{(n-1)^2}\frac{1}{\sigma^p}|K_R|\\
&\leqslant\left(\frac{n}{n-1}\right)^2\left(\frac{1-\nu_0}{1-\nu_0/2}\right)|K_R|+\frac{C}{n\sigma^p}|K_R|.
\end{aligned}$$

On the other hand,

$$\begin{aligned}
&\left|\left\{x \in K_R\colon u(x,t) > \mu_+ - \frac{\omega}{2^{n+1}}\right\}\right| \\
&\quad \leqslant \left|\left\{x \in K_{(1-\sigma)R}\colon u(x,t) > \mu_+ - \frac{\omega}{2^{n+1}}\right\}\right| + |K_R \setminus K_{(1-\sigma)R}| \\
&\quad \leqslant |S| + N\sigma|K_R|,
\end{aligned}$$

thus

$$\begin{aligned}
&\left|\left\{x \in K_R\colon u(x,t) > \mu_+ - \frac{\omega}{2^{n+1}}\right\}\right| \\
&\quad \leqslant \left\{\left(\frac{n}{n-1}\right)^2\left(\frac{1-\nu_0}{1-\nu_0/2}\right) + \frac{C}{n\sigma^p} + N\sigma\right\}|K_R|
\end{aligned}$$

for all $t \in (t^\circ, t^*)$. Choose σ so small that $N\sigma \leqslant \frac{3}{8}\nu_0^2$ and n so large that

$$\left(\frac{n}{n-1}\right)^2 \leqslant \left(1 - \frac{\nu_0}{2}\right)(1+\nu_0) \equiv \beta \quad \text{and} \quad \frac{C}{n\sigma^p} \leqslant \frac{3}{8}\nu_0^2. \tag{2.43}$$

Note that $\beta > 1$. With this choice of n, the lemma follows with $s_2 = n+1$. □

The same type of conclusion holds in an upper portion of the full cylinder $Q(a_0R^p, R)$, say for all $t \in (-\frac{a_0}{2}R^p, 0)$. Indeed, (2.21) holds for *all* cylinders of the type $(0, t^*) + Q(dR^p, R)$ so that the conclusion of the previous lemma holds true for all time levels

$$t \geqslant -(a_0 - d)R^p - \frac{\nu_0}{2}dR^p.$$

So if we choose

$$2^{(\lambda-1)(p-2)} \geqslant 2, \tag{2.44}$$

we get $\frac{a_0}{d} \geqslant 2 - \nu_0$, which is equivalent to

$$-(a_0 - d)R^p - \frac{\nu_0}{2}dR^p \leqslant -\frac{a_0}{2}R^p.$$

COROLLARY 2. *Assume* (2.21) *holds and let* (2.16) *be in force. For all* $t \in (-\frac{a_0}{2}R^p, 0)$,

$$\left|\left\{x \in K_R\colon u(x,t) > \mu_+ - \frac{\omega}{2^{s_2}}\right\}\right| \leqslant \left(1 - \left(\frac{\nu_0}{2}\right)^2\right)|K_R|.$$

The main result of this section states that in fact u is strictly below its supremum μ^+ in a smaller cylinder with the same vertex and axis as $Q(\frac{a_0}{2}R^p, R)$.

PROPOSITION 7. *Assume* (2.21) *holds and let* (2.16) *be in force. The choice of* λ *can be made so that*

$$u(x,t) \leqslant \mu_+ - \frac{\omega}{2^{\lambda+1}}, \quad a.e.\ (x,t) \in Q\left(\frac{a_0}{2}\left(\frac{R}{2}\right)^p, \frac{R}{2}\right). \tag{2.45}$$

PROOF. Define

$$R_n = \frac{R}{2} + \frac{R}{2^{n+1}}, \quad n = 0, 1, \ldots,$$

and construct the family of nested shrinking cylinders $Q(\frac{a_0}{2} R_n^p, R_n)$. Consider piecewise smooth cutoff functions $0 < \zeta_n \leqslant 1$, defined in these cylinders and satisfying the following set of assumptions:

$$\zeta_n = 1 \quad \text{in } Q\left(\frac{a_0}{2} R_{n+1}^p, R_{n+1}\right), \qquad \zeta_n = 0 \quad \text{in } \partial_p Q\left(\frac{a_0}{2} R_n^p, R_n\right),$$

$$|\nabla \zeta_n| \leqslant \frac{2^{n+1}}{R}, \qquad 0 \leqslant \partial_t \zeta_n \leqslant \frac{2^{p(n+1)}}{\frac{a_0}{2} R^p}.$$

The energy inequality (2.5) for the functions $(u - k_n)_+$, with

$$k_n = \mu_+ - \frac{\omega}{2^{\lambda+1}} - \frac{\omega}{2^{\lambda+1+n}}, \quad n = 0, 1, \ldots,$$

in the cylinders $Q(\frac{a_0}{2} R_n^p, R_n)$ and with $\zeta = \zeta_n$, reads

$$\begin{aligned}
&\sup_{-\frac{a_0}{2} R_n^p < t < 0} \int_{K_{R_n} \times \{t\}} (u - k_n)_+^2 \zeta_n^p \, dx + \int_{-\frac{a_0}{2} R_n^p}^0 \int_{K_{R_n}} |\nabla (u - k_n)_+ \zeta_n|^p \, dx\, dt \\
&\quad \leqslant C \int_{-\frac{a_0}{2} R_n^p}^0 \int_{K_{R_n}} (u - k_n)_+^p |\nabla \zeta_n|^p \, dx\, dt \\
&\qquad + C \int_{-\frac{a_0}{2} R_n^p}^0 \int_{K_{R_n}} (u - k_n)_+^2 \zeta_n^{p-1} \partial_t \zeta_n \, dx\, dt \\
&\quad \leqslant C \frac{2^{p(n+1)}}{R^p} \left\{ \int_{-\frac{a_0}{2} R_n^p}^0 \int_{K_{R_n}} (u - k_n)_+^p \, dx\, dt \right. \\
&\qquad \left. + \frac{2}{a_0} \int_{-\frac{a_0}{2} R_n^p}^0 \int_{K_{R_n}} (u - k_n)_+^2 \, dx\, dt \right\}.
\end{aligned} \tag{2.46}$$

Observe that

$$(u - k_n)_+^2 = (u - k_n)_+^{2-p} (u - k_n)_+^p \geqslant \left(\frac{\omega}{2^\lambda}\right)^{2-p} (u - k_n)_+^p.$$

Therefore, from (2.46),

$$\left(\frac{\omega}{2^\lambda}\right)^{2-p} \sup_{-\frac{a_0}{2}R_n^p<t<0} \int_{K_{R_n}\times\{t\}} (u-k_n)_+^p \zeta_n^p \,\mathrm{d}x + \int_{-\frac{a_0}{2}R_n^p}^0 \int_{K_{R_n}} \left|\nabla(u-k_n)_+\zeta_n\right|^p \mathrm{d}x\,\mathrm{d}t \leqslant C\frac{2^{p(n+1)}}{R^p}\left\{\left(\frac{\omega}{2^\lambda}\right)^p + \frac{2}{a_0}\left(\frac{\omega}{2^\lambda}\right)^2\right\} \int_{-\frac{a_0}{2}R_n^p}^0 \int_{K_{R_n}} \chi_{\{(u-k_n)_+>0\}}\,\mathrm{d}x\,\mathrm{d}t. \tag{2.47}$$

Now recall that $a_0 = (\omega/2^\lambda)^{2-p}$, and divide (2.47) by a_0 to get

$$\sup_{-\frac{a_0}{2}R_n^p<t<0} \int_{K_{R_n}\times\{t\}} (u-k_n)_+^p \zeta_n^p \,\mathrm{d}x + \frac{1}{a_0}\int_{-\frac{a_0}{2}R_n^p}^0 \int_{K_{R_n}} \left|\nabla(u-k_n)_+\zeta_n\right|^p \mathrm{d}x\,\mathrm{d}t \leqslant \frac{C2^{pn}}{R^p}\left(\frac{\omega}{2^\lambda}\right)^p \frac{1}{a_0}\int_{-\frac{a_0}{2}R_n^p}^0 \int_{K_{R_n}} \chi_{\{(u-k_n)_+>0\}}\,\mathrm{d}x\,\mathrm{d}t. \tag{2.48}$$

Next, perform a change in the time variable in (2.48), putting

$$\bar{t} = \frac{t}{a_0/2}$$

and defining

$$\bar{u}(\cdot,\bar{t}) = u(\cdot,t) \quad \text{and} \quad \bar{\zeta}_n(\cdot,\bar{t}) = \zeta_n(\cdot,t),$$

and obtain the simplified inequality

$$\left\|(\bar{u}-k_n)_+\bar{\zeta}_n\right\|_{V_0^p(Q(R_n^p,R_n))}^p \leqslant \frac{C2^{pn}}{R^p}\left(\frac{\omega}{2^\lambda}\right)^p \int_{-R_n^p}^0 \int_{K_{R_n}} \chi_{\{(\bar{u}-k_n)_+>0\}}\,\mathrm{d}x\,\mathrm{d}\bar{t}. \tag{2.49}$$

Define, for each n,

$$A_n = \int_{-R_n^p}^0 \int_{K_{R_n}} \chi_{\{(\bar{u}-k_n)_+>0\}}\,\mathrm{d}x\,\mathrm{d}\bar{t}$$

and estimate

$$\begin{aligned}
\frac{1}{2^{p(n+2)}}\left(\frac{\omega}{2^{\lambda}}\right)^{p} A_{n+1} &\leqslant |k_{n+1}-k_n|^p A_{n+1} \\
&\leqslant \left\|(\bar{u}-k_n)_+\right\|^p_{p,Q(R^p_{n+1},R_{n+1})} \\
&\leqslant \left\|(\bar{u}-k_n)_+\bar{\zeta_n}\right\|^p_{p,Q(R^p_n,R_n)} \\
&\leqslant C\left\|(\bar{u}-k_n)_+\bar{\zeta_n}\right\|^p_{V_0^p(Q(R^p_n,R_n))} A_n^{p/(N+p)} \\
&\leqslant \frac{C2^{pn}}{R^p}\left(\frac{\omega}{2^{\lambda}}\right)^{p} A_n^{1+p/(N+p)}.
\end{aligned} \tag{2.50}$$

The first three inequalities are obvious; and the fourth is a consequence of the embedding Theorem 1. Define the numbers

$$X_n = \frac{A_n}{|Q(R^p_n, R_n)|},$$

divide (2.50) by $\left|Q(R^p_{n+1}, R_{n+1})\right|$, and obtain the recursive relation

$$X_{n+1} \leqslant C4^{pn} X_n^{1+p/(N+p)}.$$

By Lemma 2 on fast geometric convergence, if

$$X_0 \leqslant C^{-(N+p)/p} 4^{-(N+p)^2/p} \equiv \nu_*, \tag{2.51}$$

then

$$X_n \to 0. \tag{2.52}$$

Thus if (2.51) holds,

$$\left|\left\{(x,t)\in Q\left(\frac{a_0}{2}\left(\frac{R}{2}\right)^p, \frac{R}{2}\right):\ u(x,t) > \mu_+ - \frac{\omega}{2^{\lambda+1}}\right\}\right| = 0$$

and the result follows.

We are left to prove (2.51). To simplify the symbolism introduce the sets

$$B_\sigma(t) = \left\{x \in K_R:\ u(x,t) > \mu^+ - \frac{\omega}{2^\sigma}\right\}$$

and

$$B_\sigma = \left\{(x,t)\in Q\left(\frac{a_0}{2}R^p, R\right):\ u(x,t) > \mu_+ - \frac{\omega}{2^\sigma}\right\}.$$

With this notation (2.51) reads

$$|B_\lambda| \leqslant \nu_* \left| Q\left(\frac{a_0}{2} R^p, R\right) \right|.$$

We will use the information contained in Corollary 2 to show that this holds, i.e., that the subset of the cylinder $Q(\frac{a_0}{2} R^p, R)$ where u is close to its supremum μ^+ can be made arbitrarily small. Consider the local energy inequalities (2.5) for the functions $(u-k)_+$ in the cylinders $Q(a_0 R^p, 2R)$, with

$$k = \mu_+ - \frac{\omega}{2^s},$$

where s will be chosen later satisfying $s_2 \leqslant s \leqslant \lambda$ (recall that s_2 was chosen on Lemma 6). Take a piecewise smooth cutoff function $0 < \zeta \leqslant 1$, defined in $Q(a_0 R^p, 2R)$, and such that

$$\zeta = 1 \quad \text{in } Q\left(\frac{a_0}{2} R^p, R\right), \qquad \zeta = 0 \quad \text{in } \partial_p Q\big(a_0 R^p, 2R\big),$$

$$|\nabla \zeta| \leqslant \frac{1}{R}, \qquad 0 \leqslant \partial_t \zeta \leqslant \frac{2}{a_0 R^p}.$$

Neglecting the first term on the left-hand side of the estimates, we obtain for the indicated choices,

$$\begin{aligned} &\int_{-\frac{a_0}{2} R^p}^{0} \int_{K_R} \left|\nabla (u-k)_+\right|^p \mathrm{d}x\,\mathrm{d}t \\ &\quad \leqslant \frac{C}{R^p} \int_{-a_0 R^p}^{0} \int_{K_{2R}} (u-k)_+^p \,\mathrm{d}x\,\mathrm{d}t + \frac{C}{a_0 R^p} \int_{-a_0 R^p}^{0} \int_{K_{2R}} (u-k)_+^2 \,\mathrm{d}x\,\mathrm{d}t. \end{aligned} \tag{2.53}$$

We estimate the two terms on the right-hand side of this inequality as follows:

$$\frac{C}{R^p} \int_{-a_0 R^p}^{0} \int_{K_{2R}} (u-k)_+^p \,\mathrm{d}x\,\mathrm{d}t \leqslant \frac{C}{R^p} \left(\frac{\omega}{2^s}\right)^p \left| Q\left(\frac{a_0}{2} R^p, R\right) \right|$$

and, recalling the definition of a_0,

$$\begin{aligned} \frac{C}{a_0 R^p} \int_{-a_0 R^p}^{0} \int_{K_{2R}} (u-k)_+^2 \,\mathrm{d}x\,\mathrm{d}t &\leqslant \frac{C}{R^p} \left(\frac{\omega}{2^\lambda}\right)^{p-2} \left(\frac{\omega}{2^s}\right)^2 \left| Q\left(\frac{a_0}{2} R^p, R\right) \right| \\ &\leqslant \frac{C}{R^p} \left(\frac{\omega}{2^s}\right)^p \left| Q\left(\frac{a_0}{2} R^p, R\right) \right| \end{aligned}$$

since $s \leqslant \lambda$. Putting this in (2.53) gives

$$\iint_{B_s} |\nabla u|^p \,\mathrm{d}x\,\mathrm{d}t \leqslant \frac{C}{R^p} \left(\frac{\omega}{2^s}\right)^p \left| Q\left(\frac{a_0}{2} R^p, R\right) \right|. \tag{2.54}$$

We next apply Lemma 1 to the function $u(\cdot, t)$, for all $-\frac{a_0}{2}R^p \leqslant t \leqslant 0$, and with

$$k = \mu_+ - \frac{\omega}{2^s}, \qquad l = \mu_+ - \frac{\omega}{2^{s+1}}, \qquad l - k = \frac{\omega}{2^{s+1}}.$$

Observing that, owing to Corollary 2,

$$\left|\left\{x \in K_R\colon u(x,t) < \mu_+ - \frac{\omega}{2^s}\right\}\right| \equiv |K_R| - \left|B_s(t)\right| \geqslant \left(\frac{\nu_0}{2}\right)^2 |K_R|,$$

we obtain

$$\frac{\omega}{2^{s+1}}\left|B_{s+1}(t)\right| \leqslant \frac{4CR^{N+1}}{\nu_0^2|K_R|}\int_{B_s(t)\setminus B_{s+1}(t)} |\nabla u|\,\mathrm{d}x\,\mathrm{d}t$$

for $t \in (-\frac{a_0}{2}R^p, 0)$. Integrating over this interval, we conclude that

$$\begin{aligned}\frac{\omega}{2^{s+1}}|B_{s+1}| &\leqslant \frac{CR}{\nu_0^2}\int\!\!\int_{B_s\setminus B_{s+1}} |\nabla u|\,\mathrm{d}x\,\mathrm{d}t \\ &\leqslant \frac{CR}{\nu_0^2}\left(\int\!\!\int_{B_s} |\nabla u|^p\,\mathrm{d}x\,\mathrm{d}t\right)^{1/p} |B_s \setminus B_{s+1}|^{(p-1)/p} \\ &\leqslant \frac{C}{\nu_0^2}\left(\frac{\omega}{2^s}\right)\left|Q\left(\frac{a_0}{2}R^p, R\right)\right|^{1/p} |B_s \setminus B_{s+1}|^{(p-1)/p},\end{aligned}$$

using also (2.54). Taking the $\frac{p}{p-1}$ power and dividing through by $(\omega/2^{s+1})^{p/(p-1)}$, we obtain

$$|B_{s+1}|^{p/(p-1)} \leqslant C(\nu_0)^{-2p/(p-1)}\left|Q\left(\frac{a_0}{2}R^p, R\right)\right|^{1/(p-1)} |B_s \setminus B_{s+1}|.$$

Since these inequalities are valid for $s_2 \leqslant s \leqslant \lambda$, we add them for

$$s = s_2, s_2 + 1, s_2 + 2, \ldots, \lambda - 1,$$

and since the sum on the right-hand side can be bounded above by $|Q(\frac{a_0}{2}R^p, R)|$, we obtain

$$(\lambda - s_2)|B_\lambda|^{p/(p-1)} \leqslant C(\nu_0)^{-2p/(p-1)}\left|Q\left(\frac{a_0}{2}R^p, R\right)\right|^{p/(p-1)},$$

that is,

$$|B_\lambda| \leqslant \frac{C}{(\lambda - s_2)^{(p-1)/p}}(\nu_0)^{-2}\left|Q\left(\frac{a_0}{2}R^p, R\right)\right|.$$

We obtain (2.51) if λ is chosen so large that

$$\frac{C}{\nu_0^2(\lambda - s_2)^{(p-1)/p}} \leqslant \nu_*.$$

We finally make the choice

$$\lambda = \max\left\{s_2 + \left(\frac{C}{\nu_0^2 \nu_*}\right)^{p/(p-1)}, 1 + \frac{1}{p-2}\right\} \tag{2.55}$$

(recall that s_2 is given through (2.43), ν_0 is given by (2.27), and ν_* is given by (2.51)) thus concluding the proof of the proposition. □

REMARK 3. Observe that the choice of λ was made so that (2.44) holds, and a larger λ is admissible. Choosing λ determines the length of the cylinder $Q(a_0 R^p, R)$, since $a_0 = (\omega/2^\lambda)^{2-p}$. The proposition has a double scope: we determine a level $\mu_+ - \omega/2^{\lambda+1}$ and a cylinder (fixing λ and consequently a_0) such that the conclusion holds in that particular cylinder.

COROLLARY 3. *Assume* (2.21) *holds and let* (2.16) *be in force. There exist constants* $\nu_0, \sigma_1 \in (0, 1)$, *depending only on the data, such that if* (2.21) *holds then*

$$\operatorname*{ess\,osc}_{Q(a_0/2(R/2)^p, R/2)} u \leqslant \sigma_1 \omega.$$

PROOF. It is similar to the proof of Corollary 1 for the choice

$$\sigma_1 = \left(1 - \frac{1}{2^{\lambda+1}}\right).$$ □

2.3.4. *The Hölder continuity.* We finally prove the Hölder continuity of weak solutions. An immediate consequence of Corollaries 1 and 3 is

PROPOSITION 8. *There exists a constant* $\sigma \in (0, 1)$, *that depends only on the data, such that*

$$\operatorname*{ess\,osc}_{Q(d(R/8)^p, R/8)} u \leqslant \sigma\omega.$$

PROOF. Assume (2.16) is in force. Then by Corollaries 1 and 3

$$\operatorname*{ess\,osc}_{Q(d(R/8)^p, R/8)} u \leqslant \sigma\omega, \quad \text{where } \sigma = \max\{\sigma_0, \sigma_1\}, \tag{2.56}$$

since

$$d\left(\frac{R}{8}\right)^p \leqslant \frac{a_0}{2}\left(\frac{R}{2}\right)^p.$$ □

We define now an iteration process that will lead to the Hölder continuity of u.

PROPOSITION 9. *There exists a positive constant C, depending only on the data, such that defining the sequences*

$$R_n = C^{-n} R \quad \textit{and} \quad \omega_n = \sigma^n \omega$$

for $n = 0, 1, 2, \ldots$, where $\sigma \in (0, 1)$ is given by (2.56), *and constructing the family of cylinders*

$$Q_n = Q\left(a_n R_n^p, R_n\right), \quad \textit{with } a_n = \left(\frac{\omega_n}{2^\lambda}\right)^{2-p},$$

where $\lambda > 1$ is given by (2.55), *we have*

$$Q_{n+1} \subset Q_n \quad \textit{and} \quad \operatorname*{ess\,osc}_{Q_n} u \leqslant \omega_n \tag{2.57}$$

for all $n = 0, 1, 2, \ldots$.

PROOF. Recall the definition of $a_0 = (\omega/2^\lambda)^{2-p}$ and the construction of the initial cylinder so that the starting relation

$$\operatorname*{ess\,osc}_{Q_0} u \leqslant \omega \tag{2.58}$$

holds. We find

$$\begin{aligned}
d\left(\frac{R}{8}\right)^p &= \left(\frac{\omega}{2}\right)^{2-p} \frac{R^p}{8^p} \\
&= \left(\frac{\omega}{2}\right)^{2-p} \left(\frac{2^\lambda}{\omega_1}\right)^{2-p} \left(\frac{\omega_1}{2^\lambda}\right)^{2-p} \frac{R^p}{8^p} \\
&= \left(\frac{\omega}{\omega_1}\right)^{2-p} \left(\frac{2^\lambda}{2}\right)^{2-p} \left(\frac{\omega_1}{2^\lambda}\right)^{2-p} \frac{R^p}{8^p} \\
&= \sigma^{p-2} 2^{(\lambda-1)(2-p)-3p} a_1 R^p \\
&= a_1 R_1^p,
\end{aligned}$$

where $R_1 = C^{-1} R$, provided C is chosen from

$$C = \sigma^{(2-p)/p} 2^{(\lambda-1)(p-2)/p+3} > 8.$$

From Proposition 8, we conclude

$$\operatorname*{ess\,osc}_{Q_1} u \leqslant \operatorname*{ess\,osc}_{Q(d(R/8)^p, R/8)} u \leqslant \sigma\omega = \omega_1$$

which puts us back to the setting of (2.58). The entire process can now be repeated inductively starting from Q_1. □

REMARK 4. The proof of Proposition 9 shows that it would have been sufficient to work with a number ω and a cylinder $Q(a_0 R^p, R)$ linked by (2.17). This relation is in general not verifiable a priori for a given cylinder, since its dimensions would have to be intrinsically defined in terms of the essential oscillation of u within it.

The role of having introduced the cylinder $Q(R^2, R)$ and having assumed (2.16) is that (2.17) holds true for the *constructed* box $Q(a_0 R^p, R)$. It is part of the proof of Proposition 9 to show that, at each step, the cylinders Q_n and the essential oscillation of u within them satisfy the intrinsic geometry dictated by (2.17).

LEMMA 7. *There exist constants $\gamma > 1$ and $\alpha \in (0,1)$, that can be determined a priori in terms of the data, such that, for all the cylinders*

$$Q(a_0\rho^p, \rho), \quad \text{with } 0 < \rho \leqslant R,$$

$$\operatorname*{ess\,osc}_{Q(a_0\rho^p,\rho)} u \leqslant \gamma\omega\left(\frac{\rho}{R}\right)^\alpha.$$

PROOF. Let $0 < \rho \leqslant R$ be fixed. There exists a nonnegative integer n such that

$$C^{-(n+1)}R \leqslant \rho \leqslant C^{-n}R.$$

So, putting $\alpha = -\frac{\ln\sigma}{\ln C}$, we deduce

$$C^{-(n+1)} \leqslant \frac{\rho}{R} \iff \sigma^{(n+1)/\alpha} \leqslant \frac{\rho}{R} \iff \sigma^{n+1} \leqslant \left(\frac{\rho}{R}\right)^\alpha.$$

Thus

$$\omega_n = \sigma^n\omega \leqslant \gamma\omega\left(\frac{\rho}{R}\right)^\alpha \quad \text{with } \gamma = \sigma^{-1}.$$

To conclude the proof, observe that the cylinder $Q(a_0\rho^p, \rho)$ is contained in the cylinder $Q_n \equiv Q(a_n R_n^p, R_n)$ since $\omega_n \leqslant \omega$ and $\rho \leqslant C^{-n}R \equiv R_n$. □

Let $\Gamma = \partial_p \Omega_T$ be the parabolic boundary of Ω_T. Introduce the degenerate intrinsic parabolic p-distance from a compact set $K \subset \Omega_T$ to Γ, by

$$p - \operatorname{dist}(K; \Gamma) \equiv \inf_{(x,t)\in K, (y,s)\in\Gamma} \left(|x-y| \wedge M^{(p-2)/p}|t-s|^{1/p}\right).$$

THEOREM 2. *Let u be a bounded local weak solution of* (2.1) *in Ω_T and $M = \|u\|_{\infty,\Omega_T}$. Then u is locally Hölder continuous in Ω_T, i.e., there exist constants $\gamma > 1$ and $\alpha \in (0,1)$, depending only upon the data, such that, for every compact subset K of Ω_T,*

$$\left|u(x_1,t_1) - u(x_2,t_2)\right| \leqslant \gamma M\left(\frac{|x_1 - x_2| + M^{(p-2)/p}|t_1 - t_2|^{1/p}}{p - \operatorname{dist}(K;\Gamma)}\right)^{\alpha}$$

for every pair of points $(x_i,t_i) \in K$, $i = 1,2$.

PROOF. Fix $(x_i,t_i) \in K$, $i = 1,2$, such that $t_2 > t_1$ and construct the cylinder $(x_2,t_2) + Q(M^{2-p}R^p, R) \subset \Omega_T$. This is realized if we choose

$$R \leqslant \inf_{x\in K, y\in\partial\Omega} |x-y| \quad \text{and} \quad M^{(2-p)/p}R \leqslant \inf_{t\in K} t^{1/p}.$$

Thus, in particular, we may choose

$$2R = p - \operatorname{dist}(K;\Gamma).$$

To prove the Hölder continuity in the t-variable assume first that $(t_2 - t_1) < M^{2-p}R^p$. Then there exists $\rho \in (0,R)$ such that $(t_2 - t_1) = M^{2-p}\rho^p$, i.e., $\rho = M^{(p-2)/p}|t_2 - t_1|^{1/p}$. The oscillation inequality of Lemma 7, applied in the cylinder $(x_2,t_2) + Q(a_0\rho^p, \rho)$ implies

$$\left|u(x_2,t_2) - u(x_2,t_1)\right| \leqslant \gamma M\left(\frac{M^{(p-2)/p}|t_2 - t_1|^{1/p}}{p - \operatorname{dist}(K;\Gamma)}\right)^{\alpha}.$$

If $(t_2 - t_1) \geqslant M^{2-p}R^p$, we have

$$\left|u(x_2,t_2) - u(x_2,t_1)\right| \leqslant 2M \leqslant 4M\left(\frac{M^{(p-2)/p}|t_2 - t_1|^{1/p}}{p - \operatorname{dist}(K;\Gamma)}\right).$$

The Hölder continuity in the space variables is proved analogously. □

REMARK 5. The theory extends to full quasilinear equations and includes statements of regularity up to the parabolic boundary of Ω_T (see [55]).

2.4. *The singular case* $1 < p < 2$

We now turn to the singular case $1 < p < 2$, still for the model equation

$$u_t - \operatorname{div}|\nabla u|^{p-2}\nabla u = 0, \quad 1 < p < 2. \tag{2.59}$$

The analysis for this case is somehow more involved, but several of the previous techniques apply. As before the Hölder continuity of u will be solely a consequence of the Caccioppoli inequalities (2.5) and the logarithmic inequalities (2.6). Throughout this section the function u is merely assumed to satisfy such inequalities. To avoid repetition of arguments, we will outline the approach to regularity emphasizing the main differences with respect to the degenerate situation.

A key point is again the choice of the appropriate intrinsic geometry in which to carry the iteration argument. Fix a point $(x_0, t_0) \in \Omega_T$ and, as before, assume $(x_0, t_0) = (0, 0)$. Consider a cylinder

$$B_{R^{p/2}}(0) \times \left(-R^p, 0\right) \equiv Q\left(R^p, R^{p/2}\right) \subset \Omega_T,$$

where $R > 0$ is taken such that the inclusion holds. Now, let

$$\mu_- := \operatorname*{ess\,inf}_{Q(R^p, R^{p/2})} u, \qquad \mu_+ := \operatorname*{ess\,sup}_{Q(R^p, R^{p/2})} u, \qquad \omega := \operatorname*{ess\,osc}_{Q(R^p, R^{p/2})} u = \mu_+ - \mu_-$$

and construct the cylinder

$$Q\left(R^p, c_0 R\right) \quad \text{with } c_0 = \left(\frac{\omega}{2^\lambda}\right)^{(p-2)/p}, \tag{2.60}$$

where λ is to be determined only in terms of the data (we use the same letter as in the case $p > 2$ but the λ's are different in the two cases).

To start the iteration, we assume

$$\left(\frac{\omega}{2^\lambda}\right)^{(p-2)/p} < R^{(p-2)/2} \tag{2.61}$$

(otherwise, $\omega \leqslant 2^\lambda R^{p/2}$ and there is nothing to prove) so that

$$Q\left(R^p, c_0 R\right) \subset Q\left(R^p, R^{p/2}\right)$$

and

$$\operatorname*{ess\,osc}_{Q(R^p, c_0 R)} u \leqslant \omega. \tag{2.62}$$

Cylinders of the type (2.60) have the space variables stretched by a factor $(\omega/2^\lambda)^{(p-2)/p}$ which is intrinsically determined by the solution. Note that, if $p = 2$, these are just the standard parabolic cylinders. The geometry chosen is not the only possible. We could have introduced, for example, a scaling with different parameters in the space and the time variables. Another possibility would be to work with a scaling formally identical to the one used in the degenerate case. Our option here was dictated only by a matter of simplicity.

The main result leading to the Hölder continuity of solutions is the following proposition.

PROPOSITION 10. *There exist constants $\eta \in (0,1)$ and $C, \lambda > 1$, that can be determined only in terms of the data, satisfying the following. Construct the sequences*

$$R_n = C^{-n} R, \qquad \omega_n = \eta^n \omega, \qquad n = 0, 1, 2, \ldots,$$

and the cylinders

$$Q_n \equiv Q(R_n^p, c_n R_n) \quad \text{with } c_n = \left(\frac{\omega_n}{2^\lambda}\right)^{(p-2)/p}, n = 0, 1, 2, \ldots.$$

Then, for all $n = 0, 1, 2, \ldots$,

$$Q_{n+1} \subset Q_n \quad \text{and} \quad \operatorname*{ess\,osc}_{Q_n} u \leqslant \omega_n.$$

This proposition implies an analogue to Lemma 7 from which the Hölder continuity follows as in the proof of Theorem 2.

The proof of the proposition is, as in the degenerate case, based on the analysis of an alternative. To begin, consider inside of $Q(R^p, c_0 R)$ subcylinders of smaller size

$$(\bar{x}, 0) + Q(R^p, d_0 R), \quad d_0 = \left(\frac{\omega}{2}\right)^{(p-2)/p} \tag{2.63}$$

These cylinders are contained in $Q(R^p, c_0 R)$ if $\bar{x}$ ranges over the cube $K_{\mathcal{R}(\omega)}$, where

$$\mathcal{R}(\omega) \equiv \left\{(2^{\lambda-1})^{(2-p)/p} - 1\right\}\left(\frac{\omega}{2}\right)^{(p-2)/p} R = L_0 d_0 R \tag{2.64}$$

for $L_0 = (2^{\lambda-1})^{(2-p)/p} - 1$.

We can regard these cylinders as boxes moving inside $Q(R^p, c_0 R)$ as the coordinates $\bar{x}$ of their centers range over the cube $K_{\mathcal{R}(\omega)}$. We may arrange L_0 to be an integer and consider the cube $K_{c_0 R}$ as the union, up to a set of measure zero, of L_0^N disjoint cubes each of them congruent to $K_{d_0 R}$. Analogously, $Q(R^p, c_0 R)$ is the disjoint union, up to a set of measure zero, of L_0^N open boxes each congruent to $Q(R^p, d_0 R)$. Then we can view $(\bar{x}, 0) + Q(R^p, d_0 R)$ as the blocks of a partition of $Q(R^p, c_0 R)$ (see [55], Figure 3.1, p. 82).

Let $\nu_0 \in (0,1)$; then either

The first alternative. There exists a cylinder of the type $(\bar{x}, 0) + Q(R^p, d_0 R)$ for which

$$\frac{|\{(x,t) \in (\bar{x}, 0) + Q(R^p, d_0 R) \colon u(x,t) < \mu_- + \omega/2\}|}{|Q(R^p, d_0 R)|} \leqslant \nu_0 \tag{2.65}$$

or

The second alternative. For every cylinder of the type $(\bar{x},0)+Q(R^p,d_0R)$,

$$\frac{|\{(x,t)\in(\bar{x},0)+Q(R^p,d_0R)\colon u(x,t)<\mu_-+\omega/2\}|}{|Q(R^p,d_0R)|}>\nu_0. \tag{2.66}$$

In both cases, we will conclude that the essential oscillation of u within a smaller cylinder, centered at the origin, decreases in a way that can be quantitatively measured.

2.4.1. *Rescaled iterations.* The study of both alternatives makes crucial use of a rescaled iteration technique which applies to any subcylinder of Ω_T. Let $m>0$ be given by

$$m=m_1+m_2,\qquad \text{where } m_1\geqslant 1 \text{ and } m_2\geqslant 0,$$

and consider the cube

$$K_{d_1R}\equiv\Bigl\{x\in\mathbb{R}^N\colon \max_{1\leqslant i\leqslant N}|x_i|<d_1R\Bigr\},\qquad d_1=\left(\frac{\omega}{2^{m_1}}\right)^{(p-2)/p}$$

and the box

$$\mathcal{Q}_R(m_1,m_2)\equiv K_{d_1R}\times\bigl(-2^{m_2(p-2)}R^p,0\bigr). \tag{2.67}$$

Fix $(\bar{x},\bar{t})\in\Omega_T$, and let $R>0$ be so small that

$$\bigl(\bar{x},\bar{t}\bigr)+\mathcal{Q}_R(m_1,m_2)\subset\Omega_T.$$

REMARK 6. Note that, for $(\bar{x},\bar{t})=(0,0)$, $m_1=\lambda$ and $m_2=0$, the cylinder $(\bar{x},\bar{t})+\mathcal{Q}_R(m_1,m_2)$ coincides with the cylinder $Q(R^p,c_0R)$. Analogously, if $m_2=0$, $m_1=1$ and $\bar{t}=0$ then, for a suitable choice of $\bar{x}$, the cylinder $(\bar{x},\bar{t})+\mathcal{Q}_R(m_1,m_2)$ coincides with one of the boxes $(\bar{x},0)+Q(R^p,d_0R)$ making up the partition of $Q(R^p,c_0R)$.

LEMMA 8. *There exists a number ν_0, that can be determined a priori only in terms of the data and independent of ω, R and m_1,m_2, such that:*

- *If u is a supersolution of* (2.59) *in $(\bar{x},\bar{t})+\mathcal{Q}_R(m_1,m_2)$ satisfying*

$$\operatorname*{ess\,osc}_{(\bar{x},\bar{t})+\mathcal{Q}_R(m_1,m_2)} u\leqslant\omega$$

and

$$\left|\left\{(x,t)\in\bigl(\bar{x},\bar{t}\bigr)+\mathcal{Q}_R(m_1,m_2)\colon u(x,t)<\mu_-+\frac{\omega}{2^m}\right\}\right|\leqslant\nu_0\bigl|\mathcal{Q}_R(m_1,m_2)\bigr|$$

then

$$u(x,t)\geqslant\mu_-+\frac{\omega}{2^{m+1}}\qquad\forall(x,t)\in\bigl(\bar{x},\bar{t}\bigr)+\mathcal{Q}_{R/2}(m_1,m_2).$$

Analogously,

- *If u is a subsolution of* (2.59) *in* $(\bar{x},\bar{t})+Q_R(m_1,m_2)$ *satisfying*

$$\operatorname*{ess\,osc}_{(\bar{x},\bar{t})+Q_R(m_1,m_2)} u \leqslant \omega$$

and

$$\left|\left\{(x,t)\in(\bar{x},\bar{t})+Q_R(m_1,m_2)\colon u(x,t)>\mu_+-\frac{\omega}{2^m}\right\}\right|\leqslant \nu_0\left|Q_R(m_1,m_2)\right|$$

then

$$u(x,t)\leqslant \mu_+-\frac{\omega}{2^{m+1}}\qquad \forall(x,t)\in(\bar{x},\bar{t})+Q_{R/2}(m_1,m_2).$$

PROOF. We only prove the statement concerning supersolutions (for subsolutions the proof is similar). For simplicity, assume $(\bar{x},\bar{t})=(0,0)$ and construct the decreasing sequences of numbers

$$R_n=\frac{R}{2}+\frac{R}{2^{n+1}},\qquad k_n=\mu_-+\frac{\omega}{2^{m+1}}+\frac{\omega}{2^{m+1+n}},\qquad n=0,1,2,\dots,$$

and the families of nested cubes and cylinders

$$K_n\equiv K_{d_1R_n},\quad d_1=\left(\frac{\omega}{2^{m_1}}\right)^{(p-2)/p}$$

$$Q_n\equiv Q_{R_n}(m_1,m_2)=K_n\times\left(-2^{(p-2)m_2}R_n^p,0\right).$$

Consider the energy estimate (2.5), written for the functions $(u-k_n)_-$ over the boxes Q_n, taking as cutoff functions ξ_n

$$\begin{cases} 0<\xi_n\leqslant 1 \quad \text{in } Q_n \quad \text{and} \quad \xi_n\equiv 1 \quad \text{in } Q_{n+1},\\ \xi_n\equiv 0 \qquad\qquad\qquad\qquad\qquad \text{on } \partial_p Q_n,\\ |\nabla\xi_n|\leqslant \frac{2^{n+2}}{R}\left(\frac{\omega}{2^{m_1}}\right)^{(2-p)/p}, \quad 0\leqslant \xi_{n,t}\leqslant 2^{(2-p)m_2}\frac{2^{p(n+2)}}{R^p}\end{cases}$$

In this context, the energy inequalities read

$$\begin{aligned}
&\sup_{-2^{(p-2)m_2}R_n^p<t<0}\int_{K_n}(u-k_n)_-^2\xi_n^p\,\mathrm{d}x+\iint_{Q_n}\left|\nabla(u-k_n)_-\xi_n\right|^p\mathrm{d}x\,\mathrm{d}t\\
&\quad\leqslant C\frac{2^{np}}{R^p}\left(\frac{\omega}{2^{m_1}}\right)^{2-p}\iint_{Q_n}(u-k_n)_-^p\,\mathrm{d}x\,\mathrm{d}t\\
&\qquad+C\frac{2^{np}}{R^p}2^{(2-p)m_2}\iint_{Q_n}(u-k_n)_-^2\,\mathrm{d}x\,\mathrm{d}t.
\end{aligned}$$

Since

$$(u-k_n)_- \leqslant \sup_{Q_n}(u-k_n)_- \leqslant \frac{\omega}{2^{m+1}} + \frac{\omega}{2^{m+1+n}} \leqslant \frac{\omega}{2^m}$$

and

$$\left(\frac{\omega}{2^{m_1}}\right)^{2-p}\left(\frac{\omega}{2^m}\right)^p = \left(\frac{\omega}{2^m}\right)^2 2^{(2-p)m_2},$$

the two terms on the right-hand side of the inequality are estimated above by

$$C\frac{2^{np}}{R^p}2^{(2-p)m_2}\left(\frac{\omega}{2^m}\right)^2 \iint_{Q_n} \chi_{\{(u-k_n)_->0\}}\,\mathrm{d}x\,\mathrm{d}t.$$

To estimate below the two integrals on the left-hand side, we introduce the level

$$\bar{k}_n \equiv \frac{k_n+k_{n+1}}{2} \in (k_{n+1}, k_n).$$

Then, for all $t \in (-2^{(p-2)m_2}R_n^p, 0)$,

$$\begin{aligned}\int_{K_n}(u-k_n)_-^2\xi_n^p\,\mathrm{d}x &\geqslant \int_{K_n}(k_n-\bar{k}_n)^{2-p}(u-\bar{k}_n)_-^p\xi_n^p\,\mathrm{d}x\\ &= \left(\frac{\omega}{2^m}\right)^{2-p}2^{(n+3)(p-2)}\int_{K_n}(u-\bar{k}_n)_-^p\xi_n^p\,\mathrm{d}x,\end{aligned}$$

and since $k_n > \bar{k}_n$,

$$\iint_{Q_n}|\nabla(u-k_n)_-\xi_n|^p\,\mathrm{d}x\,\mathrm{d}t \geqslant \iint_{Q_n}|\nabla(u-\bar{k}_n)_-\xi_n|^p\,\mathrm{d}x\,\mathrm{d}t.$$

Combining these estimates and dividing through by

$$\left(\frac{\omega}{2^m}\right)^{2-p}2^{(n+3)(p-2)}$$

we obtain

$$\begin{aligned}&\sup_{-2^{(p-2)m_2}R_n^p<t<0}\int_{K_n}(u-\bar{k}_n)_-^p\xi_n^p\,\mathrm{d}x\\ &\quad+\left(\frac{\omega}{2^m}\right)^{p-2}2^{(n+3)(2-p)}\iint_{Q_n}|\nabla(u-\bar{k}_n)_-\xi_n|^p\,\mathrm{d}x\,\mathrm{d}t\\ &\leqslant C\frac{2^{2n}}{R^p}2^{(2-p)m_2}\left(\frac{\omega}{2^m}\right)^p\iint_{Q_n}\chi_{\{(u-k_n)_->0\}}\,\mathrm{d}x\,\mathrm{d}t.\end{aligned}$$

Next consider the change of variables

$$y = d_1^{-1} x, \qquad z = 2^{(2-p)m_2} t,$$

which maps the cylinder $\mathcal{Q}_n$ into the cylinder $Q_n \equiv K_{R_n} \times (-R_n^p, 0)$, and define new functions

$$v(y, z) = u\bigl(d_1 y, 2^{(p-2)m_2} z\bigr), \qquad \hat{\xi}_n(y, z) = \xi_n\bigl(d_1 y, 2^{(p-2)m_2} z\bigr)$$

and the sets

$$A_n(z) \equiv \bigl\{y \in K_{R_n}\colon v(y, z) < k_n\bigr\} \quad \text{with } |A_n| \equiv \int_{-R_n^p}^{0} \bigl|A_n(z)\bigr|\, dz.$$

Since $1 < p < 2$ and $\omega < 1$, the coefficient

$$\left(\frac{\omega}{2^m}\right)^{p-2} 2^{(n+3)(2-p)+m_2(p-2)} = \omega^{p-2} 2^{(2-p)(m_1+n+3)} > 1,$$

and we obtain

$$\bigl\|(v - \bar{k}_n)_- \hat{\xi}_n\bigr\|_{V^p(Q_n)}^p \leqslant C \frac{2^{2n}}{R^p} \left(\frac{\omega}{2^m}\right)^p |A_n|.$$

Noting that $k_{n+1} < \bar{k}_n$, by the embedding Theorem 1,

$$\begin{aligned}
2^{-(n+3)p} \left(\frac{\omega}{2^m}\right)^p |A_{n+1}| &= (\bar{k}_n - k_{n+1})^p |A_{n+1}| \\
&= \iint_{Q_{n+1}} (\bar{k}_n - k_{n+1})^p \chi_{\{v < k_{n+1}\}}\, dy\, dz \\
&\leqslant \iint_{Q_{n+1}} (v - \bar{k}_n)_-^p\, dy\, dz \\
&\leqslant \bigl\|(v - \bar{k}_n)_- \hat{\xi}_n\bigr\|_{p, Q_n}^p \\
&\leqslant C |A_n|^{p/(N+p)} \bigl\|(v - \bar{k}_n)_- \hat{\xi}_n\bigr\|_{V^p(Q_n)}^p \\
&\leqslant C \frac{2^{2n}}{R^p} \left(\frac{\omega}{2^m}\right)^p |A_n|^{1+p/(N+p)}.
\end{aligned}$$

Thus, setting

$$Y_n = \frac{|A_n|}{|Q_n|},$$

we obtain the recursive relation

$$Y_{n+1} \leqslant C4^{np} Y_n^{1+p/(N+p)}.$$

The result now follows from Lemma 2. In fact if

$$Y_0 = \frac{|A_0|}{|Q_0|} = \frac{|v < k_0|}{|Q_0|} = \frac{|u < \mu^- + \omega/2^m|}{|Q_R(m_1, m_2)|} \leqslant C^{-1/\alpha}\left(4^p\right)^{-1/\alpha^2} \equiv \nu_0, \tag{2.68}$$

where $\alpha = \frac{p}{N+p}$, the lemma guarantees that $Y_n \to 0$ when $n \to \infty$. But this is nothing but the conclusion of the lemma and (2.68) is precisely the hypothesis. □

REMARK 7. The proof shows that ν_0 depends upon p, but it is stable as $p \nearrow 2$ in the sense that $\nu_0(p) \to \nu_0(2)$ when $p \nearrow 2$.

REMARK 8. The conclusion of Lemma 8 continues to hold for cylinders of the type

$$Q_R(m, \beta) \equiv K_r \times \left(-\beta R^p, 0\right), \quad r = \left(\frac{\omega}{2^m}\right)^{(p-2)/p} R, \beta > 0,$$

provided β is independent of ω and R. In such a case we take $m_1 = m$ and ν_0 will also depend upon β.

2.4.2. *The first alternative.* Assume that there exists a cylinder of the type $(\bar{x}, 0) + Q(R^p, d_0 R)$, making up the partition of $Q(R^p, c_0 R)$, for which (2.65) holds. Applying Lemma 8 with $m_1 = 1$ and $m_2 = 0$ we conclude that

$$u(x,t) \geqslant \mu_- + \frac{\omega}{4} \quad \forall (x,t) \in (\bar{x}, 0) + Q\left(\left(\frac{R}{2}\right)^p, d_0 \frac{R}{2}\right). \tag{2.69}$$

We view the box $(\bar{x}, 0) + Q((R/2)^p, d_0(R/2))$ as a block inside $Q(R^p, c_0 R)$. The location of $\bar{x}$ in the cube $K_{\mathcal{R}(\omega)}$, where $\mathcal{R}(\omega)$ is defined by (2.64), is only known qualitatively. However, the *positivity* of u as stated in (2.69) *spreads* over the full cube $K_{c_0 R}$ for all times

$$-\left(\frac{R}{8}\right)^p \leqslant t \leqslant 0.$$

More precisely, we will prove the following:

PROPOSITION 11. *Assume that* (2.69) *holds for some* $\bar{x} \in K_{\mathcal{R}(\omega)}$. *There exists a positive number* s_1, *that can be determined a priori only in terms of the data and the number* λ *in the definition* (2.60) *of* $Q(R^p, c_0 R)$, *such that*

$$u(x,t) \geqslant \mu_- + \frac{\omega}{2^{s_1+1}} \quad \forall (x,t) \in Q\left(\left(\frac{R}{8}\right)^p, c_0 R\right). \tag{2.70}$$

As a consequence we may rephrase the first alternative in the following way.

COROLLARY 4. *Assume* (2.65) *holds for some cylinder of the type* $(\bar{x}, 0) + Q(R^p, d_0 R)$ *making up the partition of* $Q(R^p, c_0 R)$. *There exists a positive number* s_1 *that can be determined a priori only in terms of the data and the number* λ *in the definition* (2.60) *of* $Q(R^p, c_0 R)$, *such that*

$$\operatorname*{ess\,osc}_{Q(\rho^p, c_0\rho)} u \leqslant \eta_1 \omega \quad \forall \rho \in \left(0, \frac{R}{8}\right), \tag{2.71}$$

where $\eta_1 \equiv 1 - \dfrac{1}{2^{s_1+1}}$.

To prove Proposition 11 we regard $\bar{x}$ as the center of a larger cube $\bar{x} + K_{8c_0R}$ which we may assume to be contained in $K_{R^{p/2}}$. Otherwise we would have

$$16c_0 R \geqslant R^{p/2} \quad \Longrightarrow \quad \omega \leqslant 16^{p/(2-p)} 2^{\lambda} R^{p/2}.$$

We work within the box

$$(\bar{x}, 0) + Q\left(\left(\frac{R}{2}\right)^p, 8c_0 R\right)$$

and show that the conclusion of Proposition 11 holds within the cylinder

$$(\bar{x}, 0) + Q\left(\left(\frac{R}{8}\right)^p, 2c_0 R\right).$$

This contains $Q((R/8)^p, c_0 R)$, regardless of the location of $\bar{x}$ in the cube $K_{\mathcal{R}(\omega)}$.

The proof begins by introducing of the change of variables

$$x \mapsto \frac{x - \bar{x}}{2c_0 R}, \qquad t \mapsto \frac{4^p t}{(R/2)^p},$$

which maps $(\bar{x}, 0) + Q((R/2)^p, 8c_0 R)$ into $Q_4 \equiv K_4 \times (-4^p, 0)$, and the new unknown function

$$v = (u - \mu^-)\frac{2}{\omega}. \tag{2.72}$$

Denoting again with x and t the new variables, the function v satisfies the PDE

$$v_t - c \operatorname{div} |\nabla v|^{p-2} \nabla v = 0 \quad \text{in } \mathcal{D}'(Q_4), \tag{2.73}$$

where

$$c = \frac{1}{2^{4p}} \left(\frac{2}{2^{\lambda}}\right)^{2-p} = 2^{(\lambda-1)(p-2)-4p}.$$

The information (2.69) now translates into

$$v(x,t) \geqslant \frac{1}{2}, \quad \text{a.e. } (x,t) \in Q(h_0) \equiv \{x\colon |x| < h_0\} \times (-4^p, 0), \tag{2.74}$$

where

$$h_0 = \frac{d_0}{4c_0} = \frac{1}{4}\left(\frac{2}{2^\lambda}\right)^{(2-p)/p} = 2^{(\lambda-1)(p-2)/p-2} < 1. \tag{2.75}$$

We regard $Q(h_0)$ as a thin cylinder sitting at the center of Q_4. We prove that the relative largeness of v in $Q(h_0)$ spreads sidewise over $Q_2 \equiv K_2 \times (-2^p, 0)$, thus obtaining the desired result. Indeed, we want to show that

$$u(x,t) \geqslant \mu_- + \frac{\omega}{2^{1+s_1}}, \quad (x,t) \in (\bar{x}, 0) + Q\left(\left(\frac{R}{8}\right)^p, 2c_0 R\right)$$

which, according to the change of variables and the new function, is the same as

$$v(x,t) \geqslant \frac{1}{2^{s_1}}, \quad (x,t) \in Q_1 \equiv K_1 \times (-1, 0) \subset Q_2.$$

Proposition 11 will then be a consequence of the following lemma.

LEMMA 9. *For every $\nu \in (0,1)$, there exists a positive number $\delta^* \in (0,1)$, that can be determined a priori only in terms of ν, N, p and the data, such that*

$$\left|\{x \in K_2\colon v(x,t) \leqslant \delta^*\}\right| \leqslant \nu |K_2| \tag{2.76}$$

for all time levels $t \in [-2^p, 0]$.

REMARK 9. The key feature of the lemma is that the set where v is small can be made arbitrarily small for *every* time level in $[-2^p, 0]$.

The proof of this lemma is rather technical, involving the manipulation of appropriate integral inequalities, and will be omitted; the interested reader is referred to the book [55], Chapter IV, Sections 6–9, for a detailed proof.

PROOF OF PROPOSITION 11 ASSUMING LEMMA 9. Let ν_0 be the number claimed by Lemma 8, take $\nu = \nu_0$ in Lemma 9 and determine the corresponding $\delta^* = \delta^*(\nu_0)$. Let m_2 be defined by

$$2^{-m_2} = \delta^*(\nu_0),$$

and apply Lemma 8 with $\mu_- = 0$, $\omega = 1$, $m_1 = 0$, $R = 2$, over the boxes

$$(0, \bar{t}) + K_2 \times (-2^{m_2(p-2)} 2^p, 0) \equiv (0, \bar{t}) + Q_2(0, m_2)$$

as long as they are contained in Q_2, i.e., for $\bar{t}$ satisfying

$$2^{m_2(p-2)}2^p - 2^p \leqslant \bar{t} \leqslant 0.$$

Since (2.76) holds true for *all* time levels in $t \in [-2^p, 0]$, each such box satisfies

$$\left|\left\{(x,t) \in (0,\bar{t}) + \mathcal{Q}_2(0,m_2)\colon v(x,t) \leqslant 2^{-m_2}\right\}\right| \leqslant \nu_0\left|\mathcal{Q}_2(0,m_2)\right|.$$

Therefore, by Lemma 8,

$$v(x,t) \geqslant 2^{-(m_2+1)} \quad \forall (x,t) \in (0,\bar{t}) + \mathcal{Q}_1(0,m_2),$$

for all $\bar{t} \in (2^{m_2(p-2)}2^p - 2^p, 0)$. Since

$$\left(2^{m_2(p-2)}2^p - 2^p - 2^{m_2(p-2)}, 0\right) \supset (-1,0),$$

we conclude that

$$v(x,t) \geqslant 2^{-(m_2+1)} \quad \forall (x,t) \in Q_1.$$

Returning to the original coordinates and redefining the various constants accordingly, we arrive at

$$u(x,t) \geqslant \mu_- + \frac{\omega}{2^{m_2+2}} \quad \forall (x,t) \in (\bar{x},0) + Q\left(\left(\frac{R}{8}\right)^p, 2c_0R\right),$$

and Proposition 11 follows with

$$s_1 = m_2 + 1, \quad m_2 = -\log_2\left(\delta^*(\nu_0)\right), \quad \nu_0 = \left(C4^{N+p}\right)^{-(N+p)/p}. \qquad \square$$

REMARK 10. We comment further on the expansion of positivity of Proposition 11. A crucial point in the proof of Lemma 9 (which was omitted) is the use of the information contained in (2.74) and (2.75) to apply a Poincaré inequality. But for this it is not truly necessary to know that the set $\{v \geqslant \frac{1}{2}\}$ is concentrated in a cylinder centered at the origin, it suffices to have the following information:

$$\exists \alpha_0, k_0 > 0\colon \quad \left|\{x \in K_2\colon v > k_0\}\right| \geqslant \alpha_0 \quad \forall t \in \left(-4^p, 0\right).$$

2.4.3. *The second alternative.* We will omit most of the proofs in this section; for the details see [55].

Assume that (2.66) holds for all cylinders $(\bar{x},0) + Q(R^p, d_0R)$, making up the partition of $Q(R^p, c_0R)$. Since

$$\mu_+ - \frac{\omega}{2} = \mu_- + \frac{\omega}{2},$$

we can rephrase (2.66) as

$$\frac{|\{(x,t)\in(\bar{x},0)+Q(R^p,d_0R)\colon u(x,t)>\mu_+-\frac{\omega}{2}\}|}{|Q(R^p,d_0R)|}\leqslant 1-\nu_0 \tag{2.77}$$

for *all* boxes $(\bar{x},0)+Q(R^p,d_0R)$ making up the partition of $Q(R^p,c_0R)$.

Let n be a positive number to be chosen and arrange that $2^{n(2-p)/p}$ is an integer. Then we combine $2^{(n(2-p)/p)N}$ of these cylinders to form boxes congruent to

$$Q(R^p,d_*R)\equiv K_{d_*R}\times(-R^p,0),\quad d_*=\left(\frac{\omega}{2^{n+1}}\right)^{(p-2)/p}=d_0(2^n)^{(2-p)/p}. \tag{2.78}$$

We next consider cylinders of the type $(\tilde{x},0)+Q(R^p,d_*R)$. These are contained in $Q(R^p,c_0R)$ if the abscissa $\tilde{x}$ of their vertices ranges over the cube $K_{\mathcal{R}_1(\omega)}$, where

$$\begin{aligned}\mathcal{R}_1(\omega)&=\{2^{\lambda(2-p)}-2^{(n+1)(2-p)/p}\}\omega^{(p-2)/p}R\\&=\{(2^{\lambda-(n+1)})^{(2-p)/p}-1\}\left(\frac{\omega}{2^{n+1}}\right)^{(p-2)/p}R\\&=L_1d_*R,\quad\text{where } L_1\equiv(2^{\lambda-(n+1)})^{(2-p)/p}-1.\end{aligned}$$

We will take $\lambda>n+1$ and arrange that L_1 is an integer. Then we regard $Q(R^p,c_0R)$ as the union, up to a set of measure zero, of L_1^N pairwise disjoint boxes each congruent to $Q(R^p,d_*R)$. Since each box $(\tilde{x},0)+Q(R^p,d_*R)$ is the pairwise disjoint union of boxes $(\bar{x},0)+Q(R^p,d_0R)$, each of them satisfying (2.77), we can rephrase again (2.66), this time as

$$\frac{|\{(x,t)\in(\tilde{x},0)+Q(R^p,d_*R)\colon u(x,t)>\mu_+-\omega/2\}|}{|Q(R^p,d_*R)|}\leqslant 1-\nu_0 \tag{2.79}$$

for all cylinders $(\tilde{x},0)+Q(R^p,d_*R)$ making up the partition of $Q(R^p,c_0R)$.

REMARK 11. The need of introducing a larger cylinder than the one involved in (2.66) is justified by the appearance of the factor $2^{n(2-p)}$ in the logarithmic estimates (2.6) employed in the proofs. The use of a geometry in which the space dimensions are stretched by this factor accommodates the singularity and restores the homogeneity in (2.5) and (2.6).

LEMMA 10. *Let $(\tilde{x},0)+Q(R^p,d_*R)$ be any box contained in $Q(R^p,c_0R)$ and satisfying (2.79). There exists a time level*

$$t^*\in\left(-R^p,-\frac{\nu_0}{2}R^p\right),$$

such that, for all $s \geqslant 2$,

$$\left|x \in \tilde{x} + K_{d_* R}\colon u(x, t^*) > \mu_+ - \frac{\omega}{2^s}\right| \leqslant \left(\frac{1-\nu_0}{1-\nu_0/2}\right)|K_{d_* R}|. \tag{2.80}$$

PROOF. If not,

$$\left|x \in \tilde{x} + K_{d_* R}\colon u(x, t^*) > \mu_+ - \frac{\omega}{2^s}\right| > \left(\frac{1-\nu_0}{1-\nu_0/2}\right)|K_{d_* R}|$$

for all $t \in (-R^p, -\frac{\nu_0}{2} R^p)$. Then

$$\begin{aligned}
&\left|(x,t) \in (\tilde{x}, 0) + Q(R^p, d_* R)\colon u(x,t) > \mu_+ - \frac{\omega}{2}\right| \\
&\quad = \int_{-R^p}^{0} \left|x \in \tilde{x} + K_{d_* R}\colon u(x,t) > \mu_+ - \frac{\omega}{2}\right| \mathrm{d}t \\
&\quad \geqslant \int_{-R^p}^{-\frac{\nu_0}{2} R^p} \left|x \in \tilde{x} + K_{d_* R}\colon u(x,t) > \mu_+ - \frac{\omega}{2^s}\right| \mathrm{d}t \\
&\quad > \int_{-R^p}^{-\frac{\nu_0}{2} R^p} \left(\frac{1-\nu_0}{1-\nu_0/2}\right)|K_{d_* R}|\, \mathrm{d}t \\
&\quad = (1-\nu_0)\left|Q(R^p, d_* R)\right|
\end{aligned}$$

which contradicts (2.79). □

The next lemma asserts that a property similar to (2.80) still holds for all time levels from t^* up to 0. The proof of the lemma, that we omit, will also determine the number n.

LEMMA 11. *There exists a positive integer n such that, for all $t^* < t < 0$,*

$$\left|x \in \tilde{x} + K_{d_* R}\colon u(x, t) > \mu_+ - \frac{\omega}{2^{n+1}}\right| \leqslant \left(1 - \left(\frac{\nu_0}{2}\right)^2\right)|K_{d_* R}|. \tag{2.81}$$

The information of Lemma 11 will be exploited to show that in a small cylinder about $(0, 0)$, the solution u is strictly bounded above by

$$\mu_+ - \frac{\omega}{2^m} \quad \text{for some } m > n + 1.$$

The process also determines the number λ which defines the size of $Q(R^p, c_0 R)$. To make this quantitative consider the box

$$Q(\beta R^p, c_0 R), \quad \beta = \frac{\nu_0}{2}, c_0 = \left(\frac{\omega}{2^\lambda}\right)^{(p-2)/p}.$$

We view $Q(\beta R^p, c_0 R)$ as being partitioned into subboxes $(\tilde{x}, 0) + Q(\beta R^p, d_* R)$, where $\tilde{x}$ takes finitely many points within the cube $K_{\mathcal{R}_1(\omega)}$. For each of these cylinders Lemma 11 holds.

LEMMA 12. *For every $\nu \in (0,1)$, there exists a number m, depending only on the data and independent of ω and R, such that, for all cylinders $(\tilde{x}, 0) + Q(\beta R^p, d_* R)$ making up the partition of $Q(\beta R^p, c_0 R)$,*

$$\left|(x,t) \in (\tilde{x}, 0) + Q(\beta R^p, d_* R)\colon u(x,t) > \mu_+ - \frac{\omega}{2^m}\right| \leqslant \nu \left|Q(\beta R^p, d_* R)\right|. \tag{2.82}$$

REMARK 12. The proof shows that m must be chosen so that

$$m \geqslant n + 1 + \frac{C}{\nu^{p/(p-1)}}.$$

This estimate deteriorates as $p \searrow 1$, i.e., $m \nearrow \infty$ as $p \searrow 1$. Nevertheless the choice of m is stable as $p \nearrow 2$.

To proceed we return to the box $Q(\beta R^p, c_0 R)$ and recall that it is the finite union, up to a set of measure zero, of pairwise disjoint boxes $(\tilde{x}, 0) + Q(\beta R^p, d_* R)$. Therefore, Lemma 12 implies the following:

COROLLARY 5. *For every $\nu \in (0,1)$, there exists a number m depending only upon the data and independent of ω and R such that*

$$\left|(x,t) \in Q(\beta R^p, c_0 R)\colon u(x,t) > \mu_+ - \frac{\omega}{2^m}\right| \leqslant \nu \left|Q(\beta R^p, c_0 R)\right|. \tag{2.83}$$

We finally determine the size of the cylinder $Q(\beta R^p, c_0 R)$ and consequently the number λ. First, in Corollary 5, take $\nu = \nu_0$ and determine m accordingly. Then let m_2 be given by

$$\beta = \frac{\nu_0}{2} = 2^{m_2(p-2)} \tag{2.84}$$

and assume that $m \geqslant m_2$ (if necessary take m even larger). Determine λ from

$$\lambda = m_1 \quad \text{and} \quad m = m_1 + m_2. \tag{2.85}$$

With these choices, the cylinder $Q(\beta R^p, c_0 R)$ coincides with the cylinder $\mathcal{Q}_R(m_1, m_2)$ introduced in (2.67). By Corollary 5, we have

$$\left|(x,t) \in \mathcal{Q}_R(m_1, m_2)\colon u(x,t) > \mu_+ - \frac{\omega}{2^m}\right| \leqslant \nu_0 \left|\mathcal{Q}_R(m_1, m_2)\right|$$

which implies, using Lemma 8,

$$u(x,t) \leqslant \mu_+ - \frac{\omega}{2^{m+1}} \quad \forall (x,t) \in Q_{R/2}(m_1, m_2).$$

We summarize:

PROPOSITION 12. *Assume that* (2.66) *holds for all cylinders* $(\bar{x}, 0) + Q(R^p, d_0 R)$ *making up the partition of* $Q(R^p, c_0 R)$. *Then, for all* $0 < \rho < R/2$,

$$\operatorname*{ess\,osc}_{Q(\beta\rho^p, c_0\rho)} u \leqslant \eta_0 \omega, \quad \textit{where } \eta_0 \equiv 1 - \frac{1}{2^{m+1}}. \tag{2.86}$$

2.4.4. *Proof of the main proposition.* The proof of Proposition 10 follows by combining the two alternatives. The conclusion of the first alternative is that

$$\operatorname*{ess\,osc}_{Q(\rho^p, c_0\rho)} u \leqslant \eta_1 \omega \quad \forall \rho \in \left(0, \frac{R}{8}\right), \tag{2.87}$$

where $\eta_1 = 1 - 1/(2^{m_2+2})$. The conclusion of the second alternative is that

$$\operatorname*{ess\,osc}_{Q(\beta\rho^p, c_0\rho)} u \leqslant \eta_0 \omega \quad \forall \rho \in \left(0, \frac{R}{2}\right), \tag{2.88}$$

where $\eta_0 = 1 - 1/2^{m+1}$.

Set

$$\eta = \max\{\eta_0, \eta_1\} \quad \text{and} \quad C = \frac{\beta^{1/p}}{4} = \left(\frac{\nu_0}{2^{2p+1}}\right)^{1/p}.$$

Observe that, assuming $\nu_0 \leqslant 1/2$,

$$C \leqslant \frac{1}{2^{2+2/p}} < \frac{1}{8} < \frac{1}{2} \quad \text{and} \quad C < \frac{\beta^{1/p}}{2}.$$

Define

$$R_1 = CR, \qquad \omega_1 = \eta\omega$$

and the cylinder

$$Q_1 = Q\big(R_1^p, c_1 R_1\big), \quad c_1 = \left(\frac{\omega_1}{2^\lambda}\right)^{(p-2)/p}.$$

Since $\eta < 1$,

$$c_1 R_1 = \left(\frac{\omega_1}{2^\lambda}\right)^{(p-2)/p} R_1 = \left(\frac{\omega}{2^\lambda}\right)^{(p-2)/p} \eta^{(p-2)/p} R_1 \leqslant c_0 R_1.$$

Therefore, combining both alternatives,

$$\operatorname*{ess\,osc}_{Q_1} u \leqslant \omega_1 .$$

The process can now be repeated inductively starting from such relation. This yields the Hölder continuity of u as in the degenerate case $p > 2$.

2.5. *The porous medium equation and other generalizations*

As indicated earlier, the Hölder continuity of u is solely a consequence of the Caccioppoli inequalities (2.5) and the logarithmic inequalities (2.6). For this reason the techniques just presented are rather flexible and adjust to a variety of singular and degenerate parabolic partial differential equations. The first generalization we want to mention is to equations with the full p-Laplacian type quasilinear structure

$$u_t - \operatorname{div} \mathbf{a}(x,t,u,\nabla u) = b(x,t,u,\nabla u) \quad \text{in } \mathcal{D}'(\Omega_T), \tag{2.89}$$

where $\mathbf{a} : \Omega_T \times \mathbb{R}^{N+1} \to \mathbb{R}^N$ and $b : \Omega_T \times \mathbb{R}^{N+1} \to \mathbb{R}$ are measurable and satisfy the structure assumptions:

(A_1) $\mathbf{a}(x,t,u,\nabla u) \cdot \nabla u \geqslant C_0 |\nabla u|^p - \varphi_0(x,t)$;
(A_2) $|\mathbf{a}(x,t,u,\nabla u)| \leqslant C_1 |\nabla u|^{p-1} + \varphi_1(x,t)$;
(A_3) $|b(x,t,u,\nabla u)| \leqslant C_2 |\nabla u|^p + \varphi_2(x,t)$,

for $p > 1$ and a.e. $(x,t) \in \Omega_T$. The C_i, $i = 0, 1, 2$, are given positive constants and the φ_i, $i = 0, 1, 2$, are given nonnegative functions, defined in Ω_T and subject to the integrability conditions

$$\varphi_0, \varphi_1^{p/(p-1)}, \varphi_2 \in L^{q,r}(\Omega_T)$$

with $q, r \geqslant 1$ satisfying

$$\frac{1}{r} + \frac{N}{pq} \in (0,1), \qquad 1 < p \leqslant N.$$

See [55] for the details.

Another family of equations to which the theory applies are degenerate or singular equations of porous medium type that can be cast in the form (2.89), for the structure assumptions:

(B_1) $\mathbf{a}(x,t,u,\nabla u) \cdot \nabla u \geqslant C_0 |u|^{m-1} |\nabla u|^2 - \varphi_0(x,t)$, $m > 0$;
(B_2) $|\mathbf{a}(x,t,u,\nabla u)| \leqslant C_1 |u|^{m-1} |\nabla u| + \varphi_1(x,t)$;
(B_3) $|b(x,t,u,\nabla u)| \leqslant C_2 |\nabla |u|^m|^2 + \varphi_2(x,t)$,

and the functions φ_i, $i = 0, 1, 2$, satisfy the same conditions as before with $p = 2$. We require

$$u \in L^\infty_{\mathrm{loc}}\big(0, T; L^2_{\mathrm{loc}}(\Omega)\big) \quad \text{and} \quad |u|^m \in L^2_{\mathrm{loc}}\big(0, T; W^{1,2}_{\mathrm{loc}}(\Omega)\big).$$

There is a wide literature concerning this problem. We refer the reader to the Proceedings [12,29,82,94], as well as the references therein.

Further generalizations can be obtained by replacing s^{m-1}, $s > 0$, with a function that blows up like a power when $s \searrow 0$ and is regular otherwise. To be specific, consider doubly degenerate equations of the form (2.89) with structure assumptions:

(C_1) $\mathbf{a}(x, t, u, \nabla u) \cdot \nabla u \geqslant C_0 \Phi(|u|)|\nabla u|^p - \varphi_0(x, t)$;

(C_2) $|\mathbf{a}(x, t, u, \nabla u)| \leqslant C_1 \Phi(|u|)|\nabla u|^{p-1} + \Phi^{1/p}(u)\varphi_1(x, t)$;

(C_3) $|b(x, t, u, \nabla u)| \leqslant C_2 \Phi(|u|)|\nabla u|^p + \varphi_2(x, t)$.

Here φ_i, $i = 0, 1, 2$, satisfy the same conditions as before and the function $\Phi(\cdot)$ is degenerate near the origin in the sense that

$$\exists \sigma > 0: \quad \gamma_1 s^{\beta_1} \leqslant \Phi(s) \leqslant \gamma_2 s^{\beta_2} \quad \forall s \in (0, \sigma),$$

for given constants $0 < \gamma_1 \leqslant \gamma_2$ and $0 \leqslant \beta_2 \leqslant \beta_1$. For $s > \sigma$, i.e., away from zero it is assumed that Φ is bounded above and below by given positive constants. We require that

$$u \in C_{\mathrm{loc}}\big(0, T; L^2_{\mathrm{loc}}(\Omega)\big) \quad \text{and} \quad \Phi^{1/(p-1)}(u)|\nabla u| \in L^p_{\mathrm{loc}}(\Omega_T)$$

and, denoting with $F(\cdot)$ the primitive of $\Phi^{1/(p-1)}(\cdot)$, that

$$F(u) \in L^p_{\mathrm{loc}}\big(0, T; W^{1,p}_{\mathrm{loc}}(\Omega)\big),$$

which allows for an interpretation of the equation in the weak sense. One recognizes that if $\Phi(s) \equiv 1$ the equation is of p-Laplacian type and if $\Phi(s) = s^{m-1}$ and $p = 2$ the equation is of porous medium type. The Hölder continuity of solutions was obtained independently in [96] and [147].

3. Boundedness of weak solutions

The regularity theorems of the previous section apply to *bounded* weak solutions of (2.1). The theory of local boundedness discriminates between the degenerate case $p > 2$ and the singular case $1 < p < 2$. If $p > 2$, a local bound for the solution is implicit in the notion of weak solution. If $1 < p < 2$, local or global solutions need not be bounded in general.

Another substantial difference between the two cases surfaces when studying the Dirichlet problem

$$\begin{cases} u \in C\big(0,T; L^2(\Omega)\big) \cap L^2\big(0,T; W^{1,p}(\Omega)\big) \\ \quad \equiv V^{2,p}(\Omega_T), \\ u_t - \operatorname{div}|\nabla u|^{p-2}\nabla u = 0 & \text{in } \Omega_T, \\ u(\cdot,t)|_{\partial\Omega} = g(\cdot,t) & \text{traces of functions in } V^{2,p}(\Omega_T), \\ u(\cdot,0) = u_0 & \text{in the sense of } L^2(\Omega), \end{cases} \tag{3.1}$$

or the Cauchy problem

$$\begin{cases} u \in C\big(0,T; L^1_{\mathrm{loc}}(\mathbb{R}^N)\big) \cap L^2\big(0,T; W^{1,p}_{\mathrm{loc}}(\mathbb{R}^N)\big), \\ u_t - \operatorname{div}|\nabla u|^{p-2}\nabla u = 0 & \text{in } \Sigma_T \equiv \mathbb{R}^N \times (0,T), \\ u(\cdot,0) = u_0 & \text{in the sense of } L^1_{\mathrm{loc}}\big(\mathbb{R}^N\big). \end{cases} \tag{3.2}$$

The following weak maximum principle is common to both cases.

THEOREM 3. *Let $p > 1$ and let u be a weak solution of* (3.1) *and assume that $g \in L^\infty(\partial\Omega \times (0,T))$ and $u_0 \in L^\infty(\Omega)$. Then*

$$\operatorname*{ess\,sup}_{\Omega_T} |u| \leqslant \max\Big\{\operatorname*{ess\,sup}_{\Omega} |u_0|;\ \operatorname*{ess\,sup}_{\partial\Omega\times(0,T)} |g|\Big\}.$$

THEOREM 4. *Let $p > 1$ and let u be a weak solution of* (3.2). *Then, if $u_0 \in L^\infty(\mathbb{R}^N)$,*

$$\operatorname*{ess\,sup}_{\Sigma_T} |u| \leqslant \operatorname*{ess\,sup}_{\mathbb{R}^N} |u_0|.$$

In the next sections we let u be a nonnegative weak subsolution of (2.1) and will state several upper bounds for it. The assumption that u is nonnegative is not essential and is used here only to deduce that u is locally or globally bounded. If u is a subsolution, not necessarily bounded below, our results supply a priori bounds above for u. Analogous statements hold for nonpositive local supersolutions and in particular for solutions.

3.1. *The degenerate case $p > 2$*

THEOREM 5. *Let $p > 2$. Every nonnegative, local weak subsolution u of* (2.1) *in Ω_T is locally bounded in Ω_T. Moreover, for all $\varepsilon \in (0,2]$, there exists a constant γ depending only upon N, p, Λ and ε, such that, $\forall (x_0,t_0) + Q(\tau,\rho) \subset \Omega_T$ and $\forall \sigma \in (0,1)$,*

$$\sup_{(x_0,t_0)+Q(\sigma\tau,\sigma\rho)} u \leqslant \frac{\gamma(\tau/\rho^p)^{1/\varepsilon}}{(1-\sigma)^{(N+p)/\varepsilon}} \left(\iint_{(x_0,t_0)+Q(\tau,\rho)} u^{p-2+\varepsilon}\,\mathrm{d}x\,\mathrm{d}t\right)^{1/\varepsilon} \wedge \left(\frac{\rho^p}{\tau}\right)^{1/(p-2)}.$$

REMARK 13. In the linear case $p=2$ and $\tau=\rho^2$ such an estimate holds for any positive number ε (see [134]). In our case ε is restricted in the range $(0,2]$.

It is of interest to have sup-estimates that involve "low" integral norms of the solution. The next theorem is a result in this direction.

THEOREM 6. *Let $p>2$ and let u be a nonnegative, local subsolution of* (2.1) *in Ω_T. There exists a constant $\gamma=\gamma(data)$ such that, $\forall(x_0,t_0)+Q(\tau,\rho)\subset\Omega_T$ and $\forall\sigma\in(0,1)$,*

$$\sup_{(x_0,t_0)+Q(\sigma\tau,\sigma\rho)} u$$
$$\leqslant \frac{\gamma\sqrt{\tau/\rho^p}}{(1-\sigma)^{(N(p+1)+p)/2}}\left(\sup_{t_0-\tau<t<t_0}\int_{B_\rho(x_0)} u(x,t)\,\mathrm{d}x\right)^{p/2}\wedge\left(\frac{\rho^p}{\tau}\right)^{1/(p-2)}.$$

3.1.1. *Global estimates for solutions of the Dirichlet problem.* Consider a nonnegative weak subsolution of the Dirichlet problem (3.1) and let $p>2$. If the boundary data are bounded then the weak maximum principle of Theorem 3 holds true. If, however, u_0^+ is not bounded, it is of interest to investigate how the supremum of u behaves when $t\to 0$.

THEOREM 7. *Let u be a nonnegative weak subsolution of the Dirichlet problem* (3.1). *There exists a constant $\gamma=\gamma(data)$ such that, $\forall t\in(0,T)$,*

$$\sup_\Omega u(\cdot,t)\leqslant \sup_{S_T} g+\frac{\gamma}{t^{N/\lambda}}\left(\int_0^t\int_\Omega u\,\mathrm{d}x\,\mathrm{d}s\right)^{p/\lambda},\qquad \lambda=N(p-2)+p.$$

Results of this kind could be used to construct solutions of the Dirichlet problem with initial data in $L^1(\Omega)$ or even finite measures. Indeed the regularity results of the previous section supply the necessary compactness to pass to the limit in a sequence of approximating problems.

3.1.2. *Estimates in Σ_T.* Consider a nonnegative weak subsolution u of the Cauchy problem (3.2) in the whole strip Σ_T. By this we mean that u is a *local* weak subsolution of the PDE in (3.2) in Ω_T for every *bounded* domain $\Omega\subset\mathbb{R}^N$. To derive global sup-estimates, we must impose some control on the behavior of u as $|x|\to\infty$. We assume that the quantity

$$|||u|||_{\{r,t\}}\equiv\sup_{0<s<t}\sup_{\rho\geqslant r}\int_{B_\rho}\frac{u(x,s)}{\rho^{\lambda/(p-2)}}\,\mathrm{d}x,\qquad \lambda=N(p-2)+p, \tag{3.3}$$

is finite for some $r>0$ and for all $t\in(0,T)$. This assumption is not restrictive. It is shown in [64] that it is necessary and sufficient for a nonnegative solution of (3.2) to exist in Σ_T.

The subsolution u at hand, is not necessarily bounded. However, it is locally bounded and as $|x|\to\infty$ grows no faster that $|x|^{p/(p-2)}$. This is the content of the next theorem.

THEOREM 8. *Let u be a nonnegative subsolution of* (3.2) *in Σ_T, and assume* (3.3) *holds. There exist constants γ_* and γ, depending only upon N, p and Λ, such that*

$$\|u(\cdot,t)\|_{\infty,B_\rho} \leqslant \gamma \frac{\rho^{p/(p-2)}}{t^{N/\lambda}} |||u|||_{\{r,t\}}^{p/\lambda}, \quad \lambda = N(p-2)+p,$$

for all $0 < t < \gamma_* |||u|||_{\{r,t\}}^{2-p}$ and all $\rho \geqslant r$.

Information of this kind are of interest in investigating the behavior of the solutions for t near zero and in studying the structure of the nonnegative solutions in Σ_T. In this estimate, the functional dependence as $t \searrow 0$ is sharp as it can be verified from the explicit Barenblatt solution (recall $(1.4)_p$ in the Introduction). The functional dependence as $|x| \to \infty$ is also optimal as it follows from the explicit solution

$$\mathcal{D}(x,t) = \left\{ A\left(\frac{T}{T-t}\right)^{N(p-2)/\lambda(p-1)} + \left(\frac{p-2}{p}\right)\lambda^{-1/(p-1)}\left(\frac{|x|^p}{T-t}\right)^{1/(p-1)} \right\}^{(p-1)/(p-2)}, \tag{3.4}$$

where A and T are two positive parameters.

3.2. *The singular case* $1 < p < 2$

We will give below an example of a solution with $p = 2N/(N+1)$, that is unbounded. Thus in the singular range $1 < p < 2$, the boundedness of a weak solutions is not a purely local fact and, if at all true, it must be deduced from some global information. One of them is the weak maximum principle of Theorems 3 and 4. Another is a sufficiently high order of integrability. A sharp sufficient condition can be given in terms of the numbers

$$\lambda_r = N(p-2) + rp, \quad r \geqslant 1.$$

We assume that u satisfies

$$u \in L^r_{\text{loc}}(\Omega_T) \quad \text{for some } r \geqslant 1 \text{ such that } \lambda_r > 0. \tag{3.5}$$

The global information needed here is

u can be constructed as the weak limit in $L^r_{\text{loc}}(\Omega_T)$ of a sequence of nonnegative *bounded* subsolutions of (2.1). (3.6)

The notion of weak subsolution requires u to be in the class $u \in V^{2,p}_{\text{loc}}(\Omega_T)$. This space is embedded into $L^q_{\text{loc}}(\Omega_T)$, where $q = p(N+2)/N$. Therefore if p is so close to one that $\lambda_q \leqslant 0$, the order of integrability in (3.5) is not implicit in the notion of subsolution and must be imposed.

THEOREM 9. *Let u be a nonnegative local weak subsolution of* (2.1) *in Ω_T and assume that* (3.5) *and* (3.6) *hold. There exists a constant $\gamma = \gamma(data, r)$ such that, $\forall (x_0, t_0) + Q(\tau, \rho) \subset \Omega_T$ and $\forall \sigma \in (0, 1)$,*

$$\sup_{(x_0,t_0)+Q(\sigma\tau,\sigma\rho)} u \leqslant \frac{\gamma(\rho^p/\tau)^{N/\lambda_r}}{(1-\sigma)^{p(N+p)/\lambda_r}} \left(\iint_{(x_0,t_0)+Q(\tau,\rho)} u^r \,\mathrm{d}x\,\mathrm{d}t \right)^{p/\lambda_r} \wedge \left(\frac{\tau}{\rho^p} \right)^{1/(2-p)}. \tag{3.7}$$

3.2.1. *Estimates near $t = 0$.* Fix $t \in (0, T)$ and let us rewrite (3.7) for the pair of boxes

$$B_{\sigma\rho} \times (\sigma t, t), \qquad B_\rho \times (0, t).$$

COROLLARY 6. *Let u be a nonnegative local weak subsolution of* (2.1) *in Ω_T and let* (3.5)–(3.6) *hold. There exists a constant $\gamma = \gamma(data, r)$ such that, for all $0 < t \leqslant T$ and for all $\sigma \in (0, 1)$,*

$$\sup_{B_{\sigma\rho}} u(\cdot, t) \leqslant \frac{\gamma t^{-N/\lambda_r}}{(1-\sigma)^{(N+p)/\lambda_r}} \left(\int_0^t \int_{B_\sigma} u^r \,\mathrm{d}x\,\mathrm{d}s \right)^{p/\lambda_r} \wedge \left(\frac{t}{\rho^p} \right)^{1/(2-p)}. \tag{3.8}$$

REMARK 14. Assume that (3.5) holds with $r = 1$, i.e.,

$$p > \frac{2N}{N+1}.$$

Then the behavior of the supremum of u as $t \searrow 0$ is *formally* the same as that of solutions of the Dirichlet problem (3.1) for degenerate equations as in Theorem 7.

3.2.2. *Global estimates: Dirichlet data.* A peculiar phenomenon of these equations is that, unlike their degenerate counterparts, local and global estimates take essentially the same form. This appears for example by comparing (3.8) with the next global estimate.

THEOREM 10. *Let u be a nonnegative weak subsolution of the Dirichlet problem* (3.1) *and let* (3.5)–(3.6) *hold. There exists a constant $\gamma = \gamma(data, r)$ such that, $\forall t \in (0, T)$,*

$$\sup_{\Omega} u(\cdot, t) \leqslant \sup_{S_T} g + \frac{\gamma}{t^{N/\lambda_r}} \left(\int_0^t \int_\Omega u^r \,\mathrm{d}x\,\mathrm{d}s \right)^{p/\lambda_r}.$$

3.2.3. *A counterexample.* Let $a \in (0, 1)$ be a positive constant and let B_a denote the ball of radius a in $\mathbb{R}^N$ centered at the origin. Consider the functions

$$z = \frac{(a^2 - |x|^2)_+^2}{|x|^N |\ln |x|^2|^\beta} \quad \text{and} \quad v = (1 - ht)_+ z,$$

where $\beta, h > 1$ are parameters to be chosen. One verifies that

$$z \in L^1(B_a) \quad \text{and} \quad z \notin L^{1+\varepsilon}(B_a) \quad \forall \varepsilon > 0.$$

Consider also the Cauchy problem

$$\begin{cases} u_t - \operatorname{div} |\nabla u|^{p-2} \nabla u = 0 & \text{in } \Sigma_1, \\ u(\cdot, 0) = z. \end{cases} \tag{3.9}$$

LEMMA 13. *Assume that $N(p-2) + p = 0$. The constants $a \in (0, 1)$ and $\beta, h > 1$ can be determined a priori so that v is a nonnegative, weak subsolution of* (3.9) *in Σ_1.*

Next we return to (3.9) and observe that by the comparison principle $u \geqslant v$ and therefore u is not bounded.

4. Intrinsic Harnack estimates

In this section we present some results about Harnack inequalities. More precisely, we consider nonnegative weak solutions of the type:

$$\begin{cases} u \in C_{\text{loc}}\big(0, T; L^2_{\text{loc}}(\Omega)\big) \cap L^p_{\text{loc}}\big(0, T; W^{1,p}_{\text{loc}}(\Omega)\big), & p > 1, \\ u_t - \operatorname{div} |\nabla u|^{p-2} \nabla u = 0 & \text{in } \Omega_T. \end{cases} \tag{4.1}$$

The first parabolic version of the Harnack inequality is due to Hadamard [89] and Pini [144]. Their result is the following:

Let u be a nonnegative solution of the heat equation in Ω_T. Let $(x_0, t_0) \in \Omega_T$ and assume that the cylinder $(x_0, t_0) + Q_{2\rho} \subset \Omega_T$, where $Q_\rho \equiv B_\rho \times (-\rho^2, 0)$. Then there exists a constant γ, depending only upon N, such that

$$u(x_0, t_0) \geqslant \gamma \sup_{B_\rho(x_0)} u\big(x, t_0 - \rho^2\big). \tag{4.2}$$

The proof is based on local representations by means of heat potentials. A breakthrough in the theory is due to Moser, who in his celebrated paper [134] proved that (4.2) continues to hold for nonnegative weak solution of the type

$$\begin{cases} u \in C_{\text{loc}}\big(0, T; L^2_{\text{loc}}(\Omega)\big) \cap L^2_{\text{loc}}\big(0, T; W^{1,2}_{\text{loc}}(\Omega)\big), \\ u_t - \sum_{i,j=1}^N D_i\big(a_{ij}(x, t) D_j u\big) = 0 \quad \text{in } \Omega_T, \end{cases} \tag{4.3}$$

where $a_{ij} \in L^\infty(\Omega_T)$ and satisfy the ellipticity condition

$$\sum_{i,j=1}^N a_{ij} \xi_i \xi_j \geqslant \nu |\xi|^2 \quad \forall \xi \in \mathbb{R}^N, \tag{4.4}$$

with ν a positive constant. The result of Moser can be extended (see [16,163,164]) to nonnegative weak solutions of the quasilinear parabolic equation

$$u_t - \operatorname{div} \mathbf{a}(x,t,u,\nabla u) + b(x,t,u,\nabla u) = 0 \quad \text{in } \Omega_T, \tag{4.5}$$

where the diffusion field $\mathbf{a}$ and the forcing term b are real valued and measurable over $\Omega_T \times \mathbb{R} \times \mathbb{R}^N$ and satisfy the structure conditions considered in Section 2.4 for $m = 1$ or for $p = 2$.

The proof of Moser's result is based on suitable integral estimates for powers and logarithm of the solution u; the general structure follows the same one Moser used in his earlier work on Harnack's inequality in the elliptic case [133] and it is basically articulated in three steps.

First Step (Estimates on positive powers of u). Let u be a nonnegative solution; then for all $\varepsilon > 0$ there exists a positive constant γ, depending only upon N and ε, such that, for every cylinder $Q_\rho(x_0,t_0) \subset \Omega_T$ and for every $\sigma \in (0,1)$,

$$\sup_{Q_{\sigma\rho}(x_0,t_0)} u \leqslant \frac{\gamma}{(1-\sigma)^{(N+2)/(2\varepsilon)}} \left(\frac{1}{|Q_\rho(x_0,t_0)|} \iint_{Q_\rho(x_0,t_0)} u^\varepsilon \, dx \, d\tau \right)^{1/\varepsilon}. \tag{4.6}$$

Let us remark that this estimate holds also for nonnegative *subsolutions*.

Second Step (Estimates on negative powers of u). Let u be a positive solution; then, for all $\varepsilon > 0$, there exists a positive constant γ, depending only upon N and ε, such that, for every cylinder $Q_\rho(x_0,t_0) \subset \Omega_T$ and for every $\sigma \in (0,1)$,

$$\sup_{Q_{\sigma\rho}(x_0,t_0)} \frac{1}{u} \leqslant \frac{\gamma}{(1-\sigma)^{(N+2)/(2\varepsilon)}} \left(\frac{1}{|Q_\rho(x_0,t_0)|} \iint_{Q_\rho(x_0,t_0)} \frac{1}{u^\varepsilon} \, dx \, d\tau \right)^{1/\varepsilon}. \tag{4.7}$$

Quite analogously to what happened in the first step, this estimate holds also for positive *supersolutions*.

If we consider the mean values

$$M(p,D) = \left(\frac{1}{|D|} \iint_D u^p \, dx \, d\tau \right)^{1/p},$$

it is obvious that (4.6) and (4.7) can be rewritten as

$$M(+\infty, Q_{\sigma\rho}) \leqslant \gamma_1 M(\varepsilon, Q_\rho), \qquad M(-\varepsilon, Q_\rho) \leqslant \gamma_2 M(-\infty, Q_{\sigma\rho}).$$

The main point is then to establish a so-called "crossover inequality", namely

$$M(p, D_-) \leqslant \gamma_3 M(-p, D_+),$$

for sufficiently small $p > 0$ and appropriate domains D_-,D_+. Indeed this is the result of the following:

Third Step (Crossover lemma). Let u be a positive solution, $D_+ = \{|x| < 1, \frac{1}{2} < t < 1\}$ and $D_+ = \{|x| < 1, -1 < t < -\frac{1}{2}\}$. Then there exist constants $\delta > 0$ and $C > 0$, depending only on N, such that

$$\left(\iint_{D_-} u^\delta \,\mathrm{d}x\,\mathrm{d}\tau\right)\left(\iint_{D_+} u^{-\delta} \,\mathrm{d}x\,\mathrm{d}\tau\right) \leqslant C. \tag{4.8}$$

We stated (4.8) in a normalized form just for the sake of simplicity.

It is worth saying that the third step is the most difficult part of Moser's proof. Inequality (4.8) is a straightforward consequence of an adaptation to the parabolic case of the well-known lemma of F. John and L. Nirenberg, which concerns the exponential decay of the distribution function of a function with bounded mean oscillation. Going from the elliptic to the parabolic situation the difficulty lies in the special role played by the time variable. In fact as clearly stated in (4.2), the Harnack inequality for a nonnegative solution of a parabolic equation is an *inf-bound* on the value of such a solution at a given time in terms of its value at a previous time and this necessary time lag has to be reflected in a proper parabolic John–Nirenberg lemma. This is precisely what Moser did in his main lemma in [134]. Indeed Moser's proof is hard to follow and the need for a possible simplification was immediately felt.

Moser himself published a new proof of the Harnack inequality in 1971 (see [135]), with the expressed purpose to avoid the use of his parabolic John–Nirenberg lemma; through estimates on the logarithm of the solution and via a measure lemma based on a result of Bombieri [27,28], he showed that it is possible to estimate in a quantitative way the supremum of u. Repeating the same argument for u^{-1} one gets the quantitative estimates for the infimum of u. By combining these results the Harnack estimates are proved.

A few years later Fabes and Garofalo [80] came back to Moser's main lemma and gave a simplified proof, using Calderon's proof of the original John–Nirenberg lemma, see also [81].

In Moser's approach the main feature that makes the method work is the homogeneity of the time and space terms of the equation; in fact Trudinger ([164], Section 5) shows that things run in the same way in the proof of the Harnack inequality for doubly nonlinear equations of the type

$$\left(u^{p-1}\right)_t - \operatorname{div}\left(|\nabla u|^{p-2}\nabla u\right) = 0 \quad \text{in } \Omega_T, \tag{4.9}$$

which is p-homogeneous, exactly as (4.3) is 2-homogeneous.

On the other hand, coming back to (4.1), quite surprisingly Moser's method does not work when $p \neq 2$ and this is not simply a matter of technique. As already discussed in the

Introduction, as (4.1) is invariant by the scaling $x \to hx$ and $t \to h^p t$, one would guess that Harnack estimates would hold in the cylinder $[(x_0, t_0) + B_\rho \times (-\rho^p, 0)]$, but this is not the case. Let us consider the explicit solution of (4.1) introduced by Barenblatt in [17]:

$$\mathcal{B}(x,t) = t^{-N/\lambda}\left\{1 - \gamma_p\left(\frac{|x|}{t^{1/\lambda}}\right)^{p/(p-1)}\right\}_+^{(p-1)/(p-2)}, \quad t > 0, p > 2, \tag{4.10}$$

where $\gamma_p = \lambda^{1/(1-p)}(p-2)/p$ and $\lambda = N(p-2) + p$. Let (x_0, t_0) be a point of the *free boundary* $\{t = |x|^\lambda\}$. If t_0 is large enough, the ball $B_\rho(x_0)$ taken at the time level $t_0 - \rho^p$ intersects the support of $x \to \mathcal{B}(x, t_0 - \rho^p)$ in an open set. Hence,

$$\sup_{B_\rho(x_0)} \mathcal{B}(x, t_0 - \rho^p) > 0 \quad \text{and} \quad \mathcal{B}(x_0, t_0) = 0$$

which contradicts (4.2) and we conclude that things must be more complicated.

However, a comparison between (4.4) and (4.1) suggests that one may heuristically regard (4.1) as if it were written in a time scale intrinsic to the solution itself and, loosely speaking, of the order of $t[u(x,t)]^{2-p}$. Indeed if one looks at the Barenblatt solutions once more, one realizes that for such specific functions a Harnack estimate holds with an intrinsic time scale exactly of the order $u(x_0, t_0)^{2-p}$.

A general result of this kind for (4.1) is proved in [40,54,66]:

THEOREM 11. *Let u be a nonnegative weak solution of* (4.1) *and let* $p > 2N/(N+1)$. *Fix $(x_0, t_0) \in \Omega_T$ and assume that $u(x_0, t_0) > 0$. There exists constants $\gamma > 1$ and $C > 0$, depending only upon N and p, such that*

$$u(x_0, t_0) \leqslant \gamma \inf_{B_\rho(x_0)} u(\cdot, t_0 + \theta), \tag{4.11}$$

where

$$\theta \equiv \frac{C\rho^p}{[u(x_0, t_0)]^{p-2}} \tag{4.12}$$

provided that the cylinder $(x_0, t_0) + B_{4\rho} \times (-4\theta, 4\theta)$ is contained in Ω_T.

In the next sections we will give a sketch of the proof both for the degenerate and singular cases. For the moment let us make some general remarks and point out some open questions.

REMARK 15. There is a big difference between the degenerate case ($p > 2$) and the singular case ($p < 2$) and this is due to the different behavior of the modulus of ellipticity $|Du|^{p-2}$. In the degenerate situation the modulus vanishes when Du is zero; hence, the evolution phenomenon dominates over the diffusion process and this holds more and more as p grows to infinity. We have a direct consequence in (4.11), as the constant C is larger than 1, when $p > 2$; moreover, $C \to \infty$ as $p \to \infty$. Roughly speaking the Harnack inequality states that the original positivity of u at (x_0, t_0) is spread over the full ball $B_\rho(x_0)$ and is preserved for a large time. On the other hand, when $p \to 2^+$, $\gamma(N, p)$, $C(N, p) \to \gamma(N, 2)$, $C(N, 2)$ so that, at least formally, we recover the classical Harnack inequality for nonnegative solutions of the heat equation. On the contrary, in the singular case the modulus blows up when Du vanishes, so that now the previous situation is reversed, the diffusion dominates over the evolution phenomenon and this is felt more and more as $p \to (2N)/(N+1)$. Once more we can see this clearly expressed in (4.11), since $C \in (0, 1)$ and $C \to 0^+$ as $p \to (2N)/(N+1)$: under a geometrical point of view, we can say that the original positivity of u at (x_0, t_0) spreads over the full ball $B_\rho(x_0)$ but is now preserved only for a relatively small time. Exactly as in the degenerate case, when $p \to 2^-$, we have that $\gamma(N, p), C(N, p) \to \gamma(N, 2), C(N, 2)$, so that once again we recover the classical results in the limit situation. Finally one may naturally ask about the lower bound for p: why $p > (2N)/(N+1)$ and not just $p > 1$? Indeed $p > (2N)/(N+1)$ is optimal for Harnack inequality to hold, as we will see in the next sections, discussing the phenomenon of extinction in finite time.

REMARK 16. Let apart the intrinsic height of the cylinder, inequality (4.11) is obviously equivalent to (4.2). Let us just remark that in the case of parabolic equations in nondivergence form, Krylov and Safonov gave to Harnack inequality exactly the same formulation as in (4.11).

In Theorem 11 the level θ is defined in terms of $u(x_0, t_0)$ by (4.12). Notwithstanding the previous discussion about the intrinsic scaling, it is natural to ask if an estimate holds where the geometry can be a priori prescribed independent of the solution. In [54] a positive answer is given when $p > 2$ by the following

THEOREM 12. *Let u be a nonnegative weak solution of* (4.1) *and let* $p > 2$. *There exists a constant $B = B(N, p) > 1$ such that,*

$$\forall (x_0, t_0) \in \Omega_T, \forall \rho, \theta > 0 \text{ s.t. } (x_0, t_0) + B_{4\rho} \times (-4\theta, 4\theta) \subset \Omega_T,$$

we have

$$u(x_0, t_0) \leqslant B\left\{\left(\frac{\rho^p}{\theta}\right)^{1/(p-2)} + \left(\frac{\theta}{\rho^p}\right)^{N/p}\left[\inf_{B_\rho(x_0)} u(\cdot, t)\right]^{\lambda/p}\right\}, \tag{4.13}$$

where $\lambda = N(p-2) + p$.

At first glance Theorems 11 and 12 could look markedly different, as in the second one the positivity of $u(x_0, t_0)$ is not required and $\theta > 0$ can be assumed arbitrarily. Indeed (4.13) holds trivially when $u(x_0, t_0) = 0$ and both statements are equivalent when $u(x_0, t_0) > 0$ in the sense that (4.11) $\Rightarrow$ (4.13) in any case and (4.13) $\Rightarrow$ (4.11) with a constant $\gamma(N, p)$ which may not be stable as $p \to 2^+$.

Assuming $u(x_0, t_0) > 0$, let us prove the second implication. Under the hypothesis that (4.13) is valid for all $\theta > 0$ s.t. $(x_0, t_0) + B_{4\rho} \times (-4\theta, 4\theta) \subset \Omega_T$, if we choose

$$\theta = \frac{(2B)^{p-2}\rho^p}{[u(x_0, t_0)]^{p-2}},$$

we immediately conclude that

$$u(x_0, t_0) \leqslant \gamma \inf_{B_\rho(x_0)} u(\cdot, t_0 + \theta) \quad \text{with } \gamma = 2B^{N(p-2)+\lambda}.$$

We postpone the proof of the opposite implication to the next section.

A consequence of Theorem 12 is the following:

COROLLARY 7. *Let u be a nonnegative weak solution of* (4.1) *and let $p > 2$. There exists a constant $B = B(N, p) > 1$ such that,*

$$\forall (x_0, t_0) \in \Omega_T, \forall \rho, \theta > 0 \text{ s.t. } (x_0, t_0) + B_{4\rho} \times (-4\theta, 4\theta) \subset \Omega_T,$$

we have

$$\frac{1}{\rho^N} \int_{B_\rho(x_0)} u(x, t_0)\,\mathrm{d}t \leqslant \gamma \left\{ \left(\frac{\rho^p}{\theta}\right)^{1/(p-2)} + \left(\frac{\theta}{\rho^p}\right)^{N/p} \big[u(x_0, t)\big]^{\lambda/p} \right\}. \quad (4.14)$$

Harnack inequalities like the ones stated in Theorems 11 and 12 hold for nonnegative solutions of the porous medium equation

$$\begin{cases} u \in C_{\mathrm{loc}}\big(0, T; L^2_{\mathrm{loc}}(\Omega)\big), \quad u^m \in L^2_{\mathrm{loc}}\big(0, T; W^{1,2}_{\mathrm{loc}}(\Omega)\big), \\ u_t - \Delta u^m = 0 \quad \text{in } \Omega_T, m > 1. \end{cases} \quad (4.15)$$

In particular, Theorem 11 becomes:

THEOREM 13. *Let u be a nonnegative weak solution of* (4.15) *and let $m > (N-2)_+/(N+2)$. Fix $(x_0, t_0) \in \Omega_T$ and assume that $u(x_0, t_0) > 0$. There exists constants $\gamma > 1$ and $C > 0$, depending only upon N and m, such that*

$$u(x_0, t_0) \leqslant \gamma \inf_{B_\rho(x_0)} u(\cdot, t_0 + \theta), \quad (4.16)$$

where

$$\theta \equiv \frac{C\rho^2}{[u(x_0, t_0)]^{m-1}} \quad (4.17)$$

provided that the cylinder $(x_0, t_0) + B_{4\rho} \times (-4\theta, 4\theta)$ is contained in Ω_T.

The different behavior of the constant C dependent upon $p > 2$ or $(2N)/(N+1) < p < 2$ discussed in Remark 15 comes up in this context too, where the degenerate case is given by $m > 1$ and the singular case by $(N-2)_+/(N+2) < m < 1$.

REMARK 17. Similar estimates have been proved also for the solutions of doubly nonlinear parabolic equations of the type

$$u_t = \operatorname{div}\big(|\nabla u|^{p-2}|u|^{m-1}\nabla u\big)$$

(see [177]). Equations of this type are classified as doubly nonlinear [123] or with implicit nonlinearity [103]. This class of equations have their own mathematical interest (the porous medium equation and the p-Laplacian equations belong to this larger class) and physical interest (see the review paper [103]). Also for this larger class, local Hölder continuity results hold (see [96,97,147,175]).

REMARK 18. The theory of Harnack estimates is fragmented and incomplete. The estimates for $p \neq 2$ hold only for homogeneous PDEs and this strongly depends on the method we will present in the following, which basically relies on the construction of special solutions and subsolutions. The shortcoming of such a technique is evident even in the framework of homogeneous equations since a Harnack-type estimate is not known to hold for nonnegative weak solutions of (see [123])

$$u_t = \sum_{i=1}^{N} D_i\big(|D_i u|^{p-2} D_i u\big).$$

The open question is if it is possible to extend the Harnack estimates to the case of parabolic equations with the full quasilinear structure, as it happens when $p = 2$. Results of this kind would probably require a new method independent of local representations and local subsolutions. Whenever developed, such a technique would parallel the discovery of the Moser estimates [134], based on real and harmonic analysis tools, versus the estimates by Hadamard [89] and Pini [144], based on local representations.

4.1. *Harnack estimates: the degenerate case*

First of all, let us briefly comment upon the assumption that the cylinder $(x_0, t_0) + B_{4\rho} \times (-4\theta, 4\theta)$ is contained in Ω_T. Under a geometrical point of view, this means that t_0 should be of the order of θ and this is essential. In fact if we consider the Barenblatt solution given in (4.10) with $x_0 = 0$ and t_0 arbitrarily close to the origin, it is evident that it cannot satisfy (4.13). One might think that this is due to the pointwise nature of (4.12) and (4.13), but this is not the case and the reason actually lies in the local character of the solutions we are considering. Indeed quite surprisingly also the averaged form of the Harnack inequality (4.14) does not hold without the assumption that the cylinder $(x_0, t_0) + B_{4\rho} \times (-4\theta, 4\theta)$

is contained in Ω_T. To see this, let u be the unique weak solution of the boundary value problem

$$\begin{cases} u_t - \left(|u_x|^{p-2}u_x\right)_x = 0 & \text{in } Q \equiv (0,1)\times(0,\infty), \\ u(0,t) = u(1,t) = 0 & \text{for all } t \geqslant 0, \\ u(x,0) = u_0(x) \in C_0^\infty(0,1), & \\ u_0(x) \in [0,1] \quad \forall x \in (0,1) \quad \text{and} \quad u_0(x) = 1 & \text{for } x \in \left(\frac{1}{4}, \frac{3}{4}\right). \end{cases} \tag{4.18}$$

Thanks to the results of [20], we can say that

$$u_t \geqslant -\frac{1}{p-2}\frac{u}{t} \quad \text{in } \mathcal{D}'(Q).$$

Since $0 \leqslant u \leqslant 1$, by the comparison principle we have

$$-\left(|u_x|^{p-2}u_x\right)_x \leqslant \frac{1}{(p-2)t}, \qquad t > 0.$$

At any fixed level t, the function $x \to u(x,t)$ is majorized by

$$v(x,t) = \frac{\gamma x^\delta}{t^{1/(p-1)}}, \qquad \delta \in \left(\frac{p-1}{p}, 1\right), \qquad (\gamma\delta)^{p-1}(1-\delta)(p-1) \geqslant \frac{1}{p-2},$$

as

$$-\left(|v_x|^{p-2}v_x\right)_x \geqslant \frac{1}{(p-2)t} \quad \text{and} \quad v(0,t) = 0,\ v(1,t) > 0.$$

Therefore, for every $\delta \in ((p-1)/p, 1)$, there exists a constant $C = C(\delta)$, such that

$$u\left(\frac{1}{2}, t\right) \leqslant \frac{C(\delta)}{t^{1/(p-1)}}.$$

Now if (4.14) held for $t_0 = 0$, $x_0 = \frac{1}{2}$ and $\rho = \frac{1}{4}$, for $t > 1$ we would have

$$1 \leqslant \text{const}\left(t^{-1/(p-2)} + t^{-1/p}\right) \to 0 \quad \text{as } t \to +\infty.$$

In the sequel we will see that the limitation on t_0 in (4.14) can be dropped when (4.1) is considered in the whole $\mathbb{R}^N$.

We can now finally come to the proof of Theorem 11 when $p > 2$. The technical tools used in the proof are only two: the Hölder continuity of solutions as proved in Section 2 and the comparison principle. This point of view is somehow reversed with respect to Moser's approach where the Hölder continuity is implied by the Harnack estimate. Even though not so explicitly stated, a method similar to ours is already present in the work of Krylov and Safonov [117].

We can basically recognize four steps.

First Step (Renormalization of the solution). Let $(x_0, t_0) \in \Omega_T$ and $\rho > 0$ be fixed, assume $u(x_0, t_0) > 0$ and consider the box

$$Q_{4\rho} = \{|x - x_0| < 4\rho\} \times \left\{t_0 - \frac{4C\rho^p}{[u(x_0, t_0)]^{p-2}}, t_0 + \frac{4C\rho^p}{[u(x_0, t_0)]^{p-2}}\right\},$$

where C is a positive constant to be determined later. We render the equation *dimensionless* by the change of variables

$$\xi = \frac{x - x_0}{\rho}, \qquad \tau = \frac{(t - t_0)[u(x_0, t_0)]^{p-2}}{\rho^p}, \qquad v = \frac{u}{u(x_0, t_0)}.$$

This maps $Q_{4\rho}$ into $Q = Q^+ \cup Q^-$, where $Q^+ \equiv B(4) \times [0, 4C)$, $Q^- \equiv B(4) \times (-4C, 0]$. We denote again the new variables with x and t and observe that the rescaled function v is a bounded nonnegative solution of the equation

$$v_t - \operatorname{div} |\nabla v|^{p-2} \nabla v = 0 \quad \text{in } Q,$$

with $v(0, 0) = 1$. To prove the Harnack inequality it is enough to find constants $0 < \gamma_0 < 1$ and $C > 1$, depending only upon N and p, such that, for each $x \in B(1)$, we have

$$v(x, C) \geqslant \gamma_0. \tag{4.19}$$

As a matter of fact if $u(x_0, t_0) = 0$, no rescaling is possible and we are led to consider Theorem 12, which is trivially satisfied in this case as already remarked.

Second Step (Determination of the largest value of v in Q^-). Construct the family of nested boxes $Q_\tau \equiv B_\tau \times (-\tau^p, 0]$. Define the numbers $M_\tau = \sup_{Q_\tau} v$ and $N_\tau = (1 - \tau)^{-\beta}$, where $\beta > 1$ will be chosen later. Let $0 \leqslant \tau_0 < 1$ be the largest root of the equation $M_\tau = N_\tau$. Such a root is well defined, since $M_0 = N_0$ and as $\tau \to 1^-$ M_τ remain bounded and N_τ blow up. By construction $\sup_{Q_\tau} v \leqslant N_\tau$ for all $\tau > \tau_0$. Moreover, from the continuity of v in Q, there exists at least a point $(x_1, t_1) \in N_{\tau_0}$, where $v(x_1, t_1) = (1 - \tau_0)^{-\beta}$.

Third Step (Lower bound on v at the same time-level t_1). Relying on the Hölder continuity of v [51], we can determine a small ball of radius r_0 about (x_1, t_1), where $v \geqslant (1 - \tau_0)^{-\beta}/2$. Roughly speaking, we have found a small ball $B_{r_0}(x_1)$ at time t_1, close to $(0, 0)$, where the largeness of $v(\cdot, t_1)$ is *qualitatively* determined. The proof is concluded once we choose the constant $\beta > 1$ and $C > 1$ in such a way that we come up with a quantitative lower bound on v over the full ball B_1 at a later time C: otherwise stated, we have to spread the positivity of v and this is the crucial step in the proof of the Harnack inequality.

Fourth Step (Expansion of the positivity set). The spread of positivity is achieved by means of a proper comparison function. For $t \geqslant t_1$ consider the function

$$\mathcal{B}_{k,\rho}(x, t; x_1, t_1) \equiv \frac{k\rho^n}{S^{N/\lambda}(t)} \left[1 - \left(\frac{|x - x_1|}{S^{1/\lambda}(t)}\right)^{p/(p-1)}\right]_+^{(p-1)/(p-2)},$$

where as usual $\lambda = N(p-2)+p$, $S(t) = B(N,p)k^{p-2}\rho^{N(p-2)}(t-t_1)+\rho^{\lambda}$, $b(N,p) = \lambda(p/(p-2))^{p-1}$ and choose $k = (1-\tau_0)^{-\beta}/2$ and $\rho = r_0$. By direct calculation one verifies that $\mathcal{B}_{k,\rho}(x,t;x_1,t_1)$ is a weak solution of (4.1) in $\mathbb{R}^N \times \{t_1,t\}$. This comparison function was introduced in [17,54,143].

By a proper choice of β and C, the support of $\mathcal{B}_{k,\rho}(\cdot, C; x_1, t_1)$ contains B_2 and by the comparison principle

$$\inf_{B_1} v(x,C) \geqslant \inf_{B_1} \mathcal{B}_{k,\rho}(x,C;x_1,t_1) \geqslant \gamma_0$$

for a suitable value of γ_0 and we are finished.

REMARK 19. The constant γ_0 tends to 0 as $p \to 2^+$. Therefore in order to have the constants under control as p approaches the nondegenerate case, a comparison function other than $\mathcal{B}_{k,\rho}$ is used for p close to 2.

Once we have proved Theorem 11, we can show how it implies Theorem 12 and therefore conclude about the equivalence between the two different Harnack estimates. Let $(x_0,t_0) \in \Omega_T$, $\rho > 0$ and $\theta > 0$ be fixed in such a way that the cylinder $(x_0,t_0) + B_{4\rho} \times (-4\theta, 4\theta)$ is contained in Ω_T. Without loss of generality we can assume that $(x_0,t_0) \equiv (0,0)$ and set $u_* = u(0,0)$. With C and γ as determined in Theorem 11, we can assume that

$$t^* \equiv \frac{C\rho^p}{u_*^{p-2}} \leqslant \frac{\theta}{2},$$

otherwise there is nothing to prove. Relying on this and using the comparison principle with the function $\mathcal{B}_{k,\rho}(x,t;0,t^*)$ and $k = \gamma^{-1}u_*$, we can prove that

$$u(x,\theta) \geqslant \gamma_1 u_*^{p/\lambda}\left(\frac{\rho^p}{\theta}\right)^{N/\lambda}, \quad \text{with } \gamma_1 \equiv \gamma_1(N,p).$$

We have finished once we set $B = \max\{\gamma_1^{-\lambda/p}; (2C)^{1/(p-2)}\}$.

The assumption that the cylinder $Q_{4\rho}(\theta)$ be contained in the domain of definition of the solution is essential for the Harnack estimates of Theorems 11 and 12 to hold. When the solution is defined in $\mathbb{R}^N$ any restriction on t_0 can be avoided because we do not need to impose any restriction on ρ to have that the cylinder $Q_{4\rho}(\theta)$ belongs to the domain of definition. More precisely, if we consider nonnegative weak solutions of the type

$$\begin{cases} u \in C_{\text{loc}}(0,T; L^2_{\text{loc}}(\mathbb{R}^N)) \cap L^p_{\text{loc}}(0,T; W^{1,p}_{\text{loc}}(\mathbb{R}^N)), & p > 2, \\ u_t - \operatorname{div}|\nabla u|^{p-2}\nabla u = 0 & \text{in } \Sigma_T, \end{cases} \tag{4.20}$$

where $\Sigma_T \equiv \mathbb{R}^N \times (0,T]$, we have the following:

THEOREM 14. *Let u be a nonnegative weak solution of* (4.20). *Let* $(x_0, t_0) \in \Sigma_T$, $\rho > 0$ *and* $t > t_0$. *Then*

$$\frac{1}{\rho^N}\int_{B_\rho(x_0)} u(x, t_0)\,\mathrm{d}t$$
$$\leqslant \gamma\left\{\left(\frac{\rho^p}{t-t_0}\right)^{1/(p-2)} + \left(\frac{t-t_0}{\rho^p}\right)^{N/p}\left[\inf_{B_\rho(x_0)} u(\cdot, t)\right]^{\lambda/p}\right\}, \tag{4.21}$$

where $\gamma > 1$ *is depending only upon* N, p *and* $\lambda = N(p-2)+p$.

REMARK 20. Even if we are still dealing with local estimates as with the previous Harnack inequalities, it is the switch from Ω_T to Σ_T that gives us a useful piece of *global* information and allows us to get arbitrarily close to 0 with t_0. Estimate (4.21) contains information on the initial data of (4.20). Let $x_0 \in \mathbb{R}^N$, $r > 0$ and $\varepsilon > 0$. Apply (4.21) with $(t - t_0) = T - \varepsilon$, divide by $\rho^{p/(p-2)}$ and take the supremum of both sides for $\rho > r$ and $\tau \in (0, T - \varepsilon)$. In this way, one obtains that

$$\sup_{0<\tau\leqslant T-\varepsilon}\ \sup_{\rho>r}\int_{B_\rho(x_0)} \frac{u(x,\tau)}{\rho^{\lambda/p}}\,\mathrm{d}x$$
$$\leqslant \frac{\gamma}{\varepsilon^{1/(p-2)}}\left[1 + \left(\frac{T}{r^p}\right)^{1/(p-2)} u(x_0, T-\varepsilon)\right]^{\lambda/p}. \tag{4.22}$$

The previous estimate implies that the nonnegative solutions of (4.21) are locally bounded and, as $|x| \to +\infty$ they cannot grow faster than $|x|^{p/(p-2)}$.

4.2. *Harnack estimates*: *the singular case*

The proof of the singular case is quite similar to the degenerate case, except for the last part, relative to the spread of positivity. For the sake of completeness we recall all the steps.

First Step (Renormalization of the solution). Let $(x_0, t_0) \in \Omega_T$ and $\rho > 0$ be fixed, assume $u(x_0, t_0) > 0$ and consider the box

$$Q_{4\rho} = \{|x - x_0| < 4\rho\}$$
$$\times \{t_0 - [u(x_0, t_0)]^{2-p}(4\rho)^p, t_0 + [u(x_0, t_0)]^{2-p}(4\rho)^p\}.$$

We render the equation *dimensionless* by the change of variables

$$\xi = \frac{x - x_0}{\rho}, \qquad \tau = \frac{(t-t_0)[u(x_0,t_0)]^{p-2}}{\rho^p}, \qquad v = \frac{u}{u(x_0,t_0)}.$$

This maps $Q_{4\rho}$ into $Q = Q^+ \cup Q^-$, where $Q^+ \equiv B(4) \times [0, 4^p)$, $Q^- \equiv B(4) \times (-4^p, 0]$. We denote again the new variables with x and t and observe that the rescaled function v is a bounded nonnegative solution of the equation

$$v_t - \operatorname{div} |\nabla v|^{p-2} \nabla v = 0 \quad \text{in } Q,$$

with $v(0,0) = 1$. To prove the Harnack inequality it is enough to find constants $0 < \gamma_0 < 1$ and $0 < C < 1$, depending only upon N and p, such that, for each $x \in B(1)$, we have

$$v(x, C) \geqslant \gamma_0. \tag{4.23}$$

As a matter of fact if $u(x_0, t_0) = 0$, no rescaling is possible and we are led to consider the elliptic-type Harnack inequality, we will discuss in the next section.

Second Step (Determination of the largest value of v in Q^-). Construct the family of nested boxes $Q_\tau \equiv B_\tau \times (-\delta\tau, 0]$. Define the numbers $M_\tau = \sup_{Q_\tau} v$ and $N_\tau = (1-\tau)^{-p/(2-p)}$, where $0 < \delta < 1$ will be chosen later and has the effect of making flat the boxes Q_τ. If we compare the situation with the analogous one for $p > 2$, we notice that the cylinders Q_τ are rather thin in the t-variable and the exponent for N_τ is fixed and depends only on the singularity of the equation. Let $0 \leqslant \tau_0 < 1$ be the largest root of the equation $M_\tau = N_\tau$. Such a root is well defined, since $M_0 = N_0$ and as $\tau \to 1^-$ M_τ remain bounded and N_τ blow up. By construction

$$M_{\tau_0} = (1-\tau_0)^{-p/(2-p)}, \qquad M_{(1-\tau_0)/2} \leqslant 2^{p/(2-p)} (1-\tau_0)^{-p/(2-p)}.$$

Moreover, from the continuity of v in Q, there exists at least a point $(x_1, t_1) \in Q_{\tau_0}$, where $v(x_1, t_1) = (1-\tau_0)^{-p/(2-p)}$ and

$$\sup_{|x-\bar{x}|<(1-\tau_0)/2} v(x, t_1) \leqslant 2^{p/(2-p)} (1-\tau_0)^{-p/(2-p)}.$$

Third Step (Lower bound on v at the same time-level t_1). Relying on the Hölder continuity of v, we can determine a small ball of radius $r_0 = \varepsilon(1-\tau_0)$ about (x_1, t_1) where $v \geqslant ((1-\tau_0)^{p/(p-2)}/2)$. The variable ε is a small constant that tends to 0 as $p \to 2^-$.

Fourth Step (Time expansion of positivity). All the previous arguments are independent of the quantity δ and now it is fixed. By means of a proper comparison function, the positivity of v is spread over a small time interval *without* modifying the space size of the box we are working in. More precisely, it is proved that there exist small positive numbers c_0 and δ that can be determined a priori only in terms of N and p such that

$$v(x,t) \geqslant c_0 (1-\tau_0)^{p/(p-2)} \quad \forall |x - \bar{x}| < \varepsilon(1-\tau_0), \forall \delta \leqslant t \leqslant 2\delta.$$

This step (and the next one, too) is the main difference with respect to the degenerate case when $p > 2$. Roughly speaking, in that case the positivity is spread over time and space in one stroke. Here we need to proceed one step at a time.

Fifth Step (Sidewise expansion of positivity). Using a new comparison function the positivity of v is spread over the full ball $\{|x| < 1\}$ at the time level $t = 2\delta$. This is done by showing that there exists a constant $\gamma_0 = \gamma_0(N, p)$ such that

$$v(x, 2\delta) \geqslant \gamma_0 \quad \forall |x - \bar{x}| < 2,$$

and with this we are finished.

REMARK 21. The comparison functions used in the proof were introduced in [5] and [66]. We also point out that the comparison principle is a consequence of L^1-techniques if $u_t \in L^1_{\mathrm{loc}}(\Omega_T)$. If u_t does not belong to $L^1_{\mathrm{loc}}(\Omega_T)$ the comparison principle can be proved adapting a technique introduced by Kalashnikov, Oleinik and Chzhou [104] (see also Appendix 9 of [66]).

REMARK 22. Exactly as in the degenerate case when $p > 2$, the constant γ_0 tends to 0 as $p \to 2^-$. Therefore in order to have the constants under control as p approaches the nonsingular case, a different comparison function is used for p close to 2.

REMARK 23. In the singular case too, it is possible to state an integral Harnack inequality that holds for all $1 < p < 2$: Let u be a nonnegative weak solution of (4.1). Then there exists a constant γ, depending only upon N and p, such that, for all $(x_0, t_0) \in \Omega_T$, for all $\rho > 0$ such that $B_{4\rho}(x_0) \subset \Omega$ and for all $t > t_0$,

$$\begin{aligned} &\sup_{t_0 < \tau \leqslant t} \int_{B_\rho(x_0)} u(x, \tau)\, \mathrm{d}x \\ &\quad \leqslant \gamma \inf_{t_0 < \tau \leqslant t} \int_{B_{2\rho}(x_0)} u(x, \tau)\, \mathrm{d}x + \gamma \left(\frac{t - t_0}{\rho^\lambda} \right)^{1/(2-p)} \end{aligned} \tag{4.24}$$

with $\lambda = N(p - 2) + p$. Note that λ might be of either sign. Moreover, a very important difference with respect to the degenerate case is that in the singular case the L^1 norm of $u(\cdot, t)$ over a ball bounds the L^1 norm of $u(\cdot, \tau)$ over a smaller ball for any previous or later time. We stress out that in the nonsingular case it is only possible a control for later times and NOT for previous times. Accordingly the constant γ deteriorates when $p \to 2^-$.

4.3. *Elliptic-type Harnack estimates and extinction time*

In this section we focus our attention on what is peculiar of the case $p < 2$. As discussed before, in the singular case, at the points where $|\nabla u| = 0$, the modulus of ellipticity becomes infinite. Hence, roughly speaking, the elliptic nature of the diffusion dominates the time-evolution of the process itself and this implies that the positivity of u at some point (x_0, t_0) spreads at the same time level. We have already seen a hint of this property in the fifth step of the proof of Theorem 11 but such a feature can be made quantitatively precise and assumes the form of an elliptic-type Harnack inequality, where the infimum of u over the ball $B_\rho(x_0)$ is bound by the supremum over the same ball at the *same* time level:

THEOREM 15. *Let u be a nonnegative weak solution of* (4.1) *and let* $2N/(N+1) < p < 2$. *Let* $(x_0, t_0) \in \Omega_T$, $\rho > 0$ *and* $t_0 > 0$. *Let* $\theta = C[u(x_0, t_0)]^{2-p}\rho^p$, *where the constant C is defined in* (4.12). *Assume that the cylinder* $(x_0, t_0) + B_{4\rho} \times (t_0 - 4\theta, t_0 + 4\theta)$ *is contained in* Ω_T. *Then*

$$\gamma^{-1} \sup_{B_\rho(x_0)} u(\cdot, t_0) \leqslant u(x_0, t_0) \leqslant \gamma \inf_{B_\rho(x_0)} u(\cdot, t_0), \tag{4.25}$$

where $\gamma > 1$ *depends only upon N and p and* $\gamma \to \infty$ *when either* $p \to (\frac{2N}{N+1})^+$ *or* $p \to 2^-$.

This result is proved in [67]. See also [177] for the extension of such a result to more general operators.

REMARK 24. Estimate (4.25) fails in the case of nonnegative solutions of the heat equation and also for nonnegative solutions of the p-Laplacian when $p > 2$. To verify this in the case of the heat equation, consider the fundamental solution in one space dimension

$$\Gamma(x, t) = \frac{1}{(4\pi t)^{1/2}} e^{-x^2/(4t)}.$$

If Theorem 15 were to hold, we would have, for some $\rho > 0$, that $\Gamma(n, 1) \leqslant \Gamma(n + \rho, 1)$. Letting $n \to \infty$ we obtain a contradiction. That is the reason why the constants in (4.25) deteriorate when p goes to 2.

The elliptic-like Harnack inequality holds also for the nonnegative solutions of the porous medium equation. In such a case we get sharp estimates on the solution

THEOREM 16. *Let u be a nonnegative weak solution of*

$$u_t - \Delta(u^m) = 0 \quad \text{in } \Omega_T, \tag{4.26}$$

and let $((N-2)_+)/(N+2) < m < 1$. *Let* $\theta = C[u(x_0, t_0)]^{1-m}\rho^2$ *where the constant C is defined in* (4.12). *Assume that the cylinder* $(x_0, t_0) + B_{4\rho} \times (t_0 - 4\theta, t_0 + 4\theta)$ *is contained in* Ω_T. *Then, for each multiindex* α,

$$\left|D^\alpha u(x_0, t_0)\right| \leqslant \frac{C^{|\alpha|+1}|\alpha|!}{\rho^\alpha} u(x_0, t_0) \tag{4.27}$$

and, for every nonnegative integer k,

$$\left|\frac{\partial^k}{\partial t^k} u(x_0, t_0)\right| \leqslant \frac{C^{2k+1}(k!)^2}{\rho^{2k}} u(x_0, t_0)^{1-(1-m)k}, \tag{4.28}$$

where $C > 1$ *depends only upon N and m.*

PROOF. We render the equation *dimensionless* by the change of variables

$$\xi = \frac{x - x_0}{\rho}, \qquad \tau = \frac{(t - t_0)[u(x_0, t_0)]^{m-1}}{\rho^2}, \qquad v = \frac{u}{u(x_0, t_0)}.$$

This maps $Q_{4\rho}$ into $Q = Q^+ \cup Q^-$, where $Q^+ \equiv B(4) \times [0, 4C)$, $Q^- \equiv B(4) \times (-4C, 0]$. We denote again the new variables with x and t and observe that the rescaled function v is a bounded nonnegative solution of the equation

$$v_t - \Delta\left(v^m\right) = 0 \quad \text{in } Q, \tag{4.29}$$

with $v(0,0) = 1$. By the integral Harnack inequality (4.24) and the elliptic-like Harnack inequality (4.25) we have that there exist positive constants r_0 and $\gamma > 1$, depending only upon N and m, such that $\gamma^{-1} \leqslant v(x,t) \leqslant \gamma$ for each $(x,t) \in Q(r_0^2, r_0)$. Using this new piece of information, we have that in such a cylinder we can apply classical results due to Friedman [83] and Kinderleher and Nirenberg [106] to the equation (4.29) to obtain the analyticity of the solution. More precisely we get that, for each multiindex α,

$$\left|D^\alpha v(0,0)\right| \leqslant \frac{C^{|\alpha|+1}}{|\alpha|!}$$

and, for every nonnegative integer k,

$$\left|\frac{\partial^k}{\partial t^k} v(0,0)\right| \leqslant \frac{C^{2k+1}}{(k!)^2},$$

where $C > 1$ depends only upon N and m.

By the reverse change of variables we deduce (4.27) and (4.28). □

REMARK 25. Estimate (4.27) not only implies the analyticity of the solution in the space variables but it also says that when the solution vanishes at a point then all the derivatives vanish at the same point. Therefore, for the analyticity, the solution vanishes in all the domain Ω whenever vanishes at a point of Ω. Estimate (4.27) fails in the case of nonnegative solutions of the heat equation. Indeed, consider the fundamental solution in one space dimension

$$\Gamma(x,t) = \frac{1}{(4\pi t)^{1/2}} \mathrm{e}^{-x^2/(4t)}.$$

If (4.27) were to hold, we would have, for some $C > 0$,

$$\left|\frac{\mathrm{d}}{\mathrm{d}x}\Gamma(n,1)\right| = \frac{1}{2} n \Gamma(n,1) \leqslant C\Gamma(n,1).$$

Letting $n \to \infty$ we obtain a contradiction.

REMARK 26. The previous estimates say that a bounded nonnegative solution of the singular porous medium equation is analytic in the space variables and at least Lipschitz continuous in the time variable. We stress that these estimates are optimal. Indeed let z be the nonnegative and non trivial solution of the problem $z_{xx} = \frac{1}{1-m} z^{1/m}$ in the interval $(0, 1)$, with boundary conditions $z(0) = z(1) = 0$. Then $u(x,t) = z^{1/m}(T-t)_+^{1/(1-m)}$ solves the equation $u_t = \Delta(u^m)$ and satisfies the above estimates sharply.

REMARK 27. These estimates hold also in the case of a class of quasilinear parabolic equations. More precisely for nonnegative, local weak solutions of

$$u_t = \Delta(u^m) + f(x,t,u,\nabla u),$$

with $(N-2)_+/(N+2) < m < 1$, and f locally analytic and such that $0 \leqslant f(x,t, u, \nabla u) \leqslant F u^m$ for some positive constant F (see [67]).

Another peculiarity of the case $p < 2$ is that the solution can become extinct in a finite time. The extinction profile is defined as the set $\partial[u > 0] \cap [\Omega \times (0, \infty)]$. By the elliptic-like Harnack principle the extinction profile is a portion of the hyperplane $\Omega \times \{t = T^*\}$.

Let us first consider the case of a bounded domain.

THEOREM 17. *Let Ω be a bounded domain of $\mathbb{R}^N$. Let u be the unique nonnegative weak solution of*

$$\begin{cases} u \in C(\mathbb{R}^+; L^2(\Omega)) \cap L^p(\mathbb{R}^+; W_0^{1,p}(\Omega)), & 1 < p < 2, \\ u_t - \operatorname{div} |\nabla u|^{p-2} \nabla u = 0 & \text{in } \Omega \times \mathbb{R}^+, \\ u(\cdot, 0) = u_0(x) \in L^\infty(\Omega) \quad \text{and} \quad u_0 \geqslant 0. \end{cases} \tag{4.30}$$

Then there is a finite time T^*, depending only upon N, p and u_0, such that $u(\cdot,t) \equiv 0$ for all $t \geqslant T^*$.

Moreover, if $\max(1, 2N/(N+2)) < p < 2$, then

$$0 < T^* \leqslant \gamma^* \|u_0\|_{2,\Omega}^{2-p} |\Omega|^{\frac{N(p-2)+2p}{2N}} \tag{4.31}$$

with γ^* depending only upon N and p.

If $1 < p \leqslant 2N/(N+1)$ and $N \geqslant 2$, then

$$0 < T^* \leqslant \gamma^* \|u_0\|_{s,\Omega}^{2-p} \tag{4.32}$$

with γ^* depending only upon N, p and $s = N(2-p)/p$.

REMARK 28. Note that there is an overlap in the range of p in the previous estimates. We stress that for $1 < p < 2N/(N+1)$, the upper estimate of T^* does not depend upon the measure of Ω.

PROOF. Consider first the case $\max(1, 2N/(N+2)) < p < 2$. Take u as a test function in the weak formulation of (4.30) to get

$$\frac{\mathrm{d}}{\mathrm{d}t}\|u(t)\|_{2,\Omega}^2 + 2\|\nabla u(t)\|_{p,\Omega}^p = 0.$$

By Hölder's inequality and Sobolev embedding theorem,

$$\|u(t)\|_{2,\Omega} \leqslant |\Omega|^{\frac{N(p-2)+2p}{2Np}} \|u(t)\|_{\frac{Np}{N-p},\Omega} \leqslant \gamma|\Omega|^{\frac{N(p-2)+2p}{2Np}} \|\nabla u(t)\|_{p,\Omega}.$$

In a straightforward way one may deduce that $\|u(t)\|_{2,\Omega}$ satisfies the following differential inequality

$$\frac{\mathrm{d}}{\mathrm{d}t}\|u(t)\|_{2,\Omega} + \gamma_1\|u(t)\|_{2,\Omega}^{p-1} \leqslant 0,$$

where $\gamma_1 = \gamma^{-p}|\Omega|^{-\frac{N(p-2)+2p}{2N}}$. Solving the ordinary differential equation one obtains

$$\|u(t)\|_{2,\Omega} \leqslant \|u_0\|_{2,\Omega}\left\{1 - \frac{(2-p)\gamma_1 t}{\|u_0\|_{2,\Omega}^{2-p}}\right\}_+^{1/(2-p)} \tag{4.33}$$

and

$$0 < T^* \leqslant \frac{1}{2-p}\gamma^p|\Omega|^{\frac{N(p-2)+2p}{2N}}\|u_0\|_{2,\Omega}^{2-p}.$$

Consider now the case $1 < p \leqslant 2N/(N+1)$ and $N \geqslant 2$. Let $s = N(2-p)/p$ and take u^{s-1} as a test function in the weak formulation of the equation (4.30) to get

$$\frac{1}{s}\frac{\mathrm{d}}{\mathrm{d}t}\|u(t)\|_{s,\Omega}^s + \gamma_2\|\nabla u^{(s+(p-2))/p}(t)\|_{p,\Omega}^p = 0,$$

where $\gamma_2 = (s-1)(\frac{p}{s+(p-2)})^p$. By Sobolev embedding theorem,

$$\|u(t)\|_{s,\Omega}^s \leqslant \gamma\|\nabla u^{(s+(p-2))/p}(t)\|_{p,\Omega}^{Np/(N-p)}.$$

In a straightforward way one may deduce that $\|u(t)\|_{s,\Omega}$ satisfies the following differential inequality

$$\frac{\mathrm{d}}{\mathrm{d}t}\|u(t)\|_{s,\Omega} + \gamma_3\|u(t)\|_{s,\Omega}^{p-1} \leqslant 0,$$

where $\gamma_3 = \gamma^{-p}\gamma_2$. Solving the ordinary differential equation one obtains

$$\|u(t)\|_{s,\Omega} \leqslant \|u_0\|_{s,\Omega}\left\{1 - \frac{(2-p)\gamma_3 t}{\|u_0\|_{s,\Omega}^{2-p}}\right\}_+^{1/(2-p)} \tag{4.34}$$

and

$$0 < T^* \leqslant \frac{1}{(2-p)\gamma_2}\gamma^p \|u_0\|_{s,\Omega}^{2-p}. \qquad \square$$

REMARK 29. The estimate (4.33) is *stable* as $p \to (\frac{2N}{N+1})^+$. As $p \to 2^-$ the boundary value problem (4.30) tends to the corresponding boundary value problem for the heat equation, for which the extinction in finite time does not occur. Accordingly, if $p \to 2^-$, the estimate (4.33) becomes

$$\|u(t)\|_{2,\Omega} \leqslant \|u_0\|_{2,\Omega} \exp\left\{-\frac{t|\Omega|^{2/N}}{\gamma^2}\right\},$$

where γ is the best constant of the Sobolev embedding of $W^{1,2}$ in $L^{2N/(N-2)}$.

The estimate (4.34) deteriorates as $p \to (\frac{2N}{N+1})^-$ and is *stable* as $p \to 1^+$. However, as the regularity results of the previous sections deteriorate as $p \to 1^+$, we cannot infer the convergence of (4.30) to a boundary value problem in some reasonable topology.

REMARK 30. Theorem 17 holds for solutions of variable sign. The only modification occurs in the case $1 < p < 2N/(N+1)$, $N \geqslant 2$. For this it is enough to work with the testing functions $|u|^{s-2}u$.

The Harnack principle gives also an estimate on the way the solution approaches the extinction. Let $M = \|u\|_{\infty,\Omega_\infty}$, where $\Omega_\infty \equiv \Omega \times (0,\infty)$.

THEOREM 18. *Let Ω be a bounded domain of $\mathbb{R}^N$. Let u be the unique nonnegative weak solution of* (4.30) *and let* $2N/(N+1) < p < 2$. *Then there exists a constant γ, depending only upon N and p, such that, for all* $(x,t) \in \Omega \times (T^*/2, T^*)$,

$$u(x,t) \leqslant \gamma \max\left\{M^{2-p}; \frac{T^*}{[\operatorname{dist}\{x,\partial\Omega\}]^p}\right\}^{1/(2-p)} \left(\frac{T^*-t}{T^*}\right)^{1/(2-p)}.$$

PROOF. Fix $x \in \Omega$ and $T^*/2 \leqslant t \leqslant T^*$. Assume that $u(x,t) > 0$ and set

$$4\rho = \min\left\{\operatorname{dist}\{x,\partial\Omega\}; \left(\frac{T^*}{2M^{2-p}}\right)^{1/p}\right\}. \tag{4.35}$$

Consider the cylinder $Q_{4\rho}(x,t) = B_{4\rho}(x) \times \{t - [u(x,t)]^{2-p}(4\rho)^p, t + [u(x,t)]^{2-p} \times (4\rho)^p\}$. By the choice (4.35) the cylinder is contained in Ω_∞. Apply Harnack inequality (4.11) over the ball $B_\rho(x)$ and the cylinder $Q_{4\rho}(x,t)$. We must have

$$T^* - t \geqslant C[u(x,t)]^{2-p}\rho^p,$$

otherwise, by Harnack estimate, $u(x,t) = 0$ against the assumption. $\square$

Consider now the case of the extinction in $\mathbb{R}^N$.

THEOREM 19. *Let u be the unique nonnegative weak solution of*

$$\begin{cases} u \in C(\mathbb{R}^+; L^2(\mathbb{R}^N)) \cap L^p(\mathbb{R}^+; W^{1,p}(\mathbb{R}^N)), & 1 < p < 2, \\ u_t - \operatorname{div} |\nabla u|^{p-2} \nabla u = 0 & \text{in } \mathbb{R}^N \times \mathbb{R}^+, \\ u(\cdot, 0) = u_0(x) \geqslant 0. \end{cases} \tag{4.36}$$

Assume that u_0 is continuous in a ball $B(2R)$ and vanishes outside $B(R)$. Assume $1 < p \leqslant 2N/(N+1)$ and $N \geqslant 2$, and let $s = N(2-p)/p$. Then there is a positive number T^*, depending only upon N, p and u_0, such that $u(x,t) = 0$ for each $t \geqslant T^*$. Moreover, $0 < T^* \leqslant \gamma^* \|u_0\|_{s,\Omega}^{2-p}$, with γ^* depending only upon N and p.

PROOF. The solution of (4.36) can be constructed as the uniform limit of a sequence u_n of solutions in bounded domains where n is a natural number and $n \geqslant R$. More precisely,

$$\begin{cases} u_n \in C(\mathbb{R}^+; L^2(B(n))) \cap L^p(\mathbb{R}^+; W_0^{1,p}(B(n))), \\ (u_n)_t - \operatorname{div} |\nabla u_n|^{p-2} \nabla u_n = 0 \quad \text{in } B(n) \times \mathbb{R}^+, \\ u_n(\cdot, 0) = u_0(x). \end{cases} \tag{4.37}$$

By Theorem 17, the extinction time T_n^* is independent of $B(n)$. Moreover, as $u_n \leqslant u_{n+1}$, we have that u_n converges to the solution u and T_n^* converges to T^*. □

REMARK 31. Theorem 19 holds for solutions of variable sign and for data in $L^s(\mathbb{R}^N)$ with $s = N\frac{2-p}{p}$. The only modification in the proof occurs in making precise in what sense the solutions of the approximated problems converge to the solution of (4.36).

REMARK 32. Theorem 19 implies that the range $2N/(N+1) < p < 2$ is optimal for a Harnack estimate to hold. Let $\Sigma_\infty \equiv \mathbb{R}^N \times \mathbb{R}_+$. Fix $(x_0, t_0) \in \mathbb{R}^N \times (0, T^*)$, where t_0 is so close to T^* as to satisfy

$$T^* - t_0 \leqslant \frac{C}{4^p} t_0, \tag{4.38}$$

where C is the constant appearing in (4.12). Now let $\rho > 0$ be so large that

$$C[u(x_0, t_0)]^{2-p} \rho^p = T^* - t_0. \tag{4.39}$$

By the choice (4.38) the cylinder

$$Q_{4\rho}(t, x) = B_{4\rho}(x) \times \{t_0 - [u(x_0, t_0)]^{2-p} (4\rho)^p, t_0 + [u(x_0, t_0)]^{2-p} (4\rho)^p\}$$

is contained in Σ_∞. If the Harnack inequality (4.11) were to hold for $1 < p < 2N/(N+1)$, $N \geqslant 2$, for some constants C and γ independent of ρ, it would give $0 < u(x_0, t_0) \leqslant \gamma \inf_{x \in B_\rho(x_0)} u(x, T^*) = 0$. We stress that the choice (4.39) is possible in the whole Σ_∞.

The same argument implies that no extinction in finite time can occur for solutions of (4.36) if $2N/(N+1) < p < 2$. In such a range the Harnack estimate (4.11) holds and if a finite extinction time T^* were to exists, the choices (4.38) and (4.39) would give $u(x, T^*) > 0$.

4.4. *Raleigh quotient and extinction profile*

The Harnack estimates play a fundamental role in analyzing the asymptotic behavior of solutions of singular equations. The physical motivation of such an analysis comes from the modeling of plasma assuming the Okuda–Dawson diffusion model (see [74,139], see also [22,23]). In [118] and [67] this analysis was carried out through Sobolev embedding theorem, Harnack estimates and Raleigh quotient (see also [155], where general singular operators and general boundary conditions are considered). In this section this application of Harnack inequalities is described.

Let Ω be a bounded set of $\mathbb{R}^N$. Consider the Cauchy–Dirichlet problem:

$$\begin{cases} u \in C\big(\mathbb{R}^+; L^2(\Omega)\big) \cap L^p\big(\mathbb{R}^+; W^{1,p}(\Omega)\big), & \frac{2N}{N+1} < p < 2, \\ u_t - \operatorname{div} |\nabla u|^{p-2} \nabla u = 0 & \text{in } \Omega \times \mathbb{R}^+, \\ u(\cdot, 0) = u_0(x) \in L^2(\Omega) \quad \text{and} \quad u_0 \geqslant 0, & \\ u(x, t) = 0 & \forall x \in \partial\Omega \text{ and } \forall t > 0. \end{cases} \tag{4.40}$$

The main result of this section is the following:

THEOREM 20. *Let u be the unique nonnegative weak solution of* (4.40) *and let $T^* > 0$ be the extinction time. Let $u_*(x, t) = u(x, t)(T^* - t)^{1/(p-2)}$. Then there is a sequence $t_n \to T^*$ such that $u_*(x, t) \to v(x)$, where v is a nontrivial solution of the equation*

$$\operatorname{div} |\nabla v|^{p-2} \nabla v = \frac{1}{2-p} v \quad \text{in } \Omega \tag{4.41}$$

satisfying homogeneous Dirichlet boundary conditions.

REMARK 33. The proof of Theorem 20 is heavily based upon the Sobolev embedding of $W^{1,p}$ into L^2. For this reason we assume Ω bounded and $2N/(N+1) < p < 2$. If $p \leqslant 2N/(N+1)$ the result is false. Indeed if the theorem were to hold for such a range of p, it would give the existence of a nontrivial solution of (4.41) in contradiction with known results of the elliptic theory (see [145]). We recall that the existence of a nonzero solution of (4.41) when $p \leqslant 2N/(N+1)$ is not true in general, but depends on topological properties of the set Ω (see, for instance, [30]).

Before coming to the actual proof of Theorem 20, we consider some auxiliary results. First of all let us introduce the Raleigh quotient

$$\mathcal{E}[u](t) = \left(\frac{\|\nabla u\|_{p,\Omega}}{\|u\|_{2,\Omega}} \right)^p.$$

PROPOSITION 13. *The quantity $\mathcal{E}[u](t)$ is not increasing in time.*

PROOF. Choose u as a test function in the weak form of (4.40) to get

$$\frac{1}{2}\frac{\mathrm{d}}{\mathrm{d}t}\|u(t)\|_{2,\Omega}^{2}+\|\nabla u(t)\|_{p,\Omega}^{p}=0. \tag{4.42}$$

Setting $\Delta_p(u)(t)=\operatorname{div}|\nabla u(t)|^{p-2}\nabla u(t)$, we obtain

$$\begin{aligned}\int_\Omega |\nabla u(t)|^p\,\mathrm{d}x &= -\int_\Omega u(t)\Delta_p(u)(t)\,\mathrm{d}x\\ &\leqslant \|u(t)\|_{2,\Omega}\left(\int_\Omega |\Delta_p(u)(t)|^2\,\mathrm{d}x\right)^{1/2}.\end{aligned} \tag{4.43}$$

On the other hand,

$$\begin{aligned}\frac{\mathrm{d}}{\mathrm{d}t}\int_\Omega |\nabla u(t)|^p\,\mathrm{d}x &= p\int_\Omega |\nabla u(t)|^{p-2}\nabla u(t)\nabla u_t(t)\,\mathrm{d}x\\ &= -p\int_\Omega u_t(t)\Delta_p(u)(t)\,\mathrm{d}x.\end{aligned} \tag{4.44}$$

Hence

$$\frac{\mathrm{d}}{\mathrm{d}t}\int_\Omega |\nabla u(t)|^p\,\mathrm{d}x = -p\int_\Omega |\Delta_p(u)(t)|^2\,\mathrm{d}x, \tag{4.45}$$

which gives, together with (4.43),

$$\frac{\mathrm{d}}{\mathrm{d}t}\|\nabla u(t)\|_{p,\Omega}^{p}\leqslant -p\frac{\|\nabla u(t)\|_{p,\Omega}^{2p}}{\|u(t)\|_{2,\Omega}^{2}}. \tag{4.46}$$

Directly from the equation one gets

$$\frac{1}{p}\frac{\frac{\mathrm{d}}{\mathrm{d}t}\|\nabla u(t)\|_{p,\Omega}^{p}}{\|\nabla u(t)\|_{p,\Omega}^{p}}\leqslant\frac{1}{2}\frac{\frac{\mathrm{d}}{\mathrm{d}t}\|u(t)\|_{2,\Omega}^{2}}{\|u(t)\|_{2,\Omega}^{2}}$$

and this implies that $\mathcal{E}[u](t)$ is not increasing in time. □

Let $B_{p,\Omega}$ be the best Sobolev constant of the embedding of $W^{1,p}(\Omega)$ in $L^2(\Omega)$.

PROPOSITION 14. *The inequalities*

$$B_{p,\Omega}\leqslant\mathcal{E}[u](t)\leqslant\mathcal{E}[u](0),$$

hold.

PROOF. It is a straightforward consequence of Proposition 13 and the definition of the Raleigh quotient. □

The following proposition gives a sharp estimate on the decay of the solution at the extinction time.

PROPOSITION 15. *Let u be the unique nonnegative weak solution of* (4.40) *and let* $T^* > 0$ *be the extinction time. Then*

$$\begin{aligned}&\left[(2-p)B_{p,\Omega}(T^*-t)\right]^{1/(2-p)}\\&\quad\leqslant \left\|u(t)\right\|_{2,\Omega}\leqslant \left[(2-p)\mathcal{E}[u](0)(T^*-t)\right]^{1/(2-p)}\end{aligned}$$

and

$$\begin{aligned}&B_{p,\Omega}\left[(2-p)B_{p,\Omega}(T^*-t)\right]^{1/(2-p)}\\&\quad\leqslant \left\|\nabla u(t)\right\|_{p,\Omega}\leqslant \mathcal{E}[u](0)\left[(2-p)\mathcal{E}[u](0)(T^*-t)\right]^{1/(2-p)}.\end{aligned}$$

PROOF. It is sufficient to note that $\|u(t)\|_{2,\Omega}$ solves the ODE,

$$\frac{\mathrm{d}}{\mathrm{d}t}\left\|u(t)\right\|_{2,\Omega}^{2-p}=-(2-p)\mathcal{E}[u](t).$$

Hence

$$0=\left\|u(T^*)\right\|_{2,\Omega}^{2-p}=\left\|u(t)\right\|_{2,\Omega}^{2-p}-(2-p)\int_t^{T^*}\mathcal{E}[u](s)\,\mathrm{d}s.$$

Now the statement follows from Proposition 14 and the definition of the Raleigh quotient. □

Using the previous results we can now conclude with:

PROOF OF THEOREM 20. Consider the change of variables: $t=T^*-T^*\mathrm{e}^{-\tau}$. Let

$$w(\cdot,\tau)=\frac{u(\cdot,T^*-T^*\mathrm{e}^{-\tau})}{(T^*\mathrm{e}^{-\tau})^{1/(2-p)}}.$$

The function w is a nonnegative bounded weak solution of

$$\begin{cases} w_t=\operatorname{div}|\nabla w|^{p-2}\nabla w-\frac{1}{2-p}w & \text{in } \Omega\times\mathbb{R}^+,\\ w(\cdot,0)=u_0(x)T^{*1/(p-2)}. \end{cases}\tag{4.47}$$

Consider the functional

$$F(h)=\int_{\Omega}\left(\frac{1}{p}|\nabla h|^{p}-\frac{1}{2(2-p)}|h|^{2}\right)\mathrm{d}x$$

and the function

$$g(\tau)=F\big(w(\tau)\big).$$

The function $g(\tau)$ is a nonincreasing function. Indeed, by using (4.44) and (4.47), we have

$$\begin{aligned}\frac{1}{p}\frac{\mathrm{d}}{\mathrm{d}t}\int_{\Omega}|Dw|^{p}\,\mathrm{d}x&=-\int_{\Omega}w_t^2\,\mathrm{d}x+\frac{1}{2-p}\int_{\Omega}w_t w\,\mathrm{d}x\\&=\int_{\Omega}w_t^2\,\mathrm{d}x+\frac{1}{2(2-p)}\frac{\mathrm{d}}{\mathrm{d}t}\int_{\Omega}w^2\,\mathrm{d}x.\end{aligned}$$

Moreover, by Proposition 13 $\gamma(\tau)$ is bounded from below. Therefore there exists a sequence $\tau_n\to\infty$ such that

$$\lim_{n\to\infty}\frac{\mathrm{d}}{\mathrm{d}t}F\big(w(\tau_n)\big)=0.$$

From the previous calculations, this implies that $w_t(\tau_n)\to 0$ in $L^p(\Omega)$. Hence, we get that there is a sequence $t_n\to T^*$ such that

$$\frac{u(\cdot,t_n)}{(T^*-t_n)}\to v,$$

where v is the solution of (4.41). On the other hand, by Proposition 15, there are two positive constants c_1 and c_2, such that, for each t_n,

$$0<c_1\leqslant\left\|\frac{\nabla u(\cdot,t_n)}{(T^*-t_n)}\right\|_{p,\Omega}\leqslant c_2.$$

Therefore

$$\frac{u(\cdot,t_n)}{(T^*-t_n)}\to v$$

in $W^{1,p}(\Omega)$. Applying the regularity results of the previous sections, it is easy to show that this convergence holds also in $C^{\alpha}(\overline{\Omega})$. Applying the results of the previous section, we have that v satisfies (4.25). □

REMARK 34. If the asymptotic profile of the singular porous medium equation is considered, arguing as before one can prove that the limiting solution v satisfies the sharpest estimates (4.27) and (4.28).

The approach through the Raleigh quotient can be applied also in the case of degenerate parabolic equations using similar arguments (see [125]).

THEOREM 21. *Let $p > 2$ and let u be the unique nonnegative weak solution of* (4.40). *Let $u_*(x,t) = u(x,t)t^{1/(2-p)}$. Then there is a sequence $t_n \to \infty$ such that $u_*(x,t) \to w(x)$, where w is a nontrivial solution of* (4.41), *satisfying Dirichlet boundary conditions.*

REMARK 35. In the literature there are several papers devoted to the study of the asymptotic behavior of the solutions of the porous medium equation and the p-Laplacian equation. Among them we quote [14,15,24,25,174]. In these papers the approach is different from what we followed here. Indeed they first study the elliptic equation (4.41), then using some comparison principles they analyze the asymptotic behavior of the evolution equation. With this approach things are somehow reversed: the basic properties of the evolution equation allow for the study of the asymptotic behavior and the elliptic result follows as a consequence.

REMARK 36. The proofs of Theorems 20 and 21 are based on Sobolev embedding and weak convergence arguments. For this reason one can apply this approach also in the case of initial data with variable sign and in the case of Neumann or mixed boundary conditions [155].

5. Stefan-like problems

In this section we present some results about the local and global behavior of weak solution of singular parabolic equations which model physical phenomena like transitions of phase and/or the flow of immiscible fluids in a porous medium. More precisely, let us consider parabolic inclusions of the type

$$\frac{\partial}{\partial t}\beta(u) - \operatorname{div}\mathbf{A}(x,t,u,\nabla u) + B(x,t,u,\nabla u) \ni 0 \quad \text{in } \Omega_T, \tag{5.1}$$

where β is a maximal monotone graph in $\mathbb{R} \times \mathbb{R}$. We assume the coerciveness of the graph $\beta(\cdot)$, i.e., there exists a positive constant γ_0 such that

$$\beta(s_1) - \beta(s_2) \geqslant \gamma_0(s_1 - s_2) \quad \forall s_i \in \mathbb{R}. \tag{5.2}$$

We also assume that,

$$\forall M > 0, \qquad \sup_{-M \leqslant s \leqslant M} \left|\beta(s)\right| \equiv \gamma_1 < \infty. \tag{5.3}$$

We stress that we do not assume any further assumptions on the behavior of $\beta(\cdot)$. In particular, in any finite interval, the graph might exhibit infinitely many jumps or become vertical infinitely many times with any possible growth (exponentially fast or faster). Examples of such $\beta(\cdot)$ are

$$\beta(s)=\begin{cases} s & \text{if } s<0,\\ [0,1] & \text{if } s=0,\\ 1+s & \text{if } 0<s<1,\end{cases} \tag{5.4}$$

$$\beta(s)=\begin{cases} s & \text{if } s<0,\\ [0,1] & \text{if } s=0,\\ 1+s & \text{if } 0<s<1,\\ [2,3] & \text{if } s=1,\\ 2+s & \text{if } s>1,\end{cases} \tag{5.5}$$

$$\beta(s)=|s|^{1/m}\operatorname{sign} s, \quad m>1, \tag{5.6}$$

or

$$\beta(s)=1+s^{\alpha_1}-(1-s)^{\alpha_2}, \tag{5.7}$$

where $s\in[0,1]$ and $\alpha_i\in[0,1]$.

Equation (5.4) describes the enthalpy function in the weak formulation of a Stefan-like problem modeling a transition of phase, while (5.5) could be a good prototype to model the behavior of the enthalpy in a double transition of phase. There is a wide literature concerning the classical Stefan problem. For a summary of the main results we refer the reader to the monographs of Meirmanov [129] and Visintin [178], the review papers by Danilyuk [45], DiBenedetto [57] and by Visintin, Fasano, Magenes and Verdi [179], the proceedings [29,82,94], as well as the references therein. Here, we consider only the aspects related to the local continuity of a weak solution. Equation (5.6) is the classical porous medium equation, describing the flow of a single fluid in a porous matrix, that was already mentioned in Section 2.5. Equation (5.7) is a first approximation of the flows of two immiscible fluids in a porous matrix. Once more this is a widely studied problem. For instance, Van Duijn and Zhang [173] considered a one-dimensional model in hydrology (see also [93] for an investigation from a numerical point of view). For multidimensional multiphase models we refer the reader to the monographs [18,19,34,44,156] and to the references therein.

The diffusion field $\mathbf{A}$ and the forcing term B in (5.1) are real valued and measurable over $\Omega_T\times\mathbb{R}\times\mathbb{R}^N$ and satisfy the structure conditions:

$$\left|\mathbf{A}(x,t,v,\bar p)\right|\geqslant \mu_0\big(|v|\big)|\bar p|^2-\phi_0(x,t), \tag{5.8}$$

$$\left|\mathbf{A}(x,t,v,\bar p)\right|\leqslant \mu_1\big(|v|\big)|\bar p|-\phi_1(x,t), \tag{5.9}$$

$$\left|B(x,t,v,\bar p)\right|\leqslant \mu_2\big(|v|\big)|\bar p|^2-\phi_2(x,t), \tag{5.10}$$

where $\mu_0:\mathbb{R}_+\to\mathbb{R}_+$ is a continuous and decreasing function, $\mu_1,\mu_2(\cdot):\mathbb{R}_+\to\mathbb{R}_+$ are continuous and increasing functions and ϕ_i, $i=0,1,2$, are nonnegative and satisfy

$$\|\phi_0,\phi_2\|_{\hat q,\hat r,\Omega_T}, \qquad \|\phi_1\|_{2\hat q,2\hat r,\Omega_T}\leqslant \mu_3. \tag{5.11}$$

Here μ_3 is a given constant and $\hat{q}$, $\hat{r}$ are positive numbers linked by

$$\frac{1}{\hat{r}} + \frac{N}{2\hat{q}} = 1 - \kappa_1, \quad 0 < \kappa_1 < 1, \tag{5.12}$$

with

$$\hat{q} \in \left[\frac{N}{2(1-\kappa_1)}, \infty\right], \qquad \hat{r} \in \left[\frac{1}{1-\kappa_1}, \infty\right] \quad \text{if } N \geqslant 2, \tag{5.13}$$

$$\hat{q} \in (1, \infty), \qquad \hat{r} \in \left[\frac{1}{1-\kappa_1}, \frac{1}{1-2\kappa_1}\right], \quad 0 < \kappa_1 < \frac{1}{2}, \quad \text{if } N = 1. \tag{5.14}$$

The inclusion (5.1) is in the sense of the graphs and in the weak sense. More precisely a function

$$u \in L^2_{\text{loc}}\big(0, T; W^{1,2}_{\text{loc}}(\Omega)\big) \tag{5.15}$$

is a weak solution of (5.1) if there exists a measurable selection $w \subseteq \beta(u)$, such that

$$t \to w(\cdot, t) \quad \text{is weakly continuous in } L^2_{\text{loc}}(\Omega)$$

and

$$\begin{aligned} &\int_\Omega w(x,\tau)\phi(x,\tau)\,\mathrm{d}x \Big|_{t_1}^{t_2} + \int_{t_1}^{t_2}\int_\Omega \left[-w(x,\tau)\frac{\partial}{\partial\tau}\phi(x,\tau)\right] \mathrm{d}x\,\mathrm{d}\tau \\ &\quad + \int_{t_1}^{t_2}\int_\Omega \big\{\mathbf{A}(x,\tau,u,\nabla u)\cdot\nabla\phi + B(x,\tau,u,\nabla u)\phi\big\}\,\mathrm{d}x\,\mathrm{d}\tau = 0 \end{aligned} \tag{5.16}$$

for all $\phi \in W^{1,2}_{\text{loc}}(0, T; L^2_{\text{loc}}(\Omega)) \cap L^2_{\text{loc}}(0, T; W^{1,2}_0(\Omega))$ and all intervals $[t_1, t_2] \subset (0, T]$.

In the sequel, when possible and for the sake of simplicity, we will work with the simplest example of (5.1), i.e.,

$$\frac{\partial}{\partial t}\beta(u) - \Delta u \ni 0, \quad \text{in } \Omega_T. \tag{5.17}$$

We will obviously point out the situations in which the claimed results hold only for this simplest inclusion.

5.1. *The continuity of weak solutions*

It is quite natural to investigate if locally bounded weak solutions of (5.2) are continuous and if their modulus of continuity can be estimated quantitatively. Let us state this more precisely. For the sake of simplicity, assume that u is a solution of (5.2) and that it is bounded in Ω_T. Set

$$\|u\|_{\infty,\Omega_T} = M. \tag{5.18}$$

We recall that this assumption is not restrictive if we define Ω_T as the domain of definition of u. In a similar way, we assume the following integrability assumption

$$\left\|\phi_0 + \phi_1^2 + \phi_2\right\|_{\bar{q},\bar{r},\Omega_T} = \Phi. \tag{5.19}$$

We set the numbers

$$N, \gamma_i, M, \Phi, \mu_i, \quad i = 0, 1, 2,$$

as the data. Consequently, we say that a constant $C = C(data)$ or a continuous function $\omega(\cdot) = \omega_{data}(\cdot)$ if they can be determined a priori only in term of the above parameters. Let $\Theta \subset\subset \Omega_T$ be an arbitrary subset. In the sequel we investigate the problem of the continuity of u in Θ with a modulus of continuity $\omega_{data}(\cdot)$ depending only upon the data and the distance from Θ and the parabolic boundary of Ω_T.

REMARK 37. If $\beta(\cdot)$ is the identity, we are in De Giorgi's setting so any locally bounded weak solution is Hölder continuous in Ω_T. The assumptions (5.11)–(5.14) are optimal for this result to hold (see, for instance, the monograph by Ladyzhenskaja, Solonnikov, and Ural'tzeva [120], Chapters 1, 2 and 5). In this section we want to study how the singularity of β affects the regularity of u.

REMARK 38. The assumption that u is bounded is essential. Even in the most favorable case when $\beta(\cdot)$ is the identity, weak solutions need not to be bounded (even in the elliptic case; see a counterexample by Stampacchia [159]). This is due to the critical growth of the forcing term $B(x,t,u,\nabla u)$ with respect to ∇u. We recall that in the nonsingular case, if we assume that

$$B(x,t,u,\nabla u) \leqslant \mu_2|\nabla u|^q + \phi_2(x,t),$$

where $0 \leqslant q < (N+4)/(N+2)$, instead of (5.10), then any weak solution is locally bounded. This statement follows by means of a simple adaptation of the method of [120] (see also [47] and [132]). Even when the solution is not necessarily locally bounded, if one has some a priori qualitative knowledge of the boundedness of the solution such a qualitative information can, in many cases, be turned into a quantitative one (see, for instance, [26,141,176] and the references therein).

In the sequel we assume in addition that the solutions of (5.1) can be constructed as the limit in the topology of (5.15) of a sequence of smooth local solutions of (5.1) with smooth $\beta(\cdot)$. This assumption is made in order to justify some of the calculations and to deal with equations instead of inclusions. We stress that the modulus of continuity of u

must be independent of any approximating procedure and must depend only upon the data. Such a result gives us some compactness that, in many cases, is fundamental to obtain existence of a solution. Actually if one approximates the equation with a sequence of regular ones, through the estimate of the modulus of continuity, one obtains that the approximating solutions are uniformly continuous. Then, via the Ascoli–Arzelá theorem one gets the convergence in the uniform norm of a subsequence of approximating functions to a continuous function v. Thus, in many cases, by applying the method exploited by Kinderlehrer and Stampacchia [107] and based on Minty's lemma [130], it is possible to prove that v is a weak solution of the original equation.

Lastly we recall that if

$$B(x,t,u,\nabla u) \leqslant \mu_2|\nabla u|^q + \phi_2(x,t),$$

where $0 \leqslant q < (N+4)/(N+2)$, the questions of existence and uniqueness are well understood. We refer the readers to the monographs [84,120,123], to the proceedings [29,82,94] and to the references therein.

5.2. *A bridge between singular and degenerate equations*

The understanding of the physical model requires the analysis of (5.1) in its full generality. For instance, the flow of two immiscible fluids is described by a system of two parabolic equations, written in terms of the saturations and pressures of the two fluids (see, for instance, Chapter 9 of [18], Chapter 6 of [19], Chapter 6 of [44], Chapter 10 of [156] and the article by Leverett [122]). The transformation by Kruzkov and Sukorjanski [115] reduces the above system of two equations to a system of a degenerate-elliptic equation in terms of the mean pressure and a parabolic equation of the type

$$v_t - \operatorname{div} \mathbf{a}(x,t,v,\nabla v) + b(x,t,v,\nabla v) = 0 \quad \text{in } \Omega_T \tag{5.20}$$

in terms of the saturation v of only one of the two fluids. In (5.20), the forcing term $b(x,t,v,\nabla v)$ depends essentially on the mean pressure.

The diffusion field $\mathbf{a}$ and the forcing term b in (5.20) are real valued and measurable over $\Omega_T \times \mathbb{R} \times \mathbb{R}^N$ and satisfy the structure conditions:

$$\big|\mathbf{a}(x,t,v,\bar{v})\big| \geqslant \phi\big(|v|\big)|\bar{v}|^2 - \phi_0(x,t), \tag{5.21}$$

$$\big|\mathbf{a}(x,t,v,\bar{v})\big| \leqslant \phi\big(|v|\big)|\bar{v}| - \phi_1(x,t) \tag{5.22}$$

and

$$\big|b(x,t,v,\bar{v})\big| \leqslant \phi\big(|v|\big)|\bar{v}|^2 - \phi_2(x,t), \tag{5.23}$$

where $\phi : \mathbb{R}_+ \to \mathbb{R}_+$ is a continuous function and ϕ_i, $i = 0,1,2$, are nonnegative and satisfy assumptions (5.11) and (5.12).

As the function $(x,t) \to v(x,t)$ represents the local relative saturation of one of the two fluids (see, for instance, [5,18,19,44,115,156]), it is natural to assume it is bounded, for example $v \in [0,1]$. The equation is degenerate because $\phi(\cdot)$ is allowed to vanish. More precisely we assume that

$$\phi(v) > 0 \quad \forall v \in [0,1], \qquad \phi(0) = \phi(1) = 0. \tag{5.24}$$

Obviously v satisfies (5.20) in the weak sense defined previously and we require that

$$\nabla\big(\phi(v)\big) \in L^2_{\text{loc}}(\Omega_T). \tag{5.25}$$

The problem of proving the local continuity of the saturations was studied in [5,53,170], for example. The continuity of the saturations comes from the continuity of the solution of (5.20) arguing as in [5] and [53]. Proving the continuity of the solution is difficult, not only for the double degeneracy of $\phi(\cdot)$, but also because of the lack of a precise quantitative and qualitative information on its modulus of continuity. Actually the function ϕ is related to the permeability of both fluids and the permeability vanishes as one fluid is totally replaced by the other one (that is, when $v = 0$ or $v = 1$); this is the physical origin of the degeneracy of $\phi(\cdot)$. The information on the rate of vanishing is limited because it is derived only from hydrostatic experiments (see [18,19,44,156]), dimensional analysis (see [122]) and heuristic arguments. As a matter of fact, such a limited information on the nature of the degeneracy is typical of models of flows of a mixture of fluids in a porous medium. Hence $\phi(\cdot)$ could degenerate at $v = 0$ and $v = 1$ at different rates (exponentially fast or faster). By the phenomenon of connate water it might be even completely flat in a small right neighborhood of 0 or in a small left neighborhood of 1 (see Chapter 9 of [18], Chapter 2 of [44], Chapters 3 and 10 of [156]). So the problem of the continuity of weak solutions of (5.20) consists in showing that v is continuous *whatever* the nature of the degeneracy of $\phi(\cdot)$ is.

Let $u \in [0,1]$ be a solution of (5.1) with $\beta(\cdot) \in C^1(0,1)$ and singular in 0 and 1. For example assume

$$\lim_{u\to 0^+} \beta'(u) = \lim_{u \to 1^-} \beta'(u) = +\infty.$$

Then, by setting $v = \beta(u)$ and $\phi(\cdot) = \beta^{-1'}(\cdot)$, the singular equation (5.1) in terms of u is re-casted as the degenerate equation (5.20) in terms of v. Moreover, all the assumptions on $\mathbf{a}$ and b are satisfied. Vice versa, putting

$$u = \int_0^v \phi(s)\,\mathrm{d}s, \qquad \beta(u) = v,$$

one gets the singular equation from the degenerate one. In this case, however, while the assumptions on $\mathbf{A}$ are verified, the assumption on the free term B cannot be verified in the case of a superlinear growth with respect to ∇v. We refer the reader to [5] for a detailed technical analysis of this case.

5.3. *Parabolic equations with one-point singularity*

Let us consider the case where $\beta(\cdot)$ is singular at only one point (the prototype cases are given by the examples (5.4) and (5.6)). The local continuity of the solutions was proved in the papers [31,48,50,52,149,150,184]. However the situations of (5.4) and (5.6) are very different; actually, for $\beta(\cdot)$ of the type (5.6) it is possible to repeat De Giorgi's argument (i.e., to find suitable Caccioppoli and logarithmic estimates such that an embedding in the space of Hölder continuous functions holds) while that procedure is impossible in the case of (5.4). In this case, the Caccioppoli and logarithmic estimates are more complicated than in the nonsingular case:

there exists a constant γ, only depending upon the data, such that, for each

$$(y,s)+Q(\sigma\theta\rho^2,\sigma\rho)\subset(y,s)+Q(\theta\rho^2,\rho),\quad \sigma\in(0,1),$$

$$\begin{aligned}
&\sup_{s-\theta\rho^2\leqslant t\leqslant s}\int_{y+K_{\sigma\rho}}(u-k)_\pm^2(x,t)+\iint_{(y,s)+Q(\sigma\theta\rho^2,\sigma\rho)}\left|\nabla(u-k)_\pm\right|^2\\
&\quad\leqslant\frac{\gamma}{(1-\sigma)^2\rho^2}\iint_{(y,s)+Q(\theta\rho^2,\rho)}(u-k)_\pm^2\\
&\qquad+\frac{\gamma}{(1-\sigma)\theta\rho^2}\iint_{(y,s)+Q(\theta\rho^2,\rho)}(u-k)_\pm,
\end{aligned}\tag{5.26}$$

$$\begin{aligned}
&\sup_{s-\theta\rho^2\leqslant t\leqslant s}\int_{y+K_{\sigma\rho}}(u-k)_\pm^2(x,t)+\iint_{(y,s)+Q(\sigma\theta\rho^2,\sigma\rho)}\left|\nabla(u-k)_\pm\right|^2\\
&\quad\leqslant\frac{\gamma}{(1-\sigma)^2\rho^2}\iint_{(y,s)+Q(\theta\rho^2,\rho)}(u-k)_\pm^2\\
&\qquad+\gamma\int_{y+K_\rho}(u-k)_\pm\big(x,s-\theta\rho^2\big),
\end{aligned}\tag{5.27}$$

$$\begin{aligned}
&\sup_{s-\theta\rho^2\leqslant t\leqslant s}\int_{y+K_{\sigma\rho}}\Psi^2\big(\mathcal{H}_k^\pm,(u-k)_\pm,c\big)(x,t)\\
&\quad\leqslant\frac{\gamma}{(1-\sigma^2)\rho^2}\iint_{(y,s)+Q(\theta\rho^2,\rho)}\Psi\big(\mathcal{H}_k^\pm,(u-k)_\pm,c\big)\\
&\qquad+\frac{\gamma}{c}\int_{y+K_\rho}\Psi^2\big(\mathcal{H}_k^\pm,(u-k)_\pm,c\big)\big(x,s-\theta\rho^2\big).
\end{aligned}\tag{5.28}$$

Note that, for the sake of simplicity, we have written the above estimates only in the case of the prototype equation (5.17).

If $N=2$, the existence of the modulus of continuity can be derived from the above estimates (see [72]). If $N\geqslant 3$ there are bounded discontinuous functions satisfying the previous estimates (see [158]). So in order to prove the local continuity of the solutions one has to use the structure of the parabolic equation. In [31,48,50,52,149,150,184] the proof of continuity has this common point: assume that $\beta(\cdot)$ is singular only at a point, say, for example, at 0. Fix a cylinder $(y,s)+Q_\rho$. There are two possibilities: (i) either the singularity occupies a small portion of such a cylinder (and in such a case it plays a negligible role); or (ii) the singularity occupies a large portion. In such a situation, outside the singular set, the evolution equation is uniformly parabolic so the solution remains close to 0 because it cannot grow too fast due to the classical regularity properties of non singular parabolic equations. This gives us a control on the oscillation of the solution and allows us to obtain some recursive inequalities that will imply the local continuity. As it is absolutely evident, the whole argument is based on the fact that $\beta(\cdot)$ has only a singularity. The proofs of these recursive inequalities differ in [31,48,50,149,150,184]. In [31], the authors use the local representation in terms of heat potentials (for this reason their approach works only in the case of the prototype equation (5.17)). In [48] and [50], De Giorgi's iterations are used following the setting of [120]. The shrinking technique introduced by Krylov and Safonov [117] is applied in [149] and [150]. As this method is genuinely based on the nondivergence structure of the operator, also in this case the approach works only for the prototype equation (5.17). Lastly in [184] the Harnack-type techniques of Moser are applied, following the setting of [16,132,134,164].

The singularity of $\beta(\cdot)$ changes the iterative procedure of the nonsingular case. Actually the singularity affects the sequence of the radii of the nested cylinders and the reduction of the oscillation. More precisely, let

$$\eta,\delta:(0,2M]\to(0,1),\qquad \eta(0),\delta(0)=0, \tag{5.29}$$

and define

$$\begin{aligned} \omega_0 &= \max\{2M; C\rho_0^\lambda\},\\ \rho_{n+1} &= \delta(\omega_n)\rho_n,\\ \omega_{n+1} &= \max\{(1-\eta(\omega_n))\omega_n; C\rho_n^\lambda\}, \end{aligned} \tag{5.30}$$

where $C>1$ and $\lambda\in(0,1)$ are two given constants. Define also the corresponding family of shrinking nested cylinders $(x_0,t_0)+Q_{\rho_n}$.

PROPOSITION 16. *Let u be a weak solution of* (5.1) *with $\beta(\cdot)$ a graph of Stefan type* (5.4). *Then there exist constants $C>1$ and $\lambda\in(0,1)$, and two continuous functions $\delta(\cdot)$ and $\eta(\cdot)$ as in* (5.29), *that can be determined a priori only in terms of data, such that, for every $(x_0,t_0)\in\Omega_T$ and every $n\in\mathbb{N}$,*

$$\operatorname*{ess\,osc}_{(x_0,t_0)+Q_{\rho_n}} u \leqslant \omega_n, \tag{5.31}$$

where λ is a number determined only in terms of the integrability conditions (5.8)–(5.13) *and is independent of δ and η. As a consequence u is locally continuous in Ω_T.*

PROOF. For a detailed proof we refer the reader to [48]. Here we assume that (5.31) holds so that to prove the proposition it is sufficient to show that $\{\omega_n\}\to 0$ when $n\to\infty$. From the definition, the sequences $\{\omega_n\}$ and $\{\rho_n\}$ are nonincreasing, so their limits exist when $n \to \infty$. It is clear that $\lim_{n\to\infty} \rho_n = 0$ because the function $\delta(\cdot) \in (0,1)$.

Now, assume that $\lim_{n\to\infty} \omega_n = \omega_\infty > 0$. Then

$$\omega_{n+1} = \max\{(1 - \eta(\omega_\infty))\omega_n; C\delta(\omega_\infty^{\lambda n} \rho^\lambda)\}$$

and, hence, we get that $\lim_{n\to\infty} \omega_n = 0$, which contradicts the assumption. □

REMARK 39. The constants C and λ appearing in (5.30) are due only to the functions ϕ_i appearing in the structure conditions (5.8)–(5.10), and they are zero for the prototype equation (5.17).

REMARK 40. With respect to the case of a uniformly parabolic equation, here the modulus of continuity is not explicit but it can be derived quantitatively from (5.30). More precisely, in [48], it is proved that $\delta(s)$, $\eta(s)$ have the form $K^{-h/s}$ with K and h large constants, but it is not obtained an explicit modulus of continuity for u in terms of K and h.

5.4. *Parabolic equations with multiple singularities*

Graphs $\beta(\cdot)$ that are singular at multiple points, besides their intrinsic mathematical interest, arise naturally in phenomena of multiple transitions of phase. An example is a water–ice–vapor triple point and another one is the Buckley–Leverett model of two immiscible fluids in a porous medium (see the previous sections for more details about these models).

The first attempt to prove continuity results in such a setting was made in [5], where some restrictions were made on the singularities. Actually, the authors considered the case of the Buckley–Leverett model, i.e., the case of only two singularities. They assumed that $\beta(\cdot)$ could be singular at any rate in one point, while in the second point the singularity allowed was only of logarithmic type. This result was improved in [53], by allowing the second singularity to have a power-like behavior, and in [170] where Hölder continuity was proved assuming that both degeneracies are power-like. The alternative argument is quite similar to the one of only one singularity. Let Q_ρ be the cylinder where we want to reduce the oscillation of the solution. If the nonrestricted singularity occupies a small portion of such a cylinder it means that it plays a negligible role and the continuity results for porous medium equation hold. If the singularity occupies a large portion, outside the singular set, the solution cannot grow too fast due to the regularity properties of porous medium equations. This gives a control on the oscillation of the solution and allows one to obtain some recursive inequalities that imply the local continuity. It is clear, however, that any parabolic approach cannot face the case of two unrestricted singularities. In the pioneering paper [72], the approach is based on the energy estimates and on some measure-theoretical results. The results obtained in [72] are optimal in the case $N = 2$ and hold for the prototype equation (5.17).

THEOREM 22. *Let $N = 2$. Let u be a locally bounded weak solution of* (5.1), *where $\beta(\cdot)$ is any maximal monotone graph satisfying conditions* (5.2) *and* (5.3), *and let the structure* *conditions* (5.8)–(5.14) *be satisfied. Then u is continuous in Ω_T. Moreover, for every compact subset $K \subset \Omega_T$, there exists a continuous, nonnegative, increasing function*

$$s \mapsto \omega_{data,K}(s), \qquad \omega_{data,K}(0) = 0,$$

that can be determined a priori only in terms of the data and the distance from K to the parabolic boundary of Ω_T, such that

$$\left|u(x_1, t_1) - u(x_2, t_2)\right| \leqslant \omega_{data,K}\left(|x_1 - x_2| + |t_1 - t_2|^{1/2}\right)$$

for every pair of points $(x_i, t_i) \in K$, $i = 1, 2$.

In the case $N \geqslant 3$, it is necessary to assume some conditions either on the maximal graph $\beta(\cdot)$ or on the structure of the elliptic operator. In [72] the following result is proved.

THEOREM 23. *Let $N \geqslant 3$. Let u be a locally bounded weak solution of the prototype equation* (5.17), *where $\beta(\cdot)$ is any maximal monotone graph satisfying conditions* (5.2) *and* (5.3). *Then u is continuous in Ω_T. Moreover, for every compact subset $K \subset \Omega_T$, there exists a continuous, nonnegative, increasing function*

$$s \mapsto \omega_{data,K}(s), \qquad \omega_{data,K}(0) = 0$$

that can be determined a priori only in terms of the data and the distance from K to the parabolic boundary of Ω_T, such that

$$\left|u(x_1, t_1) - u(x_2, t_2)\right| \leqslant \omega_{data,K}\left(|x_1 - x_2| + |t_1 - t_2|^{1/2}\right)$$

for every pair of points $(x_i, t_i) \in K$, $i = 1, 2$.

5.4.1. *The statement of the alternative.* For the moment we consider the general equation (5.1) in any number of dimensions. We assume that $\beta(\cdot)$ is any maximal monotone graph satisfying conditions (5.2) and (5.3) and that the structure conditions (5.8)–(5.14) hold. Only later we will point the differences between $N = 2$ and $N \geqslant 3$. Without loss of generality, we assume that the generic point (x_0, t_0) is equal to $(0, 0)$.

Consider the following coaxial cylinders with vertex in $(0, \tilde{t})$ and congruent to $Q_{4\delta\rho}$

$$(0, \tilde{t}) + Q_{4\delta\rho} = K_{4\delta\rho} \times \left(\tilde{t} - (4\delta\rho)^2, \tilde{t}\right),$$

where

$$\tilde{t} \in \left(-\left(1 - 16\delta^2\right)\rho^2, 0\right), \tag{5.32}$$

and $\delta \in (0, \frac{1}{4})$ is a positive number to be chosen.

By moving the time, one seeks to find a cylinder where one can apply the techniques developed in the previous sections. More precisely, we look for $\tilde{t}$ such that the subset

of $(0,\tilde{t})+Q_{4\delta\rho}$ where u is close to μ^+ or to μ^- is small. Denote by ω the oscillation of u in Q_ρ. Then, one of these two possible alternatives takes place:
there exists a $\tilde{t}\in(-(1-16\delta^2)\rho^2,0)$ and a positive δ such that either

$$\operatorname{Meas}\left\{(x,t)\in\left(0,\tilde{t}\right)+Q_{4\delta\rho}\colon u(x,t)\geqslant\mu^+-\frac{1}{6}\omega\right\}\leqslant\nu|Q_{4\delta\rho}| \tag{5.33}$$

or

$$\operatorname{Meas}\left\{(x,t)\in\left(0,\tilde{t}\right)+Q_{4\delta\rho}\colon u(x,t)\leqslant\mu^-+\frac{1}{6}\omega\right\}\leqslant\nu|Q_{4\delta\rho}|. \tag{5.34}$$

Otherwise,

$$\text{both (5.33) and (5.34) are violated } \forall\tilde{t}\in\left(-\left(1-16\delta^2\right)\rho^2,0\right), \tag{5.35}$$

where $\nu\in(0,1)$ is a number that will be determined a priori only in terms of the data.

Assume for the moment that one of (5.33) and (5.34) holds. By using an iterative argument based on the energy estimates, we can show that it is possible to reduce the oscillation of u in the smaller cylinder $(0,\tilde{t})+Q_{2\delta\rho}$ if one chooses ν very small and depending upon ω (see Proposition $2.5^\pm$ of [72]):

PROPOSITION 17. *There exists a number $\nu\in(0,1)$, that can be determined a priori only in terms of the data and ω, such that, if* (5.33) *holds for some $\tilde{t}\in(-(1-16\delta^2)\rho^2,0)$, then either $\omega\leqslant C\rho^\lambda$ or*

$$u(x,t)\leqslant\mu^+-\frac{1}{16}\omega,\qquad (x,t)\in\left(0,\tilde{t}\right)+Q_{2\delta\rho}. \tag{5.36}$$

Analogously, if (5.34) *holds for some $\tilde{t}\in(-(1-16\delta^2)\rho^2,0)$, then either $\omega\leqslant C\rho^\lambda$ or*

$$u(x,t)\geqslant\mu^-+\frac{1}{16}\omega,\qquad (x,t)\in\left(0,\tilde{t}\right)+Q_{2\delta\rho}. \tag{5.37}$$

We recall to the reader that C is a constant depending upon the structure conditions of equation (5.1).

By using the logarithmic estimates (see Proposition $3.2^\pm$ of [72]) it is possible to bring the information of the reduction of the oscillation of the solution up to the level $t=0$. More precisely:

PROPOSITION 18. *Assume that there is $\tilde{t}\in(-(1-16\delta^2)\rho^2,-4\delta^2\rho^2)$ such that*

$$u\left(x,\tilde{t}\right)\leqslant\mu^+-\frac{1}{16}\omega\quad\forall x\in K_{2\delta\rho}. \tag{5.38}$$

Then there are constants $\eta,\lambda\in(0,1)$ and $C>1$, depending upon the data and δ, but independent of ω and ρ, such that either $\omega\leqslant C\rho^\lambda$ or

$$u(x,t)\leqslant\mu^+-\eta\omega,\qquad (x,t)\in(0,0)+Q_{\delta\rho}. \tag{5.39}$$

Analogously, if there is $\tilde{t} \in (-(1-16\delta^2)\rho^2, -4\delta^2\rho^2)$ *such that*

$$u(x,\tilde{t}) \leqslant \mu^- + \frac{1}{16}\omega \quad \forall x \in K_{2\delta\rho} \tag{5.40}$$

then either $\omega \leqslant C\rho^\lambda$ *or*

$$u(x,t) \geqslant \mu^- + \eta\omega, \quad (x,t) \in (0,0) + Q_{\delta\rho}. \tag{5.41}$$

Summarizing, if the alternatives (5.33) and (5.34) are satisfied, then respectively (5.38) and (5.40) hold. Therefore the oscillation of u is reduced in the smaller cylinder $(0,0) + Q_{\delta\rho}$. If we are able to prove that even when (5.35) holds we have a reduction of the oscillation of u in a smaller cylinder, by repeating the arguments of the previous sections, one can deduce the local continuity of the solution. Therefore, in the sequel, we assume that, for each small δ and for each $\tilde{t} \in (-(1-16\delta^2)\rho^2, 0)$,

$$\text{Meas}\left\{(x,t) \in (0,\tilde{t}) + Q_{4\delta\rho}\colon u(x,t) \geqslant \mu^+ - \frac{1}{8}\omega\right\} \geqslant \nu|Q_{4\delta\rho}|$$

and

$$\text{Meas}\left\{(x,t) \in (0,\tilde{t}) + Q_{4\delta\rho}\colon u(x,t) \geqslant \mu^- + \frac{1}{8}\omega\right\} \geqslant \nu|Q_{4\delta\rho}|.$$

By using a lemma of measure theory, the above inequalities will imply regions where the energy is concentrated. From this fact, we are able to find a contradiction in assuming (5.35) and hence to prove the regularity of the solutions. In the next section we state and prove the lemma of measure theory. We feel that it is of intrinsic interest and that it can be applied in different fields.

5.4.2. *A lemma of measure theory*

LEMMA 14. *Let* v *be a function in* $W^{1,p}(K_\rho)$, $p > 1$, *satisfying*

$$\int_{K_\rho} |\nabla v|^p \,\mathrm{d}x \leqslant \gamma^p \rho^{N-p} \tag{5.42}$$

for a given positive constant γ, *and let*

$$\text{Meas}\{x \in K_\rho\colon v(x) < 1\} > \alpha|K_\rho| \tag{5.43}$$

for a given $\alpha \in (0,1)$. *Then, for every* $\eta \in (0,1)$ *and* $\lambda > 1$, *there exists* $x^* \in K_\rho$ *and a number* $\delta \in (0,1)$, *that can be determined a priori only in terms of* $N, p, \gamma, \alpha, \lambda, \eta$, *such that within the cube* $K_{\delta\rho}(x^*)$ *centered at* x^* *with wedge* $2\delta\rho$, *there holds*

$$\text{Meas}\{x \in K_{\delta\rho}(x^*)\colon v(x) < \lambda\} > (1-\eta)|K_{\delta\rho}(x^*)|. \tag{5.44}$$

If v were continuous, this lemma would be an easy consequence of the permanence of positivity. For the sake of simplicity, we prove this lemma assuming $N = 2$ (the case $N \geqslant 3$ can be proved using an inductive procedure) and $v \in C^1(K_\rho)$. Obviously we establish (5.44) with δ independent of the modulus of continuity of v. The assumption $v \in C^1(K_\rho)$ is made only to justify some calculations and can be removed via a limiting procedure. For more details on this proof, we refer the reader to Proposition A.1 of [72]. Before proving the lemma we stress that δ goes to 0 when either η goes to 0 or λ goes to 1.

PROOF OF LEMMA 14. Let (x, y) be the coordinates in $\mathbb{R}^2$. Denote by $Y(x)$ the cross-section of the set $[v < 1] \cap K_\rho$ with lines parallel to the y-axis, i.e.,

$$Y(x) = \{y \in (-\rho, \rho)\colon\ v(x, y) < 1\}.$$

Hence,

$$\big|[v < 1] \cap K_\rho\big| = \int_{-\rho}^{\rho} \big|Y(x)\big|\, dx.$$

As $|[v < 1] \cap K_\rho| \geqslant 2\alpha\rho$, there exists some $x_0 \in (-\rho, \rho)$ such that

$$\big|Y(x_0)\big| \geqslant 2\alpha\rho. \tag{5.45}$$

For each $y \in Y(x_0)$, consider the segment

$$I_{\delta_0\rho}(y) = [x_0 - \delta_0\rho, x_0 + \delta_0\rho] \times \{y\},$$

where δ_0 will be chosen later. Denote with $Y_{\delta_0}(x_0)$ the set of $Y(x_0)$ such that, in the corresponding intervals $I_{\delta_0\rho}(y)$, the function $v(\cdot, y)$ is less than $\frac{1}{2}(1 + \lambda)$, i.e.,

$$Y_{\delta_0}(x_0) = \left\{y \in Y(x_0)\colon\ v(x, y) < \frac{1}{2}(1 + \lambda)\ \forall x \in I_{\delta_0\rho}(y)\right\}.$$

Now, we want to prove that , for each $\eta_0 < 1$ there exists a small δ_0 such that

$$\big|Y_{\delta_0}(x_0)\big| \geqslant (1 - \eta_0)\big|Y(x_0)\big|.$$

Let $Y_{\delta_0}^C(x_0)$ be the complement of $Y_{\delta_0}(x_0)$. Fix $y \in Y_{\delta_0}^C(x_0)$ and some $x \in I_{\delta_0\rho}(y)$ such that $v(x, y) \geqslant \frac{1}{2}(1 + \lambda)$. Then

$$\frac{1}{2}(\lambda - 1) \leqslant v(x, y) - v(x_0, y) = \int_{x_0}^{x} v_x(s, y)\, ds.$$

By integrating over $Y_{\delta_0}^C(x_0)$ and majorizing the obtained result via Hölder's inequality, we get

$$\frac{1}{2}(\lambda - 1)\big|Y_{\delta_0}^C(x_0)\big| \leqslant \int_{Y(x_0)} \int_{-\delta_0\rho}^{\delta_0\rho} |\nabla v|\, dx\, dy \leqslant \big(\big|Y_{\delta_0}(x_0)\big| 2\delta_0\rho\big)^{1-1/p} \|\nabla v\|_{p, K_\rho}.$$

Therefore, using (5.42) and (5.45) one gets

$$\left|Y_{\delta_0}^C(x_0)\right| < \frac{4\gamma\delta_0^{1-1/p}}{(\lambda-1)(4\alpha)^{1/p}}\left|Y(x_0)\right|.$$

So we have that

$$\left|Y_{\delta_0}(x_0)\right| \geqslant (1-\eta_0)\left|Y(x_0)\right|,$$

choosing δ_0 such that $4\gamma\delta_0^{1-1/p}/(\lambda-1)(4\alpha)^{1/p} \leqslant \eta_0$.

Next, fix $y^* \in Y_{\delta_0}(x_0)$. We recall that $v(x, y^*) \leqslant (1+\lambda)/2$ for each $x \in I_{\delta_0\rho}(y^*)$. Consider the vertical segment

$$J_{\delta\rho}(x) = \{x\} \times \left[y^* - \delta\rho, y^* + \delta\rho\right],$$

where δ will be chosen later. Denote with $H_\delta(y^*)$ the set of $I_{\delta_0\rho}(y^*)$ such that, in the corresponding intervals $J_{\delta\rho}(x)$, the function $v(x,\cdot)$ is less than λ, i.e.,

$$H_\delta(y^*) = \left\{x \in I_{\delta_0\rho}(y^*)\colon\ v(x,y) < \lambda\ \forall y \in J_{\delta\rho}(x)\right\}.$$

Now, we want to prove that, for each $\eta < 1$, there exists a small δ such that

$$\left|H_\delta(y^*)\right| \geqslant (1-\eta)\left|I_{\delta_0\rho}(y^*)\right|.$$

Let $H_\delta^C(y^*)$ be the complement of $H_\delta(y^*)$. Fix $x \in H_\delta^C(y^*)$ and some $y \in J_{\delta\rho}(x)$ such that $v(x,y) \geqslant \lambda$. Then

$$\frac{1}{2}(\lambda-1) \leqslant v(x,y) - v(x,y^*) = \int_{y^*}^{y} v_y(x,s)\,\mathrm{d}s.$$

By integrating over $H_\delta^C(y^*)$ and majorizing the obtained result via Hölder's inequality, we get

$$\frac{1}{2}(\lambda-1)\left|H_\delta^C(y^*)\right| \leqslant \int_{I_{\delta_0\rho}(y^*)}\int_{-\delta\rho}^{\delta\rho} |\nabla v|\,\mathrm{d}y\,\mathrm{d}x \leqslant \left(4\delta_0\delta\rho^2\right)^{1-1/p}\|\nabla v\|_{p,K_\rho}.$$

Therefore, using (5.42) one obtains

$$\left|H_\delta^C(y^*)\right| < \frac{8\gamma\delta^{1-1/p}}{(\lambda-1)(4\delta_0)^{1/p}}\left|I_{\delta_0\rho}(y^*)\right|.$$

So we have that

$$\left|H_\delta(y^*)\right| \geqslant (1-\eta)\left|I_{\delta_0\rho}(y^*)\right|$$

choosing δ such that $\delta_0 \leqslant 8\gamma\delta^{1-1/p}/(\lambda-1)(4\delta_0)^{1/p} \leqslant \eta$.

Without loss of generality (assuming δ smaller) we may assume that δ_0/δ is a positive integer. Consider the interval $I_{\delta_0\rho}(y^*)$ and a partition of it with δ_0/δ intervals with length $2\delta\rho$. Let (x_i, y^*) be the centers of such intervals. Let x^* be one of such indices such that

$$\mathrm{Meas}\bigl\{(x^*-\delta\rho, x^*+\delta\rho)\cap H_\delta(y^*)\bigr\} > (1-\eta)(2\delta\rho).$$

Then, from the definition of $H_\delta(y^*)$, the cube $K_{\delta\rho}(x^*, y^*)$ satisfies the assumption of the lemma. □

5.4.3. *The geometric approach.* In this section, we will apply the previous lemma to get some geometric information about the localization of the energy of the solution. This is, in essence, the novelty of the approach in [72]. We recall that we assumed that, for each small δ and for each $\bar{t} \in (-(1-16\delta^2)\rho^2, 0)$,

$$\mathrm{Meas}\left\{(x,t)\in(0,\bar{t})+Q_{4\delta\rho}\colon u(x,t)\geqslant\mu^+-\frac{1}{8}\omega\right\}\geqslant\nu|Q_{4\delta\rho}|$$

and

$$\mathrm{Meas}\left\{(x,t)\in(0,\bar{t})+Q_{4\delta\rho}\colon u(x,t)\geqslant\mu^-+\frac{1}{8}\omega\right\}\geqslant\nu|Q_{4\delta\rho}|.$$

Roughly speaking (for more details see Sections 5–8 of [72]) it means that the region where the function u is close to μ^+ (respectively, to μ^-) is relatively large. By applying the lemma of the previous section (together with energy and logarithmic estimates) one can get that the set where u is close to μ^+ (respectively, to μ^-) must have a concentration region, even though it might be scattered in the whole cylinder. Moreover, it is possible to prove that these two concentration regions are localized at the same time levels. More precisely, it is possible to prove:

PROPOSITION 19. *Let u be a locally bounded weak solution of* (5.1) *where $\beta(\cdot)$ is any maximal monotone graph satisfying conditions* (5.2) *and* (5.3) *and suppose that the structure conditions* (5.8)–(5.14) *are satisfied. Assume that the alternative* (5.35) *holds. Then there exists a time $\bar{t} \in (-(1-16\delta^2)\rho^2, 0)$, and two points $x_1, x_2 \in K_{4\delta\rho}$ such that $[x_i + K_{\delta^2\rho}]$, $i = 1, 2$, have their cross sections mutually separated by a distance of at least $\delta^2\rho$, and*

$$u(x_1,\bar{t})\geqslant\mu^+-\frac{1}{8}\omega\quad\forall(x,t)\in(x_1,\bar{t})+Q_{\delta^2\rho}\tag{5.46}$$

and

$$u(x_2,\bar{t})\leqslant\mu^-+\frac{1}{8}\omega\quad\forall(x,t)\in(x_2,\bar{t})+Q_{\delta^2\rho}.\tag{5.47}$$

SKETCH OF THE PROOF. Suppose that (5.34) is violated. Then for some time levels $t \in (\tilde{t} - 16\delta^2\rho^2)$ we have

$$\operatorname{Meas}\left\{x \in K_{\delta^2\rho^2}\colon u(x,t) \leqslant \mu^- + \frac{1}{16}\omega\right\} \geqslant \nu |K_{\delta^2\rho^2}|.$$

By setting

$$v(x,t) = \frac{16}{\omega}\big(u(x,t) - \mu^-\big),$$

we have

$$\operatorname{Meas}\{x \in K_{\delta^2\rho^2}\colon v(x,t) \leqslant 1\} \geqslant \nu |K_{\delta^2\rho^2}|. \tag{5.48}$$

Changing the constant ν into a smaller positive constant α, it is possible to find a time τ such that the following inequalities hold:

$$\operatorname{Meas}\{x \in K_{\delta^2\rho^2}\colon v(x,t) \leqslant 1\} \geqslant \alpha |K_{\delta^2\rho^2}|,$$

$$\int_{K_{\delta^2\rho^2}} |\nabla v(x,\tau)|^2 \, dx \leqslant \gamma_{data}(\omega)(\rho\delta)^{2N-4},$$

where γ_{data} is independent of τ. Therefore by the lemma of measure theory, there exists $\eta\delta_0 \in (0,1)$ and a cube $[x^* + K_{\delta^2\rho^2\delta_0}] \subset K_{\delta^2\rho^2}$, such that

$$\operatorname{Meas}\{x \in [x^* + K_{\delta^2\rho^2\delta_0}]\colon v(x,\tau) \leqslant 2\} \geqslant (1-\eta)|K_{\delta^2\rho^2\delta_0}|.$$

It means that $u(x,\tau) \leqslant \mu^- + \frac{1}{16}\omega$ except, at most, at a set of measure less than $\eta |K_{\delta^2\rho^2\delta_0}|$. The information at the time τ is almost complete (with the exception of an arbitrarily small set). Removing this set of small measure through the energy and logarithmic estimates, it is possible to get (5.47). □

Summarizing, if the alternative (5.35) holds, then (5.46) and (5.47) are verified. Therefore

$$\frac{1}{4}\omega \leqslant u(y_1,t) - u(y_2,t) \quad \forall y_i \in [x_i + K_{\delta^2\rho}], i = 1, 2, \tag{5.49}$$

for all time levels

$$t \in (\tilde{t} - \delta^4\rho^2, \tilde{t}). \tag{5.50}$$

In such temporal range, integrate (5.49) over a path, piecewise parallel to the coordinates axes and joining $y_1 \in [x_1 + K_{\delta^2\rho}]$ and $y_2 \in [x_2 + K_{\delta^2\rho}]$. Integrate the resulting segment-integrals over the remaining $N-1$ variables, and then over the time in the range (5.50). Therefore

$$\gamma(\omega)(\delta\rho)^N \leqslant \int_{\tilde{t}-\delta^4\rho^2}^{\tilde{t}} \int_{K_{\delta\rho}\setminus K_{\delta^2\rho^2}} |\nabla u|^2 \, dx \, d\tau. \tag{5.51}$$

This inequality has been derived for all $\tilde{t}$ for which (5.33) and (5.34) are both violated. Noting that, in the temporal range, the number of cylinders of the type $(0, \tilde{t}) + Q_{4\delta\rho}$ is of order δ^{-2}, we add (5.51) over the corresponding boxes to get

$$\gamma(\omega)\delta^{N-2}\rho^N \leqslant \int_{-\rho^2}^{0} \int_{K_{\delta\rho}\setminus K_{\delta^2\rho^2}} |\nabla u|^2 \, dx \, d\tau. \tag{5.52}$$

We may repeat the argument replacing δ^2 with δ. Iterating this procedure, we have that for each $n \in \mathbb{N}$,

$$\gamma(\omega)\delta^{n(N-2)}\rho^N \leqslant \int_{-\rho^2}^{0} \int_{K_{\delta^n\rho}\setminus K_{\delta^{n+1}\rho^2}} |\nabla u|^2 \, dx \, d\tau. \tag{5.53}$$

For more details about the estimates of this section, we refer the reader to Sections 9–12 of [72].

5.4.4. *The case $N = 2$.* Adding (5.53) for $n = 1, \ldots, n_0$, one obtains

$$\gamma(\omega)n_0\rho^N \leqslant \iint_{Q_\rho} |\nabla u|^2 \, dx \, d\tau. \tag{5.54}$$

On the other hand, via a standard energy estimate, we have

$$\iint_{Q_\rho} |\nabla u|^2 \, dx \, d\tau \leqslant C\rho^N, \tag{5.55}$$

where C is a constant only depending on the data. Therefore, combining (5.54) and (5.55), one gets

$$\gamma(\omega)n_0 \leqslant C.$$

This is a contradiction if n_0 is sufficiently large depending on the data and ω. It follows that at least one of the alternatives (5.33) or (5.34) holds in the range (5.50) and for some radius $\rho_0 \in [\rho, \delta^{n_0}\rho]$. But we have already shown that if one of the alternatives (5.33) or (5.34) holds, then the oscillation of u is reduced in the cylinder $Q_{\rho_0/2}$. Hence, the local continuity of the solution is proved in the case $N = 2$.

REMARK 41. The same argument works in the case in which one has information that essentially reduces the space dimension N to 1 or 2.

REMARK 42. In the case $N=2$ the previous argument works for a more general maximal monotone graph $\beta=\beta_{AC}+\beta_s$ of bounded variation, with β_{AC} an absolutely continuous and strictly increasing function, and $\beta_s \geqslant 0$ a nondecreasing function where, roughly speaking, the jumps occur (for more details, see [77]). This method works also for more general (other than second order) operators (see [152–154]).

5.4.5. *The geometric approach continued.* In order to face the case $N \geqslant 3$, the geometric approach needs to be improved. We stress, that all the results of this section hold under the most general assumptions.

To prove the continuity of u at a point $(x,t) \in \Omega_T$ we assume that such a point coincides with the origin up to a translation. The novelty of this improved approach is that we work with a longer cylinder $Q(\theta\rho^2, \rho)$, where θ is a positive integer number to be chosen. The idea is to find a "long" cylinder where u is away from μ^- (and μ^+) and this property should allow a space extension of positivity of the solution. It is clear that the structure itself of these singular parabolic equations forces us to deal with "long" cylinders. Actually, if one tries to repeat the argument of intrinsically rescaled cylinders (used in the case of the porous medium equation) for graphs of the Stefan type, one realizes that $\beta'(\cdot)$ is the Dirac mass at the origin. As a consequence the time should be intrinsically rescaled into another one that would remain constant on the transition set $u=0$. In other words, one has to work with cylinders whose length depends upon the singularity of $\beta(\cdot)$. We recall that a similar approach was used for the first time in [52] in the context of the boundary regularity in the case of a single singularity.

Consider the cylinders

$$(0,t_i)+Q(\rho^2,\rho), \quad t_i=-i\rho^2, i=0,1,\ldots,\theta-1, \tag{5.56}$$

that make a partition of the cylinder $Q(\theta\rho^2,\rho)$. Analogously to what was made in the previous sections we can state an alternative:

there exists $i=0,1,\ldots,\theta-1$ such that either

$$\operatorname{Meas}\left\{(x,t)\in(0,t_i)+Q(\rho^2,\rho)\colon u(x,t)\geqslant \mu^+-\frac{1}{6}\omega\right\}\leqslant \nu\left|Q(\rho^2,\rho)\right| \tag{5.57}$$

or

$$\operatorname{Meas}\left\{(x,t)\in(0,t_i)+Q(\rho^2,\rho)\colon u(x,t)\leqslant \mu^-+\frac{1}{6}\omega\right\}\leqslant \nu\left|Q(\rho^2,\rho)\right|. \tag{5.58}$$

Otherwise,

$$\text{both (5.57) and (5.58) are violated } \forall i=0,1,\ldots,\theta-1. \tag{5.59}$$

If (5.57) and (5.58) hold then, as shown in the previous sections, one can deduce the reduction of the oscillation of the solution. So we assume that (5.59) holds. Analogously this assumption implies that the solution has regions of concentration in any of the cylinders $(0,t_i)+Q(\rho^2,\rho)$.

Let us state this result in a different way. Let m be an integer and define $\delta = (4m)^{-1}$. Consider a partition of the original cube K_ρ into m^N pairwise disjoint subcubes of wedge $8\delta\rho$ and centered at points $x_l \in K_\rho$. That is,

$$[x_l + K_{4\delta\rho}] \subset K_\rho \quad \forall l = 1, \dots, m^N,$$
$$[x_l + K_{4\delta\rho}] \cap [x_j + K_{4\delta\rho}] = \emptyset \quad \text{if } j \neq l,$$
$$K_\rho = \bigcup_{l=1}^{m^N} [x_l + K_{4\delta\rho}].$$

Partition any cylinder $(0, t_i) + Q(\rho^2, \rho)$ into $16m^{N+2}$ subcylinders

$$(x_l, t_i + j\delta^2\rho^2) + Q_{4\delta\rho} \quad \forall l = 1, \dots, m^N, \forall j = 1, \dots, 16m^2,$$
$$(x_l, t_i + j\delta^2\rho^2) + Q_{4\delta\rho} \cap (x_r, t_i + s\delta^2\rho^2) + Q_{4\delta\rho} = \emptyset \quad \text{if } r \neq l, \text{or } j \neq s,$$
$$Q_\rho = \bigcup_{l=1}^{m^N} \bigcup_{j=1}^{16m^2} (x_l, t_i + j\delta^2\rho^2) + Q_{4\delta\rho}.$$

If (5.59) holds then, for m large enough it is possible to show (see Proposition 23.1 of [72]) that, for each $i = 0, 1, \dots, \theta - 1$, there are $1 \leqslant l, r \leqslant m^N$ and $1 \leqslant j, s \leqslant 16m^2$ such that

$$u(x, t) \geqslant \mu^+ - \frac{1}{8}\omega \quad \forall (x, t) \in (x_l, t_i + j\delta^2\rho^2) + Q_{4\delta\rho} \tag{5.60}$$

and

$$u(x, t) \leqslant \mu^- + \frac{1}{8}\omega \quad \forall (x, t) \in (x_r, t_i + s\delta^2\rho^2) + Q_{4\delta\rho}. \tag{5.61}$$

Using the logarithmic estimates one can bring this kind of information up to the level zero (see Sections 22–29 of [72]). Let m_0 be an integer and define $\delta_0 = (4m_0)^{-1}$. Consider the thin cylinder

$$(0, 0) + Q(\rho^2, \delta_0\rho) \subset (0, 0) + Q(\rho^2, \rho)$$

and consider a partition

$$(0, t_i) + Q(\rho^2, \delta_0\rho), \quad t_i = -i\delta_0^2\rho^2, i = 0, 1, \dots, 16m_0^2 - 1.$$

PROPOSITION 20. *Let u be a locally bounded weak solution of* (5.1), *with $\beta(\cdot)$ any maximal monotone graph satisfying conditions* (5.2) *and* (5.3), *and assume that the structure conditions* (5.8)–(5.14) *are satisfied. Then either the oscillation of u is reduced (in a way* *that can be quantitatively determined) in* $Q(\delta_0^2\rho^2, \delta_0\rho)$ *or there exist* $\varepsilon > 0$, *depending only upon the data,* ω *and* δ_0, *and* $x_l, x_j \in K_{\delta_0\rho}$ *such that*

$$u(x,t) \geqslant \mu^- + \varepsilon\omega \quad \forall (x,t) \in (x_l, 0) + Q\big((1-\delta_0^2)\rho^2, \delta_0\rho\big) \tag{5.62}$$

and

$$u(x,t) \leqslant \mu^+ - \varepsilon\omega \quad \forall (x,t) \in (x_j, 0) + Q\big((1-\delta_0^2)\rho^2, \delta_0\rho\big). \tag{5.63}$$

SKETCH OF THE PROOF. If in some of the subcylinders

$$(0, t_i) + Q\big(\rho^2, \delta_0\rho\big), \quad t_i = -i\delta_0^2\rho^2, i = 0, 1, \ldots, 16m_0^2 - 1,$$

(5.57) or (5.58) holds, then, as shown in the previous sections, one can deduce the reduction of the oscillation of the solution in $Q(\delta_0\rho, \delta_0^2\rho^2)$. So we assume that (5.59) holds. Under such an assumption, (5.60) and (5.61) hold. Therefore, there are $1 \leqslant l, r \leqslant m^N$ and $1 \leqslant j, s \leqslant 16m^2$ such that

$$u(x,t) \geqslant \mu^+ - \frac{1}{8}\omega \quad \forall (x,t) \in \big(x_l, -\big(1 - j\delta^2\delta_0^2\big)\rho^2\big) + Q_{\delta_0\delta\rho}$$

and

$$u(x,t) \leqslant \mu^- + \frac{1}{8}\omega \quad \forall (x,t) \in \big(x_r, -\big(1 - s\delta^2\delta_0^2\big)\rho^2\big) + Q_{\delta_0\delta\rho}.$$

Applying the logarithmic estimates from this box up to level zero, one gets the statement. □

If one is only interested in having, at any time level, a set where the solution is away from μ^+ and μ^- (thus losing the geometrical information that this set has the shape of a very long and thin cylinder) one can prove that the constant ε does not depend on δ_0. More precisely,

PROPOSITION 21. *Let u be a locally bounded weak solution of* (5.1), *with* $\beta(\cdot)$ *any maximal monotone graph satisfying conditions* (5.2) *and* (5.3), *and assume that the structure conditions* (5.8)–(5.14) *are satisfied. Then either the oscillation of u is reduced (in a way that can be quantitatively determined) in* $Q(\delta_0^2\rho^2, \delta_0\rho)$ *or there exist* $\varepsilon, \eta > 0$, *depending only upon the data and* ω, *and* $x_l, x_j \in K_{\delta_0\rho}$ *such that, for each time* $\tau \in [0, (1-\delta_0^2)\rho^2]$,

$$\operatorname{Meas}\big\{x \in K_{\delta_0\rho}\colon u(x,\tau) \geqslant \mu^- + \varepsilon\omega\big\} \geqslant \eta|K_{\delta_0\rho}| \tag{5.64}$$

and

$$\operatorname{Meas}\big\{x \in K_{\delta_0\rho}\colon u(x,\tau) \leqslant \mu^+ - \varepsilon\omega\big\} \geqslant \eta|K_{\delta_0\rho}|. \tag{5.65}$$

SKETCH OF THE PROOF. Reasoning as in the previous proposition, we may assume that (5.60) and (5.61) hold. Therefore, for each $i = 0, \dots, 16m_0^2 - 1$, there are $1 \leqslant l, r \leqslant m^N$ and $1 \leqslant j, s \leqslant 16m^2$ such that

$$u(x,t) \geqslant \mu^+ - \frac{1}{8}\omega \quad \forall (x,t) \in \bigl(x_l, -\bigl(i - j\delta^2\bigr)\delta_0^2\rho^2\bigr) + Q_{\delta_0\delta\rho}$$

and

$$u(x,t) \leqslant \mu^- + \frac{1}{8}\omega \quad \forall (x,t) \in \bigl(x_r, -\bigl(i - s\delta^2\bigr)\delta_0^2\rho^2\bigr) + Q_{\delta_0\delta\rho}.$$

Applying the logarithmic estimates from this box up to $t = -(i-2)\delta_0^2\rho^2$ one gets the statement. □

REMARK 43. An open question is to prove under general assumptions that (5.62)–(5.65) imply a space extension of positivity. What seems to be missing is some sort of weak form of the Harnack inequality for solutions of singular parabolic equations.

REMARK 44. The estimates (5.64) and (5.65) give us the same piece of information in the interior of Ω_T that a suitable Dirichlet condition gives on the boundary in [52]. This allows to extend the techniques introduced in [52] to the case of multiple singularities (see [87]).

If one accepts to work with cylinders whose base is very small with respect to the length, the previous results can be improved. The next proposition is proved in [72] (see Proposition 24.1, p. 291) and it is based on a tricky combinatorial argument.

PROPOSITION 22. *Let u be a locally bounded weak solution of* (5.1), *with $\beta(\cdot)$ any maximal monotone graph satisfying conditions* (5.2) *and* (5.3), *and assume that the structure conditions* (5.8)–(5.14) *are satisfied. Then there exists $\varepsilon \in (0,1)$, depending only upon the data and ω, such that, for each $\theta > 0$, there exists a $\delta_0 > 0$ such that either the oscillation of u is reduced (in a way that can be quantitatively determined) in $Q(\delta_0^2\rho^2, \delta_0\rho)$ or there exists $(x_i, t_i) \in Q_{(\rho^2,\delta_0\rho)}$, with $i = 1, 2$, such that the cylinders $(x_i, t_i) + Q_{(\theta\delta_0^2\rho^2,\delta_0\rho)}$ are contained in the cylinder $Q_{(\rho^2,\delta_0\rho)}$ and*

$$u(x,t) \geqslant \mu^- + \varepsilon\omega \quad \forall (x,t) \in (x_1, t_1) + Q_{(\theta\delta_0^2\rho^2,\delta_0\rho)} \tag{5.66}$$

and

$$u(x,t) \leqslant \mu^+ - \varepsilon\omega \quad \forall (x,t) \in (x_2, t_2) + Q_{(\theta\delta_0^2\rho^2,\delta_0\rho)}. \tag{5.67}$$

Summarizing, the previous proposition says that if one wants that the concentration of the solution still has the shape of a long cylinder and does not want to pay the price of having ε depending upon the length of the cylinders, then one looses the information about the precise localization of such a cylinder.

The next proposition is based on energy estimates, a De Giorgi lemma (see Lemma 2.2 of [55]) and estimates (5.64) and (5.65). For a detailed proof see Lemma 4.10 of [87].

PROPOSITION 23. *Let u be a locally bounded weak solution of* (5.1), *with $\beta(\cdot)$ any maximal monotone graph satisfying conditions* (5.2) *and* (5.3), *and assume that the structure conditions* (5.8)–(5.14) *are satisfied. Then there exist two continuous strictly increasing functions $\delta(\cdot)$, $\eta(\cdot)$, $\delta(0)=\eta(0)=0$, that can be determined a priori only in terms of the data and ω, such that either the oscillation of u is reduced* (*in a way that can be quantitatively determined*) *in $Q(\delta^2(s)\rho^2,\delta(s)\rho)$ or*

$$\operatorname{Meas}\{(x,t)\in Q_{(\rho^2,\delta(s)\rho)}\colon u(x,t)\leqslant \mu^- + s\omega\} \geqslant \eta(s)|Q_{(\rho^2,\delta(s)\rho)}| \tag{5.68}$$

and

$$\operatorname{Meas}\{(x,t)\in Q_{(\rho^2,\delta(s)\rho)}\colon u(x,t)\geqslant \mu^+ - s\omega\} \geqslant \eta(s)|Q_{(\rho^2,\delta(s)\rho)}|. \tag{5.69}$$

Summarizing, one is able to estimate the sets where the solution is close to μ^+ and μ^- paying the price of considering "thin" cylinders.

5.4.6. *The case $N\geqslant 3$.* As already stressed, we are not able to prove the regularity in the general case. We follow the approach of [72] (see also [86] and [160]) and we prove continuity results using a suitable comparison function.

In [72], the local continuity for weak solutions of the prototype equation (5.17) is proved. The key point in the proof of the continuity theorem are some estimates on a proper function v, which is then compared with the solution u of the original singular parabolic equation. In these estimates the radial symmetry of the problem is heavily used, but a careful examination of the whole procedure shows that this assumption can be done away with, provided the maximum principle and a Harnack inequality for the corresponding elliptic operator hold for all time levels (see [86]). Let us just remark that the basic reason to use the comparison function is to mimic a parabolic Harnack inequality, whose validity is not known in this context. For a detailed proof of the results of this section we refer the reader to [72,86,160]).

Consider the singular parabolic equation

$$\beta(u)_t = \mathcal{L}u, \tag{5.70}$$

where $\mathcal{L}$ is an elliptic operator with principal part in divergence form. We assume

$$\mathcal{L}u = \sum_{ij} D_i\big(a_{ij}(x,t)D_j u + a_i(x,t)u\big) + b_i(x,t)D_i u + e(x,t)u, \tag{5.71}$$

where $a_{ij}(x,t),\ a_i(x,t),\ b_i(x,t),\ e(x,t)$ are continuous functions with respect to the time variable and are measurable functions with respect to the spatial variables, and satisfy

$$\frac{1}{\mu_1}|\xi|^2 \leqslant \sum_{ij} a_{ij}(x,t)\xi_i\xi_j \leqslant \mu_1|\xi|^2, \tag{5.72}$$

$$\left\|\sum a_i^2, \sum b_i^2, e\right\|_{q,r,\Omega_T} \leqslant \mu_2, \tag{5.73}$$

with q and r such that

$$\frac{1}{r} + \frac{N}{2q} = 1 - \kappa_1$$

and

$$q \in \left[\frac{N}{2(1-\kappa_1)}, \infty\right], \qquad r \in \left[\frac{1}{1-\kappa_1}, \infty\right], \quad 0 < \kappa_1 < 1, \quad N \geqslant 2.$$

Moreover, we suppose that,

$$\forall 0 < t < s < T, \qquad \int_t^s \int_\Omega \Big(ev - \sum_i a_i D_i v\Big)\,dx\,dt \leqslant 0 \tag{5.74}$$

for all $v \in C_0^1(\Omega \times (t,s))$, $v \geqslant 0$. Assumptions (5.72) and (5.73) mean that the structure conditions (5.8)–(5.14) are satisfied, while (5.74) is assumed in order to have the maximum principle in any parabolic cylinder contained in Ω_T. Before giving a sketch of the proof, we stress that the only properties of $\mathcal{L}(x,t,u,Du)$ on which we will rely are the following:

(1) $\mathcal{L}$ satisfies the maximum principle;

(2) the coefficients of $\mathcal{L}$ are continuous in t;

(3) in the case of time-dependent coefficients, the elliptic operator $\mathcal{L}$ satisfies a *uniform* Harnack inequality t by t.

The most important of the three assumptions is the last one and this shows once more how the Harnack inequality is crucial when proving regularity results for solution of PDEs.

For the sake of simplicity we consider only the case of time independent coefficients (the reader finds a detailed proof of the general case in [86] and [160]). Without loss of generality, assume that (5.66) holds, i.e.,

$$u(x,t) \geqslant \mu^- + \varepsilon\omega \quad \forall (x,t) \in (x_1,t_1) + Q_{(\theta\delta_0^2\rho^2, \delta_0\rho)}$$

(if not, it means that the oscillation of the solution is automatically reduced in a smaller cylinder). As in [72], apply the change of variables

$$x \to \frac{4(x-x_1)}{|x_1|}, \qquad t \to \frac{t-t_1}{\delta_0^2\rho^2}$$

and introduce the function

$$\tilde{u} \equiv \frac{u - \mu^-}{\varepsilon \omega}.$$

We have that $\tilde{u}$ solves the differential equation

$$\left(\tilde{\beta}(\tilde{u})\right)_t = \tilde{\mathcal{L}}(\tilde{u}),$$

where $\tilde{\beta}$ and $\tilde{\mathcal{L}}$ satisfy the same structural conditions of the operator (5.71) (for more details see [72]). Moreover, $\tilde{u}(x,t) \geqslant 1$ in $K_{\varepsilon_0} \times (0,\theta)$. Without loss of generality, we may assume that $\varepsilon_0 \leqslant \frac{1}{2}$. Define $\mathcal{A}_{(d_1,d_2)} \times (t_1,t_2)$ as the annulus $\{d_1 < |x| < d_2\} \times (t_1,t_2)$. Finally, we are interested in the behavior of $\tilde{u}$ in an annulus contained in $\mathcal{A}_{(\varepsilon_0,4d)} \times (0,\theta)$ where we may assume $d \geqslant 4$ (see [72]). For this reason we introduce a proper comparison function. Namely, let v solve the following boundary value problem

$$\begin{cases} \left(\tilde{\beta}(v)\right)_t = \tilde{\mathcal{L}}v & \text{in } \mathcal{A}_{(\varepsilon_0,4d)} \times (0,\theta), \\ v(x,t) = 0 & \text{on } |x| = 4d, \\ v(x,t) = 1 & \text{on } |x| = \varepsilon_0, \\ v(x,0) = 0. \end{cases} \tag{5.75}$$

By the maximum principle, $\tilde{u} \geqslant v$. Hence if we are able to prove that there is a level $t_0 \in (0,\theta)$ such that there is a $\sigma_0 > 0$, depending only upon the data, such that

$$v(x,t_0) \geqslant \sigma_0 \quad \forall 1 \leqslant |x| \leqslant 2d, \tag{5.76}$$

we have that a similar bound works for $\tilde{u}$. By returning to the original coordinates, we conclude that there exists a time level t_1 such that

$$u(x,t_1) > \mu_- + \sigma_0 \varepsilon \omega \quad \forall x \in K_{\delta^* \rho},$$

with δ^* a positive constant that can be determined a priori only in terms of the data. As in [72] (Sections 24 and 25), by using the logarithmic estimates, one can bring this piece of information up to level zero and in this way reduce the oscillation of u in the small cylinder $Q(\delta^* \rho)$.

In order to prove (5.76), let ζ be the solution of the elliptic problem

$$\begin{cases} \mathcal{L}\zeta = 0 & \text{in } \mathcal{A}_{\varepsilon_0,4d}, \\ \zeta(x) = 0 & \text{on } |x| = 4d, \\ \zeta(x) = 1 & \text{on } |x| = \varepsilon_0. \end{cases} \tag{5.77}$$

By well-known classical results, there is a unique solution ζ satisfying (5.77) (see [88], Theorems 8.1 and 8.3). Moreover, ζ is Hölder continuous (see [120]). Finally, the hypotheses on the coefficients ensure that w belongs to an elliptic De Giorgi's class, which in turn guarantees that w satisfies a Harnack inequality (under this point of view, see for example [124], Chapter 3). Therefore we may assume that there is σ_0 such that

$$\zeta(x) \geqslant 2\sigma_0 \quad \forall 1 \leqslant |x| \leqslant 2d. \tag{5.78}$$

The aim now is to transfer this information to v. With this purpose, put $z = v - \zeta$ and set $\Gamma(x, \cdot) = \tilde{\beta}(\cdot - \zeta(x))$. It is easy to see that z satisfies

$$\begin{cases} \big(\Gamma(z)\big)_t = \mathcal{L}z & \text{in } \mathcal{A}_{\varepsilon_0,4d} \times (0,k), \\ z(x,t) = 0 & \text{on } |x| = 4d, \\ z(x,t) = 0 & \text{on } |x| = \varepsilon_0, \\ z(x,0) = -\zeta(x) & \text{on } x \in \mathcal{A}_{\varepsilon_0,4d}. \end{cases} \tag{5.79}$$

By the energy estimates, one gets that

$$\int_0^k \int_{\mathcal{A}_{\varepsilon_0,4d}} |\nabla z|^2 \, dx \, dt \leqslant C. \tag{5.80}$$

There must exist a time level t^* such that

$$\int_{\mathcal{A}_{\varepsilon_0,4d}} |\nabla z|^2 \, dx \leqslant \delta_0, \tag{5.81}$$

with τ_0 a proper small quantity. In fact, if it were not so, integrating on $(0, k)$ we obtain

$$\int_0^k \int_{\mathcal{A}_{\varepsilon_0,4d}} |\nabla z|^2 \, dx \geqslant k\delta_0$$

and it suffices to choose k large enough to get a contradiction.

Now we claim that, if we choose δ_0 small enough, a consequence of (5.81) is that

$$\forall x \in \mathcal{A}_{\varepsilon_0,2d}, \quad (\zeta - v)\big(x, t^*\big) \leqslant \sigma_0. \tag{5.82}$$

In fact, if it were not true, reproducing the same argument of [72], Sections 6–8, and using the measure lemma, we conclude that there exist a $y^* \in \mathcal{A}_{\varepsilon_0,4d}$ and a small cube $K_\rho(y^*) \subset \mathcal{A}_{\varepsilon_0,4d}$ such that

$$\forall x \in K_\rho\big(y^*\big), \quad (\zeta - v)\big(x, t^*\big) > \frac{\sigma_0}{2}.$$

Connecting, through a path, the boundary $K_\rho(y^*)$ with the boundary of $\mathcal{A}_{\varepsilon_0,4d}$, i.e., with a portion of the boundary where $|x| = 4d$, and working as in Section 9 of [72] we get a lower bound for $\int_{\mathcal{A}_{\varepsilon_0,4d}} |\nabla z(x, t^*)|^2 \, dx$, thus obtaining a contradiction by choosing δ_0 small enough. Therefore (5.82) holds.

The proof of the local continuity of a bounded solution of (5.70) follows from the remark that estimate (5.76) is a direct consequence of (5.78) and (5.82).

REMARK 45. The boundary regularity for singular equations like the ones we are dealing with here still presents several open questions. In the case of β with a single jump, interior regularity is again a matter of energy and logarithmic inequalities and this allows for a complete solution of boundary regularity for variational data, as in [48].

In the case of a general β it remains an open problem to devise a technique of proof only based on energy and logarithmic estimates. Therefore, for the moment, it is only possible to consider homogeneous Dirichlet boundary conditions under mild assumptions on $\partial\Omega$, without taking into account initial conditions. Assume, for instance, the compactness and the Lipschitz continuity of the boundary. By the compactness of the boundary of Ω, we can cover $\partial\Omega$ with a finite number of neighborhoods centered at points of $\partial\Omega$. The Lipschitz continuity of the boundary allows us to find a map from every neighborhood into a half ball of $\mathbb{R}^N$. The transformed equation via this map has coefficients which are still measurable with respect to x, properly summable with respect to the pair (x,t), and continuous with respect to t. Now we reflect the operator through the entire ball and notice that this reflection does not affect the $(\beta(u))_t$ term, since it is done only with respect to the x variable. We have therefore reduced the study to a problem in the interior and we can apply the previous results to prove the boundary regularity.

5.4.7. *The case $N \geqslant 3$ continued.* In this section, we prove the regularity for particular cases of (5.1). Here we permit the maximal generality allowed from conditions (5.8)–(5.14), but we consider a very special β of the type

$$\beta(s) = \begin{cases} -s-\nu_1 & \text{if } s<0, \\ [-\nu_1, 0] & \text{if } s=0, \\ s & \text{if } 0<s<1, \\ [1, 1+\nu_2] & \text{if } s=1, \\ s+\nu_2 & \text{if } s>1. \end{cases}$$

This special β allows us to write explicitly the logarithmic and energy estimates. The use of the estimates, together with the geometrical constructions of the previous sections, will allow to prove the local continuity of the solutions. This approach is developed in a very detailed way in [87]. A similar approach can be found in [52] in a different context. We feel that the right approach to the local continuity is this one, that is, it must be based only on logarithmic and energy estimates and geometric constructions.

We only give a sketch of the ideas on which the proof is based. Without loss of generality, we may assume that $\mu^+ = 1+\varepsilon_0$ and $\mu^- = -\varepsilon_0$, with ε_0 very "small". The idea is to reduce the oscillation of u such that either $u>0$ or $u<1$. In this way, the problem has been reduced to the case of only a singularity. Without loss of generality, we can focus our attention to find a subcylinder where $u>0$. Apply Proposition 23. Choose $s_0 > \varepsilon_0$ such that either the oscillation of u is so reduced in $Q(\delta^2(s_0)\rho^2, \delta(s_0)\rho)$ that $u>0$ in such a cylinder or

$$\operatorname{Meas}\bigl\{(x,t)\in Q_{(\rho^2,\delta(s_0)\rho)}\colon u(x,t)\leqslant \mu^- + s\omega\bigr\} \geqslant \eta(s_0)\lvert Q_{(\rho^2,\delta(s_0)\rho)}\rvert. \tag{5.83}$$

Note that $\eta(s_0)$ is very small (in a quantitatively way that will be clear in the sequel) if ε_0 is chosen really small.

Partition the cylinder $Q_{(\rho^2,\delta(s_0)\rho)}$ in subcylinders $Q_{(\delta^2(s_0)\rho^2,\delta(s_0)\rho)}$. Apply the iteration method based on energy and logarithmic estimates, choosing as test function $(u - \frac{1}{4})_-$. There are two possibilities:

(i) either in one of the subcylinders the set where u is less than or equal to zero is negligible; then the iteration method works as in the nonsingular case. So one can find a small cylinder where $u \geqslant \frac{1}{8}$. Then, via the logarithmic estimates, one can bring this positivity up to the level zero if ε_0 is chosen small enough;

(ii) or, in each of the subcylinders, the set where u is less or equal to zero is not negligible. Then by (5.83) one can deduce that there exists a subcylinder where the measure of the set where u is less than $\frac{1}{8}$ is a function of η_0. So if we choose η_0 small enough, the alternative (5.34) holds. So one can find a small cylinder where $u \geqslant \frac{1}{16}$. Then, via the logarithmic estimates, one can bring this positivity up to the level zero if ε_0 is chosen small enough.

REMARK 46. The fact that $\beta(\cdot)$ has only two jumps is not essential. Any finite number of jumps would be acceptable.

REMARK 47. This approach seems to work not only with jumps but also with singularities that grow in a very fast way. It could be of some interest to start an investigation on such a direction.

Acknowledgments

The research of J.M. Urbano was supported by CMUC/FCT and Project POCTI/34471/MAT/2000.

References

[1] R. Adams, *Sobolev Spaces*, Academic Press, New York (1975).
[2] L.V. Alfors, *Lectures on Quasiconformal Mappings*, Wadsworth and Brooks/Cole, Monterey, CA (1987).
[3] H.W. Alt and E. DiBenedetto, *Nonsteady flow of water and oil through inhomogeneous porous media*, Ann. Scuola Norm. Sup. Pisa (3) **XII** (1985), 335–392.
[4] D. Andreucci, *A priori bounds for weak solutions of the filtration equation*, SIAM J. Math. Anal. **22** (1991), 138–145.
[5] D. Andreucci and E. DiBenedetto, *A new approach to initial traces in nonlinear filtration*, Ann. Inst. H. Poincaré **4** (1990), 305–334.
[6] D. Andreucci and E. DiBenedetto, *On the Cauchy problem and initial traces for a class of evolution equations with strongly nonlinear sources*, Ann. Sc. Norm. Sup. Pisa Cl. Sci. (4) **18** (1991), 363–441.
[7] S.N. Antoncev, *On the solvability of boundary value problems for degenerate two-phase porous flow equations*, Dynamika Sploshnoi Sredy Vyp. **10** (1972), 28–53 (in Russian).
[8] S.N. Antoncev, *Axially symmetric problems of gas dynamics with free boundaries*, Doklady Akad. Nauk SSSR **216** (1974), 473–476.
[9] S.N. Antoncev, A.V. Kazhikov and V.N. Monakov, *Boundary-Value Problems in the Mechanics of Nonuniform Fluids*, Studies in Mathematics and Its Applications, North-Holland, Amsterdam (1990).
[10] G. Anzellotti, *On the extremal stress in Hencky plasticity*, Duke Math. J. **551** (1984), 133–147.

[11] G. Anzellotti and M. Giaquinta, *On the existence of the fields of stress and displacement for an elasto-plastic body in static equilibrium*, J. Math. Pures Appl. **61** (1982), 219–244.
[12] D.G. Aronson, *The porous medium equation*, Nonlinear Diffusion Problems (Montecatini Terme), Lecture Notes in Math., Vol. 1224, C.I.M.E. series, Springer-Verlag, Berlin (1986), 1–46.
[13] D.G. Aronson and L.A. Caffarelli, *The initial trace of a solution of the porous medium equation*, Trans. Amer. Math. Soc. **280** (1983), 351–366.
[14] D.G. Aronson, M.G. Crandall and L.A. Peletier, *Stability of a degenerate nonlinear diffusion problem*, Nonlinear Anal. **6** (1982), 1001–1022.
[15] D.G. Aronson and L.A. Peletier, *Large time behavior of solutions of the porous medium equation in bounded domain*, J. Differential Equations **39** (1981), 378–412.
[16] D.G. Aronson and J. Serrin, *Local behavior of solutions of quasilinear parabolic equations*, Arch. Ration. Mech. Anal. **25** (1967), 81–123.
[17] G.I. Barenblatt, *On some unsteady motions of a liquid or a gas in a porous medium*, Prikl. Mat. Mech. **16** (1952), 67–78.
[18] J. Bear, *Dynamics of Fluids in Porous Media*, Elsevier, New York (1972).
[19] J. Bear, *Hydraulics of Groundwater*, McGraw-Hill, New York (1979).
[20] Ph. Bénilan and M.G. Crandall, *The continuous dependence on ϕ of solutions of $u_t - \Delta\phi(u) = 0$*, Indiana Univ. Math. J. **30** (1981), 161–177.
[21] S.N. Bernstein, *Collected Works. III. Differential Equations, Calculus of Variations and Geometry (1903–1947)*, Izdat. Akad. Nauk SSSR, Moscow (1960) (in Russian).
[22] J.G. Berryman, *Evolution of a stable profile for a class of nonlinear diffusion equations with fixed boundaries*, J. Math. Phys. **18** (1977), 2108–2115.
[23] J.G. Berryman and C.J. Holland, *Stability of the separable solutions for fast diffusion equations*, Arch. Ration. Mech. Anal. **18** (1976), 379–388.
[24] M. Bertsch, N. Nambu and L.A. Peletier, *Decay of solutions of a nonlinear diffusion equation*, Nonlinear Anal. **6** (1982), 539–554.
[25] M. Bertsch and L.A. Peletier, *The asymptotic profile of solution of degenerate diffusion equation*, Arch. Ration. Mech. Anal. **91** (1986), 207–229.
[26] L. Boccardo, F. Murat and J.P. Puel, *L^∞ estimate for some nonlinear elliptic partial differential equations and application to an existence result*, SIAM J. Math. Anal. **23** (1992), 326–333.
[27] E. Bombieri, *Theory of minimal surfaces and a counterexample to the Bernstein conjecture in high dimension*, Mimeographed Notes of Lectures Held at Courant Institut, New York University (1970).
[28] E. Bombieri and E. Giusti, *Harnack's inequality for elliptic differential equations on minimal surfaces*, Invent. Math. **15** (1972), 24–46.
[29] A. Bossavit, A. Damlamian and M. Fremond, eds, *Free Boundary Problems: Applications and Theory*, Res. Notes in Math., Vols 120, 121, Pitman, Boston (1985).
[30] H. Brézis, *Elliptic equations with limiting Sobolev exponents – The impact of topology*, Comm. Pure Appl. Math. **39** (1986), 17–39.
[31] L.A. Caffarelli and L.C. Evans, *Continuity of the temperature in the two phase Stefan problem*, Arch. Ration. Mech. Anal. **81** (1983), 199–220.
[32] L.A. Caffarelli and A. Friedman, *Continuity of the density of a gas flow in a porous medium*, Trans. Amer. Math. Soc. **252** (1979), 99–113.
[33] L.A. Caffarelli and A. Friedman, *Regularity of the free-boundary of a gas in a n-dimensional porous medium*, Indiana Univ. Math. J. **29** (1980), 361–369.
[34] G. Chavent and J. Jaffré, *Mathematical Models and Finite Elements Methods for Reservoir Simulation*, North-Holland, Amsterdam (1986).
[35] Y.Z. Chen, *Hölder estimates for solutions of uniformly degenerate quasilinear parabolic equations*, Chinese Ann. Math. Ser. B **5**(4) (1984), 661–678.
[36] Y.Z. Chen, *Hölder continuity of the gradients of solutions of nonlinear degenerate parabolic systems*, Acta Math. Sinica **2** (1986), 309–331.
[37] Y.Z. Chen and E. DiBenedetto, *On the local behavior of solutions of singular parabolic equations*, Arch. Ration. Mech. Anal. **103** (1988), 319–345.
[38] Y.Z. Chen and E. DiBenedetto, *Boundary estimates for solutions of non linear degenerate parabolic systems*, J. Reine Angew. Math. **395** (1989), 102–131.

[39] Y.Z. Chen and E. DiBenedetto, *Hölder estimates of solutions of singular parabolic equations with measurable coefficients*, Arch. Ration. Mech. Anal. **118** (1992), 257–271.

[40] Y.Z. Chen and E. DiBenedetto, *On the Harnack inequality for nonnegative solutions of singular parabolic equations*, Degenerate Diffusions (Minneapolis, 1991), IMA Vol. Math. Appl., Vol. 47, Springer-Verlag, New York (1993), 61–69.

[41] K.S. Cheng and W.M. Ni, *On the structure of the conformal Gaussian curvature equation on* $\mathbb{R}^2$, Duke Math. J. **62** (1991), 721–727.

[42] H. Choe, *Hölder regularity for the gradient of solutions of certain singular parabolic equations*, Comm. Partial Differential Equations **16**(11) (1991), 1709–1732.

[43] H. Choe, *Hölder continuity of solutions of certain degenerate parabolic systems*, Nonlinear Anal. **8** (1992), 235–243.

[44] R.E. Collins, *Flow of Fluids through Porous Materials*, Reinhold, New York (1961).

[45] I.I. Danilyuk, *The Stefan problem*, Russian Math. Surveys **40** (1985), 157–223.

[46] P. Daskalopoulos and M. del Pino, *On a singular diffusion equation*, Comm. Anal. Geom. **3** (1995), 523–542.

[47] E. De Giorgi, *Sulla differenziabilitá e l'analiticitá delle estremali degli integrali multipli regolari*, Mem. Accad. Sci. Torino Cl. Sci. Fis. Mat. Natur. (3) **3** (1957), 25–43.

[48] E. DiBenedetto, *Continuity of weak solutions to certain singular parabolic equations*, Ann. Mat. Pura Appl. (4) **CXXI** (1982), 131–176.

[49] E. DiBenedetto, *Local regularity of weak solutions of degenerate elliptic equations*, Nonlinear Anal. **7** (1983), 827–850.

[50] E. DiBenedetto, *Continuity of weak solutions to a general porous medium equation*, Indiana Univ. Math. J. **32** (1983), 83–118.

[51] E. DiBenedetto, *On the local behavior of solutions of degenerate parabolic equations with measurable coefficients*, Ann. Sc. Norm. Super. Pisa **13** (1986), 487–535.

[52] E. DiBenedetto, *A boundary modulus of continuity for a class of singular parabolic equations*, J. Differential Equations **63** (1986), 418–447.

[53] E. DiBenedetto, *The flow of two immiscible fluids through a porous medium: Regularity of saturation*, Theory and Applications of Liquid Crystals, J.L. Ericksen and D. Kinderlehrer, eds, Springer-Verlag, New York (1987), 123–141.

[54] E. DiBenedetto, *Intrinsic Harnack type inequalities for solutions of certain degenerate parabolic equations*, Arch. Ration. Mech. Anal. **100** (1988), 129–147.

[55] E. DiBenedetto, *Degenerate Parabolic Equations*, Springer-Verlag, New York (1993).

[56] E. DiBenedetto, *Degenerate and singular parabolic equations*, Recent Advances in Partial Differential Equations, El Escorial Res. Appl. Math., Vol. 30, Masson, Paris (1994), 55–84.

[57] E. DiBenedetto, *Parabolic equations with multiple singularities*, Proceedings Equadiff 9, Brno (1997), 25–48.

[58] E. DiBenedetto, *Real Analysis*, Birkhäuser, Basel (2002).

[59] E. DiBenedetto and D.J. Diller, *About a singular parabolic equation arising in thin film dynamics and in the Ricci flow for complete* $\mathbb{R}^2$, Partial Differential Equations and Applications. Collected Papers in Honor of Carlo Pucci on the Occasion of his 70th Birthday, P. Marcellini et al., eds, Lecture Notes Pure Appl. Math., Vol. 177, Dekker, New York (1996), 103–119.

[60] E. DiBenedetto and D.J. Diller, *Singular semilinear elliptic equations in* $L^1(\mathbb{R}^N)$, Topics in Nonlinear Analysis. The Herbert Amann Anniversary Volume, J. Escher et al., eds, Progr. Nonlinear Differential Equations Appl., Vol. 35, Birkhäuser, Basel (1999), 143–181.

[61] E. DiBenedetto and A. Friedman, *Regularity of solutions of nonlinear degenerate parabolic systems*, J. Reine Angew. Math. **349** (1984), 83–128.

[62] E. DiBenedetto and A. Friedman, *Hölder estimates for nonlinear degenerate parabolic systems*, J. Reine Angew. Math. **357** (1985), 1–22.

[63] E. DiBenedetto and R. Gariepy, *On the local behavior of solutions for an elliptic-parabolic equation*, Arch. Ration. Mech. Anal. **97** (1987), 1–17.

[64] E. DiBenedetto and M.A. Herrero, *On the Cauchy problem and initial traces for a degenerate parabolic equation*, Trans. Amer. Math. Soc. **314** (1989), 187–224.

[65] E. DiBenedetto and M.A. Herrero, *Non negative solutions of the evolution p-Laplacian equation. Initial traces and Cauchy problem when* $1 < p < 2$, Arch. Ration. Mech. Anal. **111** (1990), 225–290.

[66] E. DiBenedetto and Y.C. Kwong, *Intrinsic Harnack estimates and extinction profile for certain singular parabolic equations*, Trans. Amer. Math. Soc. **330** (1992), 783–811.

[67] E. DiBenedetto, Y.C. Kwong and V. Vespri, *Local space analiticity of solutions of certain parabolic equations*, Indiana Univ. Math. J. **40** (1991), 741–765.

[68] E. DiBenedetto and J. Manfredi, *On the higher integrability of the gradient of weak solutions of certain degenerate elliptic systems*, Amer. J. Math. **115** (1993), 1107–1134.

[69] E. DiBenedetto, J. Manfredi and V. Vespri, *A note on boundary regularity for certain degenerate parabolic equations*, Nonlinear Diffusion Equations and Their Equilibrium States, 3 (Gregynog, 1989), Progr. Nonlinear Differential Equations Appl., Vol. 7, Birkhäuser, Boston (1992), 177–182.

[70] E. DiBenedetto and M. O'Leary, *Three-dimensional conduction–convection problems with change of phase*, Arch. Ration. Mech. Anal. **123** (1993), 99–116.

[71] E. DiBenedetto and N.S. Trudinger, *Harnack inequalities for quasi-minima of variational integrals*, Ann. Inst. H. Poincaré **1** (1984), 295–308.

[72] E. DiBenedetto and V. Vespri, *On the singular equation* $\beta(u)_t = \Delta u$, Arch. Ration. Mech. Anal. **132** (1995), 247–309.

[73] D.J. Diller, *A remark on the continuous dependence on* ϕ *of solutions to* $u_t \Delta\phi(u) = 0$, Differential Integral Equations **11** (1998), 425–438.

[74] J.R. Drake, J.R. Greenwood, G.A. Navratil and R.S. Post, *Diffusion coefficient scaling in the Wisconsin levitated octupole*, Phys. Fluids **20** (1977), 148–175.

[75] C. Ebmeyer and J.M. Urbano, *Regularity in Sobolev spaces for doubly nonlinear parabolic equations*, J. Differential Equations **187** (2003), 375–390.

[76] C. Ebmeyer and J.M. Urbano, *The smoothing property for a class of doubly nonlinear parabolic equations*, Preprint of DMUC 02-15, submitted.

[77] L. Esposito, G. Mingione and B. Stroffolini, *On the continuity of the singular equation* $\beta(u)_t = \Delta u$, Nonlinear Anal. **36** (1999), 1037–1048.

[78] J.E. Esteban, A. Rodriguez and J.L. Vázquez, *The fast diffusion equation with logarithmic nonlinearity and the evolution of a conformal metric in the plane*, Adv. Differential Equations **1** (1996), 21–50.

[79] L.C. Evans and J. Spruck, *Motion of level sets by mean curvature. I; II; III; IV*, J. Differential Geom. **33** (1991), 635–681; Trans. Amer. Math. Soc. **330** (1992), 321–332; J. Geom. Anal. **2** (1992), 121–150; J. Geom. Anal. **5** (1995), 77–114.

[80] E.B. Fabes and N. Garofalo, *Parabolic B.M.O. and Harnack's inequality*, Proc. Amer. Math. Soc. **95** (1985), 63–69.

[81] E.B. Fabes and D.W. Stroock, *A new proof of Moser's parabolic Harnack inequality using the old ideas of Nash*, Arch. Ration. Mech. Anal. **96** (1986), 327–338.

[82] A. Fasano and M. Primicerio, eds, *Free Boundary Problems: Theory and Applications*, Pitman Res. Notes Math., Vol. 78, Longman, Harlow (1983).

[83] A. Friedman, *On the regularity of the solutions of nonlinear elliptic and parabolic systems of partial diffusion equations*, J. Math. Mech. **7** (1958), 43–60.

[84] A. Friedman, *Variational Principles and Free Boundary Problems*, Wiley–Interscience, New York (1982).

[85] A. Friedman, *Mathematics in Industrial Problems – Part* 2, IMA, Vol. 24, Math. Appl., Springer-Verlag, Heidelberg (1989).

[86] U. Gianazza, B. Stroffolini and V. Vespri, *Interior and boundary continuity of the solution of the singular equation* $(\beta(u))_t = \mathcal{L}u$, Nonlinear Anal. **56** (2004), 157–183.

[87] U. Gianazza and V. Vespri, *Continuity of weak solutions of a singular parabolic equation*, Adv. Differential Equations **8** (2003), 1341–1376.

[88] D. Gilbarg and N. Trudinger, *Elliptic Partial Differential of Second Order*, Springer-Verlag, Berlin (1983).

[89] J. Hadamard, *Extension à l'équation de la chaleur d'un théorem de A. Harnack*, Rend. Circ. Mat. Palermo (2) **3** (1954), 337–346.

[90] R.S. Hamilton, *The Ricci flow on surfaces*, Contem. Math. **71** (1988), 237–262.

[91] R. Hardt and Y. Tonegawa, *Partial regularity for evolution problems with discontinuity*, Manuscripta Math. **90** (1996), 85–103.

[92] R. Hardt and X. Zhou, *An evolution problem for linear growth functionals*, Comm. Partial Defferential Equations **19**(11/12) (1994), 1879–1907.

[93] A. Hoff, *A scheme for computing solutions and interface curves for a doubly degenerate parabolic equation*, SIAM J. Numer. Anal. **22** (1985), 687–712.

[94] K.H. Hoffman and J. Spekels, eds, *Free Boundary Problems: Application and Theory*, Res. Notes Math., Vols 185, 186, New York (1990).

[95] C. Huang, *Asymptotic drying of gelatin in a photographic film*, J. Math. Anal. Appl. **187** (1994), 663–675.

[96] A.V. Ivanov, *Uniform Hölder estimates for generalized solutions of quasilinear parabolic equations admitting a doubly degeneracy*, Algebra i Analiz **3** (1991), 139–179.

[97] A.V. Ivanov, *The class B_{ml} and Hölder estimates for quasilinear doubly degenerate parabolic equations*, Zap. Nauchn. Sem. S.-Peterburg. Otdel. Math. Inst. Steklov (LOMI) **197** (1992), 42–70.

[98] A.V. Ivanov and P.Z. Mkrtchen, *On the regularity up to the boundary of weak solutions of the first initial-boundary value problem for quasilinear doubly degenerate parabolic equations*, Zap. Nauchn. Sem. S.-Peterburg. Otdel. Math. Inst. Steklov (LOMI) **196** (1991), 3–28.

[99] T. Iwaniec, *Projections onto gradient fields and L^p-estimates for degenerate elliptic equations*, Studia Math. **75** (1983), 293–312.

[100] A.S. Kalashnikov, *Cauchy's problem in classes of increasing functions for certain quasi-linear degenerate parabolic equations of the second order*, Differentsial'nye Uravneniya **9** (1973), 682–691.

[101] A.S. Kalashnikov, *On uniqueness conditions for the generalized solutions of the Cauchy problem for a class of quasi-linear degenerate parabolic equations*, Differentsial'nye Uravneniya **9** (1973), 2207–2212.

[102] A.S. Kalashnikov, *On a heat conduction equation for a medium with non-uniformly distributed non-linear heat sources or absorbers*, Bull. Univ. Moscow Math. Mech. **3** (1983), 20–24.

[103] A.S. Kalashnikov, *Some problems of the qualitative theory of nonlinear degenerate second order parabolic equations*, Russian Math. Surveys **42** (1987), 169–222.

[104] A.S. Kalashnikov, O.A. Oleinik and Y.-L. Chzhou, *The Cauchy problem and boundary value problems for equations of the type of non-stationary filtration*, Izv. Akad. Nauk. USSR Ser. Mat. **22** (1958), 667–704.

[105] S. Kamenomostkaya, *On the Stefan problem*, Naučnye Dokl. Vysšei Školy **1** (1958), 60–62 (in Russian); Mat. Sb. **53**(95) (1961), 489–514.

[106] D. Kinderlehrer and L. Nirenberg, *Analiticity at the boundary of solutions of non linear second order parabolic equations*, Comm. Pure Appl. Math. **31** (1978), 283–338.

[107] D. Kinderlehrer and G. Stampacchia, *An Introduction to Variational Inequalities and Their Applications*, Reprint of the 1980 original, Classics Appl. Math., Vol. 31, SIAM, Philadelphia (2000).

[108] J. Kinnunen and J.L. Lewis, *Higher integrability for parabolic systems of p-Laplacian type*, Duke Math. J. **102** (2000), 253–271.

[109] J. Kinnunen and J.L. Lewis, *Very weak solutions of parabolic systems of p-Laplacian type*, Ark. Mat. **40** (2002), 105–132.

[110] N.V. Kruzkov, *On the a priori estimation of solutions of linear parabolic equations and of solutions of boundary value problems for a certain class of quasilinear parabolic equations*, Dokl. Akad. NAUK SSSR **138** (1961), 1005–1008; English transl.: Soviet Math. Dokl. **2** (1961), 764–767.

[111] N.V. Kruzkov, *A priori estimates and certain properties of the solutions of elliptic and parabolic equations of second order*, Math. Sbornik **65** (1968), 522–570; English. transl.: Amer. Math. Soc. Transl. **2**(68) (1968), 169–220.

[112] N.V. Kruzkov, *Results concerning the nature of the continuity of solutions of parabolic equations and some of their applications*, Math. Zametki **6** (1969), 97–108 (in Russian).

[113] N.V. Kruzkov, *First order quasilinear equations with several space variables*, Mat. Sbornik **123** (1970), 228–255; English transl.: Math. USSR Sb. **10** (1970), 217–273.

[114] N.V. Kruzkov, *Uniqueness of the solutions of mixed problems for the degenerate system of the theory of two-phase porous flow*, Vestnik Moskovskoga Univ. Mat. **40**(2) (1985), 28–33.

[115] N.V. Kruzkov and S.M. Sukorjanski, *Boundary value problems for systems of equations of two phase porous flow type: Statement of the problems, questions of solvability, justification and approximation methods*, Math. USSR-Sb. **44** (1977), 62–80.

[116] N.V. Krylov, *Non Linear Elliptic and Parabolic Equations of the Second Order*, Reidel, Dordrecht, The Netherlands (1987).
[117] N.V. Krylov and M.V. Safonov, *A certain property of solutions of parabolic equations with measurable coefficients*, Math. USSR-Izv. **16** (1981), 151–164.
[118] Y.C. Kwong, *Asymptotic behavior of a plasma type equation with finite extinction*, Arch. Ration. Mech. Anal. **104** (1988), 277–294.
[119] O.A. Ladyzhenskaja, *New equations for the description of motion of viscous incompressible fluids and solvability in the large of boundary value problems for them*, Proc. Steklov Inst. Math. **102** (1967), 95–118; English transl.: Trudy Math. Inst. Steklov **102** (1967), 85–104.
[120] O.A. Ladyzhenskaja, V.A. Solonnikov and N.N. Ural'ceva, *Linear and Quasilinear Equations of Parabolic Type*, Amer. Math. Soc. Transl. Math. Mono., Vol. 23, Providence, RI (1968).
[121] O.A. Ladyzhenskaja and N.N. Ural'ceva, *Linear and Quasilinear Elliptic Equations*, Academic Press, New York (1968).
[122] M.C. Leverett, *Capillary behavior in porous solids*, Trans. Amer. Inst. Mining and Metallurgical Engrs **142** (1941), 151–169.
[123] J.L. Lions, *Quelques méthodes de résolution dés problèmes aux limites nonlineaires*, Dunod, Paris (1969).
[124] J. Malý and W.P. Ziemer, *Fine Regularity of Solutions of Elliptic Partial Differential Equations*, Amer. Math. Soc. Math. Surveys Mono., Vol. 51, Providence, RI (1997).
[125] J.J. Manfredi and V. Vespri, *Large time behavior of solutions to a class of doubly nonlinear parabolic equations*, Electron. J. Differential Equations **2** (1994), 1–17.
[126] L.K. Martinson and K.B. Paplov, *The effect of magnetic plasticity in non-Newtonian fluids*, Magnit. Gidrodinamika **3** (1969), 69–75.
[127] L.K. Martinson and K.B. Paplov, *Unsteady shear flows of a conducting fluid with a rheological power law*, Magnit. Gidrodinamika **2** (1970), 50–58.
[128] V.G. Mazja, *Sobolev Spaces*, Springer-Verlag, New York (1985).
[129] A. Meirmanov, *The Stefan Problem*, de Gruyter, Berlin (1992).
[130] G.J. Minty, *Monotone operators in Hilbert spaces*, Duke Math. J. **29** (1962), 341–346.
[131] C.B. Morrey, *Multiple Integrals in the Calculus of Variations*, Springer-Verlag, New York (1966).
[132] J. Moser, *A new proof of De Giorgi's theorem concerning the regularity problem for elliptic differential equations*, Comm. Pure Appl. Math. **13** (1960), 457–468.
[133] J. Moser, *On Harnack's theorem for elliptic differential equations*, Comm. Pure Appl. Math. **14** (1961), 577–591.
[134] J. Moser, *A Harnack inequality for parabolic differential equations*, Comm. Pure Appl. Math. **17** (1964), 101–134.
[135] J. Moser, *On a pointwise estimate for parabolic differential equations*, Comm. Pure Appl. Math. **24** (1971), 727–740.
[136] J. Nash, *Continuity of solutions of parabolic and elliptic equations*, Amer. J. Math. **80** (1958), 931–954.
[137] W.M. Ni, *On the elliptic equation $\Delta u + Ke^u = 0$ and conformal metrics with prescribed Gaussian curvatures*, Invent. Math. **66** (1982), 343–352.
[138] W.M. Ni, L. Peletier and J.L. Vazquez, eds, *Degenerate Diffusion*, IMA Vol. Math. Appl., Vol. 47, Springer-Verlag, New York (1993).
[139] H. Okuda and J.M. Dawson, *Theory and numerical simulation on plasma diffusion across a magnetic field*, Phys. Fluids **16** (1973), 567–597.
[140] O.A. Oleinik, *On the equation $\Delta u + K(x)e^u = 0$*, Russian Math. Surveys **33** (1978), 243–244.
[141] L. Orsina and M.M. Porzio, *$L^\infty(Q)$-estimate and existence of solutions for some nonlinear parabolic equations*, Boll. Unione Mat. Ital. **6** (1992), 631–647.
[142] R. Ossermann, *On the inequality $\Delta u \geqslant f(u)$*, Pacific J. Math. **12** (1959), 1641–1647.
[143] R.E. Pattle, *Diffusion from an instantaneous point source with a concentration dependent coefficient*, Quart. Appl. Math. **12** (1959), 407–409.
[144] B. Pini, *Sulla soluzione generalizzata di Wiener per il primo problema di valori al contorno nel caso parabolico*, Rend. Sem. Math. Univ. Padova **23** (1954), 422–434.
[145] S.I. Pohozaev, *On the eigenfunctions of the equation $\Delta(u) + \lambda f(u) = 0$*, Dokl. Akad. Nauk SSSR **165** (1965), 36–39.

[146] M.M. Porzio, *$L^{\infty}_{\mathrm{loc}}$-estimates for degenerate and singular parabolic equations*, Nonlinear Anal. **17** (1991), 1093–1107.

[147] M.M. Porzio and V. Vespri, *Hölder estimates for local solutions of some doubly nonlinear degenerate parabolic equations*, J. Differential Equations **103** (1993), 146–178.

[148] J.F. Rodrigues, *Variational methods in the Stefan problem*, Phase Transitions and Hysteresis, A. Visintin, ed., Lecture Notes in Math., Vol. 1584, Springer-Verlag, Berlin (1994), 147–212.

[149] P. Sachs, *Continuity of solutions of singular parabolic equations*, Nonlinear Anal. **7** (1983), 387–409.

[150] P. Sachs, *The initial and boundary value problem for a class of degenerate parabolic equations*, Comm. Partial Differential Equations **8** (1983), 693–734.

[151] M.V. Safonov, *The Harnack inequality for elliptic equations and the Hölder continuity of their solutions. Boundary problems of mathematical physics and adjacent questions in the theory of functions*, Zap. Nauchn. Sem. Leningrad. Otdel. Math. Inst. Steklov (LOMI) **96** (1980), 272–287; Engl. transl.: J. Soviet Math. (LOMI) **20** (1983), 851–863.

[152] G. Sargenti, *Continuity results for solutions of certain degenerate parabolic equations*, Rend. Sem. Mat. Univ. Padova **99** (1998), 105–131.

[153] G. Sargenti, *Local boundedness and regularity results for a class of singular anisotropic parabolic equations*, Atti Sem. Mat. Fis. Univ. Modena **XLVII** (2000), 45–78.

[154] G. Sargenti and V. Vespri, *Regularity results for multiphase Stefan-like equations*, Modelling and Optimization of Distributed Parameter Systems, K. Malanowski, Z. Nahorski and M. Peszynska, eds, Chapman and Hall, New York (1996), 111–117.

[155] G. Savaré and V. Vespri, *The asymptotic profile of solutions of a certain class of doubly nonlinear equations*, Nonlinear Anal. **22** (1994), 1553–1565.

[156] A.E. Scheidegger, *The Physics of Flow through Porous Media*, 3rd Edition, Univ. Toronto Press, Toronto, Ontario (1974).

[157] J. Serrin, *Local behavior of solutions of quasilinear elliptic equations*, Acta Math. **111** (1964), 101–134.

[158] S. Spanne, *Some function spaces defined using the mean oscillation over cubes*, Ann. Scuola Norm. Sup. Pisa **19** (1965), 593–608.

[159] G. Stampacchia, *Equation elliptiques du second ordre à coefficientes discontinues*, Sém. Math. Sup., Vol. 16, Les Presses de l'Univ. de Montréal, Montréal (1966).

[160] B. Stroffolini and V. Vespri, *On the continuity of the solution of the singular equation $(\beta(u))_t = \mathcal{L}u$*, Matematiche (Catania) **55** (2001), 165–195.

[161] R. Temam, *Plasticity*, Gauthier–Villars, Paris (1983).

[162] P. Tolksdorff, *Everywhere regularity for some quasi-linear systems with lack of ellipticity*, Ann. Mat. Pura Appl. **4** (1983), 241–266.

[163] N.S. Trudinger, *On Harnack type inequalities and their application to quasilinear elliptic partial differential equations*, Comm. Pure Appl. Math. **20** (1967), 721–747.

[164] N.S. Trudinger, *Pointwise estimates and quasilinear parabolic equations*, Comm. Pure Appl. Math. **21** (1968), 205–226.

[165] K. Uhlenbeck, *Regularity for a class of nonlinear elliptic systems*, Acta Math. **138** (1977), 219–240.

[166] N.N. Ural'tceva, *Degenerate quasilinear elliptic systems*, Zap. Nauchn. Sem. Leningrad. Otdel. Math. Inst. Steklov (LOMI) **7** (1968), 184–222 (in Russian).

[167] J.M. Urbano, *A free boundary problem with convection for the p-Laplacian*, Rend. Mat. Appl. (7) **17** (1997), 1–19.

[168] J.M. Urbano, *Continuous solutions for a degenerate free boundary problem*, Ann. Mat. Pura Appl. **CLXXVIII** (2000), 195–224.

[169] J.M. Urbano, *A singular-degenerate parabolic problem: Regularity up to the Dirichlet boundary*, Free Boundary Problems – Theory and Applications I, GAKUTO Internat. Ser. Math. Sci. Appl., Vol. 13, Gakktosho, Tokyo (2000), 399–410.

[170] J.M. Urbano, *Hölder continuity of local weak solutions for parabolic equations exhibiting two degeneracies*, Adv. Differential Equations **6** (2001), 327–358.

[171] J.M. Urbano, *A free boundary problem: Contributions from modern analysis*, European Congress of Mathematics, Progress in Mathematics, Vol. 202, Birkhäuser, Basel (2001), 319–329.

[172] J.M. Urbano, *Regularity for partial differential equations: From De Giorgi–Nash–Moser theory to intrinsic scaling*, CIM Bull. **12** (2002), 8–14.

[173] J. Van Duijn and H. Zhang, *Regularity properties of a doubly degenerate equation in hydrology*, Comm. Partial Differential Equations **13** (1988), 261–319.

[174] J.L. Vázquez, *Asymptotic behavior and propagation properties of one dimensional flow of gas in a porous medium*, Trans. Amer. Math. Soc. **277** (1983), 507–527.

[175] V. Vespri, *On the local behavior of a certain class of doubly nonlinear parabolic equations*, Manuscripta Math. **75** (1992), 65–80.

[176] V. Vespri, L^∞ *estimates for nonlinear parabolic equations with natural growth conditions*, Rend. Sem. Mat. Univ. Padova **90** (1993), 1–8.

[177] V. Vespri, *Harnack type inequality for solutions of certain doubly nonlinear parabolic equations*, J. Math. Anal. Appl. **181** (1994), 104–131.

[178] A. Visintin, *Models of Phase Transitions*, Progress in Nonlinear Differential Equations and Their Applications, Vol. 28, Birkhäuser, Boston, MA (1996).

[179] A. Visintin, A. Fasano, E. Magenes and C. Verdi, *About mathematical models of phase transitions*, Boll. Unione Mat. Ital. Ser. B Artic. Ric. Mat. **8** (1998), 1–47; 49–69; 71–81; 83–108.

[180] M. Wiegner, *On* C_α*-regularity of the gradient of solutions of degenerate parabolic systems*, Ann. Mat. Pura Appl. **4** (1986), 385–405.

[181] L. Wu, *A new result for the porous medium equation, derived from the Ricci flow*, Bull. Amer. Math. Soc. **28** (1993), 90–94.

[182] L. Wu, *The Ricci flow on the complete* $\mathbb{R}^2$, Comm. Anal. Geom. **1** (1993), 439–472.

[183] X. Zhou, *An evolution problem for plastic antiplanar shear*, Appl. Math. Optimiz. **25** (1992), 263–285.

[184] W.P. Ziemer, *Interior and boundary continuity of weak solutions of degenerate parabolic equations*, Trans. Amer. Math. Soc. **271** (1982), 733–748.

CHAPTER 4

Nonlinear Hyperbolic–Parabolic Coupled Systems

Ling Hsiao

Academy of Mathematics and Systems Science, Academia Sinica, Beijing 100080, China
E-mail: hsiaol@mail.amss.ac.cn

Song Jiang

Institute of Applied Physics and Computational Mathematics, P.O. Box 8009, Beijing 100088, China
E-mail: jiang@iapcm.ac.cn

Contents

1. Introduction . . . 289
2. The general governing equations . . . 291
3. Globally defined solutions for one-dimensional continuum medium with large smooth initial data . . . 296
4. Globally defined weak solutions for one-dimensional compressible flow with large discontinuous initial data . . . 322
5. Vacuum problem for one-dimensional compressible flow . . . 331
6. The Navier–Stokes equations for multidimensional compressible fluids . . . 339
   6.1. The isentropic fluid case . . . 341
   6.2. The heat-conducting case . . . 352
7. Nonlinear one-dimensional thermoelasticity . . . 354
8. Nonlinear three-dimensional thermoelasticity . . . 365

Acknowledgment . . . 375

References . . . 375

HANDBOOK OF DIFFERENTIAL EQUATIONS
Evolutionary Equations, volume 1
Edited by C.M. Dafermos and E. Feireisl

1. Introduction

The aim of this chapter is to introduce the study on nonlinear systems of partial differential equations which are of hyperbolic–parabolic type. This kind of system widely arises from various applied sciences among which we only choose certain fundamental models from continuum physics: the nonlinear systems of partial differential equations with both hyperbolic and parabolic effects from thermoelasticity, thermoviscoelasticity and dynamics of compressible viscous fluids.

In continuum physics, material bodies are modeled as continuous media whose motion and equilibrium are governed by balance laws and constitutive relations. The list of balance laws in force identifies the theory, for example, mechanics, thermomechanics, electrodynamics, etc. The referential (Lagrangian) and the spatial (Eulerian) formulation of the typical balance laws – the balance laws of mass, momentum and energy, can be presented as certain nonlinear hyperbolic–parabolic coupled systems of partial differential equations (see, e.g., the books [17,32] by Carlson and by Dafermos for the detailed derivation). The type of constitutive relation characterizes the nature of material response. We will discuss the constitutive equations in thermoelasticity, thermoviscoelasticity and compressible viscous fluid dynamics.

The equations of thermoelasticity describe the elastic and the thermal behavior of elastic and heat conductive media, in particular, the reciprocal actions between elastic stresses and temperature differences where the dissipation is induced by thermal diffusion, while the equations of thermoviscoelasticity as well as the compressible Navier–Stokes equations, an extension of thermoelasticity, encompass materials with, besides thermal diffusion, internal dissipation induced by viscosity. In this chapter we consider the classical nonlinear thermoelastic, thermoviscoelastic systems and the compressible Navier–Stokes equations which are nonlinear hyperbolic–parabolic coupled systems.

It is well known both for hyperbolic and for parabolic nonlinear equations and systems that smooth solutions in general might not exist globally in time but develop singularities in finite time. The criteria according to which global solutions still exist are different for hyperbolic and parabolic equations. Hence for a hyperbolic–parabolic coupled system, the question arises whether the behavior of the system will be dominated by the hyperbolic or by the parabolic part. The answer will depend on the number of space dimensions, as well as the strength of the dissipation involved: the combined dissipation of viscosity and heat diffusion, or the pure heat diffusion. This also applies to the question of asymptotic behavior of solutions.

In Section 2, we give a brief derivation to the equations of thermoelasticity and thermoviscoelasticity, and the compressible Navier–Stokes equations.

Section 3 is devoted to the nonlinear hyperbolic–parabolic system in thermoviscoelasticity. We elucidate the influence of dissipation mechanism in the one-space dimensional system on qualitative behavior of solutions in this section. We first show that the combined dissipation effects of viscosity and thermal diffusion may counterbalance the destabilizing influence of nonlinearity and thus induce the existence of globally defined smooth solutions to the Cauchy problem or initial boundary value problems even for large smooth initial data. Then, the influence of dissipation on the large-time behavior of solutions is discussed, with different kind of materials or with different type of boundary conditions.

The globally defined weak solutions to the one-dimensional system with large discontinuous initial data are investigated in Section 4 in which we consider the system for isentropic viscous flow first and turn to the nonisentropic case in the second part. The results on the well-posedness and large-time behavior of solutions are introduced, as well as certain results on regularity and stability of solutions. Moreover, it is shown that neither vacuum nor concentration can occur in the solution if there is neither vacuum nor concentration initially, no matter how large the external force in the system and how large the oscillation of the initial data are. Section 5 is devoted to vacuum problems for one-dimensional isentropic viscous gas flow. We consider the compressible Navier–Stokes equations when the viscosity coefficient is a positive constant first and then discuss the case when the viscosity coefficient depends on the density. In both cases the results on the existence and behavior of weak solutions to the compressible Navier–Stokes equations with vacuum are presented.

For the equations of multidimensional nonlinear thermoviscoelasticity, to our best knowledge, there are few results on the global existence for general large data. Hence in this chapter, instead of multidimensional thermoviscoelasticity, we only discuss the Navier–Stokes equations of multidimensional compressible fluids (which can be considered as a thermoviscoelastic system for compressible fluids).

Section 6 is concerned with mathematical topics on the Navier–Stokes equations for *multidimensional* compressible fluids. The mathematical study of the compressible Navier–Stokes equations perhaps goes back to the work by Nash [173] on the local in time existence in the last sixties. Since then, significant progress has been made on the mathematical topics, such as the global existence and the time-asymptotic behavior of solutions. However, a number of important questions still remain open for large data.

Concerning the global well-posedness and the large-time behavior of solutions, the picture is more or less complete when data are sufficiently small. For large data the situation becomes quite complex. The global well-posedness for the Navier–Stokes equations of compressible heat-conducting fluids still remains open. On the other hand, the global existence and the large-time behavior of solutions have been obtained in the case of isentropic fluids, or heat-conducting fluids with symmetric data in symmetric domains without the origin in order to avoid the mathematical difficulties induced by the singularity at the origin. We will discuss this issue in more details in Section 6.

The last two sections (Sections 7 and 8) are devoted to the study of the equations of classical nonlinear thermoelasticity, which are a coupling of the equations of elasticity and of the heat equation and thus build a hyperbolic–parabolic system. For thermoelasticity, however, the dissipation is induced only by thermal diffusion, hence weaker than that for thermoviscoelasticity. We are mainly interested in proving the well-posedness in the class of smooth solutions and in describing the asymptotic behavior of the solutions as time tends to infinity. Although the well-posedness in the linear theory has been studied for years, the description of the general dynamical system with its asymptotic behavior as time tends to infinity and, in particular, the study of nonlinear systems was only started in the late sixties and the early eighties, respectively, and led to very interesting features.

In one space dimension the picture is more or less complete. Bounded or unbounded intervals representing the reference configuration can be dealt with for all the classical boundary conditions. For small initial data global smooth solutions to the nonlinear system will exist; large initial data lead to a blow-up, then weak solutions must be considered.

These results show that the dissipation induced by thermal diffusion is strong enough for small data to prevent a smooth solution from a finite-time blow-up but not for large data, and that the behavior of the nonlinear thermoelastic system in one space dimension is dominated by the heat conduction at least with respect to the existence of solutions for small data and the time-asymptotic behavior of solutions.

In more than one space dimension the local well-posedness is known in most cases, but concerning the global existence or blow-up, the situation becomes delicate. It is well known that in the case of pure elasticity there are global small solutions if the nonlinearity degenerates up to order two, i.e., if the nonlinearity is cubic (near zero values of its arguments). In the "genuinely nonlinear" case, a blow-up in finite time has to be expected. On the other hand quadratic nonlinearities in $\mathbb{R}^3$ still lead to global, small solutions of the heat equation. The question remains whether here the dissipative impact through heat conduction is strong enough to prevent solutions from blowing up at least for small data. The answer to this question will be positive for the Cauchy problem if one essentially excludes those nonlinearities which are responsible for blow-up results in pure elasticity or one considers the radially symmetric case; otherwise a smooth solution generally breaks down for small data for general (quadratic) nonlinearities in the "genuinely nonlinear" case. We will present these results in Sections 7 and 8.

Most of the material presented here has only been previously published in original papers, and some of the material has never been published until now. A long list of references has been included, although it is not intended to be exhaustive and some of the references included there are not quoted in the text. Of course the selection of the material follows personal interests and experiences.

2. The general governing equations

In this section we derive briefly the equations of thermo- and thermoviscoelasticity and the compressible Navier–Stokes equations.

In continuum physics the mathematical model of an n-dimensional body ($n = 1, 2$, or 3) is a manifold characterized by a reference configuration, i.e., an open subset $\mathcal{B}$ of the reference space $\mathbb{R}^n$. The typical point in $\mathcal{B}$ is called a material particle. A configuration of the body $\mathcal{B}$ is a Lipschitz homeomorphism $X = X(x)$ from $\mathcal{B}$ to the physical space $\mathbb{R}^n$. A motion of $\mathcal{B}$ is an evolution $X = X(t,x)$ of configurations. Thus $X(t,x)$ is the position of particle x at time t; the curve $X(\cdot, x)$ is the trajectory of particle x; $X(t, \cdot)$ is the configuration of $\mathcal{B}$ at time t. $X(t,x)$ is called deformation.

In continuum physics one seeks to determine the time evolution of the fields of various physical quantities, such as density, stress, temperature, etc., defined over the moving body. Since every configuration is homeomorphic to the reference configuration, these field quantities can be represented equally well as functions of (t,x) (Lagrangian or referential coordinates) or of (t, X) (Eulerian or spatial coordinates). To avoid cumbersome notation, we denote

$$F = \nabla_x X \quad \left(F^{ij} = \frac{\partial X_i}{\partial x_j}\right), \qquad \dot{} = \frac{\partial}{\partial t} + \mathbf{u} \cdot \nabla_X, \quad \mathbf{u} = \dot{X},$$

where F denotes the deformation gradient field and $\mathbf{u}$ is the velocity field, ∇_x and ∇_X denote the gradients of their Lagrangian and Eulerian descriptions with respect to x and to X, respectively, and analogously for div_x and div_X. In particular, since each configuration of the body is homeomorphic to its reference configuration $\mathcal{B}$ we assume

$$\det F \neq 0.$$

Continuum thermomechanics rests upon the balance laws of mass, linear momentum and energy. The field equations in Lagrangian coordinates read

$$\dot{\rho}_c = 0, \tag{2.1}$$

$$\overset{\cdot}{\overbrace{\rho_c \mathbf{u}}} = \operatorname{div}_x \widetilde{S} + \rho_c b, \tag{2.2}$$

$$\overset{\cdot}{\overbrace{\rho_c\left(\varepsilon + \frac{1}{2}|\mathbf{u}|^2\right)}} = \operatorname{div}_x\left(\mathbf{u} \cdot \widetilde{S} - Q\right) + \rho_c b \cdot \mathbf{u} + \rho_c r, \tag{2.3}$$

where ρ_c is the reference density, $\widetilde{S}$ denotes the Piola–Kirchhoff stress tensor, and ε is the specific internal energy, and Q is the heat flux, b and r denote the specific external body force and the external heat supply respectively.

The balance law of mass (2.1) simply states that the reference density ρ_c is a function of x, in particular, a constant for a homogeneous medium. The balance law of angular momentum gives $SF' = FS'$, where and in what follows a prime denotes transposition. If $\mathbf{u}$ is Lipschitz continuous, we may combine (2.1) and (2.2) to deduce (2.3) into the simpler form

$$\rho_c \dot{\varepsilon} = \operatorname{tr}\left(\widetilde{S}\dot{F}'\right) - \operatorname{div}_x Q + \rho_c r. \tag{2.4}$$

By η we denote the entropy and by

$$\psi := \varepsilon - \theta\eta \tag{2.5}$$

the Helmholtz free energy, where θ is the absolute temperature. Then, the local form of the second law of thermodynamics is expressed by the Clausius–Duhem inequality:

$$\rho_c \dot{\eta} \geqslant -\operatorname{div}_x\left(\frac{Q}{\theta}\right) + \frac{\rho_c r}{\theta}. \tag{2.6}$$

One can use (2.4) and (2.5) to write (2.6) in the form:

$$\rho_c \dot{\psi} + \rho_c \eta \dot{\theta} - \operatorname{tr}\left(\widetilde{S}\dot{F}'\right) + \frac{1}{\theta} Q \cdot \nabla_x \theta \leqslant 0. \tag{2.7}$$

Constitutive equations, which express how the fields appearing in the balance laws are related to the motion, serve to identify the material of the body. In general, constitutive relations have to comply with a number of requirements stemming from material frame indifference, material symmetry and the second law of thermodynamics. Here we outline the theory of constitutive equations for the class of thermoviscoelastic materials of the differential type and the class of thermoelastic materials.

A material is homogeneous thermoviscoelastic of the differential type when the free energy, stress, entropy and heat flux at any material particle x and time t are solely determined by the values of F, $\nabla_x \mathbf{u}$, θ, $\nabla_x \theta$ at (t, x), i.e., in Lagrangian coordinates and upon setting $W = \nabla_x \mathbf{u} = \dot{F}$, $\mathbf{g} = \nabla_x \theta$:

$$\begin{aligned} \psi &= \hat{\psi}(F, W, \theta, \mathbf{g}), \\ \widetilde{S} &= \widehat{S}(F, W, \theta, \mathbf{g}), \\ \eta &= \hat{\eta}(F, W, \theta, \mathbf{g}), \\ Q &= \widehat{Q}(F, W, \theta, \mathbf{g}). \end{aligned} \tag{2.8}$$

As mentioned before, the constitutive relations (2.8) should be consistent with the requirement that every smooth process compatible with (2.1)–(2.3) must satisfy the Clausius–Duhem inequality (2.6) (and (2.7)). Substituting (2.8) into (2.7), we obtain after a calculation that ψ is independent of the velocity and temperature gradients $\nabla_x \mathbf{u}$, $\nabla_x T$, and (cf. [32])

$$\begin{aligned} \psi &= \hat{\psi}(F, \theta), \\ \widetilde{S} &= \rho_c \frac{\partial \hat{\psi}(F, \theta)}{\partial F} + \widehat{Z}(F, W, \theta, \mathbf{g}), \\ \eta &= -\frac{\hat{\psi}(F, \theta)}{\partial \theta}, \\ Q &= \widehat{Q}(F, W, \theta, \mathbf{g}), \end{aligned} \tag{2.9}$$

where $\widehat{Z}(F, W, \theta, \mathbf{g})$ and $\widehat{Q}(F, W, \theta, \mathbf{g})$ should satisfy

$$\operatorname{tr}\big(\widehat{Z}(F, W, \theta, \mathbf{g}) W'\big) - \frac{1}{\theta} \hat{Q}(F, W, \theta, \mathbf{g}) \cdot \mathbf{g} \geqslant 0 \tag{2.10}$$

We recall that $SF' = FS'$ (the balance law of angular momentum) which reduces here to $\widehat{Z}(F, W, \theta, \mathbf{g}) F' = F \widehat{Z}'(F, W, \theta, \mathbf{g})$. Further reduction on the constitutive relations results from the principle of material frame indifference, we refer to [32] for the details.

By substituting the relations (2.8) and (2.5) into (2.2) and (2.4), one obtains a complete system of nonlinear *thermoviscoelasticity* for the unknown functions $X(t, x)$, $\theta(t, x)$. In Section 3 we will discuss some mathematical issues on the one-dimensional case.

The limiting case of a thermoviscoelasticity material with stress and heat flux independent of velocity gradient yields the thermoelastic material. In that case, the left-hand side of (2.10) is an affine function of W and so this inequality will hold identically only if $\widehat{Z}$ vanishes. Consequently, for a thermoelastic material, the constitutive equations (2.9) and the inequality (2.10) become

$$\begin{aligned}
&\psi = \hat{\psi}(F,\theta),\\
&\widetilde{S} = \rho_c \frac{\partial\hat{\psi}(F,\theta)}{\partial F},\\
&\eta = -\frac{\hat{\psi}(F,\theta)}{\partial\theta},\\
&Q = \widehat{Q}(F,\theta,\mathbf{g}),\\
&-\widehat{Q}(F,\theta,\mathbf{g})\cdot\mathbf{g} \geqslant 0.
\end{aligned} \tag{2.11}$$

Using (2.11), we may write (2.4) as

$$\rho_c\theta\left\{-\frac{\partial^2\psi}{\partial\theta^2}\dot{\theta} - \frac{\partial^2\psi}{\partial\theta\,\partial F}\dot{F}\right\} + \operatorname{div}_x Q = \rho_c r. \tag{2.12}$$

Equation (2.2) is mainly a hyperbolic system of second order for X, while (2.12) is mainly a parabolic equation for θ. Instead of X and θ, the displacement U and the temperature difference $\theta - T_0$, still denoted by θ, i.e.,

$$U := X - x, \qquad \theta := \theta - T_0 \tag{2.13}$$

are often used, where T_0 is a constant reference temperature. We shall write $\hat{\psi}(F,T) = \hat{\psi}(\nabla U,\theta)$ with the same symbol $\hat{\psi}$, analogously for the other response functions. Then, the system of *thermoelasticity* consists of (2.2) and (2.12) together with (2.11), where the unknown functions are $U(t,x)$ and $\theta(t,x)$. The problem of finding U and θ will become well posed if additionally boundary (when $\Omega \neq \mathbb{R}^n$) and initial conditions are prescribed. Mathematical issues on thermoelasticity will be discussed in Sections 7 and 8.

REMARK 2.1. For an extended derivation of the equations of thermoelasticity see Carlson [17].

As stated in the above, the balance laws of mass, momentum and energy, as well as the Clausius–Duhem inequality, admit equivalent Eulerian descriptions. In Eulerian coordinates (t,X), the balance laws (2.1)–(2.3) for a compressible viscous heat-conducting fluid with $b=0$ and $r=0$ can be written:

$$\partial_t\rho + \operatorname{div}_X(\rho\mathbf{u}) = 0, \tag{2.14}$$

$$\partial_t(\rho\mathbf{u}) + \operatorname{div}_X(\rho\mathbf{u}\otimes\mathbf{u}) + \nabla P = \operatorname{div}_X\mathbf{T}, \tag{2.15}$$

$$E_t + \operatorname{div}_X\big[(E+P)\mathbf{u}\big] = \operatorname{div}_X(\mathbf{T}\mathbf{u}) - \operatorname{div}_X\mathbf{q}. \tag{2.16}$$

Here ρ and $\mathbf{u} = (u_1,\ldots,u_n)$ denote the density and velocity of the fluid respectively, being functions of Eulerian coordinates (t,X), $E := \rho|\mathbf{u}|^2/2 + \rho e$ denotes the total energy, e is the specific internal energy, and $\mathbf{q}$ is the heat flux, P is the pressure, $\mathbf{T}$ denotes the viscous stress tensor of the form for a Newtonian fluid:

$$T_{ij} = \lambda \operatorname{div}_X \mathbf{u} \delta_{ij} + 2\mu D_{ij}, \tag{2.17}$$

λ and μ are the constant viscosity coefficients satisfying $\mu > 0$, $\lambda + 2\mu/n \geqslant 0$, and $\mathbf{D} = \{D_{ij}\}$ is the deformation tensor

$$D_{ij} = \frac{1}{2}\left(\frac{\partial u_i}{\partial X_j} + \frac{\partial u_j}{\partial X_i}\right).$$

The system (2.14)–(2.16) is also called the *Navier–Stokes equations* for compressible heat-conducting fluids in $\mathbb{R}^n$ which describe the motion of a compressible viscous heat-conducting gas. It can be seen that the relations between $\rho, \mathbf{q}, \mathbf{T}$ and $\rho_c, Q, \widetilde{S}$ are given by

$$\rho = (\det F)^{-1}\rho_c, \qquad \mathbf{q} = (\det F)^{-1} Q F', \qquad -P\,\mathrm{id} + \mathbf{T} = (\det F)^{-1}\widetilde{S}F'.$$

In view of thermodynamics (cf. the constitutive relations (2.9)) we have the thermodynamic equations of the state $e = \hat{e}(\rho, \theta)$, $P = \hat{p}(\rho, \theta)$ with the known functions $\hat{e}$ and $\hat{p}$, where θ denotes the absolute temperature. In this chapter we restrict our study to the physical situation where e, P, $\mathbf{q}$ have the form

$$\begin{aligned} &e = c_V\theta, \qquad P = R\rho\theta \quad \text{(polytropic ideal gas)},\\ &\mathbf{q} = -\kappa\nabla\theta \quad \text{(Fourier's law)}, \end{aligned} \tag{2.18}$$

where R, c_V and κ are positive constants.

Substituting (2.17) and (2.18) into (2.14)–(2.16), we obtain the complete system for the unknowns $\rho, \mathbf{u}, \theta$ in Eulerian coordinates (t, X):

$$\partial_t\rho + \operatorname{div}(\rho\mathbf{u}) = 0, \tag{2.19}$$

$$\partial_t(\rho\mathbf{u}) + \operatorname{div}(\rho\mathbf{u}\otimes\mathbf{u}) + R\nabla(\rho\theta) = \mu\Delta\mathbf{u} + (\lambda+\mu)\nabla\operatorname{div}\mathbf{u}, \tag{2.20}$$

$$\begin{aligned} &\partial_t\left[\rho\left(\frac{|\mathbf{u}|^2}{2} + c_V\theta\right)\right] + \operatorname{div}\left[\rho\left(\frac{|\mathbf{u}|^2}{2} + c_V\theta + R\theta\right)\mathbf{u}\right]\\ &\quad = \kappa\Delta\theta + \lambda\operatorname{div}(\mathbf{u}\operatorname{div}\mathbf{u}) + \mu\operatorname{div}\big[(\mathbf{u}\cdot\nabla)\mathbf{u}\big] + \frac{\mu}{2}\Delta|\mathbf{u}|^2, \end{aligned} \tag{2.21}$$

where the differential operators div, ∇ and Δ are with respect to the Eulerian coordinate X.

In the case of isentropic ideal gases, the system (2.19)–(2.21) turns to (see, e.g., [145]):

$$\partial_t\rho + \operatorname{div}(\rho\mathbf{u}) = 0, \tag{2.22}$$

$$\partial_t(\rho\mathbf{u}) + \operatorname{div}(\rho\mathbf{u}\otimes\mathbf{u}) + a\nabla(\rho^\gamma) = \mu\Delta\mathbf{u} + (\lambda+\mu)\nabla\operatorname{div}\mathbf{u}, \tag{2.23}$$

where $\gamma \geqslant 1$ is the specific heat ratio and a a positive constant. The multidimensional theory for the compressible isentropic Navier–Stokes equations (2.22) and (2.23) is the main issue in Section 6.

It can be easily seen that both (2.19)–(2.21) and (2.22), (2.23) are hyperbolic–parabolic coupled systems in which there is dissipation induced by viscosity and heat-diffusion.

Finally, we introduce some notation used throughout this chapter.

NOTATION. Let Ω be a domain in $\mathbb{R}^n$. Let m be an integer and let $1 \leqslant p \leqslant \infty$. By $W^{m,p}(\Omega)$ ($W_0^{m,p}(\Omega)$) we denote the usual Sobolev space defined over Ω. $W^{m,2}(\Omega) \equiv H^m(\Omega)$ ($W_0^{m,2}(\Omega) \equiv H_0^m(\Omega)$), $W^{0,p}(\Omega) \equiv L^p(\Omega)$ with norm $\|\cdot\|_{L^p(\Omega)}$. $L^p(I,B)$ resp. $\|\cdot\|_{L^p(I,B)}$ denotes the space of all strongly measurable, pth-power integrable (essentially bounded if $p=\infty$) functions from I to B resp. its norm, $I \subset \mathbb{R}$ an interval, B a Banach space. $C(I,B)$ (or $C^0(I,B)$) is the space of all functions which are continuous on I with values in B. $C(I, B-w)$ denotes the space of all functions which are in $L^\infty(I,B)$ and continuous on I with values in B endowed with the weak topology.

Unless there is a statement to the contrary, the Einstein summation convention (i.e., repeated indices indicate summation) is used, and a prime "$'$" denotes transposition. For simplicity, we use the abbreviations:

$$\|\cdot\| \equiv \|\cdot\|_{L^2(\Omega)}, \qquad \|\cdot\|_{L^p} \equiv \|\cdot\|_{L^p(\Omega)}, \qquad \|\cdot\|_{W^{m,p}} \equiv \|\cdot\|_{W^{m,p}(\Omega)},$$

$$\|\cdot\|_{H^m} \equiv \|\cdot\|_{H^m(\Omega)}, \qquad \langle h\rangle^2_{\partial\Omega} \equiv h^2|_{x=1} + h^2|_{x=0} \quad \text{when } \Omega=(0,1);$$

$$D \equiv (\partial_t, \nabla).$$

3. Globally defined solutions for one-dimensional continuum medium with large smooth initial data

The intent of this section is to elucidate the role of dissipation in the continuum physics, i.e., the influence of dissipation mechanism on the qualitative behavior of solutions.

Within the framework of thermomechanics, the referential (Lagrangian) description of the balance laws of mass, momentum and energy for one-dimensional materials with reference density $\rho_c = 1$ can be written, due to the discussion in Section 2, as

$$\begin{cases} v_t - u_x = 0, \\ u_t - \sigma_x = 0, \\ \left(e + \frac{1}{2}u^2\right)_t - (\sigma u)_x + q_x = 0, \end{cases} \tag{3.1}$$

while the second law of thermodynamics is expressed by the Clausius–Duhem inequality

$$\eta_t + \left(\frac{q}{\theta}\right)_x \geqslant 0, \tag{3.2}$$

where $v, u, e, \sigma, \eta, \theta$ and q ($\equiv Q$ in Section 2) denote deformation gradient (specific volume), velocity, internal energy, stress, specific entropy, temperature and heat flux, respectively. Note that v, e and θ may only take positive values.

For one-dimensional, homogeneous, thermoviscoelastic materials, internal energy, stress, entropy and heat flux are given by the constitutive relations

$$e = \hat{e}(v,\theta), \qquad \sigma = \hat{\sigma}(v,\theta,u_x), \qquad \eta = \hat{\eta}(u,\theta), \qquad q = \hat{q}(v,\theta,\theta_x),$$

which, in order to comply with (3.2), must satisfy

$$\begin{aligned} &\hat{\sigma}(v,\theta,0) = \hat{\psi}_v(v,\theta), \qquad \hat{\eta}(v,\theta) = -\hat{\psi}_\theta(v,\theta), \\ &\big(\hat{\sigma}(v,\theta,w) - \hat{\sigma}(v,\theta,0)\big)w \geqslant 0, \qquad \hat{q}(v,\theta,g)g \leqslant 0, \end{aligned} \tag{3.3}$$

where $\psi = e - \theta\eta$ is the Helmholtz free energy.

A typical problem is to consider a body with reference configuration the interval $[0, 1]$ whose boundary is stress-free and thermally insulated,

$$\begin{aligned} &\sigma(t,0) = \sigma(t,1) = 0, \\ &q(t,0) = q(t,1) = 0, \end{aligned} \qquad 0 \leqslant t < \infty, \tag{3.4}$$

and determine the thermomechanical process, described by the functions $(v(t,x), u(t,x), \theta(t,x))$, $0 \leqslant x \leqslant 1$, $t \geqslant 0$, under prescribed initial conditions

$$v(0,x) = v_0(x), \qquad u(0,x) = u_0(x), \qquad \theta(0,x) = \theta_0(x), \quad 0 \leqslant x \leqslant 1. \tag{3.5}$$

In the absence of dissipation, when the body is a thermoelastic nonconductor, i.e., $\sigma = -\hat{p}(v,\theta), e = \hat{e}(v,\theta), q \equiv 0$, (3.1) reduces to a hyperbolic system of conservation laws which does not generally possess globally defined smooth solutions, even when the initial data are very smooth.

The question is whether the combined dissipative effects of viscosity and thermal diffusion, when the material is thermoviscoelastic, may counterbalance the destabilizing influence of nonlinearity and thus induce the existence of globally defined smooth solutions to the initial-boundary value problem (3.1), (3.4), (3.5) even for large initial data.

This is first verified by Kazhikhov and Shelukhin [137,139] for the case where the material is an ideal, linearly viscous, gas with constant specific heats:

$$e = c_V\theta, \qquad \sigma = -R\frac{\theta}{v} + \mu\frac{u_x}{v}, \qquad q = -k\frac{\theta_x}{v},$$

by means of an interesting analysis which depends crucially on the special form of the above constitutive assumption. For the case of ideal gases, the global existence of smooth or strong solutions to initial boundary value problems or to the Cauchy problem, with large initial data have been investigated also by Antontsev, Kazhikhov and Monakhov in [10,137,138], Kawashima and Nishida in [134], Nagasawa in [169–171], Iskenderova, Smagulov, Durmagambetov [103,209] (also cf. the references cited therein), etc.

The corresponding results for the material of real gases have been obtained by Kawohl [140], Jiang [109,115], Luo [153,154], Pan [182], Amosov [3], Qin [186], Qin and Rivera [187], etc.

On the other hand, Dafermos and Hsiao in [34] and Dafermos in [31] considered the problem (3.1), (3.4), (3.5) for a fairly general class of solid like linearly viscous materials,

$$e = \hat{e}(v, \theta), \qquad \sigma = -\hat{p}(v, \theta) + \mu u_x, \qquad q = \hat{q}(v, \theta, \theta_x).$$

In the paper [34], shear viscosity μv is inversely proportional to density. Consequently, its dissipative effect is so weak at high density that existence of global solutions is to be expected only when the initial energy is not too large. To overcome this limitation, Dafermos [31] considers the problem (3.1), (3.4), (3.5) for linearly viscous materials

$$e = \hat{e}(v, \theta), \qquad \sigma = -\hat{p}(v, \theta) + \hat{\mu}(v) u_x, \qquad q = -\hat{k}(v, \theta)\theta_x, \tag{3.6}$$

where the viscosity $\hat{\mu}(v)v$ is uniformly positive, that is,

$$\hat{\mu}(v)v \geqslant \mu_0 > 0, \quad 0 < v < \infty. \tag{3.7}$$

REMARK 3.1. In the case when viscosity may depend on temperature, a typical problem to understand the effect, caused by the dependence of viscosity on temperature, has been studied first by Dafermos and Hsiao in [35] for adiabatic shearing of incompressible flow. See, e.g., Tzavaras [218,219], Belov and Belov [14], Bertsch, Peletier and Verduyn Lunels [15] and Jiang [111] for related results, and among others.

Assume that $\hat{e}(v, \theta)$, $\hat{p}(v, \theta)$, $\hat{\mu}(v)$ and $\hat{k}(v, \theta)$ are twice continuously differentiable on $0 < v < \infty$, $0 \leqslant \theta < \infty$, and are interrelated by

$$\hat{e}_v(v, \theta) = -\hat{p}(v, \theta) + \theta \hat{p}_\theta(v, \theta), \tag{3.8}$$

so as to comply with (3.3). Furthermore, we require that, at any temperature, the elastic part of the stress is compressive at high density and tensile at low density, i.e., there are $0 < \tilde{v} \leqslant \widetilde{V} < \infty$ such that

$$\begin{aligned} \hat{p}(v, \theta) &\geqslant 0, \qquad 0 < v < \tilde{v}, 0 \leqslant \theta < \infty, \\ \hat{p}(v, \theta) &\leqslant 0, \qquad \widetilde{V} < v < \infty, 0 \leqslant \theta < \infty. \end{aligned} \tag{3.9}$$

Employing an idea of Andrews [7], it can be shown, as a consequence of (3.7) and (3.9), that the deformation gradient is a priori confined in a bounded interval $0 < \bar{v} < v(t, x) < \overline{V}$ and, hence, no restrictions are necessary on the behavior of $\hat{e}(v, \theta)$, $\hat{p}(v, \theta)$, $\hat{k}(v, \theta)$ at $v = 0^+$ and $v = \infty$. As regards growth with respect to temperature, we assume that, for any $0 < \bar{v} < \overline{V} < \infty$, there are positive constants ν, k_0 and N, possibly depending on $\bar{v}$ and/or $\overline{V}$, such that, for any $\bar{v} < v < \overline{V}$, $0 \leqslant \theta < \infty$,

$$\hat{e}(v,0) \geqslant 0, \qquad v \leqslant \hat{e}_\theta(v,\theta) \leqslant N\big(1+\theta^{1/3}\big), \tag{3.10}$$

$$\big|\hat{p}_v(v,\theta)\big| \leqslant N\big(1+\theta^{1+1/3}\big), \qquad \big|\hat{p}_\theta(v,\theta)\big| \leqslant N\big(1+\theta^{1/3}\big), \tag{3.11}$$

$$\begin{aligned} &k_0 \leqslant \hat{k}(v,\theta) \leqslant N, \qquad \big|\hat{k}_v(v,\theta)\big| \leqslant N, \\ &\big|\hat{k}_\theta(v,\theta)\big| \leqslant N, \qquad \big|\hat{k}_{vv}(v,\theta)\big| \leqslant N. \end{aligned} \tag{3.12}$$

(An explanation on the above growth rates can be found in [31].)

The existence theorem established in [31] reads as the following:

THEOREM 3.1. *Consider the initial-boundary value problem* (3.1), (3.4), (3.5) *under conditions* (3.6)–(3.12). *Assume that* $v_0(x)$, $v_0'(x)$, $u_0(x)$, $u_0'(x)$, $u_0''(x)$, $\theta_0(x)$, $\theta_0'(x)$ *and* $\theta_0''(x)$ *are all in* $C^\alpha([0,1])$ *and* $v_0(x) > 0$, $\theta_0(x) > 0$, $0 \leqslant x \leqslant 1$. *Furthermore, let the initial data be compatible with the boundary conditions at* $(0,0)$ *and* $(1,0)$. *Then there exists a unique solution* $\{v(t,x), u(t,x), \theta(t,x)\}$ *on* $[0,1] \times [0,\infty)$ *such that, for any* $T > 0$, *the functions* v, v_x, v_t, v_{xt}, u, u_x, u_t, u_{xx}, θ, θ_x, θ_{xx} *are all in* $C^{\alpha,\alpha/2}(Q_T)$ *and* v_{tt}, v_{xt}, θ_{xt} *are in* $L^2(Q_T)$. *Moreover,* $\theta(t,x) > 0$, $\bar{v} < v(t,x) < \overline{V}$, *for* $0 \leqslant x \leqslant 1$, $0 \leqslant t < \infty$, *where* $\bar{v}$ *and* $\overline{V}$ *are positive constants depending on the initial data,* $C^\alpha([0,1])$ *denotes the Banach space of functions on* $[0,1]$ *which are uniformly Hölder continuous with exponent* α, *while* $C^{\alpha,\alpha/2}(Q_T)$ *stands for the Banach space of functions on* $Q_T = [0,T] \times [0,1]$ *which are uniformly Hölder continuous with exponent* α *in* x *and* $\alpha/2$ *in* t.

The theorem can be proved by the procedure devised by Dafermos and Hsiao in [34], namely, solutions to (3.1), (3.4), (3.5) are visualized as fixed points of a map P on the Banach space $\mathcal{B}$ of functions $\{V(t,x), U(t,x), \Theta(t,x)\}$ with V, U, U_x, Θ, Θ_x in $C^{1/3,1/6}(Q_T)$ and existence is established by means of the Leray–Schauder fixed point theorem.

The map P carries $\{V(t,x), U(t,x), \Theta(t,x)\}$ into the solution of a complicated linear "parabolic" system obtained by linearizing (3.1) about $\{V(t,x), U(t,x), \Theta(t,x)\}$. Due to the smoothing action of linear parabolic systems, P is completely continuous and its range is contained in the set of functions $\{v(t,x), u(t,x), \theta(t,x)\}$ with v, v_x, v_t, v_{xt}, u, u_x, u_t, u_{xx}, θ, θ_x, θ_t, θ_{xx}, in $C^{\alpha,\alpha/2}(Q_T)$.

The construction of P and the precise statements and proofs of its aforementioned properties can be found in [34]. The detailed discussion to show that any possible fixed point of P, i.e., any solution $\{v(t,x), u(t,x), \theta(t,x)\}$ of (3.1), (3.4), (3.5) satisfies the admissibility condition $\theta(t,x) > 0$, $0 < \bar{v} < v(t,x) < \overline{V}$ and is contained in an a priori bounded set of $\mathcal{B}$ can be found in [31] and [99]. This will complete the list of requirements for the application of the Leray–Schauder fixed point theorem.

The global existence with different boundary conditions to (3.4) have been obtained later by Jiang [108], Watson [228], Amosov [2], Guo and Zhu [76], etc.

For the study on other thermoviscoelastic systems related to (3.1) (for instance, the systems for a viscous reactive gas, and shape memory alloys, etc.), we refer to Guo and Zhu [77], Chen and Hoffmann [26], Shen, Zheng and Zhu [204], Jiang [113], Bressan [16],

Bebernes and Bressan [11], Bebernes and Eberly [12], Belov and Belov [14], Pan and Zhu [183] and the references cited therein.

The next emphasis of this section is to understand the influence of dissipation on the large-time behavior of solutions. Various phenomena may occur on the large-time behavior of solutions corresponding to different materials in consideration. For instance, with the same boundary conditions as stress-free and thermally insulated, the solution may go to infinity as $t \to \infty$ with certain rate, may approach to a unique state exponentially fast or may have phase transition phenomena, according to different kind of constitutive relations to be concerned.

Consider the problem (3.1), (3.4), (3.5). For the material of ideal gas, it has been proved by Nagasawa in [169] that the solution (v^*, u^*, θ^*) of (3.1), (3.4), (3.5) satisfies

$$v^*(t,x) \geqslant C^* \log(1+t), \qquad C^* > 0.$$

However, totally different phenomena may occur on the large-time behavior of solutions for other kind of constitutive relations. In the case of isothermal viscoelasticity, the solution may approach to a unique state exponentially fast as shown by Greenberg and MacCamy in [74], or phase transition may take place as discovered by Andrews and Ball in [8] with nonmonotone pressure. It is proved in [8] that the large-time behavior of strain is described by a Young measure whose support is confined in the set of zeros of pressure. The analysis used in [74] and [8] is extended to the nonisothermal case by Hsiao and Luo in [102] to deal with thermoviscoelastic materials.

For simplicity, we consider the kind of solid-like materials with the following constitutive relations

$$e = c_V \theta, \qquad \sigma = -f(v)\theta + \hat{\mu}(v)u_x, \qquad q = -k(v)\theta_x, \tag{3.13}$$

where c_V is a positive constant, $k(v)$ and $f(v)$ are twice continuously differentiable for $v > 0$ such that

$$\begin{aligned} &k(v) > 0 \quad \text{for } v > 0, \\ &f(v) \geqslant 0, \quad 0 < v < \tilde{v}, \qquad f(v) \leqslant 0, \quad \widetilde{V} < v < \infty, \end{aligned} \tag{3.14}$$

for some fixed $0 < \tilde{v} \leqslant \widetilde{V} < \infty$, and the viscosity $\hat{\mu}(v)v$ is uniformly positive, namely

$$\hat{\mu}(v)v \geqslant \mu_0 > 0, \quad 0 < v < \infty. \tag{3.15}$$

REMARK 3.2. The example proposed by Ericksen [52] shows that $f(v)$ is not monotone in v and satisfies (3.14). For rubber, it is known that a good model of pressure takes the form

$$\hat{p}(v,\theta) = -\gamma\theta\left(v - \frac{1}{v^2}\right), \quad \gamma \text{ is a positive constant,}$$

namely, $f(v) = -\gamma(v - 1/v^2)$ which satisfies (3.14) with $\tilde{v} = \widetilde{V} = 1$.

Turn to the initial data, we normalize $u_0(x)$, by superimposing a trivial rigid motion if necessary, so that

$$\int_0^1 u_0(x)\,\mathrm{d}x = 0. \tag{3.16}$$

The global existence of (3.1), (3.4), (3.5), under the assumptions of (3.13)–(3.16), has been obtained in [31] (also see [99]), which concerns the solid-like material with more general constitutive relations than (3.13), as stated in Theorem 3.1.

The result on large-time behavior of solutions, obtained in [102], is stated as the following theorem.

THEOREM 3.2. *Assume that* (3.13)–(3.16) *are satisfied. Let* $\{v(t,x), u(t,x), \theta(t,x)\}$, $(t,x) \in [0,1] \times [0,\infty)$ *be the globally defined smooth solution of the problem* (3.1), (3.4), (3.5) *obtained in the Theorem* 3.1. *Then, as* $t \to \infty$,

$$\text{I.}\qquad \left\|\hat{p}(v,\theta)(t,\cdot)\right\|_{L^1} = \left\|f(v)\theta(t,\cdot)\right\|_{L^1} \to 0,$$
$$\left\|f(v)(t,\cdot)\right\| \to 0,$$
$$\left\|u(t,\cdot)\right\| \to 0,$$

and

$$\bar{\theta}(t) := \int_0^1 \theta(t,x)\,\mathrm{d}x \to \frac{E_0}{c_V},$$

where $E_0 = \int_0^1 [c_V\theta_0 + \frac{1}{2}u_0^2](x)\,\mathrm{d}x$.

II. *There exists a family of probability measure* $\{\nu_x\}_{x\in[0,1]}$ *on* $\mathbb{R}$ *(depending measurably on* x*) with* $\operatorname{supp}\nu_x \subset K = \{z\colon f(z) = 0\}$ *such that if* $\Phi \in C(\mathbb{R})$ *and*

$$g_\Phi(x) := \langle \nu_x, \Phi\rangle, \quad \text{a.e.},$$

then

$$\Phi\big(v(t,\cdot)\big) \overset{*}{\rightharpoonup} g_\Phi(\cdot) \quad \text{in } L^\infty(0,1), \text{ as } t \to \infty.$$

COROLLARY 3.3. *Suppose the equation* $f(z) = 0$ *has exactly* m *roots,* $z_1, z_2, \ldots, z_m$, $m > 1$. *Then there exist nonnegative functions* $\mu_i \in L^\infty(0,1)$, $1 \leqslant i \leqslant m$, *such that*

$$\Phi\big(v(t,\cdot)\big) \overset{*}{\rightharpoonup} \sum_{i=1}^m \Phi(z_i)\mu_i(\cdot) \quad \text{in } L^\infty(0,1), \text{ as } t \to \infty,$$

for any $\Phi \in C(\mathbb{R})$. *Furthermore,* $\sum_{i=1}^m \mu_i(x) = 1$, *a.e.*

COROLLARY 3.4. *Suppose the equation* $f(z) = 0$ *possesses only one root* $z = z_1$. *Then*

$$v(t,\cdot) \to z, \quad \text{strongly in } L^q(0,1), \text{ as } t \to \infty,$$

for all $q, 1 \leqslant q < \infty$.

If $f(v)$ *is strictly monotone decreasing,*

$$f'(v) < 0 \quad \text{for } v \in [\bar{v}, \overline{V}], \tag{3.17}$$

it follows from (3.7) *that there exists a unique* $\hat{v} \in [\tilde{v}, \widetilde{V}]$ *such that* $f(\hat{v}) = 0$.

Thus, we have further results as the following theorem.

THEOREM 3.5. *Assume that* (3.13)–(3.17) *hold. Then there are positive constants* $\beta, \widehat{T}$ *and* A, *independent of* t, *such that*

$$\|v(t,\cdot) - \hat{v}\|_{H^1} + \|u(t,\cdot)\|_{H^1} + \left\|\theta(t,\cdot) - \frac{E_0}{c_V}\right\|_{H^1} \leqslant Ae^{-\beta t} \quad \text{for } t \geqslant \widehat{T}.$$

REMARK 3.3. Theorem 3.2 and Theorem 3.5 generalize the results obtained in [8] and [74], respectively, which concern the case of isothermal viscoelasticity.

We only give the proof of Theorem 3.2 next.

PROOF OF THEOREM 3.2. Let $\{u(t,x), v(t,x), \theta(t,x)\}$ denote the solution obtained in the global existence theorem from which it is known that

$$0 < \bar{v} \leqslant v(t,x) \leqslant \overline{V}, \qquad \theta(t,x) > 0, \quad x \in [0,1], t \in [0,\infty). \tag{3.18}$$

This, combined with (3.15), yields

$$0 < \mu_1 \leqslant \hat{\mu}\big(v(t,x)\big) \leqslant \mu_2, \quad x \in [0,1], t \in [0,\infty), k_1 \leqslant k(v) \leqslant k_2, \tag{3.19}$$

where μ_1, μ_2, k_1 and k_2 are positive constants, independent of t.

In the sequel, Λ will denote a generic constant, independent of t.

The conservation laws of total momentum and energy can be easily obtained, namely

$$\int_0^1 u(t,x)\,\mathrm{d}x = \int_0^1 u_0(x)\,\mathrm{d}x = 0, \quad 0 \leqslant t < \infty. \tag{3.20}$$

$$\int_0^1 \left[e(t,x) + \frac{1}{2}u^2(t,x)\right]\mathrm{d}x = \int_0^1 \left[e(0,x) + \frac{1}{2}u_0^2(x)\right]\mathrm{d}x =: E_0, \quad 0 \leqslant t < \infty. \tag{3.21}$$

The part I of Theorem 3.2 can be established by the following estimates given in the following steps.

Step 1.

$$\int_0^t \int_0^1 \left[\frac{\mu_1 u_x^2}{\theta} + \frac{k_1 \theta_x^2}{\theta^2} \right](\tau, x)\,\mathrm{d}x\,\mathrm{d}\tau \leqslant \Lambda, \quad t \in [0, \infty). \tag{3.22}$$

By substituting (3.13) to the system (3.1), we may get the equation

$$c_V \theta_t + f(v)\theta u_x - \hat{\mu}(v)u_x^2 - \big(k(v)\theta_x\big)_x = 0. \tag{3.23}$$

Multiplying (3.23) by θ^{-1}, integrating over $[0, t] \times [0, 1]$, and using (3.4) and $(3.1)_1$, we arrive at (3.22) with the help of (3.18), (3.19) and the inequality $\log\theta \leqslant \theta - 1$ for $\theta > 0$.

Due to (3.20)–(3.22), we are able to show that

$$\int_0^t \max_{[0,1]} u^2(\tau, \cdot)\,\mathrm{d}\tau \leqslant \Lambda, \quad t \in [0, \infty). \tag{3.24}$$

Step 2.

$$\int_0^1 \big(u^4 + \theta^2\big)(t, x)\,\mathrm{d}x + \int_0^t \int_0^1 \big[\theta_x^2 + u^2 u_x^2\big](\tau, x)\,\mathrm{d}x\,\mathrm{d}\tau \leqslant \Lambda, \quad t \in [0, \infty). \tag{3.25}$$

Multiplying $(3.1)_3$ with $(c_V\theta + u^2/2)$, integrating over $[0, t] \times [0, 1]$, and using (3.1), (3.2), (3.4), (3.13) and Young's inequality, we arrive at

$$\begin{aligned}
&\frac{1}{2}\int_0^1 \left[c_V\theta + \frac{u^2}{2} \right]^2 (t, x)\,\mathrm{d}x + \mu_1 \int_0^t \int_0^1 u^2 u_x^2(\tau, x)\,\mathrm{d}x\,\mathrm{d}\tau \\
&\quad + \frac{c_V k_1}{2} \int_0^t \int_0^1 \theta_x^2(\tau, x)\,\mathrm{d}x\,\mathrm{d}\tau \\
&\leqslant \Lambda + \Lambda \int_0^t \max_{[0,1]} u^2(\tau, \cdot) \int_0^1 \theta^2(\tau, x)\,\mathrm{d}x\,\mathrm{d}\tau \\
&\quad + \Lambda \int_0^t \int_0^1 u^2 u_x^2(\tau, x)\,\mathrm{d}x\,\mathrm{d}\tau.
\end{aligned} \tag{3.26}$$

To estimate the term $\int_0^t \int_0^1 u^2 u_x^2(\tau, x)\,\mathrm{d}x\,\mathrm{d}\tau$, we multiply $(3.1)_2$ by u^3, integrate it over $[0, t] \times [0, 1]$, and use (3.4), (3.18), (3.19) and the Cauchy–Schwarz inequality. It then follows that

$$\begin{aligned}
&\frac{1}{4}\int_0^1 u^4(t, x)\,\mathrm{d}x + 2\mu_1 \int_0^t \int_0^1 u^2 u_x^2(\tau, x)\,\mathrm{d}x\,\mathrm{d}\tau \\
&\leqslant \Lambda + \Lambda \int_0^t \max_{[0,1]} u^2(\cdot, \tau) \int_0^1 \theta^2(\tau, x)\,\mathrm{d}x\,\mathrm{d}\tau.
\end{aligned} \tag{3.27}$$

By using the Cauchy–Schwarz inequality with the term $(c_V\theta + \frac{u^2}{2})^2$ in (3.26), we obtain

$$\begin{aligned}
&\int_0^1 \theta^2(t,x)\,\mathrm{d}x + \int_0^t\int_0^1 u^2u_x^2(\tau,x)\,\mathrm{d}x\,\mathrm{d}\tau + \int_0^t\int_0^1 \theta_x^2(\tau,x)\,\mathrm{d}x\,\mathrm{d}\tau \\
&\quad\leqslant \Lambda + \Lambda\int_0^t\int_0^1 u^2u_x^2(\tau,x)\,\mathrm{d}x\,\mathrm{d}\tau + \Lambda\int_0^1 u^4(t,x)\,\mathrm{d}x \\
&\qquad + \Lambda\int_0^t \max_{[0,1]} u^2(\tau,\cdot)\int_0^1 \theta^2(\tau,x)\,\mathrm{d}x\,\mathrm{d}\tau. \qquad (3.28)
\end{aligned}$$

Multiplying (3.27) with a suitably large positive constant, and combing with (3.28), we get

$$\begin{aligned}
&\int_0^1 \theta^2(\tau,x)\,\mathrm{d}x + \int_0^1 u^4(t,x)\,\mathrm{d}x + \int_0^t\int_0^1 u^2u_x^2(\tau,x)\,\mathrm{d}x\,\mathrm{d}\tau \\
&\qquad + \int_0^t\int_0^1 \theta_x^2(\tau,x)\,\mathrm{d}x\,\mathrm{d}\tau \\
&\quad\leqslant \Lambda + \Lambda\int_0^t \max_{[0,1]} u^2(\tau,\cdot)\int_0^1 \theta^2(\tau,x)\,\mathrm{d}x\,\mathrm{d}\tau. \qquad (3.29)
\end{aligned}$$

Applying Gronwall's inequality to (3.29) and using (3.24), we obtain (3.25).

Step 3.

$$\int_0^t\int_0^1 u_x^2(\tau,x)\,\mathrm{d}x\,\mathrm{d}\tau \leqslant \Lambda. \qquad (3.30)$$

Multiplying $(3.1)_2$ by u and integrating over $[0,t]\times[0,1]$, it follows, with the help of (3.4), (3.18) and (3.19), that

$$\frac{1}{2}\int_0^1 u^2(t,x)\,\mathrm{d}x + \mu_1\int_0^t\int_0^1 u_x^2(\tau,x)\,\mathrm{d}x\,\mathrm{d}\tau \leqslant \Lambda + \int_0^t\int_0^1 f(v)\theta u_x\,\mathrm{d}x\,\mathrm{d}\tau.$$

It can be shown that

$$\begin{aligned}
&\int_0^t\int_0^1 f(v)\theta u_x\,\mathrm{d}x\,\mathrm{d}\tau \\
&\quad\leqslant \frac{\mu_1}{4}\int_0^t\int_0^1 u_x^2(\tau,x)\,\mathrm{d}x\,d\tau + \Lambda\int_0^t\int_0^1 (\theta-\bar\theta)^2(\tau,x)\,\mathrm{d}x\,\mathrm{d}\tau \\
&\qquad + \int_0^t\int_0^1 f(v)\bar\theta v_t(\tau,x)\,\mathrm{d}x\,\mathrm{d}\tau,
\end{aligned}$$

where

$$\bar{\theta}(t) = \int_0^1 \theta(t, x)\,\mathrm{d}x, \quad t \in [0, \infty).$$

By the mean value theorem, there exists a $z(t) \in [0, 1]$ such that $\bar{\theta}(t) = \theta(t, z(t))$. Thus

$$\begin{aligned} |\theta - \bar{\theta}|(\tau, x) &= \left| \int_{z(\tau)}^{x} \theta_x(\xi, \tau)\,\mathrm{d}\xi \right| \\ &\leqslant \left[\int_0^1 \theta_x^2(\tau, x)\,\mathrm{d}x \right]^{1/2}, \quad \tau \in [0, \infty). \end{aligned} \tag{3.31}$$

Integrating $(3.1)_3$ over $[0, 1]$ and using (3.4) and (3.13), we arrive at

$$\bar{\theta}_t(\tau) = -\frac{1}{c_V}\left[\int_0^1 \frac{1}{2} u^2(\tau, x)\,\mathrm{d}x \right]_t, \quad \tau \in [0, \infty), \tag{3.32}$$

which, together with (3.21), implies

$$\bar{\theta}(\tau) = \frac{1}{c_V}\left[E_0 - \int_0^1 \frac{1}{2} u^2(\tau, x)\,\mathrm{d}x \right], \quad \tau \in [0, \infty). \tag{3.33}$$

This, together with $(3.1)_1$, (3.18), (3.21) and (3.24), implies, upon integrating by parts and using the Cauchy–Schwarz inequality, that

$$\begin{aligned} &\int_0^t \int_0^1 \left[f(v)\bar{\theta} v_t \right](\tau, x)\,\mathrm{d}x\,\mathrm{d}\tau \\ &\quad \leqslant \Lambda - \int_0^t \int_0^1 \left\{ f'(v) v v_t \left[\frac{1}{c_V}\left(E_0 - \int_0^1 \frac{1}{2} u^2(\tau, x)\,\mathrm{d}x \right) \right] \right\}(\tau, x)\,\mathrm{d}x\,\mathrm{d}\tau \\ &\quad\quad + \int_0^t \int_0^1 \left\{ f(v) v \left[\frac{1}{c_V}\left(\int_0^1 \frac{1}{2} u^2(\tau, x)\,\mathrm{d}x \right) \right]_\tau \right\}(\tau, x)\,\mathrm{d}x\,\mathrm{d}\tau \\ &\quad \leqslant \Lambda + \frac{\mu_1}{4} \int_0^t \int_0^1 u_x^2(\tau, x)\,\mathrm{d}x\,\mathrm{d}\tau + \frac{1}{4} \int_0^1 u^2(\tau, x)\,\mathrm{d}x, \end{aligned}$$

which, combined to the estimates obtained above, implies (3.30).

Step 4.

$$\int_0^t \int_0^1 \left[f(v)\theta \right]^2(\tau, x)\,\mathrm{d}x\,\mathrm{d}\tau \leqslant \Lambda. \tag{3.34}$$

By integrating $(3.1)_2$ over $[0, x]$ for any $x \in [0, 1]$ and using the boundary conditions (3.4), we obtain

$$\begin{aligned} &[f(v)\theta](t, x) \\ &\quad = [\hat{\mu}(v)u_x](t, x) - \left(\int_0^x u(t, y)\,\mathrm{d}y\right)_t, \quad t \in [0, \infty), x \in [0, 1]. \end{aligned} \tag{3.35}$$

Multiplying (3.35) by $f(v)\theta$ and integrating it over $[0, t] \times [0, 1]$, we arrive at

$$\begin{aligned} &\int_0^t \int_0^1 [f(v)\theta]^2(\tau, x)\,\mathrm{d}x\,\mathrm{d}\tau \\ &\quad = \int_0^t \int_0^1 [\hat{\mu}(v)u_x f(v)\theta](\tau, x)\,\mathrm{d}x\,\mathrm{d}\tau \\ &\qquad - \int_0^t \int_0^1 \left\{\left[\int_0^x u(\tau, y)\,\mathrm{d}y\right]_\tau f(v)\left(\theta - \bar{\theta}\right)\right\}(\tau, x)\,\mathrm{d}x\,\mathrm{d}\tau \\ &\qquad + \int_0^t \int_0^1 \left\{\left[\int_0^x u(\tau, y)\,\mathrm{d}y\right]_\tau f(\theta)\bar{\theta}\right\}(\tau, x)\,\mathrm{d}x\,\mathrm{d}\tau. \end{aligned} \tag{3.36}$$

Next, we estimate each term separately. In view of (3.18), (3.30) and the Cauchy–Schwarz inequality,

$$\begin{aligned} &\left|\int_0^t \int_0^1 [\hat{\mu}(v)u_x f(v)\theta](\tau, x)\,\mathrm{d}x\,\mathrm{d}\tau\right| \\ &\quad \leqslant \frac{1}{4}\int_0^t \int_0^1 [f(v)\theta]^2(\tau, x)\,\mathrm{d}x\,\mathrm{d}\tau + \Lambda, \quad t \in [0, \infty). \end{aligned}$$

By (3.18), (3.25), (3.30), (3.31), (3.35) and the Cauchy–Schwarz inequality,

$$\begin{aligned} &\left|\int_0^t \int_0^1 \left\{\left[\int_0^x u(t, y)\,\mathrm{d}y\right]_\tau f(v)\left(\theta - \bar{\theta}\right)\right\}(\tau, x)\,\mathrm{d}x\,\mathrm{d}\tau\right| \\ &\quad \leqslant \frac{1}{8}\int_0^t \int_0^1 [f(v)\theta]^2(\tau, x)\,\mathrm{d}x\,\mathrm{d}\tau + \Lambda \end{aligned}$$

for all $t \in [0, \infty)$. Integrating by parts and using $(3.1)_1$, (3.21), (3.24), (3.30), (3.32) and (3.33), (3.35), the Hölder inequality and Cauchy–Schwarz inequality, we obtain

$$\begin{aligned} &\left|\int_0^t \int_0^1 \left\{\left[\int_0^x u(\tau, y)\,\mathrm{d}y\right]_\tau f(v)\bar{\theta}\right\}(\tau, x)\,\mathrm{d}x\,\mathrm{d}\tau\right| \\ &\quad \leqslant \Lambda + \frac{1}{8}\int_0^t \int_0^1 [f(v)\theta]^2(\tau, x)\,\mathrm{d}x\,\mathrm{d}\tau. \end{aligned}$$

Inequality (3.34) follows then from these estimates on each term of the right-hand side of (3.36).

Step 5.

$$\int_0^1 u^2(t,x)\,\mathrm{d}x \to 0, \quad \text{as } t \to \infty. \tag{3.37}$$

It is clear from (3.24) that $\int_0^1 u^2(t,x)\,\mathrm{d}x \in L^1(0,\infty)$.

Multiplying $(3.1)_2$ by u and integrating over $[0,1]$, we obtain, with the help of (3.13) and (3.4), that

$$\left|\int_0^1 (uu_t)(t,x)\,\mathrm{d}x\right| \leqslant \Lambda \int_0^1 \left[\left(f(v)\theta\right)^2 + u_x^2\right](t,x)\,\mathrm{d}x.$$

Combined with (3.30) and (3.34), it implies

$$\int_0^\infty \left|\frac{\mathrm{d}}{\mathrm{d}t}\int_0^1 u^2(t,x)\,\mathrm{d}x\right|\mathrm{d}t \leqslant \Lambda.$$

Thus, (3.37) follows.

Step 6. It holds that

$$\bar\theta(t) = \int_0^1 \theta(t,x)\,\mathrm{d}x \to \frac{E_0}{c_V}, \qquad \text{as } t \to \infty. \tag{3.38}$$

$$\int_0^t\int_0^1 \left[f(v)\right]^2(\tau,x)\,\mathrm{d}x\,\mathrm{d}\tau \leqslant \Lambda, \quad t \in [0,\infty). \tag{3.39}$$

$$\left\|f(v)(t,\cdot)\right\| \to 0, \qquad \text{as } t \to \infty. \tag{3.40}$$

$$\left\|\left(f(v)\theta\right)(t,\cdot)\right\|_{L^1} \to 0, \qquad \text{as } t \to \infty. \tag{3.41}$$

It is easy to see that (3.38) follows from (3.33) and (3.37) directly. It is known from (3.38) that there exists $T_0 > 0$ such that

$$\bar\theta(t) \geqslant \frac{E_0}{2c_V}, \quad \text{as } t \geqslant T_0.$$

Together with (3.18), (3.25), (3.31) and (3.34), it implies

$$\int_{T_0}^t\int_0^1 \left[f(v)\right]^2(\tau,x)\,\mathrm{d}x\,\mathrm{d}\tau \leqslant \frac{4c_V^2}{E_0^2}\int_{T_0}^t\int_0^1 \left[f(v)\bar\theta\right]^2(\tau,x)\,\mathrm{d}x\,\mathrm{d}\tau \leqslant \Lambda,$$

which and (3.18) yield (3.39).

To prove (3.40), we use $(3.1)_1$, (3.18), (3.30) and (3.39) to deduce the following estimate

$$\int_0^\infty \left| \frac{\mathrm{d}}{\mathrm{d}t} \int_0^1 \left[f(v)\right]^2(t,x)\,\mathrm{d}x \right| \mathrm{d}t \leqslant \Lambda + \Lambda \int_0^\infty \int_0^1 u_x^2(t,x)\,\mathrm{d}x\,\mathrm{d}t \leqslant \Lambda.$$

This together with (3.39) implies (3.40) directly. The limit (3.41) follows from (3.25) and (3.40) easily. Thus, the part I of Theorem 3.2 is established by the above estimates.

Next, we employ an idea of Andrews and Ball (see [8]) and the above results to prove the part II of Theorem 3.2.

Suppose $\Psi \in L^2(0,1)$ with $\Psi \geqslant 0$ and $\Phi \in C^2([\bar{v}, \overline{V}])$ satisfying

$$\Phi'(t) f(z) \geqslant 0 \quad \text{for } z \in \left[\bar{v}, \overline{V}\right], \tag{3.42}$$

where $\bar{v}$ and $\overline{V}$ are the lower and upper bounds of v.

Let

$$\varphi(t,x) := \int_0^x \Psi(y) \frac{\Phi'(v(t,y))}{\hat{\mu}(v(t,y))}\,\mathrm{d}y \quad \text{for } x \in [0,1], t \geqslant 0.$$

Multiplying $(3.1)_2$ with φ and integrating over $[0,t] \times [0,1]$ and using the boundary conditions (3.4), we get, with the help of (3.1) and integration by parts, that

$$\begin{aligned}
&\int_0^t \int_0^1 \left[\big(f(v)\theta\big) \Psi(x) \frac{\Phi'(v)}{\hat{\mu}(v)} \right](\tau,x)\,\mathrm{d}x\,\mathrm{d}\tau \\
&\quad = \int_0^1 u(t,x) \left[\int_0^x \Psi(y) \frac{\Phi'(v(t,y))}{\hat{\mu}(v(t,y))}\,\mathrm{d}y \right] \mathrm{d}x \\
&\qquad - \int_0^1 u_0(x) \left[\int_0^x \Psi(y) \frac{\Phi'(v_0(y))}{\hat{\mu}(v_0(y))}\,\mathrm{d}y \right] \mathrm{d}x \\
&\qquad - \int_0^t \int_0^1 u(\tau,x) \left[\int_0^x \Psi(x) \left(\frac{\Phi'(v)}{\hat{\mu}(v)} \right)' (\tau,y) u_x(\tau,y)\,\mathrm{d}y \right] \mathrm{d}x\,\mathrm{d}\tau \\
&\qquad + \int_0^1 \Psi(x) \Phi\big(v(t,x)\big)\,\mathrm{d}x - \int_0^1 \Psi(x) \Phi\big(v_0(x)\big)\,\mathrm{d}x,
\end{aligned} \tag{3.43}$$

where "$'$" denotes the differentiation with respect to v.

To show the existence of the limit of the left-hand side of (3.43) as $t \to \infty$, we estimate each term on the right-hand side of (3.43).

For the first term, it is easy to see that

$$\left| \int_0^1 u(t,x) \left[\int_0^x \Psi(y) \frac{\Phi'(v(t,y))}{\hat{\mu}(v(t,y))}\,\mathrm{d}y \right] \mathrm{d}x \right| \leqslant \Lambda \left\| u(t,\cdot) \right\|,$$

which tends to zero as $t \to \infty$, due to (3.37).

The third term can be treated as follows:

$$\left|\int_0^1 u(t,x)\left[\int_0^x \Psi(y)\left(\frac{\Phi'(v)}{\hat{\mu}(v)}\right)'(t,y)u_x(t,y)\,\mathrm{d}y\right]\mathrm{d}x\right|$$
$$\leqslant \Lambda\left[\left\|u(t,\cdot)\right\|^2+\left\|u_x(t,\cdot)\right\|^2\right] \quad \text{for all } t \geqslant 0.$$

Therefore, the limit of the third term as $t \to \infty$ exists by (3.24), (3.30) and the dominated convergence theorem.

It is obvious that the term $\mu_0 \int_0^1 \Psi(x)\Phi(v(t,x))\,\mathrm{d}x$ is uniformly bounded in $t \geqslant 0$ since (3.18). Thus, the above estimates imply that

$$\int_0^t \int_0^1 \left[\left(f(v)\theta\right)\Psi(x)\frac{\Phi'(v)}{\hat{\mu}(v)}\right](\tau,x)\,\mathrm{d}x\,\mathrm{d}\tau$$

is bounded uniformly in $t \geqslant 0$. This, together with (3.42), (3.2) and $\theta > 0$, yields the existence of

$$\lim_{t\to\infty}\int_0^t \int_0^1 \left[f(v)\theta\Psi(x)\Phi'(v)\right](\tau,x)\,\mathrm{d}x\,\mathrm{d}\tau.$$

Furthermore, the existence of

$$\lim_{t\to\infty}\int_0^1 \Psi(x)\Phi\left(v(t,x)\right)(\tau,x)\,\mathrm{d}x$$

for all $\Psi \in L^2(0,1)$ with $\Psi \geqslant 0$, is established since each term in (3.43), apart from $\int_0^1 \Psi(x)\Phi(v(t,x))(\tau,x)\,\mathrm{d}x$, is either independent of t or tends to a limit as $t \to \infty$. Therefore, it follows that

$$\Phi\left(v(t,\cdot)\right) \rightharpoonup g_\Phi(\cdot) \quad \text{in } L^2(0,1),$$

as $t \to \infty$, for some $g_\Phi \in L^2(0,1)$.

In view of $\|\Phi(v(t,\cdot))\|_{L^\infty} \leqslant \Lambda$, we can show that

$$g_\Phi \in L^\infty(0,1)$$

and

$$\Phi\left(v(t,\cdot)\right) \overset{*}{\rightharpoonup} g_\Phi(\cdot) \quad \text{in } L^\infty(0,1). \tag{3.44}$$

Now, let $\Phi \in C([\bar{v},\overline{V}])$ be arbitrary and $\Psi \in L^1(0,1)$. By using the same method in [8] and the following Lemma 3.6 (which can be proved by the same argument as used for Lemma 3.1 in [8]), it is easy to verify that

$$\lim_{t\to\infty}\int_0^1 \Psi(x)\Phi\left(v(t,x)\right)(\tau,x)\,\mathrm{d}x$$

exists for all $\Psi \in L^1(0,1)$ and $\Phi \in C([\bar{v}, \overline{V}])$.

LEMMA 3.6. *Let $f \in C(\mathbb{R})$ and let $0 < \bar{v} < \overline{V}$. Then the set*

$$S = \operatorname{span}\{\Phi \in C^2([\bar{v}, \overline{V}])\colon\ \Phi'(z) f(z) \geqslant 0,\ \text{if } z \in [\bar{v}, \overline{V}]\}$$

is dense in $C([\bar{v}, \overline{V}])$.

Thus, it turns that (3.44) holds for an arbitrary $\Psi \in C([\bar{v}, \overline{V}])$. The existence of probability measures ν_x follows from (3.44) and Theorem 5 in Tartar's paper [216]. To prove that $\operatorname{supp} \nu_x \subset K = \{z\colon f(z) = 0\}$ a.e., it suffices to show if Φ is zero on K, then $\langle \nu_x, \Phi \rangle = 0$, a.e. But, if Φ is zero on K, then $\Phi(v(t,\cdot)) \to 0$ in measure as $t \to \infty$ due to (3.40). Therefore, $\Phi(v(t,\cdot)) \rightharpoonup 0$ in $L^\infty(0,1)$ as $t \to \infty$, and hence $\langle \nu_x, \Phi \rangle = 0$, a.e., as required.

Thus, the proof of Theorem 3.2 is complete. □

A similar result on large-time behavior of solutions has been established by Hsiao and Jian in [100] to deal with the following boundary conditions

$$\begin{aligned} &\sigma(t,0) = \gamma u(t,0), \qquad \sigma(t,1) = -\gamma u(t,1), \\ &\theta(t,0) = \theta(t,1) = T_0, \end{aligned} \qquad t \geqslant 0,$$

where $\gamma = 0$ or $\gamma = 1$, and $T_0 > 0$ is the reference temperature. Physically, the case with $\gamma = 1$ represents that the endpoints of the interval $[0,1]$ are connected to some sort of dash pot.

Other study on the large-time behavior of solutions for thermoviscoelasticity can be found in [194,204], etc.

REMARK 3.4. There is an extensive body of work in the isothermal setting, namely the equations of viscoelasticity, which addresses the issues of unique global solvability and temporal-asymptotic behavior, see, for example, [7–9,30,68,75,130,184] and the references therein.

The last part of this section is devoted to the study on various behavior of solutions corresponding to different boundary conditions. Such understanding is quite important not only in mathematical theory but also in applications. For instance, the outer pressure problem,

$$q(t,0) = q(t,1) = 0, \qquad \sigma(t,0) = \sigma(t,1) = -R(t) < 0, \tag{3.45}$$

with the outer pressure $R(t)$ being a given function, possesses a completely different large-time behavior comparing to the problem with classical *homogeneous* boundary conditions, e.g.,

$$q(t,0) = q(t,1) = 0, \qquad \sigma(t,0) = \sigma(t,1) = 0,$$

which has been discussed before. We mention that there are a number of articles dedicating to the study of the large-time behavior of strong or smooth solutions to initial boundary value problem with the classical homogeneous boundary conditions and to the Cauchy problem, we refer, for example, to [10,19,62,116,117,172].

For the outer pressure problem, the large-time behavior is analyzed by Nagasawa [171] for an ideal gas, by Hsiao and Luo [101] for a real gas, and by Ducomet [42,45] for models of real reactive fluids arising from astrophysics.

The global existence for the outer pressure problem has been obtained by Nagasawa [170] and Luo [154], corresponding to the case of an ideal gas and real gas, respectively.

A real gas is well approximated by an ideal gas within moderate ranges of θ and v. However, it becomes inadequate for high temperature and density, since the specific heat, conductivity and viscosity vary with temperature and density.

The result on the asymptotic behavior of solutions to (3.1), (3.5), (3.45), established in [101], is for a real gas with the following constitutive relations:

$$\begin{cases} \sigma = -\hat{p}(v,\theta) + \frac{\mu(v,\theta)}{v} u_x, \\ q = -\frac{k(v,\theta)}{v} \theta_x, \\ e = \hat{e}(v,\theta), \end{cases} \tag{3.46}$$

where $\hat{e}(v,\theta)$, $\hat{p}(v,\theta)$, $\mu(v,\theta)$, and $k(v,\theta)$ are twice continuously differentiable on $0 < v < \infty$ and $0 \leqslant \theta < \infty$.

Since the internal energy may only take nonnegative values, we assume

$$\hat{e}(v,\theta) \geqslant 0, \quad 0 < v < \infty, 0 \leqslant \theta < \infty. \tag{3.47}$$

For compatibility with the second law of thermodynamics (3.2), we need $\mu(v,\theta) > 0$, $k(v,\theta) \geqslant 0$ and (3.8).

Finally, we impose upon $\hat{e}(v,\theta)$, $\hat{p}(v,\theta)$ and $k(v,\theta)$ the following growth condition: there are exponents $r \in [0,1]$, $s \geqslant 2 + 2r$ and positive constants ν, k_0, p_i $(i = 1,2,3,4)$, and there are positive constants $N(\underline{v})$ and $k_1(\underline{v})$ for any $\underline{v} > 0$ such that, for any $v \geqslant \underline{v}$ and $\theta \geqslant 0$,

$$\nu(1+\theta^r) \leqslant \hat{e}_\theta(v,\theta) \leqslant N(\underline{v})(1+\theta^r), \tag{3.48}$$

$$-\frac{p_2[l + (1-l)\theta + \theta^{1+r}]}{v^2} \leqslant \hat{p}_v(v,\theta) \leqslant -\frac{p_1[l + (1-l)\theta + \theta^{1+r}]}{v^2}, \quad l = 0 \text{ or } l = 1, \tag{3.49}$$

$$\frac{p_4[l + (1-l)\theta + \theta^{1+r}]}{v} \leqslant \hat{p}(v,\theta) \leqslant \frac{p_3[l + (1-l)\theta + \theta^{1+r}]}{v}, \quad l = 0 \text{ or } l = 1, \tag{3.50}$$

and

$$|\hat{p}_\theta(v,\theta)| \leqslant N(\underline{v})(1+\theta^r), \tag{3.51}$$

$$k_0(1+\theta^s) \leqslant k(v,\theta) \leqslant k_1(\underline{v})(1+\theta^s), \tag{3.52}$$

$$|k_v(v,\theta)| + |k_{vv}(v,\theta)| \leqslant k_1(\underline{v})(1+\theta^s). \tag{3.53}$$

The above assumptions are motivated by the facts in [13] and [236]. In these books, it is pointed out that $\hat{e}$ grows as θ^{1+r} with $r \approx 0.5$ and k increases like θ^s with $4.5 \leqslant s \leqslant 5.5$.

For simplicity, we only discuss the case of

$$\mu(v,\theta) \equiv \mu_0 > 0. \tag{3.54}$$

For the outer pressure $R(t)$, we assume that

$$m_R = \inf_{t\in[0,\infty)} R(t) > 0 \tag{3.55}$$

$$TV(R) = \int_0^\infty |R'(t)|\,\mathrm{d}t < \infty, \tag{3.56}$$

which implies that there exist $\overline{R}$ and M_R such that $\overline{R} = \lim_{t\to\infty} R(t) > 0$ and $0 < M_R = \sup_{t\in[0,\infty)} R(t) < \infty$.

The initial data are normalized as in (3.16) and compatible with the boundary conditions (3.45).

The global existence of smooth solutions to the initial-boundary problem (3.1), (3.5), (3.45) is established in [154] under the above assumptions. The theorem on the asymptotic behavior of solutions, established in [101], is stated as

THEOREM 3.7. *Under the assumptions* (3.46)–(3.56) *and* (3.16), *it holds, for the globally defined solution of* (3.1), (3.5), (3.45), *that*

$$0 < \underline{v} \leqslant v(t,x) \leqslant \bar{v} \quad \textit{for } t \in [0,\infty),\ x \in [0,1], \tag{3.57}$$

where $\underline{v}$ and $\bar{v}$ are positive constants independent of t,

$$\int_0^x u(t,y)\,\mathrm{d}y \to 0, \quad \textit{as } t \to \infty, \textit{ uniformly with respect to } x \in [0,1], \tag{3.58}$$

$$\left\{v(t,x) - \int_0^x v(t,x)\,\mathrm{d}x\right\} \to 0, \quad \textit{as } t \to \infty, \tag{3.59}$$

uniformly with respect to $x \in [0,1]$,

$$\left\{v(t,x) - \frac{\int_0^\infty R'(\tau)\int_0^1 v(\tau,x)\,\mathrm{d}x\,\mathrm{d}\tau}{\overline{R}} + \frac{\int_0^t Y(\tau)\hat{e}(v,\theta)(\tau,x)\,\mathrm{d}\tau}{\mu_0 Y(t)}\right\} \to \frac{E_1}{\overline{R}}, \tag{3.60}$$

as $t \to \infty$, *uniformly with respect to* $x \in [0, 1]$, *where*

$$E_1 = \int_0^1 \left[\hat{e}\big(v_0(x), \theta_0(x)\big) + \frac{u_0^2(x)}{2} + R(0)v_0(x) \right] \mathrm{d}x,$$

$$Y(t) = \exp\left\{ \int_0^t \frac{R(\tau)}{\mu_0} \,\mathrm{d}\tau \right\}, \quad t \geqslant 0.$$

Moreover,

$$\int_0^t \frac{Y(\tau)}{Y(t)} v(\tau, x)\big[\hat{p}(v, \theta)(\tau, x) - R(\tau)\big] \mathrm{d}\tau \to 0, \quad \textit{as } t \to \infty, \tag{3.61}$$

uniformly with respect to $x \in [0, 1]$.

REMARK 3.5. In contrast with the case of $R(t) \equiv 0$, in which the specific volume v satisfies $v(t, x) \geqslant C(\log(1 + t))^{\bar{k}}$ with some constants $C > 0$ and $\bar{k} \geqslant 1$ for the ideal gas [169], (3.57) disclose that $v(t, x)$ is uniformly bounded even for the ideal gas if the boundary conditions are (3.45) instead of (3.4).

Equation (3.60) describes the asymptotic relation between $v(t, x)$, $\theta(t, x)$ and $R(t)$ as $t \to \infty$. (3.61) can be understood as follows: With a weight function $\frac{Y(\tau)}{Y(t)} v(\tau, x)$, the weighted average of $[\hat{p}(v, \theta)(\tau, x) - R(\tau)]$ over $[0, t]$ tends to zero as $t \to \infty$, uniformly in $x \in [0, 1]$. For the weight function, it is known that

$$\underline{v} \exp\big[-M_r(t - \tau)\big] \leqslant \frac{Y(\tau)}{Y(t)} v(\tau, x) \leqslant \bar{v} \exp\big[-m_R(t - \tau)\big].$$

PROOF OF THEOREM 3.7. For the proof, we need a sequence of estimates to which we give an outline in Steps (1)–(6). The detail can be found in [101] and [99].

Step (1). It holds that

$$\begin{aligned} &\int_0^1 \left[\left(\hat{e}(v, \theta) + \frac{u^2}{2} \right)(t, x) + R(t)v(t, x) \right] \mathrm{d}x \\ &\quad = E_1 + \int_0^t R'(t) \int_0^1 v(\tau, x) \,\mathrm{d}x \,\mathrm{d}\tau, \end{aligned} \tag{3.62}$$

$$\int_0^1 \left[\hat{e}(v, \theta) + \frac{u^2}{2} + v \right](t, x) \,\mathrm{d}x \leqslant C, \tag{3.63}$$

$$\begin{aligned} &\int_0^1 \left(\theta + \theta^{1+r} + \frac{u^2}{2} \right)(t, x) \,\mathrm{d}x \\ &\quad + \int_0^t \int_0^1 \left[\frac{\mu_0 u_x^2}{v\theta} + \frac{k_0(1 + \theta^s)}{v\theta^2} \theta_x^2 \right] \mathrm{d}x \,\mathrm{d}s \leqslant C, \end{aligned} \tag{3.64}$$

where C is a generic constant independent of t.

It is easy to get (3.62) and (3.63). To establish (3.64), we need some estimates related to the Helmholtz free energy function.

Let $\psi(v,\theta)=\hat{e}(v,\theta)-\theta\eta(v,\theta)$ denote the Helmholtz free energy function. It is known from (3.3) that (see (2.9))

$$\begin{aligned}
&-\psi_\theta(v,\theta)=\eta(v,\theta),\\
&\psi_v(v,\theta)=-\hat{p}(v,\theta),\\
&\psi_{\theta\theta}(v,\theta)=-\frac{\hat{e}_\theta(v,\theta)}{\theta}.
\end{aligned}\tag{3.65}$$

Define

$$E(v,\theta)=\psi(v,\theta)-\psi(1,1)-\psi_v(1,1)(v-1)-\psi_\theta(v,\theta)(\theta-1).$$

It can be shown, on account of (3.1), (3.46), (3.65), (3.45), (3.54), (3.55) and (3.63), that

$$\begin{aligned}
&\int_0^1\left[E(v,\theta)+\frac{u^2}{2}\right](t,x)\,\mathrm{d}x\\
&\quad+\int_0^t\int_0^1\left[\frac{\mu_0 u_x^2}{v\theta}+\frac{k(v,\theta)}{v\theta^2}\theta_x^2\right](\tau,x)\,\mathrm{d}x\,\mathrm{d}\tau\leqslant C_0.
\end{aligned}\tag{3.66}$$

In view of (3.48), (3.49) and (3.65), it follows from Taylor's theorem that

$$\begin{aligned}
&E(v,\theta)-\psi(v,\theta)+\psi(v,1)+(\theta-1)\psi_\theta(v,\theta)\geqslant 0,\\
&\psi(v,\theta)-\psi(v,1)-(\theta-1)\psi_\theta(v,\theta)\\
&\quad\geqslant \nu(\theta-\log\theta-1)+\begin{cases}\nu(\theta-\log\theta-1), & r=0,\\ \frac{\theta^{r+1}}{r+1}+\frac{\frac{1}{r+1}-\theta^r}{r}, & r>0.\end{cases}
\end{aligned}$$

Thus,

$$E(v,\theta)\geqslant \nu(\theta-\log\theta-1),\quad r=0,\tag{3.67}$$

or

$$E(v,\theta)\geqslant C\theta+C\theta^{1+r}-C+\nu(\theta-\log\theta-1),\quad 0<r\leqslant 1.\tag{3.68}$$

This, together with (3.66), yields (3.64).

Step (2). Let α and β be the two positive roots of equation

$$\theta-\log\theta-1=\frac{C_0}{\nu},$$

where C_0 is the same positive constant as in (3.66). Then, there exists an $a(t) \in [0, 1]$ for each $t \geqslant 0$ such that

$$0 < \alpha \leqslant \theta\big(a(t), t\big) \leqslant \beta. \tag{3.69}$$

Furthermore, it holds, for any $t \geqslant 0$, that

$$\alpha \leqslant \int_0^1 \theta(t, x)\,\mathrm{d}x \leqslant \beta. \tag{3.70}$$

It is clear that (3.66)–(3.68) yield

$$\nu \int_0^1 \big[\theta(t, x) - \log\theta(t, x) - 1\big]\,\mathrm{d}x \leqslant C_0, \quad t \geqslant 0, \tag{3.71}$$

from which (3.69) follows. Using (3.71) and applying Jensen's inequality to the convex function $U - \log U - 1$, we deduce

$$\int_0^1 \theta(t, x)\,\mathrm{d}x - \log\int_0^1 \theta(t, x)\,\mathrm{d}x - 1 \leqslant \frac{C_0}{\nu},$$

which implies (3.71).

Step (3). There exist positive constants $\underline{v} > 0$ and $\bar{v} > 0$, independent of t, such that for any $(t, x) \in [0, \infty) \times [0, 1]$ it holds

$$\underline{v} \leqslant v(t, x) \leqslant \bar{v}. \tag{3.72}$$

To prove (3.72), we find the expression of $v(t, x)$ first, namely

$$v(t, x) = \frac{v_0(x) + \frac{1}{\mu_0}\int_0^t B(\tau, x) Y(\tau) \hat{p}(v, \theta)(\tau, x)\,\mathrm{d}\tau}{B(t, x) Y(t)}, \tag{3.73}$$

where

$$B(t, x) = \exp\left\{\frac{1}{\mu_0}\int_0^x \big[u_0(y) - u(t, y)\big]\,\mathrm{d}y\right\},$$

$$Y(t) = \exp\left\{\frac{1}{\mu_0}\int_0^t R(\tau)\,\mathrm{d}\tau\right\}.$$

Next, we establish the upper bound of v. With the help of (3.64), it is easy to estimate $B(t, x)$ as follows

$$0 < C^{-1} \leqslant B(t, x) \leqslant C \quad \text{for } (t, x) \in [0, \infty) \times [0, 1].$$

Using (3.63), the Cauchy–Schwarz inequality and the condition $s \geqslant 2+2r$, we can show that

$$\left|\theta^{(r+1)/2}(t,x)-\theta^{(r+1)/2}\big(t,a(t)\big)\right| \leqslant CV^{1/2}(t), \tag{3.74}$$

where

$$V(t)=\int_0^1\left[\frac{u_x^2}{v\theta}+\frac{(1+\theta^s)}{v\theta^2}\theta_x^2\right](t,x)\,\mathrm{d}x,$$

for which it holds

$$\int_0^t V(\tau)\,\mathrm{d}\tau \leqslant C \quad \text{for any } t \geqslant 0, \tag{3.75}$$

due to (3.64). This and (3.69) yield

$$\frac{\alpha^{r+1}}{2}-CV(t)\leqslant \theta^{r+1}(t,x)\leqslant 2\beta^{r+1}+V(t) \tag{3.76}$$

for $(t,x)\in[0,\infty)\times[0,1]$.

Using (3.50), we deal with $v\hat{p}(v,\theta)$ to get

$$\begin{aligned} p_4\big[l+(1-l)\theta+\theta^{1+r}\big] &\leqslant v\hat{p}(v,\theta)\\ &\leqslant p_3\big[l+(1-l)\theta+\theta^{1+r}\big], \quad l=0 \text{ or } l=1. \end{aligned} \tag{3.77}$$

Finally, (3.73)–(3.77) imply

$$v(t,x)\leqslant C. \tag{3.78}$$

Now, we estimate the positive lower bound of v. First, there exists $b(t)\in[0,1]$ for any $t\geqslant 0$ such that

$$\int_0^1 \theta(t,x)\,\mathrm{d}x=\theta\big(t,b(t)\big).$$

Thus, due to (3.63), (3.69) and (3.78),

$$\theta^{1/2}(t,x)=\theta^{1/2}\big(t,b(t)\big)+\int_{b(t)}^x \frac{\theta_x(t,y)}{2\theta^{1/2}(t,y)}\,\mathrm{d}y\geqslant C\big[1-V^{1/2}(t)\big],$$

which gives

$$\theta(t,x)\geqslant C\big(1-V(t)\big). \tag{3.79}$$

So, (3.73), (3.75)–(3.77) and (3.79) yield

$$v(t,x) \geqslant C \int_0^t \left\{ \exp\left[\frac{-M_R}{\mu_0}(t-\tau) \right] \right\} \left\{ C - CV(\tau) \right\} \mathrm{d}\tau. \tag{3.80}$$

In view of (3.75), we claim that

$$\lim_{t\to\infty} \int_0^t \left\{ \exp\left[\frac{-M_R}{\mu_0}(t-\tau) \right] \right\} V(\tau)\, \mathrm{d}\tau = 0. \tag{3.81}$$

It is proved in [154] that

$$v(t,x) \geqslant \underline{v}(t) > 0. \tag{3.82}$$

On account of (3.80)–(3.82), we end up with

$$v(t,x) \geqslant \underline{v} > 0 \quad \text{for any } (t,x) \in [0,\infty) \times [0,1].$$

Step (4).

$$\int_0^1 v_x^2(t,x)\,\mathrm{d}x + \int_0^t \int_0^1 \left(1+\theta^{1+r}\right) v_x^2(\tau,x)\,\mathrm{d}x\,\mathrm{d}\tau \leqslant C. \tag{3.83}$$

We rewrite $(3.1)_2$ into the form

$$\left(u - \mu_0 \frac{v_x}{v} \right)_t + \hat{p}_v(v,\theta) v_x = -\hat{p}_\theta(v,\theta)\theta_x. \tag{3.84}$$

Multiplying (3.84) by $(u - \mu_0 v_x/v)$ and integrating over $[0,t] \times [0,1]$, we get, due to (3.49), (3.51), (3.64) and the condition $s \geqslant 2+2r$, that

$$\begin{aligned} &\int_0^1 \left[\mu_0 (\log v)_x - u \right]^2 \mathrm{d}x + C_1 \int_0^t \int_0^1 \left[l + (1-l)\theta + \theta^{1+r} \right] v_x^2 \,\mathrm{d}x\,\mathrm{d}\tau \\ &\quad \leqslant \frac{C_1}{4} \int_0^t \int_0^1 \left[l + (1-l)\theta + \theta^{1+r} \right] v_x^2 \,\mathrm{d}x\,\mathrm{d}\tau + C \int_0^t \int_0^1 \left(l + \theta^{1+r} \right) v_x^2 \,\mathrm{d}x\,\mathrm{d}\tau \\ &\qquad + \frac{C_1}{8} \int_0^t \int_0^1 \left(l + \theta^{1+r} \right) v_x^2 \,\mathrm{d}x\,\mathrm{d}\tau + C. \end{aligned} \tag{3.85}$$

By virtue of (3.16), there exists $d(t) \in [0,1]$ for any $t \geqslant 0$ such that

$$u\big(t, d(t)\big) = 0.$$

Hence, $u(t,x)$ can be expressed as $u(t,x)=\int_{d(t)}^{x}u_x(t,y)\,\mathrm{d}y$. This, together with (3.64) and (3.72), yields

$$\int_0^t \max_{x\in[0,1]} u^2(\tau,x)\,\mathrm{d}\tau \leqslant C, \tag{3.86}$$

which, combined to (3.64), implies

$$\int_0^t\int_0^1 (1+\theta^r)u^2\,\mathrm{d}x\,\mathrm{d}\tau \leqslant C. \tag{3.87}$$

For the case of $l=1$ in (3.49), the desired result of (3.83) follows in view of (3.64), (3.72), (3.85) and (3.87).

The case of $l=0$ can be dealt with as follows, using (3.64) and the condition $s\geqslant 2+2r$:

$$\left|\int_0^t\int_0^1 \hat{p}_\theta(v,\theta)\theta_x\mu_0\frac{v_x}{v}\,\mathrm{d}x\,\mathrm{d}\tau\right| \leqslant \varepsilon\int_0^t\int_0^1 (\theta+\theta^{1+r})v_x^2\,\mathrm{d}x\,\mathrm{d}\tau + C, \tag{3.88}$$

where ε is a small positive constant.

By a similar approach as in (3.85), with the help of (3.87) and (3.88), we arrive at

$$\int_0^1 v_x^2\,\mathrm{d}x + \int_0^t\int_0^1 (\theta+\theta^{1+r})v_x^2\,\mathrm{d}x\,\mathrm{d}\tau \leqslant C. \tag{3.89}$$

Due to (3.79),

$$\theta(t,x)+CV(t)\geqslant C.$$

Thus,

$$v_x^2(t,x)\leqslant C\theta(t,x)v_x^2(t,x)+CV(t)v_x^2(t,x).$$

This, together with (3.64) and (3.89), gives

$$\int_0^t\int_0^1 v_x^2(t,x)\,\mathrm{d}x\,\mathrm{d}t\leqslant C,$$

which, combined to (3.89), implies (3.83).

Step (5).

$$\int_0^x u(t,y)\,\mathrm{d}y\to 0, \quad \text{as } t\to\infty, \text{ uniformly with respect to } x\in[0,1]. \tag{3.90}$$

Integrating $(3.1)_2$ over $[0, x]$ and using (3.5), we obtain

$$\left[\int_0^x u(t,y)\,\mathrm{d}y\right]_t = \left[-\hat{p}(v,\theta) + R(t)\right] + \mu_0 \frac{u_x}{v}. \tag{3.91}$$

Multiplying (3.91) by $\int_0^x u(t,y)\,\mathrm{d}y$, and integrating over [0, 1], we arrive at

$$\begin{aligned}\frac{1}{2}\left\{\int_0^1 \left(\int_0^x u\,\mathrm{d}y\right)^2 \mathrm{d}x\right\}_t = {} & \int_0^1 \left[-\hat{p}(v,\theta) + R\right]\left(\int_0^x u\,\mathrm{d}y\right)\mathrm{d}x \\ & - \mu_0 \int_0^1 \frac{u^2}{v}\,\mathrm{d}x + \mu_0 \int_0^1 \frac{u v_x}{v^2} \int_0^x u\,\mathrm{d}y\,\mathrm{d}x,\end{aligned}$$

from which we get, with the help of (3.64) and the inequality

$$\int_0^1 u^2\,\mathrm{d}x \geqslant \int_0^1 \left(\int_0^x u\,\mathrm{d}y\right)^2 \mathrm{d}x,$$

that

$$\begin{aligned}&\left[\int_0^1 \left(\int_0^x u\,\mathrm{d}y\right)^2 \mathrm{d}x\right]_t + C\int_0^1 \left(\int_0^x u\,\mathrm{d}y\right)^2 \mathrm{d}x \\ &\quad \leqslant C\left(\int_0^1 u^2\,\mathrm{d}x\right)^{1/2}\left\{1 + \int_0^1 (u^2 + v_x^2)\,\mathrm{d}x\right\}.\end{aligned} \tag{3.92}$$

Define

$$z(t) = \int_0^1 \left(\int_0^x u\,\mathrm{d}y\right)^2 \mathrm{d}x.$$

Inequality (3.92) means

$$z'(t) + Cz(t) \leqslant C\left(\int_0^1 u^2\,\mathrm{d}x\right)^{1/2}\left\{1 + \int_0^1 (u^2 + v_x^2)\,\mathrm{d}x\right\},$$

which implies

$$\begin{aligned}z(t) \leqslant {} & \mathrm{e}^{-Ct} z(0) + C\int_0^t \mathrm{e}^{-C(t-\tau)}\left(\int_0^1 u^2\,\mathrm{d}x\right)^{1/2} \mathrm{d}\tau \\ & + C\int_0^t \mathrm{e}^{-C(t-\tau)} \int_0^1 (u^2 + v_x^2)\,\mathrm{d}x\,\mathrm{d}\tau.\end{aligned}$$

With the help of (3.86), (3.83) and the following Lemma 3.8 established by Nagasawa in [171], it can be easily proved that $z(t) \to 0$ as $t \to \infty$. Namely,

$$\int_0^1 \left(\int_0^x u \,\mathrm{d}y \right)^2 \mathrm{d}x \to 0, \quad \text{as } t \to \infty. \tag{3.93}$$

LEMMA 3.8. *Let $\lambda(t)$ ($\geqslant 0$) and $w(t)$ be continuous functions satisfying that there exist positive constants C_i, $i = 1, 2, 3, 4$, such that*

$$C_1 \mathrm{e}^{C_2(t-\tau)} \leqslant \exp\left\{ \int_\tau^t w(\xi)\,\mathrm{d}\xi \right\} \leqslant C_3 \mathrm{e}^{C_4(t-\tau)} \quad \text{for } 0 \leqslant \tau \leqslant t.$$

Denote $\Lambda(t)$ by $\Lambda(t) := \int_t^{t+1} \lambda(\tau)\,\mathrm{d}\tau$. Then

$$\begin{aligned} C^{-1} \varliminf_{t\to\infty} \Lambda(t) &\leqslant \varliminf_{t\to\infty} \int_0^t \exp\left\{ -\int_\tau^t w(\xi)\,\mathrm{d}\xi \right\} \lambda(\tau)\,\mathrm{d}\tau \\ &\leqslant C \varlimsup_{t\to\infty} \Lambda(t). \end{aligned}$$

Now, using the embedding theorem $W^{1,1}(0,1) \hookrightarrow L^\infty(0,1)$, (3.93) and (3.64), we finish the proof of (3.90).

Step (6). It holds that

$$v(t,x) - \int_0^1 v(t,x)\,\mathrm{d}x \to 0, \quad \text{as } t \to \infty, \text{ uniformly in } x \in [0,1], \tag{3.94}$$

$$\begin{aligned} &v(t,x) + \frac{\int_0^t Y(\tau) \int_0^1 \hat{e}(v(\tau,x), \theta(\tau,x))\,\mathrm{d}x\,\mathrm{d}\tau}{\mu_0 Y(t)} \\ &\quad \to \frac{E_1 + \int_0^\infty R'(\tau) \int_0^1 v(\tau,x)\,\mathrm{d}x\,\mathrm{d}\tau}{\overline{R}}, \quad \text{as } t \to \infty, \text{ uniformly in } x \in [0,1], \end{aligned} \tag{3.95}$$

$$\begin{aligned} &\int_0^t \frac{Y(\tau)}{Y(t)} v(\tau,x) \big[\hat{p}(v,\theta)(\tau,x) - R(\tau) \big]\,\mathrm{d}\tau \\ &\quad \to 0, \quad \text{as } t \to \infty, \text{ uniformly in } x \in [0,1]. \end{aligned} \tag{3.96}$$

To prove (3.94), we rewrite (3.73) as

$$\begin{aligned} v(t,x) = {} & \frac{v_0}{B(t,x) Y(t)} + \frac{1}{\mu_0 Y(t)} \int_0^t \hat{p}(v,\theta) v(\tau,x) \left[\frac{B(\tau,x)}{B(t,x)} - 1 \right] Y(\tau)\,\mathrm{d}\tau \\ & + \frac{1}{\mu_0 Y(t)} \int_0^t \hat{p}(v,\theta) v(\tau,x) Y(\tau)\,\mathrm{d}\tau. \end{aligned} \tag{3.97}$$

It is easy to show that the first term on the right-hand side of (3.97) tends to zero as $t \to \infty$ uniformly in $x \in [0, 1]$. We can also prove that the second term tends to zero as $t \to \infty$, uniformly in $x \in [0, 1]$, with the help of (3.90), (3.50), (3.76), (3.75) and Lemma 3.8.

To estimate the third term, we rewrite it as

$$\begin{aligned}
&\frac{1}{\mu_0 Y(t)} \int_0^t \hat{p}(v,\theta)v(\tau,x)Y(\tau)\,\mathrm{d}\tau \\
&\quad = \frac{1}{\mu_0 Y(t)} \int_0^t \left[\hat{p}(v,\theta)v(\tau,x) - \int_0^1 \hat{p}(v,\theta)v(\tau,x)\,\mathrm{d}x\right] Y(\tau)\,\mathrm{d}\tau \\
&\qquad + \frac{1}{\mu_0 Y(t)} \int_0^t Y(\tau) \int_0^1 \hat{p}(v,\theta)v(\tau,x)\,\mathrm{d}x\,\mathrm{d}\tau.
\end{aligned} \tag{3.98}$$

For any $\tau \geqslant 0$, there exists $y(\tau) \in [0, 1)$ such that

$$\int_0^1 \hat{p}(v,\theta)v(\tau,x)\,\mathrm{d}x = \hat{p}\big(v(\tau, y(\tau)), \theta(\tau, y(\tau))\big)v\big(\tau, \eta(\tau)\big).$$

This, with the help of (3.49)–(3.51), (3.64), (3.83) and Lemma 3.8, makes possible to show that

$$\begin{aligned}
&\frac{1}{\mu_0 Y(t)} \int_0^t \left[\hat{p}(v,\theta)v(\tau,x) - \int_0^1 \hat{p}(v,\theta)v(\tau,x)\,\mathrm{d}x\right] Y(\tau)\,\mathrm{d}\tau \to 0, \\
&\text{as } t \to \infty, \text{ uniformly in } x \in [0, 1].
\end{aligned} \tag{3.99}$$

In order to deal with the second term on the right-hand side of (3.98), we multiply (3.91) by $v(t, x)$ to imply

$$\begin{aligned}
&\frac{1}{\mu_0 Y(t)} \int_0^t Y(\tau) \int_0^1 \hat{p}(v,\theta)v(\tau,x)\,\mathrm{d}x\,\mathrm{d}\tau \\
&\quad = -\frac{1}{\mu_0 Y(t)} \int_0^t Y(\tau) \int_0^1 \left[\int_0^x u(\tau,y)\,\mathrm{d}y\right]_t v(\tau,x)\,\mathrm{d}x\,\mathrm{d}\tau \\
&\qquad + \frac{1}{\mu_0 Y(t)} \int_0^t Y(\tau) \int_0^1 R(\tau)v(\tau,x)\,\mathrm{d}x\,\mathrm{d}\tau \\
&\qquad + \frac{1}{\mu_0 Y(t)} \int_0^t Y(\tau) \int_0^1 u_x(\tau,x)\,\mathrm{d}x\,\mathrm{d}\tau.
\end{aligned} \tag{3.100}$$

It can be shown, by using integrating by parts, (3.72), (3.90), (3.86) and Lemma 3.8, that the first term on the right-hand side of (3.100) tends to zero as $t \to \infty$, uniformly in $x \in [0, 1]$. With a similar approach, we can claim that the third term in (3.100) tends to zero as well as $t \to \infty$, uniformly in $x \in [0, 1]$.

From the above arguments, we end up with

$$v(t,x) - \frac{1}{\mu_0 Y(t)} \int_0^t Y(\tau)R(\tau) \int_0^1 v(\tau,x)\,dx\,d\tau \to 0,$$
$$\text{as } t \to \infty, \text{ uniformly in } x \in [0,1].$$

Therefore, (3.94) follows.

The arguments on getting (3.95) and (3.96) can be found in [101]. □

Another kind of boundary condition is discussed by Jiang in [110] and Wang in [226], namely,

$$\begin{aligned} &\theta_x(t,0) = \theta_x(t,1) = 0, \\ &u(t,0) = u(t,1) = 0, \end{aligned} \qquad t > 0. \tag{3.101}$$

In [110] the exponential stability as $t \to \infty$ of smooth solutions is obtained, while in [226] the main issue in the analysis is whether shock waves, vacuum, and concentration will be developed in the solutions in finite time without outer pressure, provided that the initial data are bounded, smooth, and do not contain vacuum. For the solutions with large initial data in H^1, it is shown that neither shock waves nor vacuum and concentration will be developed in finite time. This is also true for a free boundary condition (see [226]).

Related results are given by Ducomet, Zlotnik in a series of papers [40,42,43,45,46,49] on models with nonmonotone pressure arising from astrophysics and nuclear physics, where the global existence and time-asymptotic behavior of solutions to initial boundary value problems with classical and outer pressure boundary conditions are investigated. We also mention the work by Frid and Shelukhin [67,201,202] where the limit of zero shear viscosity for the compressible Navier–Stokes equations is studied.

4. Globally defined weak solutions for one-dimensional compressible flow with large discontinuous initial data

We discuss the one-dimensional isentropic viscous flow first in this section and turn to the nonisentropic case in the second part.

Consider the global existence of weak solutions to the classical Navier–Stokes equations with external force for one-dimensional isentropic compressible flow, which can be written, due to the discussion in Section 2, as

$$\begin{cases} \rho_t + (\rho u)_x = 0, \\ (\rho u)_t + \left(\rho u^2 + P(\rho)\right)_x = \mu u_{xx} + \rho f, \end{cases} \tag{4.1}$$

with initial data

$$\left(\rho(0,\cdot), u(0,\cdot)\right) = \left(\rho_0, u_0\right). \tag{4.2}$$

Here ρ and u denote the fluid density and velocity, μ is a positive viscosity constant, and $f = f(t,x)$ is the external force. $P = P(\rho)$ is the pressure, satisfying

$$P(\rho), \quad P'(\rho) > 0 \quad \text{for } \rho > 0. \tag{4.3}$$

Assume that there is a positive constant C such that

$$\rho + P(\rho) \leqslant C\big[1 + G(\rho, \rho')\big], \tag{4.4}$$

$$\liminf_{\rho\to 0} G(\rho, \rho') \geqslant C^{-1} \tag{4.5}$$

for ρ' between two fixed positive reference densities ρ_+ and ρ_-, where the potential energy density G is given by

$$G(\rho, \rho') = \rho \int_{\rho'}^{\rho} \left[\frac{P(s) - P(\rho')}{s^2}\right] \mathrm{d}s$$

which is nonnegative for positive ρ and ρ'.

REMARK 4.1. It is easy to check that (4.3)–(4.5) hold for the special pressure $P(\rho) = \kappa\rho^\gamma$, $\gamma \geqslant 1$, which corresponds to the polytropic fluid.

The external force f is assumed to satisfy:

$$f \in L^\infty_{\mathrm{loc}}\big([0,\infty), L^2(\mathbb{R})\big) \cap L^1_{\mathrm{loc}}\big([0,\infty), L^\infty(\mathbb{R})\big), \tag{4.6}$$

$$f_t,\ f_x \in L^2_{\mathrm{loc}}\big([0,\infty), L^2(\mathbb{R})\big). \tag{4.7}$$

The hypotheses on the initial data are:

$$\rho_0 \in L^\infty(\mathbb{R}) \quad \text{and} \quad \operatorname{ess\,inf} \rho_0 > 0, \tag{4.8}$$

$$\rho_0 - \bar\rho,\ u_0 - \bar u \in L^2(\mathbb{R}), \tag{4.9}$$

where $\bar\rho(x)$ and $\bar u(x)$ are defined as smooth monotone functions with the following property

$$\big(\bar\rho(x), \bar u(x)\big) = (\rho_\pm, u_\pm), \quad \text{when } \pm x \geqslant 1,$$

for some constants u_+, u_-, $\rho_+ > 0$, $\rho_- > 0$.

REMARK 4.2. In particular, it is allowed for the initial data to be piecewise constant with arbitrarily large jump discontinuities.

Under the assumptions (4.3)–(4.9), it is proved by Hoff [88] that:

THEOREM 4.1. *Assume the system* (4.1) *satisfies* (4.3)–(4.7) *and the initial data satisfy* (4.8), (4.9). *Then the initial value problem* (4.1), (4.2) *has a global weak solution* (ρ, u), *for which*

$$\begin{aligned}
&\rho - \bar{\rho}, \quad \rho u - \bar{\rho}\bar{u} \in C\big([0,\infty), H^{-1}(\mathbb{R})\big), \qquad u - \bar{u} \in C\big((0,\infty), L^2(\mathbb{R})\big),\\
&u(t,\cdot), \quad \mu u_x(t,\cdot) - P\big(\rho(t,\cdot)\big) + P(\bar{\rho}) \in H^1(\mathbb{R}), \quad t > 0,\\
&u_t(t,\cdot), \quad \dot{u}(t,\cdot) \in L^2(\mathbb{R}), \quad t > 0,
\end{aligned}$$

where $\dot{u}$ is the convective derivative $\dot{u} = \mathrm{d}u/\mathrm{d}t = u_t + uu_x$. Additionally, given $T > 0$, there is a positive constant $C(T)$, depending on T, μ, P, f and on upper bounds for $\|\rho_0 - \bar{\rho}\|$, $\|u_0 - \bar{u}\|$, $\|\rho_0\|_{L^\infty}$, $\|\rho_0^{-1}\|_{L^\infty}$, such that if $\sigma(t) = \min\{1, t\}$, then

$$\begin{aligned}
&C(T)^{-1} \leqslant \rho(t,\cdot) \leqslant C(T) \quad a.e.,\\
&\sup_{0<t\leqslant T} \big[\big\|\rho(t,\cdot) - \bar{\rho}\big\| + \big\|u(t,\cdot) - \bar{u}\big\| + \sigma(t)^{1/2}\big\|u_x(t,\cdot)\big\|\\
&\qquad + \sigma(t)\big(\big\|\dot{u}(t,\cdot)\big\| + \big\|\mu u_x(t,\cdot) - P\big(\rho(t,\cdot)\big) + P(\bar{\rho})\big\|\big)\big] \leqslant C(T),\\
&\int_0^T \big[\|u_x\|^2 + \sigma\|\dot{u}\|^2 + \sigma\big\|\mu u_x - P + P(\bar{\rho})\big\|^2 + \sigma^2\|\dot{u}_x\|^2\big]\,\mathrm{d}t \leqslant C(T);
\end{aligned}$$

and for $0 < \tau < T$,

$$\sigma(\tau)^{1/4}\|u\|_{L^\infty([\tau,T]\times\mathbb{R})} + \sigma(\tau)^{1/2}\langle u\rangle^{1/2,1/4}_{[\tau,T]\times\mathbb{R}} \leqslant C(T),$$

where $\|\cdot\|$ is the usual norm in $L^2(\mathbb{R})$ and $\langle u\rangle^{1/2,1/4}_{[\tau,T]\times\mathbb{R}}$ is the usual Hölder norm

$$\sup\left\{\frac{|u(t,x) - u(s,y)|}{|t-s|^{1/4} + |x-y|^{1/2}}, t, s \in [\tau, T], x, y \in \mathbb{R}, (t,x) \neq (s,y)\right\}.$$

The proof of Theorem 4.1 consists of four steps.

The first three are related to getting the a priori estimates for smooth solutions of (4.1) with smooth initial data satisfying (4.8), (4.9). That is, to give a bound for $\|\rho(u-\bar{u})^2 + G(\rho,\bar{\rho})\|_{L^1}$ first, which expresses the total energy in the fluid, at least for the special case when the functions $\bar{u}$ and $\bar{\rho}$ are constant. Namely,

$$\sup_{0\leqslant t\leqslant T} \int \left[\frac{1}{2}\rho(u-\bar{u})^2 + G(\rho,\bar{\rho})\right]\mathrm{d}x + \int_0^T \int u_x^2\,\mathrm{d}x\,\mathrm{d}t \leqslant C(T); \tag{4.10}$$

then, to derive a priori pointwise bounds for ρ from the energy estimate

$$C(T)^{-1} \leqslant \rho(t,x) \leqslant C(T), \quad t \in [0,T],\ x \in \mathbb{R}. \tag{4.11}$$

The proof is based on the hyperbolicity which serves to control the growth of ρ along particle trajectories. These pointwise bounds can be used then to exploit the parabolicity of the second equation in (4.1) to derive certain higher order regularity estimates for u:

$$\sup_{0\leqslant t\leqslant T}\int\left[(u-\bar{u})^2+(\rho-\bar{\rho})^2\right]\mathrm{d}x+\int_0^T\int u_x^2\,\mathrm{d}x\,\mathrm{d}t\leqslant C(T), \tag{4.12}$$

$$\sup_{0<t\leqslant T}\sigma(t)\int u_x^2\,\mathrm{d}x+\int_0^T\int\sigma^2\dot{u}^2\,\mathrm{d}x\,\mathrm{d}t\leqslant C(T), \tag{4.13}$$

$$\sup_{0<t\leqslant T}\sigma(t)^2\int\dot{u}^2\,\mathrm{d}x+\int_0^T\int\sigma^2\dot{u}_x^2\,\mathrm{d}x\,\mathrm{d}t\leqslant C(T), \tag{4.14}$$

$$\left\|u(t,\cdot)\right\|_{L^\infty}\leqslant C(T)\sigma(t)^{-1/4}, \tag{4.15}$$

$$\langle u\rangle^{1/2,1/4}_{[\tau,T]\times\mathbb{R}}\leqslant C(T)\sigma(\tau)^{-1/2}, \tag{4.16}$$

$$\left\|u(t_2,\cdot)-u(t_1,\cdot)\right\|\leqslant C(T)\sigma(t_1)^{-1/2}(t_2-t_1)^{1/2},\quad 0<t_1\leqslant t_2\leqslant T, \tag{4.17}$$

$$\begin{aligned}&\left\|\rho(t_2,\cdot)-\rho(t_1,\cdot)\right\|_{H^{-1}}+\left\|(\rho u)(t_2,\cdot)-(\rho u)(t_1,\cdot)\right\|_{H^{-1}}\\&\quad\leqslant C(T)(t_2-t_1)^{1/2},\quad 0<t_1\leqslant t_2\leqslant T.\end{aligned} \tag{4.18}$$

It should be pointed out that the bound given for the quantity $\mu u_x-P(\rho)$ has a special interest. It is known (see [80], for example) that initial discontinuities in ρ persist for all time, so that $\rho(t,\cdot)$ and $u_x(t,\cdot)$ may be locally integrable but no smoother. On the other hand, the fact that $\mu u_x-P(\rho)$ is in $H^1_{\mathrm{loc}}\hookrightarrow C^{1/2}$ evidently implies a cancellation of singularities. This cancellation of singularities can be shown to prevent oscillations in weakly converging sequences of solutions of (4.1). These ideas are carried out in the last step where the proof of Theorem 4.1 is completed by obtaining the solution (ρ,u) as the strong limit of approximate solutions corresponding to mollified initial data.

More precisely, let (ρ_0,u_0) be the initial data required in the theorem and define $\rho_0^\delta=j_\delta*\rho_0$, $u_0^\delta=j_\delta*u_0$, where $j_\delta(x)=\delta^{-1}j(x/\delta)$ is the standard mollifier. The estimates of (4.10)–(4.18) can show that the corresponding smooth solutions (ρ^δ,u^δ) of (4.1) with initial data $(\rho_0^\delta,u_0^\delta)$ exist for all time and satisfy all the estimates of (4.10)–(4.18) with constants $C(T)$ independent of δ. It can be easily shown that there is a subsequence $\delta\to 0$, for which

$$\begin{cases}u^\delta\to u, & \text{uniformly on compact sets in } \mathbb{R}\times(0,\infty),\\ u^\delta(t,\cdot)-u(t,\cdot)\to 0, & \text{strongly in } L^2(\mathbb{R}),\, t\geqslant 0,\\ u_x^\delta(t,\cdot)-u_x(t,\cdot)\rightharpoonup 0, & \text{weakly in } L^2(\mathbb{R}),\, t\geqslant 0.\end{cases} \tag{4.19}$$

The limiting function u keeps all the above bounds which pertain to u and u_x. Although we could obtain a subsequence for which ρ^δ converges weakly to a function ρ, but we could not guarantee that $P(\rho^\delta)$ converges, even weakly, to $P(\rho)$. Therefore we are not able to conclude that the limiting pair (ρ, u) is a weak solution of (4.1). We are required to obtain the density ρ as a strong limit of ρ^δ. In fact, it can be proved that there is a further subsequence $\delta \to 0$ and a limiting density ρ such that

$$\begin{cases} \rho^\delta(t,\cdot) - \rho(t,\cdot) \to 0, & \text{strongly in } L^2(\mathbb{R}), t \geqslant 0, \\ u_x^\delta(t,\cdot) - P\big(\rho^\delta(t,\cdot)\big) & \\ \quad \to u_x(t,\cdot) - P\big(\rho(t,\cdot)\big), & \text{strongly in } L^2_{\mathrm{loc}}(\mathbb{R}), t > 0. \end{cases} \tag{4.20}$$

The detailed analysis can be found in [88].

REMARK 4.3. An important consequence of Theorem 4.1 is that neither vacuum nor concentration can occur in this solution, no matter how large the external force or how large the oscillation of the initial data is.

The recent progress obtained in [94] is motivated by the above result. It is proved by Hoff and Smoller [94], for the system (4.1) with more general pressure P and weaker requirement on the external force, that weak solutions of the system (4.1) do not exhibit vacuum state, provided that no vacuum states are present initially. The solutions and external forces considered are quite general: the essential requirements are that the mass and energy densities of the fluid be locally integrable at each time, and that the $L^2_{\mathrm{loc}}-$ norm of the velocity gradient be locally integrable in time. It shows that if a vacuum state were to occur, the viscous force would impose an impulse of infinite magnitude on the adjacent fluid, thus violating the hypothesis that the momentum remains locally finite.

The initial-value problem of (4.1), (4.2) has been studied by many people. See, for example, Kanel [130], Serre [198] and Hoff [81,88] for the existence of solutions with constant time-asymptotic states. Also see Liu [149], Hoff and Liu [92], Liu and Xin [150], Szepessy and Xin [214], Xin [229], Matsumura and Nishihara [160] and [161] for the cases in which the time-asymptotic state contains a viscous shock or rarefaction wave, usually of small strength. Finally, there are a large number of results concerning solutions of (4.1) on a finite interval with suitable boundary conditions, for example, Hoff and Ziane [96], Matsumura and Yanagi [162], Shelukhin [200], Ducomet, Straškraba, Valli, Zlotnik [48,211–213,243, 247,248], Nishida [176, survey article] and the references contained in these papers. See also Hoff and Zarnowski [95], Hoff [87] and Hoff and Ziane [96] for uniqueness and continuous dependence results for solutions with strictly positive densities.

Let us turn to the nonisentropic case next. We consider the Navier–Stokes equations for compressible, heat-conducting flow in one space dimension with large, discontinuous initial data.

The equations under consideration describe the conservation of mass and momentum and the balance of energy, and take the following form in Lagrangian coordinates

$$\begin{cases} v_t - u_x = 0, \\ u_t + p(v,e)_x = \big(\frac{\mu u_x}{v}\big)_x, \\ \big(e + \frac{u^2}{2}\big)_t + \big(up(v,e)\big)_x = \big(\frac{\mu u u_x + \lambda e_x}{v}\big)_x, \end{cases} \tag{4.21}$$

where v, u, e and p denote the specific volume, velocity, specific internal energy and pressure, respectively; μ and λ are the fixed positive viscosity and heat conductivity parameters respectively; and x is the Lagrangian coordinate, so that $x =$ constant corresponds to a particle trajectory.

We first discuss the case with the equation of state for an ideal, polytropic fluid:

$$e = \frac{pv}{\gamma - 1} = c_V \theta, \tag{4.22}$$

where θ is the temperature, and $\gamma > 1$ and $c_V > 0$ are constants.

Consider the following initial-boundary value problem for (4.21) with boundary conditions

$$\begin{aligned} &u(t,0) = u(t,1) = 0, \\ &e_x(t,0) = e_x(t,1) = 0, \end{aligned} \qquad t > 0, \tag{4.23}$$

and initial conditions

$$(v,u,e)|_{t=0} = \big(v_0(x), u_0(x), e_0(x)\big), \quad 0 \leqslant x \leqslant 1, \tag{4.24}$$

satisfying

$$\begin{aligned} &C_0^{-1} \leqslant v_0(x) \leqslant C_0, \qquad e_0(x) \geqslant C_0^{-1}, \\ &\|u_0\|_{L^4} + \|\rho_0\| + TV(v_0) \leqslant C_0 \end{aligned} \tag{4.25}$$

for a constant $C_0 > 0$.

The result on well-posedness and large-time behavior of solutions to (4.21)–(4.25), given by Chen, Hoff and Trivisa in [22], is states as the following theorem.

THEOREM 4.2. *Given initial data (v_0, u_0, e_0) satisfying (4.25), there is a global weak solution (v,u,e) such that $v, u \in C([0,\infty), L^2(0,1))$, $e \in C([0,\infty), L^2(0,1))$ with $e(t,\cdot) \rightharpoonup e_0$ weakly in $L^2(0,1)$ as $t \to 0$. Furthermore, there is a positive constant M depending on C_0, but independent of $t > 0$, such that the following hold*:

$$M^{-1} \leqslant v(t,x) \leqslant M, \qquad M^{-1} \leqslant e(t,x) \leqslant M\sigma^{-1}(t), \tag{4.26}$$

$$TV_{[0,1]}\big(v(t,x)\big) \leqslant M, \qquad \big\|v(t',\cdot) - v(t,\cdot)\big\| \leqslant M\big|t' - t\big|^{1/2}, \tag{4.27}$$

$$\big\|u_x(t,\cdot)\big\| \leqslant M\sigma^{-1/2}(t), \qquad \big\|e_x(t,\cdot)\big\| \leqslant M\sigma^{-1}(t), \tag{4.28}$$

$$\big\|u(t,\cdot)\big\|_{L^\infty} \leqslant M\sigma^{-1/4}(t), \qquad \mathcal{E}(t) + \mathcal{F}(t) \leqslant M, \tag{4.29}$$

where $\sigma(t) = \min\{t, 1\}$, $\|\cdot\|$ denotes the norm in $L^2(0,1)$, $\mathcal{E}(t)$ and $\mathcal{F}(t)$ are functionals *for the weak solutions, defined as*

$$\mathcal{E}(t) = \sup_{0\leqslant s\leqslant t} \left(\sigma(s)\left\|u_x(s,\cdot)\right\|^2 + \sigma^2(s)\left\|e_x(s,\cdot)\right\|^2\right)$$
$$+ \int_0^t \Big(\left\|u_x(s,\cdot)\right\|^2 + \left\|e_x(s,\cdot)\right\|^2$$
$$+ \sigma(s)\left\|u_t(s,\cdot)\right\|^2 + \sigma^2(s)\left\|e_t(s,\cdot)\right\|^2\Big)\,\mathrm{d}s,$$
$$\mathcal{F}(t) = \sup_{0\leqslant s\leqslant t} \left(\sigma^2(s)\left\|u_t(s,\cdot)\right\|^2 + \sigma^3(s)\left\|e_t(s,\cdot)\right\|^2\right)$$
$$+ \int_0^t \sigma^2(s)\left\|u_{xt}(s,\cdot)\right\|^2 \mathrm{d}s + \int_0^t \sigma^3(s)\left\|e_{xt}(s,\cdot)\right\|^2 \mathrm{d}s.$$

Finally, the solution tends to a constant state as $t \to \infty$ in the sense that

$$\left\|(v - v_\infty)(t,\cdot)\right\|_{L^\infty} + \left\|(u, e - e_\infty)(t,\cdot)\right\|_{H^1} \to 0 \tag{4.30}$$

with $v_\infty = \int_0^1 v_0(x)\,\mathrm{d}x$ and $e_\infty = \int_0^1 (e_0(x) + u_0^2(x)/2)\,\mathrm{d}x$.

The proof of this theorem contains two important steps. In the first, it is assumed that v_0 is piecewise H^1, having a finite number of points of discontinuity. Solutions for this case can be obtained as the limits of approximate solutions derived from a suitable semidiscrete difference approximation to (4.21). Then in the second step, the uniform total variation estimate (4.27) is exploited to complete the solution operator for the more general case that $v_0 \in BV$. The result in [22] has improved those in [82,83] by allowing for large, discontinuous initial data, and by obtaining estimates (4.26)–(4.29) which are independent of time and can be used to show the large-time behavior of solutions. The crucial point involved here is to get pointwise upper and lower bounds for the specific volume v. The idea of Kazhikhov and Shelukhin in [139] with elaborating on it in a new way, is used in the derivation of these bounds.

REMARK 4.4. The pointwise bounds in (4.26) for the specific volume (equivalently, the density) show that neither vacuum nor concentration states occur, no matter how large the initial data is. This is one of several important differences between the Navier–Stokes equations and the inviscid Euler equations, for which vacuum states may take place for large initial data with certain equations of state (cf. [18,21]).

With the help of these pointwise bounds (4.26) on v, the higher-order regularity assertions of Theorem 4.2 can be obtained similarly as in [82,83]. The asymptotic behavior (4.30) can be derived as a consequence of the time-independent bounds (4.26)–(4.29).

The construction of solutions for the intermediate case when v_0 is piecewise H^1, used as the first step in the proof of Theorem 4.2, plays an important role. For this case some interesting observations concerning the propagation of singularities can be made.

Let v_0 be piecewise smooth, having jump discontinuities at isolated points $y_1 < \cdots < y_N$. Then we could find at the heuristic level that discontinuities in v, p, u_x and e_x occur only at $x = y_i$, and satisfy the following jump condition

$$\left[p(v,e) - \frac{\mu u_x}{v}\right] = 0, \qquad \left[\frac{e_x}{v}\right] = 0, \tag{4.31}$$

where $[w]$ denotes a jump of the function w across $x = y_i$: $[w(t, y_i)] = w(t, y_i + 0) - w(t, y_i - 0)$.

From (4.31), it follows, with the help of the first equation in (4.21) that

$$[\log v]_t = \mu^{-1}[p] = \frac{\gamma - 1}{\mu} e \left[\frac{1}{v}\right].$$

Therefore the exponential decay of the magnitudes of these discontinuities as $t \to \infty$ can be anticipated, since e is positive and $1/v$ is a decreasing function of $\log v$. In fact, this is indeed true, as proved in [22] in the following theorem.

THEOREM 4.3. *Assume, in addition to the hypotheses of Theorem* 4.2, *that* v_0 *is piecewise* H^1, *having isolated jump discontinuities at points* $y_1 < \cdots < y_N$. *Then, for* $t > 0$, *each of the quantities* $v(t,\cdot)$, $p(t,\cdot)$, $u_x(t,\cdot)$ *and* $e_x(t,\cdot)$ *has one-sided limits at each* y_i, *and the jump conditions* (4.31) *are satisfied in a strict pointwise sense. Moreover,*

$$\big[\log v(t, y_i)\big] = \big[\log v(0, y_i)\big] \exp\left(-\mu^{-1} \int_0^t \alpha_i(s)\,\mathrm{d}s\right),$$

where $\alpha_i(t) = -(\gamma - 1)e(t, y_i)[v^{-1}(t, y_i)]/[\log v(t, y_i)]$.

Finally, there is a constant M *depending on* C_0, *but independent of* t *and* N, *such that*

$$\begin{aligned}&\big|\big[(v, p, u_x, e_x)(t, y_i)\big]\big| \\ &\quad \leqslant M \min\big(\exp\{-M^{-1}t^{1/2}\}, \sigma(t)^{-3/2}\exp\{-M^{-1}t\}\big) \to 0, \quad \textit{as } t \to \infty.\end{aligned}$$

As far as uniqueness is concerned, it is a rather delicate issue for the kind of general solutions as in Theorem 4.2, due to the absence of uniform regularity in the initial layer near $t = 0$. However, we can improve the smoothing rates implicit in the definitions of $\mathcal{E}(t)$ and $\mathcal{F}(t)$ sufficiently, by proposing slightly stronger conditions on u_0 and e_0, to prove that solutions are indeed unique and depend continuously on their initial values.

THEOREM 4.4 (Regularity and stability). *Assume, in addition to the hypotheses of Theorem* 4.2, *that*

$$TV(u_0) + TV(e_0) \leqslant C_0.$$

Then there exists $M > 0$ independent of t such that the solution of Theorem 4.2 satisfies the additional estimates:

$$\begin{aligned} &\|u_x(t,\cdot)\| \leqslant M\sigma^{-1/4}(t), \qquad \|e_x(t,\cdot)\| \leqslant M\sigma^{-1/4}(t),\\ &TV_{[0,1]}\big(u(t,\cdot)\big) \leqslant M\sigma^{-1/4}(t), \qquad TV_{[0,1]}\big(e(t,\cdot)\big) \leqslant M\sigma^{-1/4}(t). \end{aligned} \tag{4.32}$$

Moreover, solutions satisfying (4.26)–(4.29) *and* (4.32) *are unique and depend continuously on their initial data in the sense that, if* (v_1, u_1, e_1) *and* (v_2, u_2, e_2) *are any two such solutions, and if* $S(t)$ *is defined by*

$$S(t) = \|(v_2 - v_1)(t,\cdot)\| + \|(u_2 - u_1)(t,\cdot)\|_{H^{-\alpha}} + \|(u_2 - u_1)(t,\cdot)\|_{H^{-\beta}},$$

where α and β are small and positive constants, then given $T > 0$, there is a constant $C(T)$ such that, for $0 \leqslant t \leqslant T$,

$$S(t) \leqslant C(T)S(0).$$

REMARK 4.5. The results of Theorems 4.2 and 4.3 can be converted to equivalent statements for the Navier–Stokes equations in Eulerian coordinates for which we still use (t, x) to denote the time and space variable

$$\begin{cases} \rho_t + (\rho u)_x = 0,\\ (\rho u)_t + \big(\rho u^2 + p\big)_x = (\mu u_x)_x,\\ (\rho E)_t + \big(u(\rho E + p)\big)_x = (\mu u u_x)_x + (\lambda e_x)_x, \end{cases} \tag{4.33}$$

where $E = e + u^2/2$ is the total specific energy.

However, the corresponding statements on uniqueness and continuous dependence are more subtle in view of the fact that the change of variables from Lagrangian to Eulerian coordinates is solution-dependent, and the solutions concerned are only minimally regular (also see [126]).

The approach in [22] can also be applied to get the global well-posedness and large-time behavior of discontinuous solutions with large initial data to the Navier–Stokes equations for a compressible reacting flow (see [21,23,29,72]).

The Navier–Stokes system (4.21) has been studied by a number of authors in a large variety of contexts. See for example the earlier results of Kazhikhov and Shelukhin [139], Serre [198], Hoff [83], also the results of Amosov and Zlotnick [4–6,244–246], Jiang and Zhang [115,125], Fujita-Yashima, Padula and Novotný [71], Liu and Zeng [152] and Wang [226], and the references contained in these papers.

In [226], the global existence of large discontinuous solutions with large discontinuous data to the Navier–Stokes system (4.21) for perfect gases is established by a different argument which is achieved by first mollifying the initial data and then using the estimates obtained for continuous solutions (see the discussion in Section 3 and a convergence argument in [215,217]). The uniqueness and the Lipschitz continuous dependence on data of weak solutions to initial boundary value problems in bounded intervals and to the Cauchy problem are established in [4,244,245] and in [126], respectively.

5. Vacuum problem for one-dimensional compressible flow

The one-dimensional isentropic viscous gas flow is governed by the compressible Navier–Stokes equations which can be written, in Eulerian coordinates, as (in the case without external force)

$$\begin{cases} \rho_t + (\rho u)_x = 0, \\ (\rho u)_t + \big(\rho u^2 + P(\rho)\big)_x = \big(\mu(\rho) u_x\big)_x, \end{cases} \tag{5.1}$$

due to the derivation given in Section 2, where $x \in \mathbb{R}$, $t > 0$; ρ, u and $P(\rho)$ denote the density, velocity and pressure, respectively; $\mu \geqslant 0$ is the viscosity coefficient. For simplicity we consider only the polytropic gas, i.e., $P(\rho) = A\rho^\gamma$ with $\gamma > 1$, $A > 0$ being constants.

We consider the case when $\mu(\rho) = \mu > 0$ is a constant first. This corresponds to the classical Navier–Stokes equations.

We are interested in the evolution of the interfaces separating the one-dimensional isentropic viscous gases from vacuum when the gases are in contact with the vacuum on a finite interval initially, with or without external force. This is a free boundary problem for which the free boundaries are the interfaces separating the gases from vacuum.

Consider (5.1) with the initial data

$$\rho(0,x) = \rho_0(x), \qquad u(0,x) = u_0(x). \tag{5.2}$$

The regularity and behavior of solutions of (5.1), (5.2) near the interfaces between the gas and vacuum are investigated in [155] where it is assumed that the entire gas initially occupies a finite interval $(a,b) \subset \mathbb{R}^1$ and is in contact with the vacuum, with the following assumptions:

(H$_1$) $\rho_0 \in C(\mathbb{R}^1)$, $\operatorname{supp}\rho_0 \subset (a,b)$, $\rho_0(x) > 0\ \forall x \in (a,b)$;

(H$_2$) $\rho_0^k \in H^1([a,b])$ for some constant $0 < k \leqslant \gamma - 1/2$; and the initial velocity $u_0(x)$ has the regularity as

(H$_3$) $\rho_0 u_0^2$, $(\partial_x u_0)^2$, $\rho_0^{-1}(\partial_{xx} u_0)^2 \in L^1([a,b])$.

Suppose $a = a(t)$ and $b = b(t)$ are two particle paths issuing from a and b, respectively, i.e.,

$$\dot a(t) = u\big(t, a(t)\big), \qquad \dot b(t) = u\big(t, b(t)\big), \tag{5.3}$$

with $a(0) = a$ and $b(0) = b$. The free boundary problem studied is

$$\begin{cases} (5.1), \quad \text{in } \big(a(t), b(t)\big) \times (0,\infty), 0 < t < \infty, \\ \rho\big(t, a(t)\big) = \rho\big(t, b(t)\big) = 0, \\ (\rho, u)(0,x) = (\rho_0, u_0)(x), \quad x \in [a,b]. \end{cases} \tag{5.4}$$

DEFINITION 5.1. A function (ρ, u) is called a weak solution to (5.4) if there exist $a(t)$, $b(t) \in C[0, \infty)$ such that

$$\rho \in C\big((0,\infty), L^2\big([a(t), b(t)]\big)\big), \qquad u \in C\big((0,\infty), H^1\big([a(t), b(t)]\big)\big),$$
$$\lim_{x \to a(t)+} \rho(t,x) = 0 = \lim_{x \to b(t)-} \rho(t,x).$$

Furthermore,

$$\int_a^b \rho_0 v(0,x)\,\mathrm{d}x + \int_0^\infty \int_{a(t)}^{b(t)} (\rho v_t + \rho u v_x)\,\mathrm{d}x\,\mathrm{d}t = 0,$$
$$\int_a^b \rho_0 u_0 w(0,x)\,\mathrm{d}x + \int_0^\infty \int_{a(t)}^{b(t)} \big(\rho u w_t + \big(\rho u^2 + P(\rho) + \mu u_x\big) w_x\big)\,\mathrm{d}x\,\mathrm{d}t = 0$$

hold for all test functions v, $w \in C_0^1(\Omega)$ with $\Omega = \{(t,x) \mid 0 \leqslant t < \infty, a(t) \leqslant x \leqslant b(t)\}$. The result on the existence and behavior of weak solutions to the free boundary problem (5.4) is stated as

THEOREM 5.1 [155]. *Let ρ_0 and u_0 satisfy* (H$_1$)–(H$_3$). *Then the free boundary problem* (5.4) *admits a globally defined weak solution with $a(\cdot)$, $b(\cdot) \in C^1[0,\infty)$. Moreover, the solution $(\rho(t,x), u(t,x))$ has the properties*:

(1) The regularity:

$$\rho > 0 \textit{ in } Q \quad \textit{with } Q =: \big\{(t,x)\colon 0 \leqslant t < \infty, a(t) < x < b(t)\big\},$$
$$\int_{a(t)}^{b(t)} \big(\rho u^2 + u_x^2\big)(t,x)\,\mathrm{d}x \leqslant C, \quad 0 \leqslant t < \infty,$$
$$\int_{a(t)}^{b(t)} \big(\big[(\rho^k)_x\big]^2 + \rho^{-1} u_{xx}^2\big)(t,x)\,\mathrm{d}x \leqslant C(T), \quad 0 \leqslant t \leqslant T, \textit{ for any } T > 0.$$

Hereafter, we use C (or $C(T)$) to denote a generic positive constant depending only on the initial data (and the given time T).

(2) Decay rate of the density and expanding rate of the interface: *There exist positive constants C_i $(i = 1, 2, \dots, 7)$, independent of the time t, such that*

$$C_1 \rho_0(x_1)\big(1 + C_2 \rho_0^\gamma(x_1) t\big)^{-1/\gamma} \leqslant \rho(t,x) \leqslant C_3 \rho_0(x_1)\big(1 + C_4 \rho_0^\gamma(x_1) t\big)^{-1/\gamma},$$

where x_1 is determined uniquely by $\int_a^{x_1} \rho_0(t)\,\mathrm{d}t = \int_{a(t)}^x \rho(t,z)\,\mathrm{d}z$ for any x,

$$C_5(1+t)^{1/\gamma} \leqslant b(t) - a(t) \leqslant C_6(1+t)^{1/\gamma}, \quad 0 \leqslant t < \infty,$$
$$\sup_{a(t) \leqslant x \leqslant b(t)} \big|u(t,x)\big| \leqslant C_7(1+t)^{1/\gamma}, \quad 0 \leqslant t < \infty.$$

(3) Behavior near the interface: *It holds, for (t,x) near the interfaces, that*

$$\rho(t,x) \leqslant C(T)\left|x-a(t)\right|^{1/2k}, \qquad \rho(t,x) \leqslant C(T)\left|x-b(t)\right|^{1/2k},$$
$$\left|u_x(t,x)-u_x\big(t,a(t)+\big)\right| \leqslant C(T)\left|x-a(t)\right|^{(1/2+1/4k)},$$
$$\left|u_x(t,x)-u_x\big(t,b(t)-\big)\right| \leqslant C(T)\left|x-b(t)\right|^{(1/2+1/4k)}$$

for any $0 \leqslant t \leqslant T$.

(4) Basic conserved quantities:

$$\int_{a(t)}^{x} (\rho u_t + \rho u u_x)(t,z)\,\mathrm{d}z + P(\rho) = \mu u_x \quad \text{for } a(t) < x < b(t) \text{ and } t > 0,$$
$$\int_{a(t)}^{b(t)} \rho u(t,x)\,\mathrm{d}x = \int_a^b \rho_0 u_0(x)\,\mathrm{d}x \quad \text{for } t > 0.$$

It is important and interesting to investigate the regularity of the solution to the above free boundary problem and the behavior near interfaces due to the degeneracy of vacuum. In general, it is not expected to have the global existence of the smooth solution because of the degeneracy. However, we may expect the smoothness of the solution in the region $\{(t,x)\colon \rho(t,x) > 0\}$ when the initial data have certain appropriate regularity and the smoothness up to the boundary when the initial density is connected to vacuum smooth enough. This is proved in [155] as the following theorem.

THEOREM 5.2 [155]. *Let (ρ,u) be the weak solution to* (5.4), *obtained in Theorem* 5.1. *It holds that*:

(1) *If $\rho_0^\alpha(\partial_x^2\rho_0)^2 \in L^1(a,b)$, then $\rho^{\alpha_1}(\partial_x^2\rho)^2(t,\cdot) \in L^1(a(t),b(t))$ with $\alpha_1 = 2k-3+\max\{2k+3+\alpha, 6k+1\}$. And the weak solution (ρ,u) to* (5.4) *is smooth in the region $Q =: \{(t,x)\colon 0 \leqslant t < \infty, a(t) < x < b(t)\}$ with $\rho(t,\cdot) \in C^{1+\lambda}(a(t),b(t))$, $u(t,\cdot) \in C^{2+\lambda}(a(t),b(t))$ for some $0 < \lambda < 1$ and any $t > 0$;*

(2) *$u_{xx}(t,a(t)+) = u_{xx}(t,b(t)-) = 0$ if $k < 1/2$; furthermore $\rho(t,\cdot) \in H^2(a(t),b(t))$, $0 \leqslant t \leqslant \infty$, if $\alpha \leqslant -4k$ and $k \leqslant 1/4$, and this implies ρ_x is Hölder continuous in the whole interval $[a(t),b(t)]$ for any $t \geqslant 0$. Also, the gradient of the pressure satisfies*

$$\begin{aligned}\left|P(\rho)_x(t,x)\right| &\leqslant C(T)\rho^{\gamma-1}\\ &\leqslant C(T)\big(\left|x-a(t)\right|^{(\gamma-1)/(2k)} + \left|x-b(t)\right|^{(\gamma-1)/(2k)}\big).\end{aligned}$$

To prove Theorem 5.1, one converts the free boundary problem to a fixed boundary problem by using Lagrangian coordinates and obtains some basic estimates on the corresponding initial-boundary value problem to which the global solution can be constructed by a slight modification of the method of lines used in [80,151,175,178,179].

Based on these estimates, the smoothness result in Theorem 5.2 can be obtained with the L^2-estimate of $\rho^{\alpha/2}\rho_{xx}$ which is crucial to the improvement of regularity of the solution.

Moreover, the uniqueness result holds.

THEOREM 5.3 [155]. *Assume* (H_1)–(H_3). *Let* (ρ_1, u_1) *and* (ρ_2, u_2) *be two weak solutions to the free boundary problem* (5.4) *in* $0 \leqslant t \leqslant T$ *as described in Definition* 5.1. *Then* $(\rho_1, u_1)(t,x) = (\rho_2, u_2)(t,x)$ *in* $a(t) < x < b(t)$ *and* $0 \leqslant t \leqslant T$.

REMARK 5.1. The above results are stated for the case without external force. However, similar results hold true also in the presence of external forces such as gravity.

The free boundary problem of the one-dimensional Navier–Stokes equations with one boundary fixed was investigated earlier by Okada in [178]; see also [175], where the global existence of the weak solutions was proved and the regularity of $(\rho^{1/2}u, u_x, \rho^{-1/2}u_{xx}) \times (t, \cdot) \in L^2$, $\rho \in BV$ was obtained. Similar results were derived in [163] and [179] for the equations of spherically symmetric motion of viscous gases. Furthermore, the free boundary problem for the one-dimensional viscous gases which expand into the vacuum has been studied a lot, see [178,179,230] and the references therein.

Other interesting results of free boundary problems related to the classical Navier–Stokes equations can be found in [80,85,93,151], also in [4,5,10,88,198], etc.

We turn to the case when the viscosity coefficient depends on the density, $\mu = \mu(\rho)$, instead of a positive constant. This modified Navier–Stokes system (5.1) was proposed by Liu, Xin and Yang in [151] to study the one-dimensional isentropic viscous gas flow with vacuum states. This is motivated by certain considerations in both mathematics and physics. When the viscosity coefficient μ is a constant, the study in [93] shows the failure of the continuous dependence on the initial data of solutions to the compressible Navier–Stokes equations with vacuum. As pointed out in [151], the main reason for this noncontinuous dependence comes from the independence of the kinetic viscosity coefficient on the density. From the physical point of view, one can find in the derivation of the Navier–Stokes equations from the Boltzmann equations through the Chapman–Enskog expansion to the second order (cf. [73]) that the viscosity is not a constant but depends on the temperature. For isentropic flow, this dependence is translated into the dependence on the density by the laws of Boyle and Gay-Lussac for ideal gases as discussed in [151].

For simplicity, we only discuss the polytropic gas

$$P(\rho) = A\rho^{\gamma}$$

and assume that

$$\mu(\rho) = B\rho^{\alpha}, \tag{5.5}$$

where $\gamma > 1$, $A > 0$, $B > 0$, $\alpha > 0$ are constants.

Now the viscosity coefficient vanishes at vacuum. This property yields the well-posedness of the Cauchy problem when the initial density is of compact support. In this aspect, the local existence of weak solutions to the Cauchy problem for the Navier–Stokes equations with vacuum was studied in [151] and [156], where the initial density was assumed to be connected to vacuum with discontinuities. This property can be maintained for some finite time. Jiang [115] studied the Navier–Stokes equations for a one-dimensional heat-conducting gas and proved the global existence of smooth solutions provided that

$0 < \alpha < 1/4$ in (5.5). By using the techniques similar to those in [115] to derive a priori estimates and the finite difference method, Okada, Matušů-Nečasová and Makino [180] obtained the existence of global weak solutions with isentropic flow for $0 < \alpha < 1/3$. This result has been improved to the case $0 < \alpha < 1/2$ by Yang, Yao and Zhu [233]. In both [180] and [233] the initial data are required to satisfy $\rho_0, \partial_x u_0 \in \mathrm{Lip}[0,1]$.

Some new progress has been made by Jiang, Xin and Zhang [121] recently under the conditions $\rho_0 \in W^{1,p}(0,1)$, $u_0 \in L^p(0,1)$ for some p and $0 < \alpha < 1$. Furthermore, the uniqueness can be established provided $u_0 \in H^1(0,1)$. This result improves those in [180,233]. First, more general pressure laws can be dealt with in [121]. For example, it is assumed in [180,233] that $1 < \gamma < 2$ for the perfect fluid to which $\alpha = (\gamma - 1)/2$, while the more general situation with $1 < \gamma < 3$ is allowed now. Second, less regularity of the initial data is required.

Consider the system (5.1) with the boundary conditions:

$$\begin{cases} \big(-P(\rho) + \mu(\rho)u_x\big)\big(t, a(t)\big) = 0, \\ \big(-P(\rho) + \mu(\rho)u_x\big)\big(t, b(t)\big) = 0, \\ \frac{\mathrm{d}a(t)}{\mathrm{d}t} = u\big(t, a(t)\big), \quad \frac{\mathrm{d}b(t)}{\mathrm{d}t} = u\big(t, b(t)\big). \end{cases} \tag{5.6}$$

To study the free boundary problem (5.1)–(5.6), it is convenient to convert the free boundaries to the fixed boundaries by using Lagrangian coordinates. We introduce the coordinate transformation

$$\xi = \int_{a(t)}^{x} \rho(t, y)\,\mathrm{d}y, \qquad \tau = t,$$

then the free boundaries $a(t)$ and $b(t)$ become

$$\xi = 0 \quad \text{and} \quad \xi = \int_{a(t)}^{b(t)} \rho(t, y)\,\mathrm{d}y = \int_a^b \rho(0, y)\,\mathrm{d}y,$$

where $\int_a^b \rho(0, y)\,\mathrm{d}y$ is the total initial mass. Without loss of generality we assume $\int_a^b \rho(0, y)\,\mathrm{d}y = 1$. Hence, in Lagrangian coordinates, the free boundary problem (5.1), (5.6) becomes, with the notation (τ, ξ)

$$\begin{aligned} &\rho_\tau + \rho^2 u_\xi = 0, \\ &u_\tau + A\big[\rho^\gamma\big]_\xi = B\big[\rho^{1+\alpha} u_\xi\big]_\xi, \quad 0 < \xi < 1, \tau > 0, \end{aligned}$$

with the boundary conditions

$$A\rho^\gamma = B\rho^{1+\alpha} u_\xi \quad \text{at } \xi = 0 \text{ and } \xi = 1, \tau \geqslant 0.$$

For convenience, we still use the notation (t, x) instead of (τ, ξ), namely we consider the following initial-boundary value problem

$$\begin{cases} \rho_t + \rho^2 u_x = 0, \\ u_t + A[\rho^\gamma]_x = B[\rho^{1+\alpha} u_x]_x, \quad 0 < x < 1, t > 0, \end{cases} \tag{5.7}$$

with the boundary conditions

$$A\rho^\gamma = B\rho^{1+\alpha} u_x \quad \text{at } x = 0 \text{ and } x = 1, t \geqslant 0 \tag{5.8}$$

and the initial conditions

$$\bigl(\rho(0,x), u(0,x)\bigr) = \bigl(\rho_0(x), u_0(x)\bigr), \quad x \in [0,1]. \tag{5.9}$$

From the boundary conditions (5.8) it is known that for $\gamma > \alpha$, $\rho(t,0)$, $\rho(t,1) = O(t^{-1/(\gamma-\alpha)})$ for large t. Hence, the density decreases as t grows. This causes the viscosity (the stabilization mechanism) decreasing to zero, so that a solution may not exist globally in time. It is shown in [121] that if $\mu(\rho)$ does not decrease to zero too rapidly, i.e., if α is not large, then a weak solution of (5.7)–(5.9) still exists globally in time. Moreover, the uniqueness of the weak solution can be obtained under the additional regularity assumption on initial data. More precisely, the result can be stated as the following theorem.

THEOREM 5.4. *Assume* $0 < \alpha < 1$, $\inf_{[0,1]} \rho_0 > 0$, $\rho_0 \in W^{1,2n}(0,1)$, $u_0 \in L^{2n}(0,1)$ *for some* $n \in \mathbb{N}$ *satisfying* $n(2n-1)/(2n^2+2n-1) > \alpha$. *Then the problem* (5.7)–(5.9) *has a global weak solution* (ρ, u) *in the sense that, for any* $T > 0$,

$$\begin{aligned} &\rho \in L^\infty\bigl([0,T], W^{1,2n}(0,1)\bigr), \qquad \rho_t \in L^2\bigl([0,T], L^2(0,1)\bigr), \\ &u \in L^\infty\bigl([0,T], L^{2n}(0,1)\bigr) \cap L^2\bigl([0,T], H^1(0,1)\bigr), \\ &\rho(t,x) \geqslant C \quad \text{on } [0,T] \times [0,1] \end{aligned}$$

for some positive constant $C = C(\|\rho_0\|_{W^{1,2n}}, \|u_0\|_{L^{2n}}, \inf_{[0,1]} \rho_0, T)$, *and the following equations hold*:

$$\begin{aligned} &\rho_t + \rho^2 u_x = 0, \qquad \rho(0,x) = \rho_0(x) \quad \text{for a.e. } x \in (0,1) \text{ and any } t \geqslant 0, \\ &\int_0^\infty \int_0^1 \bigl\{u\phi_t + \bigl(A\rho^\gamma - B\rho^{\alpha+1} u_x\bigr)\phi_x\bigr\}\,\mathrm{d}x\,\mathrm{d}t + \int_0^1 u_0(x)\phi(0,x)\,\mathrm{d}x = 0 \end{aligned}$$

for any test function $\phi(t,x) \in C_0^\infty(Q)$ *with* $Q := \{(t,x) \mid t \geqslant 0, 0 \leqslant x \leqslant 1\}$. *Moreover, if, in addition,* $u_0 \in H^1(0,1)$, *then* u *satisfies the additional estimates*:

$$\begin{aligned} &u \in L^\infty\bigl([0,T], H^1(0,1)\bigr) \cap L^2\bigl([0,T], H^2(0,1)\bigr), \\ &u_t \in L^2\bigl([0,T], L^2(0,1)\bigr), \end{aligned}$$

and furthermore, this weak solution is unique in the class:

$$\rho \in L^\infty([0,T], H^1(0,1)), \qquad \rho_t \in L^2([0,T], L^2(0,1)),$$
$$u \in L^\infty([0,T], L^2(0,1)) \cap L^2([0,T], H^1(0,1)),$$
$$\rho(t,x) > 0 \quad \text{on } [0,T]\times[0,1].$$

The proof of Theorem 5.4 is based on a priori estimates for the approximate solutions of (5.7)–(5.9) and a limit procedure. To derive the a priori estimates, the crucial step is to obtain lower and upper bounds of the density, that is, if the initial density has no vacuum and concentration of mass on [0, 1], then the same should be true for the density for all $t > 0$. By exploiting the high integrability of u, $(\rho^\alpha)_x$ (i.e., u, $(\rho^\alpha)_x \in L^\infty([0,T]$, $L^{2n}(0,1))$, $u^{n-1}u_x \in L^2([0,T], L^2(0,1))$ for any $n \in \mathbb{N}$) and the energy conservation, one gets the boundedness of the density from below and above. The detail can be found in Section 2 of [121]. As far as the uniqueness is concerned, the regularity of the weak solution constructed for smooth data is proved first, then the uniqueness of the weak solution under the additional regularity assumption on u_0 can be shown. (See Section 3 in [121] for details.)

One should notice that the analysis, used to deal with the situation when the initial density was assumed to be connected to vacuum with discontinuities, is based on the uniform positive lower bound of the density with respect to the construction of the approximate solutions. This estimate is crucial since the other estimates concerned can be obtained by standard techniques as long as the vacuum does not appear in the solutions in finite time. And this uniform positive lower bound on the density function can only be obtained when the density function connects to vacuum with discontinuities. In this situation, the density function is positive for any finite time. So that the viscosity coefficient never vanishes.

The situation is quite different if the density function connects to vacuum continuously because there is no positive lower bound for the density any more. The degeneracy in the viscosity coefficient which vanishes at vacuum gives rise to new difficulties in analysis due to the less regularity effect on the solutions. A local existence result is obtained in [234]. Another difficulty comes from the singularity at the vacuum boundary when the density function connects to the vacuum continuously. This can be found from the analysis used in [230] on the nonglobal existence of the regular solution to the classical Navier–Stokes equations with a constant viscosity coefficient when the density function is of compact support. The proof in [230] is based on the estimation of the growth rate of the support in terms of the time t. If the growth rate is sublinear, then the nonlinear functional introduced there could yield the nonglobal existence of regular solutions. It seems that this phenomenon comes from the effect of the pressure in the gas. No matter how smooth the initial data are, the pressure of the gas will build up at the vacuum boundary in finite time, so that the gas will be pushed into the vacuum region. It turns out the pressure will have the effect on the evolution of the vacuum boundary in finite time so that the density function at the interface will not be smooth. This singularity at the derivatives cause certain analytic difficulty which can be overcome by introducing some appropriate weights in the energy estimates.

This is carried out by Yang and Zhu in [235], where some weight functions, vanishing at the vacuum boundary, are introduced and some new a priori estimates on the solutions are established to prove two new global existence results. The first one is for the case when the density function is of compact support in Eulerian coordinates, and the other is when the support of gas in Eulerian coordinates is infinite but the total mass is finite. A nonglobal existence theorem for regular solutions is also given when the initial data is of compact support.

To state the result in [235] more precisely, we also choose Lagrangian coordinates on which the free boundary problem to (5.1) can be written as the following initial-boundary value problem to (5.7)

$$\begin{cases} (5.7), \quad \text{together with:} & \\ \rho(t,0)=\rho(t,1)=0, & t\geqslant 0, \\ (\rho,u)(0,x)=\big(\rho_0(x),u_0(x)\big), & 0\leqslant x\leqslant 1. \end{cases} \tag{5.10}$$

The assumptions on the initial data, γ and α proposed in [235] for the first global existence result are the following:

(A_1) For any positive integer n, $0\leqslant \rho_0(x)\leqslant C(x(1-x))^\beta$ with $0<\beta<1$, $(\rho_0(x))^{-1}\in L^1(0,1)$, for some k_1 with

$$1<k_1<\min\left\{1+(1-3\alpha)\beta, (2\gamma-3\alpha+1)\beta, \frac{5-15\alpha}{1+3\alpha}, \frac{3-5\alpha}{1+\alpha}, \beta(4-2\alpha)\right\},$$

such that

$$\big(x(1-x)\big)^{k_1}\rho_0^{2\alpha-2}\in L^1(0,1), \qquad \big(x(1-x)\big)^{1/2}(\rho_0^\alpha)_x\in L^2(0,1),$$
$$\big(\rho_0^\gamma\big)_x\in L^{2n}(0,1);$$

(A_2) $u_0\in L^\infty(0,1)$ and $(\rho_0^{1+\alpha}u_{0x})_x\in L^{2n}(0,1)$;

(A_3) $0<\alpha<2/9$, $\gamma>1$.

Under the assumptions (A_1)–(A_3), the existence of a global weak solution to the initial-boundary value problem (5.10), when the density function is of compact support in Eulerian coordinates, is proved as the following theorem.

THEOREM 5.5. *Under the conditions* (A_1)–(A_3), *the problem* (5.10) *has a global weak solution* $(\rho(t,x),u(t,x))$ *with* $\rho,u\in C^1([0,T];H^1(0,1))$ *such that it holds*

$$C(T)\big(x(1-x)\big)^{k_2/(1-2\alpha)}\leqslant \rho(t,x)\leqslant C(T)\big(x(1-x)\big)^\beta,$$

where $k_2=(1+k_1)/2$.

The definition of the weak solution obtained above is similar to the one in [180], namely:

DEFINITION 5.2. A pair of functions $(\rho(t,x), u(t,x))$ is called a global weak solution to the initial boundary value problem (5.10), if, for any $T > 0$,

$$\rho, u \in L^\infty([0,T]\times[0,1]) \cap C^1([0,T], L^2(0,1)),$$
$$\rho^{1+\alpha} u_x \in L^\infty([0,T]\times[0,1]) \cap C^{1/2}([0,T], L^2(0,1)).$$

Furthermore, the following equations hold

$$\int_0^\infty \int_0^1 (\rho\phi_t - \rho^2 u_x \phi)\,\mathrm{d}x\,\mathrm{d}t + \int_0^1 \rho_0(x)\phi(0,x)\,\mathrm{d}x = 0,$$
$$\int_0^\infty \int_0^1 \big(u\psi_t + (P(\rho) - \mu\rho u_x)\psi_x\big)\,\mathrm{d}x\,\mathrm{d}t + \int_0^1 u_0(x)\psi(0,x)\,\mathrm{d}x = 0$$

for any test functions ϕ and ψ belonging to $C_0^\infty(\Omega)$ with $\Omega = \{(t,x)\colon t \geqslant 0, 0 < x < 1\}$.

We refer the reader to [235] for the interesting proof of Theorem 5.5 and the discussion on other results there.

6. The Navier–Stokes equations for multidimensional compressible fluids

As mentioned in the Introduction, the mathematical study of the compressible Navier–Stokes equations goes back to the work by Nash [173] in the last sixties. In [173] he proved the local existence of smooth solutions to the Navier–Stokes equations of compressible heat-conducting fluids. Since then, significant progress has been made on the mathematical topics, such as the global existence and the time-asymptotic behavior of solutions. However, a number of important questions, for example, the existence of global solutions in the case of heat-conducting gases and the uniqueness of weak solutions, still remain open for large data.

Concerning the global well-posedness and the large-time behavior of solutions, the picture is more or less complete when data are sufficiently small. In this case, global smooth and weak solutions exist and converge to the corresponding stationary solutions as time tends to infinity (see, e.g., [85,158,159,224], and there are a vast amount of literature contributing to this issue, we refer the reader to the extensive bibliography in [146]) and the survey article [223].

For large data the situation becomes quite complex. The global well-posedness for the Navier–Stokes equations of compressible heat-conducting fluids still remains open. On the other hand, the global existence and the large-time behavior of solutions have been obtained in the case of isentropic fluids or heat-conducting fluids with symmetric data in symmetric domains without the origin. A global existence result for general large data was obtained first by Lions [146] for isentropic fluids (i.e., for (2.22) and (2.23)), where he used the method ofweak convergence and established delicate techniques to obtain global weak solutions provided the specific heat ratio γ is appropriately large, for example, $\gamma \geqslant 3n/(n+2)$, $n = 2, 3$. Then, by combining Lions' techniques and convex analysis to reduce the integrability requirement of the density around the origin, Jiang and Zhang [122] showed the global existence of weak solutions for any $\gamma > 1$ to the Cauchy problem when the initial data are spherically symmetric. Recently, Feireisl, Novotný and Petzeltová [60,63] delicately use the curl–div lemma to derive certain compactness, and apply Lions' idea [146] and a technique from [122] to extend Lions' existence result to the case $\gamma > n/2$ $(n = 2, 3)$. More recently, the global existence of axisymmetric weak solutions for $\gamma > 1$ and spherically symmetric weak solutions in the two-dimensional isothermal fluid case (i.e., $\gamma = 1$) was studied in [123,124] by combining different ideas from [60,63,122,146,181], and the result in [123] improves that of Hoff [84]. Further results on the compressible isentropic Navier–Stokes equations are contained in [59,61] where the existence of time-periodic weak solutions and the large-time behavior of the global weak solutions are discussed, in [222] in which strong solutions in two space dimensions are shown to exist globally in time provided that μ varies with ρ in a very specific way (i.e., $\mu = O(\rho^\beta)$ for some $\beta \geqslant 3$ as $\rho \to \infty$), and in [47,55] where self-gravitating fluids and nonmonotone pressure laws are investigated, in [57] where the motion of rigid body in a viscous compressible fluid is studied, and in [56,64,65] where the dynamics, such as bounded absorbing sets, global attractors and the ω-limit sets, are investigated.

There are related results, for example, by Desjardins, Grenier, Lions and Masmoudi in [39,89,147,157] where the incompressible limit of the compressible isentropic Navier–Stokes equations is studied, and by Min, Kazhikhov, Ukai and Vaigant in [164,221], where the potential flow equations at small Reynolds numbers and the Stokes approximation equations in $\mathbb{R}^2$ are dealt with, and by Xin and Yanagisawa in [231,232, etc.], where the zero dissipation limit and boundary layers are discussed.

The multidimensional theory for the compressible isentropic Navier–Stokes equations (2.22) and (2.23) is the main issue in this section.

For the heat-conducting gas case, the global well-posedness is known only when the initial data and the domains are spherically symmetric and the origin is excluded from the domains in order to avoid the mathematical difficulties induced by the singularities at the origin (see, e.g., [24,112,114,174]). Moreover, the global spherically symmetric solutions are asymptotically stable as time goes to infinity. On the other hand, in [230] Xin gave a nonexistence result of smooth (small) solutions when there is vacuum initially (also cf. [220]). So, one can only expect the existence of global weak solutions with the initial data containing vacuum. There are continuous developments on this issue. In [241] Zheng and Qin investigated the existence of maximal attractors for the spherically symmetric case in an annulus, while in [41,44] Ducomet studied the global existence and the large-time behavior of spherically symmetric solutions for hydrodynamical models from astrophysics, and among others.

In Section 6.1 we present a global existence theorem for the compressible isentropic Navier–Stokes equations (2.22), (2.23) and in Section 6.2 we discuss the nonisentropic fluid case. *Throughout this section* we use (t, x) to denote the Eulerian coordinates (t, X) in Section 2.

6.1. *The isentropic fluid case*

We consider the initial boundary value problem for (2.22), (2.23) with initial and nonslip boundary conditions

$$(\rho, \rho\mathbf{u})|_{t=0} = (\rho_0, \mathbf{m}_0), \quad x \in \Omega, \tag{6.1}$$

$$\mathbf{u}|_{\partial\Omega} = 0, \quad t \geqslant 0, \tag{6.2}$$

where $\Omega \subset \mathbb{R}^n$ $(n = 2, 3)$ is a bounded domain with smooth boundary $\partial\Omega$, and we have used (t, x) to denote the Eulerian coordinates (t, X). Now, let us recall the definition of weak solutions.

DEFINITION 6.1. We call $(\rho, \mathbf{u})$ a global finite energy weak solution of (2.22), (2.23), (6.1), (6.2) on the interval $[0, T]$, if

(i) $\rho \geqslant 0$, $\rho \in L^\infty((0, T), L^\gamma(\Omega))$, $\mathbf{u} \in L^2((0, T), H_0^1(\Omega))$;

(ii) the energy $E(t)$ is locally integrable on $(0, T)$ and satisfies the following inequality in $\mathcal{D}'(0, T)$:

$$\frac{\mathrm{d}}{\mathrm{d}t} E(t) + \int_\Omega \left(\mu|\nabla\mathbf{u}|^2 + (\lambda + \mu)|\operatorname{div}\mathbf{u}|^2\right)(t, x)\,\mathrm{d}x \leqslant 0, \tag{6.3}$$

where

$$E(t) = \int_\Omega \left(\frac{1}{2}\rho|\mathbf{u}|^2 + \frac{a}{\gamma - 1}\rho^\gamma\right)(t, x)\,\mathrm{d}x;$$

(iii) equations (2.22) and (2.23) are satisfied in $\mathcal{D}'((0, T) \times \Omega)$; moreover, (2.22) holds in $\mathcal{D}'((0, T) \times \mathbb{R}^n)$ provided $\rho, \mathbf{u}$ were prolonged to be zero on $\mathbb{R}^n \backslash \Omega$;

(iv) equation (2.22) is satisfied in the sense of renormalized solutions, that is, the following equation holds in $\mathcal{D}'((0, T) \times \Omega)$,

$$b(\rho)_t + \operatorname{div}\left[b(\rho)\mathbf{u}\right] + \left[b'(\rho)\rho - b(\rho)\right]\operatorname{div}\mathbf{u} = 0 \tag{6.4}$$

for any $b \in C^1(\mathbb{R})$ such that

$$\left|b'(z)z\right| + \left|b(z)\right| \leqslant C \quad \forall z \in \mathbb{R}.$$

It is easy to see from (2.22) and (2.23) that a finite energy weak solution satisfies

$$\rho \in C\left([0, T], L^\gamma(\Omega) - w\right), \qquad \rho\mathbf{u} \in C\left([0, T], L^{2\gamma/(\gamma+1)}(\Omega) - w\right), \tag{6.5}$$

and consequently, the initial conditions (6.1) make sense and are satisfied in the weak topology of $L^\gamma(\Omega) \times L^{2\gamma/(\gamma+1)}(\Omega)$.

The global existence reads then [54,58,60,146]:

THEOREM 6.1. *Assume that* $\gamma > n/2$ $(n = 2, 3)$, *and* $\rho_0 \in L^\gamma(\Omega)$, $\rho_0 \geqslant 0$, $\mathbf{m}_0 = 0$ *whenever* $\rho_0(x) = 0$, *and* $|\mathbf{m}_0|^2/\rho_0 \in L^1(\Omega)$. *Then there exists a finite energy weak solution* $(\rho, \mathbf{u})$ *to the initial boundary value problem* (2.22), (2.23), (6.1) *and* (6.2).

SKETCH OF THE PROOF. Roughly speaking, the proof of Theorem 6.1 is mainly based on the method of weak convergence, and it is very long and complicated. Here in the following, we only give the main steps of the proof for the case $n = 3$, for which $\gamma > 3/2$ should be kept in mind. The case $n = 2$ can be dealt with in the same manner.

Step I. The first step is to solve a regularized approximate system to (2.22), (2.23), (6.1) and (6.2):

$$\begin{cases} \rho_t + \operatorname{div}(\rho\mathbf{u}) = \varepsilon\Delta\rho, \\ (\rho\mathbf{u})_t + \operatorname{div}(\rho\mathbf{u}\otimes\mathbf{u}) + a\nabla(\rho^\gamma) + \delta\nabla(\rho^\beta) + \varepsilon\nabla\mathbf{u}\cdot\nabla\rho \\ \quad = \mu\Delta\mathbf{u} + (\lambda+\mu)\nabla\operatorname{div}\mathbf{u}, \end{cases} \tag{6.6}$$

with mollified initial data and boundary conditions

$$(\rho, \rho\mathbf{u})|_{t=0} = (\rho_0, \mathbf{m}_0) \in C^{2+\nu}(\overline{\Omega}) \times C^2(\overline{\Omega}), \quad 0 < \underline{\rho} \leqslant \rho_0(x) \leqslant \bar{\rho}, \tag{6.7}$$

$$\left.\frac{\partial\rho}{\partial\mathbf{n}}\right|_{\partial\Omega} = 0, \qquad \mathbf{u}|_{\partial\Omega} = 0, \quad t \geqslant 0, \tag{6.8}$$

where $\underline{\rho}, \bar{\rho}, \varepsilon, \delta, \beta$ are positive constants, ε and δ are sufficiently small and β sufficiently large, $\mathbf{n}$ denotes the unit outward normal to $\partial\Omega$.

The existence and uniqueness of solutions to (6.6)–(6.8) are proved by means of a modified Faedo–Galerkin approximation, where $(6.6)_1$ is solved directly by applying the classical theory of parabolic equations, and the solution is then inserted into $(6.6)_2$ while $(6.6)_2$ is replaced by a finite-dimensional system to give finite-dimensional approximate solutions to (6.6)–(6.8). With the help of the a priori estimates that are obtained by employing the energy estimates and the L^p-theory for parabolic equations as well as the Lions–Aubin lemma, the approximate solutions converge to a limit which solves (6.6)–(6.8). The result is summarized in the following lemma (see [60] for the details).

LEMMA 6.2. *Let* $\beta > \max\{4, \gamma\}$. *Assume that* ρ_0, $\mathbf{m}_0$ *satisfy the conditions in* (6.7). *Then there exists a weak solution* $(\rho, \mathbf{u})$ *of* (6.6)–(6.8) *such that* $\rho \in L^{\beta+1}((0,T)\times\Omega)$ *and the following estimates hold*:

$$\begin{aligned} &\sup_{t\in[0,T]} \left\{\|\rho\|_{L^\gamma}^\gamma + \delta\|\rho\|_{L^\beta}^\beta + \left\|\sqrt{\rho}\mathbf{u}\right\|^2\right\}(t) + \int_0^T \left(\|\mathbf{u}\|^2 + \|\nabla\mathbf{u}\|^2\right)(t)\,\mathrm{d}t \\ &\quad \leqslant CE_\delta(\rho_0, \mathbf{m}_0), \end{aligned} \tag{6.9}$$

$$\varepsilon\int_0^T \left\|\nabla\rho(t)\right\|^2 \mathrm{d}t \leqslant C(\beta, \delta, \rho_0, \mathbf{m}_0), \tag{6.10}$$

and

$$\frac{\mathrm{d}}{\mathrm{d}t}\left\{\int_\Omega\left(\frac{1}{2}\rho|\mathbf{u}|^2+\frac{a}{\gamma-1}\rho^\gamma+\frac{\delta}{\beta-1}\rho^\beta\right)(t,x)\,\mathrm{d}x\right\}$$
$$+\int_\Omega\left(\mu|\nabla\mathbf{u}|^2+(\lambda+\mu)|\operatorname{div}\mathbf{u}|^2\right)(t,x)\,\mathrm{d}x\leqslant 0\quad \text{in } \mathcal{D}'(0,T), \tag{6.11}$$

where

$$E_\delta(\rho_0,\mathbf{m}_0):=\int_\Omega\left\{\frac{|\mathbf{m}_0|^2}{2\rho_0}+\frac{a}{\gamma-1}\rho_0^\gamma+\frac{\delta}{\gamma-1}\rho_0^\beta\right\}\mathrm{d}x.$$

Moreover, there exists a $r_1>0$ *such that* $\rho_t,\Delta\rho\in L^{r_1}((0,T)\times\Omega)$ *and* $(6.6)_1$ *holds a.e. on* $(0,T)\times\Omega$.

Step II (Vanishing viscosity limit). In this step we let the artificial viscosity terms represented by the ε-quantities tend to zero. This is a delicate matter due to the lack of suitable estimates of the density ρ. Here the techniques developed in [60,146] are used.

Denote the solution of (6.6)–(6.8) established in Lemma 6.2 by $(\rho_\varepsilon,\mathbf{u}_\varepsilon)$. First, one uses the linear bounded operator

$$\mathcal{B}:\left\{f\in L^p(\Omega)\,\Big|\,\int_\Omega f=0\right\}\mapsto W_0^{1,p}(\Omega),\quad 1<p<\infty,$$

satisfying

$$\operatorname{div}\mathcal{B}(f)=f\quad\text{in }\Omega,\qquad \mathcal{B}(f)|_{\partial\Omega}=0,$$

to derive a higher integrability estimate of ρ_ε, i.e., taking the operator $\mathcal{B}$ on both sides of $(6.6)_2$, multiplying then by ρ_ε and integrating, one obtains after a lengthy but straightforward calculation that

$$\|\rho_\varepsilon\|_{L^{\gamma+1}((0,T)\times\Omega)}+\|\rho_\varepsilon\|_{L^{\beta+1}((0,T)\times\Omega)}\leqslant C(\delta,\rho_0,\mathbf{m}_0), \tag{6.12}$$

where $C(\delta,\rho_0,\mathbf{m}_0)$ is a positive constant independent of ε.

From (6.9), (6.10) and (6.12), we can extract a subsequence of $(\rho_\varepsilon,\mathbf{u}_\varepsilon)$, still denoted by $(\rho_\varepsilon,\mathbf{u}_\varepsilon)$, such that (also cf. Lemma C.1 of [145], Appendix C, and Lemma 5.1 of [146]), noticing that the system (6.6) gives bounds of $\partial_t\rho_\varepsilon$ and $\partial_t(\rho_\varepsilon\mathbf{u}_\varepsilon)$ in certain Sobolev spaces with a negative index,

$$\varepsilon\nabla\rho_\varepsilon\cdot\nabla\mathbf{u}_\varepsilon\to 0\quad\text{in } L^1\big((0,T)\times\Omega\big),$$
$$\varepsilon\Delta\rho_\varepsilon\to 0\quad\text{in } L^2\big((0,T),H^{-1}(\Omega)\big),$$
$$\rho_\varepsilon\to\rho\quad\text{in } C\big([0,T],L^\beta(\Omega)-w\big)\text{ and weakly in } L^{\beta+1}\big((0,T)\times\Omega\big),$$
$$\mathbf{u}_\varepsilon\rightharpoonup\mathbf{u}\quad\text{weakly in } L^2\big((0,T),H^1(\Omega)\big),$$

$$\rho_\varepsilon \mathbf{u}_\varepsilon \to \rho\mathbf{u} \quad \text{in } C\big([0,T], L^{2\gamma/(\gamma+1)}(\Omega) - w\big),$$
$$\rho_\varepsilon \mathbf{u}_\varepsilon \otimes \mathbf{u}_\varepsilon \to \rho\mathbf{u}\otimes\mathbf{u} \quad \text{in } \mathcal{D}'\big((0,T)\times\Omega\big).$$

Hence, letting $\varepsilon \to 0$ in (6.6), the limit $(\rho, \mathbf{u})$ satisfies the initial data (6.7) and

$$\begin{cases} \rho_t + \operatorname{div}(\rho\mathbf{u}) = 0, \\ (\rho\mathbf{u})_t + \operatorname{div}(\rho\mathbf{u}\otimes\mathbf{u}) + \nabla\overline{P} = \mu\Delta\mathbf{u} + (\lambda+\mu)\nabla\operatorname{div}\mathbf{u} \end{cases} \tag{6.13}$$

in $\mathcal{D}'((0,T)\times\Omega)$, where

$$a\rho_\varepsilon^\gamma + \delta\rho_\varepsilon^\beta \rightharpoonup \overline{P} \quad \text{weakly in } L^{(\beta+1)/\beta}\big((0,T)\times\Omega\big).$$

Our next goal is to show that

$$\overline{P} = a\rho^\gamma + \delta\rho^\beta, \tag{6.14}$$

which is equivalent to strong convergence of ρ_ε in $L^1((0,T)\times\Omega)$. To this end, we introduce the so-called effective viscous flux

$$a\rho^\gamma + \delta\rho^\beta - (\lambda+2\beta)\operatorname{div}\mathbf{u},$$

which enjoys many remarkable properties (see, e.g., [85,146,199]). It can be shown that as $\varepsilon \to 0$,

$$\big[a\rho_\varepsilon^\gamma + \delta\rho_\varepsilon^\beta - (\lambda+2\beta)\operatorname{div}\mathbf{u}_\varepsilon\big]\rho_\varepsilon \to \big[a\rho^\gamma + \delta\rho^\beta - (\lambda+2\beta)\operatorname{div}\mathbf{u}\big]\rho \tag{6.15}$$

in $\mathcal{D}'((0,T)\times\Omega)$, for $\beta \geqslant 6\gamma/(2\gamma-3)$. (6.15) plays a key role for the existence in [146]. The property (6.15) was proved by Serre [199] in the one-dimensional case and by Lions [146] in the multidimensional case, while in [60] Feireisl, Novotný and Petzeltová gave an elementary (simplified) proof based on the div–curl lemma and the following compactness lemma on commutators involving the Riesz transforms.

LEMMA 6.3. *Assume* $(v_n, w_n) \rightharpoonup (v, w)$ *weakly in* $L^p(\mathbb{R}^n) \times L^q(\mathbb{R}^n)$, $1/p + 1/q = 1/r < 1$. *Then*

$$v_n\mathcal{R}_{i,j}[w_n] - w_n\mathcal{R}_{i,j}[v_n] \rightharpoonup v\mathcal{R}_{i,j}[w] - w\mathcal{R}_{i,j}[v] \quad \textit{weakly in } L^r\big(\mathbb{R}^n\big),$$

where $\mathcal{R}_{i,j}[v] = \partial_{x_j}\Delta^{-1}\partial_{x_i} v$.

To show (6.14), we use (6.15) to prove the strong convergence of ρ_ε. Using the regularization procedure due to DiPerna and Lions, we see that $(\rho, \mathbf{u})$ solves $(6.13)_1$ in the sense of renormalized solutions, i.e., (6.4) holds. Taking $b(z) = z\log z$ in (6.4) which is possible by an approximation argument, we find that

$$\int_0^T \int_\Omega \rho \operatorname{div}\mathbf{u}\,\mathrm{d}x\,\mathrm{d}t = \int_\Omega \rho_0 \log\rho_0\,\mathrm{d}x - \int_\Omega \rho(T)\log\rho(T)\,\mathrm{d}x. \tag{6.16}$$

On the other hand, ρ_ε solves $(6.6)_1$, a.e. on $(0, T) \times \Omega$, in particular,

$$b(\rho_\varepsilon)_t + \operatorname{div}\big[b(\rho_\varepsilon)\mathbf{u}_\varepsilon\big] + \big[b'(\rho_\varepsilon)\rho_\varepsilon - b(\rho_\varepsilon)\big]\operatorname{div}\mathbf{u}_\varepsilon - \varepsilon\Delta b(\rho_\varepsilon) \leqslant 0$$

for any b convex and globally Lipschitz on $\mathbb{R}^+$, from which it follows

$$\int_0^T\int_\Omega \rho_\varepsilon \operatorname{div}\mathbf{u}_\varepsilon \,\mathrm{d}x\,\mathrm{d}t \leqslant \int_\Omega \rho_0\log\rho_0\,\mathrm{d}x - \int_\Omega \rho_\varepsilon(T)\log\rho_\varepsilon(T)\,\mathrm{d}x. \tag{6.17}$$

Now, combining (6.15) with (6.16) and (6.17), we are able to conclude that

$$\limsup_{\varepsilon\downarrow 0}\int_0^T \psi_m \int_\Omega \phi_m\big(a\rho_\varepsilon^\gamma + \delta\rho_\varepsilon^\beta\big)\rho_\varepsilon\,\mathrm{d}x\,\mathrm{d}t \leqslant \int_0^T\int_\Omega \overline{P}\rho\,\mathrm{d}x\,\mathrm{d}t \tag{6.18}$$

for all $m = 1, 2, \ldots$, where $\psi_m \in \mathcal{D}(0, T)$, $\psi_m \to 1$, $\phi_m \in \mathcal{D}(\Omega)$, $\phi_m \to 1$.

By terms of the monotonicity of $P(z) = az^\gamma + \delta z^\beta$,

$$\int_0^T \psi_m \int_\Omega \phi_m\big(P(\rho_\varepsilon) - P(v)\big)(\rho_\varepsilon - v)\,\mathrm{d}x\,\mathrm{d}t \geqslant 0,$$

which together with (6.18) implies

$$\begin{aligned}\int_0^T\int_\Omega \overline{P}\rho\,\mathrm{d}x\,\mathrm{d}t &+ \int_0^T \psi_m \int_\Omega \phi_m P(v)v\,\mathrm{d}x\,\mathrm{d}t \\ &\geqslant \int_0^T \psi_m \int_\Omega \phi_m\big(\overline{P}v + P(v)\rho\big)\,\mathrm{d}x\,\mathrm{d}t,\end{aligned}$$

whence, by letting $m \to \infty$,

$$\int_0^T\int_\Omega \big(\overline{P} - P(v)\big)(\rho - v)\,\mathrm{d}x\,\mathrm{d}t \geqslant 0,$$

and the choice $v = \rho + \eta\varphi$, $\eta \to 0$, φ arbitrary, yields (6.14). Summarizing the above analysis, the main result in this step reads:

LEMMA 6.4. *Let $\beta > 6\gamma/(2\gamma - 3)$ and $(\rho_0, \mathbf{m}_0)$ satisfy the conditions in* (6.7). *Then there is a finite energy weak solution $(\rho, \mathbf{u})$ to the problem*

$$\begin{cases} \rho_t + \operatorname{div}(\rho\mathbf{u}) = 0, \\ (\rho\mathbf{u})_t + \operatorname{div}(\rho\mathbf{u}\otimes\mathbf{u}) + \nabla\big(a\rho^\gamma + \delta\rho^\beta\big) = \mu\Delta\mathbf{u} + (\lambda + \mu)\nabla\operatorname{div}\mathbf{u}, \\ \mathbf{u}|_{\partial\Omega} = 0 \quad \text{and the initial conditions (6.1).} \end{cases} \tag{6.19}$$

Moreover, the following estimate holds

$$\sup_{t\in[0,T]}\big(\|\rho\|_{L^\gamma}^\gamma + \delta\|\rho\|_{L^\beta}^\beta + \big\|\sqrt{\rho}\mathbf{u}\big\|^2\big)(t) + \int_0^T \|\mathbf{u}(t)\|_{H^1}^2\,\mathrm{d}t \leqslant CE_\delta(\rho_0, \mathbf{m}_0), \tag{6.20}$$

where the constant C is independent of δ.

Step III (Passing to the limit in the artificial pressure term). Our ultimate goal is to take $\delta \to 0$ in $(6.19)_2$. We start with the construction of the initial data. Let $(\rho_0, \mathbf{m}_0)$ be the initial data in (6.1) and satisfy the conditions in Theorem 6.1. Then, it is easy to find a sequence $(\rho_0^\delta, \mathbf{m}_0^\delta)$ that satisfies the conditions in (6.7) and converges to $(\rho_0, \mathbf{m}_0)$ in $L^\gamma(\Omega) \times L^1(\Omega)$ as $\delta \to 0$; moreover, $|\mathbf{m}_0^\delta|^2/\rho_0^\delta$ is uniformly bounded in $L^1(\Omega)$. Denote the finite energy solution of (6.19) with the initial data $(\rho_0^\delta, \mathbf{m}_0^\delta)$ established in Lemma 6.4 by $(\rho_\delta, \mathbf{u}_\delta)$. Then for $(\rho_\delta, \mathbf{u}_\delta)$, the estimate (6.20) holds uniformly in δ.

Similarly to Step I, one first uses the operator $\mathcal{B}$ to derive higher integrability of ρ_δ. Let $b \in C^1$ be uniformly bounded and $\langle b(\rho_\delta)\rangle_\varepsilon$ denote the mollified function of $b(\rho_\delta)$ by the Friedrichs' mollifier. Thus, using

$$\varphi(t,x) = \psi(t)\mathcal{B}\Big[\langle b(\rho_\delta)\rangle_\varepsilon - \fint_\Omega \langle b(\rho_\delta)\rangle_\varepsilon \,\mathrm{d}x\Big], \qquad \psi \in \mathcal{D}'(0,T),$$

to test $(6.19)_2$, one obtains a representation of

$$\int_0^T \int_\Omega \psi\big(a\rho_\delta^\gamma + \delta\rho_\delta^\beta\big)\langle b(\rho_\varepsilon)\rangle_\varepsilon \,\mathrm{d}x\,\mathrm{d}t,$$

then letting $\varepsilon \to 0$ and approximating $b(z) = z^\theta$ for some $\theta > 0$ by a sequence of bounded functions b_n and a limit process, one gets an representation of $\int_0^T \int_\Omega \psi(a\rho_\delta^{\gamma+\theta} + \delta\rho_\delta^{\beta+\theta})\,\mathrm{d}x\,\mathrm{d}t$. Carefully bounding each term on the right-hand side of the representation, one gets after a lengthy but straightforward calculation that there is a $\theta > 0$ such that

$$\int_0^T \int_\Omega \big(a\rho_\delta^{\gamma+\theta} + \delta\rho_\delta^{\gamma+\theta}\big)\,\mathrm{d}x\,\mathrm{d}t \leqslant C \tag{6.21}$$

with C being a constant independent of δ.

REMARK 6.1. In fact, it can be shown that the optimal value of θ is $2\gamma/n - 1$ (cf. [146]).

Now, by the uniform estimates (6.20) and (6.21), we can extract a subsequence of $(\rho_\delta, \mathbf{u}_\delta)$, still denoted by $(\rho_\delta, \mathbf{u}_\delta)$, such that

$$\begin{aligned}
&\rho_\delta \to \rho && \text{in } C\big([0,T], L^\gamma(\Omega) - w\big),\\
&\mathbf{u}_\delta \rightharpoonup \mathbf{u} && \text{in } L^2\big((0,T), H_0^1(\Omega)\big),\\
&\rho_\delta\mathbf{u}_\delta \to \rho\mathbf{u} && \text{in } C\big([0,T], L^{2\gamma/(\gamma+1)}(\Omega) - w\big),\\
&\rho_\delta^\gamma \rightharpoonup \overline{\rho^\gamma} && \text{in } L^{(\gamma+\theta)/\gamma}\big((0,T)\times\Omega\big),\\
&\rho_\delta\mathbf{u}_\delta \otimes \mathbf{u}_\delta \to \rho\mathbf{u}\otimes\mathbf{u} && \text{in } \mathcal{D}'\big((0,T)\times\Omega\big),\\
&\delta\rho_\delta^\beta \to 0 && \text{in } L^1\big((0,T)\times\Omega\big).
\end{aligned} \tag{6.22}$$

Consequently, by taking $\delta \to 0$ in (6.19), the limit $(\rho, \mathbf{u})$ satisfies

$$\rho_t + \operatorname{div}(\rho\mathbf{u}) = 0, \tag{6.23}$$

$$(\rho\mathbf{u})_t + \operatorname{div}(\rho\mathbf{u} \otimes \mathbf{u}) + a\nabla\overline{\rho^\gamma} = \mu\Delta\mathbf{u} + (\lambda + \mu)\nabla \operatorname{div}\mathbf{u}, \tag{6.24}$$

in $\mathcal{D}'((0, T) \times \Omega)$.

So, to complete the proof of Theorem 6.1, it remains to prove $\overline{\rho^\gamma} = \rho^\gamma$, or equivalently, the strong convergence of ρ_δ in L^1. To this end, since $(\rho_\delta, \mathbf{u}_\delta)$ is a renormalized solution of (6.19) in $\mathcal{D}'((0, T) \times \mathbb{R}^n)$, we take b in (6.4) to be the cut-off function $T_k(z) = kT(z/k)$ with $T \in C^\infty(\mathbb{R})$ being concave and satisfying $T(z) = z$ for $z \leqslant 1$ and $T(z) = 2$ for $z \geqslant 3$, and let $\delta \to 0$ to arrive at

$$\partial_t \overline{T_k(\rho)} + \operatorname{div}\big(\overline{T_k(\rho)\mathbf{u}}\big) + \overline{\big(T_k'(\rho)\rho - T_k(\rho)\big)\operatorname{div}\mathbf{u}} = 0 \quad \text{in } \mathcal{D}'\big((0, T) \times \Omega\big). \tag{6.25}$$

Analogously as in Step II, we can use the div–curl lemma, Lemma 6.3 and (6.25) to derive the following identity involving the effective viscous flux:

$$\big[a\rho_\delta^\gamma - (\lambda + 2\mu)\operatorname{div}\mathbf{u}_\delta\big]T_k(\rho_\delta) \to \overline{\big[a\overline{\rho^\gamma} - (\lambda + 2\mu)\operatorname{div}\mathbf{u}\big]T_k(\rho)} \tag{6.26}$$

in $\mathcal{D}'((0, T) \times \Omega)$. In order to show $\overline{\rho^\gamma} = \rho^\gamma$, we need that the limit $(\rho, \mathbf{u})$ is a renormalized solution of (6.23), i.e., (6.4) holds. The equation (6.4) is proved by regularizing (6.25), and employing a limit procedure and the crucial observation that

$$\limsup_{\delta \downarrow 0} \big\| T_k(\rho_\delta) - T_k(\rho)\big\|_{L^{\gamma+1}((0,T)\times\Omega)} \leqslant C \tag{6.27}$$

with C being a constant independent of k. The estimate (6.27) is in the spirit of Jiang and Zhang [122], which means, roughly speaking, that neither the sequence ρ_δ nor its limit ρ is square integrable but the amplitude of possible oscillations is.

Introduce a family of functions

$$L_k(z) = \begin{cases} z\log z & \text{for } 0 \leqslant z < k, \\ z\log k + z\int_k^z \frac{T_k(s)}{s^2}\,\mathrm{d}s & \text{for } z \geqslant k. \end{cases}$$

Since $(\rho_\delta, \mathbf{u}_\delta)$ and $(\rho, \mathbf{u})$ are the renormalized solutions of $(6.19)_1$ and (6.23), respectively, we have by taking $b(z) = L_k(z)$ that

$$\partial_t L_k(\rho_\delta) + \operatorname{div}\big(L_k(\rho_\delta)\mathbf{u}_\delta\big) + T_k(\rho_\delta)\operatorname{div}\mathbf{u}_\delta = 0, \tag{6.28}$$

$$\partial_t L_k(\rho) + \operatorname{div}\big(L_k(\rho)\mathbf{u}\big) + T_k(\rho)\operatorname{div}\mathbf{u} = 0, \tag{6.29}$$

and by (6.20), we can assume

$$L_k(\rho_\delta) \to \overline{L_k(\rho)} \quad \text{in } C\big([0, T], L^\gamma(\Omega) - w\big)$$

and, approximating $z \log z \approx L_k(z)$,

$$\rho_\delta \log \rho_\delta \to \overline{\rho \log \rho} \quad \text{in } C\big([0,T], L^\alpha(\Omega) - w\big) \text{ for any } 1 \leqslant \alpha < \gamma.$$

Finally, we are able to show strong convergence of $\{\rho_\delta\}$ by comparing $L_k(\rho)$ with $\overline{L_k(\rho)}$ for $k \to \infty$. Subtracting (6.29) from (6.28) and letting $\delta \to 0$, we infer

$$\begin{aligned}
&\int_\Omega \big(\overline{L_k(\rho)} - L_k(\rho)\big)\phi \,\mathrm{d}x \\
&\quad = \int_0^T \int_\Omega \big(\overline{L_k(\rho)} - L_k(\rho)\big)\mathbf{u} \cdot \nabla\phi \,\mathrm{d}x\,\mathrm{d}t \\
&\qquad + \lim_{\delta \downarrow 0} \int_0^T \int_\Omega \big(T_k(\rho)\operatorname{div}\mathbf{u} - T_k(\rho_\delta)\operatorname{div}\mathbf{u}_\delta\big)\phi \,\mathrm{d}x\,\mathrm{d}t \quad \text{for any } \phi \in \mathcal{D}(\Omega).
\end{aligned} \tag{6.30}$$

On the other hand, making use of (6.26), the monotonicity of the pressure and (6.27), we can show that the second term on the right-hand side of (6.30) tends to zero as $k \to \infty$, and accordingly, we can pass to the limit for $k \to \infty$ in (6.30) to conclude $\overline{\rho \log \rho} = \rho \log \rho$ a.e., which implies strong convergence of $\{\rho_\delta\}$ in $L^1((0,T) \times \Omega)$ (cf., e.g., [129]). This completes the proof of Theorem 6.1. □

The same arguments as used for Theorem 6.1 can be applied and adapted to the Cauchy problem, initial boundary value problems in exterior domains, the initial boundary value problem with spatially periodic boundary condition, the problem on self-gravitating fluids and on nonmonotone pressure laws as well as the motion driven by a periodic external force, and a similar existence theorem can be proved, see [47,55,59,146].

When $1 < \gamma \leqslant n/2$, the global existence is known only for initial data with symmetry by now. Spherically symmetric solutions are investigated in [122] for any $\gamma > 1$, while axisymmetric solutions in [124] for any $\gamma > 1$. In these papers, the mathematical difficulties induced by the singularity at the symmetric axis are circumvented by combining different techniques from [60,122,146] and the concentrated compactness theory. When $\gamma = 1$, which corresponds to isothermal fluids, we have the global existence only in the case of spherically symmetric initial data. The global existence of weak solutions was obtained first by Hoff [84] for spherically symmetric BV-initial data, and the two-dimensional result in [84] was extended to L^p-initial data in [123] by exploiting certain compactness between the Orlicz and Sobolev spaces.

In the sequel, we present a existence theorem for the axisymmetric case in $\mathbb{R}^3$ for any $\gamma > 1$. For axisymmetric flow, there is no flow in the θ-direction and all θ derivatives are identically zero. So we consider only two variables, r – the radial direction and z – the axial direction, and denote by u_1 and u_2 the radial and axial components of the velocity $\mathbf{u} = (u_1, u_2)$, respectively. Then, the Cauchy problem for the system (2.22), (2.23) for axisymmetric isentropic flows becomes

$$\partial_t \rho + \frac{1}{r}\partial_r(r\rho u_1) + \partial_z(\rho u_2) = 0, \tag{6.31}$$

$$\begin{aligned} &\partial_t(\rho u_1) + \frac{1}{r}\partial_r\left(r\rho u_1^2\right) + \partial_z(\rho u_1 u_2) \\ &\quad = -a\partial_r \rho^\gamma + \mu\left[\frac{1}{r}\partial_r(r\partial_r u_1) + \partial_z^2 u_1\right] \\ &\qquad + (\lambda+\mu)\partial_r\left[\frac{1}{r}\partial_r(ru_1) + \partial_z u_2\right] - \mu\frac{u_1}{r^2}, \end{aligned} \tag{6.32}$$

$$\begin{aligned} &\partial_t(\rho u_2) + \frac{1}{r}\partial_r(r\rho u_1 u_2) + \partial_z\left(\rho u_2^2\right) \\ &\quad = -a\partial_z \rho^\gamma + \mu\left[\frac{1}{r}\partial_r(r\partial_r u_2) + \partial_z^2 u_2\right] \\ &\qquad + (\lambda+\mu)\,\partial_z\left[\frac{1}{r}\partial_r(ru_1) + \partial_z u_2\right] \end{aligned} \tag{6.33}$$

together with

$$\rho(0,r,z) = \rho_0(r,z), \qquad \rho\mathbf{u}(0,r,z) = \mathbf{m}_0(r,z), \quad (r,z) \in \mathbb{R}^+ \times \mathbb{R}, \tag{6.34}$$

$$u_1(t,0,z) = 0, \qquad \partial_r u_2(t,0,z) = 0, \quad t \geqslant 0, z \in \mathbb{R}. \tag{6.35}$$

Comparing with the two-dimensional or the spherically symmetric case, the difficulties here lie in the singularity at $r = 0$, the fact that the singularity set here is the plane $\mathbb{R}_0^+ \times \{0\} \times \mathbb{R}$ in $\mathbb{R}^3$ but not a line as in the spherically symmetric case, and the Neumann boundary condition for u_2 which could induce concentration of singularity involving u_2 at $r = 0$ in passing to the limit $r \downarrow 0$.

For the sake of simplicity of the presentation, let us assume that $\lambda + \mu \equiv 0$. It can be easily seen that the case $\lambda + \mu > 0$ will not arouse any new difficulties. Now we modify the definition of the finite energy solutions in the following way:

DEFINITION 6.2. We call $(\rho(t,r,z), \mathbf{u}(t,r,z))$ $(\mathbf{u} = (u_1, u_2))$ a finite energy weak solution of (6.31)–(6.35), if

(1) $\rho \geqslant 0$ a.e., and for any $T > 0$,

$$\begin{aligned} &\rho \in L^\infty\left([0,T], \mathcal{L}^\gamma\left(\mathbb{R}^+ \times \mathbb{R}\right)\right), \qquad \rho|\mathbf{u}|^2 \in L^\infty\left([0,T], \mathcal{L}^1\left(\mathbb{R}^+ \times \mathbb{R}\right)\right), \\ &\nabla\mathbf{u}, \quad u_1/r \in L^2\left([0,T], \mathcal{L}^2\left(\mathbb{R}^+ \times \mathbb{R}\right)\right), \\ &\rho \in C\left([0,T], \mathcal{L}^\gamma_{\mathrm{loc}}\left(\mathbb{R}_0^+ \times \mathbb{R}\right) - w\right), \\ &\rho\mathbf{u} \in C\left([0,T], \mathcal{L}^{2\gamma/(\gamma+1)}_{\mathrm{loc}}\left(\mathbb{R}_0^+ \times \mathbb{R}\right) - w\right), \\ &(\rho, \rho\mathbf{u})(0,x) = (\rho_0, \mathbf{m}_0)(x) \quad \text{weakly in } \mathcal{L}^\gamma_{\mathrm{loc}}\left(\mathbb{R}_0^+ \times \mathbb{R}\right) \times \mathcal{L}^{2\gamma/(\gamma+1)}_{\mathrm{loc}}\left(\mathbb{R}_0^+ \times \mathbb{R}\right), \end{aligned}$$

where

$$\mathcal{L}^p(\Omega) := \left\{ f \in L^p_{\mathrm{loc}}(\Omega);\ \int_\Omega |f(r,z)|^p r\,\mathrm{d}r\,\mathrm{d}z < \infty \right\}$$

with norm $\|\cdot\|_{\mathcal{L}^p(\Omega)} := (\int_\Omega |\cdot|^p r\,\mathrm{d}r\,\mathrm{d}z)^{1/p}$; $\mathcal{L}^p_{\mathrm{loc}}(\Omega)$ is defined similarly to $L^p_{\mathrm{loc}}(\Omega)$.

(2) For any $b \in C^1(\mathbb{R})$ such that $|b(s)| + |b'(s)s| \leqslant C$ for all $s \in \mathbb{R}$, there holds

$$\partial_t b(\rho) + \frac{1}{r}\partial_r\big[rb(\rho)u_1\big] + \partial_z\big[b(\rho)u_2\big] + \big[b'(\rho)\rho - b(\rho)\big]\left(\frac{u_1}{r} + \operatorname{div}\mathbf{u}\right) = 0$$

in $\mathcal{D}'((0,T)\times\mathbb{R}^+\times\mathbb{R})$, i.e. $(\rho,\mathbf{u})$ is a renormalized solution of (6.31).

(3) For any $t_2 \geqslant t_1 \geqslant 0$ and any $\psi \in C^1_0(\mathbb{R}^3)$, $\varphi \in C^1_0(\mathbb{R}^3)$ with $\varphi(t,0,z) = 0$, $\phi \in C^1_0(\mathbb{R}^3)$ with $\phi_r(t,0,z) = 0$, the following equations hold:

$$\int_{\mathbb{R}^+\times\mathbb{R}} \rho u_1 \psi r\,\mathrm{d}r\,\mathrm{d}z\Big|_{t_1}^{t_2} - \int_{t_1}^{t_2}\int_{\mathbb{R}^+\times\mathbb{R}} (\rho\psi_t + \rho u_1\psi_r + \rho u_2\psi_z) r\,\mathrm{d}r\,\mathrm{d}z\,\mathrm{d}t = 0,$$

$$\int_{\mathbb{R}^+\times\mathbb{R}} \rho u_1 \varphi r\,\mathrm{d}r\,\mathrm{d}z\Big|_{t_1}^{t_2} - \int_{t_1}^{t_2}\int_{\mathbb{R}^+\times\mathbb{R}} \big\{\rho u_1\varphi_t + \rho u_1^2\varphi_r + \rho u_1 u_2\varphi_z\big\} r\,\mathrm{d}r\,\mathrm{d}z\,\mathrm{d}t$$
$$= \int_{t_1}^{t_2}\int_{\mathbb{R}^+\times\mathbb{R}} \left\{ a\rho^\gamma\left[\frac{\varphi}{r} + \varphi_r\right] - \mu\partial_r u_1\varphi_r - \mu\frac{u_1}{r^2}\varphi \right\} r\,\mathrm{d}r\,\mathrm{d}z\,\mathrm{d}t,$$

$$\int_{\mathbb{R}^+\times\mathbb{R}} \rho u_2 \phi r\,\mathrm{d}r\,\mathrm{d}z\Big|_{t_1}^{t_2} - \int_{t_1}^{t_2}\int_{\mathbb{R}^+\times\mathbb{R}} \big\{\rho u_2\phi_t + \rho u_1 u_2\phi_r + \rho u_2^2\phi_z\big\} r\,\mathrm{d}r\,\mathrm{d}z\,\mathrm{d}t$$
$$= \int_{t_1}^{t_2}\int_{\mathbb{R}^+\times\mathbb{R}} \big\{ a\rho^\gamma\phi_z - \mu\partial_r u_2\phi_r - \mu\partial_z u_2\phi_z \big\} r\,\mathrm{d}r\,\mathrm{d}z\,\mathrm{d}t.$$

(4)

$$\int_{\mathbb{R}^+\times\mathbb{R}} \left(\rho\frac{|\mathbf{u}|^2}{2} + \frac{a\rho^\gamma}{\gamma-1}\right)(t,r,z) r\,\mathrm{d}r\,\mathrm{d}z + \mu\int_0^t\int_{\mathbb{R}^+\times\mathbb{R}} \left(|\nabla\mathbf{u}|^2 + \frac{u_1^2}{r^2}\right) r\,\mathrm{d}r\,\mathrm{d}z\,\mathrm{d}t$$
$$\leqslant \int_{\mathbb{R}^+\times\mathbb{R}} \left(\frac{|\mathbf{m}_0|^2}{2\rho_0} + \frac{a\rho_0^\gamma}{\gamma-1}\right) r\,\mathrm{d}r\,\mathrm{d}z \quad \forall t \geqslant 0.$$

Then we have the following existence theorem:

THEOREM 6.5. *Let $\gamma > 1$, $0 \leqslant \rho_0 \in \mathcal{L}^\gamma(\mathbb{R}^+\times\mathbb{R}) \cap \mathcal{L}^1(\mathbb{R}^+\times\mathbb{R})$ and $\mathbf{m}_0/\sqrt{\rho_0} \in \mathcal{L}^2(\mathbb{R}^+\times\mathbb{R})$. Then there exists a global finite energy weak solution of* (6.31)–(6.35), *such that for any $T, L > 0$ and $\alpha \in (0,1)$,*

$$\int_0^T\int_0^1\int_{-L}^L (\rho^\gamma + \rho u_1^2) r^\alpha\,\mathrm{d}r\,\mathrm{d}z\,\mathrm{d}t \leqslant C. \tag{6.36}$$

REMARK 6.2. If we define $\rho(t,\mathbf{x}) := \rho(t,r,z)$, $\mathbf{U}(t,\mathbf{x}) := (\frac{\mathbf{x}'}{r}u_1(t,r,z), u_2(t,r,z))$, where $\mathbf{x} = (x,y,z) \in \mathbb{R}^3$, $\mathbf{x}' = (x,y) \in \mathbb{R}^2$ and $r = |\mathbf{x}'|$. Then it is easy to see that $(\rho(t,\mathbf{x}), \mathbf{U}(t,\mathbf{x}))$ is a weak (finite energy) solution of the Cauchy problem for the compressible isentropic Navier–Stokes equations in $\mathbb{R}^3$ (cf. the proof of Theorem 5.7 in [84]).

Roughly speaking, Theorem 6.5 is proved by adapting the techniques from [60,122,146] and the concentration compactness principle. In order to exclude the possible concentration of singularity at the symmetric axis induced by the term $\rho(u_2)^2$, namely the possible concentration in the weak limit

$$\rho^\varepsilon\big(u_2^\varepsilon\big)^2 r\,\mathrm{d}r\,\mathrm{d}z\,\mathrm{d}t \rightharpoonup \rho(u_2)^2 r\,\mathrm{d}r\,\mathrm{d}z\,\mathrm{d}t + \sum_{i\in J} c_i\delta(t_i,0,z_i), \qquad \sum_{i\in J} c_i < \infty,$$

in the sense of measures, where $(\rho^\varepsilon, \mathbf{u}^\varepsilon)$, J and c_i are the approximate solutions, an at most countable set (possibly empty) and constants respectively, besides modifying the arguments in the proof of Theorem 6.1, we have to adapt Lions' concentration compactness method for stationary isothermal flows. Moreover, one obtains a new integrability estimate (6.36) of the density near $r = 0$ simultaneously.

Recently, the dynamics of the compressible isentropic Navier–Stokes equations has been studied by Feireisl and Petzeltová in a series of papers. In [56,61] they investigated the large-time behavior of weak solutions driven by a potential external force and the existence of global attractors by exploiting the weak convergence associated with the existence result and carefully analyzing the time-evolution of the energy. We summarize these results in Theorems 6.6 and 6.7.

THEOREM 6.6. *Let $n = 3$ and $\gamma > 3/2$, let $\Omega \subset \mathbb{R}^3$ be a domain with compact Lipschitz boundary. Assume that $G(x)$ is bounded and Lipschitz continuous on $\overline{\Omega}$ and the upper level sets $\{x \in \Omega;\ G(x) > k\}$ are connected in Ω for any $k < \sup_{x\in\Omega} G(x)$. Moreover, if Ω is unbounded, assume*

$$\lim_{R\to\infty} \operatorname{ess} \sup_{x\in\Omega, |x|\geqslant R} \big(\big|G(x)\big| + \big|\nabla G(x)\big|\big) = 0.$$

Then, for any finite energy weak solution $(\rho, \mathbf{u})$ of the problem (2.22), (2.23), (6.1), (6.2) *with the potential external force of density ∇G on $(0,\infty)\times\Omega$, there is a stationary state ρ_s determined by $a\nabla(\rho_s^\gamma) = \rho_s\nabla G$ on Ω, such that, as $t\to\infty$,*

$$\rho(t) \to \rho_s \quad \text{strongly in } L^\gamma(\Omega), \qquad \operatorname*{ess\,sup}_{\tau>t} \int_\Omega \rho(\tau,x)\big|\mathbf{u}(\tau,x)\big|^2\,\mathrm{d}x \to 0.$$

For the description of Theorem 6.7 we need some notation. Let $\mathcal{F}$ be a bounded subset of $L^\infty(\mathbb{R}\times\Omega)$ and denote

$$\mathcal{F}^+ := \Big\{\mathbf{f}\,\Big|\,\mathbf{f} = \lim_{\tau_n\to\infty} \mathbf{h}_n(\cdot+\tau_n) \text{ weak-}* \text{ in } L^\infty(\mathbb{R}\times\Omega) \\ \text{for a certain } \mathbf{h}_n \in \mathcal{F} \text{ and } \tau_n \to \infty\Big\},$$

$$\mathcal{A}[\mathcal{F}] := \big\{(\rho_0, \mathbf{m}_0) \,\big|\, (\rho_0, \mathbf{m}_0) = (\rho, \rho\mathbf{u})|_{t=0}, \text{ where } (\rho, \mathbf{u}) \text{ is a finite energy weak solution of the problem (2.22), (2.23), (6.2) on } \mathbb{R} \times \Omega \text{ with the external force of density } \mathbf{f} \in \mathcal{F}^+ \text{ in (2.23), } \Omega \in \mathbb{R}^3\big\},$$

$$\mathcal{U}[E_0, \mathcal{F}](t_0, t) := \Big\{(\rho, \rho\mathbf{u}) \,\Big|\, (\rho, \mathbf{u}) \text{ is a finite energy weak solution of the problem (2.22), (2.23), (6.2) on } \mathbb{R} \times \Omega \text{ with external force of density } \mathbf{f} \in \mathcal{F}, \text{ and such that } \limsup_{t \to t_0+} E\big(\rho(t), \mathbf{u}(t)\big) \leqslant E_0\Big\},$$

where $\Omega \subset \mathbb{R}^3$ is a bounded domain with Lipschitz boundary, E_0 is a given constant and

$$E(\rho, \mathbf{u}) := \int_\Omega \left(\frac{1}{2}\rho|\mathbf{u}|^2 + \frac{a}{\gamma - 1}\rho^\gamma\right)(t, x)\,\mathrm{d}x.$$

THEOREM 6.7. *Let $n = 3$ and $\gamma > 9/5$. Then $\mathcal{A}[\mathcal{F}]$ is compact in $L^\alpha(\Omega) \times (L^{2\gamma/(\gamma+1)}(\Omega) - w)$ and as $t \to \infty$,*

$$\sup_{(\rho,\mathbf{m}) \in \mathcal{U}[E_0,\mathcal{F}](t_0,t)} \left[\inf_{(\bar\rho,\bar{\mathbf{m}}) \in \mathcal{A}[\mathcal{F}]} \left(\|\rho - \bar\rho\|_{L^\alpha(\Omega)} + \left|\int_\Omega (\mathbf{m} - \bar{\mathbf{m}}) \cdot \Phi\,\mathrm{d}x\right|\right)\right] \to 0$$

for any $1 \leqslant \alpha < \gamma$ and any $\Phi \in L^{2\gamma/(\gamma-1)}(\Omega)$.

Theorem 6.7 shows that $\mathcal{A}[\mathcal{F}]$ is a global attractor in the sense of Foias and Temam [66]. Other related results on dynamics are contained in the articles [64,65] where bounded absorbing sets, complete bounded trajectories and the ω-limit sets are discussed.

6.2. *The heat-conducting case*

As stated in the Introduction, in the heat-conducting case we have only the global existence of spherically symmetric solutions in symmetric domains without the origin (in order to exclude the difficulties induced by the singularity at the symmetric axis). Next, we mainly present the corresponding existence result.

The spherically symmetric form of (2.19)–(2.21) in Lagrangian coordinates in the exterior domain $\{x \in \mathbb{R}^n;\ |x| > 1\}$ can be written

$$u_t = \big(r^{n-1}v\big)_y, \tag{6.37}$$

$$v_t = r^{n-1}\left[\tilde\mu\frac{(r^{n-1}v)_y}{u} - R\frac{\theta}{u}\right]_y, \quad y \in (0, \infty), t > 0, \tag{6.38}$$

$$c_V\theta_t = \kappa\left[\frac{r^{2n-2}\theta_y}{u}\right]_y + \frac{1}{u}\big[\tilde\mu\big(r^{n-1}v\big)_y - R\theta\big]\big(r^{n-1}v\big)_y - 2\mu(n-1)\big(r^{n-2}v^2\big)_y, \tag{6.39}$$

where $u \equiv u(t, y)$ is the specific volume, $v \equiv v(t, y)$ is the velocity component in the radial direction, (t, y) and (t, r) denotes the Lagrangian and Eulerian coordinates respectively, $\tilde{\mu} = \lambda + 2\mu$, and $r \equiv r(t, y)$ is determined by

$$r(t, y) = r_0(y) + \int_0^t v(y, \tau)\,\mathrm{d}\tau, \qquad y, t \geqslant 0,$$

$$r_0(y) := \left\{1 + n\int_0^y u_0(\xi)\,\mathrm{d}\xi\right\}^{1/n}.$$

As boundary conditions we consider nonslip and thermally insulated boundary, i.e.,

$$v(t, 0) = 0, \qquad \theta_y(t, 0) = 0, \qquad t \geqslant 0; \tag{6.40}$$

and initial conditions

$$\big(u(0, y), v(0, y), \theta(0, y)\big) = \big(u_0(y), v_0(y), \theta_0(y)\big), \quad y \geqslant 0. \tag{6.41}$$

For the problem (6.37)–(6.41), we have the following existence theorem [112]:

THEOREM 6.8. *Assume that*

$$u_0 - 1, v_0, \theta_0 - 1, r_0^{n-1}\partial_y u_0, r_0^{n-1}\partial_y v_0, r_0^{n-1}\partial_y \theta_0 \in L^2(0, \infty);$$
$$u_0(y), \theta_0(y) > 0 \quad \textit{for all } y \in [0, \infty),$$

and that the initial data are compatible with the boundary conditions (6.40). *Then the problem* (6.37)–(6.41) *has a unique solution* $\{u, v, \theta\}$ *with* $u, \theta > 0$ *such that, for any* $T > 0$,

$$u - 1, v, \theta - 1 \in L^\infty\big((0, T), H^1(0, \infty)\big), \qquad u_t \in L^\infty\big((0, T), L^2(0, \infty)\big),$$
$$v_t, \theta_t, u_{yt}, v_{yy}, \theta_{yy}, r^{n-1}\theta_y, r^{2n-2}\theta_{yy} \in L^2\big((0, T) \times (0, \infty)\big),$$

and $\{u, v, \theta\}$ *satisfies* (6.37)–(6.41) *almost everywhere in* $(0, T) \times (0, \infty)$ *and takes on the given boundary and initial conditions in the sense of traces. Moreover, when* $n = 3$, *there are positive constants* $\underline{u}, \bar{u}$ *independent of* t *and* y, *such that* $\underline{u} \leqslant u(t, y) \leqslant \bar{u}$ *for all* $t, y \geqslant 0$, *and for an arbitrary but fixed integer* $j \geqslant 2$, $\|v(t)\|_{L^{2j}(0,\infty)} \to 0$ *as* $t \to \infty$.

IDEA OF THE PROOF. The proof of Theorem 6.8 is essentially based on a careful examination of a priori estimates and a limit procedure. Since the domain is unbounded and the coefficients tend to infinity as $y \to \infty$, some difficulties arise; for example, from the a priori estimates we could get only $v(y, t) = o(y^{-1/2+1/(2n)})$ as $y \to \infty$; but this is not sufficient to guarantee integration by parts in the proof where $v = o(y^{-1+1/n})$ is required. To overcome such difficulties we first study an approximate problem in the bounded interval $(0, k)$ and show the a priori estimates independent of k by utilizing some cut-off function and modifying a technique of Kazhikhov [10,137] for the one-dimensional case, then letting k tend to infinity and using the obtained a priori estimates, we get a global spherically symmetric solution as the limit.

The large-time behavior is obtained by adapting an idea of Kazhikhov [137] to derive a representation of u from which the uniform boundedness of u from below and above follows, and utilizing tricky energy estimates to derive L^{2j}-bounds for v uniformly in t. □

REMARK 6.3. The same techniques work and an analogous theorem is obtained when (6.40) is replaced by the following boundary conditions:

$$v(t,0)=0, \qquad \theta(t,0)=1, \quad t\geqslant 0 \quad \text{(nonslip, constant temperature)}. \tag{6.42}$$

There are other papers contributing to the mathematical topics on spherically symmetric solutions in *bounded* annular domains. Nikolaev in [174] first studied the initial boundary value problem (6.37)–(6.39), (6.41), (6.42) in an annular domain and proved the global existence of smooth spherically symmetric solutions for strictly positive initial density and temperature, while the exponential stability as $t\to\infty$ of the corresponding spherically symmetric solutions was gotten in [114]. Yashima and Benabidallah [69,70] dealt with the case of nonnegative initial density and temperature and proved the global existence of weak spherically symmetric solutions. Using a difference scheme and a limit procedure, Chen and Kratka obtained the global spherically symmetric solutions for a free boundary problem in an annulus [24]. In [241] Zheng and Qin studied the existence of maximal attractors for the spherically symmetric smooth solutions in an annulus.

7. Nonlinear one-dimensional thermoelasticity

We shall study spatially one-dimensional models and prove the global existence of solutions for small data, i.e., for small deviations from the reference configuration, while for large data blow-up results will be presented. This shows that the dissipation induced by thermal diffusion is strong enough for small data to prevent a smooth solution from a finite-time blow-up but not for large data.

The equations for the displacement u and the temperature difference θ are those given by (2.2) and (2.12), where we shall assume without loss of generality that the medium is homogeneous and that the reference density ρ_c equals one. Write f and f_2 for the body force and for the heat supply, respectively, and for the sake of simplicity, denote $q:=Q$ in Section 2, and

$$\begin{aligned} &a:=\frac{\partial\widetilde{S}}{\partial u_x}, \qquad b:=-\frac{\partial\widetilde{S}}{\partial\theta}, \qquad c:=\frac{\partial\eta}{\partial\theta}, \\ &d:=-\frac{1}{\theta+T_0}\frac{\partial q}{\partial\theta_x}, \qquad g:=\frac{f_2}{\theta+T_0} \end{aligned} \tag{7.1}$$

(for $|\theta| < T_0$). Observing $-\partial\widetilde{S}/\partial\theta = \partial\eta/\partial u_x$ and assuming $q = q(\theta_x)$ for simplicity, we may rewrite the equations (3.1) and (2.12) as follows:

$$u_{tt} - a(u_x, \theta)u_{xx} + b(u_x, \theta)\theta_x = f \quad \text{in } (0, \infty) \times \Omega, \tag{7.2}$$

$$c(u_x, \theta)\theta_t + b(u_x, \theta)u_{tx} - d(\theta, \theta_x)\theta_{xx} = g \quad \text{in } (0, \infty) \times \Omega. \tag{7.3}$$

Additionally, we consider the following boundary and initial conditions, respectively:

$$u(t, 0) = u(t, 1) = \theta(t, 0) = \theta(t, 1) = 0 \quad \text{in } [0, \infty), \tag{7.4}$$

$$u(0, x) = u^0(x), \qquad u_t(0, x) = u^1(x), \qquad \theta(0, x) = \theta^0(x) \quad \text{in } \Omega, \tag{7.5}$$

with prescribed data u^0, u^1 and θ^0. The boundary conditions (7.4) mean that the boundary is rigidly clamped and the temperature is kept constant on the boundary.

Using the energy method and a tricky treatment of some boundary terms with techniques from the theory of boundary control, we can obtain the global existence and the exponential decay of smooth small solutions. More precisely, assume:

ASSUMPTION 7.1. *Variables a, b, c, d are C^2-functions of their arguments. There exist positive constants γ_0, γ_1 and K with $K < T_0$, such that if $|u_x| \leqslant K$, $|\theta| \leqslant K$, $|\theta_x| \leqslant K$, we have*

$$\gamma_0 \leqslant a(u_x, \theta), \qquad c(u_x, \theta), \qquad d(\theta, \theta_x) \leqslant \gamma_1, \qquad b(u_x, \theta) \neq 0.$$

f and f_2 satisfy:

$$f, f_2 \in C^2\big([0, \infty), L^2(\Omega)\big) \cap C^1\big([0, \infty), H^1(\Omega)\big).$$

And next:

ASSUMPTION 7.2. *Suppose $u^0 \in H^3(\Omega)$, $u^1 \in H^2(\Omega)$, $u^2 \in H^1(\Omega)$, $\theta^0 \in H^3(\Omega)$, $\theta^1 \in H^2(\Omega)$ and $|\theta^0(x)| < T_0$ in $\overline{\Omega}$,*

$$u^0 = u^1 = u^2 = \theta^0 = \theta^1 = 0 \quad \text{on } \partial\Omega,$$

where $u^2 := u_{tt}|_{t=0}$, $\theta^1 := \theta_t|_{t=0}$ are given formally through the differential equations, explicitly in terms of the initial data u^0, u^1, θ^0:

$$u^2 = a\big(u^0_x, \theta^0\big)u^0_{xx} + b\big(u^0_x, \theta^0\big)\theta^0_{xx} + f|_{t=0},$$

$$\theta^1 = \frac{1}{c(u^0_x, \theta^0)}\big\{d\big(\theta^0, \theta^0_x\big)\theta^0_{xx} - b\big(u^0_x, \theta^0\big)u^1_x + g|_{t=0}\big\}.$$

REMARK 7.1. Since we are looking for the solution in a K–neighborhood of the origin, we can assume without loss of generality that the functions a, b, c, d and their derivatives are bounded.

Thus, we have [193]:

THEOREM 7.1. *Let*

$$\lambda(t) := \sum_{j=0}^{1} \left\| D^j (f, f_2)(t, \cdot) \right\|^2 + \left\| \partial_t^2 (f, f_2)(t, \cdot) \right\|^2 + \left\| \partial_t \partial_x (f, f_2)(t, \cdot) \right\|^2$$

and suppose that Assumptions 7.1 *and* 7.2 *are satisfied. Then there exists a small constant* $\varepsilon_0 > 0$ *such that if*

$$\left\| u^0 \right\|_{H^2}^2 + \left\| u^1 \right\|_{H^2}^2 + \left\| u^2 \right\|_{H^1}^2 + \left\| \theta^0 \right\|_{H^2}^2 + \left\| \theta^1 \right\|_{H^2}^2 + \sup_{t \geqslant 0} \lambda(t) \leqslant \varepsilon_0,$$

then the initial boundary value problem (7.2)–(7.5) *admits a unique global solution*

$$u \in \bigcap_{j=0}^{3} C^j\big([0, \infty), H^{3-j}(\Omega)\big), \qquad \theta \in \bigcap_{j=0}^{1} C^j\big([0, \infty), H^{3-j}(\Omega)\big),$$
$$\theta \in C^2\big([0, \infty), L^2(\Omega)\big).$$

Moreover, there are constants $d_1, d_2 > 0$ *such that, for* $t \geqslant 0$,

$$\begin{aligned} &\sum_{j=0}^{3} \left\| D^j u(t) \right\|^2 + \sum_{j=0}^{2} \left\| D^j \theta(t) \right\|^2 \\ &\quad \leqslant d_1 \mathrm{e}^{-d_2 t} \bigg(\left\| u^0 \right\|_{H^2}^2 + \left\| u^1 \right\|_{H^2}^2 + \left\| u^2 \right\|_{H^1}^2 \\ &\qquad + \left\| \theta^0 \right\|_{H^2}^2 + \left\| \theta^1 \right\|_{H^2}^2 + \int_0^t \mathrm{e}^{d_2 r} \lambda(r) \, \mathrm{d}r \bigg). \end{aligned}$$

PROOF. First, for appropriately small ε_0, by using the Banach contraction mapping principle and a standard procedure, one obtains a local smooth solution $U = (u, \theta)$ defined on the time interval $[0, T_1]$ for some $T_1 > 0$ which is in the function space given in Theorem 7.1 (cf. Notes at the end of Section 7) and satisfies

$$|u_x| < K, \qquad |\theta| < K, \qquad |\theta_x| < K \quad \text{in } [0, T_1] \times \overline{\Omega}. \tag{7.6}$$

Next we derive a uniform a priori estimate for (u, θ), then with the help of this a priori estimate, one can continue the local solution globally in time.

Taking ∂_t^j $(j = 0, 1, 2)$ on both sides of (7.2) and (7.3), and multiplying by $\partial_t^j u_t$ and $\partial_t^j \theta$, respectively, adding then together, integrating with respect to x and integrating by parts, we obtain

$$\frac{1}{2}\sum_{j=0}^{2}\left\{\frac{\mathrm{d}}{\mathrm{d}t}\left[\left\|\partial_t^j u_t\right\|^2+\int_0^1 a\left(\partial_t^j u_x\right)^2\mathrm{d}x+\int_0^1 c\left(\partial_t^j \theta\right)^2\mathrm{d}x\right]+\int_0^1 d\left(\partial_t^j \theta_x\right)^2\mathrm{d}x\right\}$$

$$=R_1(t)+\int_0^1\left(\partial_t^j f\,\partial_t^j u_t+\partial_t^j g\,\partial_t^j \theta\right)\mathrm{d}x, \tag{7.7}$$

where R_1 contains the cubic or higher-order terms of U with derivatives up to third order, and can be easily bounded from above as follows, using (7.6) and Sobolev's embedding theorem ($H^1 \hookrightarrow L^\infty$).

$$\left|R_1(t)\right| \leqslant C\left(X_2^{3/2}(t)+X_2^2(t)+X_2^{1/2}(t)X_3(t)\right), \tag{7.8}$$

with X_2 and X_3 being defined by (recalling $D=(\partial_t,\nabla)$)

$$X_2(t):=\sum_{j=0}^{2}\left\|D^j u\right\|^2+\sum_{j=0}^{1}\left\|D^j\theta\right\|^2+\left\|\left(u_{ttt},u_{ttx},u_{txx},\theta_{tx},\theta_{tt}\right)\right\|^2,$$

$$X_3(t):=\left\|\left(u_{xxx},\theta_{xx},\theta_{ttx},\theta_{txx},\theta_{xxx}\right)\right\|^2.$$

To bound θ_{xx} and θ_{xxx}, we take $\partial_t^j\partial_x$ ($j=0,1$) on both sides of (7.2) and (7.3), and multiply the resulting equations by $\partial_t^j u_{tx}$ and $\partial_t^j\theta_x$, respectively, add together and integrate in x to deduce that

$$\frac{1}{2}\frac{\mathrm{d}}{\mathrm{d}t}\left(\left\|\partial_t^j u_{tx}\right\|^2+\int_0^1 a\left(\partial_t^j u_{xx}\right)^2\mathrm{d}x+\int_0^1 c\left(\partial_t^j\theta_x\right)^2\mathrm{d}x\right)+\int_0^1 d\left(\partial_t^j\theta_{xx}\right)^2\mathrm{d}x$$

$$=\sum_{j=0}^{1}\left\{\int_0^1\left(\partial_t^j f_x\,\partial_t^j u_{tx}+\partial_t^j g_x\,\partial_t^j\theta_x\right)\mathrm{d}x\right.$$

$$\left.+\,b\,\partial_t^j u_{tx}\,\partial_t^j\theta_x\big|_0^1-\partial_t^j f\,\partial_t^j u_{tx}\big|_0^1-\partial_t^j g\,\partial_t^j\theta_x\big|_0^1\right\}+R_2(t), \tag{7.9}$$

where R_2, similarly to R_1, contains the cubic or higher-order terms in (7.9) and is bounded from above by the right-hand side of (7.8); moreover, we have used the relations

$$bu_{tx}-g-d\theta_{xx}\big|_0^1=-c\theta_t\big|_0^1=0,$$

$$d\theta_{txx}-bu_{ttx}+d_t\theta_{xx}-b_t u_{tx}\big|_0^1=-g_t\big|_0^1.$$

In order to deal with the boundary terms in (7.9) we use an approach of Munõz Rivera [166] from the theory of boundary control. Differentiating (7.2) with respect to t

once and twice and multiplying by $(x-1/2)u_{tx}$ and $(x-1/2)u_{ttx}$ respectively, integrating in x, one gets after a lengthy but straightforward calculation that

$$\begin{aligned}\frac{1}{4}\sum_{j=0}^{1}\langle\sqrt{a}\partial_t^j u_{tx}\rangle_{\partial\Omega}^2 &= \sum_{j=0}^{1}\left\{\frac{\mathrm{d}}{\mathrm{d}t}\int_0^1\left(x-\frac{1}{2}\right)\partial_t^j u_{tx}\,\partial_t^j u_{tt}+\frac{1}{2}\|\partial_t^j u_{tt}\|^2+\frac{1}{2}\int_0^1(\partial_t^j u_{tx})^2\,\mathrm{d}x\right.\\ &\quad+\int_0^1\left(x-\frac{1}{2}\right)b\,\partial_t^j u_{tx}\,\partial_t^j\theta_{tx}\,\mathrm{d}x\\ &\quad\left.-\int_0^1\left(x-\frac{1}{2}\right)\partial_t^j f_t\,\partial_t^j u_{tx}\,\mathrm{d}x\right\}+R_3 \end{aligned}\tag{7.10}$$

with R_3 being bounded from above by the right-hand side of (7.8), and $\langle h\rangle_{\partial\Omega}^2=h^2|_{x=1}+h^2|_{x=0}$.

The strategy in the sequel consists of estimating u in terms of θ and its derivatives. Using (7.3) and Poincaré's inequality, we have

$$\|u_{tx}\|^2+\|u_t\|^2\leqslant C\big(\|\theta_{tx}\|^2+\|\theta_{xx}\|^2+\|g\|^2\big).\tag{7.11}$$

In order to produce the terms $\int_0^1 au_x^2\,\mathrm{d}x$ and $\int_0^1 au_{tt}^2\,\mathrm{d}x$, we employ (7.2) and (7.11) to arrive at

$$\frac{1}{2}\int_0^1 au_x^2\,\mathrm{d}x+\frac{\mathrm{d}}{\mathrm{d}t}\int_0^1 uu_t\,\mathrm{d}x\leqslant C\big\|(\theta_{tx},\theta_x,\theta_{xx},f,g)\big\|^2+R_4,\tag{7.12}$$

$$\|u_{tt}\|^2-\frac{\mathrm{d}}{\mathrm{d}t}\int_0^1 u_tu_{tt}\,\mathrm{d}x\leqslant C\big\|(\theta_{tx},\theta_{xx},g,f_t)\big\|^2+R_5,\tag{7.13}$$

where R_4, R_5 contain the cubic terms and are bounded from above by the right-hand side of (7.8). Using (7.2), (7.3) and (7.12), we get

$$\begin{aligned}&\|u_{xx}\|^2+\|u_{ttx}\|^2\\ &\quad\leqslant\frac{\mathrm{d}}{\mathrm{d}t}\int_0^1 u_tu_{tt}\,\mathrm{d}x+C\big\|(\theta_x,\theta_{tx},\theta_{xx},f,f_t,g,\theta_{tt},\theta_{txx},g_t)\big\|^2+R_6\end{aligned}\tag{7.14}$$

with R_6 being bounded from above by the right-hand side of (7.8). Similarly to (7.13), we have

$$\begin{aligned}\|u_{ttt}\|^2&\leqslant\frac{\mathrm{d}}{\mathrm{d}t}\int_0^1(u_{tt}u_{ttt}+u_tu_{tt})\,\mathrm{d}x+\int_0^1 u_{tt}f_{tt}\,\mathrm{d}x\\ &\quad+C\big\|(\theta_{tt},\theta_{txx},\theta_{tx},\theta_{xx},g,g_t,f_t)\big\|^2+R_7,\end{aligned}\tag{7.15}$$

where R_7, similar to R_1, contains the cubic terms and is bounded from above by the right-hand side of (7.8).

Now if we differentiate (7.2) and (7.3), we easily see that

$$\begin{aligned} &\|u_{txx}\|^2 + \|u_{xxx}\|^2 + \|\theta_{xxx}\|^2 \\ &\quad \leqslant C\big\|(u_{ttt}, u_{ttx}, u_{txx}, \theta_{tx}, \theta_{xx}, f_t, f_x, g_x)\big\|^2 + R_8, \end{aligned} \tag{7.16}$$

where R_8 is bounded from above by the right-hand side of (7.8). On the other hand, the boundary terms $b\,\partial_t^j u_{tx}\,\partial_t^j \theta_x|_0^1$ in (7.9) can be estimated as follows

$$\begin{aligned} &a\,\partial_t^j u_{tx}\,\partial_t^j \theta_x\big|_0^1 \\ &\quad \leqslant \frac{\varepsilon^{1/2}}{4}\big\langle\sqrt{a}\,\partial_t^j u_{tx}\big\rangle_{\partial\Omega}^2 + C\frac{1}{\varepsilon^{1/2}}\big\langle\partial_t^j \theta_x\big\rangle_{\partial\Omega}^2 \\ &\quad \leqslant \frac{\varepsilon^{1/2}}{4}\big\langle\sqrt{a}\,\partial_t^j u_{tx}\big\rangle_{\partial\Omega}^2 + C\big(\varepsilon^{1/2}\big\|\partial_t^j \theta_{xx}\big\|^2 + \varepsilon^{-3/2}\big\|\partial_t^j \theta_x\big\|^2\big) \end{aligned} \tag{7.17}$$

for any $0 < \varepsilon < 1$.

Now, define

$$\begin{aligned} X_1(t) :={}& \frac{1}{\varepsilon^2}\big\|(u_t, u_{tt}, u_{ttt})\big\|^2 + \int_0^1 a\big(u_x^2 + u_{tx}^2 + u_{ttx}^2\big)\,\mathrm{d}x \\ &+ \int_0^1 c\big(\theta^2 + \theta_t^2 + \theta_{tt}^2\big)\,\mathrm{d}x \\ &+ \big\|(u_{tx}, u_{ttx})\big\|^2 + \int_0^1 a\big(u_{xx}^2 + u_{txx}^2\big)\,\mathrm{d}x + \int_0^1 c\big(\theta_x^2 + \theta_{tx}^2\big)\,\mathrm{d}x \\ &- \varepsilon^{1/2}\int_0^1 \Big(x - \frac{1}{2}\Big)(u_{tx}u_{tt} + u_{ttx}u_{ttt})\,\mathrm{d}x \\ &+ \varepsilon^{1/4}\int_0^1 (uu_t - u_t u_{tt} - u_{tt}u_{ttt})\,\mathrm{d}x \end{aligned}$$

with a small positive constant ε.

It is easy to see that there exists $\varepsilon^* > 0$ such that, for $0 < \varepsilon \leqslant \varepsilon^*$,

$$K_1 X_2(t) \leqslant X_1(t) \leqslant K_2 X_2(t) \tag{7.18}$$

for some constants K_1, K_2, where $X_2(t)$ is defined below (7.8).

We multiply (7.7) by ε^{-2} and add to (7.9), multiply (7.10) by $\varepsilon^{1/2}$, (7.12) and (7.13) as well as (7.15) by $\varepsilon^{1/4}$. Taking ε small enough, utilizing (7.17), (7.10), (7.11) and (7.13)–(7.16), recalling (7.18), we obtain

$$\begin{aligned}
&\frac{\mathrm{d}X_1(t)}{\mathrm{d}t}+C_1X_1(t)+C_2\big\|(u_{xxx},\theta_{xx},\theta_{ttx},\theta_{txx},\theta_{xxx})\big\|^2\\
&\quad\leqslant C\Bigg(\sum_{j=1}^{8}|R_j|+\sum_{j=0}^{1}\big\|D^j(f,f_2)(t)\big\|^2\\
&\qquad+\big\|\partial_t^2(f,f_2)(t)\big\|^2+\big\|\partial_t\partial_x(f,f_2)(t)\big\|^2\Bigg)\\
&\quad\equiv C\Bigg(\sum_{j=1}^{8}|R_j|+\lambda(t)\Bigg),
\end{aligned}$$

which together with the upper bounds of R_j (cf. (7.8)) yields

$$\frac{\mathrm{d}X_1(t)}{\mathrm{d}t}+C_1X_1(t)+C_2X_3(t)\leqslant C_3\big\{X_1^{3/2}+X_1^2+X_1^{1/2}X_3+\lambda\big\}(t). \tag{7.19}$$

If we now assume a priori

$$X_1(s)\leqslant\min\left\{\left[\frac{C_2}{C_3}\right]^2,\left[\frac{C_1}{C_3}\right]^2\frac{1}{16},\frac{C_1}{4C_3}\right\}\quad\text{for }0\leqslant s\leqslant t, \tag{7.20}$$

then we conclude from (7.19) that

$$\frac{\mathrm{d}X_1(s)}{\mathrm{d}s}+\frac{C_1}{2}X_1(s)\leqslant C_3\lambda(s),$$

which gives immediately

$$X_1(s)\leqslant \mathrm{e}^{-C_1s/2}X_1(0)+C_3\mathrm{e}^{-C_1s/2}\int_0^s\mathrm{e}^{C_1r/2}\lambda(r)\,\mathrm{d}r. \tag{7.21}$$

Now let

$$\lambda_1:=\sup_{s\geqslant0}\mathrm{e}^{-C_1s/2}\int_0^s\mathrm{e}^{C_1r/2}\lambda(r)\,\mathrm{d}r.$$

If

$$X_1(0)+C_3\lambda_1<\frac{1}{2}\min\left\{\left[\frac{C_2}{C_3}\right]^2,\left[\frac{C_1}{C_3}\right]^2\frac{1}{16},\frac{C_1}{4C_3}\right\},$$

then (7.20) holds for $0 \leqslant s \leqslant t_1$, for some $t_1 > 0$ by continuity, hence, (7.21) holds for $0 \leqslant s \leqslant t_1$, consequently

$$X_1(s) \leqslant \frac{1}{2}\min\left\{\left[\frac{C_2}{C_3}\right]^2, \left[\frac{C_1}{C_3}\right]^2 \frac{1}{16}, \frac{C_1}{4C_3}\right\} \quad \text{for } 0 \leqslant s \leqslant t_1.$$

By continuity, (7.21) holds for any $s \geqslant 0$, which is the desired a priori estimate.

Now the usual combination of the local existence and uniqueness theorem with the uniform a priori estimate (7.21) concludes the proof of Theorem 7.1. Finally, the exponential decay follows easily from (7.21). □

The arguments in Theorem 7.1 can be modified and applied to the Cauchy problem to get the global existence. Moreover, one can utilize the decay of solutions to the linearized equations to obtain the time-asymptotic behavior of solutions to the nonlinear equations. To prove the global existence for the Cauchy problem we assume without loss of generality that $f = f_2 = 0$ and

$$q = -\hat{\kappa}(u_x, \theta)\theta_x \quad \text{(Fourier's law)},$$

and introduce the variables $w := u_x$, $v := u_t$ to write the system (7.2), (7.3) as a first-order system:

$$w_t = v_x, \tag{7.22}$$

$$v_t = a(w,\theta)w_x - b(w,\theta)\theta_x \quad \left(\Longleftrightarrow v_t = \widetilde{S}(w,\theta)_x\right), \tag{7.23}$$

$$c(w,\theta)\theta_t = \frac{[\hat{\kappa}(w,\theta)\theta_x]_x}{(\theta + T_0)} - b(w,\theta)v_x \tag{7.24}$$

with the prescribed initial data

$$(w, v, \theta)(t = 0) = \left(w^0, v^0, \theta^0\right) \equiv V^0 \quad \text{in } \mathbb{R}. \tag{7.25}$$

Thus, we have the following theorem.

THEOREM 7.2. *Let $V^0 \in H^2(\mathbb{R})$ and $\hat{\kappa}$ be a C^3-function of its arguments. Assume that Assumption* 7.1 *with d replaced by $\hat{\kappa}$ is satisfied. Then there exists an $\varepsilon > 0$ such that if $\|V^0\|_{H^2} \leqslant \varepsilon$, then the Cauchy problem* (7.22)–(7.25) *has a unique global solution $V = (w, v, \theta)$ satisfying $\|V(t)\|_{L^\infty} \leqslant K < T_0$ for all $t \geqslant 0$ and*

$$w, v \in C^0\left([0,\infty), H^2(\mathbb{R})\right) \cap C^1\left([0,\infty), H^1(\mathbb{R})\right),$$

$$\theta \in C^0\left([0,\infty), H^2(\mathbb{R})\right) \cap C^1\left([0,\infty), L^2(\mathbb{R})\right),$$

$$\theta_t, \theta_{xx} \in L^2\left((0,\infty), H^1(\mathbb{R})\right).$$

Moreover, if $V^0 \in H^3(\mathbb{R}) \cap L^1(\mathbb{R})$ *and* $\|V^0\|_{H^3\cap L^1}$ *is sufficiently small, then*

$$\|V(t)\|_{H^1} \leqslant C(1+t)^{-1/4}\|V^0\|_{H^3\cap L^1} \quad \text{for all } t \geqslant 0. \tag{7.26}$$

SKETCH OF THE PROOF. To obtain the global existence, as in Theorem 7.1, the key point is to derive a uniform a priori estimate for V. Let (w, v, θ) be a solution of (7.22)–(7.25) on $[0, T]$ with the regularities given in Theorem 7.2 and satisfies $\|V(t)\|_{L^\infty} \leqslant K$ for $t \in [0, T]$.

To derive the a priori estimate, comparing with the proof in Theorem 7.1, the main difference lies in the estimate of V in $L^\infty((0,\infty), L^2(\mathbb{R}))$-norm which cannot be derived directly by applying the arguments used in Theorem 7.1 because of the unboundedness of the domain and the L^1 nonintegrability with respect to t of $\|V(t)\|^2$. The estimate of V in $L^\infty((0,\infty), L^2(\mathbb{R}))$ is obtained by exploiting some relations associated with the 2nd law of thermodynamics, and this idea is due to Kawashima [131,132] (also cf. [98]).

Setting

$$\Lambda(w,\theta) := \psi(w,\theta) - \psi(0,0) - \theta\psi_\theta(w,\theta)$$

with ψ being the Helmholtz free energy, and using (7.22)–(7.24), one obtains after a straightforward calculation that

$$\left[\Lambda(w,\theta) + \frac{v^2}{2}\right]_t + \frac{T_0\hat{\kappa}(w,\theta)\theta_x^2}{(\theta+T_0)^2} = \left[\widetilde{S}v + \frac{\hat{\kappa}(w,\theta)\theta\theta_x}{(\theta+T_0)}\right]_x. \tag{7.27}$$

Recalling $\|V(t)\|_{L^\infty} \leqslant K$, we use a Taylor expansion and the assumptions of the theorem to infer

$$\frac{\gamma_1^2}{2}\left(w^2+\theta^2\right) \geqslant \Lambda(w,\theta) \geqslant \frac{\gamma_0^2}{2}\left(w+\theta^2\right). \tag{7.28}$$

Integrating (7.27) over $(0,t)\times\mathbb{R}$ and utilizing (7.28), we find that

$$\|V(t)\|^2 + \int_0^t \|\theta_x(s)\|^2\,ds \leqslant C\|V^0\|^2 \quad \forall t \geqslant 0. \tag{7.29}$$

Once (7.29) is shown, the desired a priori estimate can be obtained, using the same arguments as in Theorem 7.1, and therefore, the global existence of a unique solution follows. Moreover, the large-time behavior of the global solution (7.26) can be proved by using the smallness of the solution and the decay-rates of the solutions to the linearized equations which are obtained by applying the Fourier transform and spectral analysis. It should be pointed out that the decay of $V(t)$ in other norms can also be given. We refer to [131,132,240,242] for the detailed proof of Theorem 7.2. □

We can combine and adapt the techniques in the proof of Theorems 7.1 and 7.2 to deal with various initial boundary value problems in bounded or unbounded domains, such as the problem (7.2), (7.3), (7.5) with one of the following boundary conditions:

$$\begin{aligned}
&\widetilde{S}(u_x,\theta)\big|_{\partial\Omega}=0, \qquad q(u_x,\theta_x,\theta)\big|_{\partial\Omega}=0 \quad \text{(traction free, thermally insulated);}\\
&\widetilde{S}(u_x,\theta)\big|_{\partial\Omega}=0, \qquad \theta|_{\partial\Omega}=0 \quad \text{(traction free, constant temperature);}\\
&u|_{\partial\Omega}=0, \qquad q(u_x,\theta_x,\theta)|_{\partial\Omega}=0 \quad \text{(rigidly clamped, thermally insulated),}
\end{aligned} \tag{7.30}$$

where $\Omega=(0,1)$ or $\Omega=(0,\infty)$. In the same way, we can obtain a similar global existence theorem for small smooth initial data, see, e.g., [104,106,107,131,135,136,192,208, 238,240,242], and [191,203,239] on thermoelasticity with dissipation, also see the monograph [120]. In the case of $\Omega=(0,1)$, the exponential stability of the global solution is proved [106,208], while in the case of $\Omega=(0,\infty)$, the polynomial decay rates of the global solution for the boundary conditions $(7.30)_2$, or $(7.30)_3$ or $u|_{\partial\Omega}=\theta|_{\partial\Omega}=0$ is obtained by using the Fourier sine and cosine transforms, the method of spectral analysis and the Laplace transform (cf. [104,131,242]). The decay of the global solution for the boundary conditions $(7.30)_1$ in the half line $\Omega=(0,\infty)$ still remains open.

In the sequel we show by an example that for large data a smooth solution will break down in finite time. In other words, the dissipation induced by thermal diffusion is too weak to prevent a smooth solution from blow-up for large data.

We consider the special constitutive equations of the form

$$\begin{cases}
\widetilde{S}(w,\theta):=p(w)+\gamma\theta,\\
\varepsilon(w,\theta):=P(w)+\delta\theta-\gamma T_0 w,\\
q(w,\theta,\theta_x):=-\kappa\theta_x,
\end{cases} \tag{7.31}$$

where $T_0,\kappa,\delta>0$, $\gamma\neq 0$ are constants, T_0 is the reference temperature; $p:(-1,\infty)\to\mathbb{R}$ is a given function,

$$P(w):=\int_0^w p(\xi)\,d\xi, \qquad w>-1. \tag{7.32}$$

It is easy to see that these constitutive equations satisfy $\varepsilon_w=\widetilde{S}-(\theta+T_0)\widetilde{S}_\theta$ and $g\,q(w,\theta,g)\leqslant 0$ and that

$$\psi(w,\theta)=P(w)+\gamma w\theta+\delta(\theta+T_0)\log\frac{T_0}{(\theta+T_0)}+\delta\theta \tag{7.33}$$

is a Helmholtz free energy; moreover, the corresponding entropy is given by

$$\eta(w,\theta)=-\gamma w-\delta\log\frac{T_0}{(\theta+T_0)}.$$

It should be pointed out that we do not claim that the constitutive equations (7.31)–(7.33) are appropriate to any real material; however, they do satisfy the assumptions used to establish the global existence of smooth solutions (i.e., Theorem 7.2) when the initial data are sufficiently small. Moreover, they are fully compatible with the second law of thermodynamics, in particular, there is a free energy.

Under the constitutive equations (7.31), the system (7.22)–(7.24) becomes

$$w_t = v_x, \tag{7.34}$$

$$v_t = p'(w)w_x + \gamma\theta_x, \tag{7.35}$$

$$\delta\theta_t = \kappa\theta_{xx} + \gamma(\theta + T_0)v_x. \tag{7.36}$$

Thus, the blow-up result on the Cauchy problem (7.34)–(7.36), (7.25) reads as the theorem.

THEOREM 7.3. *Assume that* $p \in C^4(-1,\infty)$, $p(0) = 0$, $p'(\xi) > 0$ *for* $\xi > -1$ *and* $p''(0) \neq 0$. *Let* β, L, $J > 0$ *be given. Then there exist* ε, $M > 0$ *(depending on* β, L, J*) with the following property*: *For each* $w^0, v^0, \theta^0 \in H^3(\mathbb{R})$,

$$|w^0(x)|, |v^0(x)|, |\theta^0(x)| \leqslant \varepsilon, \qquad |\partial_x\theta^0(x)| \leqslant J \quad \text{for } x \in \mathbb{R},$$

$$\|(w^0, v^0, \theta^0)\|^2 + \|\theta^0\|_{H^1} \leqslant \varepsilon^2,$$

$$\min_{x\in\mathbb{R}}\{\partial_x v^0 + p'(w_0)^{1/2}\partial_x w^0\}(x) + \min_{x\in\mathbb{R}}\{\partial_x v^0 - p'(w_0)^{1/2}\partial_x w^0\} \geqslant -J$$

and

$$\max_{x\in\mathbb{R}}\{\partial_x v^0 + p'(w_0)^{1/2}\partial_x w^0\}(x) + \max_{x\in\mathbb{R}}\{\partial_x v^0 - p'(w_0)^{1/2}\partial_x w^0\}(x) \geqslant M,$$

the length T_m *of the maximal interval of existence of a smooth solution* (w, v, θ) *of the Cauchy problem* (7.34)–(7.36), (7.25) *is less than (or equal to)* L; *moreover, the local solution satisfies* $|w(t,x)|, |v(t,x)|, |\theta(x,t)| \leqslant \beta$ *for all* $x \in \mathbb{R}$, $t \in [0, T_m)$.

PROOF. The proof is based on the combination of the techniques for blow-up results in one-dimensional nonlinear elasticity and a delicate estimate of the temperature gradient in (7.35). It is well known that a smooth solution of the purely elastic part in (7.34), (7.35) breaks down in finite time. On the other hand, the behavior of the elastic part is influenced only by the temperature gradient θ_x in (7.35). Using the semilinear heat equation (7.36) for θ, we may represent θ in terms of v_x. With the help of this representation, one can control θ_x by w_x and v_x, that is, for large data, $|\theta_x|$ grows like $O(|w_x| + |v_x|)$. Consequently, the growth of $|\theta_x|$ in the case of large data is slower than that of the elastic part. Therefore, the elastic behavior dominates in the equations (7.34), (7.35) for large data and singularities of the smooth solution develop in finite time. See [97] for the details of the proof of Theorem 7.3. □

The first blow-up result on the thermoelastic system was given by Dafermos and Hsiao [36]. They studied a slightly different class of thermoelastic materials in [36].

They assume that the stress and the heat flux are given by $(7.31)_1$ and $(7.31)_3$, and that $\varepsilon(w,\theta) = P(w) + \delta\theta$. As they point out their constitutive relations comply with $\varepsilon_w = \widetilde{S} - (\theta + T_0)\widetilde{S}_\theta$ only if $T_0 = 0$ (or $\gamma = 0$). For some initial boundary value problems in a bounded interval or the half line, we have a similar blow-up result, see [78] for the details.

The blow-up result shows the nonexistence of global classical solutions for large data, in general. It is not only mathematically interesting and natural to ask for global, large solutions with weaker regularity, but also from the physical and modeling point of view.

There exists only little work on this subject due to the inherent mathematical difficulties. In [50] Durek proved the global existence of weak solutions to the Cauchy problem under the simplifying assumption that $\widetilde{S}$ satisfies $\widetilde{S}(u_x, \theta) = \sigma(u_x) + \theta$ for a function σ and that the equation for θ in (7.3) is linear, i.e., $\delta\theta_t - \kappa\theta_{xx} + \gamma u_{tx} = 0$. The basic idea in [50] is first to represent θ in terms of u_{tx} by employing the linear heat equation for θ and to insert this representation into the momentum equation leading to a (perturbed) nonlinear elastic equation for u, and then to apply the method of compensated compactness to the perturbed elastic equation to get a global weak solution. In [197] Rieger and Zimmer investigated a nonlinear one-dimensional thermoelastic model with nonconvex energy. Using the Young measure method and the technique of vanishing regularization terms, they proved the global existence of a Young measure solution.

Further results for the one-dimensional systems are contained in the papers [53,193] on periodic small solutions and in [20] on weak solutions for the systems that do not conduct heat.

NOTES ON THE LOCAL EXISTENCE. There are many papers devoted to the local existence in $\mathbb{R}^n$. For systems of hyperbolic–parabolic type which are first-order hyperbolic the local existence for the Cauchy problem was studied by Vol'pert and Hudjaev [225] and by Kawashima [131], see also [165]. For initial boundary value problems in $\mathbb{R}^3$ there are results for systems with first-order hyperbolic part for noncharacteristic boundaries, see [237], or special characteristic boundaries and admissible boundary conditions in the Friedrichs sense for bounded domains, see [144]. In [133] Kawashima and Matsumura studied hyperbolic–parabolic systems in both bounded and exterior domains with first-order hyperbolic part and their results may be applied to nonlinear thermoelasticity after some necessary additional consideration. With a different method, Shibata [207] and Dan [37] proved a local existence result for the Neumann problem. The Dirichlet problem was investigated by Jiang and Racke [119]. A similar result was given by Chrzęszczyk [28], using the methods from Dafermos and Hrusa [33].

8. Nonlinear three-dimensional thermoelasticity

The models for one space dimension discussed in the previous section demonstrated that the behavior of the nonlinear thermoelastic system is dominated by the heat conduction at least with respect to the existence of solutions for small data and the time-asymptotic behavior of solutions. In more than one space dimension the situation becomes delicate. It is known (cf., e.g., [190] and the references therein) that in the case of pure elasticity there are global small solutions if the nonlinearity degenerates up to order two, i.e., if the nonlinearity is cubic (near zero values of its arguments). In the "genuinely nonlinear" case, a blow up in finite time has to be expected; this was proved for plane waves and for radial solutions, cf. [127,128,190]. On the other hand quadratic nonlinearities in $\mathbb{R}^3$ still lead to global, small solutions of the heat equation, cf. [190]. The question remains whether here the dissipative impact through heat conduction is strong enough to prevent solutions from blowing up at least for small data.

The answer to this question will be positive for the Cauchy problem if one essentially excludes those nonlinearities already responsible for blow-up results in pure elasticity; this will first be shown in this section, and is completed by proving a blow-up result also for small data for general (quadratic) nonlinearities, in the "genuinely nonlinear" case. Then, we present a global existence theorem for an initial boundary value problem in $\mathbb{R}^n$ ($n = 2, 3$) for the radially symmetric case.

Without loss of generality, we restrict ourselves to three space dimensions. The system (2.2), (2.12) for a homogeneous medium with unit reference density in $\mathbb{R}^3$ can be written as

$$\frac{\partial^2 U_i}{\partial t^2} = C_{ijkl}(\nabla U, \theta)\frac{\partial^2 U_k}{\partial x_j \partial x_l} + \widetilde{C}_{ij}(\nabla U, \theta)\frac{\partial \theta}{\partial x_j}, \quad i = 1, 2, 3, \tag{8.1}$$

$$a(\nabla U, \theta)\theta_t = -\frac{1}{\chi(\theta)} \operatorname{div} q(\nabla U, \theta, \nabla\theta) + \widetilde{C}_{ij}(\nabla U, \theta)\frac{\partial^2 U_i}{\partial x_j \partial t}, \tag{8.2}$$

where

$$\begin{aligned} C_{ijkl} &= \frac{\partial \widetilde{S}^{ij}}{\partial(\partial_l U_k)}, & \widetilde{C}_{ij} &= \frac{\partial \widetilde{S}^{ij}}{\partial \theta}, \\ \widetilde{S}^{ij} &= \frac{\partial \psi}{\partial(\partial_j U_i)}, & a &= -\frac{\partial^2 \psi}{\partial \theta^2} \geqslant a_0 > 0 \end{aligned} \tag{8.3}$$

for some positive constant a_0. Here $U = (U_1, U_2, U_3)'$, $q = (q_1, q_2, q_3)'$, χ is a C^∞-function such that $\chi(\theta) = \theta + T_0$ for $|\theta| \leqslant T_0/2$ and $0 < \chi_1 \leqslant \chi(\theta) \leqslant \chi_2 < \infty$, χ_1, χ_2 are constants, $-\infty < \theta < \infty$. Equations (8.1) and (8.2) are derived for small values of $|\theta|$, i.e., for $|\theta| \leqslant T_0/2$, which is a posteriori justified by the smallness of the solutions obtained later on.

We assume that $C_{ijkl}(\nabla U, \theta)$ and $q(\nabla U, \theta, \nabla\theta)$ satisfy the usual ellipticity, i.e.,

$$\begin{aligned} &C_{ijkl}(P, \mu)\xi_j \xi_l \eta_i \eta_k \geqslant \kappa_0 |\xi|^2 |\eta|^2, \\ &-\frac{\partial q_i(P, \mu, \nu)}{\partial \nu_j}\xi_i \xi_j \geqslant \kappa_0 |\xi|^2, \quad \kappa_0 \leqslant a(P, \mu) \leqslant \kappa_1, \\ &\frac{\partial q_i(P, \mu, \nu)}{\partial \nu_j} = \frac{\partial q_j(P, \mu, \nu)}{\partial \nu_i} \quad \forall P \in \mathbb{R}^{3\times 3}, \xi = \{\xi_i\}, \eta = \{\eta_i\}, \nu \in \mathbb{R}^3, \mu \in \mathbb{R}, \end{aligned}$$

for some constants $\kappa_0, \kappa_1 > 0$, and that the medium is *initially isotropic*, i.e.,

$$
\begin{aligned}
&C_{ijkl}(0,0) = \lambda\delta_{ij}\delta_{kl} + \mu(\delta_{ik}\delta_{jl} + \delta_{jk}\delta_{il}), \qquad \widetilde{C}_{ij}(0,0) = -\gamma\delta_{ij}, \\
&\frac{\partial q_i(0,0,0)}{\partial(\partial\theta/\partial x_k)} = -\kappa T_0\delta_{ik}, \qquad a(0,0) = \delta, \quad \lambda, \mu\text{: the Lamé moduli,}
\end{aligned}
\tag{8.4}
$$

and that

$$
\frac{\partial q_i(0,0,0)}{\partial\theta} = 0, \qquad \frac{\partial q_i(0,0,0)}{\partial(\partial U_j/\partial x_l)} = 0, \quad 1 \leqslant i,k,j,l \leqslant 3. \tag{8.5}
$$

Using (8.4) we can write (8.1) and (8.2) in the form:

$$
U_{tt} - \mu\Delta U - (\mu+\lambda)\nabla\operatorname{div} U + \gamma\nabla\theta = f^1\big(\nabla U, \nabla^2 U, \theta, \nabla\theta\big), \tag{8.6}
$$

$$
\delta\theta_t - \kappa\Delta\theta + \gamma\operatorname{div} U_t = f^2\big(\nabla U, \nabla^2 U, \nabla U_t, \theta, \nabla^2\theta\big), \tag{8.7}
$$

where $f^1 = (f_1, f_2, f_3)'$, and

$$
\begin{aligned}
f_i^1 := {} & \big(C_{ijkl}(\nabla U, \theta) - C_{ijkl}(0,0)\big)\frac{\partial^2 U_k}{\partial x_j\,\partial x_l} \\
& + \big(\widetilde{C}_{ij}(\nabla U, \theta) - \widetilde{C}_{ij}(0,0)\big)\frac{\partial\theta}{\partial x_j},
\end{aligned}
\tag{8.8}
$$

$$
\begin{aligned}
f^2 := {} & \delta\left(\frac{1}{a(0,0)\chi(0)}\frac{\partial q_i(0,0,0)}{\partial(\partial\theta/\partial x_k)} - \frac{1}{a(\nabla U,\theta)\chi(\theta)}\frac{\partial q_i(\nabla U,\theta,\nabla\theta)}{\partial(\partial\theta/\partial x_k)}\right)\frac{\partial^2\theta}{\partial x_i\,\partial x_k} \\
& + \delta\left(\frac{\widetilde{C}_{ij}(\nabla U,\theta)}{a(\nabla U,\theta)} - \frac{\widetilde{C}_{ij}(0,0)}{a(0,0)}\right)\frac{\partial^2 U_i}{\partial x_j\,\partial t} \\
& - \frac{\delta}{a(\nabla U,\theta)\chi(\theta)}\frac{\partial q_i(\nabla U,\theta,\nabla\theta)}{\partial\theta}\frac{\partial\theta}{\partial x_i} \\
& - \frac{\delta}{a(\nabla U,\theta)\chi(\theta)}\frac{\partial q_i(\nabla U,\theta,\nabla\theta)}{\partial(\partial U_j/\partial x_l)}\frac{\partial^2 U_j}{\partial x_i\,\partial x_l}.
\end{aligned}
\tag{8.9}
$$

To establish the global existence for the Cauchy problem of (8.1), (8.2) with initial conditions

$$
(U, U_t, \theta)|_{t=0} = \big(U^0, U^1, \theta^0\big) \quad \text{in } \mathbb{R}^3, \tag{8.10}
$$

we assume that the nonlinearity satisfies:

$$\begin{array}{l}\text{There are no purely quadratic terms only involving } \nabla U, \nabla U_t, \nabla^2 U\\ \text{and additionally one of the following two cases is given:}\\ \text{Case I: Only quadratic terms appear.}\\ \text{Case II: Only at least cubic terms appear and}\\ \qquad\qquad \text{one quadratic term of the type } \theta\Delta\theta.\end{array} \tag{8.11}$$

Then the global existence for (8.1), (8.2), (8.10) reads:

THEOREM 8.1. *Let the nonlinearity satisfy* (8.11). *Then there exist an integer* s_0 *and a* $\delta > 0$ *such that the following holds: If* $(\nabla U^0, U^1, \theta^0) \in H^s(\mathbb{R}^3) \cap W^{s,1}(\mathbb{R}^3)$ *with* $s \geqslant s_0$ *and*

$$\|(\nabla U^0, U^1, \theta^0)\|_{H^s} + \|(\nabla U^0, U^1, \theta^0)\|_{W^{s,1}} < \delta,$$

then there is a unique global solution (U, θ) *of the Cauchy problem* (8.1), (8.2), (8.10) *with*

$$(\nabla U, U_t) \in C^0([0,\infty), H^s(\mathbb{R}^3)) \cap C^1([0,\infty), H^{s-1}(\mathbb{R}^3)),$$
$$\theta \in C^0([0,\infty), H^s(\mathbb{R}^3)) \cap C^1([0,\infty), H^{s-2}(\mathbb{R}^3)).$$

Moreover, the solution decays polynomially as time goes to infinity.

SKETCH OF THE PROOF. The basic idea of the proof is to derive a uniform a priori estimate for (U, θ) by exploiting the decay of solutions to the related linearized system and delicate calculations. With the help of the a priori estimate, one can continue a local solution globally in time. The proof is technical and very long, we refer to [185,188] for the detail. In the following we give the main Steps A–E.

We first transform (8.6) and (8.7) to a suitable first-order system by introducing

$$V(t) := (S\mathcal{D}U, U_t, \theta)'(t) = (V^1, V^2, V^3)'(t), \tag{8.12}$$

where $\mathcal{D}$ is an abbreviation for a generalized gradient,

$$\mathcal{D} := \begin{pmatrix} \partial_1 & 0 & 0\\ 0 & \partial_2 & 0\\ 0 & 0 & \partial_3\\ 0 & \partial_3 & \partial_2\\ \partial_3 & 0 & \partial_1\\ \partial_2 & \partial_1 & 0 \end{pmatrix}, \qquad S = \begin{pmatrix} 2\mu+\lambda & \lambda & \lambda & 0 & 0 & 0\\ \cdot & 2\mu+\lambda & \lambda & 0 & 0 & 0\\ \cdot & \cdot & 2\mu+\lambda & 0 & 0 & 0\\ \cdot & \cdot & \cdot & \mu & 0 & 0\\ \cdot & \cdot & \cdot & \cdot & \mu & 0\\ \cdot & \cdot & \cdot & \cdot & \cdot & \mu \end{pmatrix}.$$

To recover ∇U from a known function $V^1 = S\mathcal{D}U$, we define the operator $\widetilde{B}^j$ by

$$\widetilde{B}^j : S\mathcal{D}H^1(\mathbb{R}^3) \to L^2, \qquad \widetilde{B}^j Z := \partial_j (S\mathcal{D})^{-1} Z, \quad j = 1, 2, 3,$$

where $S\mathcal{D}H^1(\mathbb{R}^3) := \{S\mathcal{D}Z | Z \in H^1(\mathbb{R}^3)\}$.

By Korn's first inequality (cf. [143]) $\widetilde{B}^j$ can be continuously extended to a bounded operator $B^j : \overline{S\mathcal{D}H^1(\mathbb{R}^3)} \to L^2$.

Let $B^\nabla := (B^1, B^2, B^3)$. Observe that ∇^k, $k \in \mathbb{N}$, commutes with B^∇. Under the transformation (8.12) the system of equations (8.6), (8.7), (8.10) now turns into (note $\mu\Delta + (\mu + \lambda)\nabla \operatorname{div} = \mathcal{D}' S\mathcal{D}$)

$$\begin{cases} V_t + AV = F\big(V, \nabla V, \nabla^2 V^3, B^\nabla V^1, \nabla B^\nabla V^1\big), \\ V|_{t=0} = V^0, \end{cases} \tag{8.13}$$

with the nonlinearity

$$F\big(V, \nabla V, \nabla^2 V^3, B^\nabla V^1, \nabla B^\nabla V^1\big) := \begin{pmatrix} 0 \\ f^1(B^\nabla V^1, \nabla B^\nabla V^1, V^3, \nabla V^3) \\ f^2(B^\nabla V^1, \nabla V^2, \nabla B^\nabla V^1, V^3, \nabla V^3, \nabla^2 V^3) \end{pmatrix},$$

and A is the differential operator formally given by

$$A = \begin{pmatrix} 0 & -S\mathcal{D} & 0 \\ -\mathcal{D}' & 0 & \gamma\nabla \\ 0 & \gamma \operatorname{div} & -\kappa\Delta \end{pmatrix}.$$

The operator $-A$ is the generator of a contraction semigroup in the Hilbert space:

$$\mathcal{H} := \overline{SDH^1(\mathbb{R}^3)} \times L^2(\mathbb{R}^3) \times L^2(\mathbb{R}^3) \quad (6 + 3 + 1 \text{ components})$$

with domain $D(A) := \{V \in \mathcal{H} \mid AV \in L^2\}$; the inner product in $\mathcal{H}$ is a weighted L^2-inner product:

$$(W, Z)_{\mathcal{H}} := \big(S^{-1} W^1, Z^1\big) + \big(W^2, Z^2\big) + \big(W^3, Z^3\big).$$

Step A (Decay for $F \equiv 0$). The decay of solutions to the linearized equations $V_t + AV = 0$ can be obtained by applying the Fourier transform and careful spectral analysis.

Step B (Local existence and uniqueness). The local existence is proved by applying the Banach contraction mapping principle and a standard procedure, cf. the Notes at the end of the last subsection. The local solution in some time interval $[0, T]$ satisfies $\|DU(t)\|_{L^\infty} + \|\theta(t)\|_{W^{2,\infty}} < \eta < 1$ for all $t \in [0, T]$ (η arbitrary, $\delta = \delta(\eta)$).

Step C (High energy estimates).

$$\begin{aligned}&\big\|(DU,\theta)(t)\big\|_{H^s}\\&\quad\leqslant C\big\|\big(\nabla U^0,U^1,\theta^0\big)\big\|_{H^s}\\&\qquad\times\exp\left\{C\int_0^t\big(\|DU\|^2_{W^{1,\infty}}+\|\theta\|_{W^{2,\infty}}+\|\theta_t\|_{L^\infty}\big)(r)\,\mathrm{d}r\right\},\\&\quad t\in[0,T],\ C=C(s),\end{aligned}$$

for $s\geqslant 4$, provided η is small enough.

The crucial point here is to obtain the quadratic term $\|DU\|^2_{W^{1,\infty}}$ in the exponent; a linear term $\|DU\|_{W^{1,\infty}}$, which is excluded by the assumption (8.11), would not be sufficient because of the weak decay rates for DU. The above estimate is proved by using the energy estimates and the assumption (8.11), and by estimating delicately the nonlinear terms.

Step D (Weighted a priori estimates). Exploiting the decay rates established in Step A and the assumption (8.11), one gets

$$M(T)\leqslant C\delta+C\delta\big(M^2(T)+M(T)\big)\exp\big\{C\big(M^2(T)+M(T)\big)\big\}. \tag{8.14}$$

Here for Case I in (8.11),

$$\begin{aligned}M(T):=\sup_{0\leqslant t\leqslant T}\Big\{&(1+t)^{3/4-\varepsilon}\big\|\nabla^3\theta(t)\big\|_{H^k};\ (1+t)^{3/2-2\varepsilon}\big\|\nabla^3\theta(t)\big\|_{H^{k'}};\\&(1+t)^{5/4}\big\|\nabla\theta(t)\big\|_{H^1};\ (1+t)^{3/4}\big\|\theta(t)\big\|_{H^2};\\&(1+t)^{3/2-\varepsilon}\big\|\theta(t)\big\|_{W^{2,\infty}};\ (1+t)^{5/4}\big\|\theta_t(t)\big\|_{L^\infty};\\&(1+t)^{3/4-\varepsilon}\big\|DU(t)\big\|_{W^{l,\infty}}\Big\}\end{aligned}$$

with $l\geqslant 6$, $k'\geqslant l+1$, $k\geqslant k'+7$, $0<\varepsilon<1/8$ arbitrary, $s_0\geqslant k+7$; and for Case II in (8.11)

$$\begin{aligned}M(T):=\sup_{0\leqslant t\leqslant T}\Big\{&(1+t)^{5/9}\big\|DU(t)\big\|_{W^{s_1,9/2}};\\&(1+t)^{27/26}\\&\times\Big(\big\|\theta(t)\big\|_{W^{s_3,13/2}}+\big\|\nabla^2\theta(t)\big\|_{W^{s_5,26/11}}+\big\|\nabla^2\theta(t)\big\|_{W^{s_5,18/7}}\Big)\Big\}\end{aligned}$$

for some s_1,s_3,s_5 being large enough, and V^0 satisfies $\|V^0\|_{H^{s_0}}+\|V^0\|_{W^{s_0,1}}<\delta$, where s_0 has to be sufficiently large.

From (8.14), it follows that

$$M(T)\leqslant M_0 \tag{8.15}$$

provided δ is sufficiently small, where the constant M_0 is independent of T.

Step E (Final energy estimate). Combining (8.15) with Step C, we obtain

$$\|V(t)\|_{H^s} \leqslant K\|V^0\|_{H^s} \quad \text{for all } 0 \leqslant t \leqslant T,$$

this allows to continue a local solution up to infinity. □

REMARK 8.1. The case of two space dimensions can be dealt with similarly in the sense that appropriate assumptions on the nonlinearities will lead to global solutions. But, since the decay is weaker, the assumptions would have to be stronger.

Next, we give an example to show that one has to expect the development of singularities in finite time if there are purely quadratic nonlinear terms only involving derivatives of the displacement, not involving the temperature, such as $\nabla U \nabla^2 U$, $\nabla U \nabla U_t$.

The idea of proving a blow-up is that we decompose (U, θ) into $(U^\sigma, 0) + (U^\pi, \theta)$ with the divergence-free U^σ and the curl-free U^π, and choose an appropriate Helmholtz free energy ψ, so that U^σ is no longer coupled to θ and satisfies a genuinely nonlinear strictly hyperbolic system. Then the general results from [127,148] will imply that there are nonlinearities which on the one hand satisfy at least the basic (physical) properties, and on the other hand are such that for sufficiently small smooth data, a plane-wave solution cannot be of class C^2 for all positive t.

We consider a plane wave $U(t,x) = \widetilde{U}(t, \tau \cdot x)$, $\tau = (1,0,0)'$ and assume $\theta(t,x) = \widetilde{\theta}(t, \tau \cdot x)$, i.e., U and θ become functions of x_1 (and t) only. We may decompose U into the divergence-free part U^σ and the curl-free part U^π:

$$U = (0, U_2, U_3)' + (U_1, 0, 0)' \equiv U^\sigma + U^\pi.$$

Let P^σ denote the corresponding projection $U \mapsto P^\sigma U := U^\sigma$. As stated above, by taking the projection P^σ to (8.1), the aim is to obtain the relation

$$\begin{aligned} 0 &= \partial_t^2 U^\sigma - P^\sigma\{C_{ijkl}(\nabla U, \theta)\,\partial_j\partial_l U_k + \widetilde{C}_{ij}(\nabla U, \theta)\,\partial_j\theta\}_{i=1,2,3} \\ &= \partial_t^2 U^\sigma - \{C_{ijkl}(\nabla U^\sigma, 0)\,\partial_j\partial_l U_k^\sigma\}_{i=1,2,3}, \end{aligned} \tag{8.16}$$

where the coefficient $C_{ijkl}(\nabla U, 0)$ does not depend on θ. For this purpose we require first

$$\begin{aligned} &P^\sigma\{\widetilde{C}_{ij}(\nabla U, \theta)\,\partial_j\theta\}_{i=1,2,3} = 0, \\ &P^\sigma\{C_{ijkl}(\nabla U, \theta)\,\partial_j\partial_l U_k\}_{i=1,2,3} = \{C_{ijkl}(\nabla U^\sigma, 0)\,\partial_j\partial_l U_k^\sigma\}_{i=1,2,3}. \end{aligned} \tag{8.17}$$

In order to have a nonlinear dependence of C_{ijkl} in θ we additionally require

$$\frac{\partial C_{ijkl}}{\partial\theta} \neq 0 \quad \text{at least for one quadruple } (i,j,k,l). \tag{8.18}$$

Since we are considering plane waves, the final system for U^σ should be

$$\begin{cases} \partial_t^2 U_i^\sigma - B_{ik}(\partial_1 U^\sigma)\,\partial_1^2 U_k^\sigma = 0, \quad i = 1,2,3, \\ U^\sigma(t=0) = U^{0,\sigma}, \quad U_t^\sigma(t=0) = U^{1,\sigma}, \end{cases} \tag{8.19}$$

where

$$B_{ik}(\partial_1 U^\sigma) = C_{i1k1}(\nabla U^\sigma, 0), \quad i,k = 1,2,3.$$

Let

$$V(\alpha) := \psi\left(\begin{pmatrix} \alpha_1 & 0 & 0 \\ \alpha_2 & 0 & 0 \\ \alpha_3 & 0 & 0 \end{pmatrix}, 0\right), \quad \alpha \in \mathbb{R}^3,$$

$$V_{ik}(\alpha) := B_{ik}\left(\begin{pmatrix} \alpha_1 & 0 & 0 \\ \alpha_2 & 0 & 0 \\ \alpha_3 & 0 & 0 \end{pmatrix}\right).$$

Then the system (8.19) is *strictly hyperbolic* if

$$\text{the matrix } \big(V_{ik}(\alpha)\big)_{ik} \text{ has only positive distinct eigenvalues} \tag{8.20}$$

and it is *genuinely nonlinear* if

$$\left.\frac{\mathrm{d}^3(V(s\beta))}{\mathrm{d}s^3}\right|_{s=0} \neq 0 \quad \text{for any right eigenvector } \beta \text{ of } \big(V_{ik}(\alpha)\big)_{ik}. \tag{8.21}$$

Now define a Helmholtz free energy function

$$\begin{aligned} \psi(\nabla U, \theta) := {} & 3(\partial_1 U_1)^2 + (\partial_1 U_2)^2 + (\partial_1 U_3)^2 + (\partial_1 U_2)(\partial_1 U_3) + a_{111}(\partial_1 U_1)^3 \\ & + a_{222}(\partial_1 U_2)^3 + a_{333}(\partial_1 U_3)^3 + a_{223}(\partial_1 U_2)^2(\partial_1 U_3) \\ & + a_{233}(\partial_1 U_2)(\partial_1 U_3)^2 + (\partial_1 U_1)^2\theta + \gamma\theta \sum_{k=1}^{3} (\partial_k U_k) - \theta^2 \end{aligned}$$

with the coefficients satisfying

$$a_{111}, a_{222}, a_{333}, a_{223}, a_{233}, \gamma \in \mathbb{R} \setminus \{0\},$$

$$a_{222} + a_{333} + a_{223} + a_{233} \neq 0, \qquad a_{222} - a_{333} - a_{223} + a_{233} \neq 0.$$

Then, a straightforward calculation implies that for the above defined ψ, (8.17), (8.18), (8.20) and (8.21) hold in a neighborhood of zero. Moreover, one has $a(\nabla U, \theta) \geqslant a_0 > 0$

for some constant a_0 and $(\widetilde{C}_{ij})_{ij} \neq 0$ ("really coupled"!). Therefore, an application of the general result in [127,148] to the system (8.19) gives the nonexistence of a global C^2-solution to (8.19). Thus, we have proved [189]:

THEOREM 8.2. *There exist nonlinearities such that for compactly supported non-vanishing smooth data* $(U^{0,\sigma}, U^{1,\sigma})$ *which are sufficiently small, i.e.,*

$$\sup_{x_1 \in \mathbb{R}} \left| \partial_1 \left(\partial_1 U_2^0, \partial_1 U_3^0, U_2^1, U_3^1 \right)(x_1) \right|$$

is sufficiently small, a plane-wave solution of the Cauchy problem (8.1), (8.2), (8.10) *cannot be of class* C^2 *for all positive* t.

Concerning initial boundary value problems in bounded domains, the work by Henry [79], Lebeau and Zuazua [142], Koch [141] shows that in general there is no decay rate of solutions to the associated linearized equations due to reflecting rays. So, the techniques in the proof of Theorem 8.1 cannot be directly applied to the nonlinear problems in bounded domains. However, in the case of radially symmetric solutions, the divergence-free part of solutions to the linearized equations vanishes, and consequently, the solutions decay exponentially as time tends to infinity. This allows us to prove the global existence for the initial boundary value problem with radially symmetric initial data in radially symmetric domains. (Compare this to Theorem 8.2, where the divergence-free part U^σ induces singularities.)

Consider the initial boundary value problem (8.6), (8.7) with initial and boundary conditions

$$(U, U_t, \theta)|_{t=0} = \left(U^0, U^1, \theta^0\right), \quad x \in \Omega, \tag{8.22}$$

$$U|_{\partial\Omega} = 0, \qquad \theta|_{\partial\Omega} = 0, \quad t \geqslant 0. \tag{8.23}$$

We assume that for $(\widetilde{U}, \tilde{\theta})$ with $\operatorname{rot} \widetilde{U} = 0$, $C_{ijkl}(\nabla U, \theta)$ satisfy

$$C_{ijkl}\left(\nabla \widetilde{U}, \tilde{\theta}\right) \frac{\partial^2 \widetilde{U}_k}{\partial x_j \, \partial x_l} \equiv A_{ik}\left(\nabla \widetilde{U}, \tilde{\theta}\right) \Delta \widetilde{U}_k, \quad i = 1, 2, 3. \tag{8.24}$$

Then we have [118]:

THEOREM 8.3. *Let* $\Omega \subset \mathbb{R}^n$ $(n = 2, 3)$ *be a ball or an annulus. Assume that* $U^j \in H^{4-j}(\Omega)$ $(j = 0, \ldots, 4)$, $\theta^j \in H^{4-j}(\Omega)$ $(j = 0, 1, 2)$, $\theta^3 \in L^2(\Omega)$, *and* U^0, U^1, θ^0 *are radially symmetric. Also assume that* (8.24) *is satisfied and the initial data are compatible with the boundary conditions. Then there is a constant* $\varepsilon > 0$ *such that if*

$$\sum_{j=0}^{4} \left\| U^j \right\|_{H^{4-j}}^2 + \sum_{j=0}^{2} \left\| \theta^j \right\|_{H^{4-j}}^2 + \left\| \theta^3 \right\|^2 \leqslant \varepsilon^2,$$

then there exists a unique solution (U,θ) *of* (8.6), (8.7), (8.22), (8.23) *on* $[0,\infty)$ *such that*

$$U \in \bigcap_{j=0}^{4} C^j\big([0,\infty), H^{4-j}(\Omega)\big), \qquad \theta \in \bigcap_{j=0}^{2} C^j\big([0,\infty), H^{4-j}(\Omega)\big),$$

$$\theta_{ttt} \in C^0\big([0,\infty), L^2(\Omega)\big) \cap L^2\big([0,\infty), H^1(\Omega)\big),$$

$$\forall (t,x) \in [0,\infty) \times \overline{\Omega}: \quad \big|\nabla U(t,x)\big|, \big|\theta(t,x)\big|, \big|\nabla\theta(t,x)\big| < \min\{1, T_0/2\}.$$

Moreover, $\|U(t)\|_{H^4}$, $\|\theta(t)\|_{H^4}$ *decay to zero exponentially as* $t \to \infty$.

OUTLINE OF THE PROOF. The first step is to prove the exponential decay of solutions to the linearized equations of (8.6) and (8.7), i.e., (8.6) and (8.7) with $f^1 = 0$ and $f^2 = 0$. Note that under the assumptions of the theorem, the local smooth solution is radially symmetric due to the invariance of (8.6) and (8.7) under rotation and uniqueness. For the radially symmetric solution U, the H^1-norm of U is equivalent to $\|\operatorname{div} U\|$ $(= \|\nabla U\|)$. On the one hand, the exponential decay of θ is expected due to parabolicity of (8.7) with $f^2 = 0$ for θ; on the other hand, in view of (8.7) with $f^2 = 0$, $\operatorname{div} U_t$ (a "good" term) is controlled by θ, and that is, ∇U_t is bounded in L^2 by θ. Exploiting boundedness of ∇U_t in L^2 and applying (delicate) energy estimates for the elastic equations (8.6) with $f^1 = 0$, employing the techniques from boundary control to bound the "ill" behaved terms as in the one-dimensional case (compare to Theorem 7.1), one obtains the exponential decay for the linearized equations of (8.6) and (8.7). The next step is to exploit the exponential decay for the linearized equations and employ (tricky) energy estimates to derive a uniform *a priori* estimate for (8.6), (8.7), (8.22) and (8.23), with which one can continue a local radially symmetric solution globally in time. □

Other boundary conditions were studied by Rieger [196]. However, there are not yet any results for unbounded domains with boundary.

NOTES ON THE RELATED RESULTS. There are a number of related results for nonlinear evolution equations of thermoelastic type. Fully dynamical and quasi-static thermoelastic contact problems were studied, for example, in [1,38,51,167,168,205,206] and among others. The corresponding system models the evolution of temperature and displacement in an elastic body that may come into contact with a rigid foundation, and consists of the linearized equations together with nonlinear boundary conditions which reflect the contact situation. Under certain conditions the global well-posedness and the large-time behavior of solutions are discussed.

In [25,195,227] the propagation of singularities in nonlinear thermoelasticity was investigated using techniques from the theory of pseudo-differential operators and microlocal analysis.

The far-field behavior of solutions to the equations of nonlinear thermoelasticity is useful for numerical simulations in unbounded domains, and can be utilized to construct "appropriate boundary conditions" on the artificial boundary. The far-field behavior in one and three space dimensions was investigated with the help of the Fourier transform and the use of suitable weighted function spaces (cf. [105] and the references cited therein).

Acknowledgment

Research was supported by the Special Funds of State Major Basic Research Projects (Grant No. 1999075107, G1999032801), NSFC (Grant No. 10225105), NSAF (Grant No. 10276036) and the CAEP.

References

[1] K.A. Ames and L.E. Payne, *Uniqueness and continuous dependence of solutions to a multidimensional thermoelastic contact problem*, J. Elasticity **34** (1994), 139–148.

[2] A.A. Amosov, *Existence of global generalized solutions to the equations for a viscous one-dimensional real gas and for one-dimensional nonlinear thermoviscoelasticity*, Dokl. Math. **60** (1999), 347–350.

[3] A.A. Amosov, *The existence of global generalized solutions of the equations of one-dimensional motion of a real viscous gas with discontinuous data*, Differ. Equ. **36** (2000), 540–558.

[4] A.A. Amosov and A.A. Zlotnik, *Global generalized solutions of the equations of the one-dimensional motion of a viscous heat-conducting gas*, Soviet Math. Dokl. **38** (1989), 1–5.

[5] A.A. Amosov and A.A. Zlotnik, *Solvability "in the large" of a system of equations of the one-dimensional motion of an inhomogeneous viscous heat-conducting gas*, Math. Notes **52** (1992), 753–763.

[6] A.A. Amosov and A.A. Zlotnik, *Semidiscrete method for solving equations of a one-dimensional motion of viscous heat-conductive gas with non-smooth data*, Russian Math. (Iz. VUZ) **41** (1997), 3–19.

[7] G. Andrews, *On the existence of solutions to the equation* $u_{tt} = u_{xxt} + \sigma(u_x)_x$, J. Differential Equations **35** (1980), 200–231.

[8] G. Andrews and J.M. Ball, *Asymptotic behavior and change of phase in one-dimensional nonlinear viscoelasticity*, J. Differential Equations **44** (1982), 306–341.

[9] S.S. Antman and T.I. Seidman, *Quasilinear hyperbolic–parabolic equations of one-dimensional viscoelasticity*, J. Differential Equations **124** (1996), 132–185.

[10] S.N. Antontsev, A.V. Kazhikhov and V.N. Monakhov, *Boundary Value Problems in Mechanics of Nonhomogeneous Fluids*, North-Holland, Amsterdam (1990).

[11] J. Bebernes and A. Bressan, *Global a priori estimates for a viscous reactive gas*, Proc. Roy. Soc. Edinburgh Sect. A **101** (1985), 321–333.

[12] J. Bebernes and D. Eberly, *Mathematical Problems from Combustion Theory*, Appl. Math. Sci., Vol. 83, Springer-Verlag, New York–Heidelberg (1989).

[13] E. Becker, *Gasdynamik*, Teubner-Verlag, Stuttgart (1966).

[14] S. Belov and V. Belov, *On the solvability of a boundary value problem for model equations of linearly viscous materials*, J. Math. Anal. Appl. **200** (1996), 635–652.

[15] M. Bertsch, L.A. Peletier and S.M. Verduyn Lunels, *The effect of temperature dependent viscosity on shear flow of incompressible fluids*, SIAM J. Math. Anal. **22** (1991), 328–343.

[16] A. Bressan, *Global solutions for the one-dimensional equations of a viscous reactive gas*, Boll. Unione Mat. Ital. B **5** (1985), 291–308.

[17] D.E. Carlson, *Linear thermoelasticity*, Handbuch der Physik **VIa/2**, Springer-Verlag, Berlin–New York (1972), 297–346.

[18] T. Chang and L. Hsiao, *The Riemann Problem and Interaction of Waves in Gas Dynamics*, Longman, Harlow (1989).

[19] G.Q. Chen, *Global solutions to the compressible Navier–Stokes equations for a reacting mixture*, SIAM J. Math. Anal. **23** (1992), 609–634.

[20] G.Q. Chen and C.M. Dafermos, *The vanishing viscosity method in one-dimensional thermoelasticity*, Trans. Amer. Math. Soc. **347** (1995), 531–541.

[21] G.Q. Chen and J. Glimm, *Global solutions to the compressible Euler equations with geometrical structure*, Comm. Math. Phys. **180** (1996), 153–193.

[22] G.Q. Chen, D. Hoff and K. Trivisa, *Global solutions of the compressible Navier–Stokes equations with large discontinuous initial data*, Comm. Partial Differential Equations **25** (2000), 2233–2257.

[23] G.Q. Chen, D. Hoff and K. Trivisa, *Global solutions to a model for exothermically data*, Arch. Ration. Mech. Anal. **166** (2003), 321–358.

[24] G.Q. Chen and M. Kratka, *Global solutions to the Navier–Stokes equations for compressible heat-conducting flow with symmetry and free boundary*, Comm. Partial Differential Equations **27** (2002), 907–943.

[25] S. Chen and Y. Wang, *Propagation of singularities of solutions to hyperbolic–parabolic coupled systems*, Math. Nachr. **242** (2002), 46–60.

[26] Z. Chen and K.H. Hoffmann, *On a one dimensional thermoviscoelastic model for structural phase transitions in shape memory alloys*, J. Differential Equations **112** (1994), 325–350.

[27] H.J. Choe and H. Kim, *Strong solutions of the Navier–Stokes equations for isentropic compressible fluids*, J. Differential Equations **190** (2003), 504–523.

[28] A. Chrzęszczyk, *Some existence results in dynamical thermoelasticity. Part I. Nonlinear case*, Arch. Mech. **39** (1987), 605–617.

[29] P. Collela, A. Majda and V. Roytbund, *Theoretical and numerical structure for reacting shock waves*, SIAM J. Sci. Statist. Comput. **7** (1986), 1059–1080.

[30] C.M. Dafermos, *The mixed initial-boundary value problem for the equations of nonlinear one-dimensional viscoelasticity*, J. Differential Equations **6** (1969), 71–86.

[31] C.M. Dafermos, *Global smooth solutions to the initial-boundary value problem for the equations of one-dimensional nonlinear thermoviscoelasticity*, SIAM J. Math. Anal. **13** (1982), 397–408.

[32] C.M. Dafermos, *Hyperbolic Conservation Laws in Continuum Physics*, Springer-Verlag, Berlin (2000).

[33] C.M. Dafermos and W.J. Hrusa, *Energy methods for quasilinear hyperbolic initial-boundary value problems. Applications to elastodynamics*, Arch. Ration. Mech. Anal. **87** (1985), 267–292.

[34] C.M. Dafermos and L. Hsiao, *Global smooth thermomechanical processes in one-dimensional nonlinear thermoviscoelasticity*, Nonlinear Anal. **6** (1982), 435–454.

[35] C.M. Dafermos and L. Hsiao, *Adiabatic shearing of incompressible fluids with temperature-dependent viscosity*, Quart. Appl. Math. **41** (1983), 45–58.

[36] C.M. Dafermos and L. Hsiao, *Development of singularities in solutions of the equations of nonlinear thermoelasticity*, Quart. Appl. Math. **44** (1986), 463–474.

[37] W. Dan, *On a local in time solvability of the Neumann problem of quasilinear hyperbolic parabolic coupled systems*, Math. Methods Appl. Sci. **18** (1995), 1053–1082.

[38] W.A. Day, *Heat Conduction within Linear Thermoelasticity*, Springer Tracts Nat. Philos., Vol. 30, Springer-Verlag, New York (1985).

[39] B. Desjardins, E. Grenier, P.-L. Lions and N. Masmoudi, *Incompressible limit for solutions of the isentropic Navier–Stokes equations with Dirichlet boundary conditions*, J. Math. Pures Appl. **78** (1999), 461–471.

[40] B. Ducomet, *Evolution of a self-gravitating Rosseland gas in one dimension*, Math. Models Meth. Appl. Sci. **5** (1995), 999–1012.

[41] B. Ducomet, *Hydrodynamical models of gaseous stars*, Rev. Math. Phys. **8** (1996), 957–1000.

[42] B. Ducomet, *On the stability of a stellar structure in one dimension. II. The reactive case*, RAIRO Model. Math. Anal. Numer. **31** (1997), 381–407.

[43] B. Ducomet, *A model of thermal dissipation for a one-dimensional viscous reactive and radiative gas*, Math. Meth. Appl. Sci. **22** (1999), 1323–1349.

[44] B. Ducomet, *Some asymptotics for a reactive Navier–Stokes–Poisson system*, Math. Models Methods Appl. Sci. **9** (1999), 1039–1076.

[45] B. Ducomet, *Asymptotic behaviour for a non-monotone fluid in one dimension: The positive temperature case*, Math. Methods Appl. Sci. **24** (2001), 543–559.

[46] B. Ducomet, *Global existence for a nuclear fluid in one dimension: The $T > 0$ case*, Appl. Math. **47** (2002), 45–75.

[47] B. Ducomet, E. Feireisl, H. Petzeltová and I. Straškraba, *Existence globale pour un fluide barotrope autogravitant*, C. R. Math. Acad. Sci. Paris Ser. I **332** (2001), 627–632.

[48] B. Ducomet and A.A. Zlotnik, *Remark on the stabilization of a viscous barotropic medium with a nonmonotonic equation of state*, Appl. Math. Lett. **14** (2001), 921–926.

[49] B. Ducomet and A.A. Zlotnik, *Stabilization for viscous compressible heat-conducting media equations with nonmonotone state functions*, C. R. Math. Acad. Sci. Paris Ser. I **334** (2002), 119–124.

[50] U. Durek, *Ein Existenzsatz für nichtlineare Thermoelastizitätsgleichungen zu großen Daten*, Ph.D. Thesis, University of Konstanz (1997).

[51] C.M. Elliott and Q. Tang, *A dynamic contact problem in thermoelasticity*, Nonlinear Anal. **23** (1994), 883–898.

[52] J.L. Ericksen, *Equilibrium of bars*, J. Elasticity **5** (1975), 191–201.

[53] E. Feireisl, *Forced vibrations in one-dimensional nonlinear thermoelasticity as a local coercive-like problem*, Comment. Math. Univ. Carolin. **31** (1990), 243–255.

[54] E. Feireisl, *On compactness of solutions to the compressible isentropic Navier–Stokes equations when the density is not integrable*, Comment. Math. Univ. Carolin. **42** (2001), 83–98.

[55] E. Feireisl, *Compressible Navier–Stokes equations with a non-monotone pressure law*, J. Differential Equations **184** (2002), 97–108.

[56] E. Feireisl, *Propagation of oscillations, complete trajectories and attractors for compressible flows*, NoDEA Nonlinear Differential Equations Appl. **10** (2003), 33–55.

[57] E. Feireisl, *On the motion of rigid bodies in a viscous compressible fluid*, Arch. Ration. Mech. Anal. **167** (2003), 281–308.

[58] E. Feireisl, *Dynamics of Viscous Compressible Fluids*, Oxford Science Publications, Clarendon Press, Oxford (2003).

[59] E. Feireisl, Š. Matušů-Nečasová, H. Petzeltová and I. Straškraba, *On the motion of a viscous compressible flow driven by a time-periodic external force*, Arch. Ration. Mech. Anal. **149** (1999), 69–96.

[60] E. Feireisl, A. Novotný and H. Petzeltová, *On the existence of globally defined weak solutions to the Navier–Stokes equations of isentropic compressible fluids*, J. Math. Fluid Mech. **3** (2001), 358–392.

[61] E. Feireisl and H. Petzeltová, *Large-time behaviour of solutions to the Navier–Stokes equations of compressible flow*, Arch. Ration. Mech. Anal. **150** (1999), 77–96.

[62] E. Feireisl and H. Petzeltová, *Unconditional stability of stationary flows of compressible heat-conducting fluids driven by large external forces*, J. Math. Fluid Mech. **1** (1999), 168–186.

[63] E. Feireisl and H. Petzeltová, *On compactness of solutions to the Navier–Stokes equations of compressible flow*, J. Differential Equations **163** (2000), 57–75.

[64] E. Feireisl and H. Petzeltová, *Bounded absorbing sets for the Navier–Stokes equations of compressible fluid*, Comm. Partial Differential Equations **26** (2001), 1133–1144.

[65] E. Feireisl and H. Petzeltová, *Asymptotic compactness of global trajectories generated by the Navier–Stokes equations of compressible fluid*, J. Differential Equations **173** (2001), 390–409.

[66] C. Foias and R. Temam, *The connection between the Navier–Stokes equations, dynamical systems and turbulence*, Directions in Partial Differential Equations, Academic Press, Boston, MA (1987), 55–73.

[67] H. Frid and V. Shelukhin, *Vanishing shear viscosity in the equations of compressible fluids for the flows with the cylinder symmetry*, SIAM J. Math. Anal. **31** (2000), 1144–1156.

[68] G. Friesecke and J.B. McLeod, *Dynamics as a mechanism preventing the formation of finer and finer microstructure*, Arch. Ration. Mech. Anal. **133** (1996), 199–247.

[69] H. Fujita-Yashima and R. Benabidallah, *Unicite' de la solution de l'équation monodimensionnelle ou a' symétrie sphérique d'un gaz visqueux et calorifère*, Rend. Circ. Mat. Palermo (2) **XLII** (1993), 195–218.

[70] H. Fujita-Yashima and R. Benabidallah, *Equation à symétrie sphérique d'un gaz visqueux et calorifère avec la surface libre*, Ann. Mat. Pura Appl. **CLXVIII** (1995), 75–117.

[71] H. Fujita-Yashima, M. Padula and A. Novotný, *Equation monodimensionnelle d'un gaz visqueux et calarifere avec des conditions initiales moins restrictives*, Richerche Mat. **XLII** (1983), 199–248.

[72] J. Glimm, *The continuous structure of discontinuities*, Lecture Notes in Phys., Vol. 344, Springer-Verlag, Berlin (1989), 177–186.

[73] H. Grad, *Asymptotic theory of the Boltzmann equation II*, Rarefied Gas Dynamics, 1, J. Laurmann, ed., Academic Press, New York (1963), 26–59.

[74] J.M. Greenberg and R.C. MacCamy, *On the exponential stability of solutions of* $E(u_x)u_{xx} + \lambda u_{xtx} = \rho u_{tt}$, J. Math. Anal. Appl. **31** (1970), 406–417.

[75] J.M. Greenberg, R.C. MacCamy and V.J. Mizel, *On the existence, uniqueness and stability of solutions of the equation* $\sigma'(u_x)u_{xx} + \lambda u_{xtx} = \rho_0 u_{tt}$, J. Math. Mech. **17** (1968), 707–728.

[76] B.L. Guo and P.C. Zhu, *Global existence of smooth solution to nonlinear thermoviscoelastic system with clamped boundary conditions in solid-like materials*, Comm. Math. Phys. **203** (1999), 365–383.

[77] B.L. Guo and P.C. Zhu, *Asymptotic behavior of the solution to the system for a viscous reactive gas*, J. Differential Equations **155** (1999), 177–202.

[78] I. Hansen, *Lebensdauer von klassischen Lösungen nichtlinearer Thermoelastizitätsgleichungen*, Diploma Thesis, Univ. of Bonn, Germany (1994).

[79] D.B. Henry, *Topics in analysis*, Publ. Sec. Mat. Univ. Autònoma Barselona **31** (1987), 29–84.

[80] D. Hoff, *Construction of solutions for compressible, isentropic Navier–Stokes equations in one space dimension with nonsmooth initial data*, Proc. Roy. Soc. Edinburgh Sect. A **103** (1986), 301–315.

[81] D. Hoff, *Global existence for 1-D, compressible, isentropic Navier–Stokes equations with large initial data*, Trans. Amer. Math. Soc. **303** (1987), 169–181.

[82] D. Hoff, *Discontinuous solutions of the Navier–Stokes equations for compressible flow*, Arch. Ration. Mech. Anal. **114** (1991), 15–46.

[83] D. Hoff, *Global well-posedness of the Cauchy problem for nonisentropic gas dynamics with disconinuous initial data*, J. Differential Equations **95** (1992), 33–74.

[84] D. Hoff, *Spherically symmetric solutions of the Navier–Stokes equations for compressible, isothermal flow with large discontinuous initial data*, Indiana Univ. Math. J. **41** (1992), 1225–1302.

[85] D. Hoff, *Strong convergence to global solutions for multidimensional flows of compressible, viscous fluids with polytropic equations of state and discontinuous initial data*, Arch. Ration. Mech. Anal. **132** (1995), 1–14.

[86] D. Hoff, *Global solutions of the Navier–Stokes equations for multidimensional, compressible flow with discontinuous initial data*, J. Differential Equations **120** (1995), 215–254.

[87] D. Hoff, *Continuous dependence on initial data for discontinuous solutions of the Navier–Stokes equations for one -dimensional compressible flow*, SIAM J. Math. Anal. **27** (1996), 1193–1211.

[88] D. Hoff, *Global solutions of the equations of one-dimensional, compressible flow with large data and forces, and with differing end states*, Z. Angew. Math. Phys. **49** (1998), 774–785.

[89] D. Hoff, *The zero-mach limit of compressible flows*, Comm. Math. Phys. **192** (1998), 543–554.

[90] D. Hoff, *Dynamics of singularity surfaces for compressible, viscous flows in two space dimensions*, Comm. Pure Appl. Math. **55** (2002), 1365–1407.

[91] D. Hoff and H.K. Jenssen, *Symmetric nonbarotropic flows with large data and forces*, Preprint (2003).

[92] D. Hoff and T.P. Liu, *The inviscid limit for Navier–Stokes equations of compressible, isentropic flow with shock data*, Indiana Univ. Math. J. **38** (1989), 861–915.

[93] D. Hoff and D. Serre, *The failure of continuous dependence on initial data for the Navier–Stokes equations of compressible flow*, SIAM J. Appl. Math. **51** (1991), 887–898.

[94] D. Hoff and J. Smoller, *Non-formation of vacuum states for compressible Navier–Stokes equations*, Comm. Math. Phys. **216** (2001), 255–276.

[95] D. Hoff and R. Zarnowski, *Continuous dependence in L^2 for discontinuous solutions of the viscous p-system*, Ann. Inst. H. Poincaré Anal. Non Linéaire **11** (1994), 159–187.

[96] D. Hoff and M. Ziane, *The global attractor and finite determining nodes for the Navier–Stokes equations of compressible flow with singular initial data*, Indiana Univ. Math. J. **49** (2000), 843–889.

[97] W.J. Hrusa and S.A. Messaoudi, *On formation of singularities in one-dimensional nonlinear thermoelasticity*, Arch. Ration. Mech. Anal. **111** (1990), 135–151.

[98] W.J. Hrusa and M.A. Tarabek, *On smooth solutions of the Cauchy problem in one-dimensional nonlinear thermoelasticity*, Quart. Appl. Math. **47** (1989), 631–644.

[99] L. Hsiao, *Quasilinear Hyperbolic Systems and Dissipative Mechanics*, World Scientific, Singapore (1998).

[100] L. Hsiao and H.Y. Jian, *Asymptotic behavior of solutions to the system of one-dimensional nonlinear thermoviscoelasticity*, Chinese Ann. Math. Ser. B **19** (1998), 143–152.

[101] L. Hsiao and T. Luo, *Large-time behavior of solutions for the outer presser problem of a viscous heat-conductive one-dimensional real gas*, Proc. Roy. Soc. Edinburgh Sect. A **126** (1996), 1277–1296.

[102] L. Hsiao and T. Luo, *Large-time behavior of solutions to the equations of one-dimensional nonlinear thermoviscoelasticity*, Quart. Appl. Math. **56** (1998), 201–219.

[103] D.A. Iskenderova and Sh.S. Smagulov, *The Cauchy problem for the equations of a viscous heat-conducting gas with degenerate density*, Comput. Math. Math. Phys. **33** (1993), 1109–1117.

[104] S. Jiang, *Global existence of smooth solutions in one-dimensional nonlinear thermoelasticity*, Proc. Roy. Soc. Edinburgh Sect. A **115** (1990), 257–274.

[105] S. Jiang, *Rapidly decreasing behaviour of solutions in nonlinear 3-D-thermoelasticity*, Bull. Austral. Math. Soc. **43** (1991), 89–99.

[106] S. Jiang, *Global solutions of the Neumann problem in one-dimensional nonlinear thermoelasticity*, Nonlinear Anal. **19** (1992), 107–121.

[107] S. Jiang, *On global smooth solutions to the one-dimensional equations of nonlinear inhomogeneous thermoelasticity*, Nonlinear Anal. **20** (1993), 1245–1256.

[108] S. Jiang, *Global large solutions to initial boundary value problem in one-dimensional nonlinear thermoviscoelasticity*, Quart. Appl. Math. **51** (1993), 731–744.

[109] S. Jiang, *On the initial boundary value problem for a viscous, heat-conducting, one-dimensional real gas*, J. Differential Equations **110** (1994), 157–181.

[110] S. Jiang, *On the asymptotic behavior of the motion of a viscous, heat-conducting, one-dimensional real gas*, Math. Z. **216** (1994), 317–336.

[111] S. Jiang, *Asymptotic stability of shearing flows of incompressible thermoviscous fluids*, Nonlinear Anal. **24** (1995), 1207–1220.

[112] S. Jiang, *Global spherically symmetric solutions to the equations of a viscous polytropic ideal gas in an exterior domain*, Comm. Math. Phys. **178** (1996), 339–374.

[113] S. Jiang, *Global smooth solutions to a one-dimensional nonlinear thermoviscoelastic model*, Adv. Math. Sci. Appl. **7** (1997), 771–787.

[114] S. Jiang, *Large-time behavior of solutions to the equations of a viscous polytropic ideal gas*, Ann. Mat. Pura Appl. **175** (1998), 253–275.

[115] S. Jiang, *Global smooth solutions of the equations of a viscous, heat-conducting, one-dimensional gas with density-dependent viscosity*, Math. Nachr. **190** (1998), 169–183.

[116] S. Jiang, *Large-time behavior of solutions to the equations of a one-dimensional viscous polytropic ideal gas in unbounded domains*, Comm. Math. Phys. **200** (1999), 181–193.

[117] S. Jiang, *Remarks on the asymptotic behavior of solutions to the compressible Navier–Stokes equations in the half-line*, Proc. Roy. Soc. Edinburgh Sect. A **132** (2002), 627–638.

[118] S. Jiang, J.E. Muñoz Rivera and R. Racke, *Asymptotic stability and global existence in thermoelasticity with symmetry*, Quart. Appl. Math. **56** (1998), 259–275.

[119] S. Jiang and R. Racke, *On some quasilinear hyperbolic–parabolic initial boundary value problems*, Math. Methods Appl. Sci. **12** (1990), 315–339.

[120] S. Jiang and R. Racke, *Evolution Equations in Thermoelasticity*, Chapman and Hall/CRC, Boca Raton–London–New York (2000).

[121] S. Jiang, Z.P. Xin and P. Zhang, *Global weak solutions to 1D compressible isentropic Navier–Stokes equations with density-dependent viscosity*, Preprint (1999).

[122] S. Jiang and P. Zhang, *Global spherically symmetric solutions of the compressible isentropic Navier–Stokes equations*, Comm. Math. Phys. **215** (2001), 559–581.

[123] S. Jiang and P. Zhang, *Remarks on weak solutions to the Navier–Stokes equations for 2-D compressible isothermal fluids with spherically symmetric initial data*, Indiana Univ. Math. J. **51** (2002), 345–355.

[124] S. Jiang and P. Zhang, *Axisymmetric solutions of the 3-D Navier–Stokes equations for compressible isentropic fluids*, J. Pures Appl. Math. **82** (2003), 949–973.

[125] S. Jiang and P. Zhang, *Global weak solutions to the Navier–Stokes equations for a 1D viscous polytropic ideal gas*, Quart. Appl. Math. **61** (2003), 435–449.

[126] S. Jiang and A.A. Zlotnik, *Global well-posedness of the Cauchy problem for the equations of a 1D viscous heat-conducting gas with the Lebesgue initial data*, Proc. Roy. Soc. Edinburgh Sect. A, to appear.

[127] F. John, *Formation of singularities in one-dimensional nonlinear wave propagation*, Comm. Pure Appl. Math. **27** (1974), 377–405.

[128] F. John, *Formation of singularities in elastic waves*, "Trends and applications of pure mathematics to mechanics", Proceedings Palaiseau (1983), P.G. Ciarlet and M. Roseau, eds, Lecture Notes in Phys., Vol. 195, Springer-Verlag, Berlin (1984), 194–210.

[129] J.L. Joly, G. Métivier and J. Rauch, *Focusing at a point and absorption of nonlinear oscillations*, Trans. Amer. Math. Soc. **347** (1995), 3921–3969.

[130] Y. Kanel, *On a model system of equations of one dimensional gas motion*, Differ. Equ. **4** (1968), 374–380.

[131] S. Kawashima, *Systems of a hyperbolic–parabolic composite type, with applications to the equations of magnetohydrodynamics*, Ph.D. Thesis, Kyoto University (1983).

[132] S. Kawashima, *Large-time behaviour of solutions to hyperbolic–parabolic systems of conservation laws and applications*, Proc. Roy. Soc. Edinburgh Sect. A **106** (1987), 169–194.

[133] S. Kawashima and A. Matsumura, *Mixed problems for symmetric hyperbolic–parabolic systems*, Manuscript (1989).

[134] S. Kawashima and T. Nishida, *Global solutions to initial value problem for the equations of one-dimensional motion of viscous polytropic gases*, J. Math. Kyoto Univ. **21** (1981), 825–837.

[135] S. Kawashima and M. Okada, *Smooth global solutions for the one-dimensional equations in magnetohydrodynamics*, Proc. Japan Acad. Ser. A **58** (1982), 384–387.

[136] S. Kawashima and Y. Shibata, *On the Neumann problem of one-dimensional nonlinear thermoelasticity with time-independent forces*, Czechoslovak Math. J. **45** (1995), 39–67.

[137] A.V. Kazhikhov, *On the theory of initial-boundary value problems for the equations of one-dimensional nonstationary motion of a viscous heat-conducting gas*, Dinamika Sploshn. Sredy **50** (1981), 37–62 (in Russian).

[138] A.V. Kazhikhov, *Cauchy problem for viscous gas equations*, Siberian Math. J. **23** (1982), 44–49.

[139] A. Kazhikhov and V. Shelukhin, *Unique global solutions in time of initial boundary value problems for one-dimensional equations of a viscous gas*, J. Appl. Math. Mech. **41** (1977), 273–282.

[140] B. Kawohl, *Global existence of large solutions to initial boundary value problem for a viscous, heat-conducting one-dimensional real gas*, J. Differential Equations **58** (1985), 76–103.

[141] H. Koch, *Slow decay in linear thermoelasticity*, Quart. Appl. Math. **58** (2000), 601–612.

[142] G. Lebeau and E. Zuazua, *Decay rates for the three-dimensional linear system of thermoelasticity*, Arch. Ration. Mech. Anal. **148** (1999), 179–231.

[143] R. Leis, *Initial Boundary Value Problems in Mathematical Physics*, Teubner-Verlag, Stuttgart; Wiley, Chichester (1986).

[144] D. Li, *The nonlinear initial-boundary value problem and the existence of multi-dimensional shock wave for quasilinear hyperbolic–parabolic coupled systems*, Chinese Ann. Math. Ser. B **8** (1987), 252–280.

[145] P.-L. Lions, *Mathematical Topics in Fluid Mechanics, Vol. I, Incompressible Models*, Oxford Science Publications, Clarendon Press, Oxford (1996).

[146] P.-L. Lions, *Mathematical Topics in Fluid Mechanics, Vol. II, Compressible Models*, Oxford Science Publications, Clarendon Press, Oxford (1998).

[147] P.-L. Lions and N. Masmoudi, *Incompressible limit for a viscous compressible fluid*, J. Math. Pures Appl. **77** (1998), 585–627.

[148] T.P. Liu, *Development of singularities in the nonlinear waves for quasi-linear hyperbolic partial differential equations*, J. Differential Equations **33** (1979), 92–111.

[149] T.P. Liu, *Shock waves for compressible Navier–Stokes equations are nonlinearly stable*, Comm. Pure Appl. Math. **39** (1986), 565–594.

[150] T.P. Liu and Z. Xin, *Nonlinear stability of rarefaction waves for compressible Navier–Stokes equations*, Comm. Math. Phys. **118** (1988), 451–465.

[151] T.P. Liu, Z. Xin and T. Yang, *Vacuum states of compressible flow*, Discrete Contin. Dynam. Systems **4** (1998), 1–32.

[152] T.P. Liu and Y. Zeng, *Large Time Behavior of Solutions for General Quasilinear Hyperbolic–Parabolic Systems of Conservation Laws*, Mem. Amer. Math. Soc., No. 599, Amer. Math. Soc., Providence, RI (1997).

[153] T. Luo, *Global smooth solutions to the Cauchy problem for a viscous heat-conductive one-dimensional real gas*, Acta Math. Sinica (N.S.) **11** (1995), 201–214.

[154] T. Luo, *On the outer pressure problem of a viscous heat-conductive one-dimensional real gas*, Acta Math. Appl. Sinica (English Ser.) **13** (1997), 251–264.

[155] T. Luo, Z. Xin and T. Yang, *Interface behavior of compressible Navier–Stokes equations with vacuum*, SIAM J. Math. Anal. **31** (2000), 1175–1191.

[156] T. Makino, *On a local existence theorem for the evolution equations of gaseous stars*, Patterns and Waves Qualitative Analysis of Nonlinear Differential Equations, North-Holland, Amsterdam (1986), 459–479.

[157] N. Masmoudi, *Incompressible, inviscid limit of the compressible Navier–Stokes system*, Ann. Inst. H. Poincaré Anal. Non Linéaire **18** (2001), 199–224.

[158] A. Matsumura and T. Nishida, *The initial value problem for the equations of motion of viscous and heat-conductive gases*, J. Math. Kyoto Univ. **20** (1980), 67–104.

[159] A. Matsumura and T. Nishida, *Initial boundary value problems for the equations of motion of compressible viscous and heat-conductive fluids*, Comm. Math. Phys. **89** (1983), 445–464.

[160] A. Matsumura and K. Nishihara, *On the stability of travelling wave solutions of a one-dimensional model system for compressible viscous gas*, Japan J. Appl. Math. **2** (1985), 17–25.

[161] A. Matsumura and K. Nishihara, *Global stability of the rarefaction waves of a one-dimensional model system for compressible viscous gas*, Comm. Math. Phys. **144** (1992), 325–335.

[162] A. Matsumura and S. Yanagi, *Uniform boundedness of the solutions for one-dimensional isentropic model system of compressible viscous gas*, Comm. Math. Phys. **175** (1996), 259–274.

[163] Š. Matušů-Nečasová, M. Okada and T. Makino, *Free boundary value problems for the equation of one-dimensional motion of viscous gas (II)*, Japan J. Indust. Appl. Math. **12** (1995), 195–203.

[164] L. Min, A.V. Kazhikhov and S. Ukai, *Global solutions to the Cauchy problem of the Stokes approximation equations for two-dimensional compressible flows*, Comm. Partial Differential Equations **23** (1998), 985–1006.

[165] T. Mukoyama, *On the Cauchy problem for quasilinear hyperbolic–parabolic coupled systems in higher dimensional spaces*, Tsukuba J. Math. **13** (1989), 363–386.

[166] J.E. Muñoz Rivera, *Energy decay rates in linear thermoelasticity*, Funkcial. Ekvac. **35** (1992), 19–30.

[167] J.E. Muñoz Rivera and M. de Lacerda Oliveira, *Exponential stability for a contact problem in thermoelasticity*, IMA J. Appl. Math. **58** (1997), 71–82.

[168] J.E. Muñoz Rivera and R. Racke, *Multidimensional contact problems in thermoelasticity*, SIAM J. Appl. Math. **58** (1998), 1307–1337.

[169] T. Nagasawa, *On the one-dimensional motion of the polytropic ideal gas no-fixed on the boundary*, J. Differential Equations **65** (1986), 49–67.

[170] T. Nagasawa, *On the outer pressure problem of the one-dimensional polytropic ideal gas*, Japan J. Appl. Math. **5** (1988), 53–85.

[171] T. Nagasawa, *Global asymptotics of the outer pressure problem with free boundary*, Japan J. Appl. Math. **5** (1988), 205–224.

[172] T. Nagasawa, *On the asymptotic behavior of the one-dimensional motion of the polytropic ideal gas with stress-free condition*, Quart. Appl. Math. **46** (1988), 665–679.

[173] J. Nash, *Le problème de Cauchy pour les équations différentielles d'un fluide général*, Bull. Soc. Math. France **90** (1962), 487–497.

[174] V.B. Nikolaev, *On the solvability of mixed problem for one-dimensional axisymmetrical viscous gas flow*, Dinamicheskie Zadachi Mekhaniki Sploshnoj Sredy Sibirsk. Otd. Acad. Nauk SSSR, Inst. Gidrodinamiki **63** (1983), 136–141 (in Russian).

[175] T. Nishida, *Equations of fluid dynamics-free surface problems*, Comm. Pure. Appl. Math. **39** (1986), 221–238.

[176] T. Nishida, *Equations of motion of compressible viscous fluids, pattern and waves*, T. Nishida, M. Mimura and H. Fujii, eds, Kinokuniya/North-Holland, Amsterdam–Tokyo (1996), 97–128.

[177] A. Novotný and I. Straškraba, *Convergence to equilibria for compressible Navier–Stokes equations with large data*, Ann. Mat. Pura Appl. **CLXXIX** (2001), 263–287.

[178] M. Okada, *Free boundary value problems for the equation of one-dimensional motion of viscous gas*, Japan J. Appl. Math. **6** (1989), 161–177.

[179] M. Okada and T. Makino, *Free boundary problem for the equations of spherically symmetrical motion of viscous gas*, Japan J. Indust. Appl. Math. **10** (1993), 219–235.

[180] M. Okada, Š. Matušů-Nečasová and T. Makino, *Free boundary problem for the equation of one-dimensional motion of compressible gas with density-dependent viscosity*, Preprint (1999).

[181] M. Padula, *Existence of global solutions for two-dimensional viscous compressible flows*, J. Funct. Anal. **69** (1986), 1–20.

[182] R.H. Pan, *Global smooth solutions and the asymptotic behavior of the motion of a viscous, heat-conductive one-dimensional real gas*, J. Partial Differential Equations **11** (1998), 273–288.

[183] R. Pan and P. Zhu, *Asymptotic behavior of the solution to a model system in linearly viscous materials*, J. Partial Differential Equations **12** (1999), 251–261.

[184] R.L. Pego, *Phase transitions in one-dimensional nonlinear viscoelasticity: Admissibility and stability*, Arch. Ration. Mech. Anal. **97** (1987), 353–394.

[185] G. Ponce and R. Racke, *Global existence of small solutions to the initial value problem for nonlinear thermoelasticity*, J. Differential Equations **87** (1990), 70–83.

[186] Y.M. Qin, *Global existence and asymptotic behavior for a viscous, heat-conductive, one-dimensional real gas with fixed and thermally insulated endpoints*, Nonlinear Anal. **44** (2001), 413–441.

[187] Y.M. Qin and J.E. Muñoz Rivera, *Universal attractors for a nonlinear one-dimensional heat-conductive viscous real gas*, Proc. Roy. Soc. Edinburgh Sect. A **132** (2002), 685–709.

[188] R. Racke, *On the Cauchy problem in nonlinear 3-d-thermoelasticity*, Math. Z. **203** (1990), 649–682.

[189] R. Racke, *Blow-up in nonlinear three-dimensional thermoelasticity*, Math. Methods Appl. Sci. **12** (1990), 267–273.

[190] R. Racke, *Lectures on Nonlinear Evolution Equations. Initial Value Problems*, Aspects of Math., Vol. E19, Vieweg, Braunschweig/Wiesbaden (1992).

[191] R. Racke, *Exponential decay for a class of initial boundary value problems in thermoelasticity*, Comput. Math. Appl. **12** (1993), 67–80.

[192] R. Racke and Y. Shibata, *Global smooth solutions and asymptotic stability in one-dimensional nonlinear thermoelasticity*, Arch. Ration. Mech. Anal. **116** (1991), 1–34.

[193] R. Racke, Y. Shibata and S. Zheng, *Global solvability and exponential stability in one-dimensional nonlinear thermoelasticity*, Quart. Appl. Math. **51** (1993), 751–763.

[194] R. Racke and S. Zheng, *Global existence and asymptotic behavior in nonlinear thermoviscoelasticity*, J. Differential Equations **134** (1997), 46–67.

[195] M. Reissig and Y.-G. Wang, *Propagation of mild singularities in higher dimensional thermoelasticity*, J. Math. Anal. Appl. **240** (1999), 398–415.

[196] M.O. Rieger, *Asymptotic Behaviour of Radially Symmetric Solutions in Thermoelasticity*, Konstanzer Schriften Math. Inf., Vol. 73, Hartung-Gorre, Konstanz (1998).

[197] M.O. Rieger and J. Zimmer, *Global existence for nonconvex thermoelasticity*, Preprint (2002).

[198] D. Serre, *Sur l'équation monodimensionall d'un fluids visgueux, compressible et conducteur de chaleur*, C. R. Acad. Sci. Paris Sér I **303** (1986), 703–706.

[199] D. Serre, *Variation de grande amplitude pour la densité d'un fluid visqueux compressible*, Phys. D **48** (1991), 113–128.

[200] V.V. Shelukhin, *On the structure of generalized solutions of the one-dimensional equations of a polytropic viscous gas*, Prikl. Mat. Mech. **48** (1984), 665–672.

[201] V.V. Shelukhin, *The limit of zero shear viscosity for compressible fluids*, Arch. Rat. Mech. Anal. **143** (1998), 357–374.

[202] V.V. Shelukhin, *Vanishing shear viscosity in a free-boundary problem for the equations of compressible fluids*, J. Differential Equations **167** (2000), 73–86.

[203] W. Shen and S. Zheng, *Global smooth solutions to the system of one-dimensional thermoelasticity with dissipation boundary conditions*, Chinese Ann. Math. Ser. B **7** (1986), 303–317.

[204] W. Shen, S. Zheng and P. Zhu, *Global existence and asymptotic behavior of weak solutions to nonlinear thermoviscoelastic system with clamped boundary conditions*, Quart. Appl. Math. **57** (1999), 93–116.

[205] P. Shi and M. Shillor, *Existence of a solution to the n-dimensional problem of thermoelastic contact*, Comm. Partial Differential Equations **17** (1992), 1597–1618.

[206] P. Shi and Y. Xu, *Decoupling of the quasistatic system of thermoelasticity on the unit disk*, J. Elasticity **31** (1993), 209–218.

[207] Y. Shibata, *On a local existence theorem for some quasilinear hyperbolic–parabolic coupled system with Neumann type boundary condition*, Manuscript (1989).

[208] M. Slemrod, *Global existence, uniqueness, and asymptotic stability of classical smooth solutions in one-dimensional non-linear thermoelasticity*, Arch. Ration. Mech. Anal. **76** (1981), 97–133.

[209] Sh.S. Smagulov, A.A. Durmagambetov and D.A. Iskenderova, *The Cauchy problem for equations of magnetic gas dynamics*, Differentsial'nye Uravneniya **29** (1993), 337–348 (in Russian).

[210] J. Smoller, *Shock Waves and Reaction–Diffusion Equations*, Springer-Verlag, New York–Heidelberg (1994).

[211] I. Straškraba, *Asymptotic development of vacuums for one-dimensional Navier–Stokes equations of compressible flows*, Nonlinear World **3** (1996), 519–535.

[212] I. Straškraba and A. Valli, *Asymptotic behaviour of the density for one-dimensional Navier–Stokes equations*, Manuscripta Math. **62** (1988), 401–416.

[213] I. Straškraba and A.A. Zlotnik, *On a decay rate for 1D-viscous compressible barotropic fluid equations: The case of positive stationary density*, Preprint No. 137, Acad. Sci. Czech Rep. (2000).

[214] A. Szepessy and Z. Xin, *Nonlinear stability of viscous shock waves*, Arch. Ration. Mech. Anal. **122** (1993), 53–103.

[215] L. Tartar, *Topics in Nonlinear Analysis*, Publ. Math. Orsay, Vol. 78, Univ. de Paris-Sud, Orsay (1978).

[216] L. Tartar, *Compensated Compactness and Partial Differential Equations*, Nonlinear Analysis and Mechanics, Heriort-Watt Symposium, Vol. IV, R.J. Knops, ed., Pitman, London (1979).

[217] R. Temam, *Navier–Stokes Equations: Theory and Numerical Analysis*, 3rd Edition, Elsevier, Amsterdam (1984).

[218] A.E. Tzavaras, *Shearing of materials exhibiting thermal softening or temperature dependent viscosity*, Quart. Appl. Math. **44** (1986), 1–12.

[219] A.E. Tzavaras, *Effect of thermal softening in shearing of strain-rate dependent materials*, Arch. Ration. Mech. Anal. **99** (1987), 349–374.

[220] V.A. Vaigant, *An example of nonexistence globally in time of a solution of the Navier–Stokes equations for a compressible viscous barotropic fluid*, Dokl. Math. **50** (1995), 397–399.

[221] V.A. Vaigant and A.V. Kazhikhov, *Global solutions to the potential flow equations for a compressible viscous fluid at small Reynolds numbers*, Differ. Equ. **30** (1994), 935–947.

[222] V.A. Vaigant and A.V. Kazhikhov, *On existence of global solutions to the two-dimensional Navier–Stokes equations for a compressible viscous fluid*, Siberian J. Math. **36** (1995), 1283–1316.

[223] A. Valli, *Mathematical results for compressible flows*, Mathematical Topics in Fluid Mechanics, J.F. Rodrigues and A. Sequeira, eds, Pitman Research Notes in Math. Ser., Vol. 274, Longman, Harlow (1992), 193–229.

[224] A. Valli and W.M. Zajączkowski, *Navier–Stokes equations for compressible fluids: Global existence and qualitative properties of the solutions in the general case*, Comm. Math. Phys. **103** (1986), 259–296.

[225] A.I. Vol'pert and S.I. Hudjaev, *On the Cauchy problem for composite systems of nonlinear differential equations*, Math. USSR-Sb. **16** (1972), 517–514.

[226] D. Wang, *Global solutions of the Navier–Stokes equations for viscous compressible flows*, Nonlinear Anal. **52** (2003), 1867–1890.

[227] Y. Wang, *Microlocal analysis in nonlinear thermoelasticity*, Nonlinear Anal. **54** (2003), 683–705.

[228] S.J. Watson, *Unique global solvability for initial-boundary value problems in one-dimensional nonlinear thermoviscoelasticity*, Arch. Ration. Mech. Anal. **153** (2000), 1–37.

[229] Z. Xin, *Zero dissipation limit to rarefaction waves for the one-dimensional Navier–Stokes equations of compressible isentropic gases*, Comm. Pure Appl. Math. **46** (1993), 621–665.

[230] Z. Xin, *Blow-up of smooth solutions to the compressible Navier–Stokes equations with compact density*, Comm. Pure Appl. Math. **51** (1998), 229–240.

[231] Z. Xin, *Viscous boundary layers and their stability*, J. Partial Differential Equations **11** (1998), 97–124.

[232] Z. Xin and T. Yanagisawa, *Zero-viscosity limit of the linearized Navier–Stokes equations for a compressible viscous fluid in the half-plane*, Comm. Pure Appl. Math. **52** (1999), 479–541.

[233] T. Yang, Z.A. Yao and C.J. Zhu, *Compressible Navier–Stokes equations with density-dependent viscosity and vacuum*, Comm. Partial Differential Equations **26** (2001), 965–981.

[234] T. Yang and H.J. Zhao, *A vacuum problem for the one-dimensional compressible Navier–Stokes equations with density-dependent viscosity*, J. Differential Equations **184** (2002), 163–184.

[235] T. Yang and C.J. Zhu, *Compressible Navier–Stokes equations with degenerate viscosity coefficient and vacuum*, Comm. Math. Phys. **230** (2002), 329–363.

[236] Y.B. Zel'dovich and Y.P. Raizer, *Physics of Shock Waves and High-Temperature Hydrodynamic Phenomena, Vol. II*, Academic Press, New York (1967).

[237] S. Zheng, *Initial boundary value problems for quasilinear hyperbolic–parabolic coupled systems in higher dimensional spaces*, Chinese Ann. Math. Ser. B **4** (1983), 443–462.

[238] S. Zheng, *Global solutions and applications to a class of quasilinear hyperbolic–parabolic coupled systems*, Sci. Sinica Ser. A **27** (1984), 1274–1286.

[239] S. Zheng, *Global smooth solutions to the Cauchy problem of nonlinear thermoelastic equations with dissipation*, Chinese Ann. Math. Ser. B **8** (1987), 142–155.

[240] S. Zheng, *Nonlinear parabolic equations and hyperbolic–parabolic coupled systems*, Pitman Monographs Surveys Pure Appl. Math., Vol. 76, Longman, New York (1995).

[241] S. Zheng and Y. Qin, *Universal attractors for the Navier–Stokes equations of compressible and heat-conductive fluid in bounded annular domains in* $\mathbb{R}^n$, Arch. Ration. Mech. Anal. **160** (2001), 153–179.

[242] S. Zheng and W. Shen, *Global solutions to the Cauchy problem of quasilinear hyperbolic–parabolic coupled systems*, Sci. Sinica Ser. A **30** (1987), 1133–1149.

[243] A.A. Zlotnik, *Global behavior of* 1-*D viscous compressible barotropic flows with free boundary and self-gravitation*, Math. Methods Appl. Sci. **26** (2003), 671–690.

[244] A.A. Zlotnik and A.A. Amosov, *On stability of generalized solutions to the equations of one-dimensional motion of a viscous heat-conducting gas*, Siberian Math. J. **38** (1997), 663–684.

[245] A.A. Zlotnik and A.A. Amosov, *Stability of generalized solutions to equations of one-dimensional motion of viscous heat conducting gases*, Math. Notes **63** (1998), 736–746.

[246] A.A. Zlotnik and A.A. Amosov, *Weak solutions to viscous heat-conducting gas* 1*D-equations with discontinuous data: Global existence, uniqueness, and regularity*, The Navier–Stokes Equations: Theory and Numerical Methods, R. Salvi, ed., Lecture Notes in Pure and Appl. Math., Vol. 223, Dekker, New York–Basel (2002), 141–158.

[247] A.A. Zlotnik and Z.B. Nguyen, *Properties and asymptotic behavior of solutions of some problems of one-dimensional motion of a viscous barotropic gas*, Math. Notes **55** (1994), 471–482.

[248] A.A. Zlotnik and I. Straškraba, *Global behavior of solutions to the one-dimensional equations of viscous compressible barotropic flow with variable viscosity*, Dokl. Math. **63** (2001), 404–409.

CHAPTER 5

Nonlinear Parabolic Equations and Systems

Alessandra Lunardi

Dipartimento di Matematica, Università di Parma, via D'Azeglio 85/A, 43100 Parma, Italy
E-mail: lunardi@unipr.it, www: http://math.unipr.it/~lunardi

Contents

1. Introduction 387
2. Abstract parabolic problems 388
   2.1. Optimal regularity in linear problems 390
   2.2. The nonlinear problem 392
   2.3. Applications and drawbacks 397
3. Equations and systems in Hölder spaces 401
4. Existence at large and stability 411
   4.1. Asymptotic behavior in linear problems 415
   4.2. Principle of linearized stability and local invariant manifolds 419
5. The fully nonlinear approach to free boundary problems 424
   5.1. Hele–Shaw models 424
   5.2. Models from combustion theory 429
References 435

HANDBOOK OF DIFFERENTIAL EQUATIONS
Evolutionary Equations, volume 1
Edited by C.M. Dafermos and E. Feireisl

1. Introduction

The theory of nonlinear parabolic problems is so widely developed that it is impossible to give an overview in a few pages. Therefore in this chapter we consider only a specific class of equations and systems with a high degree of nonlinearity, that are called fully nonlinear because the nonlinearities involve the highest-order derivatives of the unknowns appearing in the problems. For instance, a simple significant example is the Cauchy problem for a second-order equation,

$$\begin{cases} u_t(t,x) = \Phi\big(D^2u(t,x)\big), & t \geqslant 0, x \in \overline{\Omega}, \\ \Psi\big(Du(t,x)\big) = 0, & t \geqslant 0, x \in \partial\Omega, \\ u(0,x) = u_0(x), & x \in \overline{\Omega}, \end{cases} \tag{1.1}$$

where the matrix $\partial\Phi/\partial q_{ij}(D^2u_0)$ is positive definite at each $x \in \overline{\Omega}$, and the vector with components $\partial\Psi/\partial p_i(Du_0)$ is nontangential at each $x \in \partial\Omega$. Ω is an open set in $\mathbb{R}^N$ with sufficiently smooth boundary $\partial\Omega$, D and D^2 denote the gradient and the matrix of the second-order derivatives with respect to the space variables x.

The theory of fully nonlinear parabolic equations and systems is not new. After a few papers in the sixties and seventies dealing mainly with local existence and uniqueness of regular solutions [23,25,26,38], big improvements came in the eighties with the papers of Krylov about a priori estimates and existence large for second-order equations (see, e.g., the book [27]), and of Da Prato and Grisvard who initiated the theory of abstract fully nonlinear parabolic equations in Banach spaces, in [12]. The latter paper was followed by a series of works about geometric theory of fully nonlinear parabolic equations, an account of which up to 1995 may be found in Chapters 8 and 9 of the book [33].

In the last years, a further impulse to the general theory was given by the study of multidimensional parabolic free boundary problems that can be transformed to fixed boundary ones by suitable changes of coordinates, and the resulting final systems are fully nonlinear. See [6–10,16,18–20,31].

Let us describe the contents of this chapter.

In Section 2 we give an overview on the theory of fully nonlinear parabolic equations in Banach spaces, including the discussion about stability, instability, and invariant manifolds of stationary solutions. Problems like (1.1) with the nonlinear boundary condition $\Psi(Du(t,x)) = 0$ replaced by a linear one, such as $u = 0$ or $\partial u/\partial\nu = 0$, may be turned into evolution equations in Banach spaces in a standard way. The function u is seen as a function of the only variable t with values in a Banach space X of functions defined in Ω, i.e., setting $U(t) = u(t,\cdot)$ we rewrite (1.1) as

$$\begin{cases} U'(t) = F\big(U(t)\big), & t \geqslant 0, \\ U(0) = u_0, \end{cases}$$

where $F(U) = \Phi(D^2U)$ is defined in (an open subset of) a Banach space $D \subset X$, and the linear operator $A = F'(u_0) : D \mapsto X$ is the generator of an analytic semigroup. The difference between the above problem and the more popular semilinear problems treated for instance in [22] is that the nonlinearity is defined in the domain of A instead of in some intermediate space between X and $D(A)$. This creates several technical difficulties that will be described in Section 2.

In Section 3 we see in detail second-order equations in smooth domains of $\mathbb{R}^N$ with fully nonlinear boundary conditions such as (1.1) and their generalizations that cannot be treated as immediate applications of the abstract theory. In Section 4 we discuss the principle of linearized stability and the construction of invariant manifolds near stationary solutions of these equations.

In Section 5 we show how the general theory may be applied to different parabolic free boundary problems. As model problems we consider the free boundary heat equation, arising in combustion theory,

$$\begin{cases} u_t = \Delta u, & t > 0, x \in \Omega_t, \\ u = 0, \quad \frac{\partial u}{\partial n} = -1, & t > 0, x \in \partial\Omega_t, \end{cases} \tag{1.2}$$

and Stefan-type problems like the Hele–Shaw flow,

$$\begin{cases} \Delta u = 0, & t > 0, x \in \Omega_t, \\ u = 0, \quad V = -\frac{\partial u}{\partial n}, & t > 0, x \in \Gamma_t, \\ \frac{\partial u}{\partial n} = b, & t > 0, x \in J, \end{cases} \tag{1.3}$$

where the unknowns are the open sets $\Omega_t \subset \mathbb{R}^N$ for $t > 0$ and the function $u(t,x)$ for $t > 0$, $x \in \Omega_t$. In the boundary conditions, $n = n(t,x)$ is the exterior normal vector to $\partial\Omega_t$ at $x \in \partial\Omega_t$. In the case of problem (1.3) we have $N = 2$ and the boundary of Ω_t is made by a fixed known interior component J and a moving unknown exterior component Γ_t, V represents the normal velocity of the free boundary Γ_t in such a way that V is positive for expanding curves, and $b \geqslant 0$.

Our procedure is to reduce the free boundary problems to fixed boundary ones by suitable changes of coordinates, and then to eliminate one of the unknowns (either u or the free boundary), expressing it in terms of the other unknown, to get a final problem for only one unknown. In both cases the final problem will be fully nonlinear, and nonlocal. In the case of problem (1.2) we eliminate the free boundary and the final problem is studied with the methods of Sections 3 and 4. In problem (1.3) we eliminate u, and the final problem is studied with the methods of parabolic evolution equations in Banach spaces of Section 2.

2. Abstract parabolic problems

Let D, X be Banach spaces with respective norms $\|\cdot\|_D$, $\|\cdot\|$, and such that D is continuously embedded in X. We shall discuss the problem

$$\begin{cases} u'(t) = F\big(t, u(t)\big), & t > 0, \\ u(0) = u_0, \end{cases} \tag{2.1}$$

where $F:[0,T]\times\mathcal{O}\mapsto X$ is a sufficiently smooth function, $T>0$, and $\mathcal{O}$ is a neighborhood of the initial datum $u_0\in D$.

The abstract parabolicity assumption near u_0 is that the operator $A:D(A)=D\mapsto X$, $A=F_u(0,u_0)$ is sectorial.

DEFINITION 2.1. A linear operator $A:D(A)\mapsto X$ is called sectorial if the resolvent set $\rho(A)$ contains a sector $\Sigma=\{\lambda\in\mathbb{C}\colon\ \lambda\neq\omega,\ |\arg(\lambda-\omega)|<\theta\}$ with $\omega\in\mathbb{R}$, $\theta\in(\pi/2,\pi)$, and there is $M>0$ such that $\|(\lambda-\omega)R(\lambda,A)\|_{L(X)}\leqslant M$ for each $\lambda\in\Sigma$.

We recall that if A is sectorial, the analytic semigroup generated by A is defined by

$$\mathrm{e}^{0A}=I,\qquad \mathrm{e}^{tA}=\frac{1}{2\pi\mathrm{i}}\int_{\omega+\gamma}\mathrm{e}^{t\lambda}R(\lambda,A)\,\mathrm{d}\lambda,\quad t>0,\tag{2.2}$$

where $r>0$, $\eta\in(\pi/2,\theta)$, and γ is the curve $\{\lambda\in\mathbb{C}\colon\ |\arg\lambda|=\eta,\ |\lambda|\geqslant r\}\cup\{\lambda\in\mathbb{C}\colon\ |\arg\lambda|\leqslant\eta,\ |\lambda|=r\}$, oriented counterclockwise. A summary of the general theory of sectorial operators and analytic semigroups in Banach spaces may be found in Chapter 2 of [33].

It is easy to see that if $A:D(A)=D\mapsto X$ is sectorial and $B\in L(D,X)$ is small enough, then $A+B:D(A+B)=D\mapsto X$ is still sectorial. So we may assume without much loss of generality that $F_u(t,x):D\mapsto X$ is sectorial for each $t\in[0,T]$ and $x\in\mathcal{O}$.

As a first step we look for a local solution $u\in C([0,a];D)\cap C^1([0,a];X)$ for some $a\in(0,T]$. The most natural way to solve problem (2.1), at least locally, is by linearization near u_0. Setting $G(t,x)=F(t,x)-Ax=F(t,x)-F_u(0,u_0)x$ for $t\in[0,T]$, $x\in\mathcal{O}$, we rewrite problem (2.1) in the form

$$\begin{cases} u'(t)=Au(t)+G\big(t,u(t)\big), & t>0,\\ u(0)=u_0,\end{cases}\tag{2.3}$$

and we try to solve it by a fixed point argument, i.e., we look at a solution as a fixed point of the operator Γ defined by $\Gamma u=v$ where v is the solution to the linear problem

$$\begin{cases} v'(t)=Av(t)+G\big(t,u(t)\big), & 0\leqslant t\leqslant a,\\ v(0)=u_0.\end{cases}\tag{2.4}$$

The simplest space where to set the fixed point would be (a closed ball in) $C([0,a];D)$. In this case, for every u in the ball, the function $f(t)=G(t,u(t))$ is in $C([0,a];X)$, and unless X and D are special spaces, well-known counterexamples show that in general v does not belong to $C([0,a];D)$, and Γ cannot map a ball of $C([0,a];D)$ into itself. Therefore, we turn to Hölder spaces where optimal regularity results and estimates for linear problems are available.

Such optimal regularity results, needed to solve locally (2.1) and to describe the properties of the solution, are stated in the next section.

Note that these difficulties do not arise in semilinear equations, i.e., equations of the type (2.4) where G is defined in $[0,a]\times Y$, Y being an intermediate space between

X and $D(A)$. This is because if $f(t) = G(t, u(t))$ is in $C([0,a]; X)$ and $u_0 \in D$, the solution v of (2.4) belongs to $C([0,a]; Y)$, provided Y satisfies the interpolatory embedding property

$$\|y\|_Y \leqslant C(\|y\|_X)^{1-\alpha}(\|y\|_D)^{\alpha}, \quad y \in Y,$$

for some $C > 0$, $\alpha \in (0,1)$. In this case we have in fact $v \in C^{1-\alpha}([0,a]; Y)$; see [33], Chapter 4.

2.1. *Optimal regularity in linear problems*

Let us consider the problem

$$\begin{cases} u'(t) = Au(t) + f(t), & 0 < t < a, \\ u(0) = u_0, \end{cases} \tag{2.5}$$

where A is a linear sectorial operator in general Banach space X and $f : [0,a] \mapsto X$ is (at least) continuous.

DEFINITION 2.2. A classical solution to problem (2.5) is a function $u \in C([0,a]; X) \cap C((0,a]; D(A)) \cap C^1((0,a]; X)$ that satisfies $u'(t) = Au(t) + f(t)$ for $0 < t \leqslant a$ and $u(0) = u_0$. A strict solution is a function $u \in C([0,a]; D) \cap C^1([0,a]; X)$ that satisfies $u'(t) = Au(t) + f(t)$ for $0 \leqslant t \leqslant a$ and $u(0) = u_0$.

It is well known that if problem (2.5) has a classical solution, then it is unique and is given by the variation of constants formula,

$$u(t) = e^{tA}u_0 + \int_0^t e^{(t-s)A} f(s)\, ds, \quad 0 \leqslant t \leqslant a.$$

We state below two optimal regularity results in Hölder spaces, whose proofs are due to [32] and [36], respectively, and may be found in [33], Chapter 4.

We need to introduce a class of real interpolation spaces between X and $D(A)$. For $0 < \alpha < 1$, the real interpolation space $D_A(\alpha, \infty) := (X, D(A))_{\alpha,\infty}$ is characterized by

$$\begin{aligned} &D_A(\alpha,\infty) = \{x \in X : t \mapsto v(t) = \|t^{1-\alpha} A e^{tA} x\| \in L^\infty(0,1)\}, \\ &\|x\|_{D_A(\alpha,\infty)} = \|x\| + [x]_{D_A(\alpha,\infty)} = \|x\| + \|v\|_\infty. \end{aligned} \tag{2.6}$$

The weighted Hölder space $C^\alpha_\alpha((a,b]; X)$ is defined as the set of all bounded functions $f : (a,b] \mapsto X$ such that $t \mapsto (t-a)^\alpha f(t)$ is α-Hölder continuous in $(a,b]$. The norm is $\|f\|_{C^\alpha_\alpha((a,b];X)} = \|f\|_\infty + \|(\cdot - a)^\alpha f(\cdot)\|_{C^\alpha((a,b];X)}$.

THEOREM 2.3. *Let $0 < \alpha < 1$, $f \in C^\alpha_\alpha((0, T]; X)$, $u_0 \in D(A)$. Then problem* (2.5) *has a classical solution u such that u' and Au belong to $C^\alpha_\alpha((0, T]; X)$, $t \mapsto t^\alpha u'(t)$ is bounded with values in $(X, D(A))_{\alpha,\infty}$, and there is $C = C(T) > 0$, increasing in T, such that*

$$\|u'\|_{C^\alpha_\alpha((0,T];X)} + \|Au\|_{C^\alpha_\alpha((0,T];X)} + \sup_{0<t<T} \|t^\alpha u'(t)\|_{(X,D(A))_{\alpha,\infty}} \leqslant C\big(\|f\|_{C^\alpha_\alpha((0,T];X)} + \|u_0\|_{D(A)}\big). \tag{2.7}$$

If in addition $f \in C([0, T]; X)$ and $Au_0 + f(0) \in \overline{D(A)}$, then u', $Au \in C([0, T]; X)$, and u is a strict solution to problem (2.5).

THEOREM 2.4. *Let $0 < \alpha < 1$, $f \in C^\alpha([0, T], X)$, $u_0 \in D(A)$ be such that $Au_0 + f(0) \in (X, D(A))_{\alpha,\infty}$. Then both u' and Au belong to $C^\alpha([0, T], X)$, u' is bounded with values in $(X, D(A))_{\alpha,\infty}$, and there is $C = C(T) > 0$, increasing in T, such that*

$$\|u\|_{C^{1+\alpha}([0,T],X)} + \|Au\|_{C^\alpha([0,T],X)} + \sup_{0<t<T} \|u'(t)\|_{(X,D(A))_{\alpha,\infty}} \leqslant C\big(\|f\|_{C^\alpha([0,T],X)} + \|u_0\|_{D(A)} + \|Au_0 + f(0)\|_{(X,D(A))_{\alpha,\infty}}\big). \tag{2.8}$$

We emphasize that we need Hölder spaces because of the lack of optimal regularity in spaces of continuous functions, in the sense that if $f : [0, T] \mapsto X$ is continuous, in general it is not true that the u' and Au in (2.5) are continuous with values in X. However, we have optimal regularity in spaces of continuous functions if we replace X and $D(A)$ by the continuous interpolations spaces $E_0 = D_A(\alpha)$ and $E_1 = D_A(\alpha + 1)$, respectively.

For $0 < \alpha < 1$, the space $D_A(\alpha) := (X, D(A))_\alpha$ is defined as the closure of $D(A)$ in $(X, D(A))_{\alpha,\infty}$. It may be characterized by

$$(X, D(A))_\alpha = \Big\{x \in (X, D(A))_{\alpha,\infty}\colon \lim_{t\to 0} t^{1-\alpha} A e^{tA} x = 0\Big\},$$

and it is a Banach space under the norm of $(X, D(A))_{\alpha,\infty}$. The space $D_A(\alpha + 1)$ is the domain of the part of A in $(X, D(A))_\alpha$:

$$D_A(\alpha + 1) = \big\{x \in D(A)\colon\ Ax \in D_A(\alpha)\big\},$$

and it is endowed with the graph norm $\|x\|_{D_A(\alpha+1)} = \|x\|_{D_A(\alpha)} + \|Ax\|_{D_A(\alpha)}$.

The following is a well-known result due to Da Prato and Grisvard [12].

THEOREM 2.5. *Let $0 < \alpha < 1$, and let $f \in C([0, T], D_A(\alpha))$, $u_0 \in D_A(\alpha + 1)$. Then the solution u of problem* (2.5) *belongs to $C([0, T], D_A(\alpha + 1)) \cap C^1([0, T], D_A(\alpha))$, and there is C such that*

$$\|u'\|_{C([0,T];D_A(\alpha,\infty))} + \|Au\|_{C([0,T];D_A(\alpha,\infty))} + \|Au\|_{C^\alpha([0,T];X)} \leqslant C\big(\|f\|_{C([0,T];D_A(\alpha,\infty))} + \|u_0\|_{D_A(\alpha+1,\infty)}\big). \tag{2.9}$$

2.2. *The nonlinear problem*

Here we collect several results about problem (2.1), that are proved using Theorems 2.3 and 2.4 as main tools.

The minimal assumptions on $F:[0,T]\times \mathcal{O} \mapsto X$, $\mathcal{O}$ being an open set in D, are the following:

(H1) The function $(t,u)\mapsto F(t,u)$ is continuous with respect to (t,u) and it is Fréchet differentiable with respect to u. There exists $\alpha \in (0,1)$ such that, for all $\bar{u}\in \mathcal{O}$, there are $R=R(\bar{u})$, $L=L(\bar{u})$, $K=K(\bar{u})>0$ verifying

$$\begin{aligned} &\|F_u(t,v)-F_u(t,w)\|_{L(D,X)}\leqslant L\|v-w\|_D,\\ &\|F(t,u)-F(s,u)\|+\|F_u(t,u)-F_u(s,u)\|_{L(D,X)}\leqslant K|t-s|^\alpha \end{aligned}\tag{2.10}$$

for all $t,s\in[0,T]$, $u,v,w\in B(\bar{u},R)\subset D$.

(H2) For every $t\in[0,T]$ and $v\in\mathcal{O}$, the Fréchet derivative $F_v(t,v)$ is sectorial in X and its graph norm is equivalent to the norm of D.

We now state the main local existence theorem.

THEOREM 2.6. *Let $\mathcal{O}\subset D$ be an open set. Let $F:[0,T]\times\mathcal{O}\mapsto X$ satisfy assumptions* (H1) *and* (H2). *Let $u_0\in\mathcal{O}$ be such that $F(0,u_0)\in\overline{D}$. Then there is a maximal $\tau=\tau(u_0)>0$ such that problem* (2.1) *has a solution $u\in C([0,\tau);D)\cap C^1([0,\tau);X)$ with the following properties.*

(i) *For every $\varepsilon\in(0,\tau)$, $u\in C^\alpha_\alpha((0,\tau-\varepsilon];D)$, $u'\in C^\alpha_\alpha((0,\tau-\varepsilon];X)$ and $t^\alpha u'(t)$ is bounded in $(0,\tau-\varepsilon]$ with values in $(X,D)_{\alpha,\infty}$.*

(ii) *u is the unique solution of* (2.1) *belonging to*

$$\bigcup_{0<\beta<1} C^\beta_\beta\big((0,\tau-\varepsilon];D\big)\cap C\big([0,\tau-\varepsilon];D\big)$$

for each $\varepsilon\in(0,\tau)$.

(iii) *u depends continuously on u_0, in the sense that, for each $\bar{u}\in\mathcal{O}$ such that $F(0,\bar{u})\in\overline{D}$ and for each $\bar{\tau}\in(0,\tau(\bar{u}))$, there are $\epsilon=\epsilon(\bar{u},\bar{\tau})>0$, $H=H(\bar{u},\bar{\tau})>0$ such that if*

$$u_0\in\mathcal{O},\quad F(0,u_0)\in\overline{D},\qquad \|u_0-\bar{u}\|_D\leqslant\epsilon,$$

then $\tau(u_0)\geqslant\bar{\tau}$ and

$$\begin{aligned} &\big\|u(\cdot;u_0)-u(\cdot;\bar{u})\big\|_{C^\alpha_\alpha((0,\bar{\tau}];D)}+\big\|u_t(\cdot;u_0)-u_t(\cdot;\bar{u})\big\|_{C^\alpha_\alpha((0,\bar{\tau}];X)}\\ &\qquad+\sup\big\{t^\alpha\big\|u_t(t,u_0)-u_t(t;\bar{u})\big\|_{(X,D)_{\alpha,\infty}}:0<t\leqslant\bar{\tau}\big\}\\ &\quad\leqslant H\|u_0-\bar{u}\|_D. \end{aligned}$$

(iv) *If in addition $F(0,u_0)\in(X,D)_{\alpha,\infty}$, then u is more regular up to $t=0$, precisely $u\in C^\alpha([0,\tau-\varepsilon];D)\cap C^{1+\alpha}([0,\tau-\varepsilon];X)$ for each $\varepsilon\in(0,\tau)$.*

It is possible to obtain several further regularity results, as well as results of dependence on parameters, stability of stationary solutions and of periodic orbits. See [33], Chapter 8.

The uniqueness part of the statement of Theorem 2.6 is not completely satisfactory. In a sense it is natural, because we get uniqueness in the same space where we prove existence of the solution. But a theorem of uniqueness of the strict solution, i.e., uniqueness in $C([0,a]; D) \cap C^1([0,a]; D)$, is not available (except in special cases, of course), and uniqueness of the strict solution is still an open problem.

Applying Theorem 2.5 gives well-posedness results for fully nonlinear problems in continuous interpolation spaces.

Let $E_1 \subset E_0 \subset X$ be Banach spaces, let $\mathcal{O}$ be an open subset of E_1, and let $0 < \theta < 1$, $T > 0$. $F : [0,T] \times \mathcal{O} \mapsto E_0$ is a nonlinear function such that

(H3) F and F_x are continuous in $[0,T] \times \mathcal{O}$, for every $(\bar{t}, \bar{u}) \in [0,T] \times \mathcal{O}$, the operator $F_x(\bar{t}, \bar{u}) : E_1 \mapsto E_0$ is the part in E_0 of a sectorial operator $L : D(L) \subset X \mapsto X$, such that $D_L(\theta) = E_0$, $D_L(\theta + 1) = E_1$ with equivalence of the respective norms.

THEOREM 2.7. *Let F satisfy assumption* (H3), *and let $u_0 \in \mathcal{O}$. Then there is a maximal $\tau = \tau(u_0) > 0$ such that problem* (2.1) *has a unique solution $u \in C([0,\tau); E_1) \cap C^1([0,\tau); E_0)$.*

The solution depends continuously on u_0, in the sense that, for each $\bar{u} \in \mathcal{O}$ and for each $\bar{\tau} \in (0, \tau(\bar{u}))$, there are $\epsilon = \epsilon(\bar{u}, \bar{\tau}) > 0$, $H = H(\bar{u}, \bar{\tau}) > 0$ such that if

$$u_0 \in \mathcal{O}, \quad \|u_0 - \bar{u}\|_{E_1} \leqslant \epsilon,$$

then $\tau(u_0) \geqslant \bar{\tau}$ and

$$\big\|u(t; u_0) - u(t; \bar{u})\big\|_{E_1} + \big\|u_t(t; u_0) - u_t(t, \bar{u})\big\|_{E_0} \leqslant H\|u_0 - \bar{u}\|_{E_1}.$$

The original proof due to Da Prato and Grisvard was simplified and clearly written in [5]. See also [33], Chapter 8.

A geometric theory of fully nonlinear abstract evolution equations may be developed, see [33], Chapter 9. Here we quote the principle of linearized stability and the construction of stable, unstable, and center manifolds of stationary solutions made in [13], that will be used in the applications to free boundary problems of Section 5.

Without loss of generality, we assume that the stationary solution is 0. In next Theorems 2.8–2.11 we shall assume that $F : \mathcal{O} \mapsto E_0$ satisfies assumption (H3), $\mathcal{O}$ being a neighborhood of 0 in E_1, and that $F(0) = 0$. Moreover, we set

$$A = F'(0).$$

THEOREM 2.8. *The following statements hold true.*

(i) *If $\omega_A := \sup\{\operatorname{Re}\lambda : \lambda \in \sigma(A)\} < 0$, then, for every $\omega \in (0, \omega_A)$, there are r, M such that if $\|u_0\|_{E_1} \leqslant r$ then the solution u of* (2.1) *is defined in* $[0, +\infty)$, *and*

$$\big\|u(t)\big\|_{E_1} + \big\|u'(t)\big\|_{E_0} \leqslant Me^{-\omega t}\|u_0\|_{E_1}, \quad t \geqslant 0.$$

(ii) *If* $\omega_A > 0$ *and* $\inf\{\mathrm{Re}\,\lambda\colon \lambda \in \sigma(A), \mathrm{Re}\,\lambda > 0\} > 0$, *then the null solution of* $u' = F(u)$ *is unstable in* E_1. *Specifically, there exist nontrivial backward solutions of* $u' = F(u)$ *going to* 0 *as* t *goes to* $-\infty$.

In the case where A is hyperbolic, i.e.,

$$\sigma(A) \cap \mathrm{i}\mathbb{R} = \emptyset, \tag{2.11}$$

a saddle point theorem may be shown. We denote by P the spectral projection associated to the subset $\sigma^+(A)$ of $\sigma(A)$ with positive real part,

$$P = \frac{1}{2\pi \mathrm{i}} \int_C R(\lambda, A)\,\mathrm{d}\lambda,$$

where C is any closed simple regular curve in $\{\mathrm{Re}\,\lambda > 0\}$ surrounding $\sigma^+(A)$.

THEOREM 2.9. *Assume that* (2.11) *holds. Then there are positive numbers* r_0, r_1 *such that*

(i) *There exist* $R_0 > 0$ *and a Lipschitz continuous function*

$$\varphi : B(0, r_0) \subset P(E_0) \mapsto (I - P)(E_1),$$

differentiable at 0 *with* $\varphi'(0) = 0$, *such that for every* u_0 *belonging to the graph of* φ *problem* (2.1) *has a unique backward solution* v *in* $C((-\infty, 0]; E_1)$, *such that*

$$\sup_{t<0} \|v(t)\|_{E_1} \leqslant R_0. \tag{2.12}$$

Moreover, $\|v(t)\mathrm{e}^{-\omega t}\|_{E_1} \to 0$ *as* $t \to -\infty$ *for every* $\omega \in (0, \inf\{\mathrm{Re}\,\lambda\colon \lambda \in \sigma^+(A)\})$.

Conversely, if problem (2.1) *has a backward solution* v *which satisfies* (2.12) *and* $\|Pv(0)\|_{E_0} \leqslant r_0$, *then* $v(0) \in \operatorname{graph}\varphi$.

(ii) *There exist* $R_1, r_1 > 0$ *and a Lipschitz continuous function*

$$\psi : B(0, r_1) \subset (I - P)(E_1) \mapsto P(E_0),$$

differentiable at 0 *with* $\psi'(0) = 0$, *such that, for every* u_0 *belonging to the graph of* ψ, *problem* (2.1) *has a unique solution* u *in* $C([0, +\infty); E_1)$, *such that*

$$\sup_{t>0} \|u(t)\|_{E_1} \leqslant R_1. \tag{2.13}$$

Moreover, $\|u(t)\mathrm{e}^{\omega t}\|_{E_1} \to 0$ *as* $t \to +\infty$ *for every* $\omega \in (0, -\sup\sigma^-(A))$, *where* $\sigma^-(A) := \{\lambda \in \sigma(A), \mathrm{Re}\,\lambda < 0\}$.

Conversely, if problem (2.1) *has a solution* u *which satisfies* (2.13) *and* $\|(I - P)u(0)\|_{D_A(\theta+1,\infty)} \leqslant r_1$, *then* $u(0) \in \operatorname{graph}\psi$.

(iii) *If in addition* $F \in C^k(\mathcal{O}; E_0)$ *and* $F^{(k)}$ *is Lipschitz continuous for some* $k \in \mathbb{N}$, *then* ψ *and* φ *are k times differentiable, with Lipschitz continuous kth order derivatives.*

As in the case of ordinary differential equations, the construction of center manifolds, or center-unstable manifolds, is more delicate. In addition to (H3) and to $F(0) = 0$ we shall assume that the set $\{\lambda \in \sigma(A)\colon \operatorname{Re}\lambda \geqslant 0\}$ consists of a finite-number of isolated eigenvalues with finite algebraic multiplicity. We shall denote by P_0 the spectral projection associated to it. The fact that the range of P_0 is finite-dimensional is of fundamental importance in the proofs.

Applying P_0 and $I - P_0$ we see that problem

$$u'(t) = F\big(u(t)\big), \quad t \geqslant 0,$$

is equivalent to the system

$$\begin{cases} x'(t) = A_+x(t) + P_0F\big(x(t) + y(t)\big), & t \geqslant 0, \\ y'(t) = A_-y(t) + (I - P_0)F\big(x(t) + y(t)\big), & t \geqslant 0, \end{cases} \tag{2.14}$$

with $x(t) = P_0u(t)$, $y(t) = (I - P_0)u(t)$, $A_+ = A_{|P_0(E_0)}\colon P_0(E_0) \mapsto P_0(E_0)$, $A_- = A_{|(I-P_0)(E_1)}\colon (I - P_0)(E_1) \mapsto (I - P_0)(E_0)$.

We modify F by introducing a smooth cutoff function $\rho\colon P_0(E_0) \mapsto \mathbb{R}$ such that

$$0 \leqslant \rho(x) \leqslant 1, \quad \rho(x) = 1 \quad \text{if } \|x\|_0 \leqslant 1/2, \qquad \rho(x) = 0 \quad \text{if } \|x\|_0 \geqslant 1.$$

Since $P_0(E_0)$ is finite dimensional, such a ρ does exist. For small $r > 0$, we consider the system

$$\begin{cases} x'(t) = A_+x(t) + f\big(x(t), y(t)\big), & t \geqslant 0, \\ y'(t) = A_-y(t) + g\big(x(t), y(t)\big), & t \geqslant 0, \end{cases} \tag{2.15}$$

with initial data

$$x(0) = x_0 \in P_0(E_0), \qquad y(0) = y_0 \in (I - P_0)(E_0), \tag{2.16}$$

where

$$f(x, y) = P_0F\big(\rho(x/r)x + y\big), \qquad g(x, y) = (I - P_0)F\big(\rho(x/r)x + y\big).$$

System (2.15) coincides with (2.14) if $\|x(t)\|_{E_0} \leqslant r/2$, and it is possible to show that if r and the initial data are small enough, then the solution of (2.15) and (2.16) exists in the large.

A finite-dimensional invariant manifold $\mathcal{M}$ for system (2.15) with small r may be constructed as the graph of a bounded, Lipschitz continuous function $\gamma\colon P_0(E_0) \mapsto (I - P_0)(E_1)$.

THEOREM 2.10. *Under the above assumptions, there exists $r_1 > 0$ such that, for $r \leqslant r_1$, there is a Lipschitz continuous function $\gamma : P_0(E_0) \mapsto (I - P_0)(E_1)$ such that the graph of γ is invariant for system* (2.15). *If in addition F is k times continuously differentiable, with $k \geqslant 2$, then there exists $r_k > 0$ such that if $r \leqslant r_k$ then $\gamma \in C^{k-1}$, $\gamma^{(k-1)}$ is Lipschitz continuous, and*

$$\gamma'(x)\big(A_{-}x + f\big(x, \gamma(x)\big)\big) = A_{+}\gamma(x) + g\big(x, \gamma(x)\big), \quad x \in P_0(X). \tag{2.17}$$

Then it is possible to see that the graph $\mathcal{M}$ of γ attracts exponentially all the orbits which start from an initial datum sufficiently close to $\mathcal{M}$. Moreover, each one of these orbits decays exponentially to an orbit in $\mathcal{M}$, in the sense specified by the next theorem.

THEOREM 2.11. *Let F be twice continuously differentiable. For every $\omega > 0$ such that $\omega < -\sup\{\operatorname{Re}\lambda\colon \lambda \in \sigma(A), \operatorname{Re}\lambda < 0\}$ there are $r(\omega)$, $M(\omega)$ such that if $\|x_0\|_{E_0}$ and $\|y_0\|_{E_1}$ are sufficiently small, then the solution of* (2.15) *and* (2.16) *exists in the large and satisfies*

$$\big\|y(t) - \gamma\big(x(t)\big)\big\|_{E_1} \leqslant M(\omega)e^{-\omega t}\big\|y_0 - \gamma(x_0)\big\|_{E_1}, \quad t \geqslant 0. \tag{2.18}$$

Moreover, there is $C(\omega) > 0$ such that if $\|x_0\|_{E_0}$ and $\|y_0\|_{E_1}$ are small enough there exists $\bar{x} \in P_0(E_0)$ such that

$$\begin{aligned} &\big\|x(t) - \bar{z}(t)\big\|_{E_1} + \big\|y(t) - \gamma\big(\bar{z}(t)\big)\big\|_{E_1} \\ &\quad \leqslant C(\omega)e^{-\omega t}\big\|y_0 - \gamma(x_0)\big\|_{E_0}, \quad t \geqslant 0, \end{aligned} \tag{2.19}$$

where $\bar{z}(t) = z(t; \gamma, \bar{x})$ is the solution of

$$z' = A_{+}z + f\big(z + \gamma(z)\big), \quad z(0) = \bar{x}. \tag{2.20}$$

As a consequence, the problem of the stability of the null solution to (2.1) in the critical case

$$\sup\big\{\operatorname{Re}\lambda\colon \lambda \in \sigma(A)\big\} = 0 \tag{2.21}$$

is reduced to the stability of the null solution to a finite-dimensional system.

COROLLARY 2.12. *Let $\mathcal{O}$ be a neighborhood of 0 in E_1, and let $F : \mathcal{O} \mapsto E_0$ be a C^2 function satisfying* (H3), *with $F(0) = 0$. Assume that $A = F'(0)$ satisfies* (2.21) *and that $\sigma(A) \cap i\mathbb{R}$ consists of a finite number of isolated eigenvalues with finite algebraic multiplicity.*

Then the null solution of (2.1) *is stable* (*respectively, asymptotically stable, unstable*) *in E_1 if and only if the null solution of the finite-dimensional system* (2.20) *is stable* (*respectively, asymptotically stable, unstable*).

These stability results (precisely, Theorems 2.8–2.11 and Corollary 2.12) may be extended to the case where E_0 and E_1 are real interpolation spaces $D_L(\theta, \infty)$, $D_L(\theta + 1, \infty)$ instead of continuous interpolation spaces. See [33], Chapter 9.

2.3. *Applications and drawbacks*

Let us describe the applicability of the abstract theory in a simple significant example, Ω being a bounded open set in $\mathbb{R}^N$ with regular boundary $\partial\Omega$:

$$\begin{cases} u_t(t,x) = \Phi\big(D^2u(t,x)\big), & t \geqslant 0, x \in \Omega, \\ \mathcal{B}u(t,x) = 0, & t \geqslant 0, x \in \partial\Omega, \\ u(0,x) = u_0(x), & x \in \Omega. \end{cases} \tag{2.22}$$

Here $\mathcal{B}$ is a first-order differential operator with regular coefficients,

$$\mathcal{B}u = \sum_{i=1}^{N} \beta_i(x) D_i u(x) + \gamma(x)u(x),$$

satisfying the nontangentiality condition

$$\sum_{i=1}^{N} \beta_i(x)\nu_i(x) \neq 0, \quad x \in \partial\Omega, \tag{2.23}$$

where $\nu(x)$ is the unit exterior vector normal to $\partial\Omega$ at x. We may consider also a Cauchy–Dirichlet problem,

$$\begin{cases} u_t(t,x) = \Phi\big(D^2u(t,x)\big), & t \geqslant 0, x \in \Omega, \\ u(t,x) = 0, & t \geqslant 0, x \in \partial\Omega, \\ u(0,x) = u_0(x), & x \in \Omega. \end{cases} \tag{2.24}$$

The initial datum u_0 is a regular (at least, C^2) function satisfying the natural compatibility condition $\mathcal{B}u_0 = 0$ at $\partial\Omega$ for problem (2.22), or $u_0 = 0$ at $\partial\Omega$ for problem (2.24). Moreover, Φ is a regular nonlinear function defined in a neighborhood of the range of D^2u_0 in $\mathbb{R}^{N^2}$, satisfying an ellipticity assumption

$$\sum_{i,j=1}^{N} \frac{\partial\Phi}{\partial q_{ij}}(Q)\xi_i\xi_j \geqslant \nu|\xi|^2, \quad x \in \overline{\Omega}, \tag{2.25}$$

and the symmetry condition

$$\Phi(Q) = \Phi\big(Q^*\big) \tag{2.26}$$

for every matrix Q with entries close to the range of D^2u_0.

Let us see how we can apply Theorem 2.6 to problem (2.22). The choice $X = L^p(\Omega)$, $D = \{\varphi \in W^{2,p}(\Omega)\colon \mathcal{B}\varphi = 0 \text{ at } \partial\Omega\}$ does not work, because the function

$$F(\varphi)(x) = \Phi\big(D^2\varphi(x)\big), \quad x \in \Omega, \tag{2.27}$$

does not map D into X, unless Φ has (not more than) linear growth. For instance, if Φ is a quadratic polynomial then F maps D into $L^{p/2}(\Omega)$. Much worse, even if Φ has linear growth, F is not differentiable unless Φ is linear.

After L^p and $W^{2,p}$, the simplest choice for the spaces D and X seems to be $X = C(\overline{\Omega})$, the space of the continuous functions from $\overline{\Omega}$ to $\mathbb{R}$, and $D = \{\varphi \in C^2(\overline{\Omega})\colon \mathcal{B}\varphi = 0 \text{ at } \partial\Omega\}$. If Φ is smooth enough, assumption (H1) is easily checked for the function F defined in (2.27).

But assumption (H2) does not hold, unless $N = 1$. Indeed, $F'(u_0)$ is the realization of the elliptic operator $\mathcal{A}$ in $C(\overline{\Omega})$ with the above boundary condition, where

$$(\mathcal{A}\varphi)(x) = \sum_{i,j=1}^{N} \frac{\partial \Phi}{\partial q_{ij}} \big(D^2 u_0(x)\big) D_i\varphi(x) D_j\varphi(x), \quad x \in \Omega, \tag{2.28}$$

which is sectorial in $C(\overline{\Omega})$ thanks to the Stewart's theorems [33], Chapter 3, and [39,40], but whose domain contains properly $C^2(\overline{\Omega})$. This is not due to the lack of regularity of the coefficients, or to the boundary condition, but it is a structural well-known difficulty, shared by all the elliptic operators including the Laplacian: if φ and $\Delta\varphi$ are continuous in some open set, φ is not necessarily a C^2 function. Of course, this difficulty disappears in dimension 1. So, we may apply Theorem 2.6 with $X = C(\overline{\Omega})$ either in dimension $N = 1$, or for special nonlinearities, for example, $F(u)(x) = \Psi(\Delta u(x))$, where we can take D as the domain of the Laplacian in X. For the details, see [33], Chapter 8.

If we replace $C(\overline{\Omega})$ by its subspace $C^\theta(\overline{\Omega})$ of the bounded and uniformly θ-Hölder continuous functions, and we take $D = \{\varphi \in C^{2+\theta}(\overline{\Omega})\colon \mathcal{B}\varphi = 0 \text{ at } \partial\Omega\}$, the function F defined in (2.27) satisfies (H1) if Φ is smooth enough, and the classical Schauder-type theorems plus generation theorems in Hölder spaces (see, e.g., [33], Chapter 3) show that also assumption (H2) is satisfied. So, Theorem 2.6 may be applied, and if the compatibility condition $F(u_0) \in \overline{D}$ holds, we get a local existence and uniqueness result. It is well known that D is not dense in X, and that the closure of D in X is a set of little-Hölder continuous functions: $\overline{D} = h^\theta(\overline{\Omega})$.

The space $h^\theta(\overline{\Omega})$ may be characterized as the subset of $C^{2+\theta}(\overline{\Omega})$ consisting of the functions φ such that

$$\lim_{h\to 0} \sup_{x,y\in\Omega, 0<|x-y|\leqslant h} \frac{|\varphi(x) - \varphi(y)|}{|x-y|^\theta} = 0,$$

and it coincides with the closure of $C^\infty(\overline{\Omega})$ in $C^\theta(\overline{\Omega})$ if $\partial\Omega$ is C^∞. Similarly, $h^{2+\theta}(\overline{\Omega})$ is the subset of $C^{2+\theta}(\overline{\Omega})$ consisting of the functions with second-order derivatives in $h^\theta(\overline{\Omega})$, and it coincides with the closure of $C^\infty(\overline{\Omega})$ in $C^{2+\theta}(\overline{\Omega})$ if $\partial\Omega$ is C^∞.

The assumption $F(u_0) \in \overline{D}$ holds provided $u_0 \in h^{2+\theta}(\overline{\Omega})$. This extra regularity assumption on the initial datum is preserved throughout the evolution because Theorem 2.6 implies that $u'(t) = F(u(t))$ belongs to $(X, D)_{\alpha,\infty} \subset \overline{D}$ for $t > 0$. Therefore, the choice of working in Hölder spaces leads naturally to little-Hölder spaces, and we can choose $X = h^\theta(\overline{\Omega})$, $D = \{\varphi \in h^{2+\theta}(\overline{\Omega})\colon \mathcal{B}\varphi = 0 \text{ at } \partial\Omega\}$, from the very beginning. Indeed, a Schauder-type theorem and a generation of analytic semigroups theorem hold in the space of the little-Hölder continuous functions, as follows (see, e.g., [33], Chapter 3):

THEOREM 2.13. *Let* $0 < \theta < 1$ *and let* $\partial\Omega$ *be of class* $h^{2+\theta}$. *Assume that the coefficients* a_{ij}, b_i, c *are in* $h^\theta(\overline{\Omega})$ *and satisfy the ellipticity condition*

$$\sum_{i,j=1}^{N} a_{ij}(x)\xi_i\xi_j > 0, \quad x \in \overline{\Omega},$$

and that the coefficients β_i, γ *are in* $h^{1+\theta}(\partial\Omega)$ *and satisfy the nontangentiality condition* (2.23). *Let* $\mathcal{A}$, $\mathcal{B}$ *be the differential operators defined by* $\mathcal{A} = \sum_{i,j=1}^{N} a_{ij}(x)D_{ij} + \sum_{i=1}^{N} b_i D_i + c$ *and* $\mathcal{B} = \sum_{i=1}^{N} \beta_i D_i + \gamma$.

If $\varphi \in \bigcap_{p>1} W^{2,p}(\Omega)$ *is such that* $\mathcal{A}\phi \in h^\theta(\overline{\Omega})$, $\mathcal{B}\varphi = 0$ *at* $\partial\Omega$, *then* $\varphi \in h^{2+\theta}(\overline{\Omega})$, *and there exists* $C > 0$, *independent of* ϕ, *such that*

$$\|\varphi\|_{C^{2+\theta}(\overline{\Omega})} \leqslant C\big(\|\mathcal{A}\varphi\|_{C^\theta(\overline{\Omega})} + \|\varphi\|_\infty\big).$$

The same conclusion holds if the boundary operator $\mathcal{B}$ *is replaced by a trace operator.*

THEOREM 2.14. *Let the assumptions of Theorem* 2.13 *hold. Then the realization of the operator* $\mathcal{A}$ *in* $h^\theta(\overline{\Omega})$, *with domain* $\{\varphi \in h^{2+\theta}(\overline{\Omega})\colon\ \mathcal{B}\varphi = 0 \text{ at } \partial\Omega\}$, *is sectorial in* $h^\theta(\overline{\Omega})$.

The application of Theorem 2.6 gives the following result.

THEOREM 2.15. *Let* $0 < \theta < 1$. *Assume that* $\partial\Omega$ *is of class* $h^{2+\theta}$, *let the coefficients* β_i, $\gamma \in h^{1+\theta}(\partial\Omega)$ *satisfy the nontangentiality condition* (2.23), *and let* $u_0 \in h^{2+\theta}(\overline{\Omega})$ *satisfy the compatibility condition*

$$\sum_{i=1}^{N} \beta_i(x)D_i u_0(x) + \gamma(x)u_0(x) = 0, \quad x \in \partial\Omega.$$

Let Φ *be a* C^3 *function defined in a neighborhood of the range of* D^2u_0, *satisfying the ellipticity condition* (2.25) *and the symmetry condition* (2.26).

Then there exists a maximal $\tau > 0$ *such that problem* (2.22) *has a solution* $u(t,x)$ *such that* $t \mapsto u(t,x)$ *belongs to* $C([0,\tau); h^{2+\theta}(\overline{\Omega})) \cap C^1([0,\tau); h^\theta(\overline{\Omega}))$. *For every* $\varepsilon \in (0,\tau)$ *and* $\beta \in (0,1)$, $u(t,\cdot) \in C^\beta_\beta((0,\tau-\varepsilon]; h^{2+\theta}(\Omega))$ *and* $u_t(t,\cdot) \in C^\beta_\beta((0,\tau-\varepsilon]; h^\theta(\overline{\Omega}))$. *u is the unique solution to* (2.22) *with such regularity properties. Moreover, it depends continuously on the initial datum* u_0 *in the sense specified by statement* (iv) *of Theorem* 2.6.

A further difficulty arises if the boundary operator $\mathcal{B}$ is replaced by the trace operator, i.e., if we consider problem (2.24) instead of (2.22). Simple counterexamples in dimension 1 show that the realizations of second-order elliptic operators with smooth coefficients and Dirichlet boundary condition in Hölder and little-Hölder spaces are not sectorial in general. The difficulty is due to the Dirichlet boundary condition, and it may be avoided replacing $C^\theta(\overline{\Omega})$ or $h^\theta(\overline{\Omega})$ by their subspaces consisting of functions that vanish at the boundary. See [33], Chapter 3, for a discussion. In the case of the choice

$X = h_0^\theta(\overline{\Omega}) = \{\phi \in h^\theta(\overline{\Omega})\colon \phi = 0 \text{ at } \partial\Omega\}$, the domain of $F'(u_0)$ is the subset of $h^{2+\theta}(\overline{\Omega})$ consisting of the functions φ such that φ and $F'(u_0)\varphi$ vanish at $\partial\Omega$; in general it does not coincide with the domain of $F'(u_1)$ for $u_1 \neq u_0$. Therefore we are not able to find a common domain D to apply Theorem 2.6, unless Φ is of a special type. For instance, if instead of a function $\Phi = \Phi(D^2u)$ we have $\Phi = \Phi(D^2u, Du, u)$ and $\Phi(q, p, 0) = 0$ for each q and p, we are done, and a theorem similar to (2.15) holds.

Now let us see how we can apply Theorem 2.7. The space E is still $C(\overline{\Omega})$, $F'(u_0)$ is the realization of the operator $\mathcal{A}$ defined in (2.28) with homogeneous boundary condition, and

$$D_{F'(u_0)}(\alpha) = h^{2\alpha}(\overline{\Omega})$$

for $\alpha < 1/2$, and

$$D_{F'(u_0)}(\alpha) = \{\varphi \in h^{2\alpha}(\overline{\Omega})\colon \mathcal{B}\varphi = 0 \text{ at } \partial\Omega\}$$

for $\alpha > 1/2$. See [33], Chapter 3.

If u_0 and $\partial\Omega$ are smooth enough, that is $\partial\Omega \in h^{2+2\alpha}$, $u_0 \in h^{2+2\alpha}(\overline{\Omega})$, Theorem 2.13 yields

$$D_{F'(u_0)}(\alpha + 1) = \{\varphi \in h^{2+2\alpha}(\overline{\Omega})\colon \mathcal{B}\varphi = 0 \text{ at } \partial\Omega\}$$

for $\alpha < 1/2$, and

$$D_{F'(u_0)}(\alpha + 1) = \{\varphi \in h^{2+2\alpha}(\overline{\Omega})\colon \mathcal{B}\varphi = 0, \mathcal{B}(F'(u_0)\varphi) = 0 \text{ at } \partial\Omega\}$$

for $\alpha > 1/2$.

With $\theta \in (0, 1)$ fixed, we take $\alpha = \theta/2 \in (0, 1/2)$ and we may apply Theorem 2.7, with $E_0 = h^\theta(\overline{\Omega})$, $E_1 = \{\varphi \in h^{2+\theta}(\overline{\Omega})\colon \mathcal{B}\varphi = 0 \text{ at } \partial\Omega\}$. It is easy to see that the regularity assumption in (H3) is satisfied if Φ is a C^2 function, and the other assumptions in (H3) are satisfied thanks to Theorems 2.13 and 2.14. The final result is the following.

THEOREM 2.16. *Let $0 < \theta < 1$. Assume that $\partial\Omega$ is of class $h^{2+\theta}$, let the coefficients β_i, $\gamma \in h^{1+\theta}(\partial\Omega)$ satisfy the nontangentiality condition* (2.23), *and let $u_0 \in h^{2+\theta}(\overline{\Omega})$ satisfy the compatibility condition*

$$\sum_{i=1}^N \beta_i(x) D_i u(x) + \gamma(x) u(x) = 0, \qquad x \in \partial\Omega.$$

Let Φ be a C^2 function defined in a neighborhood of the range of D^2u_0, satisfying the ellipticity condition (2.25) *and the symmetry condition* (2.26).

Then there exists a maximal $\tau > 0$ such that problem (2.22) *has a unique solution $u(t, x)$ such that $t \mapsto u(t, x)$ belongs to $C([0, \tau); h^{2+\theta}(\overline{\Omega})) \cap C^1([0, \tau); h^\theta(\overline{\Omega}))$. It depends continuously on the initial datum u_0 in the sense specified in Theorem 2.7.*

So, there is not much difference between this theorem and Theorem 2.15. Here we do not get extra time regularity, but Φ can be taken of class C^2 instead of C^3. Substantially, the applications of Theorems 2.6 and 2.7 to problem (2.22) give the same results.

In any case, problems with nonlinear boundary condition of the type $G(Du(t,x)) = 0$ for $x \in \partial\Omega$, with nonlinear smooth G, cannot be treated by a direct application of Theorems 2.6 and 2.7, even in the case of linear Φ. First, the boundary condition has to be incorporated in the domain D (if we use Theorem 2.6) or in the domain E_1 (if we use Theorem 2.7), but D and E_1 have to be linear spaces and the boundary condition is nonlinear. Second, and more important, even if we rewrite the boundary condition as a linear boundary condition plus the rest and then try to get rid of the rest using suitable trace theorems, the domain of the linearized operator changes with u_0. Take for instance $\Phi(D^2u) = \Delta u$, $X = h^\theta(\overline{\Omega})$. Assuming that the linearized operator

$$\varphi \mapsto \sum_{i=1}^{N} D_i G\big(Du_0(x)\big) D_i\varphi(x), \quad x \in \partial\Omega,$$

is nontangential, the realization of the Laplace operator in $h^\theta(\overline{\Omega})$, with domain $D = \{\varphi \in h^{2+\theta}(\overline{\Omega})\colon \sum_{i=1}^{N} D_i G(Du_0(x)) D_i\varphi(x) = 0 \text{ at } \partial\Omega\}$, is in fact sectorial but its domain strongly depends on u_0.

For this type of problems, a direct approach in Hölder spaces seems to be simpler and more fruitful than applying abstract results. This approach is described in the next section.

3. Equations and systems in Hölder spaces

Throughout this section we shall consider an open set $\Omega \subset \mathbb{R}^N$ with uniform $C^{2+\theta}$ boundary, $0 < \theta < 1$. This means that there is $r > 0$ such that for each $x_0 \in \partial\Omega$ there is a $C^{2+\theta}$ diffeomorphism φ from the open ball $B(x_0, r)$ centered at x_0 and with radius r to the unit open ball in $\mathbb{R}^N$ with the property that $\varphi(\Omega \cap B(x_0, r)) = \{(x, y) \in \mathbb{R} \times \mathbb{R}^{N-1}\colon |(x, y)| < 1, x < 0\}$; moreover, the $C^{2+\theta}$ norm of the diffeomorphisms and their inverses are bounded by a constant independent of x_0.

For $k \in \mathbb{N}$, $C_b^k(\overline{\Omega})$ denotes the space of the functions with continuous bounded derivatives in $\overline{\Omega}$, and for $k \in \mathbb{N}$, $0 < \theta < 1$, $C^{k+\theta}(\overline{\Omega})$ denotes the subspace of $C_b^k(\overline{\Omega})$ consisting of the functions with uniformly θ-Hölder continuous kth order derivatives.

We shall use the parabolic Hölder spaces $C^{\theta/2,\theta}(I \times \overline{\Omega})$, $C^{1/2+\theta/2,1+\theta}(I \times \overline{\Omega})$, $C^{1+\theta/2,2+\theta}(I \times \overline{\Omega})$, I being a real interval, $0 < \theta < 1$, with the usual meanings and norms.

We recall that a function w belongs to $C^{\theta/2,\theta}(I \times \overline{\Omega})$ if and only if w is bounded and, moreover,

$$[w]_{C^{\theta/2,\theta}(I\times\overline{\Omega})} = \sup_{x\in\overline{\Omega}} \big[w(\cdot, x)\big]_{C^{\theta/2}(I)} + \sup_{t\in I} \big[w(t, \cdot)\big]_{C^{\theta}(\overline{\Omega})} < \infty.$$

In this case, we set

$$\|w\|_{C^{\theta/2,\theta}(I\times\overline{\Omega})} = \|w\|_\infty + [w]_{C^{\theta/2,\theta}(I\times\overline{\Omega})}.$$

The function $w \in C^{1/2+\theta/2,1+\theta}(I \times \overline{\Omega})$ if and only if w and its first-order space derivatives $D_i w$ are bounded and moreover,

$$[w]_{C^{1/2+\theta/2,1+\theta}(I\times\overline{\Omega})} = \sup_{x\in\overline{\Omega}} \big[w(\cdot,x)\big]_{C^{1/2+\theta/2}(I)} + \sum_{i=1}^{N} \sup_{t\in I} \big[D_i w(t,\cdot)\big]_{C^{\theta}(\overline{\Omega})} < \infty.$$

In this case we set

$$\|w\|_{C^{1/2+\theta/2,1+\theta}(I\times\overline{\Omega})} = \|w\|_\infty + \sum_{i=1}^{N} \|D_i w\|_\infty + [w]_{C^{1/2+\theta/2,1+\theta}(I\times\overline{\Omega})}.$$

The space $C^{1+\theta/2,2+\theta}(I \times \overline{\Omega})$ is defined similarly; w belongs to $C^{1+\theta/2,2+\theta}(I \times \overline{\Omega})$ if and only if w is bounded, there exist the derivatives w_t, $D_{ij}w$ for i, $j = 1,\dots N$, and they belong to $C^{\theta/2,\theta}(I \times \overline{\Omega})$. It is easy to see that in this case w belongs also to $C^{1/2+\theta/2,1+\theta}(I \times \overline{\Omega})$. The norm is

$$\|w\|_{C^{1+\theta/2,2+\theta}(I\times\overline{\Omega})} = \|w\|_{C^{1/2+\theta/2,1+\theta}(I\times\overline{\Omega})} + \|w_t\|_{C^{\theta/2,\theta}(I\times\overline{\Omega})} + \sum_{i,j=1}^{N} \|D_{ij} w\|_{C^{\theta/2,\theta}(I\times\overline{\Omega})}.$$

The spaces $C^{\theta/2,\theta}(I \times \partial\Omega)$ and $C^{1/2+\theta/2,1+\theta}(I \times \partial\Omega)$ are defined similarly, by means of a smooth ($C^{2+\theta}$) atlas for $\partial\Omega$. There exists $C_\theta > 0$ such that, for every $w \in C^{\theta/2,\theta}(I \times \overline{\Omega})$, we have $\|w_{|I\times\partial\Omega}\|_{C^{\theta/2,\theta}(I\times\partial\Omega)} \leqslant C_\theta \|w\|_{C^{\theta/2,\theta}(I\times\overline{\Omega})}$, and similarly, for every $w \in C^{1/2+\theta/2,1+\theta}(I \times \overline{\Omega})$, we have $\|w_{|I\times\partial\Omega}\|_{C^{1/2+\theta/2,1+\theta}(I\times\partial\Omega)} \leqslant C_\theta \|w\|_{C^{1/2+\theta/2,1+\theta}(I\times\overline{\Omega})}$.

Let us return to problem

$$\begin{cases} u_t(t,x) = \Phi\big(D^2u(t,x)\big), & t \geqslant 0, x \in \overline{\Omega}, \\ \Psi\big(Du(t,x)\big) = 0, & t \geqslant 0, x \in \partial\Omega, \\ u(0,x) = u_0(x), & x \in \overline{\Omega}, \end{cases}$$

where u_0 is a regular function defined in $\overline{\Omega}$, and Φ and Ψ are regular functions defined in a neighborhood of the range of D^2u_0 and of Du_0, respectively. We need also symmetry and ellipticity assumptions of the type (2.26), (2.25) on Φ, as well as a nontangentiality assumption on Ψ. Precisely, we assume that there are open sets $\mathcal{O}_1 \subset \mathbb{R}^{N^2}$, $\mathcal{O}_2 \subset \mathbb{R}^N$ such that

$$\sum_{i,j=1}^{N} \frac{\partial \Phi}{\partial q_{ij}}(Q)\xi_i\xi_j > 0, \quad Q \in \mathcal{O}_1, \xi \in \mathbb{R}^N, \tag{3.1}$$

$$\Phi(Q) = \Phi\big(Q^*\big), \quad Q \in \mathcal{O}_1, \tag{3.2}$$

and

$$\sum_{i=1}^{N} \frac{\partial \Psi}{\partial p_i}(p)\nu_i(x) \neq 0, \quad p \in \mathcal{O}_2, x \in \partial\Omega, \tag{3.3}$$

where $\nu(x)$ is the unit exterior vector normal to $\partial\Omega$ at x.

Under these conditions, problem (1.1) is the simplest significant example of a fully nonlinear parabolic problem with fully nonlinear boundary condition. We give a complete proof of the local existence theorem for (1.1) because it exhibits the typical difficulties of fully nonlinear problems, but the technical points are reduced to the minimum and it is easy to see to which extent the proof itself may be generalized.

Also, in this case we need an optimal regularity theorem for linear equations, the popular Ladyzhenskaja–Solonnikov–Ural'ceva theorem [28], Chapter 4. In next Theorem 3.1, $\mathcal{A}$ is a linear second-order differential operator,

$$\begin{aligned}(\mathcal{A}v)(\xi) = &\sum_{i,j=1}^{N} a_{ij}(\xi) D_{ij} v(\xi) \\ &+ \sum_{i=1}^{N} b_i(\xi) D_i v(\xi) + c(\xi)v(\xi), \quad \xi \in \overline{\Omega},\end{aligned} \tag{3.4}$$

satisfying the ellipticity condition

$$\sum_{i,j=1}^{N} a_{ij}(\xi)\eta_i\eta_j \geqslant \nu|\eta|^2, \quad \xi \in \overline{\Omega}, \ \eta \in \mathbb{R}^N, \tag{3.5}$$

for some $\nu > 0$, and $\mathcal{B}$ is a linear first-order differential operator,

$$(\mathcal{B}v)(\xi) = \gamma(\xi)v(\xi) + \sum_{i=1}^{N} \beta_i(\xi) D_i v(\xi), \quad \xi \in \partial\Omega, \tag{3.6}$$

satisfying the nontangentiality condition

$$\sum_{i=1}^{N} \beta_i(\xi)\nu_i(\xi) \neq 0, \quad \xi \in \partial\Omega. \tag{3.7}$$

THEOREM 3.1. *Fix $\theta \in (0,1)$ and $T > 0$. Let Ω be an open set in $\mathbb{R}^N$ with uniform $C^{2+\theta}$ boundary. Let $a_{ij}, b_i, c \in C^\theta(\overline{\Omega})$ satisfy* (3.5), *and let $\beta_i, \gamma \in C^{1/2+\theta/2, 1+\theta}([0,T] \times \partial\Omega)$ satisfy* (3.7). *Define the operators $\mathcal{A}$ and $\mathcal{B}$ by* (3.4) *and* (3.6), *respectively. Then, for every $f \in C^{\theta/2,\theta}([0,T] \times \overline{\Omega})$, $g \in C^{1/2+\theta/2, 1+\theta}([0,T] \times \partial\Omega)$, satisfying the compatibility condition*

$$\mathcal{B}w_0(0,\cdot) = g(0,\cdot) \quad \text{in } \partial\Omega, \tag{3.8}$$

the problem

$$\begin{cases} w_t = \mathcal{A}w + f, & 0 \leqslant t \leqslant T,\ \xi \in \overline{\Omega}, \\ \mathcal{B}w = g, & 0 \leqslant t \leqslant T,\ \xi \in \partial\Omega, \\ w(0,x) = w_0, & \xi \in \overline{\Omega}, \end{cases} \tag{3.9}$$

has a unique solution $w \in C^{1+\theta/2,2+\theta}([0,T] \times \overline{\Omega})$. *Moreover, there exists* $C = C(T) > 0$, *increasing with respect to* T, *such that*

$$\begin{aligned} &\|w\|_{C^{1+\theta/2,2+\theta}([0,T]\times\overline{\Omega})} \\ &\quad \leqslant C\big(\|f\|_{C^{\theta/2,\theta}([0,T]\times\overline{\Omega})} + \|g\|_{C^{1/2+\theta/2,1+\theta}([0,T]\times\partial\Omega)} + \|w_0\|_{C^{2+\theta}(\overline{\Omega})}\big). \end{aligned} \tag{3.10}$$

Now we are ready for the proof of the local existence and uniqueness theorem.

THEOREM 3.2. *Let* Ω *be an open set in* $\mathbb{R}^N$ *with uniform* $C^{2+\theta}$ *boundary,* $0 < \theta < 1$. *Let* $\Phi : \mathcal{O}_1 \mapsto \mathbb{R}$ *be a* C^2 *function satisfying* (3.2) *and* (3.1), *and let* $\Psi : \mathcal{O}_2 \mapsto \mathbb{R}$ *be a* C^3 *function satisfying* (3.3).

Then for each $u_0 \in C^{2+\theta}(\overline{\Omega})$ *such that the range of* D^2u_0 *is contained in* $\mathcal{O}_1$, *the range of* Du_0 *is contained in* $\mathcal{O}_2$, *and satisfying the compatibility condition*

$$\Psi\big(Du_0(x)\big) = 0, \quad x \in \partial\Omega,$$

there exists for $T > 0$ *a unique* $u \in C^{1+\theta/2,2+\theta}([0,T] \times \overline{\Omega})$ *that solves* (1.1) *in* $[0,T] \times \overline{\Omega}$.

PROOF. Let $\mathcal{A}$ and $\mathcal{B}$ be the operators defined by

$$\mathcal{A}\varphi(x) = \sum_{i,j=1}^{N} \frac{\partial\Phi}{\partial q_{ij}}\big(D^2u_0(x)\big)D_{ij}\varphi(x), \quad x \in \overline{\Omega},$$

$$\mathcal{B}\varphi(x) = \sum_{i=1}^{N} \frac{\partial\Psi}{\partial p_i}\big(Du_0(x)\big)D_i\varphi(x), \quad x \in \overline{\Omega}.$$

Moreover, set

$$\begin{aligned} Y = \big\{&u \in C^{1+\theta/2,2+\theta}\big([0,T] \times \overline{\Omega}\big): \\ &u(0,\cdot) = u_0,\ \|u - u_0\|_{C^{1+\theta/2,2+\theta}([0,T]\times\overline{\Omega})} \leqslant R\big\}, \end{aligned}$$

where the positive numbers T and R have to be chosen later.

The solution to (1.1) is sought as a fixed point of the operator Γ defined in Y by $\Gamma u = w$, w being the solution of

$$\begin{cases} w_t(t,x) & \\ \quad = \mathcal{A}w(t,x) + \Phi\big(D^2u(t,x)\big) - \mathcal{A}u(t,x), & 0 \leqslant t \leqslant T, x \in \overline{\Omega}, \\ \mathcal{B}w(t,x) = \mathcal{B}u(t,x) - \Psi\big(Du(t,x)\big), & 0 \leqslant t \leqslant T, x \in \partial\Omega, \\ w(0,x) = u_0(x), & x \in \overline{\Omega}. \end{cases} \tag{3.11}$$

We have to choose T and R in such a way that Γ is well defined, it maps Y into itself, it is a contraction with constant less than 1, and the unique fixed point of Γ in Y is in fact the unique solution to (1.1).

For Γ be well defined, for every $u \in Y$ the ranges of $D^2u(t,\cdot)$ and of $Du(t,\cdot)$ need to be contained in $\mathcal{O}_1$ and in $\mathcal{O}_2$. If $\mathcal{O}_1$ contains the closure of the neighborhood of the range of D^2u_0 with radius r_1 and $\mathcal{O}_2$ contains the closure of the neighborhood of the range of Du_0 with radius r_2, we take T, R such that

$$T^{\theta/2}R \leqslant \frac{r_1}{2}, \qquad T^{1/2+\theta/2}R \leqslant \frac{r_2}{2}. \tag{3.12}$$

So, the compositions $\Phi(D^2u)$ and $\Psi(Du)$ are well defined for each $u \in Y$, and they belong to $C^{\theta/2,\theta}([0,T] \times \overline{\Omega})$ and to $C^{1/2+\theta/2,1+\theta}([0,T] \times \overline{\Omega})$, respectively. The compatibility condition (3.8) holds, and then by Theorem 3.1 problem (3.11) has a unique solution $w \in C^{1+\theta/2,2+\theta}([0,T] \times \overline{\Omega})$.

Let us prove that Γ is a 1/2-contraction for suitable T and R. The constants $C(T)$ given by Theorem 3.1 increase with T. So we take

$$T \leqslant 1. \tag{3.13}$$

For each $u, v \in Y$, $\Gamma u - \Gamma v$ is the solution to (3.9) with $w_0 = 0$, and

$$\begin{aligned} f(t,x) &= \Phi\big(D^2u(t,x)\big) - \Phi\big(D^2v(t,x)\big) - \mathcal{A}(u-v)(t,x) \\ &= \int_0^1 \big\langle D\Phi\big(\sigma D^2u(t,x) + (1-\sigma)D^2v(t,x)\big) - D\Phi\big(D^2u_0(x)\big), \\ &\qquad D^2u(t,x) - D^2v(t,x)\big\rangle\,\mathrm{d}\sigma, \\ g(t,x) &= \mathcal{B}(u-v)(t,x) - \Psi\big(Du(t,x)\big) + \Psi\big(Dv(t,x)\big) \\ &= \int_0^1 \big\langle D\Psi\big(Du_0(x)\big) - D\Psi\big(\sigma Du(t,x) + (1-\sigma)Dv(t,x)\big), \\ &\qquad Du(t,x) - Dv(t,x)\big\rangle\,\mathrm{d}\sigma. \end{aligned}$$

Theorem 3.1 gives now

$$\begin{aligned}
&\|\Gamma u - \Gamma v\|_{C^{1+\theta/2,2+\theta}([0,T]\times\overline{\Omega})} \\
&\quad \leqslant C\big(\|f\|_{C^{\theta/2,\theta}([0,T]\times\overline{\Omega})} + \|g\|_{C^{1/2+\theta/2,1+\theta}([0,T]\times\overline{\Omega})}\big),
\end{aligned}$$

where $C = C(1)$.

Let us estimate $\|f\|_{C^{\theta/2,\theta}([0,T]\times\overline{\Omega})}$. Since $u(0,\cdot) = v(0,\cdot) = u_0$, for each $t \in [0,T]$, we have

$$\begin{aligned}
&\big\|D^2u(t,\cdot) - D^2u_0\big\|_{L^\infty(\Omega)} \leqslant T^{\theta/2}R, \\
&\big\|D^2v(t,\cdot) - D^2u_0\big\|_{L^\infty(\Omega)} \leqslant T^{\theta/2}R
\end{aligned}$$

and

$$\big\|D^2u(t,\cdot) - D^2v(t,\cdot)\big\|_{L^\infty(\Omega)} \leqslant T^{\theta/2}\|u - v\|_{C^{1+\theta/2,2+\theta}([0,T]\times\overline{\Omega})}.$$

Therefore, for $t, s \in [0,T]$ and $x \in \overline{\Omega}$, we have

$$\begin{aligned}
&f(t,x) - f(s,x) \\
&\quad = \int_0^1 \big\langle D\Phi\big(\sigma D^2u(t,x) + (1-\sigma)D^2v(t,x)\big) - D\Phi\big(D^2u_0(x)\big), \\
&\qquad\qquad D^2u(t,x) - D^2v(t,x)\big\rangle\, d\sigma \\
&\qquad - \int_0^1 \big\langle D\Phi\big(\sigma D^2u(s,x) + (1-\sigma)D^2v(s,x)\big) - D\Phi\big(D^2u_0(x)\big), \\
&\qquad\qquad D^2u(s,x) - D^2v(s,x)\big\rangle\, d\sigma \\
&\quad = \int_0^1 \big\langle D\Phi\big(\sigma D^2u(t,x) + (1-\sigma)D^2v(t,x)\big) \\
&\qquad\qquad - D\Phi\big(\sigma D^2u(s,x) + (1-\sigma)D^2v(s,x)\big), D^2u(t,x) - D^2v(t,x)\big\rangle\, d\sigma \\
&\qquad + \int_0^1 \big\langle D\Phi\big(\sigma D^2u(s,x) + (1-\sigma)D^2v(s,x)\big) - D\Phi\big(D^2u_0(x)\big), \\
&\qquad\qquad D^2u(t,x) - D^2v(t,x) - D^2u(s,x) + D^2v(s,x)\big\rangle\, d\sigma.
\end{aligned}$$

Let L_1 be the supremum of $|D^2\Phi|$ in the neighborhood of the range of D^2u_0 with radius r_1. Then we have

$$\begin{aligned}
&\big|f(t,x)-f(s,x)\big| \\
&\quad\leqslant \int_0^1 L_1\big(\sigma(t-s)^{\theta/2}+(1-\sigma)(t-s)^{\theta/2}\big)\big\|D^2u(t,\cdot)-D^2v(t,\cdot)\big\|_{L^\infty(\Omega)}\,\mathrm{d}\sigma \\
&\qquad+\int_0^1 L_1\big(\sigma T^{\theta/2}R+(1-\sigma)T^{\theta/2}R\big)\,\mathrm{d}\sigma\,(t-s)^{\theta/2}. \\
&\qquad\qquad\times\big[D^2u(\cdot,x)-D^2v(\cdot,x)\big]_{C^{\theta/2}([0,T])} \\
&\quad\leqslant 2L_1T^{\theta/2}R(t-s)^{\theta/2}\|u-v\|_{C^{1+\theta/2,2+\theta}([0,T]\times\overline{\Omega})}.
\end{aligned}$$

Recalling that $f(0,x)=0$, so that $\|f(\cdot,x)\|_\infty \leqslant T^{\theta/2}[f(\cdot,x)]_{C^{\theta/2}([0,T])}$, we get, for each $x\in\overline{\Omega}$,

$$\big\|f(\cdot,x)\big\|_{C^{\theta/2}([0,T])}\leqslant 2L_1\big(T^{\theta/2}+T^{\theta}\big)R\|u-v\|_{C^{1+\theta/2,2+\theta}([0,T]\times\overline{\Omega})}.$$

To evaluate $[f(t,\cdot)]_{C^\theta(\overline{\Omega})}$ we recall that

$$[\varphi\psi]_{C^\theta(\overline{\Omega})}\leqslant\|\varphi\|_\infty[\psi]_{C^\theta(\overline{\Omega})}+[\varphi]_{C^\theta(\overline{\Omega})}\|\psi\|_\infty.$$

Therefore

$$\begin{aligned}
&\big[f(t,\cdot)\big]_{C^\theta(\overline{\Omega})} \\
&\quad\leqslant\int_0^1\big\|D\Phi\big(\sigma D^2u(t,\cdot)+(1-\sigma)D^2v(t,\cdot)\big)-D\Phi\big(D^2u_0\big)\big\|_\infty\,\mathrm{d}\sigma \\
&\qquad\quad\times\big[D^2u(t,\cdot)-D^2v(t,\cdot)\big]_{C^\theta(\overline{\Omega})} \\
&\qquad+\int_0^1\big[D\Phi\big(\sigma D^2u(t,\cdot)+(1-\sigma)D^2v(t,\cdot)\big)-D\Phi\big(D^2u_0\big)\big]_{C^\theta(\overline{\Omega})}\,\mathrm{d}\sigma \\
&\qquad\qquad\times\big\|D^2u(t,\cdot)-D^2v(t,\cdot)\big\|_\infty \\
&\quad\leqslant\int_0^1 L_1\big(\sigma t^{\theta/2}R+(1-\sigma)t^{\theta/2}R\big)\,\mathrm{d}\sigma\,\|u-v\|_{C^{1+\theta/2,2+\theta}([0,T]\times\overline{\Omega})} \\
&\qquad+\int_0^1 L_1\big(\sigma R+(1-\sigma)R\big)\,\mathrm{d}\sigma\,T^{\theta/2}\|u-v\|_{C^{1+\theta/2,2+\theta}([0,T]\times\overline{\Omega})} \\
&\quad\leqslant 2L_1T^{\theta/2}R\|u-v\|_{C^{1+\theta/2,2+\theta}([0,T]\times\overline{\Omega})}.
\end{aligned}$$

Summing up,

$$\|f\|_{C^{\theta/2,\theta}([0,T]\times\overline{\Omega})} \leqslant L_1\big(4T^{\theta/2}+2T^{\theta}\big)R\|u-v\|_{C^{1+\theta/2,2+\theta}([0,T]\times\overline{\Omega})}. \tag{3.14}$$

Let us estimate $\|g\|_{C^{1/2+\theta/2,1+\theta}([0,T]\times\partial\Omega)}$. We recall that there exists $C_\theta>0$ such that $\|g\|_{C^{1/2+\theta/2,1+\theta}([0,T]\times\partial\Omega)} \leqslant C_\theta\|g\|_{C^{1/2+\theta/2,1+\theta}([0,T]\times\overline{\Omega})}$.

Let L_2, L_3 be the suprema of $|D^2\Psi|$, $|D^3\Psi|$ in the neighborhood of the range of Du_0 with radius r_2. For each $x\in\overline{\Omega}$, the estimate for the seminorm $\|g(\cdot,x)\|_{C^{1/2+\theta/2}([0,T])}$ is obtained as the estimate for $\|f(\cdot,x)\|_{C^{\theta/2}([0,T])}$. We get

$$\big\|g(\cdot,x)\big\|_{C^{1/2+\theta/2}([0,T])} \leqslant 2L_2\big(T^{1/2+\theta/2}+T^{1+\theta}\big)R\|u-v\|_{C^{1+\theta/2,2+\theta}([0,T]\times\overline{\Omega})}.$$

To estimate $\|g(t,\cdot)\|_{C^{1+\theta}(\overline{\Omega})}$ we write down the first-order derivatives of g:

$$\begin{aligned}
D_i g = {} & \int_0^1 \sum_{k=1}^N \big(D_k\Psi(Du_0) \\
& \qquad - D_k\Psi\big(\sigma Du(t,x)+(1-\sigma)Dv\big)(D_{ik}u-D_{ik}v)\big)\,\mathrm{d}\sigma \\
& - \int_0^1 \sum_{j,k=1}^N \big(D_{kj}\Psi\big(\sigma Du+(1-\sigma)Dv\big)D_{ji}\big(\sigma u+(1-\sigma)v\big) \\
& \qquad - D_{kj}\Psi(Du_0)D_{ji}u_0\big)(D_k u-D_k v)\,\mathrm{d}\sigma \\
:= {} & h_i(t,x)+m_i(t,x).
\end{aligned}$$

The functions h_i are estimated like f, and we get

$$\begin{aligned}
&\|h_i\|_\infty \leqslant NL_2T^{1/2+\theta}R\|u-v\|_{C^{1+\theta/2,2+\theta}([0,T]\times\overline{\Omega})}, \\
&[h_i]_{C^\theta(\overline{\Omega})} \leqslant 2NL_2\big(T^{\theta/2}+T^{1/2+\theta/2}\big)R\|u-v\|_{C^{1+\theta/2,2+\theta}([0,T]\times\overline{\Omega})}.
\end{aligned}$$

Concerning the functions m_i, we recall that

$$\|D_k u-D_k v\|_\infty \leqslant T^{1/2+\theta/2}\|u-v\|_{C^{1+\theta/2,2+\theta}([0,T]\times\overline{\Omega})},$$

and from the inequality

$$[\varphi]_{C^\theta(\overline{\Omega})} \leqslant K\|\varphi\|_\infty^{1-\theta}\|D\varphi\|_\infty^{\theta}$$

we get

$$\begin{aligned}
&\big\|D_k u(t,\cdot)-D_k v(t,\cdot)\big\|_{C^\theta(\overline{\Omega})} \\
&\quad \leqslant K\big\|Du(t,\cdot)-Dv(t,\cdot)\big\|_\infty^{1-\theta}\big\|D^2u(t,\cdot)-D^2v(t,\cdot)\big\|_\infty^{\theta} \\
&\quad \leqslant KT^{1/2}\|u-v\|_{C^{1+\theta/2,2+\theta}([0,T]\times\overline{\Omega})}.
\end{aligned}$$

Since

$$\begin{aligned}&\big\| D_{kj}\Psi\big(\sigma Du + (1-\sigma)Dv\big)D_{ji}\big(\sigma u + (1-\sigma)v\big) - D_{kj}\Psi(Du_0)D_{ji}u_0\big\|_{C^\theta(\overline{\Omega})}\\&\quad\leqslant C\big(L_3, R, \|u_0\|_{C^{2+\theta}(\overline{\Omega})}\big),\end{aligned}$$

with C increasing in all its arguments, we get

$$\|m_i\|_{C^\theta(\overline{\Omega})} \leqslant KC\big(L_3, R, \|u_0\|_{C^{2+\theta}(\overline{\Omega})}\big)T^{1/2}\|u-v\|_{C^{1+\theta/2,2+\theta}([0,T]\times\overline{\Omega})}.$$

Summing up, we get

$$\begin{aligned}&\|g\|_{C^{1/2+\theta/2,1+\theta}([0,T]\times\overline{\Omega})}\\&\quad\leqslant T^{\theta/2}K\big(L_2, L_3, T, R, \|u_0\|_{C^{2+\theta}(\overline{\Omega})}\big)\|u-v\|_{C^{1+\theta/2,2+\theta}([0,T]\times\overline{\Omega})},\end{aligned} \tag{3.15}$$

with $K(L_2, L_3, T, R, \|u_0\|_{C^{2+\theta}(\overline{\Omega})})$ positive and increasing with respect to all its arguments.

Taking into account (3.14) and (3.15) we obtain that Γ is a $1/2$-contraction provided

$$C\big[L_1\big(4T^{\theta/2} + 2T^\theta\big)R + T^{\theta/2}K\big(L_2, L_3, T, R, \|u_0\|_{C^{2+\theta}(\overline{\Omega})}\big)\big] \leqslant \frac{1}{2}. \tag{3.16}$$

Now we check that Γ maps Y into itself if T, R are suitably chosen. For each $u \in Y$, we write $\Gamma u = \Gamma(u-u_0) + \Gamma u_0$. We already know that if (3.12), (3.13), (3.16) hold then

$$\big\|\Gamma(u-u_0)\big\|_{C^{1+\theta/2,2+\theta}([0,T]\times\overline{\Omega})} \leqslant \frac{1}{2}\|u-u_0\|_{C^{1+\theta/2,2+\theta}([0,T]\times\overline{\Omega})} \leqslant \frac{R}{2}.$$

Therefore, Γ maps Y into itself provided $\|\Gamma u_0 - u_0\|_{C^{1+\theta/2,2+\theta}([0,T]\times\overline{\Omega})} \leqslant R/2$. The function $w = \Gamma(u_0) - u_0$ is the solution to (3.9) with $f = \Phi(D^2u_0)$, $g = -\Psi(Du_0)$, $w_0 = 0$, and its $C^{1+\theta/2,2+\theta}([0,T]\times\overline{\Omega})$ norm is not small in general if T is small. We only have, by estimate (3.10),

$$\begin{aligned}&\|\Gamma u_0 - u_0\|_{C^{1+\theta/2,2+\theta}([0,T]\times\overline{\Omega})}\\&\quad\leqslant C\big(\big\|\Phi\big(D^2u_0\big)\big\|_{C^{\theta/2,\theta}([0,T]\times\overline{\Omega})} + \big\|\Psi(Du_0)\big\|_{C^{1/2+\theta/2,1+\theta}([0,T]\times\partial\Omega)}\big)\\&\quad= C\big(\big\|\Phi\big(D^2u_0\big)\big\|_{C^\theta(\overline{\Omega})} + \big\|\Psi(Du_0)\big\|_{C^{1+\theta}([0,T]\times\partial\Omega)}\big),\end{aligned}$$

with $C = C(1)$. Then Γ maps Y into itself if

$$R \geqslant 2C\big(\big\|\Phi\big(D^2u_0\big)\big\|_{C^\theta(\overline{\Omega})} + \big\|\Psi(Du_0)\big\|_{C^{1+\theta}([0,T]\times\partial\Omega)}\big). \tag{3.17}$$

In conclusion, if (3.12), (3.13), (3.16) and (3.17) hold, Γ is a $1/2$-contraction that maps Y into itself, so that it has a unique fixed point u in Y.

To finish the proof we have to show that u is the unique solution to (1.1) in $C^{1+\theta/2,2+\theta}([0,T]\times\overline{\Omega})$. This is done in a standard way.

If (1.1) has two solutions u and v, set $t_0=\sup\{t\in[0,T]:\ u=v \text{ in } [0,t]\times\overline{\Omega}\}$. If $t_0=T$ then $u=v$ in the whole $[0,T]\times\overline{\Omega}$ and the proof is finished; if $t_0<T$, we consider the initial-boundary value problem

$$\begin{cases} w_t(t,x)=\Phi\big(D^2w(t,x)\big), & t\geqslant t_0,\ x\in\overline{\Omega},\\ \Psi\big(Dw(t,x)\big)=0, & t\geqslant t_0,\ x\in\partial\Omega,\\ w(t_0,x)=w_0(x), & x\in\overline{\Omega}, \end{cases} \tag{3.18}$$

where $w_0(x)=u(t_0,x)=v(t_0,x)$. The above proof shows that (3.18) has a unique solution in the set $Y'=\{w\in C^{1+\theta/2,2+\theta}([t_0,t_0+T']\times\overline{\Omega}):\ w(t_0,\cdot)=w_0,\ \|w-w_0\|_{C^{1+\theta/2,2+\theta}([t_0,t_0+T']\times\overline{\Omega})}\leqslant R'\}$, provided R' is large enough and T' is small enough. Taking R' larger than $\|u-u(t_0,\cdot)\|_{C^{1+\theta/2,2+\theta}([t_0,T]\times\overline{\Omega})}$ and than $\|v-v(t_0,\cdot)\|_{C^{1+\theta/2,2+\theta}([t_0,T]\times\overline{\Omega})}$ we get $u=v$ in $[t_0,t_0+T']\times\overline{\Omega}$, and this contradicts the definition of t_0. Therefore, $t_0=T$ and $u\equiv v$. □

With a little extra effort it is possible to prove that the solution depends continuously on the initial datum.

COROLLARY 3.3. *Under the assumptions of Theorem* 3.2, *fix any $u_0\in C^{2+\theta}(\overline{\Omega})$ such that the range of D^2u_0 is contained in $\mathcal{O}_1$, the range of Du_0 is contained in $\mathcal{O}_2$, and $\Psi(Du_0(x))=0$ at $\partial\Omega$. Then there exist $r>0$, $K>0$ such that, for each $v_0\in C^{2+\theta}(\overline{\Omega})$ with $\|v_0-u_0\|_{C^{2+\theta}(\overline{\Omega})}\leqslant r$ and satisfying the compatibility conditions $\Psi(Dv_0(x))=0$ at $\partial\Omega$, the solution $v(t,x)$ of problem* (1.1) *with initial datum v_0 is defined in $[0,T]\times\overline{\Omega}$, where $T>0$ is given by Theorem* 3.2, *and*

$$\|u-v\|_{C^{1+\theta/2,2+\theta}([0,T]\times\overline{\Omega})}\leqslant K\|u_0-v_0\|_{C^{2+\theta}(\overline{\Omega})}.$$

PROOF. We follow the notation of the proof of Theorem 3.2. If we take $r<r_1/2$, $r<r_2/2$ and define

$$Y_1=\big\{u\in C^{1+\theta/2,2+\theta}\big([0,T]\times\overline{\Omega}\big):\\ u(0,\cdot)=v_0,\ \|u-v_0\|_{C^{1+\theta/2,2+\theta}([0,T]\times\overline{\Omega})}\leqslant R\big\},$$

with T and R chosen as in the proof of Theorem 3.2. Then, for each $u\in Y_1$, the ranges of D^2u and of Du are contained in the neighborhoods of the ranges of D^2u_0 and of Du_0 with radii r_1, r_2, respectively, according to (3.12). Since $u-v$ is a solution to problem (3.9) with $f(t,x)=\Phi(D^2u(t,x))-\Phi(D^2v(t,x))-\mathcal{A}(u-v)(t,x)$, $g(t,x)=\mathcal{B}(u-v)(t,x)-\Psi(Du(t,x))+\Psi(Dv(t,x))$, $w_0=u_0-v_0$, combining estimate (3.10) with the estimates of the proof of Theorem 3.2 we get

$$\|u-v\|_{C^{1+\theta/2,2+\theta}([0,T]\times\overline{\Omega})}\\ \leqslant C\bigg(\frac{1}{2}\|u-v\|_{C^{1+\theta/2,2+\theta}([0,T]\times\overline{\Omega})}+\|u_0-v_0\|_{C^{2+\theta}(\overline{\Omega})}\bigg),$$

and the statement follows with $K=2C$. □

The proofs of Theorem 3.2 and of its corollary indicate that they may be extended to more general situations.

First of all, we may allow more general nonlinearities, such as $\Phi=\Phi(t,x,u,Du,D^2u)$ and $\Psi=\Psi(t,x,u,Du)$. See, e.g., [33], Chapter 8. The nonlinear boundary condition may be replaced by a Dirichlet boundary condition, $u(t,x)=g(t,x)$ with $g\in C^{1+\theta/2,2+\theta}([0,T]\times\partial\Omega)$, and the proof comes out shorter.

Second, the nonlinearities may also be nonlocal: the essential property of Φ that we used in the proof was just that the function $F(u)=\Phi(D^2u)-\mathcal{A}u$ is differentiable near u_0, with locally Lipschitz continuous (null at u_0) derivative, as a function from $C^2_{\mathrm{b}}(\overline{\Omega})$ to $C_{\mathrm{b}}(\overline{\Omega})$, and from $C^{2+\theta}(\overline{\Omega})$ to $C^\theta(\overline{\Omega})$, and moreover, that

$$\begin{aligned}\|F'(u)v\|_{C^\theta(\overline{\Omega})} \leqslant{}& \|F'(u)\|_{L(C^2_{\mathrm{b}}(\overline{\Omega}),C_{\mathrm{b}}(\overline{\Omega}))}\|v\|_{C^{2+\theta}(\overline{\Omega})}\\ &+\|F'(u)\|_{L(C^{2+\theta}(\overline{\Omega}),C^\theta(\overline{\Omega}))}\|v\|_{C^2_{\mathrm{b}}(\overline{\Omega})}.\end{aligned}$$

Third, the proof is not confined to a single second-order equation but it works as well for higher-order equations and systems. This is because optimal regularity theorems in parabolic Hölder spaces similar to Theorem 3.1 are available for higher-order equations and systems [34,37]. A detailed proof for a general class of second-order systems with Dirichlet boundary condition is in [1].

A completely different approach is in [24].

4. Existence at large and stability

Existence at large for arbitrary initial data is a hard task in the fully nonlinear case. The results available up to now concern only second-order equations. The difficulty is due to the fact that we need a priori estimates in a very high norm, substantially in a $C^{1+\theta/2,2+\theta}$ a norm, to get existence at large; therefore the nonlinearities have to satisfy severe restrictions. See [27,30] for further detailed discussion and comments.

On the other hand, existence in the large and stability for initial data close to stationary solutions or more generally to established given solutions, is a quite developed subject.

For initial data close to stationary solutions, the proof of the local existence Theorem 3.2 is easier, and it can be extended to a very general class of perturbations. We quote a result from [7], concerning problem

$$\begin{cases} u_t(t,\xi)=\mathcal{A}u+F\big(u(t,\cdot)\big)(\xi), & \xi\in\overline{\Omega},\\ \mathcal{B}u=G\big(u(t,\cdot)\big)(\xi), & \xi\in\partial\Omega,\\ u(0,\xi)=u_0(\xi), & \xi\in\overline{\Omega}.\end{cases} \tag{4.1}$$

Here the stationary solution is $u\equiv 0$. In [7] a bounded Ω is taken into consideration, but the proofs are easily extended to unbounded open sets. The assumptions on F and G are the following.

(H4) $F : B(0, R) \subset C^2_{\rm b}(\overline{\Omega}) \to C_{\rm b}(\overline{\Omega})$ is continuously differentiable with Lipschitz continuous derivative, $F(0) = 0$, $F'(0) = 0$ and the restriction of F to $B(0, R) \subset C^{2+\theta}(\overline{\Omega})$ has values in $C^\theta(\overline{\Omega})$ and is continuously differentiable; $G : B(0, R) \subset C^1_{\rm b}(\overline{\Omega}) \to C_{\rm b}(\partial\Omega)$ is continuously differentiable with Lipschitz continuous derivative, $G(0) = 0$, $G'(0) = 0$ and the restriction of G to $B(0, R) \subset C^{2+\theta}(\overline{\Omega})$ has values in $C^{1+\theta}(\partial\Omega)$ and is continuously differentiable.

THEOREM 4.1. *Let Ω and the operators $\mathcal{A}$, $\mathcal{B}$ defined in* (3.4), (3.6), *satisfy the assumptions of Theorem* 3.1. *If* (H4) *holds, for every $T > 0$, there are $r, \rho > 0$ such that* (4.1) *has a solution $u \in C^{1+\theta/2,2+\theta}([0, T] \times \overline{\Omega})$ provided $\|u_0\|_{C^{2+\theta}(\overline{\Omega})} \leqslant \rho$. Moreover, u is the unique solution in $B(0, r) \subset C^{1+\theta/2,2+\theta}([0, T] \times \overline{\Omega})$.*

PROOF. Let $0 < r \leqslant R$, and set

$$K(r) = \sup\bigl\{\|F'(\varphi)\|_{L(C^{2+\theta}(\overline{\Omega}),C^{\theta}(\overline{\Omega}))}\colon \varphi \in B(0, r) \subset C^{2+\theta}(\overline{\Omega})\bigr\},$$
$$H(r) = \sup\bigl\{\|G'(\varphi)\|_{L(C^{2+\theta}(\overline{\Omega}),C^{1+\theta}(\partial\Omega))}\colon \varphi \in B(0, r) \subset C^{2+\theta}(\overline{\Omega})\bigr\}.$$

Since $F'(0) = 0$ and $G'(0) = 0$, $K(r)$ and $H(r)$ tend to 0 as $r \to 0$. Let $L > 0$ be such that, for all $\varphi, \psi \in B(0, r) \subset C^2_{\rm b}(\overline{\Omega})$ with small r,

$$\|F'(\varphi) - F'(\psi)\|_{L(C^2_{\rm b}(\overline{\Omega}),C_{\rm b}(\overline{\Omega}))} \leqslant L\|\varphi - \psi\|_{C^2_{\rm b}(\overline{\Omega})},$$
$$\|G'(\varphi) - G'(\psi)\|_{L(C^1_{\rm b}(\overline{\Omega}),C_{\rm b}(\partial\Omega))} \leqslant L\|\varphi - \psi\|_{C^1_{\rm b}(\overline{\Omega})}.$$

For every $0 \leqslant s \leqslant t \leqslant T$ and for every $w \in B(0, r) \subset C^{1+\theta/2,2+\theta}([0, T] \times \overline{\Omega})$ with r so small that $K(r), H(r) < \infty$, we have

$$\|F(w(t,\cdot))\|_{C^{\theta}(\overline{\Omega})} \leqslant K(r)\|w(t,\cdot)\|_{C^{2+\theta}(\overline{\Omega})},$$
$$\begin{aligned}\|F(w(t,\cdot)) - F(w(s,\cdot))\|_{C_{\rm b}(\overline{\Omega})} &\leqslant Lr\|w(t,\cdot) - w(s,\cdot)\|_{C^2_{\rm b}(\overline{\Omega})}\\ &\leqslant Lr|t - s|^{\theta/2}\|w\|_{C^{1+\theta/2,2+\theta}([0,T]\times\overline{\Omega})},\end{aligned}$$

and similarly,

$$\begin{aligned}\|G(w(t,\cdot))\|_{C^{1+\theta}(\partial\Omega)} &\leqslant H(r)\|w(t,\cdot)\|_{C^{2+\theta}(\overline{\Omega})}\\ &\leqslant H(r)\|w\|_{C^{1+\theta/2,2+\theta}([0,T]\times\overline{\Omega})},\end{aligned}$$

$$\begin{aligned}\|G(w(t,\cdot)) - G(w(s,\cdot))\|_{C_{\rm b}(\partial\Omega)} &\leqslant Lr\|w(t,\cdot) - w(s,\cdot)\|_{C^1_{\rm b}(\overline{\Omega})}\\ &\leqslant Lr|t - s|^{1/2+\theta/2}\|w\|_{C^{1+\theta/2,2+\theta}([0,T]\times\overline{\Omega})}.\end{aligned}$$

Therefore, $(t,\xi) \to F(w(t,\cdot))(\xi)$ belongs to $C^{\theta/2,\theta}([0,T]\times\overline{\Omega})$, $(t,\xi) \to G(w(t,\cdot))(\xi)$ belongs to $C^{1/2+\theta/2,1+\theta}([0,T]\times\partial\Omega)$, and

$$\begin{aligned} &\big\|F(w)\big\|_{C^{\theta/2,\theta}([0,T]\times\overline{\Omega})} \leqslant \big(K(r)+Lr\big)\|w\|_{C^{2+\alpha,1+\alpha/2}([0,T]\times\overline{\Omega})},\\ &\big\|G(w)\big\|_{C^{1/2+\theta/2,1+\theta}([0,T]\times\partial\Omega)} \leqslant \big(2H(r)+Lr\big)\|w\|_{C^{1+\theta/2,2+\theta}([0,T]\times\overline{\Omega})}. \end{aligned} \tag{4.2}$$

So, if $\|w_0\|_{C^{2+\theta}(\overline{\Omega})}$ is small enough, we define a nonlinear map

$$\begin{aligned} &\Gamma:\big\{w\in B(0,r)\subset C^{1+\theta/2,2+\theta}\big([0,T]\times\overline{\Omega}\big): w(\cdot,0)=w_0\big\}\\ &\quad\mapsto C^{1+\theta/2,2+\theta}\big([0,T]\times\overline{\Omega}\big), \end{aligned}$$

by $\Gamma w = v$, where v is the solution of

$$\begin{cases} v_t(x,t) = \mathcal{A}v + F\big(w(t,\cdot)\big)(x), & 0\leqslant t\leqslant T,\ x\in\overline{\Omega},\\ \mathcal{B}v = G\big(w(t,\cdot)\big)(x), & 0\leqslant t\leqslant T,\ x\in\partial\Omega,\\ v(0,x) = w_0(x), & x\in\overline{\Omega}. \end{cases}$$

Actually, thanks to the compatibility condition $\mathcal{B}w_0 = G(w_0)$ and the regularity of $F(w)$ and $G(w)$, the range of Γ is contained in $C^{1+\theta/2,2+\theta}([0,T]\times\overline{\Omega})$. Moreover, Theorem 3.1 gives the estimate

$$\begin{aligned} &\|v\|_{C^{1+\theta/2,2+\theta}([0,T]\times\overline{\Omega})}\\ &\quad\leqslant C\big(\|w_0\|_{C^{2+\theta}(\overline{\Omega})} + \big\|F(w)\big\|_{C^{\theta/2,\theta}([0,T]\times\overline{\Omega})} + \big\|G(w)\big\|_{C^{1/2+\theta/2,1+\theta}([0,T]\times\partial\Omega)}\big), \end{aligned}$$

with $C=C(T)$, so that

$$\begin{aligned} &\big\|\Gamma(w)\big\|_{C^{1+\theta/2,2+\theta}([0,T]\times\overline{\Omega})}\\ &\quad\leqslant C\big(\|w_0\|_{C^{2+\theta}(\overline{\Omega})} + \big(K(r)+2Lr+2H(r)\big)\|w\|_{C^{1+\theta/2,2+\theta}([0,T]\times\overline{\Omega})}\big). \end{aligned}$$

Therefore, if r is so small that

$$C\big(K(r)+2Lr+2H(r)\big) \leqslant \frac{1}{2}, \tag{4.3}$$

and w_0 is so small that

$$\|w_0\|_{C^{2+\theta}(\overline{\Omega})} \leqslant \frac{Cr}{2},$$

Γ maps the ball $B(0,r)$ into itself. Let us check that Γ is a 1/2-contraction. Let

$w_1, w_2 \in B(0,r)$, $w_i(\cdot,0) = w_0$. Taking $w_i(t,\cdot) = w_i(t)$, $i = 1,2$, we have

$$
\begin{aligned}
&\|\Gamma w_1 - \Gamma w_2\|_{C^{1+\theta/2,2+\theta}([0,T]\times\overline{\Omega})} \\
&\quad \leqslant C\Big(\big\|F(w_1) - F(w_2)\big\|_{C^{\theta/2,\theta}([0,T]\times\overline{\Omega})} \\
&\qquad + \big\|G(w_1) - G(w_2)\big\|_{C^{1/2+\theta/2,1+\theta}([0,T]\times\partial\Omega)}\Big),
\end{aligned}
$$

and, arguing as above, for $0 \leqslant t \leqslant T$,

$$
\begin{aligned}
\big\|F\big(w_1(t,\cdot)\big) - F\big(w_2(t,\cdot)\big)\big\|_{C^{\theta}(\overline{\Omega})} &\leqslant K(r)\big\|w_1(t,\cdot) - w_2(t,\cdot)\big\|_{C^{2+\theta}(\overline{\Omega})} \\
&\leqslant K(r)\|w_1 - w_2\|_{C^{1+\theta/2,2+\theta}([0,T]\times\overline{\Omega})}, \\
\big\|G\big(w_1(t,\cdot)\big) - G\big(w_2(t,\cdot)\big)\big\|_{C^{1+\theta}(\partial\Omega)} &\leqslant H(r)\big\|w_1(t,\cdot) - w_2(t,\cdot)\big\|_{C^{2+\theta}(\overline{\Omega})} \\
&\leqslant H(r)\|w_1 - w_2\|_{C^{1+\theta/2,2+\theta}([0,T]\times\overline{\Omega})},
\end{aligned}
$$

while for $0 \leqslant s \leqslant t \leqslant T$,

$$
\begin{aligned}
&\big\|F\big(w_1(t,\cdot)\big) - F\big(w_2(t,\cdot)\big) - F\big(w_1(s,\cdot)\big) - F\big(w_2(s,\cdot)\big)\big\|_{C_b(\overline{\Omega})} \\
&\quad = \bigg\|\int_0^1 F'\big(\sigma w_1(t,\cdot) + (1-\sigma)w_2(t,\cdot)\big)\big(w_1(t,\cdot) - w_2(t,\cdot)\big) \\
&\qquad - F'\big(\sigma w_1(s,\cdot) + (1-\sigma)w_2(s,\cdot)\big)\big(w_1(s,\cdot) - w_2(s,\cdot)\big)\,\mathrm{d}\sigma\bigg\|_{C_b(\overline{\Omega})} \\
&\quad \leqslant \int_0^1 \big\|F'\big(\sigma w_1(t,\cdot) + (1-\sigma)w_2(t,\cdot)\big) - F'\big(\sigma w_1(s,\cdot) + (1-\sigma)w_2(s,\cdot)\big) \\
&\qquad\quad \times \big(w_1(t,\cdot) - w_2(t,\cdot)\big)\,\mathrm{d}\sigma\big\|_{C_b(\overline{\Omega})} \\
&\qquad + \int_0^1 \big\|F'\big(\sigma w_1(s,\cdot)\big)\big(w_1(t,\cdot) - w_2(t,\cdot) - w_1(s,\cdot) + w_2(s,\cdot)\big)\big\|_{C_b(\overline{\Omega})} \\
&\quad \leqslant \frac{L}{2}\Big(\big\|w_1(t,\cdot) - w_1(s,\cdot)\big\|_{C_b^2(\overline{\Omega})} + \big\|w_2(t,\cdot) - w_2(s,\cdot)\big\|_{C^2(\overline{\Omega})}\Big) \\
&\qquad \times \big\|w_1(t,\cdot) - w_2(t,\cdot)\big\|_{C_b^2(\overline{\Omega})} \\
&\qquad + Lr\big\|w_1(t,\cdot) - w_2(t,\cdot) - w_1(s,\cdot) + w_2(s,\cdot)\big\|_{C_b^2(\overline{\Omega})} \\
&\quad \leqslant 2Lr(t-s)^{\theta/2}\|w_1 - w_2\|_{C^{1+\theta/2,2+\theta}([0,T]\times\overline{\Omega})},
\end{aligned}
$$

and similarly,

$$\begin{aligned}&\|G(w_1(t,\cdot)) - G(w_2(t,\cdot)) - G(w_1(s,\cdot)) - G(w_2(\cdot,s))\|_{C_b(\partial\Omega)}\\&\quad\leqslant 2Lr(t-s)^{1/2+\theta/2}\|w_1 - w_2\|_{C^{1+\theta/2,2+\theta}([0,T]\times\overline{\Omega})}.\end{aligned}$$

Therefore,

$$\begin{aligned}&\|\Gamma w_1 - \Gamma w_2\|_{C^{1+\theta/2,2+\theta}([0,T]\times\overline{\Omega})}\\&\quad\leqslant C\big(K(r) + 2Lr + 2H(r)\big)\|w_1 - w_2\|_{C^{1+\theta/2,2+\theta}([0,T]\times\overline{\Omega})}\\&\quad\leqslant \frac{1}{2}\|w_1 - w_2\|_{C^{1+\theta/2,2+\theta}([0,T]\times\overline{\Omega})},\end{aligned}$$

the last inequality being a consequence of (4.3). The statement follows. □

In our example (4.1), assumption (H4) is satisfied with $F(\phi) = \Phi(D^2\phi) - \mathcal{A}\phi$, $G(\phi) = \Psi(D\phi) - \mathcal{B}\phi$, if Φ, Ψ satisfy the assumptions of Theorem 3.2, where $\mathcal{O}_1$, $\mathcal{O}_2$ are neighborhoods of 0 in $\mathbb{R}^{N^2}$, $\mathbb{R}^N$, respectively, and $\Phi(0) = 0$, $\Psi(0) = 0$, and $\mathcal{A}$, $\mathcal{B}$ are the operators

$$\mathcal{A}\varphi(x) = \sum_{i,j=1}^{N} \frac{\partial\Phi}{\partial q_{ij}}(0)D_{ij}\varphi(x), \quad x \in \overline{\Omega},$$

$$\mathcal{B}\varphi(x) = \sum_{i=1}^{N} \frac{\partial\Psi}{\partial p_i}(0)D_i\varphi(x), \quad x \in \overline{\Omega}.$$

Theorem 4.1 states that if 0 is a stationary solution then the solution to (4.1) is defined in an arbitrary large time interval provided the initial datum is small enough. The next natural question is now the stability of the null solution. We shall see that the principle of linearized stability holds, and that in the hyperbolic case local stable and unstable manifolds may be constructed, just like in the case of ordinary differential equations. To do this we shall see again our nonlinear problem as a perturbation of a linear one, and the main tools will be optimal regularity/asymptotic behavior results for the linear case, stated in the next section.

4.1. *Asymptotic behavior in linear problems*

Let us consider again the operators $\mathcal{A}$ and $\mathcal{B}$ defined in (3.4) and (3.6). The realization A of $\mathcal{A}$ with homogeneous boundary conditions in $X = C(\overline{\Omega})$, i.e., the operator with domain

$$D(A) = \Big\{\varphi \in C_b(\overline{\Omega}) \bigcap_{p>1} W^{2,p}_{\mathrm{loc}}(\Omega)\colon \mathcal{A}\varphi \in C_b(\overline{\Omega}), \mathcal{B}\varphi(x) = 0, x \in \partial\Omega\Big\}$$

$$A\varphi = \mathcal{A}\varphi, \quad \varphi \in D(A),$$

is a sectorial operator in $C_b(\overline{\Omega})$, according to [40]. We define

$$\sigma^-(A) = \{\lambda \in \sigma(A)\colon \operatorname{Re}\lambda < 0\}, \qquad \sigma^+(A) = \{\lambda \in \sigma(A)\colon \operatorname{Re}\lambda > 0\}.$$

We shall consider the assumptions

$$\begin{aligned} &\text{(i)} \quad \sup\{\operatorname{Re}\lambda\colon \lambda \in \sigma^-(A)\} < 0, \\ &\text{(ii)} \quad \inf\{\operatorname{Re}\lambda\colon \lambda \in \sigma^+(A)\} > 0, \end{aligned} \tag{4.4}$$

which are always true if Ω is bounded, because in this case the domain of A is compactly embedded in $C(\overline{\Omega})$, the resolvent operators $(\lambda I - A)^{-1}$ are compact, and the spectrum consists of a sequence of eigenvalues.

If (4.4)(ii) holds, $\sigma^+(A)$ is closed. We denote by P^+ the spectral projection associated to $\sigma^+(A)$, i.e.,

$$P^+ = \frac{1}{2\pi i}\int_C R(\lambda, A)\,d\lambda,$$

where C is any closed simple regular curve in $\{\operatorname{Re}\lambda > 0\}$ surrounding $\sigma^+(A)$.

If (4.4)(i) holds, $\sigma^-(A)$ is closed, and we denote by P^- the spectral projection associated to $\sigma^+(A)$, i.e.,

$$P^- = I - P, \quad P = \frac{1}{2\pi i}\int_C R(\lambda, A)\,d\lambda,$$

where now C is any closed simple regular curve surrounding $\sigma(A) \setminus \sigma^-(A)$ with index 0 with respect to all points in $\sigma^-(A)$. If the spectrum of A does not intersect the imaginary axis, then $P^- = I - P^+$.

We also need a deeper insight into the solution to (3.9). We shall consider a representation formula for w, that is an extension of the well-known Balakrishnan formula (see, e.g., [33], page 200):

$$\begin{aligned} w(t,\cdot) &= e^{tA}\bigl(w_0 - \mathcal{N}g(0,\cdot)\bigr) + \int_0^t e^{(t-s)A}\bigl[f(s,\cdot) + A\mathcal{N}g(s,\cdot)\bigr]\,ds \\ &\quad - A\int_0^t e^{(t-s)A}\bigl[\mathcal{N}g(s,\cdot) - \mathcal{N}g(0,\cdot)\bigr]\,ds + \mathcal{N}g(0,\cdot) \\ &= e^{tA}u_0 + \int_0^t e^{(t-s)A}\bigl[f(s,\cdot) + A\mathcal{N}g(s,\cdot)\bigr]\,ds \\ &\quad - A\int_0^t e^{(t-s)A}\mathcal{N}g(s,\cdot)\,ds, \quad 0 \leqslant t \leqslant T. \end{aligned} \tag{4.5}$$

Here $\mathcal{N}$ is any lifting operator such that

$$\begin{cases} \mathcal{N} \in L\big(C^{\alpha}(\partial\Omega), C^{\alpha+1}(\overline{\Omega})\big), & 0 \leqslant \alpha \leqslant \theta + 1, \\ \mathcal{B}\mathcal{N}g = g, & g \in C_b(\partial\Omega). \end{cases} \tag{4.6}$$

For instance, we can take as $\mathcal{N}$ the operator given in Theorem 0.3.2 of [33]. Later we will need an explicit expression of $\mathcal{N}$, so we give some details.

LEMMA 4.2. *Let Ω be an open set in $\mathbb{R}^N$ with uniform $C^{2+\theta}$ boundary, and let $\mathcal{B}$ satisfy the assumptions of Theorem* 3.1. *Then there exists a lifting operator satisfying* (4.6).

PROOF. As a first step we construct $\mathcal{N}$ for $\Omega = \mathbb{R}^N_- = \{(x, y)\colon x < 0, y \in \mathbb{R}^{N-1}\}$. Fix a function $\varphi \in C^{\infty}(\mathbb{R}^{N-1})$, with compact support, and such that $\int_{\mathbb{R}^{N-1}} \varphi(\xi)\,d\xi = 1$. Fix, moreover, $\delta > 0$ and $\eta \in C^{\infty}((-\infty, 0])$ such that $\eta \equiv 0$ for $x \leqslant -2\delta$, $\eta \equiv 1$ for $-\delta \leqslant x \leqslant 0$.

For each $k \in \mathbb{N}$ and $g \in L^{\infty}(\mathbb{R}^{N-1})$, set

$$Ng(x, y) = x\eta(x) \int_{\mathbb{R}^{N-1}} \varphi(\xi) g(y + \xi x)\,d\xi, \quad x \leqslant 0, y \in \mathbb{R}^{N-1}. \tag{4.7}$$

Then $N \in L(C^{\alpha}(\mathbb{R}^{N-1}); C^{1+\alpha}(\mathbb{R}^N_-))$ for each $k \in \mathbb{N}$, $\alpha \geqslant 0$, and for every $y \in \mathbb{R}^{N-1}$, it holds

$$\begin{cases} Ng(0, y) = 0, \\ \frac{\partial}{\partial x} Ng(0, y) = \psi(y). \end{cases} \tag{4.8}$$

If $\mathcal{B}$ is the normal derivative, we are done. If $\mathcal{B}v = \beta(y)\,\partial/\partial x$ plus derivatives with respect to y, we define $(\mathcal{N}g)(x, y) = (Ng)(x, y)/\beta(y)$.

The case of a general open set with uniform $C^{2+\theta}$ boundary is reduced to this one in a standard way, by locally stretching the boundary and using partitions of unity. See for instance [34], where more general lifting operators were constructed for systems of m boundary conditions. □

The following theorems were proved in [7].

THEOREM 4.3. *Let assumption* (4.4)(i) *hold, and fix $\omega > 0$ such that $\omega < -\max\{\operatorname{Re}\lambda\colon \lambda \in \sigma^-(A)\}$. Let f be such that $(t, \xi) \to e^{\omega t} f(t, \xi) \in C^{\theta/2,\theta}([0, \infty) \times \overline{\Omega})$, let g be such that $(t, \xi) \to e^{\omega t} g(t, \xi) \in C^{1+\theta,1/2+\theta/2}([0, \infty) \times \partial\Omega)$ and let $w_0 \in C^{2+\theta}(\overline{\Omega})$ satisfy the compatibility condition* (3.8). *Then $v(t, \xi) = e^{\omega t} w(t, \xi)$ is bounded in $[0, +\infty) \times \overline{\Omega}$ if and only if*

$$\begin{aligned} (I - P^-)w_0 = {} & -\int_0^{+\infty} e^{-sA}(I - P^-)\big[f(s, \cdot) + \mathcal{A}\mathcal{N}g(s, \cdot)\big]\,ds \\ & + A \int_0^{\infty} e^{-sA}(I - P^-)\mathcal{N}g(s, \cdot)\,ds. \end{aligned} \tag{4.9}$$

In this case, w is given by

$$\begin{aligned} w(t,\cdot) = {} & \mathrm{e}^{tA}P^- w_0 + \int_0^t \mathrm{e}^{(t-s)A}P^-\big[f(s,\cdot)+\mathcal{A}\mathcal{N}g(s,\cdot)\big]\,\mathrm{d}s \\ & - A\int_0^t \mathrm{e}^{(t-s)A}P^-\mathcal{N}g(s,\cdot)\,\mathrm{d}s \\ & - \int_t^{+\infty} \mathrm{e}^{(t-s)A}\big(I-P^-\big)\big[f(s,\cdot)+\mathcal{A}\mathcal{N}g(s,\cdot)\big]\,\mathrm{d}s \\ & + A\int_t^{+\infty} \mathrm{e}^{(t-s)A}\big(I-P^-\big)\mathcal{N}g(s,\cdot)\,\mathrm{d}s, \end{aligned} \tag{4.10}$$

and the function $v=\mathrm{e}^{\omega t}w$ belongs to $C^{1+\theta/2,2+\theta}([0,\infty)\times\overline{\Omega})$, with the estimate

$$\begin{aligned} & \|v\|_{C^{1+\theta/2,2+\theta}([0,\infty)\times\overline{\Omega})} \\ & \quad \leqslant C\big(\|w_0\|_{C^{2+\theta}(\overline{\Omega})} + \big\|\mathrm{e}^{\omega t}f\big\|_{C^{\theta/2,\theta}([0,\infty)\times\overline{\Omega})} + \big\|\mathrm{e}^{\omega t}g\big\|_{C^{1/2+\theta/2,1+\theta}([0,\infty)\times\partial\Omega)}\big). \end{aligned}$$

Theorem 4.3 has an important corollary in the stable case when $\sigma(A)=\sigma^-(A)$.

COROLLARY 4.4. *Assume that $\omega_A := \sup\{\mathrm{Re}\,\lambda\colon \lambda\in\sigma(A)\}<0$, and fix $\omega\in(0,\omega_A)$. Let f be such that $(t,\xi)\to \mathrm{e}^{\omega t}f(t,\xi)\in C^{\theta/2,\theta}([0,\infty)\times\overline{\Omega})$, let g be such that $(t,\xi)\to \mathrm{e}^{\omega t}g(t,\xi)\in C^{1+\theta,1/2+\theta/2}([0,\infty)\times\partial\Omega)$ and let $w_0\in C^{2+\theta}(\overline{\Omega})$ satisfy the compatibility condition* (3.8). *Then $v(t,\xi)=\mathrm{e}^{\omega t}w(t,\xi)$ belongs to $C^{1+\theta/2,2+\theta}([0,\infty)\times\overline{\Omega})$ and*

$$\begin{aligned} & \|v\|_{C^{1+\theta/2,2+\theta}([0,\infty)\times\overline{\Omega})} \\ & \quad \leqslant C\big(\|w_0\|_{C^{2+\theta}(\overline{\Omega})} + \big\|\mathrm{e}^{\omega t}f\big\|_{C^{\theta/2,\theta}([0,\infty)\times\overline{\Omega})} + \big\|\mathrm{e}^{\omega t}g\big\|_{C^{1/2+\theta/2,1+\theta}([0,\infty)\times\partial\Omega)}\big). \end{aligned}$$

Let us now consider the backward problem

$$\begin{cases} w_t = \mathcal{A}w + f(t,\xi), & t\leqslant 0,\ \xi\in\overline{\Omega}, \\ \mathcal{B}w = g(t,\xi), & t\leqslant 0,\ \xi\in\partial\Omega, \\ w(0,\xi)=w_0(\xi), & \xi\in\overline{\Omega}. \end{cases} \tag{4.11}$$

THEOREM 4.5. *Let assumption* (4.4)(ii) *hold, with $\sigma^+(A)\neq\emptyset$, and fix $\omega>0$ such that $\omega<\min\{\mathrm{Re}\,\lambda\colon \lambda\in\sigma^+(A)\}$. Let f be such that $(t,\xi)\to \mathrm{e}^{-\omega t}f(t,\xi)\in C^{\theta/2,\theta}((-\infty,0]\times\overline{\Omega})$ and let g be such that $(t,\xi)\to \mathrm{e}^{-\omega t}g(t,\xi)\in C^{1/2+\theta/2,1+\theta}((-\infty,0]\times\partial\Omega)$, $w_0\in C^{2+\theta}(\overline{\Omega})$.*

Then problem (4.11) *has a solution w such that $v(t,\xi)=\mathrm{e}^{-\omega t}w(t,\xi)$ is bounded* *in* $(-\infty, 0] \times \overline{\Omega}$ *if and only if*

$$\begin{aligned}(I - P^+)w_0 = \int_{-\infty}^{0} \mathrm{e}^{-sA}(I - P^+)\big[f(s,\cdot) + A\mathcal{N}g(s,\cdot)\big]\,\mathrm{d}s \\ - A\int_{-\infty}^{0} \mathrm{e}^{-sA}(I - P^+)\mathcal{N}g(s,\cdot)\,\mathrm{d}s.\end{aligned} \tag{4.12}$$

In this case, w is given by

$$\begin{aligned}w(t,\cdot) = \mathrm{e}^{tA}P^+w_0 + \int_0^t \mathrm{e}^{(t-s)A}P^+\big[f(s,\cdot) + A\mathcal{N}g(s,\cdot)\big]\,\mathrm{d}s \\ - A\int_0^t \mathrm{e}^{(t-s)A}P^+\mathcal{N}g(s,\cdot)\,\mathrm{d}s \\ + \int_{-\infty}^{t} \mathrm{e}^{(t-s)A}(I - P^+)\big[f(s,\cdot) + A\mathcal{N}g(s,\cdot)\big]\,\mathrm{d}s \\ - A\int_{-\infty}^{t} \mathrm{e}^{(t-s)A}(I - P^+)\mathcal{N}g(s,\cdot)\,\mathrm{d}s, \quad t \leqslant 0.\end{aligned} \tag{4.13}$$

Moreover, $v = \mathrm{e}^{-\omega t}w$ *belongs to* $C^{1+\theta/2,2+\theta}((-\infty,0] \times \overline{\Omega})$ *and*

$$\begin{aligned}\|v\|_{C^{1+\theta/2,2+\theta}((-\infty,0]\times\overline{\Omega})} \leqslant C\Big(\|w_0\|_{C(\overline{\Omega})} + \big\|\mathrm{e}^{-\omega t}f\big\|_{C^{\theta/2,\theta}((-\infty,0]\times\overline{\Omega})} \\ + \big\|\mathrm{e}^{-\omega t}g\big\|_{C^{1/2+\theta/2,1+\theta}((-\infty,0]\times\partial\Omega)}\Big).\end{aligned}$$

4.2. *Principle of linearized stability and local invariant manifolds*

With the aid of Theorems 4.3, 4.5 and Corollary 4.4 we may show similar behavior for the solutions to fully nonlinear problems with initial data close to stationary solutions, provided the linearized operator near the stationary solution under consideration satisfies assumption (4.4)(i) or (4.4)(ii). For the proofs see [7].

THEOREM 4.6. *Let* Ω *be an open set in* $\mathbb{R}^N$ *with uniform* $C^{2+\theta}$ *boundary,* $0 < \theta < 1$, *let the operators* $\mathcal{A}$ *and* $\mathcal{B}$ *satisfy the assumptions of Theorem* 3.1, *and let* F, G, *satisfy assumption* (H4).

(i) *If* $\sup\{\mathrm{Re}\,\lambda\colon \lambda \in \sigma(A)\} < 0$ *then the stationary solution* $u = 0$ *of problem* (4.1) *is stable with respect to the* $C^{2+\theta}(\overline{\Omega})$ *norm. More precisely, for every* $\omega \in (0, -\sup\{\mathrm{Re}\,\lambda\colon \lambda \in \sigma(A)\})$, *there are* $C, r > 0$ *such that for every* u_0 *satisfying* (3.8) *and* $\|u_0\|_{C^{2+\theta}(\overline{\Omega})} \leqslant r$, *the solution of* (4.1) *with initial datum* u_0 *exists at large and satisfies*

$$\big\|u(t,\cdot)\big\|_{C^{2+\theta}(\overline{\Omega})} \leqslant C\mathrm{e}^{-\omega t}\|u_0\|_{C^{2+\theta}(\overline{\Omega})}, \quad t \geqslant 0.$$

(ii) *If* $\sigma(A)$ *contains elements with positive real part and* (4.4)(ii) *holds, then* $u=0$ *is unstable in* $C^{2+\theta}(\overline{\Omega})$.

We recall that if Ω is bounded then (4.4)(ii) is satisfied, and Theorem 4.6 looks like the usual principle of linearized stability for ordinary differential equations. If Ω is unbounded, it may happen that (4.4)(ii) is not satisfied. However, it is still possible to give an instability result, relying on the next theorem taken from Henry's book [22].

THEOREM 4.7. *Let* X *be a real Banach space. Let* T *be a map from a neighborhood of the origin in* X *with* $T(0)=0$, *let* M *be a bounded linear operator on* X *with spectral radius* r *greater than* 1, *and let*

$$T(x)=Mx+O(\|x\|^p) \quad \text{as } x\to 0,$$

for some constant $p>1$. *Then the origin is unstable for the iterates of* T, *i.e., there exists a constant* $C>0$ *and there exists* x_0 *arbitrarily close to* 0 *such that, if* $x_{n+1}=T(x_n)=T^{n+1}(x_0)$ *for* $n\in\mathbb{N}$, *then for some* N *(depending on* x_0*), the sequence* $x_1,x_2,\dots,x_N$ *is well defined and* $\|x_N\|\geqslant C$.

If $G\equiv 0$ in problem (4.1), we can apply Theorem 4.7 with $X=\{u_0\in C^{2+\theta}(\overline{\Omega})\colon\ \mathcal{B}u_0=0;$ at $\partial\Omega\}$, $T(u_0)=u(1;u_0)$ is the solution of (4.1) with initial datum u_0, evaluated at time $t=1$ (the lifetime of the solution is bigger than 1 provided $\|u_0\|_{C^{2+\theta}(\overline{\Omega})}$ is small enough, by Theorem 4.1). Then $T^n u_0=u(n;u_0)$, $M=e^A$, $p=2$, and the spectral radius of M is equal to $\exp(\omega_A)$ where $\omega_A=\sup\{\mathrm{Re}\lambda\colon\ \lambda\in\sigma(A)\}$ because the spectral mapping theorem holds for analytic semigroups. So, if $\omega>0$, Theorem 4.7 implies that the null solution of (4.1) is unstable in $C^{2+\theta}(\overline{\Omega})$.

For a nonvanishing function G the set $\mathcal{I}$ of admissible initial data for problem (4.1), $\mathcal{I}=\{u_0\in C^{2+\theta}(\overline{\Omega})\colon\ \|u_0\|_{C^{2+\theta}(\overline{\Omega})}\leqslant r,\ \mathcal{B}u_0=G(u_0)\}$, is not a neighborhood of 0 in a linear space. However, we shall see in Lemma 4.8 that it is the graph of a regular function defined in a neighborhood of 0 in $D(A_\theta)$, where $D(A_\theta)$ is the domain of the part of A in $C^{2+\theta}(\overline{\Omega})$:

$$D(A_\theta)=\{u_0\in C^{2+\theta}(\overline{\Omega})\colon\ \mathcal{B}u_0=0\}.$$

The already mentioned lifting operator $\mathcal{N}$ is a right inverse of the function $C^{2+\theta}(\overline{\Omega})\mapsto C^{1+\theta}(\partial\Omega)$, $u\mapsto \mathcal{B}u$, so that $C^{2+\theta}(\overline{\Omega})$ is the direct sum $D(A_\theta)\oplus(I-\Pi)(C^{2+\theta}(\overline{\Omega}))$, where Π is the projection on $D(A_\theta)=\operatorname{Ker}\mathcal{B}$ given by

$$\Pi u=u-\mathcal{N}\mathcal{B}u.$$

LEMMA 4.8. *There is a neighborhood* $\mathcal{O}$ *of* 0 *in* $C^{2+\theta}(\overline{\Omega})$ *such that* $\mathcal{I}\cap\mathcal{O}$ *is the graph of a smooth function*

$$H: B(0,\rho)\subset D(A_\theta)\mapsto (I-\Pi)\big(C^{2+\theta}(\overline{\Omega})\big)$$

with $\rho>0$. *Moreover,* $H'(0)=0$.

PROOF. Define $J : B(0, r) \subset C^{2+\theta}(\overline{\Omega}) \mapsto C^{1+\theta}(\partial\Omega)$,

$$J(\varphi) = \mathcal{B}\varphi - G(\varphi),$$

with $r < R$, see assumption (H4). Then J is smooth and $J'(0) = \mathcal{B}$ is an isomorphism from $(I - \Pi)(C^{2+\theta}(\overline{\Omega}))$ to $C^{1+\theta}(\partial\Omega)$. Moreover, $\mathcal{B}_{|D(A_\theta)} = 0$. It is sufficient now to apply the implicit function theorem. □

COROLLARY 4.9. *Under the assumptions of Theorem 4.6, if $\sigma(A)$ contains elements with positive real part then the origin is unstable in $C^{2+\theta}(\overline{\Omega})$.*

PROOF. Theorem 4.7 is applied to the map

$$T : B(0, \rho) \subset D(A_\theta) \mapsto D(A_\theta), \qquad T(x_0) = \Pi\big(u\big(1; x_0 + H(x_0)\big)\big),$$

where ρ is given by Theorem 4.1 with $T = 1$, $u(1, x_0 + H(x_0))$ is the solution of (4.1) with initial condition $u(0) = x_0 + H(x_0)$.

We shall show that the derivative of T at $x_0 = 0$ is $M = \mathrm{e}^{A_\theta}|_{D(A_\theta)}$, and that

$$\big\|T(x_0) - \mathrm{e}^{A_\theta} x_0\big\|_{C^{2+\theta}(\overline{\Omega})} \leqslant C\|x_0\|^2_{C^{2+\theta}(\overline{\Omega})}, \tag{4.14}$$

so that the assumptions of Theorem 4.7 are satisfied with $p = 2$ and $r = \mathrm{e}^{\omega_A}$. Applying Theorem 4.7 gives immediately instability of the null solution to (4.1).

Estimate (4.14) is a consequence of the construction of the solution to (4.1), as a fixed point of the operator Γ, see Theorem 4.1. Indeed, from the representation formula (4.5) and estimates (4.2) it follows that

$$\begin{aligned} &\big\|\Gamma u - \mathrm{e}^{tA}\big(u_0 - \mathcal{N}G(u_0)\big)\big\|_{C^{1+\theta/2,2+\theta}([0,1]\times\overline{\Omega})} \\ &\quad \leqslant \frac{1}{2}\|u\|^2_{C^{1+\theta/2,2+\theta}([0,1]\times\overline{\Omega})} \end{aligned} \tag{4.15}$$

for every $u_0 \in \mathcal{I} \cap B(0, \rho)$ with ρ small enough, which implies for the fixed point u,

$$\|u\|_{C^{1+\theta/2,2+\theta}([0,T]\times\overline{\Omega})} \leqslant C\|u_0\|_{C^{2+\theta}(\overline{\Omega})}.$$

Replacing this in (4.15) and then taking $t = 1$ we obtain

$$\big\|u(1, \cdot) - \mathrm{e}^{A_\theta}\big(u_0 - \mathcal{N}G(u_0)\big)\big\|_{C^{2+\theta}(\overline{\Omega})} \leqslant C\|u_0\|^2_{C^{2+\theta}(\overline{\Omega})}$$

which implies, for $u_0 = x_0 + H(x_0)$,

$$\big\|\Pi u(1, \cdot) - \mathrm{e}^{A_\theta} x_0\big\|_{C^{2+\theta}(\overline{\Omega})} \leqslant C'\big\|x_0 + H(x_0)\big\|^2_{C^{2+\theta}(\overline{\Omega})} \leqslant C''\|x_0\|^2_{C^{2+\theta}(\overline{\Omega})},$$

and (4.14) follows. □

Corollary 4.9 improves part (ii) of Theorem 4.6; however, it is not completely satisfactory because the $C^{2+\theta}(\overline{\Omega})$ norm is very strong and consequently the instability result is rather weak. We can improve the instability result using a refinement of Theorem 4.7, see proof in [9].

THEOREM 4.10. *Let the conditions of Theorem* 4.7 *be satisfied, and assume in addition that the spectral radius r of M is an eigenvalue. Let $\bar{u} \in X$ be an eigenfunction and $x' \in X'$ (the space of all linear continuous functions from X to $\mathbb{R}$) be such that $x'(\bar{u}) \neq 0$. Then there are $C' > 0$ and initial data x_0 arbitrarily close to 0 such that if $x_{n+1} = T(x_n) = T^{n+1}(x_0)$ for $n \in \mathbb{N}$, then, for some N (depending on x_0), the sequence $x_1, x_2, \dots, x_N$ is well defined, $x'(x_N)$ has the same sign $x'(\bar{u})$, and $|x'(x_N)| \geqslant C'$.*

COROLLARY 4.11. *Under the assumptions of Theorem* 4.6, *suppose moreover that ω_A is an eigenvalue of A, and let $\bar{u}$ be an eigenvector. If $x' \in (C^{2+\theta}(\overline{\Omega}))'$ is such that $x'(\bar{u}) \neq 0$, then there is $C' > 0$ such that for every $\delta > 0$ there are $u_0 \in \mathcal{I}$ with norm less or equal to δ, and $N \in \mathbb{N}$ such that the corresponding solution u of* (4.1) *is defined at $T = N$ and $|x'(\Pi u(N, \cdot))| \geqslant C'$.*

The element $x' \in (C^{2+\theta}(\overline{\Omega}))'$ may be, for instance, the evaluation of φ or of some first- or second-order derivative of φ at some point. In this case Corollary 4.11 gives pointwise instability. Let us show this in a simple example.

EXAMPLE 4.12. Consider the problem

$$\begin{cases} u_t(t,x) = \Delta u(t,x) + au(t,x) + \mathcal{F}(D^2u(t,x)), & t \geqslant 0, x \in \overline{\Omega}, \\ \frac{\partial u}{\partial n}(t,x) = \mathcal{G}(Du(t,x)), & t \geqslant 0, x \in \partial\Omega, \\ u(0,x) = u_0(x), & x \in \overline{\Omega}, \end{cases} \tag{4.16}$$

where Ω is either a bounded open set with $C^{2+\theta}$ boundary, or a halfplane, $\mathcal{F}$, $\mathcal{G}$ are smooth functions defined in a neighborhood of 0 in $\mathbb{R}^{N^2}$, $\mathbb{R}^N$, respectively, vanishing at 0 with all their first order derivatives, and $a > 0$.

Then $\omega_A = a > 0$, and Corollary 4.9 implies that the null solution is unstable in $C^{2+\theta}(\overline{\Omega})$.

We get a much better instability result using Corollary 4.11. ω_A is an eigenvalue of A with constant eigenfunctions. Therefore for each $x_0 \in \overline{\Omega}$ the mapping $\varphi \mapsto \varphi(x_0)$ is an element of $(C^{2+\theta}(\overline{\Omega}))'$ that does not vanish on the eigenfunction 1. Corollary 4.11 implies that there is $C' > 0$ such that, for every $\delta > 0$, there are $u_0 \in \mathcal{I}$ with norm less or equal to δ, and $N \in \mathbb{N}$ such that the solution u of (4.1) is defined at $T = N$ and $|(\Pi u(N, \cdot))(x_0)| \geqslant C'$. Since $\Pi u = u - \mathcal{N}\mathcal{B}u$, if $\mathcal{N}$ satisfies $(\mathcal{N}g)(x_0) = 0$ for each g, we have $(\Pi u(N, \cdot))(x_0) = u(N, x_0)$, and hence

$$|u(N, x_0)| \geqslant C'. \tag{4.17}$$

From the construction of $\mathcal{N}$ in Lemma 4.2 we know that $\mathcal{N}g$ vanishes at $\partial\Omega$ for every g. If $x_0 \in \Omega$, taking δ small enough in the proof of Lemma 4.2 we may let $(\mathcal{N}g)(x_0) = 0$ for each g, and (4.17) follows.

Now we go further in the description of the behavior of the solutions for small initial data, showing the existence of the local stable and unstable invariant manifolds. The following results were proved in [7].

THEOREM 4.13. *Let the assumptions of Theorem* 4.6 *hold.*

(i) *Assume that $\sigma^+(A) \neq \emptyset$ has positive distance from the imaginary axis, and fix $\omega \in (0, \min\{\mathrm{Re}\,\lambda\colon \lambda \in \sigma_+(A)\})$. Then there exist $R_0, r_0 > 0$ and a Lipschitz continuous function*

$$\varphi : B(0, r_0) \subset P^+\big(C_b(\overline{\Omega})\big) = P^+\big(D(A_\theta)\big) \to \big(I - P^+\big)\big(C^{2+\theta}(\overline{\Omega})\big),$$

differentiable at 0 *with $\varphi'(0) = 0$, such that for every u_0 belonging to the graph of φ, problem* (4.1) *has a unique backward solution v such that $\tilde{v}$ defined by $\tilde{v}(t,\xi) = \mathrm{e}^{-\omega t} v(t,\xi)$ belongs to $C^{1+\theta/2,2+\theta}((-\infty, 0] \times \overline{\Omega})$ and satisfies*

$$\|\tilde{v}\|_{C^{1+\theta/2,2+\theta}((-\infty,0]\times\overline{\Omega})} \leqslant R_0. \tag{4.18}$$

Moreover, for every $\omega' \in (0, \min\{\mathrm{Re}\,\lambda\colon \lambda \in \sigma^+(A)\})$, we have $(t,\xi) \to \mathrm{e}^{-\omega' t} v(t,\xi) \in C^{1+\theta/2,2+\theta}((-\infty, 0] \times \overline{\Omega}))$. Conversely, if problem (4.1) *has a backward solution v which satisfies* (4.18) *and $\|P^+ v(0,\cdot)\| \leqslant r_0$, then $v(0,\cdot) \in \mathrm{graph}\,\varphi$.*

(ii) *Assume that $\sigma^-(A)$ has positive distance from the imaginary axis, and fix $\omega \in (0, -\max\{\mathrm{Re}\,\lambda\colon \lambda \in \sigma^-(A)\})$. Then there exist R_1, $r_1 > 0$ and a Lipschitz continuous function*

$$\psi : B(0, r_1) \subset P^-\big(D(A_\theta)\big) \to \big(I - P^-\big)\big(C^{2+\theta}(\overline{\Omega})\big),$$

differentiable at 0 *with $\psi'(0) = 0$, such that for every u_0 belonging to the graph of ψ, problem* (4.1) *has a unique solution w such that $\tilde{w}$ defined by $\tilde{w}(t,\xi) = \mathrm{e}^{\omega t} w(t,\xi)$ belongs to $C^{1+\theta/2,2+\theta}([0,\infty) \times \overline{\Omega})$ and*

$$\|\tilde{w}\|_{C^{1+\theta/2,2+\theta}([0,\infty)\times\overline{\Omega})} \leqslant R_1. \tag{4.19}$$

Moreover, for every $\omega' \in (0, -\max\{\mathrm{Re}\,\lambda\colon \lambda \in \sigma^-(A)\})$, we have $(t,\xi) \to \mathrm{e}^{\omega' t} w(t,\xi) \in C^{1+\theta/2,2+\theta}([0,\infty) \times \overline{\Omega})$. Conversely, if problem (4.1) *has a forward solution w which satisfies* (4.19) *and $\|P^- w(0,\cdot)\|_{C^{2+\theta}(\overline{\Omega})} \leqslant r_1$, then $w(0,\cdot) \in \mathrm{graph}\,\psi$.*

The graph of φ is called local unstable manifold. The graph of ψ is called local stable manifold.

In the case where the operator A is hyperbolic, i.e., (2.11) holds, both the local stable manifold and the local unstable manifold exist, and Theorem 4.13 is a saddle point theorem.

What is missing up to now is a center manifold theory for problems with fully nonlinear boundary condition. Boundary conditions of the type $\mathcal{B}u = G(u)$ with nonlinear G may be treated; see [35]. But existence of a center manifold in the case where G depends nonlinearly on the gradient is still an open problem.

5. The fully nonlinear approach to free boundary problems

A class of free boundary problems of parabolic type may be reduced to abstract evolution equations of the type treated in Section 2, or to evolution equations in Hölder spaces of the type treated in Sections 3 and 4.

The prototypes of these problems are (1.2) and (1.3). The different physical nature of these problems is reflected in their different mathematical nature. Equation (1.3) is a Stefan-type problem, where the velocity of the free boundary is explicit, while in (1.2) it is implicit.

However, the initial step to reduce the free boundary problems to fixed boundary ones by natural changes of coordinates that involve one of the unknowns is the same for both models. After that, the procedures are different: we eliminate the free boundary and arrive at a final problem of the type (4.1) for (1.2), we eliminate u and we arrive at a final problem for the free boundary of the type (2.1) for (1.3).

Equations (1.2) and (1.3) are the simplest significant examples of a wide class of parabolic free boundary problems, that can be studied with the same methods. We mention the papers [14,15,19,20] for problems of the type (1.3), arising in several fields, such as flow of viscous fluids through porous media, the injection molding process, diblock copolymer melts; and [6–10,31] for problems of the type (1.2), arising in combustion theory.

5.1. *Hele–Shaw models*

Let $\Omega \subset \mathbb{R}^N$ be a bounded open set with smooth boundary, consisting of two disjoint nonempty parts J and Γ. For each t, the free boundary Γ_t will be sought as the range of an unknown function $s(t,\cdot) \in h^{2+\alpha}(\Gamma;\mathbb{R}^N)$ with small C^1 norm, in such a way that the mapping $\xi \mapsto \xi + s(t,\xi)\nu(\xi)$ is an $h^{2+\alpha}$ diffeomorphism between Γ and Γ_t, and Γ_t and J are disjoint. Here and in what follows we denote by $\nu = \nu(\xi)$ the exterior normal vector to $\partial\Omega$ at $\xi \in \partial\Omega$. Ω_t will be the open set diffeomorphic to Ω with boundary $J \cup \Gamma_t$.

Problem (1.3) is associated with an initial condition

$$s(0,\xi) = s_0(\xi), \quad \xi \in \Gamma. \tag{5.1}$$

Now we transform the free boundary problem (1.3) into a fixed boundary one.

For $a > 0$, we define a map

$$X : \Gamma \times [-a,a] \to \mathbb{R}^N, \qquad X(\xi',r) = \xi' + r\nu(\xi'). \tag{5.2}$$

If a is sufficiently small, then (5.2) is a diffeomorphism to a compact neighborhood $\mathcal{R}$ of Γ. In $\mathcal{R}$ every ξ can be written in a unique way as $\xi = X(\xi',r)$ with $\xi' \in \Gamma$ and

$r \in [-a, a]$. So, $\xi' = \xi'(\xi)$ is the nearest point to ξ in Γ, and $r = r(\xi)$ is the signed distance from ξ to Γ.

We will look for Ω_t close to Ω in some time interval I in the sense that its free boundary Γ_t will be given by

$$\Gamma_t = \{x = \xi' + s(t, \xi')\nu(\xi'), \xi' \in \Gamma\}, \tag{5.3}$$

where $s : \Gamma \times I \to [-a, a]$ is a smooth function which is one of the unknowns of the problem. In other words, Γ_t is the zero level set of the function

$$\mathcal{R} \mapsto \mathbb{R}, \qquad \xi \mapsto N(t, \xi) = r(\xi) - s(t, \xi'(\xi)).$$

It follows that the exterior normal vector at Γ_t is given by $\nu = DN/|DN|$, and the normal velocity V at $x \in \Gamma_t$ is

$$V(t, x) = -\frac{\partial/\partial t N(t, x)}{DN(t, x)} = \frac{\partial s(t, \xi'(x))/\partial t N(t, x)}{DN(t, x)}.$$

The equation $V = -\partial u/\partial \nu$ in (1.3) may be rewritten as

$$N_t(t, x) = \langle Du(t, x), DN(t, x)\rangle, \quad t > 0, x \in \Gamma_t.$$

It will be convenient to extend the vector field

$$\Phi(t, \xi) = s(t, \xi)\nu(\xi), \quad \xi \in \Gamma, \tag{5.4}$$

to the whole of $\mathbb{R}^N$, by setting

$$\Phi(t, \xi) = \begin{cases} \alpha(r)s(t, \xi')\nu(\xi') & \text{if } \xi \in \mathcal{R}, \\ 0 & \text{otherwise}, \end{cases} \tag{5.5}$$

where $r = r(\xi)$, $\xi' = \xi'(\xi)$, and $\alpha : \mathbb{R} \mapsto [0, 1]$ is a smooth mollifier which is equal to 1 near 0 and has compact support in $(-a, a)$.

The extension Φ is used now to transform (1.3) to a problem on the fixed domain Ω. We define the coordinate transformation

$$x = \Theta(t, \xi) = \xi + \Phi(t, \xi). \tag{5.6}$$

Note that $\Theta(t, \cdot)$ differs from the identity only in a small neighborhood of Γ, and it maps Ω onto Ω_t.

Denoting by $\tilde{u}$ the unknown u in the new variables, i.e., $\tilde{u}(t, \xi) = u(t, \xi + \Phi(t, \xi))$, the couple (s, u) satisfies (1.3)–(5.1) if and only if $(s, \tilde{u})$ satisfies

$$\begin{cases} \mathcal{A}\tilde{u} = 0, & t > 0, \xi \in \overline{\Omega}, \\ \tilde{u} = 0, \quad s_t + \mathcal{B}\tilde{u} = 0, & t > 0, \xi \in \Gamma, \\ \frac{\partial}{\partial \nu}\tilde{u} = b, & t > 0, \xi \in J, \\ s(0, \cdot) = s_0, & \xi \in \Gamma, \end{cases} \tag{5.7}$$

where $\mathcal{A}$ is the Laplacian expressed in the new variables, i.e., setting $\Xi(x) = \Theta(t,\cdot)^{-1}(x)$,

$$\mathcal{A} = \sum_{i,j,h=1}^{N} \frac{\partial \Xi_j}{\partial x_i}(\Theta(t,\xi)) \frac{\partial \Xi_h}{\partial x_i}(\Theta(t,\xi)) D_{jh} + \sum_{i,j=1}^{N} \frac{\partial^2 \Xi_j}{\partial x_i^2}(\Theta(t,\xi)) D_j, \quad (5.8)$$

and $\mathcal{B}$ is the normal derivative expressed in the new variables. Since

$$n(t, \xi + \Phi) = \frac{(I +^t D\Phi)^{-1} \nu(\xi)}{|(I +^t D\Phi)^{-1} \nu(\xi)|}$$

and

$$Du(t, \xi + \Phi) = \left(I +^t D\Phi\right)^{-1} D\tilde{u}(t, \xi),$$

we get

$$\mathcal{B}v = \frac{\langle (I +^t D\Phi)^{-1} \nu, (I +^t D\Phi)^{-1} Dv \rangle}{|(I +^t D\Phi)^{-1} \nu|}. \quad (5.9)$$

Note that $\mathcal{A} = \mathcal{A}(s)$, $\mathcal{B} = \mathcal{B}(s)$ depend on s through Φ.

Now we are able to decouple the system (5.7), expressing $\tilde{u}$ in terms of s.

If the function b is smooth enough ($b \in h^{1+\alpha}(\Gamma)$), for each $\sigma \in h^{2+\alpha}(\Gamma)$ with small C^1 norm, there is a unique $v \in h^{2+\alpha}(\overline{\Omega})$ such that

$$\begin{cases} \mathcal{A}(\sigma)v = 0, & \xi \in \overline{\Omega}, \\ v(\xi) = 0, & \xi \in \Gamma, \\ \frac{\partial}{\partial \nu} v = b, & \xi \in J. \end{cases} \quad (5.10)$$

This is because $\mathcal{A}(\sigma)$ is a second-order elliptic operator with h^α coefficients and without zero-order terms.

We denote by $\mathcal{F}$ the function

$$\mathcal{F}(\sigma) = \mathcal{B}(\sigma)v, \quad (5.11)$$

where v is the solution to (5.10), and we rewrite (5.7) as a final problem for s,

$$\begin{cases} s_t(t,\xi) + \mathcal{F}\big(s(t,\cdot)\big)(t,\xi) = 0, & t \geqslant 0, \xi \in \Gamma, \\ s(0,\xi) = s_0(\xi), & \xi \in \Gamma. \end{cases} \quad (5.12)$$

This problem will be seen as an evolution equation in the space $E_0 = h^{1+\alpha}(\Gamma)$, for which the assumptions of Theorem 2.7 are satisfied. Indeed, the following statements have been proved in [18].

THEOREM 5.1. *If $a > 0$ is small enough for each $\beta \in (0,1)$, the function $\mathcal{F}: \mathcal{V}_\beta := \{s \in h^{2+\beta}(\Gamma): \|h\|_{C^1(\Gamma)} < a\} \mapsto h^{1+\beta}(\Gamma)$ is smooth.*

Assume that $b(x) \geqslant 0$ for each $x \in J$, and that $b \not\equiv 0$. Then, for each $s \in \mathcal{V}_\beta$, the operator $A_\beta = -\mathcal{F}'(s): h^{2+\beta}(\Gamma) \mapsto h^{1+\beta}(\Gamma)$ is a sectorial operator in $h^{1+\beta}(\Gamma)$.

Theorem 5.1 allows to apply the theory of Section 2.

We recall that the little-Hölder spaces $h^\beta(\Gamma)$ are stable by continuous interpolation, in the sense that for nonintegers $\beta_1 < \beta_2$ we have

$$\left(h^{\beta_1}(\Gamma), h^{\beta_2}(\Gamma)\right)_\theta = h^{\beta_1+\theta(\beta_2-\beta_1)}(\Gamma)$$

for each $\theta \in (0,1)$ such that $\beta_1 + \theta(\beta_2 - \beta_1)$ is not integer. In particular, for $0 < \beta < \alpha < 1$ we have

$$h^{1+\alpha}(\Gamma) = \left(h^{1+\beta}(\Gamma), h^{2+\beta}(\Gamma)\right)_\theta \tag{5.13}$$

with $\theta = \alpha - \beta$.

We fix $0 < \beta < \alpha < 1$ and we set $E_0 = h^{1+\alpha}(\Gamma)$, $E_1 = h^{2+\alpha}(\Gamma)$. By Theorem 5.1, the function $F = -\mathcal{F}$ is defined and smooth in an open set $\mathcal{O} = \mathcal{V}_\alpha \subset E_1$ with values in E_0, for each $s \in \mathcal{O}$, $F'(s) = A_\alpha$ is sectorial, and it may be seen as the part in E_0 of the sectorial operator $A_\beta: h^{2+\beta}(\Gamma) \mapsto h^{1+\beta}(\Gamma)$. By (5.13) we have $D_{A_\beta}(\theta) = h^{1+\alpha}(\Gamma) = E_0$, and by Theorem 5.1 we have $D_{A_\beta}(\theta+1) = E_1$, provided $\theta = \alpha - \beta$. Therefore, assumption (H3) is satisfied and Theorem 2.7 is applicable. We arrive at a final well-posedness theorem; see [18].

THEOREM 5.2. *Let $\Omega \subset \mathbb{R}^N$ be a bounded smooth domain with boundary $J \cup \Gamma$, $J \neq \emptyset$ interior to Γ and with positive distance from Γ. Let $b \in h^{2+\alpha}(J)$ $(0 < \alpha < 1)$ be such that $b(x) \geqslant 0$ for each $x \in J$, and $b \not\equiv 0$.*

Then for each $s_0 \in h^{2+\alpha}(\Gamma)$ with small C^1 norm there are $T > 0$ and a classical solution (s, u) of (1.3)–(5.1) in $[0, T]$. The function s is such that $t \mapsto s(t,\cdot)$ is in $C([0,T]; h^{2+\alpha}(\Gamma)) \cap C^1([0,T]; h^{1+\alpha}(\Gamma))$, and $\|s(t,\cdot)\|_{C^1(\Gamma)} \leqslant a$. Denoting by Ω_t the bounded open set with boundary $J \cup \mathrm{Range}\, s(t,\cdot)$, u is continuous in $\{(t,x): 0 \leqslant t \leqslant T,$ $x \in \overline{\Omega_t}\}$ and $u(t,\cdot)$ belongs to $h^{2+\alpha}(\overline{\Omega_t})$ for each $t \in [0,T]$. The couple (s,u) is the unique solution to (1.3)–(5.1) enjoying these regularity properties.

Problem (1.3) is the simplest example of a class of free boundary problems that can be treated similarly. Among them we quote one- and two-phase Hele–Shaw models with surface tension, also called Mullins–Sekerka models. See [14,15,18–20].

Let us describe the two-phase problem. Ω is again a bounded open set in $\mathbb{R}^N$ with smooth boundary $\partial\Omega$. For $t \geqslant 0$, Γ_t is a compact connected hypersurface which is the boundary of an open set $\Omega_t \subset \Omega$. The normal velocity of Γ_t and its mean curvature are denoted by $V(t,\cdot)$ and $\kappa(t,\cdot)$, respectively. Again, V is taken to be positive for expanding hypersurfaces, and κ is positive for uniformly convex hypersurfaces. $\Omega_1(t)$ and $\Omega_2(t)$ are the open subsets of Ω separated by Γ_t, $\Omega_1(t)$ being the interior region; $n(t,\cdot)$ is the unit exterior vector normal to $\partial\Omega_1(t)$. We consider the system

$$\begin{cases} \Delta u = 0, & t>0,\ x\in\Omega_1(t)\cup\Omega_2(t),\\ u=\kappa, \quad V=-\left[\frac{\partial u}{\partial n}\right], & t>0,\ x\in\Gamma_t,\\ \frac{\partial u}{\partial n}=0, & t>0,\ x\in\partial\Omega, \end{cases} \tag{5.14}$$

where the brackets denote the jump across the free boundary: $[\partial u/\partial n] = \partial u_1/\partial n - \partial u_2/\partial n$, u_j being the restrictions of u on $\Omega_j(t)$. As before, the unknowns are the free boundary Γ_t for $t>0$ and the function $u(t,x)$ for $t>0$, $x\in\overline{\Omega}$, while the initial hypersurface Γ_0 is given.

The procedure described for the Hele–Shaw flow gives local existence and uniqueness of a regular solution to (5.14) for each regular (precisely: $h^{2+\alpha}$) initial hypersurface Γ_0.

Any smooth hypersurface $\Gamma\subset\Omega$ fixed, for initial data close to Γ we look for Γ_t as the graph of an unknown function $s(t,\cdot)$ defined on Γ, and we proceed as before. We eliminate the unknown u, expressing it in terms of s. More precisely, $u(t,\cdot)$ has to be the solution v to

$$\begin{cases} \mathcal{A}(\sigma)v = 0, & \xi\in\Omega_1\cup\Omega_2,\\ v(\xi)=K(\sigma), & \xi\in\Gamma,\\ \frac{\partial}{\partial\nu}v=0, & \xi\in\partial\Omega, \end{cases} \tag{5.15}$$

with $\sigma = s(t,\cdot)$, the fixed open sets Ω_1 and Ω_2 are the images $\Omega_1(t)$ and $\Omega_2(t)$ under the change of coordinates, and $K(\sigma)$ is the transformed mean curvature operator after the change of coordinates. The jump condition $V=-[\partial u/\partial n]$ is transformed to $s_t+\mathcal{B}(s)v=0$, where now $\mathcal{B}(s)$ is the transformed jump of the normal derivative across Γ_t. So we arrive at a final equation for s of the type (5.12), where now $\mathcal{F}$ is a third-order nonlocal operator with quasilinear structure. This is because the curvature depends on s through its derivatives up to the second-order, in a quasilinear way since it depends linearly on the second-order derivatives of s. The character of (5.12) is still parabolic and $\mathcal{F}$ is still smooth; more precisely, it was proved in [19] that if $a>0$ is small enough, for each $\beta\in(0,1)$ the function $\mathcal{F}:\mathcal{U}_\beta := \{s\in h^{3+\beta}(\Gamma)\colon \|h\|_{C^1(\Gamma)}<a\}\mapsto h^\beta(\Gamma)$ is C^∞, and the operator $A_\beta = -\mathcal{F}'(s): h^{3+\beta}(\Gamma)\mapsto h^\beta(\Gamma)$ is a sectorial operator in $h^{1+\beta}(\Gamma)$.

So, as far as local existence, uniqueness and regularity are concerned, we are not forced to use Theorem 2.7 but we can use other theories of abstract quasilinear evolution equations in Banach spaces, that allow less regular initial data. In the paper [19] the theory developed in [2,3] was used to arrive at a final result of existence and uniqueness of a solution to (5.12), $s\in C^\infty((0,T]\times\Gamma)$, such that $t\mapsto s(t,\cdot)\in C([0,T]; h^{2+\beta}(\Gamma))\cap C((0,T]; h^{3+\beta}(\Gamma))$, for each initial datum $s_0\in h^{2+\beta+\varepsilon}(\Gamma)$, $\varepsilon>0$, with small C^1 norm. The number $T>0$ depends on s_0. If in addition $s_0\in h^{3+\beta}(\Gamma)$, then $t\mapsto s(t,\cdot)\in C([0,T]; h^{3+\beta}(\Gamma))$. (The last statement comes by applying Theorem 2.7 with $E_0 = h^\beta(\Gamma)$, $E_1 = h^{3+\beta}(\Gamma)$.)

Returning to (5.14) we get a unique regular local solution (Γ_t, u).

It is easy to see that (5.14) admits spheres as stationary solutions, and it is of interest to study their stability. Any sphere $S\subset\Omega$ fixed, we need to know some spectral properties of the linearized operator $A = -\mathcal{F}'(0) : h^{3+\beta}(\mathcal{S}) \mapsto h^{\beta}(\mathcal{S})$. It was proved in [20] that the spectrum of A consists of a sequence of negative eigenvalues plus the semisimple isolated eigenvalue 0 with multiplicity $N+1$. Therefore, this is a critical case of stability. The center manifold theory of Section 2, with $E_0 = h^{\beta}(\Gamma)$, $E_1 = h^{3+\beta}(\Gamma)$, may be applied, and it gives the existence of an $(N+1)$-dimensional locally invariant manifold $\mathcal{M} \subset h^{3+\beta}(\mathcal{S})$ which attracts all the small orbits. Going further in the analysis, it was proved in [20] that the center manifold itself consists of spheres, and for each small initial datum the solution exists at large and converges to one of these spheres exponentially as $t \to \infty$. The convergence is in the $h^{3+\beta}(\mathcal{S})$ norm, even for initial data in $h^{2+\beta+\varepsilon}(\mathcal{S})$. See [20] for the details.

5.2. *Models from combustion theory*

Perhaps surprisingly, well-posedness of the Cauchy problem for (1.2) is still an open problem in dimension $N \geqslant 2$. The Cauchy problem consists of (1.2) supplemented by an initial condition,

$$u(0,x) = u_0(x), \quad x \in \overline{\Omega}_0, \tag{5.16}$$

where Ω_0 is a given open set in $\mathbb{R}^N$. There are results of existence of weak solutions without uniqueness [11], of existence and uniqueness of regular classical solutions with loss of regularity with respect to the inital data [6], of local well-posedness for initial data close to special solutions such as traveling waves or self-similar solutions [7] and for special geometries [4,21,29]. But none of them can be considered a standard well-posedness result.

We describe here the approach leading to the fully nonlinear evolution equations discussed in Section 3. The change of coordinates used here is the same as in the previous section, taking as reference set Ω the initial set Ω_0. Let us assume that Ω_0 is a nonempty open set in $\mathbb{R}^N$ with $C^{3+\alpha}$ boundary Γ. The boundary Γ_t of Ω_t is sought again as the range of an unknown function $s(t,\cdot) : \Gamma \mapsto \mathbb{R}^N$. The coordinate transformation (5.6) transforms Ω_t into the fixed domain Ω_0 and it leads to a Cauchy problem for the couple $(s, \tilde{u})$ where $\tilde{u}(t,\xi) = u(t,x)$ is again the function u in the new coordinates:

$$\begin{cases} \tilde{u}_t - \langle D\tilde{u}, (I + {}^t D\Phi)^{-1}\Phi_t \rangle & \\ \quad = \mathcal{A}(s)\tilde{u}, & t > 0,\ x \in \overline{\Omega}_0, \\ \tilde{u} = 0, \quad \mathcal{B}(s)\tilde{u} = -1, & t > 0,\ x \in \Gamma_0, \\ s(0,\cdot) = 0, \quad \tilde{u}(0,\cdot) = u_0, & x \in \Gamma_0, \\ \tilde{u}(0,\cdot) = u_0, & x \in \overline{\Omega}_0, \end{cases} \tag{5.17}$$

where, as before, $\mathcal{A}(s)$ is the Laplacian in the new coordinates and $\mathcal{B}(s)$ is the normal derivative in the new coordinates, see (5.8) and (5.9).

System (5.17) still has to be decoupled. Instead of proceeding like in problem (1.3), we introduce a new unknown w by splitting $\tilde{u}$ as

$$\tilde{u}(t,\xi) = u_0(\xi) + \langle Du_0(\xi), \Phi(t,\xi)\rangle + w(t,\xi). \tag{5.18}$$

At $t=0$ we have $\tilde{u}(0,\xi) = u_0(\xi)$, $\Phi(0,\cdot) \equiv 0$, so that

$$w(0,\xi) = 0, \quad \xi \in \overline{\Omega}_0. \tag{5.19}$$

Equation (5.18) allows to get s in terms of w thanks to the boundary condition $u=0$ at Γ_t, which gives

$$s(t,\xi) = w(t,\xi), \quad t \geqslant 0, \xi \in \Gamma_0, \tag{5.20}$$

so that

$$\Phi(t,\xi) = w(t,\xi')\tilde{\nu}(\xi), \quad t \geqslant 0, \xi \in \overline{\Omega}_0, \tag{5.21}$$

where $\tilde{\nu}(\xi)$ is the extension of the normal vector field in formula (5.5): $\tilde{\nu}(\xi) = \alpha(r)\nu(\xi')$ if $\xi \in \mathcal{R}$ and $\tilde{\nu}(\xi) = 0$ otherwise. Replacing (5.21) in (5.17) we get

$$w_t = \mathcal{F}_1(\xi, w, Dw, D^2w) + \mathcal{F}_2(\xi, w, Dw)s_t, \quad t \geqslant 0, \xi \in \overline{\Omega}_0, \tag{5.22}$$

where $\mathcal{F}_1$, $\mathcal{F}_2$ are obtained respectively from

$$\mathcal{A}(s)\big(u_0 + \langle Du_0, \Phi\rangle + w\big) \tag{5.23}$$

and from

$$-\big\langle Du_0 - (I + D\Phi)^{-1} D\big(u_0 + \langle Du_0, \Phi\rangle + w\big), \tilde{\nu}\big\rangle, \tag{5.24}$$

replacing $\Phi = w(t,\xi')\tilde{\nu}(\xi)$.

Equation (5.22) still contains s_t, that we eliminate using again the identity $s = w$ at the boundary which gives $s_t = w_t$. Replacing in (5.22) for $\xi \in \Gamma_0$, we get

$$s_t\big(1 - \mathcal{F}_2(\xi, w, Dw)\big) = \mathcal{F}_1\big(\xi, w, Dw, D^2w\big), \quad t \geqslant 0, \xi \in \Gamma_0.$$

At $t=0$ we have $w \equiv 0$, and $\mathcal{F}_2$ vanishes at $(\xi, 0, 0)$, so that, at least for t small, $\mathcal{F}_2(\cdot, w(t,\cdot), Dw(t,\cdot))$ is different from 1 and we get s_t in terms of w,

$$s_t = \mathcal{F}_3\big(\xi, w, Dw, D^2w\big) = \frac{\mathcal{F}_1(\xi, w, Dw, D^2w)}{1 - \mathcal{F}_2(\xi, w, Dw)}, \quad t \geqslant 0, \xi \in \Gamma_0, \tag{5.25}$$

which, replaced in (5.22), gives the final equation for w,

$$w_t = \mathcal{F}(w)(\xi), \quad t \geqslant 0, \xi \in \overline{\Omega}_0, \tag{5.26}$$

where

$$\mathcal{F}(w)(\xi) = \mathcal{F}_1\big(\xi, w, Dw, D^2w\big) + \mathcal{F}_2(\xi, w, Dw)\mathcal{F}_3\big(\xi, w, Dw, D^2w\big). \quad (5.27)$$

Note that $\mathcal{F}(0)(\xi) = \Delta u_0(\xi)$ and $\mathcal{F}(w)(\xi) = \Delta w + \Delta u_0(\xi)$ if ξ is far from the boundary Γ_0.

The function $\mathcal{F}(v)$ is defined for $v \in C^2(\overline{\Omega}_0)$ with small C^1 norm; precisely, it is defined for $v \in C^2(\overline{\Omega}_0)$ such that $1 - \mathcal{F}_2(\cdot, v(\cdot), Dv(\cdot)) \neq 0$. From formulas (5.20) and (5.23)–(5.25) we see that $\mathcal{F}(v)(\xi)$ depends smoothly on v, Dv, D^2v and their traces at the boundary; therefore the function $v \mapsto \mathcal{F}(v)$ is continuously differentiable from $\mathcal{O} = \{v \in C^2(\overline{\Omega}_0)\colon \|v\|_{C^1(\overline{\Omega}_0)} \leqslant r\}$ to $C(\overline{\Omega}_0)$, and from $\mathcal{O}_\alpha = \{v \in C^{2+\alpha}(\overline{\Omega}_0)\colon \|v\|_{C^1(\overline{\Omega}_0)} \leqslant r\}$ to $C^\alpha(\overline{\Omega}_0)$ if r is small.

The boundary condition for w comes from the boundary condition $\partial u/\partial n = -1$ in (1.2). Using (5.9) we get

$$\big\langle \big(I +^t D\Phi\big)^{-1}\nu, \big(I +^t D\Phi\big)^{-1} D\big(u_0 + \langle Du_0, \Phi\rangle + w\big)\big\rangle + \big|\big(I +^t D\Phi\big)^{-1}\nu\big| = 0, \quad (5.28)$$

which gives

$$\mathcal{G}\big(\xi, w(t,\xi), Dw(t,\xi)\big) = 0, \quad t \geqslant 0, \xi \in \Gamma_0, \quad (5.29)$$

when we replace $\Phi = w(t, \xi')\tilde{\nu}(\xi)$ in (5.28). The function $\mathcal{G}(\xi, u, p)$ is smooth with respect to u and p_i, $i = 1, \ldots, N$, and its derivatives are continuous in (ξ, u, p) and $C^{1/2+\alpha/2}$ in ξ. It follows that $v \mapsto \mathcal{G}(\cdot, v, Dv)$ is smooth from a neighborhood of 0 in $C^1(\overline{\Omega}_0)$ to $C(\Gamma_0)$, and from a neighborhood of 0 in $C^{1+\alpha}(\overline{\Omega}_0)$ to $C^\alpha(\Gamma_0)$. Moreover, $\mathcal{G}(\xi, 0, 0) = 0$.

Concerning the linear parts of $\mathcal{F}$ and $\mathcal{G}$ near 0, the following lemma was proved in [6].

LEMMA 5.3. *$\mathcal{F}_v(0)$ is the sum of the Laplacian plus a nonlocal differential operator of order* 1. *Moreover,*

$$\mathcal{G}_v(0)w = \mathcal{B}w := \frac{\partial w}{\partial \nu} + \frac{\partial^2 u_0}{\partial \nu^2} w.$$

The final problem for the only unknown w may be rewritten in the form discussed in Section 3:

$$\begin{cases} w_t = \mathcal{A}w + F(w), & t \geqslant 0, \xi \in \overline{\Omega}_0, \\ \mathcal{B}w = G(w), & t \geqslant 0, \xi \in \Gamma_0, \\ w(0,\cdot) = 0, & \xi \in \overline{\Omega}_0. \end{cases} \quad (5.30)$$

The difference between (5.30) and (1.1) is that the linear operator $\mathcal{A}$ and the nonlinearities F, G contain nonlocal terms. The nonlocal part of $\mathcal{A}$ concerns only first-order derivatives, so that it may be considered as a nonimportant perturbation. F depends nonlocally also on the second-order derivatives of w, but it is (at least) quadratic near $w = 0$.

The arguments used in the proof of Theorem 3.2 work also for (5.30), and give a local existence and uniqueness result for w. Precisely, there is $R_0 > 0$ such that for every $R \geqslant R_0$ and for every sufficiently small $T > 0$ problem (5.30) has a unique solution in the ball $B(0,R) \subset C^{1+\alpha/2,2+\alpha}([0,T] \times \overline{\Omega}_0)$. For the details of the proof see [6].

Note that we cannot use Theorem 4.1 to get local existence for w because $F(0) \neq 0$.

Now we return to the original problem (1.2). Recalling that $s(t,\xi) = w(t,\xi)$ for each $t \in [0,T]$, $\xi \in \partial\Omega$, we can define Γ_t. Of course, s has the same regularity of w, i.e., it is in $C^{1+\alpha/2,2+\alpha}([0,T] \times \Gamma_0)$. Then we define $\tilde{u}$ through (5.18), where Φ is given by (5.5). Again, $\tilde{u}$ has the same regularity of w. As a last step we define u through the change of coordinates, $u(t,x) = \tilde{u}(t,\xi)$, where $x = \xi + \Phi(t,\xi)$. This leads to loss of regularity: starting with initial data in $C^{3+\alpha}$ we get a local solution with $C^{2+\alpha}$ space regularity. The final result (see [6]) is the following.

THEOREM 5.4. *Let $\Omega_0 \subset \mathbb{R}^N$ be a nonempty bounded open set with $C^{3+\alpha}$ boundary Γ_0, and let $u_0 \in C^{3+\alpha}(\overline{\Omega}_0)$ satisfy the compatibility conditions $u_0 = 0$, $\partial u_0/\partial n = -1$ at Γ_0. Then there is $T > 0$ such that problem* (1.2) *has a solution (Ω_t, u) such that the $(N+1)$-dimensional hypersurface $S = \{(t,x)\colon 0 \leqslant t \leqslant T, x \in \Gamma_t\}$ and each $\Gamma_t = \partial\Omega_t$ are of class $C^{1+\alpha/2,2+\alpha}$, and the function $u : \{(t,x); 0 \leqslant t \leqslant \delta, x \in \overline{\Omega}_t\} \mapsto \mathbb{R}$ is of class $C^{1+\alpha/2,2+\alpha}$.*

If in addition Γ_0 and u_0 are in $C^{4+\alpha}$, and the further compatibility condition $\mathcal{B}(\Delta u_0) = 0$ at Γ_0 holds, then S and each Γ_t are of class $C^{3/2+\alpha/2,3+\alpha}$, and the function u is of class $C^{3/2+\alpha/2,3+\alpha}$. Moreover, the couple (Ω_t, u) is the unique solution with such regularity properties.

If the initial data are close to the initial datum (Ω, U) of a given regular solution, the same method gives existence and uniqueness of a local classical solution without loss of regularity. Moreover, we can go further in the investigation of the stability properties of the established solution.

The free boundary problem is transformed into a fixed boundary problem in Ω by the change of coordinates (5.6), so that the unknown s is the signed distance from $\Gamma = \partial\Omega$. The splitting of (5.18) is replaced by

$$\tilde{u}(t,\xi) = U(\xi) + \langle DU(\xi), \Phi(t,\xi)\rangle + w(t,\xi), \quad \xi \in \overline{\Omega}. \tag{5.31}$$

This gives again $s(t,\xi) = w(t,\xi)$ for $\xi \in \Gamma$. The final problem for w has initial datum $w_0 = \tilde{u}_0 - U - \langle DU, \Phi(0,\xi)\rangle$ which does not vanish, but is small.

Let us show how to apply the results of Section 4. The simplest situation would be to have initial data close to stationary solutions, but (1.2) has no bounded stationary solutions.

So we consider the case where (Ω_0, u_0) is close to the initial datum of a self-similar solution. Existence and properties of self-similar solutions have been discussed in [11]; they are solutions of the type (Ω_t, u) where

$$u(t,x) = (T-t)^{\alpha} f\left(\frac{|x|}{(T-t)^{\beta}}\right), \qquad \Omega_t = \{|x| < r(T-t)^{\beta}\}, \quad 0 \leqslant t < T,$$

with $T, \alpha, \beta, r > 0$.

It is easy to see that there exist self-similar solutions only for $\alpha = \beta = 1/2$, and that the function $g(x) = f(|x|)$ has to be an eigenfunction of an Ornstein–Uhlenbeck type operator in the ball $B(0, r)$,

$$\Delta g - \frac{1}{2}\langle x, Dg(x)\rangle + \frac{1}{2}g = 0,$$

with two boundary conditions, $g = 0$, $\partial g/\partial \nu = -1$. In other words, f has to solve

$$\begin{cases} f''(\eta) + \frac{N-1}{\eta} f'(\eta) + \frac{1}{2} f(\eta) = \frac{1}{2}\eta f'(\eta) & \text{for } 0 < \eta \leqslant r, \\ f'(0) = f(r) = 0, & f'(r) = -1, \end{cases} \tag{5.32}$$

which looks overdetermined, but it is not, because r is also an unknown. In [11] it is proved that there exist a unique $r > 0$ and a unique C^2 function $f : [0, r] \mapsto \mathbb{R}$ satisfying (5.32) and such that $f(\eta) > 0$ for $0 \leqslant \eta < r$. Moreover, f is analytic.

It is convenient now to transform the problem to self-similar variables

$$\hat{x} = \frac{x}{(T - t)^{1/2}}, \qquad \hat{t} = -\log(T - t), \tag{5.33}$$

and to set

$$\hat{u}(\hat{x}, \hat{t}) = \frac{u(x, t)}{(T - t)^{1/2}}, \qquad \widehat{\Omega}_{\hat{t}} = \{\hat{x} \colon x \in \Omega_t\}. \tag{5.34}$$

Omitting the hats, we arrive at

$$\begin{cases} u_t = \Delta u - \frac{1}{2}\langle x, Du\rangle + \frac{1}{2}u, & t > 0, x \in \Omega_t, \\ u = 0, \quad \frac{\partial u}{\partial n} = 1, & t > 0, x \in \partial\Omega_t. \end{cases} \tag{5.35}$$

The self-similar solution is transformed by (5.33) and (5.34) into a stationary solution

$$U(x) = f(|x|), \qquad \Omega = \{x \in \mathbb{R}^N;\ |x| < r\} \tag{5.36}$$

of (5.35). From now on we proceed as before: we change variables through the isomorphism (5.6), we set $\tilde{u}(t, \xi) = u(t, x) - U(x)$, we define w by the splitting of (5.19) and we arrive at a final equation for w in the fixed domain $\Omega = B(0, r)$,

$$\begin{cases} w_t = \Delta w - \frac{1}{2}\langle \xi, Dw\rangle + \frac{w}{2} + \phi(w, Dw, D^2 w), & t \geqslant 0, \xi \in \overline{\Omega}, \\ \frac{\partial w}{\partial \nu} + \left(\frac{N-1}{r} - \frac{r}{2}\right)w = \psi(w, Dw), & t \geqslant 0, \xi \in \partial\Omega, \\ w(0, \xi) = w_0(\xi), & \xi \in \overline{\Omega}, \end{cases} \tag{5.37}$$

where ϕ and ψ are smooth and quadratic near $w = 0$.

This problem fits into the theory discussed in Section 4. Theorem 4.1 implies that for every $T > 0$ and $\alpha \in (0,1)$ there are R, $\rho > 0$ such that (5.37) has a solution $w \in C^{1+\theta/2,2+\theta}([0,T] \times \overline{\Omega})$ provided $\|w_0\|_{C^{2+\theta}(\overline{\Omega})} \leqslant \rho$ and $\partial w_0/\partial \nu + ((N-1)/r - (r/2))w_0 = \psi(w_0, Dw_0)$. Moreover, u is the unique solution in $B(0,R) \subset C^{1+\theta/2,2+\theta}([0,T] \times \overline{\Omega})$.

To proceed in the analysis, we define the operator A by

$$D(A) = \left\{ v \in \bigcap_{p>1} W^{2,p}(\Omega) : \Delta v \in C(\overline{\Omega}), \mathcal{B}v = 0 \text{ on } \partial\Omega \right\},$$

$$Av = \mathcal{A}v,$$

where

$$\mathcal{A}v = \Delta v - \frac{1}{2}\langle x, Dv\rangle + \frac{1}{2}v,$$

$$\mathcal{B}v = \frac{\partial v}{\partial \nu} + \left(\frac{N-1}{r} - \frac{r}{2}\right)v.$$

It has been proved in [7] that the spectrum of A consists of the semisimple eigenvalues 1, 1/2, plus a sequence of negative eigenvalues; moreover, the eigenspace with eigenvalue 1 is one-dimensional, the eigenspace with eigenvalue 1/2 has dimension N.

The principle of linearized stability as stated in Theorem 4.6(ii) shows that the null solution of (5.37) is unstable in $C^{2+\theta}(\overline{\Omega})$, and therefore the self-similar solution of the original problem (1.2) is unstable. This is not surprising, because the original problem is invariant under translations in x and t; if we apply a small shift to (5.36), we obtain another self-similar solution which is transformed by (5.33) and (5.34) into a solution which starts close to (5.36) but moves far from it. Therefore, the local unstable manifold of (5.36) given by Theorem 4.13(i) must contain the images under (5.33) of shifts in space and time of (5.36), that are given by

$$\sqrt{1+\epsilon_2 e^t}\, U\left(\frac{x - \epsilon_1 e^{t/2}}{\sqrt{\epsilon_2 e^t + 1}}\right), \tag{5.38}$$

with $\epsilon_1 \in \mathbb{R}^N$ and $\epsilon_2 \in \mathbb{R}$. Since the local unstable manifold has to be $(N+1)$-dimensional, then it consists only of the images of (5.38) under the transformation (5.33). However, all the orbits in the unstable manifold have the same self-similar profile, so that the equilibrium (5.36) looks stable even if it is unstable. Roughly speaking, the profile itself is stable.

The above discussion is taken from [7]. A study of the stability of the (planar) traveling wave solutions to (1.2) is in the paper [8]. Stability questions for more complicated multi-dimensional free boundary equations and systems arising in combustion theory have been studied by these methods in the papers [9,10,31] and in the papers quoted therein.

References

[1] P. Acquistapace and B. Terreni, *Fully nonlinear parabolic systems*, Recent Advances in Nonlinear Elliptic and Parabolic Problems, Nancy 1988, Ph. Bénilan, M. Chipot, L. Evans and M. Pierre, eds, Pitman Res. Notes in Math. Ser., Vol. 208, Longman, Harlow (1989), 97–111.

[2] H. Amann, *Nonhomogeneous linear and quasilinear elliptic and parabolic boundary value problems*, Function Spaces, Differential Operators and Nonlinear Analysis, H.J. Schmeisser and H. Triebel, eds, Teubner, Stuttgart (1993), 9–126.

[3] H. Amann, *Linear and Quasilinear Parabolic Problems*, Vol. 1, Birkhäuser, Basel (1995); Vols 2, 3, in preparation.

[4] D. Andreucci and R. Gianni, *Classical solutions to a multidimensional free boundary problem arising in combustion theory*, Comm. Partial Differential Equations **19** (1994), 803–826.

[5] S. Angenent, *Nonlinear analytic semiflows*, Proc. Royal Soc. Edinburgh Ser. A **115** (1990), 91–107.

[6] O. Baconneau and A. Lunardi, *Smooth solutions to a class of free boundary parabolic problems*, Trans. Amer. Math. Soc. **17** (2004), 987–1005.

[7] C.-M. Brauner, J. Hulshof and A. Lunardi, *A general approach to stability in free boundary problems*, J. Differential Equations **164** (2000), 16–48.

[8] C.-M. Brauner, J. Hulshof and A. Lunardi, *A critical case of stability in a free boundary problem*, J. Evol. Equ. **1** (2001), 85–113.

[9] C.-M. Brauner and A. Lunardi, *Instabilities in a two-dimensional combustion model with free boundary*, Arch. Ration. Mech. Anal. **154** (2000), 157–182.

[10] C.-M. Brauner, A. Lunardi and Cl. Schmidt-Lainé, *Stability analysis in a multidimensional interface problem*, Nonlinear Anal. **44** (2001), 263–280.

[11] L.A. Caffarelli and J.L. Vazquez, *A free boundary problem for the heat equation arising in flame propagation*, Trans. Amer. Math. Soc. **347** (1995), 411–441.

[12] G. Da Prato and P. Grisvard, *Equations d'évolution abstraites non linéaires de type parabolique*, Ann. Mat. Pura Appl. (4) **120** (1979), 329–396.

[13] G. Da Prato and A. Lunardi, *Stability, instability, and center manifold theorem for fully nonlinear autonomous parabolic equations in Banach space*, Arch. Ration. Mech. Anal. **101** (1988), 115–141.

[14] J. Escher and U.F. Mayer, *Loss of convexity for a modified Mullins–Sekerka model arising in diblock copolymer melts*, Arch. Math. **77** (2001), 434–448.

[15] J. Escher and Y. Nishiura, *Classical solutions for a modified Mullins–Sekerka model arising in diblock copolymer melts*, Hokkaido Math. J. **31** (2002), 137–149.

[16] J. Escher and G. Simonett, *Maximal regularity for a free boundary problem*, NoDEA Nonlinear Differential Equations Appl. **2** (1995), 463–510.

[17] J. Escher and G. Simonett, *Analyticity of the interface in a free boundary problem*, Math. Ann. **305** (1996), 439–459.

[18] J. Escher and G. Simonett, *Classical solutions of multidimensional Hele–Shaw models*, SIAM J. Math. Anal. **28** (1997), 1028–1047.

[19] J. Escher and G. Simonett, *Classical solutions for Hele–Shaw models with surface tension*, Adv. Differential Equations **2** (1997), 619–642.

[20] J. Escher and G. Simonett, *A center manifold analysis for the Mullins–Sekerka model*, J. Differential Equations **143** (1998), 267–292.

[21] V.A. Galaktionov, J. Hulshof and J.L. Vazquez, *Extinction and focusing behaviour of spherical and annular flames described by a free boundary problem*, J. Math. Pures Appl. **76** (1997), 563–608.

[22] D. Henry, *Geometric theory of parabolic equations*, Lecture Notes in Math., Vol. 840, Springer-Verlag, Berlin (1980).

[23] S.I. Hudjaev, *The first boundary value problem for non-linear parabolic equations*, Dokl. Akad. Nauk SSSR **149** (1963), 535–538 (in Russian); English transl.: Soviet Math. Dokl. **4** (1963), 441–445.

[24] A.G. Kartsatos and I.V. Skrypnik, *A global approach to fully nonlinear parabolic problems*, Trans. Amer. Math. Soc. **352** (2000), 4603–4640.

[25] S.N. Kruzhkov, A. Castro and M. Lopes, *Schauder type estimates and theorems on the existence of the solution of fundamental problem for linear and nonlinear parabolic equations*, Dokl. Akad. Nauk SSSR **20** (1975), 277–280 (in Russian); English transl.: Soviet Math. Dokl. **16** (1975), 60–64.

[26] S.N. Kruzhkov, A. Castro and M. Lopes, *Mayoraciones de Schauder y teorema de existencia de las soluciones del problema de Cauchy para ecuaciones parabolicas lineales y no lineales*, (I) Ciencias Matemáticas **1** (1980), 55–76; (II) Ciencias Matemáticas **3** (1982), 37–56.

[27] N.V. Krylov, *Nonlinear Elliptic and Parabolic Equations of the Second Order*, Nauka, Moscow (1985); English transl.: *Mathematics and Its Applications*, Reidel, Dordrecht (1987).

[28] O.A. Ladyzhenskaja, V.A. Solonnikov and N.N. Ural'ceva, *Linear and Quasilinear Equations of Parabolic Type*, Nauka, Moskow (1967) (in Russian); English transl.: Transl. Math. Monographs, Amer. Math. Soc., Providence, RI (1968).

[29] C. Lederman, J.L. Vázquez and N. Wolanski, *Uniqueness of solution to a free boundary problem from combustion*, Trans. Amer. Math. Soc. **353** (2001), 655–692.

[30] G.M. Lieberman, *Second Order Parabolic Differential Equations*, World Scientific, Singapore (1996).

[31] L. Lorenzi, *A free boundary problem stemmed from combustion theory, Part 1: Existence, uniqueness and regularity results*, J. Math. Anal. Appl. **274** (2002), 505–535; *Part 2: Stability, instability and bifurcation results*, J. Math. Anal. Appl. **275** (2002), 131–160.

[32] A. Lunardi, *On the local dynamical system associated to a fully nonlinear parabolic equation*, Nonlinear Analysis and Applications, V. Lakshmikantham, ed., Dekker, New York (1987), 319–326.

[33] A. Lunardi, *Analytic Semigroups and Optimal Regularity in Parabolic Problems*, Birkhäuser, Basel (1995).

[34] A. Lunardi, E. Sinestrari and W. Von Wahl, *A semigroup approach to the time-dependent parabolic initial boundary value problem*, Differential Integral Equations **5** (1992), 1275–1306.

[35] G. Simonett, *Center manifolds for quasilinear reaction–diffusion systems*, Differential Integral Equations **8** (1995), 753–796.

[36] E. Sinestrari, *On the abstract Cauchy problem in spaces of continuous functions*, J. Math. Anal. Appl. **107** (1985), 16–66.

[37] V.A. Solonnikov, *On the Boundary Value Problems for Linear Parabolic Systems of Differential Equations of General Form*, O.A. Ladyzenskaja, ed., Proc. Steklov Inst. Math. **83** (1965), Amer. Math. Soc. Providence, RI (1967).

[38] N.N. Sopolov, *The first boundary value problem for nonlinear parabolic equations of arbitrary order*, C. R. Acad. Bulgare Sci. **23** (1970), 899–902 (in Russian).

[39] H.B. Stewart, *Generation of analytic semigroups by strongly elliptic operators*, Trans. Amer. Math. Soc. **199** (1974), 141–162.

[40] H.B. Stewart, *Generation of analytic semigroups by strongly elliptic operators under general boundary conditions*, Trans. Amer. Math. Soc. **259** (1980), 299–310.

CHAPTER 6

Kinetic Formulations of Parabolic and Hyperbolic PDEs: From Theory to Numerics

Benoît Perthame

Département de Mathématiques et Applications, CNRS UMR 8553, École Normale Supérieure, 45, rue d'Ulm, 75230 Paris Cedex 05, France

E-mail: perthame@ens.fr

Contents

1. Introduction . . . 439
2. Kinetic formulation of parabolic–hyperbolic conservation laws . . . 440
   2.1. Nondegenerate diffusions . . . 440
   2.2. Degenerate diffusions . . . 443
   2.3. Applications of the kinetic formulation . . . 446
3. Hyperbolic scalar laws: Regularizing effects . . . 447
   3.1. Regularity and compactness . . . 447
   3.2. Averaging lemmas for kinetic equations . . . 449
4. Kinetic formulation of isentropic gas dynamics . . . 453
   4.1. Entropy structure of isentropic gas dynamics . . . 453
   4.2. Kinetic formulation . . . 455
   4.3. Applications of the kinetic formulation . . . 457
5. Kinetic schemes of isentropic gas dynamics . . . 458
   5.1. Kinetic representation . . . 458
   5.2. Finite volumes . . . 462
   5.3. Kinetic solvers . . . 463
   5.4. Numerical comparisons . . . 466
6. Conclusion . . . 469
References . . . 469

HANDBOOK OF DIFFERENTIAL EQUATIONS
Evolutionary Equations, volume 1
Edited by C.M. Dafermos and E. Feireisl

1. Introduction

The method of *kinetic formulations* can now be understood as a method to handle conservative Partial Differential Equations with a large family of *entropies*. By *entropies* we mean here conservation laws built on nonlinear functions of the unknown. When these entropies can be parametrized, the corresponding parameters can be considered as extra variables and, after simple manipulations, the equation can be linearized or at least reduced to a 'less nonlinear' equation. Why is this useful: simply because in the best cases it rises a linear equation where linear methods such as convolution, Fourier transform can be used. For the same reason it also allows to derive numerical schemes in a quite natural way with good stability properties. From this point of view the systems of gas dynamics (with temperature or with barotropic pressure) are particularly interesting.

Indeed, the motivation behind the study of kinetic formulations is to understand the derivation of the compressible Euler equation from Boltzmann equation. The Boltzmann equation for dilute gases is written

$$\frac{\partial}{\partial t} f_\varepsilon + \xi \cdot \nabla_x f_\varepsilon = \frac{1}{\varepsilon} Q(f_\varepsilon, f_\varepsilon), \tag{1}$$

where the constant ε represents the mean free path of particles and $f_\varepsilon(t, x, \xi)$ is the density of particles which at time t and at position x move with the velocity ξ. Also $Q(f, f)$ is Boltzmann's quadratic collisional operator; see Cercignani [9], Cercignani, Illner and Pulvirenti [10], Glassey [28], Lions [43], Villani [63], and for general introductions to kinetic equations see also Dautray and Lions [14] and Perthame [50]. Equation (1) formally provides the Euler system of compressible flows as $\varepsilon \to 0$, the so-called *hydrodynamic limit*. For strong solutions, some rigorous results can be found in the above references. It is an open problem to understand the functional analysis behind this limit for weak solutions (with shocks) although the case of incompressible limit is now well understood [31]. The idea behind the kinetic formulation is to pass to the limit as simply as possible in (1). With some natural a priori bounds on f_ε, we can extract a subsequence still denoted ε such that $f_\varepsilon \to f$ as $\varepsilon \to 0$, then we obtain, passing to the limit in (1),

$$\frac{\partial}{\partial t} f + \xi \cdot \nabla_x f = \mathcal{T}(t, x, \xi) \in \mathcal{D}'. \tag{2}$$

What is this object $\mathcal{T} := \lim_{\varepsilon \to 0} \frac{1}{\varepsilon} Q(f_\varepsilon, f_\varepsilon)$, what are its properties, how do they characterize the limit and what do they imply on the solution f? Answering these questions would provide a kinetic formulation for gas dynamics equations. A question which is still open. For this reason only simpler examples have been carried out completely.

The idea to use a kinetic variable in quasilinear hyperbolic scalar conservation laws goes back to two independent works by Brenier [7] and Giga and Miyakawa [27] where a time discretization and some kind of linearization were introduced where the kinetic formalism appears. The full kinetic formulation (passing from a time discretization to a continuous equation with the correct right-hand side) is introduced by Lions, Perthame and Tadmor [42] in the case of scalar conservation laws, this means that the right-hand side in (2) is identified for this case. Bounded domains, and the definition of traces was treated by Vasseur [62] using compactness arguments derived form the kinetic formulation. The case of the 2×2 system of barotropic (isentropic) gas dynamics is treated in [46] and the way to use its kinetic formulation for proving global existence of weak (entropy) solutions is due to [44] (with an improvement on the possible pressures in [11]). Another but related approach can also be found in [4] and the references therein. General 2×2 systems are treated in [53]. A recent overview of the subject is given in [49], with more details and other examples.

This presentation begins with the case of *parabolic–hyperbolic equations* in the case of a nonisotropic degenerate diffusion (Section 2). In the case of hyperbolic equations, we show how to arrive to a regularity result for the solution. We also present a weaker concept *kinetic representation* and illustrate it with a class of numerical schemes for gas dynamics equations, the so-called *kinetic schemes*.

2. Kinetic formulation of parabolic–hyperbolic conservation laws

Although the *kinetic formulation* arises naturally as the hydrodynamic limit of kinetic equations; see (1) and (2), it can also be seen as a mathematical object related to nonlinear parabolic–hyperbolic scalar conservation laws, this is the point of view we adopt here. We also consider degenerate nonisotropic diffusions as they were treated in [12]. We begin with the nondegenerate case and then consider the degenerate case as a vanishing viscosity limit.

2.1. *Nondegenerate diffusions*

To explain this concept, we consider first the nondegenerate case, a case where existence of a unique family of smooth solutions with decay at infinity is standard to establish. Then the problem is to find a smooth, vanishing at infinity, real valued function $u(t,x)$ defined for $t \geqslant 0$, $x = (x_1, x_2, \ldots, x_d) \in \mathbb{R}^d$, and which solves the quasilinear parabolic equation

$$
\begin{aligned}
&\frac{\partial}{\partial t} u + \sum_{i=1}^{d} \frac{\partial}{\partial x_i} A_i(u) - \sum_{i,j=1}^{d} \frac{\partial^2}{\partial x_i \, \partial x_j} A_{ij}(u) = 0, \\
&u(t=0,x) = u^0(x).
\end{aligned} \tag{3}
$$

We define and assume the following:

$$
\begin{aligned}
&a_i(\cdot) = A_i'(\cdot) \in L^\infty_{\text{loc}}(\mathbb{R}), \qquad a_{ij}(\cdot) = A_{ij}'(\cdot) \in L^\infty_{\text{loc}}(\mathbb{R}), \\
&a(\xi) = \big(a_1(\xi), a_2(\xi), \ldots, a_d(\xi)\big) \quad \big(\text{a mapping}: \mathbb{R} \to \mathbb{R}^d\big), \\
&a_{ij} \text{ is a symmetric matrix}, \quad a_{ij}(\cdot) \geqslant \nu I_{d\times d}, \quad \nu > 0.
\end{aligned} \tag{4}
$$

This equation satisfies a simple so-called *entropy property*. That is, for all smooth function $S(\cdot)$, after multiplying (3) by $S'(u)$, we obtain

$$\frac{\partial}{\partial t}S(u)+\sum_{i=1}^{d}\frac{\partial}{\partial x_i}\eta_i^S(u)-\sum_{i,j=1}^{d}\frac{\partial^2}{\partial x_i\,\partial x_j}B_{ij}^S(u)=-S''(u)a_{ij}(u)\frac{\partial u}{\partial x_i}\frac{\partial u}{\partial x_j}, \tag{5}$$

where the entropy fluxes η_i, B_{ij} are defined (up to an additive constant) by

$$\left(\eta_i^S\right)'(\cdot)=a_i(\cdot)S_i'(\cdot), \qquad \left(B_{ij}^S\right)'(\cdot)=a_{ij}(\cdot)S_i'(\cdot). \tag{6}$$

Notice also that a priori bounds follow from the entropy inequalities. We can choose $S(u)=|u|^p$, $1\leqslant p<\infty$, to obtain

$$\left\|u(t)\right\|_{L^p(\mathbb{R}^d)}\leqslant\left\|u^0\right\|_{L^p(\mathbb{R}^d)}, \quad 1\leqslant p<\infty,$$

and $S(u)=(u-K)_+^2$ with an appropriate choice of K yields

$$\min u^0\leqslant u(t,x)\leqslant\max u^0.$$

We can also use the entropy dissipation, for $S\geqslant 0$ and $S''\geqslant 0$, we then have

$$\int_0^\infty\int_{\mathbb{R}^d}S''(u)a_{ij}(u)\frac{\partial u}{\partial x_i}\frac{\partial u}{\partial x_j}\,\mathrm{d}x\,\mathrm{d}t\leqslant\frac{1}{2}\int_{\mathbb{R}^d}S\left(u^0\right)\mathrm{d}x. \tag{7}$$

In particular, the choice $S(u)=u^2/2$ also leads to the energy estimate

$$\int_0^\infty\int_{\mathbb{R}^d}a_{ij}(u)\frac{\partial u}{\partial x_i}\frac{\partial u}{\partial x_j}\,\mathrm{d}x\,\mathrm{d}t\leqslant\frac{1}{2}\left\|u^0\right\|_{L^2(\mathbb{R}^d)}^2. \tag{8}$$

The choice of Kruzkov's entropies,

$$S_\xi(u)=\begin{cases}\max(0,u-\xi) & \text{for } \xi>0,\\ \max(0,\xi-u) & \text{for } \xi<0,\end{cases} \tag{9}$$

leads to $S_\xi''(u)=\delta(u=\xi)$ and gives the (more interesting although not so clearly defined) a priori estimate

$$\begin{aligned}&\int_0^\infty\int_{\mathbb{R}^d}\delta\big(\xi=u(t,x)\big)a_{ij}(\xi)\frac{\partial u}{\partial x_i}\frac{\partial u}{\partial x_j}\,\mathrm{d}x\,\mathrm{d}t\\&\quad\leqslant\mu(\xi):=\begin{cases}\left\|\left(u^0-\xi\right)_+\right\|_{L^1(\mathbb{R}^d)} & \text{for } \xi\geqslant 0,\\ \left\|\left(\xi-u^0\right)_+\right\|_{L^1(\mathbb{R}^d)} & \text{for } \xi\leqslant 0.\end{cases}\end{aligned} \tag{10}$$

At least this inequality makes sense as a measure in ξ, i.e., testing it against nonnegative continuous functions of ξ.

To proceed further in analyzing the consequences of these entropy inequalities, we introduce the function (from $\mathbb{R}^2$ into $\mathbb{R}$),

$$\chi(\xi; u) = \begin{cases} +1 & \text{for } 0 < \xi < u, \\ -1 & \text{for } u < \xi < 0, \\ 0 & \text{otherwise.} \end{cases}$$

This function is related to several deep problems especially to weak limits of oscillating bounded sequences $u_n(x)$ and to Young measures (see [47,49]). Indeed, it gives a representation of any function $S(u)$ as follows

$$\int_{\mathbb{R}} S'(\xi)\chi(\xi; u)\,\mathrm{d}\xi = S(u) - S(0). \tag{11}$$

Therefore, when a sequence of functions u_n converges in $L^\infty - w*$, it allows us to study the limit of all the limits $S(u_n)$ for all smooth functions S.

We claim that the family of equalities (5) is equivalent to write in $\mathcal{D}'((0,\infty) \times \mathbb{R}^{d+1})$ the so-called *kinetic formulation* (recall that ξ is a real-valued variable here).

PROPOSITION 2.1. *Equation* (5) *is equivalent to the problem of finding a function* $u(t,x)$ *such that, in the sense of distributions,*

$$\frac{\partial}{\partial t}\chi\big(\xi, u(t,x)\big) + a(\xi)\cdot\nabla_x\chi\big(\xi, u(t,x)\big)$$

$$- a_{ij}(\xi)\sum_{i,j=1}^{d}\frac{\partial^2}{\partial x_i\,\partial x_j}\chi\big(\xi, u(t,x)\big)$$

$$= \frac{\partial}{\partial \xi}n(t,x,\xi), \tag{12}$$

$$n(t,x,\xi) = \delta\big(\xi = u(t,x)\big)a_{ij}(\xi)\frac{\partial u}{\partial x_i}\frac{\partial u}{\partial x_j}. \tag{13}$$

In fact, n is a bounded nonnegative measure thanks to (8), also, thanks to (10) and usual continuity arguments,

$$n \in C_0\big(\mathbb{R}_\xi; w - M^1\big((0,+\infty)\times\mathbb{R}^d\big)\big),$$

where M^1 denotes the Banach space of bounded Radon measures.

We now indicate the reason why the *kinetic formulation* holds true.

DERIVATION OF THE KINETIC FORMULATION (12). Using the chain rule in (12) we have

$$\frac{\partial}{\partial t}\chi(\xi; u) = \delta(\xi = u)\frac{\partial u}{\partial t},$$

therefore (12) also writes

$$\delta\big(\xi = u(t,x)\big)\left[\frac{\partial u}{\partial t} + a(u)\cdot\nabla_x u\right] - \sum_{i,j=1}^{d} \frac{\partial}{\partial x_i}\left[a_{ij}\big(u(t,x)\big)\delta(\xi = u)\frac{\partial u}{\partial x_j}\right]$$

$$= \frac{\partial}{\partial \xi}\left[\delta\big(\xi = u(t,x)\big)\sum_{i,j=1}^{d} a_{ij}\big(u(t,x)\big)\frac{\partial u}{\partial x_i}\frac{\partial u}{\partial x_j}\right].$$

On the other hand, we have

$$\frac{\partial}{\partial x_i}\left[a_{ij}\big(u(t,x)\big)\delta(\xi = u)\frac{\partial u}{\partial x_j}\right]$$

$$= \delta'(\xi = u)a_{ij}(u)\frac{\partial u}{\partial x_i}\frac{\partial u}{\partial x_j} + \delta(\xi = u)\frac{\partial}{\partial x_i}\left[a_{ij}(u)\frac{\partial u}{\partial x_j}\right]$$

and therefore (12) is also equivalent to

$$\delta\big(\xi = u(t,x)\big)\left[\frac{\partial u}{\partial t} + a(u)\cdot\nabla_x u - \sum_{i,j=1}^{d}\frac{\partial}{\partial x_i}\left(a_{ij}(u)\frac{\partial u}{\partial x_j}\right)\right] = 0;$$

an equation which is equivalent to the parabolic conservation law (3) for smooth solutions.

Another way to see the equivalence is to multiply (12) by $S'(\xi)$ (a test function) and integrate $d\xi$. Then one still recovers the entropy equalities (5) which are equivalent to parabolic equation (3). □

2.2. *Degenerate diffusions*

The interest of this formalism appears for degenerate diffusions ($\nu = 0$) and especially in the hyperbolic case $A_{ij} = 0$. Then possible singularities of the solution (shock waves for instance) make that the chain rule is no longer available. The kinetic formulation, however, holds true with the only difference that the measure in the right-hand side is no longer defined explicitly by (13). It is replaced by a bounded measure $\tilde{n}(t,x,\xi) \geqslant n(t,x,\xi)$. Degenerate diffusions correspond to the assumption,

$$a_{ij}(u) \text{ is a symmetric matrix.} \tag{14}$$

They can be obtained passing to the limit as ε vanishes in the family of solutions u_ε associated with the nondegenerate diffusion matrix $\varepsilon I + (a_{ij})$. It is possible to prove that this is a strong limit in any L^p spaces, $1 \leqslant p < \infty$, when we assume that $u^0 \in L^1 \cap L^\infty(\mathbb{R}^d)$ is fixed independently of ε. In order to state correctly the limit problem we need the notations

$\Sigma_{ik}(u)$ and $\Sigma_{ik}^{\psi}(u)$ for $\psi \in C_0(\mathbb{R})$ with $\psi \geqslant 0$:

$$
\begin{aligned}
&a_{ij}(\cdot) = \sum_{k=1}^{d} \sigma_{ik}(\cdot)\sigma_{jk}(\cdot), \\
&\Sigma_{ik}'(\cdot) = \sigma_{ik}(\cdot), \qquad \left(\Sigma_{ik}^{\psi}\right)'(\cdot) = \sqrt{\psi(\cdot)}\sigma_{ik}(\cdot).
\end{aligned}
\tag{15}
$$

Then, we end up with the two equivalent definitions:

$$
n_\varepsilon^{\psi}(t,x) := \sum_{k=1}^{d} \left(\sum_{i=1}^{d} \frac{\partial}{\partial x_i} \Sigma_{ik}^{\psi}(u^\varepsilon) \right)^2 = \sum_{k=1}^{d} \psi(u^\varepsilon) \left(\sum_{i=1}^{d} \frac{\partial}{\partial x_i} \Sigma_{ik}(u^\varepsilon) \right)^2. \tag{16}
$$

A fundamental remark is that this equality still holds in the limit $\varepsilon \to 0$, and this allows to define, for weak solutions, the nonlinear term arising in the entropy relation. Then, we end up with the following:

DEFINITION 2.2. An entropy solution is a function $u(t,x) \in L^\infty([0,\infty); L^\infty \cap L^1(\mathbb{R}^d))$ such that

(i) $\sum_{i=1}^{d} \frac{\partial}{\partial x_i} \Sigma_{ik}(u) \in L^2([0,\infty) \times \mathbb{R}^d)$ for any $k \in \{1,\dots,d\}$;

(ii) for any function $\psi \in C_0(\mathbb{R})$ with $\psi(u) \geqslant 0$ and any $k \in \{1,\dots,d\}$, the chain rules hold

$$
\sum_{i=1}^{d} \frac{\partial}{\partial x_i} \Sigma_{ik}^{\psi}(u) = \sqrt{\psi(u)} \sum_{i=1}^{d} \frac{\partial}{\partial x_i} \Sigma_{ik}(u) \in L^2\big([0,\infty) \times \mathbb{R}^d\big), \tag{17}
$$

$$
\begin{aligned}
n^{\psi}(t,x) &:= \psi\big(u(t,x)\big) \sum_{k=1}^{d} \left(\sum_{i=1}^{d} \frac{\partial}{\partial x_i} \Sigma_{ik}\big(u(t,x)\big) \right)^2 \\
&= \sum_{k=1}^{d} \left(\sum_{i=1}^{d} \frac{\partial}{\partial x_i} \Sigma_{ik}^{\psi}\big(u(t,x)\big) \right)^2, \quad \text{a.e.}
\end{aligned}
\tag{18}
$$

(iii) there exists an entropy dissipation measure $m(t,x,\xi)$ such that, for any smooth function $S(u)$, we have in $\mathcal{D}'(\mathbb{R}^+ \times \mathbb{R}^d)$,

$$
\begin{aligned}
&\frac{\partial}{\partial t} S(u) + \sum_{i=1}^{d} \frac{\partial}{\partial x_i} \eta_i^S(u) - \sum_{i,j=1}^{d} \frac{\partial^2}{\partial x_i\, \partial x_j} B_{ij}^S(u) = -\big(m^{S''} + n^{S''}\big), \\
&S\big(u(t=0)\big) = S(u_0),
\end{aligned}
\tag{19}
$$

$$
m^{S''}(t,x) = \int_{\mathbb{R}} S''(\xi) m(t,x,\xi)\, \mathrm{d}\xi, \quad \text{with } m(t,x,\xi) \text{ a nonnegative measure.} \tag{20}
$$

Before we give consequences of this definition, we give some explanations. The entropy dissipation measure m appears here because we have to take into account the weak limit process when passing to the limit as $\varepsilon \to 0$ in the quadratic terms. It accounts for the inequality

$$\left(\sum_{i=1}^{d} \frac{\partial}{\partial x_i} \Sigma_{ik}\big(u(t,x)\big)\right)^2 \leqslant \mathrm{w} - \lim_{\varepsilon \to 0} \left(\sum_{i=1}^{d} \frac{\partial}{\partial x_i} \Sigma_{ik}\big(u_\varepsilon(t,x)\big)\right)^2$$

because, weakly in L^2, we have

$$\sum_{i=1}^{d} \frac{\partial}{\partial x_i} \Sigma_{ik}\big(u_\varepsilon(t,x)\big) \to \sum_{i=1}^{d} \frac{\partial}{\partial x_i} \Sigma_{ik}\big(u(t,x)\big).$$

Another way to express the same result on the decomposition $m^{S''} + n^{S''}$ of the entropy defect measure consists in writing

$$n_\varepsilon \to m + n \quad \text{in } \mathrm{w} - M^1_{\mathrm{loc}}\big(\mathbb{R}^+ \times \mathbb{R}^{d+1}\big),$$

with

$$n(t,x,\xi) = \delta\big(\xi - u(t,x)\big) \sum_{k=1}^{d} \left(\sum_{i=1}^{d} \frac{\partial}{\partial x_i} \Sigma_{ik}\big(u(t,x)\big)\right)^2. \tag{21}$$

Therefore the following total mass control is satisfied

$$\int_0^\infty \int_{\mathbb{R}^d} (m+n)(t,x,\xi)\,\mathrm{d}t\,\mathrm{d}x \leqslant \mu(\xi) \quad \forall \xi \in \mathbb{R}^d, \tag{22}$$

recalling the definition of μ in (10).

A consequence of this definition is the following theorem.

THEOREM 2.3. *Entropy solutions to the degenerate parabolic–hyperbolic equation are equivalent to the kinetic formulation*

$$\begin{aligned} &\frac{\partial}{\partial t}\chi(\xi;u) + a(\xi)\cdot\nabla_x\chi(\xi;u) - \sum_{i,j=1}^{d} a_{ij}(\xi)\frac{\partial^2}{\partial x_i\,\partial x_j}\chi(\xi;u) \\ &\quad = \frac{\partial}{\partial_\xi}(m+n)(t,x,\xi), \\ &\chi(\xi;u)|_{t=0} = \chi(\xi;u_0), \end{aligned} \tag{23}$$

in $\mathcal{D}'(\mathbb{R}^+ \times \mathbb{R}^{d+1})$ with $m(t,x,\xi)$ an unknown nonnegative measure and n given by (21).

This is a linear kinetic equation, with analogies with the Boltzmann equation (1), which acts on the function $\chi(\xi : u)$. The nonlinearity is hidden in the nonlinearity of this function χ and also in the singular right-hand side $m + n$.

2.3. *Applications of the kinetic formulation*

Several applications of the kinetic formulation exist. We list them without further details on the proofs and simply give references for the proofs.

Regularity of the solutions is a real issue in the degenerate case. This kind of equations are L^1 contractions and thus they propagate initial regularity because, for all $h \in \mathbb{R}^d$, we have

$$\|u(t, x + h) - u(t, x)\|_{L^1(\mathbb{R}^d)} \leqslant \|u^0(x + h) - u^0(x)\|_{L^1(\mathbb{R}^d)}.$$

A more interesting question is therefore the possibility of regularizing effects. Such effects can be deduced from the method in Section 3 for the hyperbolic case under a nondegeneracy condition on the hyperbolic flux $a(\cdot)$. The parabolic part helps in the arguments and thus the same results hold in the degenerate parabolic case. However, more elaborate nondegeneracy conditions can be found that combine both the hyperbolic flux $a_i(\cdot)$ and the diffusion matrix $a_{ij}(\cdot)$. They can be proved using tools as averaging lemmas. See [45] and, for more recent and elaborate results, Tao and Tadmor [57].

Another range of applications is to uniqueness of the solution. This question has raised much interest since Kruzkov's [39] variable doubling method in the hyperbolic case which was extended to parabolic–hyperbolic equations by Carillo [8] for isotropic diffusions (see also Eymard et al. [24]). Another method in the hyperbolic case [49] is based on the kinetic formulation and differs technically from Kruzkov's original method. In [12] it was extended to the entropy solutions to the parabolic–hyperbolic nonisotropic case.

The kinetic formulation also introduces a natural tool to treat L^1 datas withdrawing the L^∞ assumption; in Theorem 2.4 we call them kinetic solutions. Indeed, (23) makes a perfect sense when $u(t, x) \in L^\infty([0, \infty); L^1(\mathbb{R}^d))$ while quantities like $a_i(u)$ or $a_{ij}(u)$ do not belong to L^1_{loc}, when a_i or a_{ij} have superlinear growth at infinity (although regularizing effects in L^p can be derived related to the growth of the flux function at infinity; see [49]). As a summary, the following result holds:

THEOREM 2.4 [12]. *Assume that* $a(\cdot) \in L^\infty_{\mathrm{loc}}(\mathbb{R}; \mathbb{R}^d)$ *and* (14), (15) *hold for some matrices* $\sigma \in L^\infty_{\mathrm{loc}}(\mathbb{R}; \mathbb{R}^{d\times d})$. *For* $u_0 \in L^1(\mathbb{R}^d)$, *there exists a unique kinetic solution* $u \in C([0, \infty); L^1(\mathbb{R}^d))$ *for the Cauchy problem* (3). *If* $u_0 \in L^\infty \cap L^1(\mathbb{R}^d)$, *then the kinetic solution is the unique entropy solution and* $|u(t, x)| \leqslant \|u_0\|_{L^\infty(\mathbb{R}^d)}$.

As another application of kinetic formulations, we mention that the construction of solutions can also be performed by a kinetic approximation. This is a rather simple and efficient approximation, again related to the basically linear structure of the kinetic formulation. The idea is to use a semilinear relaxation method of the type

$$
\begin{aligned}
&\frac{\partial}{\partial t} f_\varepsilon + a(\xi)\cdot\nabla_x f_\varepsilon - \sum_{i,j=1}^{d} a_{ij}(\xi)\frac{\partial^2}{\partial x_i\,\partial x_j} f_\varepsilon = \frac{1}{\varepsilon}\big(\chi(\xi;u_\varepsilon) - f_\varepsilon\big),\\
&u_\varepsilon(t,x) = \int_{\mathbb{R}} f_\varepsilon(t,x,\xi)\,\mathrm{d}\xi,\\
&f(t=0,x,\xi) = \chi\big(\xi;u^0(x)\big).
\end{aligned}
\tag{24}
$$

The convergence of this kind of approximation can be studied by methods à la Kruzkov; see [49] and Belhadj et al. [3] for a nonisotropic diffusion.

Related to this relaxation method is the time splitting method that motivated Brenier [7] and Giga and Miyakawa [27] in their earlier work (see also [61,62]). These authors propose the following procedure in order to approximate (5) (actually, these papers deal with the hyperbolic cases $a_{ij}=0$). Being given the initial data $u^0(x)$ and a time step $\Delta t>0$, one solves successively (for $n=0,1,\dots$) the linear kinetic equations

$$
\begin{aligned}
&\frac{\partial}{\partial t} f^n + a(\xi)\cdot\nabla_x f^n - \sum_{i,j=1}^{d} a_{ij}(\xi)\frac{\partial^2}{\partial x_i\,\partial x_j} f^n = 0, \qquad t\in\big[n\Delta t,(n+1)\Delta t\big],\\
&f^n(t=n\Delta t,x,\xi) = \chi\big(\xi;u^n(x)\big),\\
&u^{n+1}(x) = \int_{\mathbb{R}} f^n\big((n+1)\Delta t,x,\xi\big)\,\mathrm{d}\xi.
\end{aligned}
$$

This method can be viewed as a first step towards a full space–time discretization. The L^1 contraction argument allows to show that this method converges as $\Delta t\to 0$ towards a solution to (5). Extensions to systems is a much harder topic which has been carried out by Vasseur [59,60].

3. Hyperbolic scalar laws: Regularizing effects

3.1. *Regularity and compactness*

In this section we restrict ourselves to the purely hyperbolic case of equation

$$
\begin{aligned}
&\frac{\partial}{\partial t}u + \sum_{i=1}^{d}\frac{\partial}{\partial x_i}A_i(u) = 0,\\
&u(t=0,x) = u^0(x).
\end{aligned}
\tag{25}
$$

Such an equation admits several weak solutions because of the nonuniqueness related to appearance of singularities (shocks) in finite time. The continuation of a solution after the shock can be performed in different way, leading to several possible solutions. The unique entropy solution is a natural one but entropy dissipation is not true for all solutions; cf. Dafermos [13], Serre [56], Hwang and Tzavaras [32] and Godlewski and Raviart [29]. For instance, one could consider the limit $\varepsilon \to 0$ of diffusion–dispersion equations as (say in one space dimension to simplify)

$$\frac{\partial}{\partial t}u + \frac{\partial}{\partial x}A(u) = \varepsilon\frac{\partial^2}{\partial x^2}u + \beta\varepsilon^2\frac{\partial^3}{\partial x^3}u,$$

which limit depends on the parameter β because nonclassical, i.e., nonentropic shocks are obtained in the limit (see Freistühler [26] and Lefloch [41]).

This consideration leads us to assume that the entropies can be controlled but in a weaker way that we did it in Section 2. Namely, we assume a condition on the family of Kruzkov's entropies (9)

$$\frac{\partial}{\partial t}S_\xi(u) + \sum_{i=1}^{d}\frac{\partial}{\partial x_i}\eta_\xi(u) = -m(t,x,\xi), \tag{26}$$

where $m(t,x,\xi)$ is a bounded measure, with the bound

$$\int_0^\infty \int_{\mathbb{R}^d} |m|(t,x,\xi)\,\mathrm{d}t\,\mathrm{d}x \leqslant \mu(\xi) \quad \forall \xi \in \mathbb{R}^d,\ \ \mu \in L_0^\infty(\mathbb{R}), \tag{27}$$

where $L_0^\infty(\mathbb{R})$ denotes the Banach space of bounded functions that vanish at infinity. In the entropy case we recall that m is nonnegative and that the above bound is available as we proved it in (22) and (10). This leads, following the arguments of previous section to a kinetic formulation

$$\frac{\partial}{\partial t}\chi(\xi;u) + a(\xi)\cdot\nabla_x\chi(\xi;u) = \frac{\partial}{\partial_\xi}m(t,x,\xi). \tag{28}$$

Actually, such a situation arises passing to the limit in several types of examples. Multidimensional relaxation of hyperbolic systems or diffusive–dispersive limits, see Hwang and Tzavaras [32]. Another, and unexpected such situation also arises in a model for thin ferromagnetic films. More precisely several constrained variational problem arising in Ginzburg–Landau theory lead, in a singular limit, to a hyperbolic equation similar to (25) in one space dimension (Ambrosio et al. [1] and Desimone et al. [18]). A kinetic formulation thus holds true which can be used to analyze the problem; see the papers [15,33,34,39]. We also refer to [23] for another nonentropic situation.

Below we present a typical example of application of the kinetic formulation and we derive regularity in Sobolev spaces. We have the following:

THEOREM 3.1 [34]. *Let* $u(t,x) \in L^\infty(\mathbb{R}^+ \times \mathbb{R}^d)$ *satisfy* (26), (27) *and assume the nondegeneracy condition, for all* $(k,\tau) \in \mathbb{R}^d \times \mathbb{R}$ *with* $|k|^2 + \tau^2 = 1$,

$$\operatorname{meas}\{\xi \text{ such that } |\xi| \leqslant R \text{ and } |a(\xi)\cdot k - \tau| \leqslant \varepsilon\} \leqslant C\varepsilon. \tag{29}$$

Then locally we have

$$u \in W^{s,r}_{t,x} \quad \text{for all } s < \frac{1}{3},\ r < \frac{3}{2}.$$

REMARK 3.2. A recent counterexample by De Lellis and Westdickenberg [16] shows that the exponent $1/3$ is sharp even in dimension $d = 1$. Therefore weak solutions to (25) which are not entropy solutions may lack BV_{loc} regularity. Their argument is based on the construction of a family of solutions with nonentropic shocks in which the size of the discontinuity, and the distance in between, are controlled in a clever asymptotic way. It is also related, but not so directly, to another example given in [1] on the regularity of minimizers in the above mentioned micromagnetism problems.

REMARK 3.3. Kinetic formulation is also a step in order to provide another type of regularity which is not expressed in terms of Sobolev spaces. More precisely the paper [15] actually shows that the singularity set of a solution $u(t,x)$ is d-dimensional in Haussdorf sense. The proof relies on the geometric theory of measures. A similar characterization of the singularity set can be found in [2] for the related variational problem.

We do not detail here the proof of Theorem 3.1. It is an improvement of a similar conclusion in [45] for the entropy solution (with a proof using strongly the contraction property of entropy solutions). The idea is that ξ regularity of the function χ can improve the exponents of averaging lemmas as explained in Section 3.2.

Another related result goes in the direction of reducing as much as possible the nondegeneracy condition on the fluxes $A(\cdot)$. Of course we still have to discard the linear case where no regularizing effect can hold true because initial oscillations propagate. The next statement extends a previous result of Tartar [58] in dimension 1 based on compensated compactness [48] or the books [13,56,47].

THEOREM 3.4 [45]. *Let u_n be a bounded sequence in $L^\infty(\mathbb{R}^+ \times \mathbb{R}^d)$ of distributional solutions to* (25). *Assume also the nondegeneracy condition,*

$$\operatorname{meas}\bigl\{\xi \text{ such that } |\xi| \leqslant R \text{ and } \bigl|a(\xi)\cdot k - \tau\bigr| = 0\bigr\} = 0\ \forall \tau \in \mathbb{R}, k \in S^{d-1}, \tag{30}$$

with $R = \sup_n \|u_n^0\|_{L^\infty}$. Then, u_n is locally relatively compact in $L^p(\mathbb{R}_+ \times \mathbb{R}^d)$ for all $1 \leqslant p < +\infty$.

The proof is also based on the theory of averaging lemmas for kinetic equations and we present it now.

3.2. *Averaging lemmas for kinetic equations*

In order to show the various ways one can work in order to arrive at the above results we would like to present several stages of averaging lemmas. First, what averaging lemmas say, second why the derivative structure of the right-hand side looses some regularity, and how one can use the regularity of χ to arrive to Theorem 3.1.

We restrict our presentation to the L^2 case for the simplicity of notations and to the case of usual kinetic equations where the transport velocity is denoted by ξ. We refer to DiPerna et al. [21,52] and Bouchut [5,34] for other cases of interest. Especially to work in L^p spaces requires interpolation arguments that cannot be used directly on the solution itself but has to be realized at the level of Fourier multipliers that appear naturally in the proof. Interpolation between L^2 and L^1 lead to understand Calderòn–Zygmund theory in the framework of kinetic equations and, to be optimal, this uses special Hardy spaces (product $\mathcal{H}^1$ spaces). This is one of the most technical issues behind averaging lemmas. Recently, Jabin and Vega [35] introduced another method for averaging based on $T \circ T^*$ type of arguments and which improve the L^p spaces in which regularizing effects hold true.

The most elementary averaging lemma is as follows:

THEOREM 3.5 [30]. *Let $\psi \in L^\infty(\mathbb{R}^d)$, $\operatorname{supp}\psi \subset B(R)$ (ball of radius R centered at origin). Let $f, g \in L^2(\mathbb{R}^{1+2d})$ satisfy*

$$\frac{\partial}{\partial t} f + \xi \cdot \nabla_x f = g, \tag{31}$$

and set

$$\rho_\psi(t,x) := \int_{\mathbb{R}^d} \psi(\xi) f(t,x,\xi)\,\mathrm{d}\xi. \tag{32}$$

Then we have

$$\|\rho_\psi\|_{\dot{H}^{1/2}(\mathbb{R}^{d+1})} \leqslant C\big(d, R, \|\psi\|_{L^\infty}\big)\|f\|^{1/2}_{L^2(\mathbb{R}^{2d+1})}\|g\|^{1/2}_{L^2(\mathbb{R}^{2d+1})}. \tag{33}$$

REMARK 3.6. As it is, this statement does not apply to the Cauchy problem because of the lack of global L^2 integrability in time. But after time truncation we recover the situation (33) and thus the result can be applied in usual situations.

In order to give an idea of the ingredients behind averaging lemma, we give a proof of this simple case.

PROOF OF THEOREM 3.5. The proof is based on Fourier transform in space and time. We define

$$\hat{f}(\tau,k,\xi) = \int_{\mathbb{R}^d} f(t,x,\xi)\mathrm{e}^{\mathrm{i}(t\tau + x\cdot\xi)}\,\mathrm{d}x\,\mathrm{d}t,$$

and similarly $\hat{g}$, $\hat{\rho}_\psi$. We use below the property that

$$\hat{\rho}_\psi(\tau,k) = \int_{\mathbb{R}^d} \psi(\xi)\hat{f}\,\mathrm{d}\xi.$$

Equation (31) becomes very simple after Fourier transform and reads

$$\mathrm{i}(\tau + k \cdot \xi)\hat{f} = \hat{g}, \quad (\tau, k, \xi) \in \mathbb{R}^{2d+1}.$$

This allows us to invert the symbol and in order to avoid the singular hyperplane $\tau + \xi \cdot k = 0$, we add $\beta \hat{f}$ on each side of the equality, with β a positive real number to be chosen later. We obtain

$$\hat{f}(\tau, k) = \frac{\hat{g} + \beta \hat{f}}{\beta + \mathrm{i}(\tau + \xi \cdot k)}.$$

After using Cauchy–Schwarz inequality, we deduce

$$|\hat{\rho}|^2 \leqslant 2 \int_{\mathbb{R}^d} \left(|\hat{g}|^2 + \beta^2 |\hat{f}|^2\right) \mathrm{d}\xi \int_{\mathbb{R}^d} \frac{\psi^2}{\beta^2 + |\tau + \xi \cdot k|^2}\, \mathrm{d}\xi.$$

After using Lemma 3.7 we deduce

$$|\hat{\rho}|^2 \leqslant \int_{\mathbb{R}^d} \left(|\hat{g}|^2 + \beta^2 |\hat{f}|^2\right) \mathrm{d}\xi \frac{C}{\beta(|\tau| + |k|)},$$

and after integrating in τ, ξ,

$$\left\| \left(|\tau| + |k|\right)\hat{\rho} \right\|^2_{L^2(\mathbb{R}^{d+1})} \leqslant C \frac{\|\hat{g}\|^2_{L^2(\mathbb{R}^{d+1})} + \beta^2 \|\hat{f}\|^2_{L^2(\mathbb{R}^{d+1})}}{\beta}.$$

After choosing $\beta = \|\hat{g}\|_{L^2(\mathbb{R}^{d+1})} / \|\hat{f}\|_{L^2(\mathbb{R}^{d+1})}$, we obtain

$$\left\| \left(|\tau| + |k|\right)\hat{\rho} \right\|^2_{L^2(\mathbb{R}^{d+1})} \leqslant C \left\| \hat{g} \right\|_{L^2(\mathbb{R}^{d+1})} \left\| \hat{f} \right\|_{L^2(\mathbb{R}^{d+1})}.$$

This is exactly the definition of the homogeneous $\dot{H}^{1/2}(\mathbb{R}^{d+1})$ seminorm

$$\|\rho\|_{\dot{H}^{1/2}(\mathbb{R}^{d+1})} = \left\| \left(|\tau| + |k|\right)\hat{\rho} \right\|^2_{L^2(\mathbb{R}^{d+1})},$$

and thus Theorem 3.5 is proved. Notice simply that the Hilbert space $H^{1/2}$ is defined by the norm $\|\rho\|_{\dot{H}^{1/2}} + \|\rho\|_{L^2}$. □

LEMMA 3.7. *With the above notations we have, for all $\beta > 0$,*

$$\int_{\mathbb{R}^d} \frac{\psi^2}{\beta^2 + |\tau + \xi \cdot k|^2}\, \mathrm{d}\xi \leqslant \frac{C(\psi)}{\beta(|\tau| + |k| + \beta)},$$

with (*we recall that* $\operatorname{supp}\psi \subset B(R)$),

$$C(\psi) = CR \sup_{\omega\in\mathbb{S}^{d-1}} \int_{\omega\cdot\xi=0} \psi(\xi)^2\,\mathrm{d}\xi',$$

$d\xi'$ *the Lebesgue measure on* $\{\omega\cdot\xi = 0\}$.

We leave this lemma without proof.

Considering the kinetic formulation (28), we are in a situation to apply this result because

$$u(t,x) = \int_{\mathbb{R}} \chi(\xi; u)\,\mathrm{d}\xi,$$

which is equivalent to a truncated integral in (32) because u is bounded and thus $\chi(\xi; u)$ has compact support in ξ. Here, let us only mention that a difficulty arises from the only M^1 regularity of the kinetic defect measure m. Because averaging lemma do not win any derivative in L^1 this case is the worst and the L^2 case provides much more regularity. Additionally, Theorem 3.5 cannot be used directly because, in particular, of the ξ derivative that appears in the right-hand side of (28). It can be handled by the following result:

THEOREM 3.8 [30]. *Let* $\psi \in C^1(\mathbb{R}^d)$, $\operatorname{supp}\psi \subset B(R)$ (*ball of radius* R *centered at origin*). *Let* $f, g \in L^2(\mathbb{R}^{1+2d})$ *satisfy*

$$\frac{\partial}{\partial t} f + \xi\cdot\nabla_x f = \frac{\partial}{\partial \xi} g. \tag{34}$$

Then, using notation (32), *we have*

$$\|\rho_\psi\|_{\dot H^{1/4}(\mathbb{R}^{d+1})} \leqslant C(d,\psi)\|f\|_{L^2(\mathbb{R}^{2d+1})}^{1/4}\|g\|_{L^2(\mathbb{R}^{2d+1})}^{3/4}. \tag{35}$$

As a consequence, the regularity of averages is not lost but just weaker in presence of a singular right-hand side. Also, the gain of regularity is already of only weak $1/4$ derivative in L^2 while we want to prove $1/3$ in the real situation. To improve the regularity result in case of the kinetic formulation, the idea consists in taking advantage of the ξ regularity of the function χ (denoted by f in the above general averaging lemmas).

THEOREM 3.9 [34]. *Let* $\psi \in \mathcal{D}(\mathbb{R}^d)$ *and* f, g *satisfy*

$$\begin{aligned} &g \in L^p\big(\mathbb{R}^d, W_\xi^{\beta,p}(\mathbb{R}^d)\big), \quad 1 < p \leqslant 2,\ \beta \leqslant \frac{1}{2},\\ &f \in L^q\big(\mathbb{R}^d, W_\xi^{\gamma,q}(\mathbb{R}^d)\big), \quad 1 < q \leqslant 2,\ 1 - \frac{1}{q} < \gamma \leqslant \frac{1}{2}. \end{aligned} \tag{36}$$

$$\frac{\partial}{\partial t} f + \xi\cdot\nabla_x f = \Delta_x^{\alpha/2} g. \tag{37}$$

Then we have

$$\|\rho_\psi\|_{W^{s',r'}_{\mathrm{loc}}} \leqslant C\big(\|g\|_{L^p_x W^{\beta,p}_\xi} + \|f\|_{L^q_x W^{\gamma,q}_\xi}\big)$$

for $s' < s = \theta(1-\alpha)$, $0 \leqslant \alpha < 1$ *and* $r' < r$ *with* $\frac{1}{r} = \frac{\theta}{p} + \frac{1-\theta}{q}$ *and*

$$\theta = \frac{1+\gamma-1/q}{1+\gamma-\beta+1/p-1/q}. \tag{38}$$

This theorem may be applied to derive Theorem 3.1, we just use the parameters $\gamma \approx \frac{1}{2}$, $q \approx 2$, $\alpha \approx 0$, $p \approx 1$, $\beta \approx -1$, and notice it still holds with the more general structure of a transport velocity $a(\xi)$ which satisfies the nondegeneracy condition (29).

4. Kinetic formulation of isentropic gas dynamics

In this section we present an example of a system with a kinetic formulation. We consider one-dimensional isentropic gas dynamics, a hyperbolic system which admits a full family of entropies.

The system of isentropic gas dynamics consists in two equations that express physically the conservation of mass and the conservation of momentum for a barotropic gas (this refers to compressible gases where the pressure only depends on the density and thus the temperature effects do not act directly on the dynamics). It reads

$$\begin{cases} \frac{\partial}{\partial t}\rho + \frac{\partial}{\partial x}(\rho u) = 0, \quad t \geqslant 0,\ x \in \mathbb{R}, \\ \frac{\partial}{\partial t}(\rho u) + \frac{\partial}{\partial x}\big(\rho u^2 + p(\rho)\big) = 0. \end{cases} \tag{39}$$

The density ρ should be nonnegative, u denotes the velocity of the gas and p is the pressure. Here we consider the case of a polytropic gas, where the equation of state is simply given by

$$p(\rho) = \kappa\rho^\gamma, \quad 1 < \gamma < +\infty, \tag{40}$$

and we normalize the constant κ as follows (for the simplicity of formulas below)

$$\kappa = \frac{\theta^2}{\gamma}, \quad \theta = \frac{\gamma-1}{2}.$$

4.1. *Entropy structure of isentropic gas dynamics*

This system falls in the class of strictly hyperbolic systems but it degenerates at the vacuum state (it looses hyperbolicity), $\rho(t,x) = 0$. Vacuum may arise even when the initial data satisfies $\rho^0 > 0$ in a simple Riemann problem [13,56]. To deal with general global solutions to the system of isentropic gas dynamics, we face the difficulty of dealing with the vacuum state. This implies especially that the velocity u, for instance, is not defined as $\rho(t,x)=0$, but more generally all quantities depending upon ρ and u that make sense should vanish at the vacuum state.

Again the method is based on using a family of entropies. Let us recall that in general 2×2 systems admit two families of entropies. But the above mentioned degeneracy of isentropic gas dynamics at vacuum makes that only a single family can be used. For non-degenerate 2×2 systems it is possible to find the kinetic structure and for instance the case of elasticity is completely treated in [53].

A first step is to find the families of entropy inequalities. We call an *entropy pair* (or also a pair: entropy/entropy flux) and we denote it by $(S(\rho,u),\ \eta(\rho,u))$, functions of the unknowns ρ and u such that the equality

$$\frac{\partial}{\partial t}S\big(\rho(t,x),u(t,x)\big)+\frac{\partial}{\partial x}\eta\big(\rho(t,x),u(t,x)\big)=0$$

holds for all smooth solutions (ρ,u). Simple algebraic manipulations show that such entropies are characterized by the differential relations

$$\begin{cases} S_{\rho\rho}-\frac{p'(\rho)}{\rho^2}S_{uu}=0,\\[2ex] \eta_u=\rho S_\rho+uS_u.\end{cases}$$

The first equation can be seen as a wave equation in the variables $\rho\geqslant 0$ which plays the role of time and $u\in\mathbb{R}$ which plays the role of one-dimensional space. The first step consists in concentrating on the initial value problem for this wave equation. Recalling that all quantities we consider should vanish at vacuum state, we consider the so-called *weak entropies family* defined by

$$\begin{cases} S_{\rho\rho}-\frac{p'(\rho)}{\rho^2}S_{uu}=0,\\[2ex] S(\rho=0,u)=0 \quad\text{and}\quad S_\rho(\rho=0,u)=g(u).\end{cases}\tag{41}$$

The solution operator to this problem can be analyzed completely and yields the following lemma which relies on the power law assumed for $p(\rho)$ (otherwise see [11]).

LEMMA 4.1. *We set* $\lambda=\frac{3-\gamma}{2(\gamma-1)}$.

(i) *There is a fundamental solution, corresponding to* $g=\delta$ *the Dirac mass in* (41), *given by the explicit formula with* $\alpha(\lambda)$ *a normalizing constant,*

$$\chi(\rho,u)=\begin{cases}\alpha(\lambda)\big(\rho^{2\theta}-u^2\big)^\lambda & \text{for } |u|<\rho^\theta,\\ 0 & \text{otherwise.}\end{cases}$$

(ii) *Therefore the weak entropies in* (41) *are given by the formula*

$$S(\rho, u) = \int_{\mathbb{R}} g(\xi)\chi(\rho, u - \xi)\,\mathrm{d}\xi.$$

(iii) *The corresponding entropy flux is given by*

$$\eta(\rho, u) = \int_{\mathbb{R}} g(\xi)\big[\theta\xi + (1-\theta)u\big]\chi(\rho, u - \xi)\,\mathrm{d}\xi.$$

(iv) *The entropy* S *is strictly convex in the variables* $(\rho, \rho u)$ *if and only if* g *is a strictly convex function.*

As an example, we mention for future reference that the physical entropy is the energy of the system given by

$$S_E(\rho, u) = \frac{1}{2}\rho u^2 + \frac{\kappa}{\gamma - 1}\rho^\gamma. \tag{42}$$

It enters the family of weak entropies and can be recovered as

$$S_E(\rho, u) = \int_{\mathbb{R}} \frac{\xi^2}{2}\chi(\rho, u - \xi)\,\mathrm{d}\xi.$$

Also density and momentum can be recovered as

$$\rho = \int_{\mathbb{R}} \chi(\rho, u - \xi)\,\mathrm{d}\xi, \qquad \rho u = \int_{\mathbb{R}} \xi\chi(\rho, u - \xi)\,\mathrm{d}\xi.$$

We refer to [46,49] for a proof of the above lemma.

As we have seen in the scalar case (see Section 2.2), for degenerate problems the vanishing viscosity method yields inequalities in the entropy relations which take into account possible singularities as shock waves. Therefore we now complete the system (39) by the full family of entropy inequalities for weak and convex in $(\rho, \rho u)$ entropies deduced from Lemma 4.1

$$\frac{\partial}{\partial t}S\big(\rho(t,x), u(t,x)\big) + \frac{\partial}{\partial x}\eta\big(\rho(t,x), u(t,x)\big) \leqslant 0. \tag{43}$$

4.2. *Kinetic formulation*

We are now ready to state the *kinetic formulation* for the system of isentropic gas dynamics. We call an *entropy solution* to the system (39), a couple $(\rho(t,x),\ \rho u(t,x))$, such that the inequality (43) holds for all convex entropy pairs given in the Lemma 4.1.

We claim that *entropy solutions* $(\rho(t,x),\ \rho u(t,x))$ to isentropic gas dynamics are equivalently described by the *single and singular* kinetic equation

$$\frac{\partial}{\partial t}\chi(\rho,u-\xi)+\frac{\partial}{\partial x}\big[\big[\theta\xi+(1-\theta)u\big]\chi(\rho,u-\xi)\big]=\frac{\partial^2}{\partial\xi^2}m(t,x,\xi), \tag{44}$$

for some nonpositive entropy dissipation measure $m(t,x,\xi)$. Moreover, we have the bound

$$-\int_0^\infty\int_{\mathbb{R}\times\mathbb{R}} m(t,x,\xi)\leqslant\int_{\mathbb{R}} S_E(t=0,x)\,\mathrm{d}x$$

(with the notation (42)).

DERIVATION OF THE CLAIM. Indeed, this statement is equivalent, after testing the ξ derivative against a function $g(\xi)$, to the equation in (t,x)

$$\frac{\partial}{\partial t}\int_{\mathbb{R}} g(\xi)\chi(\rho,u-\xi)\,\mathrm{d}\xi+\frac{\partial}{\partial x}\int_{\mathbb{R}} g(\xi)\big[\big[\theta\xi+(1-\theta)u\big]\chi(\rho,u-\xi)\big]\,\mathrm{d}\xi$$
$$=\int_{\mathbb{R}} g''(\xi)m(t,x,\xi)\,\mathrm{d}\xi,$$

which is equivalent, using Lemma 4.1(ii) and (iii), to the equation (to be understood also in distribution sense)

$$\frac{\partial}{\partial t}S\big(\rho(t,x),u(t,x)\big)+\frac{\partial}{\partial x}\eta\big(\rho(t,x),u(t,x)\big)=\int_{\mathbb{R}} g''(\xi)m(t,x,\xi)\,\mathrm{d}\xi,$$

and the right-hand side is nonpositive for all g convex (i.e., all convex weak entropy S by Lemma 4.1(iv)) if and only if m is nonpositive (and is therefore a measure).

The bound on the measure can be obtained using $g(\xi)=\frac{|\xi|^2}{2}$ thanks to the kinetic definition of the energy after (42). Indeed, we remark that the above equality also gives

$$\frac{\mathrm{d}}{\mathrm{d}t}\int_{\mathbb{R}} S_E\big(\rho(t,x),u(t,x)\big)\,\mathrm{d}x=\int_{\mathbb{R}\times\mathbb{R}} m(t,x,\xi)\,\mathrm{d}x\,\mathrm{d}\xi,$$

and it remains to integrate in time since $S_E\geqslant 0$. □

By opposition to the scalar case, (44) for the function $\chi(t,x,\xi)$ is not anymore a purely kinetic equation because the advection velocity is now $(\theta\xi+(1-\theta)u)$ which contains explicitly the macroscopic velocity $u(t,x)$. Therefore linear operations are not allowed here. This class of semikinetic equation is not as handy as purely kinetic equations although they arise in many physical contexts.

4.3. *Applications of the kinetic formulation*

As a first application, we mention the existence of a solution to the system of isentropic gas dynamics including vacuum. It can be obtained by the vanishing viscosity method considering the viscous system of two equations on the density $\rho_\varepsilon(t,x)$ and on the momentum $q_\varepsilon(t,x) = \rho_\varepsilon u_\varepsilon(t,x)$ given by the equations:

$$\begin{cases} \frac{\partial}{\partial t}\rho_\varepsilon + \frac{\partial}{\partial x}(\rho_\varepsilon u_\varepsilon) = \varepsilon \frac{\partial^2}{\partial x^2}\rho_\varepsilon, \quad t \geqslant 0,\ x \in \mathbb{R}, \\ \frac{\partial}{\partial t}(\rho_\varepsilon u_\varepsilon) + \frac{\partial}{\partial x}\big(\rho_\varepsilon u_\varepsilon^2 + p(\rho_\varepsilon)\big) = \varepsilon \frac{\partial^2}{\partial x^2}(\rho_\varepsilon u_\varepsilon). \end{cases} \tag{45}$$

Here the pressure $p(\rho)$ is still assumed to be a polytropic law (40). Also, we assume that the initial datum satisfy

$$\begin{aligned} &\rho_\varepsilon^0, u_\varepsilon^0 \text{ are smooth, uniformly bounded and uniformly of finite energy,} \\ &\rho_\varepsilon^0(x) \geqslant \varepsilon \exp\big(-|x|^2\big). \end{aligned} \tag{46}$$

Then, it is proved in [19,46,49], that the system (45) admits a smooth solution with a lower bound on the density

$$\rho_\varepsilon(t,x) \geqslant \varepsilon \exp\big(-C(\varepsilon,t)|x|^2\big),$$

where the constant $C(\varepsilon,t)$ can be computed explicitly (this is a delicate point; see [49] for instance).

The convergence result as ε vanishes is stated in the following theorem.

THEOREM 4.2 [46,44]. *Let $1 < \gamma < \infty$ and assume $\rho_\varepsilon^0 \leqslant C$, $|u_\varepsilon^0| \leqslant C$. Then, the solutions $(\rho_\varepsilon, u_\varepsilon)$ to the system* (45) *are uniformly bounded and we may extract a subsequence, such that $(\rho_\varepsilon, \rho_\varepsilon u_\varepsilon)$ converges almost everywhere and thus in $L^p_{\mathrm{loc}}((0,\infty)\times\mathbb{R})$ for all $1 \leqslant p < +\infty$, to an entropy solution to* (39).

The proof of the strong compactness for this subsequence uses the method of compensated compactness [48,58,20] combined with its kinetic formulation.

As other applications, we can mention a priori bounds for the hyperbolic system (in L^∞) and more unusual regularizing effects in $L^p_{(t,x)}$ spaces (see [49] for details). This last result can be stated as follows. For all $y \in \mathbb{R}$, we have, for some constant which only depends on γ,

$$\int_0^\infty \big(\rho|u|^3 + \rho^{\gamma+\theta}\big)(t,y)\,\mathrm{d}t \leqslant C_\gamma \int_{\mathbb{R}} S_E^0(x)\,\mathrm{d}x$$

(recall the definition of the energy S_E in (42)). An open question is however to derive from the kinetic formulation some regularizing effect on the solutions in Sobolev spaces as this was done in the scalar case, see Section 3.1.

We also mention that Bouchut and Berthelin [4] (see also the related works in the references) handled the relaxation approximation to isentropic gas dynamics. It is not possible to prove directly the convergence for the relaxation equation

$$\begin{aligned}&\frac{\partial}{\partial t} f(t,x,\xi)\\&\quad+\frac{\partial}{\partial x}\big[\big[\theta\xi+(1-\theta)u\big] f(t,x,\xi)+\frac{1}{\varepsilon}\big[f(t,x,\xi)-\chi(\rho,u-\xi)\big]\\&=0,\end{aligned}$$

as an approximation of (44). Indeed, the entropy structure is not fulfilled by this equation because it is not true that $\chi(\rho,u-\xi)$ minimizes the $\int_{\mathbb{R}} \xi^2 f(\xi)\,d\xi$ subject to the constraints that $\int_{\mathbb{R}} f(\xi)\,d\xi=\rho$ and $\int_{\mathbb{R}} \xi f(\xi)\,d\xi=\rho u$. To achieve a relaxation approximation Bouchut and Berthelin introduced a model with two kinetic variables and which provides a pure kinetic relaxation approximation. They also prove, using all the entropies and compensated compactness, that this model converges to the system of isentropic gas dynamics. Another approximation is by time splitting, and the convergence proof is given by Vasseur [59,60].

5. Kinetic schemes of isentropic gas dynamics

In this section we wish to illustrate some numerical applications of the kinetic approach to macroscopic conservation laws. Namely we show how ideas from kinetic theory allow to derive finite volume schemes for macroscopic equations. For the sake of simplicity, we restrict our presentation to isentropic gas dynamics. We first present a formalism to represent the system of isentropic gas dynamics which is simpler than the kinetic formulation in Section 4 (first subsection). It is used to build finite volume solvers in a second subsection and numerical results are presented in a third subsection.

5.1. *Kinetic representation*

We consider again the system of isentropic gas dynamics treated in previous section,

$$\begin{cases}\frac{\partial}{\partial t}\rho+\frac{\partial}{\partial x}(\rho u)=0, & t\geqslant 0,\ x\in\mathbb{R},\\ \frac{\partial}{\partial t}(\rho u)+\frac{\partial}{\partial x}\big(\rho u^2+p(\rho)\big)=0,\end{cases}\tag{47}$$

for a polytropic gas, i.e., when the pressure p is given by the barotropic and polytropic equation of state. For our present purpose, we restrict ourselves to the values (cf. with (40))

$$p(\rho)=\kappa\rho^\gamma,\quad 1<\gamma\leqslant 3,\tag{48}$$

with $\kappa>0$ a given constant. We now complete this system with a single entropy inequality for the total energy of the system (total energy = kinetic energy + potential energy).

This is,

$$E(t,x) := E(\rho,u) = \frac{1}{2}\rho u^2 + \frac{\kappa}{\gamma-1}\rho^\gamma, \tag{49}$$

$$\frac{\partial}{\partial t}E + \frac{\partial}{\partial x}\big[(E+p)u\big] \leqslant 0. \tag{50}$$

A *kinetic representation* of this system can be given. It only uses the single entropy given by energy. This representation has the advantage to lead to a purely kinetic equation (cf. with (44)), but the concept is much weaker than that of kinetic formulation, and especially the singular right-hand side carries little information. Also, this representation is based on a minimization problem which only holds for $\gamma \leqslant 3$; this explains this limitation on γ that does not appear in Section 4.

The *kinetic representation* is an equation equivalent to the original hyperbolic system (47)–(48), and is based on the representation formula

$$\begin{pmatrix} \rho \\ \rho u \\ \rho u^2 + \kappa\rho^\gamma \end{pmatrix} = \int_{\mathbb{R}} \begin{pmatrix} 1 \\ \xi \\ \xi^2 \end{pmatrix} M(\rho, u-\xi)\,\mathrm{d}\xi. \tag{51}$$

It is fulfilled when the function M is given, for instance, as

$$\begin{aligned} &M(\rho,\xi) = \rho^{1-\theta}\chi\left(\frac{\xi}{\rho^\theta}\right), \qquad \theta = \frac{\gamma-1}{2}, \\ &\chi \text{ is an even nonnegative function, and} \\ &\int_{\mathbb{R}} \chi(w)\,\mathrm{d}w = 1, \qquad \int_{\mathbb{R}} w^2\chi(w)\,\mathrm{d}w = \kappa. \end{aligned} \tag{52}$$

With these notations, one can readily check by ξ integrations against the weights 1 and ξ, that the solutions to (47)–(48) are also the solutions to the kinetic equation

$$\frac{\partial}{\partial t}M(\rho, u-\xi) + \xi\frac{\partial}{\partial x}M(\rho, u-\xi) = Q(t,x,\xi), \tag{53}$$

for some unknown "collision term" $Q(t,x,\xi)$ (a distribution) which satisfies,

$$\int_{\mathbb{R}} Q\,\mathrm{d}\xi = 0, \qquad \int_{\mathbb{R}} \xi Q\,\mathrm{d}\xi = 0. \tag{54}$$

Indeed, integrating (53) in ξ against the weights 1 and ξ, we obtain (47)–(48) just using the property (54) for Q. Conversely, being given a solution (ρ,u) to (47)–(48), we obtain an explicit right-hand side Q (which we do not compute here) and which satisfies (54).

Again, by opposition to the *kinetic formulations*, the right-hand side Q does not vanish for smooth solutions here, whatever is the choice of χ.

A possible motivation for this kinetic representation comes from a relaxation model. Equation (53) can be considered as the limit as $\lambda \to \infty$ (but it is an open problem to give a rigorous proof) of the relaxation model

$$\frac{\partial}{\partial t} f(t,x,\xi) + \xi \frac{\partial}{\partial x} f(t,x,\xi) + \lambda\big[f(t,x,\xi) - M(\rho, u-\xi)\big] = 0,$$

$$\rho(t,x) = \int_{\mathbb{R}} f(t,x,\xi)\,\mathrm{d}\xi = \int_{\mathbb{R}} M(\rho, u-\xi)\,\mathrm{d}\xi,$$

$$\rho u(t,x) = \int_{\mathbb{R}} \xi f(\xi)\,\mathrm{d}\xi = \int_{\mathbb{R}} \xi M(\rho, u-\xi)\,\mathrm{d}\xi.$$

From this we infer that $Q(t,x,\xi) = \lim_{\lambda\to\infty} \lambda[f(t,x,\xi) - M(\rho, u-\xi)]$.

Next, we indicate the relation to the entropy inequality (50). This is more involved and is related to a Gibbs principle for the microscopic energy

$$\mathcal{E}(\xi, f) = \int_{\mathbb{R}} \left[\frac{\xi^2}{2} f(\xi) + \tilde{\kappa} f(\xi)^{(\gamma+1)/(3-\gamma)}\right] \mathrm{d}\xi.$$

PROPOSITION 5.1. *We fix a real number $\tilde{\kappa} > 0$ and set $\lambda = \frac{3-\gamma}{2(\gamma-1)}$. Then, the minimization problem,*

$$E(\rho,u) = \min\left\{\int_{\mathbb{R}} \left[\frac{\xi^2}{2} f(\xi) + \tilde{\kappa} f(\xi)^{(\gamma+1)/(3-\gamma)}\right] \mathrm{d}\xi;\right.$$
$$\left. f \geqslant 0,\ \int_{\mathbb{R}} f(\xi)\,\mathrm{d}\xi = \rho,\ \int_{\mathbb{R}} \xi f(\xi)\,\mathrm{d}\xi = \rho u\right\},$$

admits a unique minimizer given by the formula

$$M_E(\rho, u-\xi) = \alpha \rho^{(3-\gamma)/2} \left(1 - \frac{(u-\xi)^2}{\beta \rho^{\gamma-1}}\right)_+^{\lambda},$$
$$\chi_E = \alpha \left(1 - \frac{w^2}{\beta}\right)_+^{\lambda}, \tag{55}$$

with $\beta = \frac{2\kappa\gamma}{\gamma-1}$, $\alpha^{-1} = \int_{\mathbb{R}} (1 - \frac{\xi^2}{\beta})_+^{\lambda}\,\mathrm{d}\xi$, $\alpha^{1/\lambda}\tilde{\kappa}(\gamma^2 - 1) = \kappa\gamma(3-\gamma)$. Hence, the compatibility relations with the γ-law isentropic gas dynamics system, (51), *are fulfilled. The former definition* (49) *of the macroscopic energy E is also satisfied.*

A consequence of this microscopic representation of energy is an additional property of the collision term Q in the kinetic equation (53). For this choice of M_E, we have (in distributional sense)

$$\int_{\mathbb{R}} \left[\frac{\xi^2}{2} + \tilde{\kappa} \frac{\gamma+1}{3-\gamma} M_E(\rho, u-\xi)^{1/\lambda} \right] Q(t,x,\xi)\, d\xi \leqslant 0. \tag{56}$$

It is derived as follows. We formally deduce from (53) the relation

$$\begin{aligned} &\frac{\partial}{\partial t}\left[\frac{\xi^2}{2} M_E(\rho,u-\xi) + \tilde{\kappa} M_E(\rho,u-\xi)^{(\gamma+1)/(3-\gamma)}\right] \\ &\quad + \xi \frac{\partial}{\partial x}\left[\frac{\xi^2}{2} M_E(\rho,u-\xi) + \tilde{\kappa} M_E(\rho,u-\xi)^{(\gamma+1)/(3-\gamma)}\right] \\ &= \left[\frac{\xi^2}{2} + \tilde{\kappa}\frac{\gamma+1}{3-\gamma} M_E(\rho,u-\xi)^{1/\lambda}\right] Q(t,x,\xi), \end{aligned}$$

and (56) follows by ξ integration since this equation reduces then to the entropy inequality (50) thanks to the relations (deduced from Proposition 5.1 and easy calculations)

$$E = \int_{\mathbb{R}} \left[\frac{\xi^2}{2} M_E(\rho,u-\xi) + \tilde{\kappa} M_E(\rho,u-\xi)^{(\gamma+1)/(3-\gamma)}\right] d\xi,$$

$$(E+p)u = \int_{\mathbb{R}} \xi \left[\frac{\xi^2}{2} M_E(\rho,u-\xi) + \tilde{\kappa} M_E(\rho,u-\xi)^{(\gamma+1)/(3-\gamma)}\right] d\xi.$$

Of course, the multiplication of the distribution Q by $M_E(\rho, u-\xi)^{1/\lambda}$ does not make a real sense. It can be made rigorous by a convolution argument. We denote by $M_{E,\varepsilon}$ the space and time regularization of $M_E(t,x,\xi)$ and we notice that the kinetic equation (53) (by opposition to the hyperbolic system itself) can be regularized. We obtain the equation

$$\frac{\partial}{\partial t} M_{E,\varepsilon}(\rho,u-\xi) + \xi \frac{\partial}{\partial x} M_{E,\varepsilon}(\rho,u-\xi) = Q_\varepsilon(t,x,\xi),$$

which holds in a classical sense; the terms in the left-hand side are regular enough in (t,x,ξ), therefore Q_ε is regular also. Hence, we can perform the nonlinear operation and obtain

$$\begin{aligned} &\frac{\partial}{\partial t}\left[\frac{\xi^2}{2} M_{E,\varepsilon}(\rho,u-\xi) + \tilde{\kappa} M_{E,\varepsilon}(\rho,u-\xi)^{(\gamma+1)/(3-\gamma)}\right] \\ &\quad + \xi \frac{\partial}{\partial x}\left[\frac{\xi^2}{2} M_{E,\varepsilon}(\rho,u-\xi) + \tilde{\kappa} M_{E,\varepsilon}(\rho,u-\xi)^{(\gamma+1)/(3-\gamma)}\right] \\ &= \left[\frac{\xi^2}{2} + \tilde{\kappa}\frac{\gamma+1}{3-\gamma} M_{E,\varepsilon}(\rho,u-\xi)^{1/\lambda}\right] Q_\varepsilon(t,x,\xi). \end{aligned}$$

Since $M_{E,\varepsilon} \to M_E$ strongly, we may pass to the limit in the left-hand side and obtain

$$\begin{aligned}
&\frac{\partial}{\partial t}\left[\frac{\xi^2}{2} M_E(\rho, u-\xi) + \tilde{\kappa} M_E(\rho, u-\xi)^{(\gamma+1)/(3-\gamma)}\right] \\
&\quad + \xi \frac{\partial}{\partial x}\left[\frac{\xi^2}{2} M_E(\rho, u-\xi) + \tilde{\kappa} M_E(\rho, u-\xi)^{(\gamma+1)/(3-\gamma)}\right] \\
&= R(t, x, \xi).
\end{aligned}$$

This gives the rigorous meaning to (56); there is a distribution R such that, whatever is the regularization kernel, we have

$$\left[\frac{\xi^2}{2} + \tilde{\kappa} \frac{\gamma+1}{3-\gamma} M_{E,\varepsilon}(\rho, u-\xi)^{1/\lambda}\right] Q_\varepsilon(t, x, \xi) \rightharpoonup R(t, x, \xi)$$

and

$$\int_{\mathbb{R}} R(t, x, \xi)\, \mathrm{d}\xi \leqslant 0.$$

5.2. *Finite volumes*

We consider again the system of gas dynamics for a barotropic gas, that we complete with the entropy inequality for the total energy

$$\begin{aligned}
&\frac{\partial}{\partial t}\rho + \frac{\partial}{\partial x}(\rho u) = 0, \\
&\frac{\partial}{\partial t}(\rho u) + \frac{\partial}{\partial x}\big(\rho u^2 + p(\rho)\big) = 0, \qquad (57) \\
&\frac{\partial}{\partial t}E + \frac{\partial}{\partial x}\big[(E+p)u\big] \leqslant 0,
\end{aligned}$$

$$p(\rho) = \kappa \rho^\gamma, \quad 1 < \gamma \leqslant 3, \qquad (58)$$

with a given constant $\kappa > 0$, and the energy is

$$E(t, x) := E(\rho, u) = \frac{1}{2}\rho u^2 + \frac{\kappa}{\gamma - 1}\rho^\gamma. \qquad (59)$$

Finite volumes methods consist in a discretized version of system (57) under a form which preserves the conservation laws

$$\begin{aligned}
&\rho_i^{n+1} - \rho_i^n + \sigma_i\big(A_{i+1/2}^{\rho,n} - A_{i-1/2}^{\rho,n}\big) = 0, \\
&q_i^{n+1} - q_i^n + \sigma_i\big(A_{i+1/2}^{q,n} - A_{i-1/2}^{q,n}\big) = 0, \quad i \in \mathbb{Z},\ n \in \mathbb{N},
\end{aligned}$$

with the following definitions: we are given cells $C_i = (x_{i+1/2}, x_{i-1/2})$ of size $h_i = x_{i+1/2} - x_{i-1/2} > 0$ and we define $\sigma_i = \frac{\Delta t}{h_i}$ for some time step Δt which is chosen small enough using a Courant–Friedrichs–Levy (CFL) condition (see Theorem 5.2). We set $h = \min_{i \in \mathbb{Z}} h_i$. Also, the principle of finite volume methods is to use approximation in L^1 sense, namely we have in mind that

$$\rho_i^n \approx \frac{1}{h_i} \int_{x_{i-1/2}}^{x_{i+1/2}} \rho(n\Delta t, x)\,\mathrm{d}x, \qquad q_i^n \approx \frac{1}{h_i} \int_{x_{i-1/2}}^{x_{i+1/2}} \rho u(n\Delta t, x)\,\mathrm{d}x.$$

It is natural here to consider that the continuous solution is approximated by the piecewise constant functions

$$\begin{cases} \rho_h(t, x) = \rho_i^n & \text{for } x \in C_i,\ n\Delta t \leqslant t < (n+1)\Delta t, \\ q_h(t, x) = q_i^n & \text{for } x \in C_i,\ n\Delta t \leqslant t < (n+1)\Delta t. \end{cases} \tag{60}$$

Notice that the discrete conservation laws for mass and momentum are now, for all $n \in \mathbb{N}$,

$$\sum_{i \in \mathbb{Z}} h_i \rho_i^n = \sum_{i \in \mathbb{Z}} h_i \rho_i^0, \qquad \sum_{i \in \mathbb{Z}} h_i q_i^n = \sum_{i \in \mathbb{Z}} h_i q_i^0.$$

We refer to [29] for a general presentation of finite volume schemes in fluid dynamics and [24,38,42] for general mathematical and numerical introductions.

The difficulty in finite volume methods is to find stable interpolation formulas called *solvers* to compute the numerical fluxes $A_{i+1/2}^{\rho,n}$ and $A_{i+1/2}^{q,n}$ from the cell values ρ_i^n, ρ_{i+1}^n, q_i^n and q_{i+1}^n. Especially the intuitive centered formulas are unstable and cannot be used in practice (as for instance, $A_{i+1/2}^{\rho,n} = \frac{1}{2}[\rho u_i^n + \rho u_{i+1}^n]$).

The idea to use kinetic theory in order to derive solvers for fluid equations is very natural and has been used by many authors, let us mention, for instance, [17,36,54,55,64], and additional references in [49].

5.3. *Kinetic solvers*

A family of first-order approximation solvers $A_{i+1/2}^{\rho,n}$, $A_{i+1/2}^{q,n}$ in (60) can be derived from the kinetic representation of the system of isentropic gas dynamics. Using the notations (51)–(52), they correspond to an upwind scheme for the kinetic-transport equation (53) where the collision terms Q is not kept (it is only used implicitly in order to provide from f_i^{n+1} an equilibrium M_i^{n+1} at the next time step):

$$\begin{aligned} &f_i^{n+1}(\xi) - M_i^n(\xi) + \sigma_i \xi \big(M_{i+1/2}^n(\xi) - M_{i-1/2}^n(\xi)\big) = 0, \\ &M_i^n(\xi) = M\big(\rho_i^n, u_i^n - \xi\big), \quad u_i^n = \frac{q_i^n}{\rho_i^n}, \\ &M_{i+1/2}^n(\xi) = \begin{cases} M_i^n(\xi) & \text{for } \xi \geqslant 0, \\ M_{i+1}^n(\xi) & \text{for } \xi \leqslant 0. \end{cases} \end{aligned} \tag{61}$$

We now define the macroscopic variables as usual by ξ integration,

$$\rho_i^{n+1} = \int_{\mathbb{R}} f_i^{n+1}(\xi)\,\mathrm{d}\xi, \qquad q_i^{n+1} = \int_{\mathbb{R}} \xi f_i^{n+1}(\xi)\,\mathrm{d}\xi.$$

Next, we deduce from (61) a finite volume scheme (60) with fluxes under the form of a flux splitting. Namely, using vector notations for the fluxes we obtain:

$$\begin{aligned} A_{i+1/2}^n &= \mathcal{A}_+(\rho_i^n, u_i^n) + \mathcal{A}_-(\rho_{i+1}^n, u_{i+1}^n), \\ \mathcal{A}_+(\rho, u) &= \int_{\xi \geqslant 0} \xi \begin{pmatrix} 1 \\ \xi \end{pmatrix} M(\rho, u-\xi)\,\mathrm{d}\xi, \\ \mathcal{A}_-(\rho, u) &= \int_{\xi \leqslant 0} \xi \begin{pmatrix} 1 \\ \xi \end{pmatrix} M(\rho, u-\xi)\,\mathrm{d}\xi. \end{aligned} \tag{62}$$

We notice that such a scheme is always consistent, i.e.,

$$\mathcal{A}_+(\rho, u) + \mathcal{A}_-(\rho, u) = \begin{pmatrix} \rho u \\ \rho u^2 + p \end{pmatrix},$$

thanks to relations (51).

Depending on the choice of the χ function used to define M, we obtain different properties which we explain in the next results.

THEOREM 5.2. *Assume that, in* (52), *we have*

$$\chi(w) = 0 \quad \text{for } |w| \geqslant w_M.$$

Then, under the CFL condition

$$\Delta t \sup_{i\in\mathbb{Z}} \Big[|u_i^n| + \sqrt{w_M (\rho_i^n)^{(\gamma-1)/2}} \Big] \leqslant h := \min_{i\in\mathbb{Z}} h_i, \tag{63}$$

the method (61), (62) *satisfies for all* $i \in \mathbb{Z}$, $n \in \mathbb{N}$,

(i) (positivity) $\rho_i^n \geqslant 0$,

(ii) (discrete entropy inequality) *for the choice* $\chi = \chi_E$ *in* (55), *the inequality holds*

$$E_i^{n+1} - E_i^n + \sigma_i \big(A_{i+1/2}^{E,n} - A_{i-1/2}^{E,n} \big) \leqslant 0,$$

with discrete entropy flux given by

$$\begin{aligned} A_{i+1/2}^{E,n} &= \mathcal{A}_+^E(\rho_i^n, q_i^n) + \mathcal{A}_-^E(\rho_{i+1}^n, q_{i+1}^n), \\ \mathcal{A}_+^E(\rho, u) &= \int_{\xi \geqslant 0} \xi \Big(\frac{\xi^2}{2} M_E(\rho, u-\xi) + \tilde{\kappa} M_E(\rho, u-\xi)^{(\gamma+1)/(3-\gamma)} \Big)\,\mathrm{d}\xi, \\ \mathcal{A}_-^E(\rho, u) &= \int_{\xi \leqslant 0} \xi \Big(\frac{\xi^2}{2} M_E(\rho, u-\xi) + \tilde{\kappa} M_E(\rho, u-\xi)^{(\gamma+1)/(3-\gamma)} \Big)\,\mathrm{d}\xi. \end{aligned}$$

Since the kinetic representation only provides a single entropy inequality, we are unable to prove the convergence of these schemes. We recall that the only known argument for proving convergence is based on using compensated compactness on the full family of weak entropy inequalities (see [49]). The above theorem gives a weak stability criteria, in the case of the choice M_E, since it provides l^γ bounds for ρ_i^n and l^2 bounds for q_i^n. Indeed, we notice the mass conservation law

$$\sum_{i\in\mathbb{Z}} \rho_i^n = \sum_{i\in\mathbb{Z}} \rho_i^0 \quad \forall n \in \mathbb{N}^*,$$

and the inequality (deduced from (ii))

$$\sum_{i\in\mathbb{Z}} E_i^{n+1} \leqslant \sum_{i\in\mathbb{Z}} E_i^n \leqslant \cdots \leqslant \sum_{i\in\mathbb{Z}} E_i^0$$

also gives the l^γ bounds for ρ_i^n: $\forall n \in \mathbb{N}^*$,

$$\sum_{i\in\mathbb{Z}} (\rho_i^n)^\gamma \leqslant \sum_{i\in\mathbb{Z}} E_i^n \leqslant \sum_{i\in\mathbb{Z}} E_i^0,$$

and the l^2 bounds for q_i^n:

$$\Big(\sum_{i\in\mathbb{Z}} q_i^{n+1}\Big)^2 \leqslant \sum_{i\in\mathbb{Z}} \rho_i^{n+1} \sum_{i\in\mathbb{Z}} \rho_i^{n+1} (u_i^{n+1})^2 \leqslant \sum_{i\in\mathbb{Z}} \rho_i^0 \sum_{i\in\mathbb{Z}} E_i^0.$$

This stability, based on a single entropy, seems enough to guarantee convergence of the numerical scheme.

REMARK 5.3. The natural speeds of propagation $u \pm \sqrt{\kappa\gamma\rho^{\gamma-1}}$ for system (57) are usually related to CFL conditions like (63). Here we notice that we always have $w_M \geqslant \kappa$, with equality for $\chi(w) = \frac{1}{2}[\delta(w = \sqrt{\kappa}) + \delta(w = -\sqrt{\kappa})]$. This proves that some choices of χ should not converge and the statement (i) is too weak. The choice of χ_E leads to a compatible CFL condition, then we have indeed

$$w_M = \sqrt{\frac{2\kappa\gamma}{\gamma - 1}} \geqslant \sqrt{\kappa\gamma}.$$

PROOF OF THEOREM 5.2. We use the kinetic representation (61) of the method (60), (62), that we write

$$f_i^{n+1}(\xi) = M_i^n(\xi)\big(1 - \sigma_i|\xi|\big) + \sigma_i \xi_+ M_{i-1}^n(\xi) + \sigma_i \xi_- \, M_{i+1}^n(\xi).$$

From the assumption on the support of the function χ, all the terms in the right-hand side vanish for $|\xi| > \sup_{i\in\mathbb{Z}}[|u_i^n| + \sqrt{w_M(\rho_i^n)^{(\gamma-1)/2}}]$. Therefore, under the CFL condition, on the complementary interval, $f_i^{n+1}(\xi)$ is a convex combination of $M_i^n(\xi)$, $M_{i-1}^n(\xi)$ and $M_{i+1}^n(\xi)$. Therefore, since M is nonnegative, we deduce that f_i^{n+1} is nonnegative too and thus, by ξ integration, we deduce that ρ_i^{n+1} is nonnegative too. This proves statement (i).

As for statement (ii), we use again the convex combination to deduce, since $\frac{\gamma+1}{3-\gamma} > 1$, that

$$\begin{aligned}
&\frac{\xi^2}{2} f_i^{n+1}(\xi) + \tilde{\kappa} f_i^{n+1}(\xi)^{(\gamma+1)/(3-\gamma)} \\
&\quad\leqslant \bigl(1 - \sigma_i|\xi|\bigr)\left[\frac{\xi^2}{2} M_i^n(\xi) + \tilde{\kappa} M_i^n(\xi)^{(\gamma+1)/(3-\gamma)}\right] \\
&\qquad + \sigma_i \xi_+ \left[\frac{\xi^2}{2} M_{i-1}^n(\xi) + \tilde{\kappa} M_{i-1}^n(\xi)^{(\gamma+1)/(3-\gamma)}\right] \\
&\qquad + \sigma_i \xi_- \left[\frac{\xi^2}{2} M_{i+1}^n(\xi) + \tilde{\kappa} M_{i+1}^n(\xi)^{(\gamma+1)/(3-\gamma)}\right].
\end{aligned}$$

By the Gibbs principle in Proposition 5.1, we deduce that

$$\begin{aligned}
E_i^{n+1} &\leqslant \int_{\mathbb{R}} \left[\frac{\xi^2}{2} f_i^{n+1}(\xi) + \tilde{\kappa} f_i^{n+1}(\xi)^{(\gamma+1)/(3-\gamma)}\right] \mathrm{d}\xi \\
&\leqslant \int_{\mathbb{R}} \bigl(1 - \sigma_i|\xi|\bigr)\left[\frac{\xi^2}{2} M_i^n(\xi) + \tilde{\kappa} M_i^n(\xi)^{(\gamma+1)/(3-\gamma)}\right] \mathrm{d}\xi \\
&\quad + \int_{\mathbb{R}} \xi_+ \left[\frac{\xi^2}{2} M_{i-1}^n(\xi) + \tilde{\kappa} M_{i-1}^n(\xi)^{(\gamma+1)/(3-\gamma)}\right] \mathrm{d}\xi \\
&\quad + \int_{\mathbb{R}} \xi_- \left[\frac{\xi^2}{2} M_{i+1}^n(\xi) + \tilde{\kappa} M_{i+1}^n(\xi)^{(\gamma+1)/(3-\gamma)}\right] \mathrm{d}\xi.
\end{aligned}$$

This inequality is equivalent to (ii), and Theorem 5.2 is proved. □

5.4. *Numerical comparisons*

In this section we compare the results given by two different choices of the function χ in the construction of kinetic schemes for Saint-Venant system which consists in the case $\gamma = 2$, $\kappa = \frac{g}{2}$ in previous section. The different tests we have performed show that the various choices of χ lead to rather similar results as long as its support is 'large enough'. We have selected two functions to illustrate this.

Firstly, we have used Kaniel's function [37] which speed of propagation is exactly the correct eigenvalues of the hyperbolic system. The function χ is

$$\chi_K(w) = \frac{1}{g}|w|\mathbf{1}_{\{|w|\leqslant\sqrt{g}\}}. \tag{64}$$

A narrower support leads to an inconsistency with the limiting system.

Secondly, we have used the very simple, piecewise constant, function

$$\chi_C(w) = \sqrt{\frac{1}{6g}}\mathbf{1}_{\{|w|\leqslant\sqrt{3g/2}\}}. \tag{65}$$

Notice that the entropic choice χ_E in (55) has a larger support and is used in [51] for nonflat bottoms (and the results are very close to those given by χ_C).

Both functions lead to very simple formulas and the codes are efficient. We denote by $Fr = u/\sqrt{gh}$ the Froud number (no dimension), then the fluxes are given by

$$A_\pm^\rho = Ch^{3/2}\big[Fr\, I^{\pm,1}(Fr) + I^{\pm,2}(Fr)\big],$$
$$A_\pm^q = Dh^2\big[Fr^2\, I^{\pm,1}(Fr) + 2FrI^{\pm,2}(Fr) + I^{\pm,3}(Fr)\big].$$

For Kaniel's choice the values are $C = \sqrt{g}$, $D = g$, and with

$$Frm = \max\big(-1, \min(1, Fr)\big),$$
$$I^{\pm,1}(Fr) = \big(1 \pm Frm\,|Frm|\big)/2,$$
$$I^{\pm,2}(Fr) = \pm\big(1 - |Frm|^3\big)/3,$$
$$I^{\pm,3}(Fr) = \big(1 \pm Frm^3|Frm|\big)/4.$$

For the piecewise constant choice, the values are $C = \sqrt{g/6}$, $D = g/\sqrt{6}$, and with $Frm = \max(-\sqrt{3/2}, \min(\sqrt{3/2}, Fr))$

$$I^{\pm,1}(Fr) = \left(\sqrt{\frac{3}{2}} \pm Frm\right),$$
$$I^{\pm,2}(Fr) = \pm\left(\frac{3}{2} - Frm^2\right)\Big/2,$$
$$I^{\pm,3}(Fr) = \left(\left(\frac{3}{2}\right)^{3/2} \pm Frm^3\right)\Big/3.$$

The numerical tests we present now have been performed with $g = 2$.

Figure 1 gives the computed Riemann decomposition for initial values $u = 0$, $h_{\text{left}} = 2$, $h_{\text{right}} = 0.5$ at time $t = 0.18$ with 100 grid points. We use a CFL number of 0.95. We draw the height of water h (left) and the water velocity u (right). The continuous line represents the exact solution and the crosses the computed solution. Here we have used the piecewise constant equilibrium $\chi_C(w)$ in the fluxes formulae. Kaniel's choice leads to very close results $\chi_K(w)$.

Figures 2 and 3 give the computed Riemann decomposition for initial values $u = 0$, $h_{\text{left}} = 2$, $h_{\text{right}} = 0.000005$. This right state is equivalent to vacuum and shows the ability

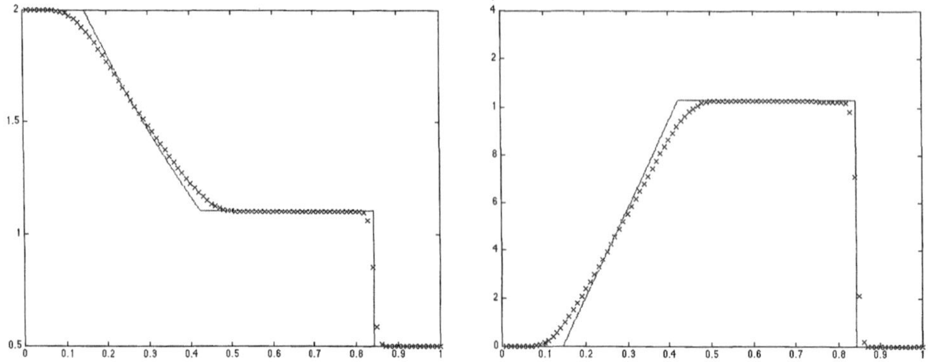

Fig. 1. A standard Riemann decomposition. A kinetic scheme based on piecewise constant equilibrium χ_C. Left: height of water, right: water velocity. Continuous line – exact solution, crosses – computed solution.

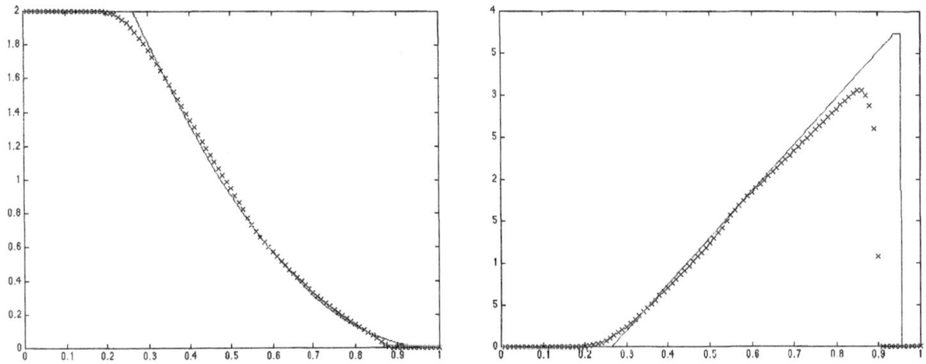

Fig. 2. Riemann decomposition including a vacuum state (dry soil). A kinetic scheme based on piecewise constant equilibrium χ_C. Left: height of water, right: water velocity. Continuous line – exact solution, crosses – computed solution.

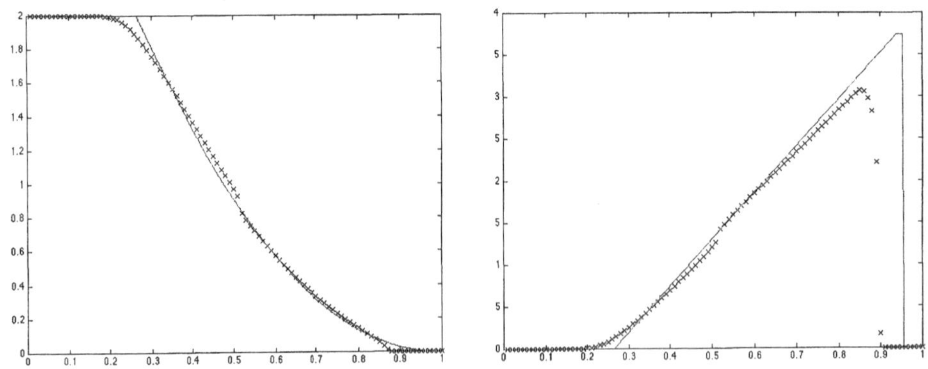

Fig. 3. The same Riemann decomposition and output as in Figure 2. Kinetic scheme based on Kaniel's equilibrium χ_K.

of the schemes to handle smoothly the case $h = 0$ at time $t = 0.12$ with 100 grid points. We use again a CFL number of 0.95. In Figure 2 we have used $\chi_C(w)$. In Figure 3 we have used $\chi_K(w)$.

Both schemes give a usual first-order accurate result. The main difference is that the results obtained with χ_K exhibit a slight jump at the sonic point for the particularly sharp Riemann problem with vacuum.

6. Conclusion

The kinetic approach to macroscopic conservation laws relies firstly on the physical description of a gas at the kinetic level thanks to Boltzmann's formalism. However, it goes much beyond in two directions.

The first direction is that of *kinetic formulations*, restricted to systems with many entropies, which systematically provides a way to handle full families of entropies by single equations. For scalar hyperbolic–parabolic equations it is a way to prove strong (and optimal) regularizing effects using the method of kinetic averaging lemmas which use the Fourier transform (it may be applied here because the kinetic formulation is a linear equation on nonlinear quantities).

The second direction is that of *kinetic representations*, which only uses one entropy, is much less demanding than the kinetic formulation. It allows to represent a hyperbolic system by integrating the underlying kinetic equation and maybe uses, for instance, to produce finite volumes solvers.

References

[1] L. Ambrosio, C. De Lellis and C. Mantegazza, *Line energies for gradient vector fields in the plane*, Calc. Var. Partial Differential Equations **9** (1999), 327–355.

[2] L. Ambrosio, B. Kircheim, M. Lecumberry and T. Rivière, *Rectifiability of Defect Measures Arising in Micromagnetics Domains*, Volume Dedicated to the 80th Birthday of O. Ladyzhenskaya, Kluwer Academic, New York (2002).

[3] M. Belhadj, J.-F. Gerbeau and B. Perthame, *A multiscale colloid transport model with anisotropic degenerate diffusion*, Asymptot. Anal. **34**(1) (2003), 41–54.

[4] F. Berthelin and F. Bouchut, *Kinetic invariant domains and relaxation limit from a BGK model to isentropic gas dynamics*, Asymptot. Anal. **31**(2) (2002), 153–176.

[5] F. Bouchut, *Entropy satisfying flux vector splittings and kinetic BGK models*, Numer. Math. **94**(4) (2003), 623–672.

[6] F. Bouchut, F. Golse and M. Pulvirenti, *Kinetic equations and asymptotic Theories*, L. Desvillettes and B. Perthame, eds, Series in Appl. Math., Vol. 4, Elsevier, Paris (2000), 1–40.

[7] Y. Brenier, *Résolution d'équations d'évolution quasilinéaires en dimensions N d'espace à l'aide d'équations linéaires en dimensions* $N+1$, J. Differential Equations **50** (1982), 375–390.

[8] J. Carillo, *Entropy solutions for nonlinear degenerate problems*, Arch. Ration. Mech. Anal. **147** (1999), 269–361.

[9] C. Cercignani, *The Boltzmann Equation and Its Application*, Appl. Math. Sci., Vol. 67, Springer-Verlag, Berlin (1988).

[10] C. Cercignani, R. Illner and M. Pulvirenti, *The Mathematical Theory of Dilute Gases*, Appl. Math. Sci., Vol. 106, Springer-Verlag, Berlin (1994).

[11] G.-Q. Chen and P.G. LeFloch, *Compressible Euler equations with general pressure law*, Arch. Ration. Mech. Anal. **153**(3) (2000), 221–259.

[12] G.-Q. Chen and B. Perthame, *Uniqueness for degenerate parabolic scalar conservation laws*, Ann. Inst. H. Poincaré Anal. Non Linéaire **20** (2003), 645–668.

[13] C.M. Dafermos, *Hyperbolic Conservation Laws in Continuum Physics*, Springer-Verlag, GM 325 (1999).

[14] R. Dautray and J.-L. Lions, *Mathematical Analysis and Numerical Methods for Science and Technology*, Springer-Verlag, Berlin (1992).

[15] C. De Lellis, F. Otto and M. Westdickenberg, *Structure of entropy solutions: Application to variational problems*, Preprint no 95 (2002); available at http://www.mis.mpg.de/preprints/2002.

[16] C. De Lellis and M. Westdickenberg, *On the optimality of velocity averaging lemmas*, Ann. Inst. H. Poincaré Anal. Non Linéaire **20**(6) (2003), 1075–1085.

[17] S.M. Deshpande, *A second order accurate, kinetic-theory based, method for inviscid compressible flows*, NASA Langley Techn. Paper **2613** (1986).

[18] A. Desimone, R.W. Kohn, S. Müller and F. Otto, *Magnetic microstructures, a paradigm of multiscale problems*, ICIAM 99 (Edinburgh), Oxford Univ. Press, Oxford (2000), 175–190.

[19] R.J. DiPerna, *Convergence of the viscosity method for isentropic gas dynamics*, Comm. Math. Phys. **91** (1983), 1–30.

[20] R.J. DiPerna, *Compensated compactness and general systems of conservation laws*, Trans. Amer. Math. Soc. **292** (1985), 383–420.

[21] R.J. DiPerna, P.-L. Lions and Y. Meyer, *L^p regularity of velocity averages*, Ann. Inst. H. Poincaré Anal. Non Linéaire **8**(3/4) (1991), 271–287.

[22] L.C. Evans, *Partial Differential Equations*, Graduate Studies in Mathematics, Vol. 19, American Mathematical Society (1998).

[23] R. Eymard, T. Gallouët and R. Herbin, *Finite volume methods*, Handbook of Numerical Analysis, Vol. VIII, P.G. Ciarlet and J.-L. Lions, eds, North-Holland, Amsterdam (2001), 713–1020.

[24] R. Eymard, T. Gallouët, R. Herbin and A. Michel, *Convergence of a finite volume scheme for nonlinear degenerate parabolic equations*, Numer. Math. **92** (2002), 41–82.

[25] E. Feireisl and H. Petzeltovà, *Singular kernels and compactness in nonlinear conservation laws*, J. Differential Equations **142**(2) (1998), 291–304.

[26] H. Freistühler, C. Fries and C. Rohde, *Existence, bifurcation, and stability of profiles for classical and non-classical shock waves*, Ergodic Theory, Analysis, and Efficient Simulation of Dynamical Systems, Vol. 814, Springer-Verlag, Berlin (2001), 287–309.

[27] Y. Giga and T. Miyawaka, *A kinetic construction of global solutions of first order quasilinear equations*, Duke Math. J. **50** (1983), 505–515.

[28] R.T. Glassey, *The Cauchy Problem in Kinetic Theory*, SIAM, Philadelphia, PA (1996).

[29] E. Godlewski and P.-A. Raviart, *Numerical Approximation of Hyperbolic Systems of Conservation Laws*, Appl. Math. Sci., Vol. 118, Springer-Verlag (1996).

[30] F. Golse, P.-L. Lions, B. Perthame and R. Sentis, *Regularity of the moments of the solution of a transport equation*, J. Funct. Anal. **76** (1988), 110–125.

[31] F. Golse and L. Saint-Raymond, *The Navier–Stokes Limit of Boltzmann Equation: Convergence Proof*, Math. Inventiones **155**(1) (2004), 81–161.

[32] S. Hwang and A. Tzavaras, *Kinetic decomposition of approximate solutions to conservation laws: Applications to relaxation and diffusion–dispersion approximations*, Comm. Partial Differential Equations **27** (2002), 1229–1254.

[33] P.E. Jabin, F. Otto and B. Perthame, *Line-energy Ginzburg–Landau models: Zero–energy states*, Ann. Sci. Norm. Sup. Pisa. Cl. Sci. **31**(1) (2002), 187–202.

[34] P.E. Jabin and B. Perthame, *Regularity in kinetic formulations via averaging lemmas*, ESAIM: Control Optim. Calc. Var. **8** (2002), 761–774.

[35] P.-E. Jabin and L. Vega, *A real space method for averaging lemmas*, Preprint (2003).

[36] M. Junk, *A new perspective in kinetic schemes*, SIAM J. Numer. Anal. **38** (2000), 1603–1625.

[37] S. Kaniel, *A kinetic model for the compressible flow equations*, Indiana Univ. Math. J. **37**(3) (1988), 537–563.

[38] D. Kröner, *Numerical Schemes for Conservation Laws*, Wiley and Teubner, Stuttgart (1997).

[39] S.N. Kruzkov, *First order quasilinear equations in several independent variables*, Math. USSR Sb. **10** (1970), 217–243.
[40] M. Lecumberry and T. Rivière, *The rectifiability of shock waves for the solutions of genuinely non-linear scalar conservation laws in dimension* $1+1$, Séminaire X-EDP (2003).
[41] P.G. LeFloch, *An introduction to nonclassical shocks of systems of conservation laws*, An Introduction to Recent Developments in Theory and Numerics for Conservation Laws, Lecture Notes in Comput. Sci., Vol. 5, D. Kröner, M. Ohlberger and C. Rohde, eds, Springer-Verlag, Berlin (1999).
[42] R.J. LeVeque, *Finite Volume Methods for Hyperbolic Problems*, Cambridge Texts Appl. Math., Cambridge Univ. Press, Cambridge (2002).
[43] P.-L. Lions, *Compactness in Boltzmann equation via Fourier integral operators and applications*, Parts I and II, J. Math. Kyoto Univ. **34**(2) (1994), 1–61; Part III, J. Math. Kyoto Univ. **34** (1994), 539–584.
[44] P.-L. Lions, B. Perthame and P.E. Souganidis, *Existence of entropy solutions for the hyperbolic systems of isentropic gas dynamics in Eulerian and Lagrangian coordinates*, Comm. Pure Appl. Math. **49** (1996), 599–638.
[45] P.-L. Lions, B. Perthame and E. Tadmor, *A kinetic formulation of multidimensional scalar conservation laws and related equations*, J. Amer. Math. Soc. **7** (1994), 169–191.
[46] P.-L. Lions, B. Perthame and E. Tadmor, *Existence of entropy solutions to isentropic gas dynamics system in Eulerian and Lagrangian variables*, Comm. Math. Phys. **163** (1994), 415–431.
[47] J. Malek, J. Necas, M. Rokyta and M. Ruzicka, *Weak and Measure Valued Solutions to Evolutionary PDEs*, Chapman and Hall, London (1996).
[48] F. Murat, *Compacité par compensation*, Ann. Sci. Norm. Sup. Pisa Cl. Sci. **5** (1978), 489–507.
[49] B. Perthame, *Kinetic Formulation of Conservation Laws*, Oxford Lecture Ser. Math. Appl., Vol. 21, Oxford Univ. Press, New York (2002).
[50] B. Perthame, *Mathematical tools for kinetic equations*, Bull. Amer. Math. Soc. **41**(2) (2004), 205–244.
[51] B. Perthame and C. Simeoni, *A kinetic scheme for the Saint-Venant system with a source term*, Calcolo **38**(4) (2001), 201–231.
[52] B. Perthame and P.E. Souganidis, *A limiting case of velocity averaging lemmas*, Ann. Sci. École Norm. Sup. (4) **31** (1998), 591–598.
[53] B. Perthame and Th. Tzavaras, *Kinetic formulation for hyperbolic systems of two conservation laws and elastodynamics*, Arch. Ration. Mech. Anal. **155** (2000), 1–48.
[54] D.I. Pullin, *Direct simulation methods for compressible inviscid ideal-gas flow*, J. Comput. Phys. **34** (1980), 231–244.
[55] R.D. Reitz, *One-dimensional compressible gas dynamics calculations using the Boltzmann equations*, J. Comput. Phys. **34** (1980), 231–244.
[56] D. Serre, *Systèmes hyperboliques de lois de conservation*, Parties I et II, Diderot, Paris (1996).
[57] E. Tadmor and T. Tao, *Averaging lemmas revisited*, Work under preparation.
[58] L. Tartar, *Compensated compactness and applications to partial differential equations*, Research Notes in Mathematics, Nonlinear Analysis and Mechanics, Herriot–Watt Symposium, Vol. 4, R.J. Knops, ed., Pitman Press, Boston, MA–London (1979), 136–212.
[59] A. Vasseur, *Time regularity for the system of isentropic gas dynamics with* $\gamma = 3$, Comm. Partial Differential Equations **24**(11/12) (1999), 1987–1997.
[60] A. Vasseur, *Convergence of a semi discrete kinetic scheme for the system of isentropic gas dynamics with* $\gamma = 3$, Indiana Univ. Math. J. **48**(1) (1999), 347–364.
[61] A. Vasseur, *Kinetic semidiscretization of scalar conservation laws and convergence by using averaging lemmas*, SIAM J. Numer. Anal. **36**(2) (1999), 465–474.
[62] A. Vasseur, *Strong traces for solutions to multidimensional scalar conservation laws*, Arch. Ration. Mech. Anal. **160**(3) (2001), 181–193.
[63] C. Villani, *A review of mathematical topics in collisional kinetic theory*, Handbook of Mathematical Fluid Dynamics, S. Friedlander and D. Serre, eds, Tome I, chapitre 2, Elsevier.
[64] K. Xu, *Unsplitting BGK-type schemes for the shallow water equations*, Internat. J. Modern Phys. C **10**(4) (1999), 505–516.

CHAPTER 7

L^1-stability of Nonlinear Waves in Scalar Conservation Laws

Denis Serre

ENS Lyon, UMPA (UMR 5669 CNRS), 46, allée d'Italie, F-69364 Lyon Cedex 07, France

Contents

Notations . . . 475
Introduction . . . 475
Scalar conservation laws with dissipation . . . 475
1. The nonlinear semigroup . . . 479
1.1. The viscous conservation law . . . 479
1.2. The inviscid conservation law . . . 482
1.3. The relaxation model . . . 487
1.4. The Rosenau model . . . 491
2. Shocks . . . 500
2.1. Inviscid shocks . . . 501
2.2. Viscous shocks . . . 503
2.3. Relaxation shocks . . . 504
2.4. Radiative shocks . . . 504
3. Shock stability (I) . . . 507
3.1. Preliminary observations . . . 509
3.2. The dynamical system approach . . . 512
3.3. Sketch of proof of Lemma 4 . . . 515
3.4. Data with values beyond $[u_+, u_-]$; inviscid case . . . 518
3.5. Data with values beyond $[u_+, u_-]$; general case . . . 523
4. The L^1-stability of constants . . . 525
4.1. The viscous case . . . 526
4.2. The Rosenau model . . . 528
5. Multidimensional stability . . . 530
5.1. Formal asymptotics . . . 531
5.2. Stability results . . . 535
6. Discrete shock profiles . . . 537
6.1. Shock profiles . . . 539
6.2. The function V . . . 540
6.3. Existence of discrete shock profiles; irrational case . . . 542

HANDBOOK OF DIFFERENTIAL EQUATIONS
Evolutionary Equations, volume 1
Edited by C.M. Dafermos and E. Feireisl

6.4. The case of upwind scheme 544
6.5. Stability of discrete shock profiles 546
7. Initial-boundary value problems 546
7.1. Stability of constants 547
7.2. Stability of boundary layers 549
Acknowledgments 550
References 550

Abstract

We consider various scalar conservation laws with dissipation. The Cauchy problem generates a (nonlinear) semigroup that satisfies a maximum principle, a contraction property with respect to the L^1-distance, and preserves the total mass (hence the word "conservation law").

A crucial role is played by traveling waves, also called shock profiles, that tend to distinct constants as $x \cdot \nu \to \pm\infty$ ($\nu \in S^{d-1}$ a direction of propagation). In particular, their stability is a good criterion for the relevance of the underlying inviscid shock waves. The precise question that we address in this article is their asymptotic stability with respect to the L^1-distance. This topology turns out to be the most natural one in the context of conservation laws because of the properties mentioned above. A related question concerns steady solutions of the IBVP, usually called "boundary layers".

In one space dimension, the stability of shock profiles is, up to some extent, a matter of dynamical systems theory. However, a complete picture needs the more analytical proof that constants are also asymptotically stable, with respect to zero-mass initial disturbances. The latter may or may not hold, depending on the strength of the dissipation involved by a given model. For instance, it does not in the inviscid case. The multidimensional stability of shock waves is significantly more involved, and is still widely open.

Numerical approximation by conservative and monotone finite difference schemes provides another kind of dissipative process. We investigate the existence and stability of discrete shock profiles. We fill a gap in Jennings' proof of existence and prove that tails are exponentially small.

Throughout this text, we present nine Open Problems that have their own interest.

Notations

Spaces $L^p(\mathbb{R}^d)$ are defined with respect to the Lebesgue measure. Their norms are denoted $\|\cdot\|_p$. The conjugate exponent of $p \in [1, \infty]$ is given by

$$\frac{1}{p'} = 1 - \frac{1}{p}.$$

Sometimes, it may be useful to remember the equality $p' = p/(p-1)$. Partial derivatives with respect to t or x_α are denoted ∂_t and ∂_α (∂_x in one space dimension). The space of C^∞-functions with compact support in Ω is $\mathcal{D}(\Omega)$. If E is a functional space, E^+ denotes the cone of nonnegative elements of E; for instance, $\mathcal{D}^+(\Omega)$ is the set of nonnegative test functions. The space of functions with bounded variations is designated by BV; the total variation of a function a is $\mathrm{TV}(a)$. The *sign* function is defined on real numbers by $s \mapsto \mathrm{sgn}(s) = s/|s|$ is $s \neq 0$ and $\mathrm{sgn}(0) = 0$. The positive part s_+ of a real number s is $\max(s, 0)$. Its negative part is $s_- := \min(s, 0)$.

Introduction

Scalar conservation laws with dissipation

We are interested in scalar conservation laws of the form

$$\partial_t u + \sum_{\alpha=1}^{d} \partial_\alpha f^\alpha(u) = Lu. \tag{1}$$

The unknown $u(x,t)$ depends on time $t > 0$ and a space variable $x = (x_1, \ldots, x_d) \in \mathbb{R}^d$. In most of this chapter, we focus on the case of a single space variable ($d = 1$). The fluxes f^α are given smooth (at least C^2) functions. They represent the nonlinear convection of some physical problem governed by (1). The linear operator L models the dissipation process present in the physics. It is the infinitesimal generator of a Markov semigroup e^{tL} on $L^\infty(\mathbb{R}^d)$. The Markov property satisfied by any operator of the from $T = \mathrm{e}^{tL}$ is summarized by the following list:

Comparison. If $a, b \in L^\infty$ are such that $a \leqslant b$ almost everywhere, then $Ta \leqslant Tb$ holds almost everywhere.

Contraction. If $a, b \in L^\infty$ are such that $b - a \in L^1$, then $Tb - Ta \in L^1$ and

$$\|Tb - Ta\|_1 \leqslant \|b - a\|_1.$$

Conservation. If $a, b \in L^\infty$ are such that $b - a \in L^1$, then $Tb - Ta \in L^1$ and

$$\int_{\mathbb{R}^d} (Tb - Ta)\,\mathrm{d}x = \int_{\mathbb{R}^d} (b - a)\,\mathrm{d}x.$$

In addition, since the flux f does not explicitly depend on x and thus commutes with translations, we shall ask that L (or T as well) share this property. In particular:

Constants. If a is a constant, then $Ta \equiv a$.

The four properties described above, whose names (Comparison, Contraction, Conservation, Constant) begin with "Co", will be referred to as the *Co-properties* in the sequel.

Such linear semigroups have long been identified, see, for instance, [50]. Their generator have the general form

$$Lu = \sum_{\alpha,\beta} a_{\alpha\beta} \partial_\alpha \partial_\beta u + \sum_{\alpha} b_\alpha \partial_\alpha u + K * u - \mu u,$$

where the differential part is a weakly elliptic second-order operator with constant coefficients:

$$0_d \leqslant A = (a_{\alpha\beta})_{1 \leqslant \alpha\beta \leqslant d},$$

and the second part is a convolution by a nonnegative integrable kernel K, while the last term is a multiplication by the constant

$$\mu := \int_{\mathbb{R}^d} K(y)\,\mathrm{d}y.$$

The first-order derivatives may be incorporated in the convective term, and thus will be ignored: $b = 0$.

An important aspect of a Markov semigroup is its decay, or dispersion, property. For instance, if L is the Laplacian Δ, we have

$$\left\| \mathrm{e}^{t\Delta} a \right\|_p \leqslant c_p \frac{\|a\|_1}{t^{d/2p'}}.$$

This decay still holds when we add a convolution term to the Laplacian, but it cannot hold in the lack of differential terms. As a matter of fact, $u \mapsto Lu = K * u - \mu u$ is a bounded operator on every L^p and thus generate a *group* of transformations that are automorphisms within L^p. In particular, $\mathrm{e}^{tL} a$ belongs to L^p if and only if a does.

The reason that motivates us in considering the perturbation of the so-called *inviscid conservation law*

$$\partial_t u + \sum_{\alpha=1}^{d} \partial_\alpha f^\alpha(u) = 0 \tag{2}$$

by such dissipation terms is two-fold. First of all, conservation laws in physics have the general form $\partial_t u + \operatorname{div} \mathbf{q} = 0$, where the relation between the flux $\mathbf{q}$ and the state u may be rather complex. In a coarse approximation (large scales), the flux may be represented by a nonlinear local term $f(u)$, but at smaller scales, additional terms must be taken in account. For instance, gas dynamics is described at large scales by the Euler equations, while a refined description needs the Navier–Stokes system. Secondly, the scalar inviscid conservation law generates a semigroup (see [39]) which displays the four Co-properties above. In some instances, these properties are inherent to the physics under consideration, and should not be destroyed by the addition of the dissipative operator L. Considering small amplitude data, we see that e^{tL} must fill the four Co-properties, hence be a Markov semigroup.

Other motivations have been given in the physics literature. For instance, the choice (here $d = 1$)

$$K_0(x) = \frac{1}{2} e^{-|x|}, \qquad \mu_0 = 1$$

represents a coupling with an elliptic equation, since then $K_0 * u - u = \partial_x q$, where q is the solution of $-\partial_x^2 q + q = \partial_x u$. Such terms have been considered in radiative gas dynamics; see [33,34] and the references herein. Of course, the present context of a scalar equation is not relevant for radiative gas dynamics, but it provides a simplified model, in the same spirit than when the Burgers equation was proposed as a simplification of full gas dynamics. Rosenau [52] considered the kernel K_0 in a regularized version of the Chapman–Enskog expansion in hydrodynamics. The study of the Cauchy problem for (1) with Rosenau's operator began with [56].

A less simple, at a first glance, modification of (2) for which we still expect that the four Co-properties hold in a suitable sense, is the relaxation model designed by Jin and Xin [32]:

$$\partial_t u + \partial_x v = 0, \tag{3}$$

$$\partial_t v + \alpha^2 \partial_x u = f(u) - v. \tag{4}$$

The first equation is the conservation law, where the flux v is related to u through the evolution equation (4). If we ignore the $\partial_x u$ term, the latter looks like an ODE $v' = f(u) - v$, which forces v to relax to the equilibrium value $f(u)$; then at equilibrium, we recover (2). Actually, for stability reasons, we may not afford to ignore the $\partial_x u$ term, where α must be large enough. This model looks much different from scalar ones. For instance, it requires two data, one for each unknown. That it shares a lot of properties of scalar equations, due to a weak decoupling, was first observed by Hanouzet and Natalini [27].

Given a nonlinear model as above, the proof that the Cauchy problem is well posed in L^∞ is more or less well known, although it is never a triviality. We reproduce it in Section 1, emphasizing that the proof of uniqueness depends heavily on which equation we are dealing with. As a matter of fact, the well-posedness of the convolution model in L^∞ (if $d = 1$) and in BV (if $d \geqslant 2$) are new, and that in L^∞ remains an open question when $d \geqslant 2$.

We show in the course of the proofs that the semigroups defined by these Cauchy problem satisfy the Co-properties. It has long been remarked that the L^1-contraction allows to extend them continuously to the larger space $L^1 + L^\infty$, in a unique way. However, it is hard to say whether this extension gives weak solutions of the models, since the nonlinear term $\operatorname{div} f(u)$ might not be a distribution if f is super-linear.

Besides the PDEs mentioned above, we also consider the discretization of the inviscid conservation law by finite difference schemes. They provide an other kind of dissipation that, in academic cases like the Lax–Friedrichs and Godunov schemes, gives rise to the Co-properties. The corresponding shock profiles are often named "discrete shock profiles", a rather inappropriate word since these profiles usually depend on the real variable.

The next step (Section 2) is the characterization of shock waves for each model. In general, these are unique up to a space translation. Their existence is more or less trivial, except in two cases. The question remains essentially open for the convolution model, apart for the kernel K_0 studied in [34,44]. The situation was uncomfortable for discrete shock profiles: Their existence was claimed by Jennings [31], whose proof covered only the so-called "rational case". We fix the state of the art in Section 6 by providing a proof of the irrational case. Our argument may be viewed as a stability analysis of profiles with respect to various data: The shock parameters as well as the scheme itself. Hence, it can be used to extend Jennings' result to schemes that are monotone but not in a strict sense, like Godunov's. In the strictly monotonous situation, with the additional assumption that the shock is noncharacteristic, we also obtain an exponential decay of the tail of the profile.

We begin the stability analysis of shock profiles in Section 3. When the values taken by the initial data lie between the states u_- and u_+, the L^1-stability is proved by arguments taken form dynamical systems theory: compactness of trajectories, ω-limit set, Lasalle's invariance principle. Only when exploiting the latter have we to make a case-by-case study. The general idea is due to Osher and Ralston [51].

The situation is more involved when the data takes values outside the shock interval. We show that L^1-stability may fail in the borderline inviscid case when the shock is characteristic on at least one side. This result is optimal, since we also prove the stability whenever the shock is noncharacteristic, that is Lax' shock inequalities are strict.

For models with $L \neq 0$, Freistühler and the author found that the L^1-stability of shock profiles is implied by that of constant states under zero-mass disturbance. The latter property is analyzed in details in Section 4. In some sense, it is the most technical part since we really have to work on a case-by-case basis. The stability is induced by the dispersion properties of the linear semigroup e^{tL}, which may be pretty strong ($L = \Delta$) or weak (convolution model, relaxation).

Section 5 is devoted to the multidimensional viscous case. Since the shift of a given profile does not belong to the same class modulo $L^1(\mathbb{R}^d)$ as the profile itself, the total mass of the initial disturbance becomes a meaningful invariant of the problem. Its nonvanishing causes a disturbance of the shock front that can be represented at the leading order by a solution of the heat equation in $\mathbb{R}^{d-1}$, which enhanced diffusion. A pretty good result has been established by Hoff and Zumbrun [29], following a preliminary study by Goodman and Miller [22]. Remaining at a qualitative level, we construct the approximate solution to which the actual one is asymptotic. Of course, because of the lack of compactness, the stability result is only known under additional restrictions such as smallness in weighted spaces. Thus a lot of interesting questions remain open.

Our last Section 7 concerns the Initial-boundary value problem. The waves under consideration are the so-called "boundary layers". Their stability analysis follows more or less the same guidelines than that of shock profiles. We give a new proof of the stability of constant in the outgoing case, by using a simple comparison argument with respect to the characteristic case. This idea actually works in several space dimensions, where the result seems to be new.

1. The nonlinear semigroup

We are interested in the Cauchy problem for (1), where the initial datum a belongs to L^∞. We wish to construct a semigroup for (1), which enjoys the four Co-properties. The situation may differ a lot when we vary the choice of the operator L.

1.1. *The viscous conservation law*

We begin with the viscous equation:

$$\partial_t u + \operatorname{div} f(u) = \Delta u. \tag{5}$$

The choice of the Laplace operator Δ is not restrictive at all, since every elliptic operator of the form

$$\sum_{\alpha,\beta} a_{\alpha\beta}\partial_\alpha\partial_\beta$$

with constant coefficients may be transformed into Δ via a suitable linear change of coordinates. Also, such a transformation does not affect the Co-properties.

In (5), the flux term $\operatorname{div} f(u)$ may be treated as a lower order perturbation of the heat equation $\partial_t u = \Delta u$. The latter generates a (linear) C^0-semigroup $(H_t)_{t\geqslant 0}$ on L^p $(1 \leqslant p \leqslant \infty)$, which satisfies the four Co-properties. This semigroup is nothing but the convolution by the *Heat kernel*;

$$H_t a = K^t * a, \quad K^t(x) := \frac{1}{(2\pi t)^{d/2}} \exp\left(-\frac{\|x\|^2}{4t}\right).$$

This representation immediately gives dispersion estimates, such as

$$\|H_t a\|_p \leqslant c_{p,q} t^{\frac{d}{2}(\frac{1}{p}-\frac{1}{q})} \|a\|_q, \quad p \geqslant q, \tag{6}$$

and

$$\|\nabla_x H_t a\|_p \leqslant c'_{p,q} t^{\frac{d}{2}(\frac{1}{p}-\frac{1}{q})-\frac{1}{2}} \|a\|_q, \quad p \geqslant q. \tag{7}$$

Notice that $c_{p,p} = 1$. The semigroup of the heat equation is weakly contractive in every L^p.

The theory of the Cauchy problem for (5) is that of *mild* solutions. One rewrites it as an integral equation, using the Duhamel's principle for the heat equation,

$$\begin{aligned} u(t) &= K^t * a - \int_0^t K^{t-s} * \big(\operatorname{div} f\big(u(s)\big)\big)\,\mathrm{d}s \\ &= K^t * a - \int_0^t \big(\nabla_x K^{t-s}\big) * \big(f\big(u(s)\big)\big)\,\mathrm{d}s. \end{aligned} \tag{8}$$

This formulation is viewed as a fixed point problem for the map

$$u \mapsto N[u] := \left(t \mapsto K^t * a - \int_0^t \big(\nabla_x K^{t-s}\big) * \big(f\big(u(s)\big)\big)\,\mathrm{d}s\right).$$

Using the estimates (7), one easily finds that N is a contraction of the ball of radius $2\|a\|_\infty$ in $L^\infty(\mathbb{R}^d \times (0,T))$, provided $T = T(\|a\|_\infty)$ is small enough. Then Picard's fixed point theorem tells that there is a unique local solution. Using again dispersion inequalities, one finds that u is smoother and smoother; for instance, u is C^∞ for positive times if f is C^∞. Then the Maximum principle implies that $\|u(t)\|_\infty \leqslant \|a\|_\infty$. Hence, the local existence can be iterated: The solution exists on $(0, 2T(\|a\|_\infty)), \ldots$. At the end, we obtain a unique bounded solution, which is global in time and smooth for positive times.

Since the solution is unique, the nonlinear map $a \mapsto V_t a := u(t)$ defines a semigroup on L^∞. Given two data a and b whose difference is integrable, denote $v(t) := V_t b$ and $u(t) = V_t a$. For the sake of simplicity, we work at the level of solutions, although the rigorous analysis consists in doing the same estimate at the level of the approximate solutions generated by Picard's fixed point argument. We have

$$v(t) - u(t) = K^t * (b-a) - \int_0^t \big(\nabla_x K^{t-s}\big) * \big(f\big(v(s)\big) - f\big(u(s)\big)\big)\,\mathrm{d}s.$$

Taking the L^1-norm, we have

$$\big\|v(t) - u(t)\big\|_1 \leqslant \|b-a\|_1 + c'_{1,1} M \int_0^t \big\|v(s) - u(s)\big\|_1 \frac{\mathrm{d}s}{\sqrt{t-s}},$$

where M is the Lipschitz constant of f on the compact interval containing $\pm\|a\|_\infty$ and $\pm\|b\|_\infty$. Defining

$$Y(t) := \sup\big\{\big\|v(s) - u(s)\big\|_1;\ s \in [0,t]\big\},$$

we obtain

$$Y(t) \leqslant Y(0) + 2c'_{1,1} M \sqrt{t}\, Y(t),$$

which gives $Y(t) \leqslant 2Y(0)$, whenever $4c'_{1,1}M\sqrt{t} < 1$. Since M depends only on the L^∞ norms, which remain bounded, this estimate may be iterated. We find that $v(t) - u(t)$ is integrable for every positive t, with a rather poor bound

$$\|v(t) - u(t)\|_1 \leqslant c e^{\omega t} \|b - a\|_1.$$

The first consequence is the norm conservation. We may write

$$\int_{\mathbb{R}^d} \big(v(t) - u(t)\big)\,\mathrm{d}x = \int_{\mathbb{R}^d} (b - a)\,\mathrm{d}x$$

because

$$\int_{\mathbb{R}^d} g * h\,\mathrm{d}x = \int_{\mathbb{R}^d} g\,\mathrm{d}x \cdot \int_{\mathbb{R}^d} h\,\mathrm{d}x$$

for every integrable functions g and h, while

$$\int_{\mathbb{R}^d} K^t\,\mathrm{d}x = 1, \qquad \int_{\mathbb{R}^d} \nabla_x K^{t-s}\,\mathrm{d}x = 0.$$

To prove the contraction property, we work with C^2 convex functions $u \mapsto \eta(u)$ (the "entropies") and vector fields ψ (their "entropy fluxes") defined up to a constant by $\psi' = \eta' f'$. The smoothness of solutions allows us to writing

$$\partial_t \eta(u) + \operatorname{div} \psi(u) = \eta'(u)\Delta u = \Delta\eta(u) - \eta''(u)|\nabla_x u|^2 \leqslant \Delta\eta(u). \tag{9}$$

Given a real number k, the convex function

$$u \mapsto \eta_k(u) := |u - k|$$

is a so-called *Kruzkhov entropy*. Approaching uniformly η_k by C^2 entropies, we observe that the fluxes converge to the flux

$$\psi_k(u) := \operatorname{sgn}(u - k)\big(f(u) - f(k)\big).$$

Moreover, one may pass to the limit in (9), whence

$$\partial_t \eta_k(u) + \operatorname{div} \psi_k(u) \leqslant \Delta\eta_k(u), \tag{10}$$

in the distributional sense. Actually, using the symmetry $\eta_k(u) = \eta_u(k)$, $\psi_k(u) = \psi_u(k)$, we may as well derive the following inequality for two solutions u and v as above:

$$\partial_t |v - u| + \operatorname{div}\big(\operatorname{sgn}(v - u)\big(f(v) - f(u)\big)\big) \leqslant \Delta|v - u|. \tag{11}$$

Writing the Duhamel's form of (11) and using the maximum principle, we have

$$\begin{aligned}\left|v(t)-u(t)\right| &\leqslant K^t * |b-a| \\ &\quad - \int_0^t \left(\nabla_x K^{t-s}\right) * \left[\operatorname{sgn}\left(v(s)-u(s)\right)\left(f\left(v(s)\right)-f\left(u(s)\right)\right)\right] \mathrm{d}s.\end{aligned}$$

Since every term is integrable on $\mathbb{R}^d$, we receive the contraction property

$$\left\|v(t)-u(t)\right\|_1 \leqslant \|b-a\|_1.$$

We conclude that the semigroup V_t enjoys the four Co-properties.

Dispersion property. A rather strange fact, noticed by Bénilan and Abourjaily [1] is that (5) exhibits a dispersion property which is independent of the flux f, even at a quantitative level. For instance, let us consider the decay of the L^2 norm of $u(t)$ for integrable initial data. The standard energy estimate, and the fact that $u \operatorname{div} f(u) = \operatorname{div} g(u)$ for $g'(s) = sf'(s)$ yield

$$\frac{\mathrm{d}}{\mathrm{d}t}\|u\|_2^2 + 2\|\nabla_x u\|_2^2 = 0.$$

Using the Nash inequality

$$\|u\|_2 \leqslant c_d \|u\|_1^{1-\theta} \|\nabla_x u\|_2^{\theta}, \quad \theta = 1 + \frac{2}{d},$$

and using the decay of $t \mapsto \|u(t)\|_1$, we obtain

$$\|a\|_1^{4/d} \frac{\mathrm{d}}{\mathrm{d}t}\|u\|_2^2 + c_d' \|u\|_2^{2+4/d} \leqslant 0.$$

This differential inequality immediately gives

$$\left\|u(t)\right\|_2 \leqslant c_d'' \frac{\|a\|_1}{t^{d/4}}. \tag{12}$$

A generalization of this idea yields the following decay estimates, where again the constants c_{pqd} do not depend on the particular shape of the flux:

$$\|V_t a\|_p \leqslant c_{pqd}\, t^{d\frac{q-p}{2pq}} \|a\|_q, \quad 1 \leqslant q \leqslant p \leqslant 2q < +\infty. \tag{13}$$

1.2. *The inviscid conservation law*

The inviscid conservation law

$$\partial_t u + \operatorname{div} f(u) = 0, \tag{14}$$

faces the problems of shock formation and nonuniqueness of distributional solutions. Shock formation is caused by the blow-up in the solution of the Ricatti equation along characteristic

$$\big(\partial_t + f'(u)\partial_x\big)(\partial_x u) + f''(u)(\partial_x u)^2 = 0.$$

This equation, supplemented by the fact that u is constant along characteristics $\mathrm{d}x/\mathrm{d}t = f'(u)$, shows that most of C^1-solutions blow up in finite time.

For this reason, we must consider distributional solutions, which are assumed bounded, at least locally, in order that all the terms in the equation make sense. Typically, solutions exhibit discontinuities across hypersurfaces. Piecewise smooth functions satisfy (14) in the weak sense if and only if they are classical solutions away from their discontinuities, the latter satisfying the *Rankine–Hugoniot* condition

$$\big(f(u_+) - f(u_-)\big) \cdot \nu = s(u_+ - u_-). \tag{15}$$

Here, $u_\pm$ are the limits of u at $x(t)$ from both sides of a discontinuity locus of $\Gamma(t)$. The unit normal to this hypersurface is oriented from the minus side towards the plus side, and is denoted by ν. Last, s is the normal velocity of $\Gamma(t)$: it is defined by

$$s\big(x(t)\big) = \nu\big(x(t)\big)\frac{\mathrm{d}x}{\mathrm{d}t}, \qquad x(t) \in \Gamma(t).$$

One often writes $[g(u)]$ for an expression $g(u_+) - g(u_-)$. For instance, (15) reads

$$\big[f(u)\big] \cdot \nu = s[u].$$

The Rankine–Hugoniot condition (15) does not depend on the orientation of Γ. In one space dimension, we choose the natural orientation $\nu = +1$, meaning that

$$u_\pm(x,t) := \lim_{y \to x \pm 0} u(y,t).$$

As it has been well known since the seminal work by Kruzkhov [39], weak solutions are far from being unique, and one must select a solution through the *entropy criterion*. The rigorous formulation of this criterion is that the solution u satisfies the integral inequalities

$$\int_{\mathbb{R}}\int_{\mathbb{R}^d} \big(|u-k|\partial_t\phi + \operatorname{sgn}(u-k)\big(f(u) - f(k)\big) \cdot \nabla_x\phi\big)\,\mathrm{d}x\,\mathrm{d}t + \int_{\mathbb{R}^d} \big|a(x) - k\big|\phi(x,0)\,\mathrm{d}x \geqslant 0 \tag{16}$$

for every constant $k \in \mathbb{R}$ and every nonnegative test function $\phi \in \mathcal{D}^+(\mathbb{R}^d \times \mathbb{R})$. Of course, if we are interested in local-in-time solutions, say on a time interval $(0, T)$, then the integral inequalities should be considered for test functions with support in $\mathbb{R}^d \times (-\infty, T)$ only. The solutions satisfying (16) are called *entropy solutions*.

An important remark is that, whenever u is bounded, at least locally in space and time, then (16) contains the fact that u solves (1) in the distributional sense (take $k > \|u\|_\infty$, and next $k < -\|u\|_\infty$). But of course, it tells much more. We shall only sketch the proof of the following well-known result.

THEOREM 1 (Kruzkhov [39]). *Let $a \in L^\infty(\mathbb{R}^d)$ and $T > 0$ be given. Then there exists a unique locally bounded function u in $\mathbb{R}^d \times \mathbb{R}$, satisfying* (16) *and*

$$u \in C\big([0,\infty); L^1_{\mathrm{loc}}(\mathbb{R}^d)\big).$$

In addition, this solution is bounded and the map $a \mapsto S_t a := u(t)$ defines a nonlinear semigroup which satisfies the four Co-properties mentioned in the Introduction.

COMMENTS.

1. The fact that the semigroup is L^1-contractive had already been noticed by Volpert [73] in the context of BV data. This fact motivated Kruzkhov in his study.
2. The L^1-contractivity, when an equation exhibits a scaling invariance, is the source of a BV regularization, as shown by Bénilan and Crandall [4]. For instance, the Burgers equation $\partial_t u + \partial_x(u^2/2) = 0$, in one space dimension, has the property that $u(t)^2$ is of bounded variation whenever the datum a is integrable, with

$$\mathrm{TV}\big(u(t)^2\big) \leqslant \frac{2\|a\|_1}{t}.$$

3. Other selection criteria have been considered for (14) by LeFloch and co-workers (see [37] and references herein). They may be relevant for specific problems, mostly in systems rather than scalar equations. Significant examples include nonstrictly hyperbolic systems and cases where elliptic regions are enclosed in domains of hyperbolicity. In the scalar case, such nonclassical solutions may not enjoy the four Co-properties. Their uniqueness has not been proved yet. In particular, we still ignore whether they define a semigroup in any sense. The most important feature for us is that they are not expected to satisfy the L^1-contraction; this is why we shall not consider them in the sequel.

SKETCH OF PROOF OF THEOREM 1.

Existence. We begin with integrable data $a \in L^1(\mathbb{R}^d) \cap L^\infty(\mathbb{R}^d)$. Let u^ε be the unique solution of the viscous model

$$\partial_t u^\varepsilon + \operatorname{div} f\big(u^\varepsilon\big) = \varepsilon \Delta u^\varepsilon$$

with data $u^\varepsilon(0) = a$. Then $\|u^\varepsilon\|_\infty = \|a\|_\infty$. Let h be a vector and denote τ_h the space shift by this vector: $\tau_h g = g(\cdot + h)$. Since $\tau_h u^\varepsilon$ is the solution associated to the data $\tau_h a$, the L^1-contraction (see the previous subsection) gives the uniform estimate

$$\big\|\tau_h u^\varepsilon - u^\varepsilon\big\|_1 \leqslant \|\tau_h a - a\|_1 =: \omega(h).$$

Since a is integrable, it is known that

$$\lim_{\|h\|\to 0} \omega(h) = 0.$$

Hence, the sequence u^ε is uniformly (in t and ε) integrable with respect to the space variable. Besides, integrating the equation on $\mathbb{R}^d \times (t, t+h)$ against a test function $\phi(x)$ gives

$$\int_{\mathbb{R}^d} \phi\big(u^\varepsilon(t+h) - u^\varepsilon(t)\big)\,\mathrm{d}x = \int_t^{t+h} \int_{\mathbb{R}^d} \big(\nabla_x \phi \cdot f\big(u^\varepsilon(s)\big) + \varepsilon u^\varepsilon(s)\Delta\phi\big)\,\mathrm{d}x\,\mathrm{d}s.$$

Using a mollifier $\rho_\theta(y) = \rho(y/\theta)$, let us choose

$$\phi = \rho_\theta * \operatorname{sgn} w, \qquad w := u^\varepsilon(t+h) - u^\varepsilon(t).$$

Without loss of generality, we may assume[1] $f(0) = 0$. Therefore we obtain, on the one hand,

$$\int_{\mathbb{R}^d} \phi w\,\mathrm{d}x \leqslant c\|h\|\big(\alpha^{-1} + \varepsilon\alpha^{-2}\big)\|a\|_1$$

for some constant c. On the other hand, the mollifying has the consequence

$$\|w\|_1 \leqslant \int_{\mathbb{R}^d} \phi w\,\mathrm{d}x + 4\Omega(\alpha),$$

with $\Omega(\alpha)$ the supremum of $\omega(y)$ in the ball $\|y\| < \alpha$. Finally, choosing for instance $\alpha = \|h\|^{1/3}$, we obtain

$$\big\|u^\varepsilon(t+h) - u^\varepsilon(t)\big\|_1 \leqslant c\big(\|h\|^{2/3} + \varepsilon\|h\|^{1/3}\big)\|a\|_1 + 4\Omega\big(\|h\|^{1/3}\big).$$

This gives uniform equiintegrability with respect to space–time variables. Given the L^∞-bound, the sequence $(u^\varepsilon)_{\varepsilon>0}$ is pre-compact in L^1_{loc}. Thus there exists a subsequence, with $\varepsilon \to 0+$, which converges boundedly almost everywhere. From (10) (with $\varepsilon\Delta$ instead of Δ) u^ε satisfies an integral inequality for every nonnegative test function ϕ. Passing to the limit as $\varepsilon \to 0$ with the help of the Dominated convergence theorem, we obtain (16). Last, by virtue of the above estimate, we have

$$\big\|u(t+h) - u(t)\big\|_1 \leqslant c\|h\|^{2/3}\|a\|_1 + 4\Omega\big(\|h\|^{1/3}\big).$$

Therefore, an entropy solution does exist.

---

[1] Replacing f by $f - f(0)$ does not change the equation.

Uniqueness. That part does not follow from the estimates above. It is needed in order to have a well-defined semigroup, and to prove the convergence of the whole sequence $(u^\varepsilon)_{\varepsilon>0}$. It follows directly from Kruzkhov's estimate: Let u and v be two entropy solutions associated with bounded data a and b. Using the symmetry between u and k in (16), a tedious and technical computation yields the inequality

$$\int_{\mathbb{R}}\int_{\mathbb{R}^d}\big(|u-v|\partial_t\phi+\operatorname{sgn}(u-v)\big(f(u)-f(v)\big)\cdot\nabla_x\phi\big)\,\mathrm{d}x\,\mathrm{d}t \\ +\int_{\mathbb{R}^d}\big|a(x)-b(x)\big|\phi(x,0)\,\mathrm{d}x\geqslant 0 \tag{17}$$

for every nonnegative test function. Then a careful choice of ϕ gives the result (uniqueness and finite propagation speed)

$$\int_{B(z;R)}\big|v(x,t)-u(x,t)\big|\,\mathrm{d}x\leqslant\int_{B(z;R+Mt)}\big|b(x)-a(x)\big|\,\mathrm{d}x \tag{18}$$

for every $z\in\mathbb{R}^d$, $t>0$ and $R>0$. The number M is as above, the Lipschitz constant of f over the interval containing $\pm\|a\|_\infty$ and $\pm\|b\|_\infty$.

Besides giving uniqueness of an entropy solution, (18) allows us to define a unique solution for every bounded data, not only integrable ones. For if $a\in L^\infty$, then truncated data $\chi_D a$, with $D=B(0;R)$ the centered ball of radius R, is integrable. Let u_R be the corresponding (unique) entropy solution. If $S>R$, then (18) tells that u_R and u_S coincide on the cone defined by $\|x\|+Mt<R$. Therefore the sequence $(u_R)_{R>1}$ converges as $R\to\infty$, to a bounded function $u\in\mathcal{C}([0,\infty);L^1_{\mathrm{loc}}(\mathbb{R}^d))$ (because each u_R belongs to $\mathcal{C}([0,\infty);L^1(\mathbb{R}^d))$), obviously an entropy solution. By (18) again, this solution is unique.

We now check the Co-properties in the general case of bounded data. Constants are obvious entropy solutions. Next, (18) together with Fatou's lemma gives the contraction property when $b-a\in L^1(\mathbb{R}^d)$. The comparison holds true for L^1 data, because it does at the viscous level and the viscous solution converges a.e. to the entropy solution. By the procedure above, the comparison property extends immediately to every bounded data. Last, given a and b with $b-a\in L^1(\mathbb{R}^d)$, we integrate the equation against a test function $\phi_R(x):=\phi(x/R)$, where ϕ has compact support and equals 1 on the unit ball. From

$$\int_{\mathbb{R}^d}\phi_R(x)(v-u)(x,t)\,\mathrm{d}x-\int_{\mathbb{R}^d}\phi_R(x)(b-a)\,\mathrm{d}x \\ =\int_0^t\int_{\mathbb{R}^d}\big(f(v)-f(u)\big)\cdot\nabla\phi_R\,\mathrm{d}x\,\mathrm{d}s,$$

we obtain

$$\left|\int_{\mathbb{R}^d}\phi_R(x)(v-u)(x,t)\,\mathrm{d}x-\int_{\mathbb{R}^d}\phi_R(x)(b-a)\,\mathrm{d}x\right|\leqslant\frac{M}{R}\|b-a\|_1.$$

Letting $R \to \infty$, conservation comes from dominated convergence

$$\int_{\mathbb{R}^d} (v-u)(x,t)\,\mathrm{d}x = \int_{\mathbb{R}^d} (b-a)\,\mathrm{d}x.$$

1.3. *The relaxation model*

We turn towards Cauchy problem for the system (3), (4). Now, the initial datum is a pair (a, b), where $u = a$ and $v = b$ at initial time. The solution of the linear problem

$$\partial_t u + \partial_x v = g, \qquad \partial_t v + \alpha^2 \partial_x u = h,$$

can be diagonalized as

$$D^+ w = g + \frac{1}{\alpha} h, \qquad D^- z = g - \frac{1}{\alpha} h,$$

where $w, z := u \pm v/\alpha$ and $D^\pm$ are the transport operators $\partial_t \pm \alpha \partial_x$. This system admits a unique solution in any reasonable sense, given by the explicit formulas

$$w(x,t) = w(x - \alpha t, 0) + \int_0^t \left(g + \frac{1}{\alpha} h\right)\bigl(x - \alpha(t-s), s\bigr)\,\mathrm{d}s,$$

$$z(x,t) = z(x + \alpha t, 0) + \int_0^t \left(g - \frac{1}{\alpha} h\right)\bigl(x + \alpha(t-s), s\bigr)\,\mathrm{d}s.$$

This allows to express (u, v) in a closed form, reminiscent to the Duhamel's formula. Such a formula may be used to state regularity results. For instance, in the homogeneous case $(g \equiv h \equiv 0)$, the function $a \pm b/\alpha$ is just propagated along lines of slope $\pm\alpha$.

When the right-hand side $(g \equiv 0, h = f(u) - v)$ is given in terms of the unknown itself, the Duhamel's formula puts the system in the form of a fixed point problem

$$w(x,t) = \left(a + \frac{1}{\alpha} b\right)(x - \alpha t) + \int_0^t G(w,z)\bigl(x - \alpha(t-s), s\bigr)\,\mathrm{d}s, \tag{19}$$

$$z(x,t) = \left(a - \frac{1}{\alpha} b\right)(x + \alpha t) - \int_0^t G(w,z)\bigl(x + \alpha(t-s), s\bigr)\,\mathrm{d}s, \tag{20}$$

where

$$G(w,z) := \frac{1}{\alpha} f\left(\frac{w+z}{2}\right) + \frac{z-w}{2}.$$

The technique employed in Section 1.1 (Picard iteration for a contractive map) applies here and yields a local existence theorem for bounded initial data. The existence time is bounded by below by a time T_0 depending only on $\|(a,b)\|_\infty$.

In order to extend the solution for all times, we need an a priori estimate in the Sup-norm. This one must be obtained through the maximum principle.[2] Following [9], such a maximum principle must hold simultaneously for the principal part

$$\partial_t u + \partial_x v = 0, \qquad \partial_t v + \alpha^2 \partial_x u = 0,$$

and the ODE part

$$\partial_t u = 0, \qquad \partial_t v = f(u) - v.$$

Let K be a domain in the plane that is positively invariant for both subsystems. The PDE part imposes (see [9]) that K is a rectangle $I_w \times I_z = [w_{\min}, w_{\max}] \times [z_{\min}, z_{\max}]$ in the characteristic coordinates (w, z). In the latter, the ODE part becomes

$$\partial_t w = G(w, z), \qquad \partial_t z = -G(w, z).$$

Hence, the invariance of K amounts to the inequalities

$$G(w_{\max}, z) \leqslant 0, \qquad G(w_{\min}, z) \geqslant 0, \quad z \in I_z, \tag{21}$$

and

$$G(w, z_{\min}) \leqslant 0, \qquad G(w, z_{\max}) \geqslant 0, \quad w \in I_z. \tag{22}$$

In particular, G must vanish at upper-left and lower-right corners, meaning that the vertices $(u_{\min}, v_{\min})$ and $(u_{\max}, v_{\max})$ belong to the equilibrium curve

$$\Gamma := \{(u, f(u)); u \in \mathbb{R}\}.$$

Whenever (21) and (22) hold and the initial datum (a, b) takes its values in the domain K, the local solution remains in K and is therefore bounded. Since its life time was bounded by below by a constant T_0 depending only on the Sup-norm of (a, b), the solution may be extended for every time $t > 0$; the solution is global in time.

When dealing with the relaxation model, which is not exactly what we think immediately about when we speak of *scalar* conservation laws, we need to interpret the Co-properties in terms of solutions with data in a box K as above only. Moreover, the order relation in $(L^\infty)^2$ and the L^1 structure have to be understood in the following sense, introduced by Hanouzet and Natalini [27]:

Comparison. We say that a pair (a, b) of bounded functions is smaller than an other one (A, B) if $w(a, b) \leqslant w(A, B)$ and $z(a, b) \leqslant z(A, B)$ almost everywhere. Let us write $(a, b) \prec (A, B)$. In that case, the comparison property for the relaxation model is

$$\{(a, b), (A, B) : \mathbb{R} \to K \text{ and } (a, b) \prec (A, B)\} \quad \Longrightarrow \quad \{(u, v) \prec (U, V)\}, \tag{23}$$

---

[2] We are not aware of such estimates, obtained by other means.

where (u, v) and (U, V) are the corresponding solutions.

In terms of w and z, which satisfy transport equations coupled by nonlinear source terms, this amounts to saying that, within the box K, G is nondecreasing with respect to z, while nonincreasing with respect to w. This exactly require the so-called *subcharacteristic condition*, which limits by below the choice of the parameter α:

$$\alpha \geqslant \sup\{|f'(u)|; u \in [u_{\min}, u_{\max}]\}. \tag{24}$$

Because borderline cases often rise major difficulties or pathologies, we usually assume the *strict* subcharacteristic condition, where the inequality is strict. Then the equilibrium curve can be written, within K, as the graph $\{z = \theta(w)\}$ of a smooth function such that $\theta' > 0$. That relation can be rewritten in the similar form $\{w = \theta^{-1}(z)\}$. Since G increases along horizontal lines (in variables (w, z)) and decreases along vertical ones, we obtain that G is positive on the lower-left side of Γ and is negative on the upper-right side. Hence, the inequalities (21) and (22) may be viewed as consequences of the subcharacteristic condition.

Contraction. The integral form of the nonhomogeneous linear problem show immediately that if a, b, g and h are integrable, then $u(t)$ and $z(t)$ are so, with

$$\frac{\mathrm{d}}{\mathrm{d}t}\|w\|_1 = \int_{\mathbb{R}} \operatorname{sgn}(w)\left(g + \frac{1}{\alpha}h\right)\mathrm{d}x, \qquad \frac{\mathrm{d}}{\mathrm{d}t}\|z\|_1 = \int_{\mathbb{R}} \operatorname{sgn}(z)\left(g - \frac{1}{\alpha}h\right)\mathrm{d}x.$$

Applying these differential equations to the difference of two solutions (w, z) and (W, Z) of the relaxation model, we obtain

$$\begin{aligned}&\frac{\mathrm{d}}{\mathrm{d}t}\left(\|W - w\|_1 + \|Z - z\|_1\right)\\ &\quad = \int_{\mathbb{R}} \left(\operatorname{sgn}(W - w) - \operatorname{sgn}(Z - z)\right)\left(G(W, Z) - G(w, z)\right)\mathrm{d}x.\end{aligned}$$

For the integrand to be nonzero, it is necessary that either $w \leqslant W$ and $Z \leqslant z$, or $W \leqslant w$ and $z \leqslant Z$. In the first case (the second is treated in the same way), the monotonicities of G in the box K provide $G(W, Z) \leqslant G(W, z) \leqslant G(w, z)$. Hence, the integrand is nonpositive and we may conclude to the specific contraction property:

$$\|W(t) - w(t)\|_1 + \|Z(t) - z(t)\|_1 \leqslant \|W(0) - w(0)\|_1 + \|Z(0) - z(0)\|_1. \tag{25}$$

Conservation. The relaxation model contains only one conservation law, namely $\partial_t u + \partial_x v = 0$, yielding the conservation property

$$\int_{\mathbb{R}} \left(U(t) - u(t)\right)\mathrm{d}x = \int_{\mathbb{R}} \left(U(0) - u(0)\right)\mathrm{d}x, \tag{26}$$

for two solutions (u, v) and (U, V). In terms of characteristic variables, (26) rewrites

$$\int_{\mathbb{R}} \big(W(t) - w(t) + Z(t) - z(t)\big)\,\mathrm{d}x \leqslant \int_{\mathbb{R}} \big(W(0) - w(0) + Z(0) - z(0)\big)\,\mathrm{d}x. \tag{27}$$

Notice the important fact that the conserved integral is a Lipschitz map with unit Lipschitz constant with respect to the distance considered in the Contraction property.

Constants. The only constants preserved by the relaxation model are the equilibrium constants (a, b), that is those belonging to Γ.

Well-posedness. As mentioned above, an initial datum taking values in a rectangle K yields a global-in-time solution that remains in K. The fact that the solution is unique in the class $L^\infty_{\mathrm{loc}}(\mathbb{R}^+; L^\infty)$ comes from the L^1-contraction property. The only point that is not obvious is the following fact: given two bounded solutions (w, z) and (W, Z) of the Cauchy problem, such that $W(0) - w(0)$ and $Z(0) - z(0)$ are integrable, then $W(t) - w(t)$ and $Z(t) - z(t)$ remain integrable forever. To do so, we remark that

$$\int_{-X}^{X} \big(\big|W(x,t) - w(x,t)\big| + \big|Z(x,t) - z(x,t)\big|\big)\,\mathrm{d}x$$

is bounded by

$$\int_{-X-\alpha t}^{X+\alpha t} \big(\big|W(x,0) - w(x,0)\big| + \big|Z(x,0) - z(x,0)\big|\big)\,\mathrm{d}x,$$

and therefore by

$$\big\|W(0) - w(0)\big\|_1 + \big\|Z(0) - z(0)\big\|_1.$$

Passing to the limit as X goes to infinity, we obtain the desired result. In particular, we justify the contraction property.

To prove the conservation property, we play the same game, knowing from above that $W(t) - w(t)$ and $Z(t) - z(t)$ remain integrable. The difference

$$\int_{-X}^{X} \big(U(x,T) - u(x,T)\big)\,\mathrm{d}x - \int_{-X-\alpha T}^{X+\alpha T} \big(U(0,t) - u(0,t)\big)\,\mathrm{d}x$$

equals the sum of the integrals of $\pm(G(W, Z) - G(w, z))$ on the triangles $\mathcal{T}_\pm$ of vertices $(\pm X, T)$ with slopes $\pm\alpha$. Since $W(T) - w(t)$ and $Z(T) - z(t)$ are space integrable, we know that the integrals along the horizontal sections of $\mathcal{T}_\pm$ tend to zero as $X \to +\infty$. Besides, this section integrals are uniformly bounded. Applying the Dominated convergence theorem, we infer that the integrals over $\mathcal{T}_\pm$ tend to zero. Hence the conservation.

COMMENT. When the equilibrium model is a system of $n \geqslant 2$ equations, most of the features of the scalar case drop out. For instance, the comparison and the contraction properties do not hold anymore. A maximum principle persists in a few specific case, mostly when $n = 2$. It was studied in details in [9] at the level of systems of viscous conservation laws. When it holds true, it extends, in a nontrivial way, to the Jin–Xin relaxation of the system [63], and allows to extend the solution of the Cauchy problem to all positive times. A thorough account of the theory of general relaxation models is contained in [49].

1.4. *The Rosenau model*

We finish this section by the study of the Cauchy problem for the generalized Rosenau model

$$\partial_t u + \operatorname{div} f(u) = K * u - u, \tag{28}$$

where the nonnegative kernel K has been normalized so that

$$\int_{\mathbb{R}^d} K(y)\,\mathrm{d}y = 1, \tag{29}$$

giving formally the mass conservation[3] for L^1-solutions.

Loss of smoothness. In this rather complex model, the diffusion term $K * u - u$ does not prevent from shock formation in finite time, for general data, though it does for small and smooth data. We refer to [33,34,41,44,56] for a general study of the Cauchy problem, including shock formation and existence of shock profiles. To explain what is going on, we first remark that $L := K * -1 = (K - \delta)*$ is a bounded operator on every Lebesgue or Sobolev space. Therefore, it generates a uniformly continuous (semi)group e^{tL}. In some sense, the diffusion is a lower order term, which cannot influence the qualitative feature of the principal part $\partial_t u + \operatorname{div} f(u)$.

Let us give a convincing example of blow-up, by choosing $d = 1$ and the Burgers' flux $f(u) = u^2/2$. Let u be a smooth solution of

$$\partial_t u + u\partial_x u = K * u - u.$$

Writing v for $\partial_x u$, and differentiating, we obtain

$$D_t v + v^2 + v = K * v,$$

where $D_t := \partial_t + u\partial_x$ is the convective derivative. If $a = u(0)$ is integrable, then $u(t)$ remains so, as we shall see in a moment, and therefore $m(t) := \inf_x v(x,t) \leqslant 0 \leqslant$

[3] We shall see later on that the vanishing of the first moments must be imposed when dealing with L^∞-solutions.

$\sup_x v(x,t) =: S(t)$. We easily obtain the following differential inequalities for both m and S:

$$m' + m^2 + m \leqslant S, \quad S' + S^2 \leqslant 0.$$

Let M be the solution of the Ricatti equation $M' + M^2 = 0$ with data $S(0)$. We have $0 \leqslant S \leqslant M \leqslant S(0)$. Therefore,

$$m' + m^2 + m \leqslant M$$

and

$$(m - M)' + (m - M)^2 + (2M + 1)(m - M) \leqslant 0.$$

Since $m - M \leqslant m - S \leqslant 0$, we deduce

$$(m - M)' + (m - M)^2 + \big(2S(0) + 1\big)(m - M) \leqslant 0,$$

where $(m - M)(0) = m(0) - S(0)$. It is well known that if this initial data is less than the smallest root $-2S(0) - 1$ of the polynomial $X^2 + (2S(0) + 1)X$, then $m - M$ must blow-up, towards $-\infty$, in finite time. Hence, if

$$\inf_x \partial_x a + \sup_x \partial_x a < -1,$$

the space derivative of u blows-up in finite time.

Global smooth solutions. On the contrary, smallness of smooth initial data ensures the existence of a global and smooth solution, as we show now. The above analysis showed that the supremum of $\partial_x u$ remains bounded above, by $1/t$ for instance. Anticipating on the maximum principle, we have $\|u\|_\infty \leqslant \|a\|_\infty$. Besides, the estimate

$$\|K * \partial_x u\|_\infty = \big\|(\partial_x K) * u\big\|_\infty \leqslant \|\partial_x K\|_1 \|u\|_\infty$$

yields a uniform bound

$$\|K * v\|_\infty \leqslant \|\partial_x K\|_1 \|a\|_\infty,$$

provided that $\partial_x K$ is integrable. There follows

$$m' + m^2 + m \geqslant -\|\partial_x K\|_1 \|a\|_\infty.$$

Let us make the smallness assumptions that

$$\|\partial_x K\|_1 \|a\|_\infty < \frac{1}{4}, \qquad \inf_x \partial_x a > m_-,$$

where m_- is the smallest root of equation

$$X^2 + X + \|\partial_x K\|_1 \|a\|_\infty = 0.$$

Then the differential inequality ensures that

$$\inf_{x,t} \partial_x u \geqslant m_-.$$

Hence, the space derivative of u remains bounded. Equation (28) shows that the time derivative is bounded as well. At last, one can prove, as in the diffusionless case, that the boundedness of the first derivatives allows to propagate the smoothness of the initial datum.

Construction of the semigroup. Because of the possibility of shock formation, a semigroup cannot be defined within spaces like $H^1(\mathbb{R}^d)$, $C^1(\mathbb{R}^d)$. Reasonable spaces are those already considered in the diffusionless case: L^p-spaces or $\mathrm{BV}(\mathbb{R}^d)$. Because we are interested in the Co-properties, we shall work within L^1, L^∞ and their sum $L^1 + L^\infty$. A special interest is given to piecewise smooth solutions, since they encode some of the most significant features. In particular, discontinuities of the solutions obey exactly the same Rankine–Hugoniot condition as those of (2).

Because of the lack of smoothing effect of the Rosenau diffusion, it seems possible to construct solutions which behave locally in the same way as piecewise smooth solutions of (2). In particular, (28) must admit unphysical solutions whose discontinuities, though satisfying the Rankine–Hugoniot condition, are not admissible with respect to (2). Therefore, we restrict the notion of solution in exactly the same way as in the diffusionless case. We say that a locally bounded measurable function $u(x,t)$ is an admissible (or an *entropy*) solution of the Cauchy problem for (28) if it satisfies the integral inequalities

$$\begin{aligned}
&\int_{\mathbb{R}}\int_{\mathbb{R}^d} \big(|u-k|\partial_t\phi + \operatorname{sgn}(u-k)\big(f(u)-f(k)\big)\cdot\nabla_x\phi\big)\,\mathrm{d}x\,\mathrm{d}t \\
&\quad + \int_{\mathbb{R}}\int_{\mathbb{R}^d} \big(\operatorname{sgn}(u-k)K*(u-k) - |u-k|\big)\phi\,\mathrm{d}x\,\mathrm{d}t \\
&\quad + \int_{\mathbb{R}^d} \big|a(x)-k\big|\phi(x,0)\,\mathrm{d}x \geqslant 0 \qquad (30)
\end{aligned}$$

for every constant $k \in \mathbb{R}$ and every nonnegative test function $\phi \in \mathcal{D}^+(\mathbb{R}^d \times \mathbb{R})$. Again, choosing large positive or negative values of k, we obtain that admissible solutions do satisfy (28) in the distributional sense.

Uniqueness. The uniqueness of an entropy solution of the Rosenau model is proved in a way similar to that followed in Section 1.2, but with extra arguments. One derives from (30) (see [56]) a formula similar[4] to (17):

$$\int_{\mathbb{R}}\int_{\mathbb{R}^d}\big(|u-v|\partial_t\phi+\operatorname{sgn}(u-v)\big(f(u)-f(v)\big)\cdot\nabla_x\phi\big)\,\mathrm{d}x\,\mathrm{d}t$$
$$+\int_{\mathbb{R}^d}\big(\operatorname{sgn}(u-v)K*(u-v)-|u-v|\big)\phi\,\mathrm{d}x\,\mathrm{d}t$$
$$+\int_{\mathbb{R}^d}\big|a(x)-b(x)\big|\phi(x,0)\,\mathrm{d}x\geqslant 0 \tag{31}$$

for every nonnegative test function.

If $v(t)-u(t)$ is known to be integrable in space for every t (a fact that is certainly true for solutions given by the viscosity method when a and b are integrable), the choice $\phi(x,t)=\rho(t)\theta(x/R)$, where $\theta(\cdot/R)$ tends to the constant 1, gives

$$\big\|v(t)-u(t)\big\|_1-\|b-a\|_1$$
$$\leqslant\int_0^t\int_{\mathbb{R}^d}\big(\operatorname{sgn}(v-u)K*(v-u)-|v-u|\big)\,\mathrm{d}x\,\mathrm{d}t,$$

which implies the contraction property

$$\big\|v(t)-u(t)\big\|_1\leqslant\|b-a\|_1.$$

The mass conservation is proved in the same way, from the identity

$$\int_{\mathbb{R}}\int_{\mathbb{R}^d}\big((u-v)\partial_t\phi+\big(f(u)-f(v)\big)\cdot\nabla_x\phi\big)\,\mathrm{d}x\,\mathrm{d}t$$
$$+\int_{\mathbb{R}^d}\big(K*(u-v)-(u-v)\big)\phi\,\mathrm{d}x\,\mathrm{d}t+\int_{\mathbb{R}^d}\big(a(x)-b(x)\big)\phi(x,0)\,\mathrm{d}x=0.$$

We are now reduced to proving the integrability of $v(t)-u(t)$, when $b-a$ is integrable. It has not yet been addressed, up to our knowledge. Several authors content themselves with the construction of a semigroup, extending by continuity the L^1-semigroup, thanks to its contraction property.

A first important case is that of integrable data, because the vanishing viscosity method (see below) gives integrable solutions. Therefore, we obtain the uniqueness for integrable data, but only within the restricted class of entropy solutions that are integrable in space. This is a bit weaker than expected.

When the space dimension equals one, the situation is nevertheless favorable. Since K has unit mass, we may introduce a function $H(x)$ with the properties that $H(\pm\infty)=0$ and

$$\frac{\mathrm{d}H}{\mathrm{d}x}=K-\delta.$$

[4] We warn the reader that an inequality such as (18) does not hold, because the model experiences an infinite speed of propagation, like the viscous model, due to the convolution term.

Then $K * w - w = H' * w$. We shall assume that $H \in L^1(\mathbb{R})$, which amounts to saying that K has a finite first moment. Following Kruzkhov's paper for conservation laws with source terms, we have the inequality

$$\begin{aligned}
&\int_{-R}^{R} \bigl|v(x,t) - u(x,t)\bigr| \, dx \\
&\quad \leqslant \int_{-R-tM}^{R+tM} |b - a| \, dx + \int_0^t ds \int_{-R-(t-s)M}^{R+(t-s)M} H' * |v - u| \, dx \\
&\quad = \int_{-R-tM}^{R+tM} |b - a| \, dx + \int_0^t \bigl[H * |v - u|\bigr]_{-R-(t-s)M}^{R+(t-s)M} \, ds \\
&\quad \leqslant \|b - a\|_1 + 2\|H\|_1 \|v - u\|_\infty \\
&\quad \leqslant \|b - a\|_1 + 2\|H\|_1 \bigl(\|a\|_\infty + \|b\|_\infty\bigr).
\end{aligned}$$

Since the right-hand side does not depend on R, we deduce that $v(t) - u(t)$ is integrable.

This argument does not extend to several space dimensions because the bound grows in general like R^{d-1}. However, if we assume that a and b have finite total variation, then there is a way to get home. We first remark that the solutions provided by the viscosity method have finite total variation, with

$$\mathrm{TV}\bigl(u(t)\bigr) \leqslant \mathrm{TV}(a).$$

We shall be considering only entropy solutions that have finite total variation. We introduce a vector field $\mathbf{H}$, vanishing at infinity and satisfying as above

$$\operatorname{div} \mathbf{H} = K - \delta.$$

In practice, it is enough to choose $\mathbf{H} = \nabla\theta$, where $\Delta\theta = K - \delta$. Then we write as above $K * |v - u| - |v - u| = \mathbf{H} \dot{*} \nabla|v - u|$. Then we have

$$\begin{aligned}
\int_{B_R} \bigl|v(x,t) - u(x,t)\bigr| \, dx &\leqslant \int_{B_{R+tM}} |b - a| \, dx + \int_0^t ds \int_{B_{R+(t-s)M}} \mathbf{H} * \nabla|v - u| \, ds \\
&\leqslant \|b - a\|_1 + \int_0^t \|\mathbf{H}\|_1 \mathrm{TV}\bigl(\bigl|v(s) - u(s)\bigr|\bigr) \, ds. \\
&\leqslant \|b - a\|_1 + \int_0^t \|\mathbf{H}\|_1 \bigl(\mathrm{TV}(a) + \mathrm{TV}(b)\bigr) \, ds.
\end{aligned}$$

If $\mathbf{H}$ is integrable,[5] the above estimate shows that $v(t) - u(t)$ is integrable, provided that $b - a \in L^1(\mathbb{R}^d)$.

[5] Remark that the singularity of $\mathbf{H}$ at he origin, of order $c_d r^{1-d}$, is integrable.

In practice, $K = K(r)$ is a radial kernel, and θ must be radial, given by

$$\theta'(r) = -\frac{1}{r^{d-1}} \int_r^{+\infty} s^{d-1} K(s)\,\mathrm{d}s.$$

Since $H(x) = \theta'(r)x/r$, we immediately see that H is integrable if, and only if $(1+|x|)K$ is integrable.

A fundamental example is that of the kernel defined by $\widehat{K}(\xi) = (1+|\xi|^2)^{-1}$. In other words, $K * u = z$, where z is the solution of the elliptic equation $-\Delta z + z = u$. Then K decays exponentially at infinity, and thus H is integrable.

Existence. We use the vanishing viscosity method. We begin by constructing the unique solution u^ε of the Cauchy problem for the perturbed equation

$$\partial_t u + \operatorname{div} f(u) = K * u - u + \varepsilon \Delta u, \tag{32}$$

where $\varepsilon > 0$ is a small number. A local solution is obtained by a fixed point argument applied to an integral form of the Cauchy problem, written through the Duhamel's principle for the heat equation (see Section 1.1). This only requires $a \in L^\infty$. The dispersion properties of the heat semigroup imply that u^ε is C^∞ whenever $t > 0$. Thus one may apply the maximum principle, giving the comparison property for (32). In particular,

$$\left\| u^\varepsilon(t) \right\|_\infty \leqslant \|a\|_\infty.$$

This uniform estimate allows us to extend u^ε to all positive times.

Given a nonnegative test function ϕ and a C^2 convex function η, we easily obtain the inequality

$$\begin{aligned}
&\int_{\mathbb{R}} \int_{\mathbb{R}^d} \left(\eta(u^\varepsilon)(\partial_t \phi + \varepsilon \Delta \phi) + \psi(u^\varepsilon) \cdot \nabla_x \phi \right) \mathrm{d}x\,\mathrm{d}t \\
&\quad + \int_{\mathbb{R}} \int_{\mathbb{R}^d} \left(K * \eta(u^\varepsilon) - \eta(u^\varepsilon) \right) \phi \,\mathrm{d}x\,\mathrm{d}t + \int_{\mathbb{R}^d} \eta(a)\phi(x,0)\,\mathrm{d}x \geqslant 0,
\end{aligned} \tag{33}$$

where ψ is the entropy flux, $\psi' = \eta' f'$. By a continuity argument, (33) holds true for Kruzkhov entropies $\eta_k = |\cdot - k|$. For two solutions, we have instead

$$\begin{aligned}
&\int_{\mathbb{R}} \int_{\mathbb{R}^d} \left\{ \left| u^\varepsilon - v^\varepsilon \right| (\partial_t \phi + \varepsilon \Delta \phi) + \operatorname{sgn}(u^\varepsilon - v^\varepsilon)\left(f(u^\varepsilon) - f(v^\varepsilon) \right) \cdot \nabla_x \phi \right\} \mathrm{d}x\,\mathrm{d}t \\
&\quad + \int_{\mathbb{R}} \int_{\mathbb{R}^d} \left(K * \left| u^\varepsilon - v^\varepsilon \right| - \left| u^\varepsilon - v^\varepsilon \right| \right) \phi \,\mathrm{d}x\,\mathrm{d}t \\
&\quad + \int_{\mathbb{R}^d} |a-b| \phi(x,0)\,\mathrm{d}x \geqslant 0,
\end{aligned} \tag{34}$$

when v^ε is the solution associated to the initial datum b. When both $b-a$ is integrable, $v^\varepsilon(t)-u^\varepsilon(t)$ is too (same proof as for the viscous model) and (34) readily gives an L^1-contraction:

$$\left\|u^\varepsilon(t)-v^\varepsilon(t)\right\|_1 \leqslant \|a-b\|_1. \tag{35}$$

In particular, taking $a \in L^1(\mathbb{R}^d)$ and $b=\tau_h a$, a shift of a by a vector h, we have a bound

$$\left\|\tau_h u^\varepsilon(t)-u^\varepsilon(t)\right\|_1 \leqslant \omega(\|h\|) \overset{\|h\|\to 0}{\longrightarrow} 0, \tag{36}$$

uniform in t and ε. Here, we only assume that

$$\lim_{h\to 0}\|\tau_h a-a\|_1=0, \tag{37}$$

a fact that holds true when $a\in L^1+\mathrm{BV}$ for instance. Using this estimate, one obtains a uniform bound of $\|u^\varepsilon(t+h)-u^\varepsilon(t)\|_1$ by the same argument as in Section 1.2. Therefore, pre-compactness holds within L^1_{loc} and we may extract a subsequence of $(u^\varepsilon)_{\varepsilon>0}$ that converges boundedly a.e. From Dominated convergence theorem, its limit is an entropy solution of the Rosenau model; the only delicate point is to pass to the limit in the term

$$\int_{\mathbb{R}}\int_{\mathbb{R}^d}\left(\operatorname{sgn}\left(u^\varepsilon-k\right)K*\left(u^\varepsilon-k\right)\right)\phi\,\mathrm{d}x\,\mathrm{d}t.$$

To do this, we only have to prove that $K*(u^\varepsilon-k)$ tends to $K*(u-k)$ boundedly a.e. The uniform bound is obvious, while the pointwise convergence follows from L^1_{loc} convergence, after extraction of a subsequence. The latter is derived from the (in)equalities

$$\int_{\mathbb{R}^d}|K*\theta|\phi\,\mathrm{d}x \leqslant \int_{\mathbb{R}^d}K*|\theta|\phi\,\mathrm{d}x=\int_{\mathbb{R}^d}|\theta|\check{K}*\phi\,\mathrm{d}x,$$

applied to $\theta=u^\varepsilon-u$.

We summarize our analysis in the following statement.

THEOREM 2. *Let K be a nonnegative, integrable kernel, with unit mass. The Cauchy problem for the Rosenau model* (28) *is well posed in the following spaces*:

- *in $L^1(\mathbb{R}^d)\cap L^\infty(\mathbb{R}^d)$, in every space dimension*;
- *in $(L^1+\mathrm{BV})\cap L^\infty(\mathbb{R}^d)$ if H is integrable. For instance, a radial K satisfying $K\in L^1((1+|x|)\,\mathrm{d}x)$ works*;
- *in $L^\infty(\mathbb{R}^d)$ if $d=1$ and $K\in L^1((1+|x|)\,\mathrm{d}x)$.*

In the sequel, we denote by $(R_t)_{\geqslant 0}$ the semigroup generated by the Rosenau model.

General data

- When the data is bounded but does not belong to $L^1+\mathrm{BV}$, the above procedure does not work. Also, the procedure employed in the diffusionless case is inefficient because of the lack of finite propagation velocity. One way to construct an entropy solution for general L^∞ data would be to use DiPerna's theory of measure valued solutions (see [14]), together with the improvement by Szepessy [69] to pass to the limit in the viscosity method. We leave this question open, with the remark that the uniqueness is already an unsolved problem.

- Notice the comparisons

$$\eta'(u)(K * u - u) \leqslant \eta(K * u) - \eta(u) \leqslant K * \big(\eta(u)\big) - \eta(u) \tag{38}$$

for every convex function η, which implies at a formal level the decay of

$$\int_{\mathbb{R}^d} \eta\big(u(t)\big)\,\mathrm{d}x$$

for entropy solutions. Examining the Picard iteration to the viscous approximation, then using the uniqueness of the limit in one of the well-posed contexts described in Theorem 2, one proves easily that if u is an entropy solution, and if $\eta \circ a$ is integrable for some convex function η with $\eta(0) = 0$ and $\eta \geqslant 0$ everywhere, then $\eta \circ u(t)$ is integrable and

$$t \mapsto \int_{\mathbb{R}^d} \eta\big(u(t)\big)\,\mathrm{d}x$$

is nonincreasing. For instance, all L^p-norms decay as time increases.

A dispersion inequality. The last comment, and the example provided by the viscous model (see Section 1.1) lead us to explore dispersion properties of the semigroup $(R_t)_{t>0}$. For that purpose, we assume that $a \in L^1 \cap L^\infty$. Hence, $a \in L^2$ and we have seen that $u(t)$ decays in L^2. We actually have a little bit more, that is,

$$\frac{\mathrm{d}}{\mathrm{d}t}\|u\|_2^2 + 2N[u]^2 \leqslant 0, \tag{39}$$

where

$$N[u] := \left(\int_{\mathbb{R}^d} \big(u^2 - u(K * u)\big)\,\mathrm{d}x\right)^{1/2}.$$

Using Fourier transform, we have

$$N[u]^2 = \int_{\mathbb{R}^d} m(\xi)\big|\hat{u}(\xi)\big|^2\,\mathrm{d}x,$$

with[6]

$$m(\xi) := 1 - \Re\widehat{K}(\xi) = 1 - \int_{\mathbb{R}^d} (\cos x \cdot \xi) K(x)\,\mathrm{d}x.$$

[6]Notice that $\overline{\hat{w}(\xi)} = \hat{w}(-\xi)$ for every real function w.

Since K is nonnegative and has unit mass, we have $m(\xi) > 0$ for $\xi \neq 0$. In addition, we know that m tends to 1 at infinity. Last, if K belongs to $L^1((1+|x|^2)\,\mathrm{d}x)$, then m is twice differentiable, and

$$D^2 m(0) = \int_{\mathbb{R}^d} K(x) x \otimes x \,\mathrm{d}x$$

is a positive definite matrix. Hence, there exists a constant $c > 0$ such that

$$m(\xi) \geqslant c\frac{|\xi|^2}{1+|\xi|^2} \quad \forall \xi \in \mathbb{R}^d. \tag{40}$$

We now estimate the L^2-norm in terms of N and the L^1-norm. We shall obtain a Nash-like inequality. Choosing a positive α, we have:

$$\|u\|_2^2 = \|\hat{u}\|_2^2 = \left(\int_{|\xi|>\alpha} + \int_{|\xi|<\alpha}\right)|\hat{u}|^2\,\mathrm{d}x.$$

We bound each term:

$$\|u\|_2^2 \leqslant c\left(\alpha^d \|\hat{u}\|_\infty^2 + \left(1+\alpha^{-2}\right)N[u]^2\right).$$

Last, we recall $\|\hat{u}\|_\infty \leqslant \|u\|_1$ and we choose

$$\alpha = \left(\frac{N[u]}{\|u\|_1}\right)^{2/(d+2)},$$

to obtain

$$\|u\|_2^2 \leqslant c\left(N[u]^2 + N[u]^{2d/(d+2)}\|u\|_1^{4/(d+2)}\right). \tag{41}$$

Apply this inequality to an entropy solution of (28) with initial datum a. Since $\|u(t)\|_1 \leqslant \|a\|_1$, we deduce from (41) the majorization

$$\left(\frac{\|u\|_2}{\|a\|_1}\right)^2 \leqslant F\left(\left(\frac{N[u]}{\|a\|_1}\right)^2\right), \qquad F(\tau) := c\left(\tau + \tau^{d/(d+2)}\right).$$

Let G denote the inverse function F^{-1}. Inserting in (39), we obtain

$$\frac{\mathrm{d}}{\mathrm{d}t}\left(\frac{\|u\|_2}{\|a\|_1}\right)^2 + 2G\left(\left(\frac{\|u\|_2}{\|a\|_1}\right)^2\right) \leqslant 0.$$

Define

$$H(y) := \int_1^y \frac{\mathrm{d}s}{G(s)},$$

there follows the bound

$$Y(t) \leqslant H^{-1}\big(H\big(Y(0)\big) - t/2\big), \quad Y(t) := \left(\frac{\|u(t)\|_2}{\|a\|_1}\right)^2. \tag{42}$$

Since H behaves like $cy^{-2/d}$ as y goes to zero, we deduce from (42) that, for every $a \in L^1 \cap L^\infty$, there holds

$$\|u(t)\|_2 = \mathcal{O}\big(t^{-d/4}\big), \quad t \to +\infty. \tag{43}$$

From (43) and the decays of the L^1 and L^∞ norms, we deduce also the weaker decays

$$\|u(t)\|_p = \mathcal{O}\big(t^{-d/(2\max(p,p'))}\big), \quad t \to +\infty, 1 < p < \infty. \tag{44}$$

Notice that these decays are weaker than those encountered in the viscous model for $p > 2$. Also, the bound of $\|u(t)\|_2$ involves both $\|a\|_1$ and $\|a\|_2$, though it involved only $\|a\|_1$ in the viscous case.

2. Shocks

In the most general acception of this word, a planar shock is a bounded traveling wave of the form

$$u(x,t) = U(x \cdot \nu - st),$$

having finite limits on each sides:

$$U(-\infty) =: u_-, \qquad U(+\infty) =: u_+.$$

The number s is the *normal velocity* of the shock while $\nu \in S^{d-1}$ is its direction of propagation. A *standing shock* corresponds to $s = 0$. It is always possible to restrict to standing shocks by using a moving frame, that is by changing variables as $(x,t) \mapsto (x + st\nu, t)$, yielding

$$(\partial_t, \nabla_x) \mapsto (\partial_t - s\nu \cdot \nabla_x, \nabla_x).$$

We shall see later on that every shock must satisfy the *Rankine–Hugoniot relations* (15). Therefore, the corresponding function

$$u_{\text{inv}}(x,t) := \begin{cases} u_-, & x \cdot \nu < st, \\ u_+, & x \cdot \nu > st \end{cases}$$

is a discontinuous solution of the inviscid model. We call it an inviscid discontinuity and denote it shortly $(u_-, u_+; s\nu)$.

The *shock profile* U obeys to a functional equation. When the latter is an ordinary differential equation and is resolved in terms of the highest derivative of U, the profile is smooth and provides an internal structure of the inviscid discontinuity.

A common feature of shock profiles for scalar equations satisfying the Co-properties is the monotony. This monotony must be read at the level of the Riemann invariants w and z in the relaxation model. The monotony of the profile is intimately related to the underlying dissipation structure. In general, the dissipation property is due to the particular way the model approaches the inviscid equation; for instance, it is present in the viscosity $\partial_x^2 u$, or in the Rosenau operator $K * u - u$. However, in the inviscid case, the dissipation cannot come from an "external term", and thus must be present in the shock itself, through the Lax shock inequality. This is why we shall meet counterexamples to the monotony in the inviscid case. As we shall see in the next section, such "bad" shocks may have poor stability properties.

2.1. *Inviscid shocks*

We begin with a characterization of the traveling waves in the inviscid case. Denoting $f(v;\nu) := \nu \cdot f(v)$ and $g(v) := f(v;\nu) - sv$, the profile is a standing wave of the one-dimensional conservation law

$$\partial_t U + \partial_x g(U) = 0.$$

In particular, $g \circ U$ is constant (say $g \circ U \equiv \bar{g}$) and, since U must satisfy the Kruzkhov entropy inequalities,

$$x \mapsto \big(\bar{g} - g(k)\big)\operatorname{sgn}\big(U(x) - k\big)$$

is nonincreasing for every real number k. For instance, if a (respectively b) is an essential value of u at some point x (respectively y), and if $b \neq a$, we obtain an Oleinik-type information: If $(b-a)(y-x) > 0$ (respectively < 0), then $g \geqslant \bar{g}$ (respectively $g \leqslant \bar{g}$) between a and b.

Assume now that U takes three distinct essential values a, b, c in this order. If b is not in the interval between a and c, say $b > \max(a, c)$, then $g \geqslant \bar{g}$ on (a, b) while $g \leqslant \bar{g}$ on (c, b), hence $g \equiv \bar{g}$ on $(\max(a, c), b)$. Therefore, making the generic assumption that g is not constant on any nontrivial interval, we see that U must be monotonous:

PROPOSITION 1. *Assume that $f(\cdot;\nu)$ is not affine on any nontrivial interval. Then the profiles of planar shock waves are (not strictly) monotonous. In particular, if $u_- = u_+$, then U is constant.*

Assuming without loss of generality that $u_- \leqslant u_+$, the graph of $f(\cdot;\nu)$ lies above the chord between u_- and u_+.

REMARKS.

- The generic assumption on the flux f is reminiscent to various approaches in the study of scalar conservation laws. In one space dimension, it ensures the strong convergence in the viscosity method, thanks to a compensated-compactness argument (see

Tartar [71,72]). Given in a quantitative form, it yields regularity properties through the well-known kinetic formulation (see Lions et al. [43]).

- As a matter of fact, the inviscid model is reversible at the level of smooth solutions, since the change of variables $(x,t) \mapsto (-x,-t)$ preserves the equation. However, it is crucial to observe that this model becomes irreversible at the level of discontinuous solutions. If a discontinuity $(u_-, u_+; s\nu)$ *and* the reverse one $(u_+, u_-; s\nu)$ are admissible, then the Oleinik criterion tells that $f(\cdot;\nu)$ is affine along the interval between u_- and u_+.
- More generally, if $f(\cdot;\nu)$ is affine along some interval J, then any measurable function U valued in J yields a traveling wave. In this rather trivial situation, nothing deep can be said.
- On the contrary, if $f(\cdot;\nu)$ is convex between u_- and u_+, then the discontinuity $(u_-, u_+; s\nu)$ is a shock if and only if it satisfies both $u_+ \leqslant u_-$ and the Rankine–Hugoniot relation.

In generic situations, the monotonicity of U and the constancy of $g \circ U$ imply that U is piecewise constant with finitely many discontinuities. Each discontinuity in U is a shock by itself, and we could consider only those which take two values only, thus named u_- and u_+. Even with this restriction, we may encounter, as soon as $f(\cdot;\nu)$ has inflexion points, degenerate situations where the graph of $f(\cdot;\nu)$ meets the chord somewhere (say at c) between a and b. In such a situation, the shock must be understood as two shocks of equal velocities s glued together, the first one linking a to c, the second one linking c to b. Finally, the only situation where U cannot be decomposed in more elementary blocks is when U takes only the values u_- and u_+ and the graph of $f(\cdot;\nu)$ does not meet the chord. When $u_+ \neq u_-$ (to avoid triviallity), this is what we shall call a *strict* shock in the sequel. Hence we have

$$U(x) = \begin{cases} u_-, & x < x_0, \\ u_+, & x > x_0, \end{cases}$$

with the following cases, symmetric to each other:

- either $u_- < u_+$ and the graph of $f(\cdot;\nu)$ in the interval (u_-, u_+) lies strictly above the chord,
- or $u_- > u_+$ and the graph of $f(\cdot;\nu)$ in the interval (u_-, u_+) lies strictly below the chord.

The Oleinik criterion implies immediately the *Lax shock inequalities*:

$$f_u(u_+;\nu) \leqslant s \leqslant f_u(u_-;\nu). \tag{45}$$

When these inequalities are strict, one speaks of a *noncharacteristic shock*, or a *Lax shock*. Otherwise, it is characteristic. Many people use the term "shock wave" in the case of noncharacteristic shocks only. But in the sequel, we shall not need to distinguish the characteristic case from the noncharacteristic one.

2.2. *Viscous shocks*

The viscous case is extremely simple. With the notations of the previous paragraph, the profile must satisfy the nonlinear ODE

$$U'' = (g \circ U)'$$

in the distributional sense. Integrating once, we obtain

$$U' = g \circ U - cst.$$

By induction, U is a smooth function, say of class C^{k+1} if f is C^k. Since U has limits $u_\pm$ at $\pm\infty$, we deduce that the constant of integration equals both of $g(u_\pm)$, hence

$$g(u_+) = g(u_-).$$

This is nothing but the Rankine–Hugoniot relation (15).

Since $g - g(u_-)$ has a constant sign between two consecutive zeroes, and since $u_\pm$ are zeroes of this function, we see that an inviscid planar discontinuity admits a *viscous profile* if, and only if, u_- and u_+ are consecutive zeroes, and if the sign of $g - g(u_-)$ equals that of $u_+ - u_-$ in the corresponding interval. This is the Oleinik inequality for is a strict shock. In conclusion, we have the following:

PROPOSITION 2. *An inviscid discontinuity admits a viscous profile if and only if it is a strict shock.*

When it exists, a viscous profile is monotonous and is unique up to a shift. Its behavior at infinity depends of the nature of the shock. For a noncharacteristic shock, the profile U tends exponentially fast to its limits $u_\pm$. But if it is characteristic, for instance at right ($f_u(u_+; \nu) = s$), then the decay of $U - u_+$ is at best algebraic; it is of order $1/x$ if $f_{uu}(u_+; \nu) \neq 0$.

In other words, one may say that, up to the decomposition of a discontinuity in elementary ones, the admissibility condition for a shock can be interpreted as the requirement that it be realized as the limit of traveling wave solutions in the viscous approximation

$$\partial_t u + \operatorname{div} f(u) = \varepsilon \Delta u,$$

As a matter of fact, a shock profile U yields the following solution

$$u^\varepsilon(x,t) = U\left(\frac{x \cdot \nu - st}{\varepsilon}\right)$$

which is a good approximation of the inviscid shock. It is clear that u^ε tends pointwise to $u_\pm$, according to the sign of $x \cdot \nu - st$.

2.3. *Relaxation shocks*

In the case of a relaxation model, say the Jin–Xin model, the profile consists in a pair (U, V), while the space dimension is one. The differential equation that it satisfies is

$$V' = sU', \qquad \alpha^2 U' = sV' + f(U) - V.$$

Eliminating V, we obtain the same equation than in the viscous case, up to a scaling,

$$\left(\alpha^2 - s^2\right)U'' = \left(f(U) - sU\right)'. \tag{46}$$

Notice the importance of the subcharacteristic condition (24), which must be fulfilled on the whole interval $[u_-, u_+]$ (or $[u_+, u_-]$ if needed). From the Rankine–Hugoniot condition, $s = f'(\bar{u})$ for some $\bar{u}$ between u_- and u_+. When (24) fails on some interval I, there exist shocks with $u_\pm \in I$, for which $\alpha^2 < s^2$. Such shocks do not admit relaxation profiles, while the nonadmissible discontinuities $(u_+, u_-; s)$ do! Notice also the degenerate case $\alpha^2 = s^2$, where both admissible and nonadmissible discontinuities may coexist; to avoid it, we shall reinforce the subcharacteristic condition into

$$\alpha > \sup\left\{\left|f'(u)\right|; u \in [u_-, u_+]\right\}. \tag{47}$$

Under condition (47), an inviscid planar discontinuity admits a *relaxation profile* if, and only if, it is a strict shock. Moreover, V is given by the formula

$$V = sU + f(u_-) - su_- = sU + f(u_+) - su_+.$$

Again, the relaxation profile is unique up to a shift. Moreover, the Riemann invariants are monotonous since

$$U \pm \frac{1}{\alpha} V = \left(1 \pm \frac{s}{\alpha}\right) U + cst,$$

and U itself is monotonous.

2.4. *Radiative shocks*

We shall call *radiative profile* the profile U of a discontinuity $(u_-, u_+; s)$ for the generalized Rosenau model, because the exact Rosenau model, whose kernel K is given by $\widehat{K}(\xi) = 1/(1 + |\xi|^2)$, may serve as a baby model for radiative gases.

There is no general theory for the existence of radiative profiles. Up to our knowledge, the literature consists in the papers [56] by Schochet and Tadmor, and [34] by Kawashima and Nishibata. In both papers, the kernel is Rosenau's one and the existence is studied through the differential equation

$$f(U)' - sU' = Q', \qquad Q'' - Q + U' = 0.$$

In the first paper, the flux g is any convex function but the amplitude $[u]$ of the shock is limited by some threshold. In the second paper, the flux is the simplest one $u^2/2$, but the amplitude is arbitrary. It is proven that whenever the Rankine–Hugoniot relation and the Oleinik condition $u_+ < u_-$ hold, there exists a unique (up to a shift) radiative profile. The profile cannot be of class $\mathcal{C}^\infty$; it is of class $\mathcal{C}^k$ where $k = k(|[u]|)$ is a decreasing function. As a matter of fact, the profile is *discontinuous* as $[u]$ is large enough. This is the origin of the restriction of the shock strength in [56], since Schochet and Tadmor did not consider discontinuous profiles. The lack of regularity is due to the lack of invertibility of the map $u \mapsto f(u) - su$ between u_- and u_+.

It must be emphasized that discontinuous profiles are perfectly relevant for this model, whose well-posedness holds in spaces of discontinuous functions (see Section 1.4.) Such profiles have at most one discontinuity, which turns out to satisfy the Rankine–Hugoniot relation and the Oleinik condition. Hence, the discontinuity of a radiative profile is itself an inviscid shock! This phenomenon reminds us the case of shock profiles in full gas dynamics, in presence of heat conduction but without viscosity; in this situation, Gilbarg [20] showed that strong shocks admit discontinuous profiles, where the discontinuity is an admissible shock of the isothermal gas dynamics.

Consistency of Rosenau approximation. The natural way to approach an inviscid scalar conservation law by a Rosenau model is to take a diffusion operator L_ε with

$$L_\varepsilon u := \frac{1}{\varepsilon}(K_\varepsilon * u - u), \qquad K_\varepsilon(x) := \frac{1}{\varepsilon^d} K\left(\frac{x}{\varepsilon}\right).$$

With this choice, a radiative profile furnishes a sequence

$$u^\varepsilon(x,t) := U\left(\frac{x \cdot \nu - st}{\varepsilon}\right)$$

of the radiative model

$$\partial_t u^\varepsilon + \operatorname{div} f\left(u^\varepsilon\right) = L_\varepsilon u^\varepsilon.$$

As in the viscous and relaxation cases, such a sequence converges pointwise to the inviscid shock $(u_-, u_+; s\nu)$.

At a formal level, which may be justified rigorously, the operator L_ε is given by the Fourier multiplier

$$\frac{1}{\varepsilon}\left(\widehat{K}(\varepsilon\xi) - 1\right).$$

Since $\widehat{K}(0) = 1$, this multiplier tends to the linear function $\xi \cdot \nabla_\xi \widehat{K}(0)$. Hence, L_ε tends to the drift

$$L_0 = \mathbf{m} \cdot \nabla_x, \quad \mathbf{m} := \int_{\mathbb{R}^d} K(x) x \, \mathrm{d}x.$$

Therefore, consistency of the Rosenau approximation requires the vanishing of the first moments of K (recall that we have assumed the existence of first moments):

$$\int_{\mathbb{R}^d} K(x)x_\alpha \,\mathrm{d}x = 0, \quad \alpha = 1, \ldots, d. \tag{48}$$

For instance, every even kernel K satisfies (48).

Naturally, we encounter condition (48) when searching for radiative profiles, as a compatibility with the Rankine–Hugoniot relation. For if there exists a radiative profile, then it satisfies

$$(g \circ U)' = K_\nu * U - U = (H_\nu * U)',$$

where K_ν and H_ν are one-dimensional kernels obtained from K and $H \cdot \nu$, after integration along the hyperplanes normal to ν. We thus obtain

$$g \circ U = H_\nu * U + cst,$$

or in other words $g(u_+) - g(u_-) = [H_\nu * U]$. Since the convolution term $H_\nu * U$ tends to $(\mathbf{m} \cdot \nu)u_\pm$ as $y \to \pm\infty$, we deduce

$$g(u_+) - g(u_-) = (\mathbf{m} \cdot \nu)(u_+ - u_-). \tag{49}$$

Hence, a necessary condition for the existence of a radiative profile for every inviscid shock is that the first moments of K vanish.

Assuming (48), the Rankine–Hugoniot relation becomes a necessary condition for the existence of a radiative profile, thanks to (49). Similarly, applying the entropy inequalities (30) to a radiative profile, we obtain

$$\begin{aligned}\big(\big(g(U) - g(k)\big)\operatorname{sgn}(U-k)\big)' &\leqslant \operatorname{sgn}(U-k)K_\nu * (U-k) - |U-k| \\ &\leqslant K_\nu * |U-k| - |U-k| \\ &= \big(H_\nu * |U-k|\big)'\end{aligned}$$

for every $k \in \mathbb{R}$. Integrating, we get

$$\big[\big(g(u) - g(k)\big)\operatorname{sgn}(u-k)\big] \leqslant \big[H_\nu * |U-k|\big].$$

Since $H_\nu * |U-k|$ vanishes at infinity, the right-hand side equals zero. Hence, the Oleinik condition appears as a necessary condition for the existence of a radiative profile.

OPEN PROBLEM 1. Assume that the kernel K satisfies (29) and (48). Prove that every strict inviscid shock $(u_-, u_+; s\nu)$ admits a radiative profile.

We give here a partial result, generalizing that was observed in the viscous and relaxation approximations.

PROPOSITION 3. *Let K be a nonnegative kernel satisfying* (29) *and* (48). *Then*
(i) *radiative shock profiles are monotonous;*
(ii) *given an inviscid shock, there exists at most one radiative profile, up to a shift.*

PROOF. We may assume that the space dimension is one, and that the shock is stationary. Let U be a radiative profile. For any real number h, the shifted $\tau_h U$ is another profile. Since both U and $\tau_h U$ are admissible stationary solutions of the Rosenau model, (30) gives

$$\begin{aligned}
&\partial_x\{\operatorname{sgn}(\tau_h U - U)(f(\tau_h U) - f(U))\}\\
&\quad\leqslant \operatorname{sgn}(\tau_h U - U)K * (\tau_h U - U) - |\tau_h U - U|\\
&\quad\leqslant K * |\tau_h U - U| - |\tau_h U - U|\\
&\quad= \partial_x(H * |\tau_h U - U|).
\end{aligned}$$

In other words, the function

$$x \mapsto \operatorname{sgn}(\tau_h U - U)(f(\tau_h U) - f(U)) - H * |\tau_h U - U|$$

is nonincreasing. Since it vanishes at infinity, it must therefore vanish identically. Hence, the above inequalities are equalities. In particular,

$$\operatorname{sgn}(\tau_h U - U)K * (\tau_h U - U) = K * |\tau_h U - U|,$$

which implies that $\tau_h U - U$ has a constant sign. Since its integral equals $h[u]$, we deduce that $h^{-1}(\tau_h U - U)$ has the sign of $[u]$. Letting h going to zero, we obtain the monotony.

The same argument can be used for uniqueness. Replacing $\tau_h U$ by any profile V in the computation above, we see that $V - U$ has a constant sign. Thus $V - \tau_h U$ has a constant sign for every $h \in \mathbb{R}$. The uniqueness follows immediately. □

3. Shock stability (I)

We now raise the main question of this paper, namely that of L^1-stability of the shock profiles. Let $U(x \cdot \nu - st)$ be a traveling wave of a scalar diffusive models, as considered above. Given an initial datum a, being an integrable perturbation of $U(\cdot\nu)$, that is,[7] $a \in U + L^1(\mathbb{R}^d)$, we already know that

$$t \mapsto \|S_t a - U(\cdot\nu - st)\|_1$$

is nonincreasing. This strongly suggests the following stability property, which was first considered by Osher and Ralston [51] in the context of one-dimensional viscous shock fronts:

$$\lim_{t\to+\infty} \|S_t a - U(\cdot\nu - st)\|_1 = 0? \tag{50}$$

[7]This assumption must interpreted in terms of the model under consideration. For instance, in the Rosenau model with $d \geqslant 2$, $\mathrm{TV}(a)$ should also be finite.

We shall give a complete answer to this question in the one-dimensional case, the multidimensional problem being essentially open.

Warning. In the sequel, we try to remain at a general level as long as possible, without distinguishing the model under consideration. It must be understood that in the Shi–Jin relaxation model, whose unknown is a pair (u, v), the L^1-norm is defined in terms of the Riemann invariants (w, z), since the corresponding distance is that which is contracted along the evolution. Similarly, a relaxation profile is a pair (U, V), and when we speak of "the mass" of the initial disturbance $(a - U, b - V)$, we mean the integral of $a - U$, since that is the conserved quantity along the evolution. At last, the comparison principle, denoted above by $\prec$, does not concern (u, v), but (w, z).

Other stability results. Our goal is not to treat stability in other functional spaces, like Hilbert ones (L^2 or other Sobolev spaces), with or without weights. As mentioned in the Introduction, we are motivated by the important physical significance of the L^1-norm and the fact that it is the norm for which the semigroup of the equation is nonexpansive. On an abstract level, Bruck and Baillon have long been studying nonexpansive semigroups and mappings in Banach spaces (see [2,3] for instance), establishing numerous deep theorems. But their results more or less assume that the norm has some kind of uniform convexity, a fact that is not true at all in L^1. On a more explicit side, one would not underestimate the voluminous literature where shock stability is obtained in Hilbert spaces through energy estimates and/or spectral analysis. One may date its beginning with the famous paper by Sattinger [55]. This method always assumes some kind of smallness of the initial disturbance, some decay rate of the data as $x \to \pm\infty$, and often a special structure of the nonlinear flux f. This flaw is a cheap price to pay for much more precise results, including accurate decay rates as time go to infinity. Also, the method has enough versatility to be applied in multidimensional situations, for which our approach remains inefficient (see Section 5 for a convincing example). Let us mention that significant difficulties arise when one considers *systems* of conservation laws instead of a scalar equation. Some of them may be observed at the level of the linearized operator $\mathcal{L}$ around the shock profile U. Even in the simplest case considered by Sattinger (strict Lax shock, exponentially decaying data), it is not any more possible to shift the essential spectrum of $\mathcal{L}$ strictly to the left of the imaginary axis. As a by-product, the null eigenvalue $\lambda = 0$ associated to the equation $\mathcal{L}\,dU/dx = 0$ remains embedded in the continuous spectrum, meaning that the translational invariance cannot be decoupled from the dynamics. This fact reflects at the level of the mass transport: the excess mass $\int_{\mathbb{R}}(a - U)\,dx$ cannot be absorbed any more by a shift of the profile. Therefore, it becomes necessary to take in account diffusion waves associated to the characteristics that emerge from the shock. Liu [45] gave a neat description of the expected asymptotics, which was rigorously proved under less and less restrictive assumptions by himself, Goodman and Xin [23] and Szepessy and Xin [70]. A refined analysis was performed by Kreiss and Kreiss [38] when the excess mass is null, and by Fries [18,19] in the general case, as long as the shock is small. A rather new fact in the case of systems is that a profile of large amplitude may be dynamically unstable. The fact that all scalar shock are stable can be viewed as a consequence of the Krein–Rutman theorem, applied to the eigenvalue $\lambda = 0$ of $\mathcal{L}$, noticing that the eigenfunction U' has a constant sign. Two mechanisms may be responsible for a dynamical instability. First of all, the instability of the underlying inviscid shock is an obstacle to the stability of any reasonable kind of profile (see Zumbrun and Serre [74]). But, even when the inviscid shock is stable, a viscosity induces a kind of Reynolds number, as noticed by Grenier and Guès [24] and Serre ([62], Chapter 15.2), and a large *Re* could cause a spectral instability. A first such example in a conservative setting is given in [64], while more realistic ones were built in Serre and Zumbrun [68]. In the stability analysis of large amplitude shock wave, one thus must assume the spectral stability. The strategy that has been developed by Howard and Zumbrun [30] and followed by many authors including Rousset [53] is to convert the spectral into linear stability (that is estimates for the semigroup $\exp(t\mathcal{L})$) and then into nonlinear stability. This is an outstandingly difficult program.

At last, let us mention a recent development that uses Logarithmic Sobolev inequalities and relative entropy, which applies in the L^1 context to the viscous Burgers' equation. With such techniques, Francesco and Markovitch prove a trend to self-similar diffusion waves.

3.1. *Preliminary observations*

To begin with, we obviously have

$$\left\| S_t a - U(\cdot \nu - st) \right\|_1 \geqslant \left| \int_{\mathbb{R}^d} \big(S_t a - U(\cdot \nu - st)\big)\, \mathrm{d}x \right|,$$

while the integral in the right-hand side equals (conservation of mass)

$$\int_{\mathbb{R}^d} \big(a - U(\cdot \nu)\big)\, \mathrm{d}x.$$

Hence,

$$\left\| S_t a - U(\cdot \nu - st) \right\|_1 \geqslant \left| \int_{\mathbb{R}^d} \big(a - U(\cdot \nu)\big)\, \mathrm{d}x \right|$$

and a necessary condition for (50) to hold is that the mass of the initial disturbance vanish:

$$\int_{\mathbb{R}^d} \big(a - U(\cdot \nu)\big)\, \mathrm{d}x = 0.$$

We shall write this restriction in the form

$$a \in U(\cdot \nu) + L_0^1(\mathbb{R}^d), \tag{51}$$

where $L_0^1(\mathbb{R}^d)$ denotes the set of integrable functions whose integral over $\mathbb{R}^d$ vanishes.

In a single space dimension, (51) is not a significant restriction, because of the following fact. Since U is monotonous, it has finite total variation. Hence, $\tau_h U - U$ is integrable and

$$\int_{\mathbb{R}} (\tau_h U - U)\,dx = \lim_{A\to+\infty} \int_{-A}^{A} (\tau_h U - U)\,dx = \lim_{A\to+\infty} \left(\int_{A}^{A+h} - \int_{-A}^{-A+h} \right) U\,dx$$

gives

$$\int_{\mathbb{R}} (\tau_h U - U)\,dx = h(u_+ - u_-). \tag{52}$$

Therefore, our initial datum a is an integrable perturbation of $\tau_h U$, and the mass of the corresponding perturbation is

$$\int_{\mathbb{R}^d} \big(a - U(\cdot\nu)\big)\,dx - h(u_+ - u_-).$$

Thus it is always possible to choose h in such a way that $a \in \tau_h U + L^1_0$. Specifically, we must take

$$h = \frac{1}{[u]} \int_{\mathbb{R}} (a - U)\,dx.$$

For this reason, given an initial datum $a \in U + L^1$, it is always possible to replace U by some $\tau_h U$, in such a way that $a \in U + L^1_0$. Hence, the one-dimensional problem reduces to the following:

QUESTION. Let $d = 1$ and $a \in U + L^1_0$ be given. Prove (or disprove) the convergence (50).

This problem is solved in this paper. Again, asking the question assumes that the Cauchy problem is well posed and that the semigroup satisfies the Co-properties. For instance, in the relaxation model (3), (4), we also assume that the initial data takes values in a characteristic box where the subcharacteristic condition (24) is fulfilled.

On the contrary, whenever $d \geqslant 2$, the difference $\tau_h U - U$ is not integrable (obvious) for nonzero vectors h. Hence, $a \in U + L^1$ is not compatible with $a \in \tau_h U + L^1$. Hence it is not possible, starting from a general integrable perturbation of U, to restrict to a zero-mass disturbance. Either the initial disturbance had zero-mass, or it had not. In the latter case, the answer to the question is definitively *no*. Let us mention however the very interesting results by Goodman and Miller [22] and Hoff and Zumbrun [29] about the viscous shocks under small perturbations of nonzero mass. These will be describe in Section 5.

OPEN PROBLEM 2. Let $d \geqslant 2$ and $a \in U + L^1_0$ be given. Prove (or disprove) the convergence (50).

Density arguments. The contraction property offers for free a few useful inequalities. Define first the L^1-distance $d(u,v) := \|v-u\|_1$ between two functions defined on $\mathbb{R}^d$. Also, given a traveling wave $U(\cdot\nu - st)$ and a data $a \in U(\cdot\nu) + L^1_0$, define the limit

$$\ell_0(a) := \lim_{t\to+\infty} d\big(S_t a, U(\cdot\nu - st)\big).$$

Since $S_\tau a - U(\cdot\nu - s\tau)$ is integrable with a zero total mass, $\ell_0(S_\tau a)$ makes sense, and we have on the one hand

$$\ell_0(S_\tau a) = \ell_0(a). \tag{53}$$

On the other hand, the contraction property and the triangle inequality show that ℓ_0 is Lipschitzian on the affine space $U(\cdot\nu) + L^1_0$, with unit constant:

$$\big|\ell_0(b) - \ell_0(a)\big| \leqslant \|b-a\|_1. \tag{54}$$

The first applications of (54) concerns the attraction basin of the traveling wave, that is the set $\mathcal{A}_0(U)$ of data such that (50) holds.

LEMMA 1. *The attraction basin $\mathcal{A}_0(U)$ of a traveling wave U is closed under the L^1-distance d. Furthermore, for every $a \in U(\cdot\nu) + L^1_0$, we have*

$$\ell_0(a) \leqslant d\big(a; \mathcal{A}_0(U)\big). \tag{55}$$

In a single space dimension, we may compare a data $a \in U + L^1$ to the shifted profiles $\tau_h U$. The distance $d(S_t a; \Upsilon(U))$, to the set $\Upsilon(U)$ of these profiles, is nonincreasing and we call $\ell(a)$ its limit. The function ℓ enjoys the same properties (53) and (54) as ℓ_0. Similarly, we define the attraction basin $\mathcal{A}(U)$ of the set $\Upsilon(U)$. Because $S_t a$ may be asymptotic only to the wave whose initial profile $\tau_h U$ is such that $a \in \tau_h U + L^1_0$, we see that $\mathcal{A}(U)$ is exactly the union of the sets $\mathcal{A}_0(\tau_h U)$ as $h \in \mathbb{R}$.

LEMMA 2. *Assume $d = 1$, and let $a \in U + L^1_0$ be an initial datum, where U is a shock profile. Then*

$$\ell_0(a) \leqslant 2d\big(a; \mathcal{A}(U)\big). \tag{56}$$

PROOF. Given $b \in \mathcal{A}(U)$, we have $b \in \tau_h U + L^1_0$ for

$$h = \frac{1}{[u]}\int_{\mathbb{R}} (b-U)\,dx = \frac{1}{[u]}\int_{\mathbb{R}} (b-a)\,dx.$$

Hence, we have, on the one hand,

$$\big|h[u]\big| \leqslant d(a,b). \tag{57}$$

On the other hand, $d(S_t b, U(\cdot - h - st))$ tends to zero. Using the triangle inequality, we have

$$d\big(S_t a, U(\cdot - st)\big) \leqslant d(S_t a, S_t b) + d\big(S_t b, U(\cdot - h - st)\big) \\ + d\big(U(\cdot - h - st), U(\cdot - st)\big).$$

The first term in the right-hand side is bounded by $d(a, b)$ (contraction) and the last one equals

$$\left| \int_{\mathbb{R}} \big(U(x - h - st) - U(x - st)\big)\, \mathrm{d}x \right| = \big|h[u]\big|.$$

Using (57) and passing to the limit, we deduce

$$\ell_0(a) \leqslant 2d(a, b).$$

There remains to take the infimum over $\mathcal{A}(U)$. □

3.2. *The dynamical system approach*

From now on, we restrict to the one-dimensional case. In a first instance, we focus on initial data that lie between two shifts of the profile U. Following Osher and Ralston [51], this case can be treated by means of tools from dynamical systems theory: ω-limit sets and Lasalle's invariance principle. Theorem 3 and Corollary 1 are due to Osher and Ralston in the viscous case, under the unnecessary assumption that the underlying inviscid shock satisfied the strict Lax inequalities.

Thus let U be a shock profile and let a be an initial datum such that there exist two real numbers j, k with

$$\tau_j U \leqslant a \leqslant \tau_k U. \tag{58}$$

Since $\tau_k U - \tau_j U$ is integrable, (58) ensures that $a \in U + L^1$. Without loss of generality, we may assume that $a \in U + L^1_0$. Up to the choice of a moving frame, we also may assume that U is a standing wave, that is a steady solution. In the relaxation case, such a choice changes slightly the system itself, but what is important is that it does not affect the Co-properties.

The comparison principle ensures that $S_t a$ still satisfies (58) for every positive time. In other words, $S_t a - U$ remains pointwise between the integrable functions $\tau_j U - U$ and $\tau_k U - U$. Hence, there exists an integrable function F such that

$$\big|S_t a(x) - U(x)\big| \leqslant F(x) \quad \forall t \in \mathbb{R}. \tag{59}$$

Besides, the semigroup commutes with the space shifts. Hence, $\tau_h S_t a = S_t \tau_h a$ is another solution, with initial datum $\tau_h a$, and the contraction principle gives

$$\|\tau_h S_t a - S_t a\|_1 \leqslant \|\tau_h a - a\|_1 =: \delta_a(h).$$

Since $a - U$ is integrable, the upper bound $\delta_a(h)$ tends to zero with h. In other words, the family of integrable functions $(S_t a - U)_{t\geqslant 0}$ is *equiintegrable*. Together with (59) that constitutes the hypotheses of the Fréchet–Kolmogorov (see Brézis [5]) compactness theorem: This family is relatively compact in $L^1(\mathbb{R})$. In other words, the solution trajectory $(S_t a)_{t\geqslant 0}$ is relatively compact in $U + L^1(\mathbb{R})$, equipped with the distance d.

The asymptotic behavior of the solution, with respect to d, is thus entirely encoded in the *ω-limit set*, defined as

$$\Omega(a) := \bigcap_{T\geqslant 0} \overline{\{S_t a;\, t > T\}}.$$

In other words, Ω_a is the set of the strong limits of subsequences $(S_{t_m} a)_{t_m \to +\infty}$. Important properties of the ω-limit set are the following:

- Ω_a is a nonvoid compact subset of $U + L^1$.
- Ω_a is invariant under the semigroup, forward *and* backward, meaning that

$$S_t \Omega_a = \Omega_a.$$

- Ω_a is a connected set.
- The elements b of Ω_a satisfy

$$\tau_j U \leqslant b \leqslant \tau_k U. \tag{60}$$

Proving (50) amounts to showing that

$$\Omega_a = \{U\}.$$

To do that, we apply the Lasalle's invariance principle (see [40]), which we recall now. Let $J : U + L^1 \to \mathbb{R}$ be a continuous Liapunov function for our semigroup. Then J takes a constant value c_J on Ω_a. Since Ω_a consists in trajectories, these must be rather special ones, for which the Liapunov functions are just constant. In general, there is no way to compute c_J, and the only information that this principle gives is the constancy. For instance, if there is some differentiability (but this will not be the case for us), that is, if

$$\frac{\mathrm{d}}{\mathrm{d}t} J[S_t a] = -R[S_t a]$$

for some dissipation rate R, then every element b of Ω_a must satisfy[8] $R[b] = 0$. The only situation where c_J is explicit is when J is a conserved quantity, in which case we have $c_J = J[a]$.

In our context, we have precisely one conservation law, namely the total mass of $u - U$. Since $\int_{\mathbb{R}} (a - U)\,\mathrm{d}x = 0$, we deduce

$$\int_{\mathbb{R}} (b - U)\,\mathrm{d}x = 0 \quad \forall b \in \Omega_a. \tag{61}$$

[8] We warn the reader that this property is a bit weaker than the constancy of J in general. See a striking example in [66].

Besides, there is a one-parameter family of Liapunov functions, given by the contraction principle and the fact that each $\tau_h U$ is a steady solution:

$$J_h[u] := \|u - \tau_h U\|_1.$$

Therefore we may state the following intermediate result.

LEMMA 3. *Let U be a profile for a steady shock, and $a \in U + L^1_0$ be given. Then, for every $h \in \mathbb{R}$ and every $b \in \Omega_a$, the function of time*

$$t \mapsto \|S_t b - \tau_h U\|_1$$

is constant.

We have reached a point where we have to treat each model separately, by exploiting Lemma 3. We shall skip technical details and retain only the main ideas. Though they differ from model to model, we always get the following conclusion.

LEMMA 4. *There holds*

$$\Omega_a \subset \Upsilon(U).$$

Then (61) tells that

$$\Omega_a \subset \Upsilon(U) \cap \left(U + L^1_0\right) = \{U\}.$$

Since Ω_a is nonvoid, we obtain that $\Omega_a = \{U\}$, which implies (50), hence the conclusion:

THEOREM 3. *Let $d = 1$ and $U(x - st)$ be a traveling wave (a shock for the diffusion model). In the inviscid case, assume that U is a strict shock. Let a be an initial datum satisfying*

$$\tau_j U \leqslant a \leqslant \tau_k U$$

for some $j, k \in \mathbb{R}$. Then $a \in \mathcal{A}(U)$. More precisely,

$$d\big(S_t a, U(\cdot - h - st)\big) \to 0$$

as $t \to +\infty$, where h is given by

$$h = \frac{1}{u_+ - u_-} \int_{\mathbb{R}} (a - U)\,\mathrm{d}x.$$

Using Lemma 1, we have a straightforward extension of Theorem 3.

COROLLARY 1. *Let $d = 1$ and $U(x - st)$ be a traveling wave (a shock for the diffusion model). In the inviscid case, assume that U is a strict shock. Let $a \in U + L^1$ be an initial datum satisfying*

$$u_+ \leqslant a(x) \leqslant u_-, \qquad a.e.\ x \in \mathbb{R}.$$

Then $a \in \mathcal{A}(U)$. More precisely,

$$d\big(S_t a, U(\cdot - h - st)\big) \to 0$$

as $t \to +\infty$, where h is given by

$$h = \frac{1}{u_+ - u_-} \int_{\mathbb{R}} (a - U)\, \mathrm{d}x.$$

3.3. *Sketch of proof of Lemma* 4

Without loss of generality, we may assume[9] that $u_- > u_+$. Thus the Oleinik condition for a strict shock tells that $f(y) < f(u_\pm)$ for every $y \in (u_+, u_-)$. Given an element b of Ω_a, denote $v(t) = S_t b$.

Inviscid case. From (60), we have $u_+ \leqslant b \leqslant u_-$. Since $U(x) = u_\pm$ for $\pm x > 0$, we have

$$J_h[v] = \int_{-\infty}^{h} (u_- - v)\, \mathrm{d}x + \int_{h}^{+\infty} (v - u_+)\, \mathrm{d}x.$$

Integrating the entropy inequality

$$\partial_t |v - \tau_h U| + \partial_x \big(\operatorname{sgn}(v - \tau_h U)\big(f(v) - f(\tau_h U)\big)\big) \leqslant 0$$

on a domain defined by $t_1 < t < t_2$, $|x| + Mt \leqslant R$ and then letting $R \to +\infty$, we obtain

$$J_h\big[v(t_2)\big] - J_h\big[v(t_1)\big] + \int_{t_1}^{t_2} \big(f(u_+) + f(u_-) - 2f \circ v(h, t)\big)\, \mathrm{d}t \leqslant 0.$$

Notice that the right-hand side makes sense, since the conservation law implies the existence of traces of the flux $f \circ v$ along vertical[10] lines. These traces are bounded measurable functions.

From the constancy of J_h along a trajectory in Ω_a, we conclude that

$$\int_{t_1}^{t_2} \big(f(u_+) + f(u_-) - 2f \circ v(h, t)\big)\, \mathrm{d}t \leqslant 0,$$

[9] Otherwise, change f into $-f$ and x into $-x$.

[10] More generally, $f \circ v - \sigma v$ has a well-defined trace on lines of slopes σ.

whenever $t_1 < t_2$. Hence,

$$f(u_+) + f(u_-) - 2f \circ v(h,t) \leqslant 0 \tag{62}$$

almost everywhere. However, since v takes values in $[u_+, u_-]$, and since $f \leqslant f(u_\pm)$ on this interval, the traces are less than or equal to $f(u_\pm)$. Hence, (62) implies that $f \circ v \equiv f(u_\pm)$, and therefore $\partial_t v = 0$. Thus $v \equiv b$ is a steady solution, in the class $U + L^1_0$. The general study of steady shocks, made in Section 2.1, shows that $b = U$.

Viscous case. Recall that Ω_a is backward invariant under the evolution, and that the semigroup has a smoothing property. We infer that Ω_a is made of smooth functions only. For instance, if $f \in C^\infty$, then $b \in C^\infty$. Even with a limited amount of regularity, we always may write that the decay rate of $J_h[v]$ vanishes. At initial time, that means

$$\int_{\mathbb{R}} \operatorname{sgn}(b - \tau_h U)\partial_x^2(b - \tau_h U)\,\mathrm{d}x = 0. \tag{63}$$

It is a bit technical, though not too much, to prove that (63) amounts to the following property (see the details in [61,65]): Every vanishing point of $b - \tau_h U$ is a vanishing point of its derivative $(b - \tau_h U)'$. Varying the value of h, that means exactly that, for every points $x, y \in \mathbb{R}$,

$$b(x) = U(y) \quad \Longrightarrow \quad b'(x) = U'(y). \tag{64}$$

Since $U : \mathbb{R} \to (u_+, u_-)$ is monotonous, it is a diffeomorphism. Since b lies between $\tau_j U$ and $\tau_k U$, it takes values in (u_+, u_-). Thus it can be written as $b = U \circ \phi$ for some smooth function ϕ. Applying (64) to the pair $(x, y = \phi(x))$, we obtain on the one hand $b' = U' \circ \phi$. On the other hand, simple differentiation would have given $b' = \phi' U' \circ \phi$. Since U' does not vanish, we deduce from both equalities that $\phi' \equiv 1$, that is, $\phi(x) = x - h$ for some h, or $b = \tau_h U$.

At last, the fact that $b \in L^1_0$ gives $h = 0$. This ends the proof of Lemma 4 in this case.

Shi–Jin relaxation. Theorem 3 was proved, in this case, by Mascia and Natalini [48].

The initial datum, being a pair of functions, is denoted by (a, b). Our assumption is that $a - U$ and $b - V$ are integrable, and that the total mass of $a - U$ is zero.

Denote by W, Z the Riemann invariants of the shock front (U, V), and by (w, z) those of a solution (u, v) under consideration. The Liapunov function J_h is defined by

$$J_h[u, v] := \|w - \tau_h W\|_1 + \|z - \tau_h Z\|_1.$$

Its rate of decay is

$$R_h[u, v] = \int_{\mathbb{R}} \big(G(w, z) - G(\tau_h W, \tau_h Z)\big)\big(\operatorname{sgn}(z - \tau_h Z) - \operatorname{sgn}(w - \tau_h W)\big)\,\mathrm{d}x.$$

This integral involves only the domain where

$$(z - \tau_h Z)(w - \tau_h W) \leqslant 0, \quad |z - \tau_h Z| \cdot |w - \tau_h W| \neq 0. \tag{65}$$

From the strict subcharacteristic condition (47), G is strictly increasing with respect to z and decreasing with respect to w. Therefore, (65) implies that $G(w, z) - G(\tau_h W, \tau_h Z)$ be positive. Hence $R_h[u, v] \geqslant 0$ (the contraction property), and the equality holds if, and only if $z - \tau_h Z$ and $w - \tau_h W$ have strictly the same sign (both positive, both negative, or both zero), almost everywhere.

If (c, d) belongs to $\Omega_{a,b}$, we must have $R_h[c, d] = 0$. Therefore, for every h, $w - \tau_h W$ and $z - \tau_h Z$ have strictly the same sign. Moving h, this implies that, given two points $x, y \in \mathbb{R}$, the following equivalence holds:

$$w(x) = W(y) \iff z(x) = Z(y). \tag{66}$$

Since, as in the viscous case, $W : \mathbb{R} \to (w_-, w_+)$ and $Z : \mathbb{R} \to (z_-, z_+)$ are diffeomorphisms, while w, z take values in (w_-, w_+) and (z_-, z_+) respectively, there exist functions ϕ and ψ such that

$$w = W \circ \phi, \qquad z = Z \circ \psi.$$

However, property (66) tells us that $\psi \equiv \phi$, hence

$$(w, z) = (W, Z) \circ \phi, \qquad x \in \mathbb{R}.$$

We now consider[11] the solution (u, v) of the relaxation model, whose initial data is (c, d). We still denote by (w, z) its Riemann invariants. Since $(u(t), v(t)) \in \Omega_{a,b}$ at each time, there exists a function $\phi(x, t)$ such that

$$(w, z) = (W, Z) \circ \phi, \qquad (x, t) \in \mathbb{R}^2.$$

Let us write that (w, z) is a solution of the coupled transport equations:[12]

$$\delta^+ w = G(w, z), \qquad \delta^- z = -G(w, z).$$

Using the fact that (W, Z) is a steady solution, namely

$$G(W, Z) = \alpha_+ W' = \alpha_- Z',$$

there comes

$$(\delta^+ \phi - \alpha_+) W' \circ \phi = 0, \qquad (\delta^- \phi + \alpha_-) Z' \circ \phi = 0.$$

[11]The following argument is distinct from that employed by Mascia and Natalini. It is somehow simpler, as it does not need to establish some regularity result for the elements of Ω_a.

[12]The differential operators $\delta^\pm = \partial_t \pm \alpha_\pm \partial_x$ differ from $D^\pm$ because of the moving frame, but we still have $\alpha_\pm > 0$.

Since W' and Z' do not vanish, we infer that

$$\partial_t \phi \pm \alpha_\pm (\partial_x \phi - 1) = 0,$$

or equivalently

$$\partial_t \phi = 0, \qquad \partial_x \phi = 1,$$

since $\alpha_\pm > 0$. Thus there exists a real number h such that $\phi \equiv x - h$. This proves that $(c, d) = \tau_h(U, V)$. At last, the property that $c - U$ has zero total mass implies $h = 0$. This ends the proof of the lemma.

Radiative model. The rate of decay of J_h is

$$R_h[u] = \int_{\mathbb{R}} \big(|u - \tau_h U| - \operatorname{sgn}(u - \tau_h U) K * (u - \tau_h U)\big)\,\mathrm{d}x.$$

Since $(\operatorname{sgn} v) K * v \leqslant K * |v|$ with equality if, and only if, v has a constant sign, $R_h[u]$ is nonnegative (the contraction property) and vanishes only when $u - \tau_h U$ has a constant sign. Hence, every b in Ω_a is such that $b - \tau_h U$ has a constant sign for every $h \in \mathbb{R}$. As above, this implies that b equals one of the shifted profiles. At last, the fact that $\Omega_a \subset U + L^1_0$ tells us that this profile is U itself. Whence the Lemma.

3.4. *Data with values beyond* $[u_+, u_-]$; *inviscid case*

Corollary 1 does not cover the case of general data $a \in U + L^1_0$, taking some values outside the interval $[u_+, u_-]$. There does not seem possible to reach the same conclusion by the use of dynamical systems tools only. As a matter of fact, we shall give an example, with an inviscid equation, that the L^1-stability does not hold. Such a phenomenon is actually due to the lack of diffusion, and will not occur in other models. We shall then show that the inviscid stability does hold when the shock is noncharacteristic. At last, we shall describe a general strategy, based on the diffusion properties of the linear operator L, to show the L^1-stability in other models; we shall explain why this method does not work in the inviscid case.

In the sequel, it will be useful to denote by $A(U)$ the set of initial data to which Corollary 1 applies. It consists in the functions $a \in U + L^1$, taking values in the interval[13] $[u_+, u_-]$.

Lack of stability for inviscid characteristic shocks. Let $U = (u_-, u_+; 0)$ be a steady inviscid strict shock with, say, $u_+ < u_-$. Assume that it is characteristic at left, that is,

$$f'(u_-) = 0.$$

[13] Exercise: Write down the correct definition of $A(U, V)$ in the case of Shi–Jin relaxation.

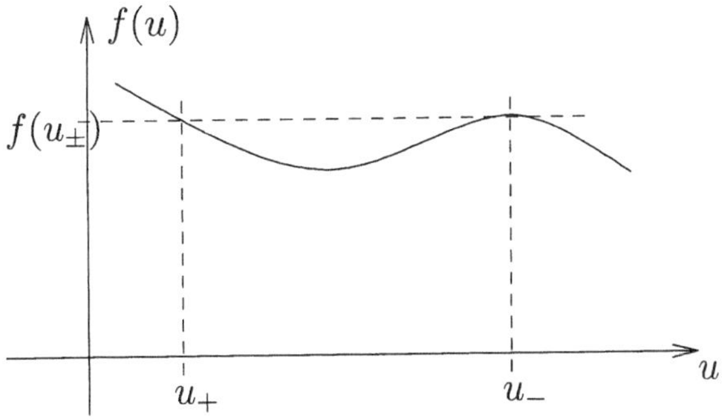

Fig. 1. A shock that is characteristic at left.

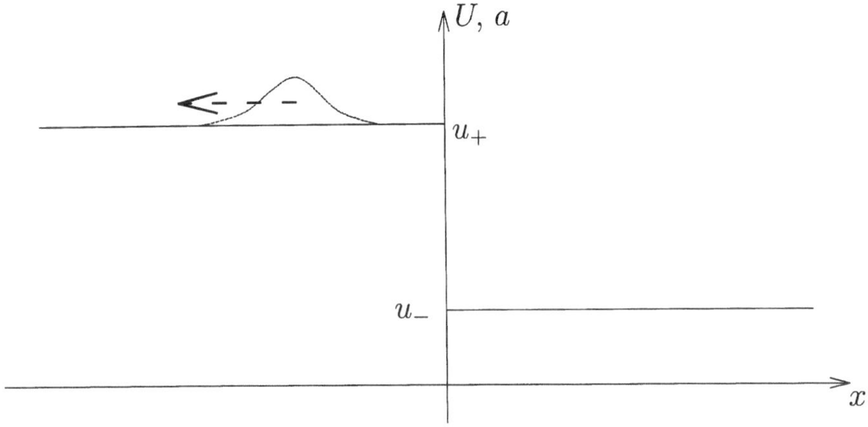

Fig. 2. The shock is characteristic at left. The disturbance travels to the left with unchanged mass.

The Oleinik criterion implies that

$$f''(u_-) \leqslant 0$$

and, generically, the inequality is strict. Thus, assume

$$f''(u_-) < 0. \tag{67}$$

Then there is some nontrivial interval $(u_-, u_- + \delta)$ on which f' is negative. These assumptions are displayed in Figure 1.

Choose an initial datum a in the following way: It coincides with U, except on some compact interval $I \subset (-\infty, 0)$, on which $U \equiv u_-$ and a takes values in $(u_-, u_- + \delta)$. We may see a as the superposition $U + b - u_-$ of two distinct data, namely U, and a data $b : \mathbb{R} \to [u_-, u_- + \delta]$, which equals u_- outside of J. We claim that the solution $S_t a$ equals the superposition of $S_t U = U$ and $S_t b$, that is, $S_t a = U + S_t b - u_-$ (see Figure 2). For, by the maximum principle, $u_- \leqslant S_t b \leqslant u_- + \delta$. Hence, because of $f' < 0$

on $(u_-, u_- + \delta)$, the support of $S_t b - u_-$ moves to the left, implying that $S_t b \equiv U$ at the right of I. Therefore $U + S_t b - u_-$ coincides with $S_t b$ on $(-\infty, 0)$ and with U on some $(-\varepsilon, +\infty)$, and thus is an entropy solution. Since its initial value is $U + b - u_- = a$, uniqueness gives the claim.

Let us compute the distance from $S_t a$ to $\tau_h U$, where h is the value given by mass conservation:

$$h = \frac{1}{u_+ - u_-} \int_{\mathbb{R}} \big(b(x) - u_-\big)\,\mathrm{d}x.$$

It is known (Dafermos [10]) that, because of (67), $S_t b$ evolves as an N-wave as time tends to infinity: The mass of $S_t b - u_-$ is obviously conserved, while its sup-norm decays as $t^{-1/2}$ and its support spreads. In particular, since h is a constant,

$$\begin{aligned} d(S_t a, \tau_h U) &= \int_{-\infty}^{0} (S_t b - \tau_h U)\,\mathrm{d}x \sim \int_{-\infty}^{\min(h,0)} (S_t b - u_-)\,\mathrm{d}x \\ &= \int_{\mathbb{R}} (b - u_-)\,\mathrm{d}x. \end{aligned}$$

Therefore,

$$\ell(a) = \int_{\mathbb{R}} (b - u_-)\,\mathrm{d}x$$

and the stability does not hold if b is not trivial.

L^1-stability of noncharacteristic inviscid shocks. Since the situation would be similar, if the shock was characteristic at right, we know assume that the shock is noncharacteristic:

$$f'(u_+) < s = 0 < f'(u_-). \tag{68}$$

Given an initial datum $a \in U + L^1$, we let

$$b(x) := \max\big(a(x), u_-\big), \qquad c := \min\big(a(x), u_+\big)$$

and

$$v, u, w := S_t c, S_t a, S_t b.$$

From the comparison principle, we have $v \leqslant u \leqslant w$, implying that

$$\begin{aligned} d\big(u(t), A(U)\big) &= \big\|u(t) - u_-\big\|_1 + \big\|u_+ - u(t)\big\|_1 \\ &\leqslant \big\|w(t) - u_-\big\|_1 + \big\|u_+ - v(t)\big\|_1. \end{aligned} \tag{69}$$

Define

$$E(t) := \{x > 0;\ u(x,t) > u_-\},$$
$$\rho := \max\left(\|(a - u_-)_+\|_\infty, \|(a - u_+)_-\|_\infty\right).$$

From the maximum principle, we have $u_+ - \rho \leqslant u \leqslant u_- + \rho$. There holds

$$\int_0^{+\infty} \left(u(t) - u_-\right)_+ \mathrm{d}x \leqslant \rho\left|E(t)\right|$$

and

$$|u_- - u_+| \cdot \left|E(t)\right| \leqslant \int_{E(t)} (u - u_+)\,\mathrm{d}x \leqslant \left\|u(t) - U\right\|_1 \leqslant \|a - U\|_1.$$

Both inequalities imply

$$\int_0^{+\infty} \left(u(t) - u_-\right)_+ \mathrm{d}x \leqslant \frac{\rho}{u_- - u_+}\|a - U\|_1.$$

The assumption ensures that there is a $\delta > 0$ such that $f' > 0$ on $[u_-, u_- + \delta]$ and $f' < 0$ on $[u_+ - \delta, u_+]$. Assuming that $\rho < \delta$, we therefore see that the support of $w(t) - u_-$ moves uniformly to the right: $w(t) \equiv u_-$ on $(-\infty, R + \varepsilon t)$ for some positive ε. There eventually happens a time T beyond which this support is contained in $\mathbb{R}^+$, implying

$$\left\|\left(u(t) - u_-\right)_+\right\|_1 = \int_0^{+\infty} \left(u(x,t) - u_-\right)_+ \mathrm{d}x \leqslant \frac{\rho}{u_- - u_+}\|a - U\|_1, \quad t > T.$$

The norm $\|(u(t) - u_+)_-\|_1$ admits the same bound, and therefore we have

$$d\left(u(t), A(U)\right) \leqslant \frac{2\rho}{u_- - u_+}\|a - U\|_1, \quad t > T.$$

Since $A(U) \subset \mathcal{A}(U)$, Lemma 2 allows to derive

$$\ell_0(a) \leqslant \frac{4\rho}{u_- - u_+}\|a - U\|_1. \tag{70}$$

The hypotheses that constrain a are also satisfied by $u(t)$ for $t > 0$, because of the Co-properties. Hence, we may apply (70) to $u(t)$ instead. By virtue of (53), we obtain

$$\ell_0(a) \leqslant \frac{4\rho}{u_- - u_+}\left\|u(t) - U\right\|_1.$$

Passing to the limit as $t \to +\infty$, there comes

$$\ell_0(a) \leqslant \frac{4\rho}{u_- - u_+}\ell_0(a).$$

Therefore, assuming that ρ is less than $(u_- - u_+)/4$, we conclude to stability:

LEMMA 5. *Assume that the shock* $(u_-, u_+; s = 0)$ *is noncharacteristic. There exists a* $\delta_1 > 0$ *such that, for every initial data* $a \in U + L^1_0$, *satisfying*

$$a(x) \in [u_+ - \delta_1, u_- + \delta_1], \quad a.e.\ x \in \mathbb{R}, \tag{71}$$

there holds $\ell_0(a) = 0$.

One only has to take $\delta_1 < \min(\delta, (u_- - u_+)/4)$.

We now get rid of the hypothesis (71). Given $a \in U + L^1_0$, we let R be the smallest integer such that a takes values in the interval $[u_+ - R, u_- + R]$. If $R \leqslant \delta_1$, Lemma 5 gives the positive answer. Otherwise, define $p := \pi \circ a$, where $\pi : \mathbb{R} \to [u_+ - \delta_1, u_- + \delta_1]$ is the projection. From contraction, we have $\ell(a) \leqslant \ell(p) + \|a - p\|_1$. In view of Lemma 5, we obtain, on the one hand,

$$\ell(a) \leqslant \|a - p\|_1. \tag{72}$$

On the other hand, p and a differ only on a set B, where p equals either $u_- + \delta_1$ or $u_+ - \delta_1$. This implies that

$$\|a - p\|_1 \leqslant \min\bigl(d\bigl(a; A(U)\bigr) - \delta_1 |B|, (R - \delta_1)|B|\bigr). \tag{73}$$

We conclude from (73) that

$$\|a - p\|_1 \leqslant \left(1 - \frac{\delta_1}{R}\right) d\bigl(a; A(U)\bigr).$$

Again, we may apply this inequality to $u(t)$ instead. By virtue of (72) and (53), we obtain

$$\ell(a) \leqslant \left(1 - \frac{\delta_1}{R}\right) d\bigl(u(t); A(U)\bigr).$$

We finish by passing to the limit:

$$\ell(a) \leqslant \left(1 - \frac{\delta_1}{R}\right) \lim_{t \to +\infty} d\bigl(u(t); A(U)\bigr) \leqslant \left(1 - \frac{\delta_1}{R}\right) \ell(a). \tag{74}$$

Since $\delta_1 < R$, this ensures $\ell(a) = 0$, that is, $\ell_0(a) = 0$ since the mass of $a - U$ is zero. We have thus proved:

THEOREM 4. *Let* $U = (u_-, u_+; s)$ *be an inviscid shock in one space dimension, satisfying the strict Lax inequalities*

$$f'(u_+) < s < f'(u_-).$$

Let a be a bounded function, with $a \in U + L^1$. Then

$$\lim_{t \to +\infty} \| S_t a - U(\cdot - h - st) \|_1 = 0,$$

where

$$h := \frac{1}{u_+ - u_-} \int_{\mathbb{R}} (a - U)\, dx.$$

Up to our knowledge, Theorem 4 is new.

3.5. *Data with values beyond $[u_+, u_-]$; general case*

The inviscid case studied above is rather specific, since it displays counter-examples to the L^1-stability of shock waves. It turns out that a small amount of dissipation is always enough to ensure the stability. The proof of this fact is pretty long, because it has to be done in the case-by-case. To begin with, there is a general argument, due to Freistühler and the author [16], which reduces the question to that of the stability of constant states under zero-mass initial disturbances.[14] The latter is an interesting problem in itself, and its treatment highly depends on the nature of the dissipation process. It will be addressed in Section 4.

The general argument. For definiteness, we assume $s = 0$, $u_+ < u_-$, and that U is a steady shock for a given diffusive model. Let $a \in U + L^1_0$ be an initial datum and u be the corresponding solution to the Cauchy problem. Our goal is to prove that the L^1-distance of $u(t)$ to the set

$$A(U) := \{ z \in U + L^1(\mathbb{R});\ z(x) \in [u_+, u_-] \text{ a.e.} \}$$

tends to zero as $t \to +\infty$. Denote

$$\hat{a}(x) := \max(a(x), u_-), \qquad \check{a}(x) := \min(a(x), u_+).$$

The functions $\hat{a} - u_-$ and $\check{a} - u_+$ are integrable. Since U tends to u_+ as $x \to +\infty$, we have

$$\int_{\mathbb{R}} (u_- - U)^+ \, dx = +\infty.$$

Hence, there is enough room between the horizontal line of level u_- and the graph of U, in order to select a function $b \in u_- + L^1$, with the property that

$$\int_{\mathbb{R}} (b - u_-)\, dx = 0, \qquad a(x) \leqslant b(x) \quad \text{a.e.} \tag{75}$$

[14] We recall that the L^1-stability of constants require that the disturbance be of zero mass.

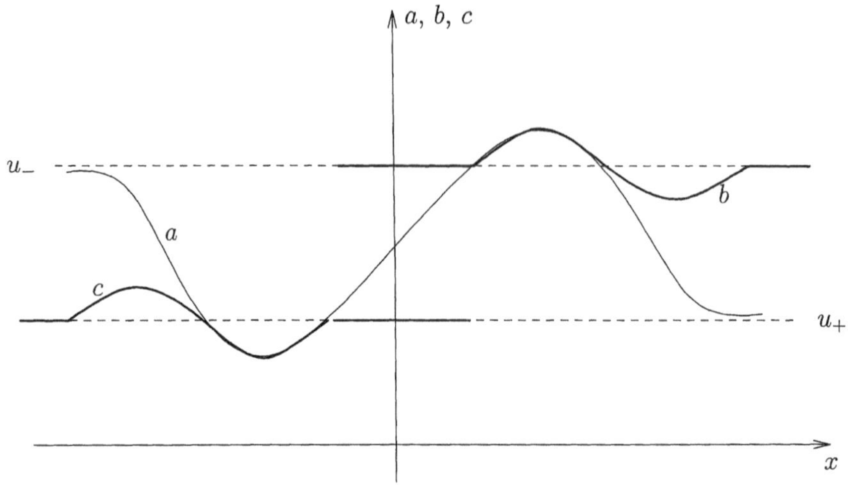

Fig. 3. The data a, with bounding functions b and c of zero relative total masses.

Figure 3 shows the construction of the bounding functions b and c. Similarly, there exists a function $c \in u_+ + L^1$ that satisfies

$$\int_{\mathbb{R}} (c - u_+)\,dx = 0, \qquad c(x) \leqslant a(x) \quad \text{a.e.} \tag{76}$$

Let v, w be the solutions of the Cauchy problem associated to the data b, c. From the comparison property, we have

$$w(x,t) \leqslant u(x,t) \leqslant v(x,y). \tag{77}$$

With the same notation as above, the distance to $A(U)$ is given by

$$d\big(u(t); A(U)\big) = \|\hat{u}(t) - u_-\|_1 + \|u_+ - \check{u}(t)\|_1.$$

From (77), we have

$$\begin{aligned} d\big(u(t); A(U)\big) &\leqslant \|\hat{v}(t) - u_-\|_1 + \|u_+ - \check{w}(t)\|_1 \\ &\leqslant \|v(t) - u_-\|_1 + \|u_+ - w(t)\|_1. \end{aligned} \tag{78}$$

Assume now that the following property holds true (Stability of constant states).

CLAIM 1. *Given a constant $u_0 \in \mathbb{R}$ and a data $b \in u_0 + L^1_0(\mathbb{R})$, the corresponding solution z of the Cauchy problem satisfies*

$$\lim_{t \to +\infty} \|z(t) - u_0\|_1 = 0. \tag{79}$$

Then (79) and (78) give $d(u(t); A(U)) \to 0$ and therefore $d(u(t); \mathcal{A}(U)) \to 0$. Using Lemma 2, we obtain

$$\ell_0(a) = \ell_0\big(u(t)\big) \leqslant 2d\big(u(t); \mathcal{A}(U)\big),$$

and thus $\ell_0(a) = 0$. Hence the stability.

THEOREM 5. *Assume that the diffusive model (in one space dimension) is such that Claim 1 holds true. Given a shock wave U and a bounded initial datum $a \in U + L_0^1$, one has*

$$\lim_{t \to +\infty} \big\| S_t a - U(\cdot - st) \big\|_1 = 0.$$

COMMENTS.

- For the Jin–Xin relaxation, the data and the shock wave are pairs (a, b) and (U, V). The assumptions are that (a, b) takes its values in a characteristic domain where the subcharacteristic condition (47) holds strictly, and that $a - U \in L_0^1$.
- Claim 1 does not hold true in the inviscid case. When $f''(u_0) \neq 0$, zero-mass initial disturbances develop in the form of "N-waves", that are pairs of bumps of opposite signs, going to infinity from both sides. Their heights are of order $1/\sqrt{t}$, while their supports have typical length $\sqrt{t}$. They carry nonzero masses, which may be computed explicitly from the initial datum when f is either convex or concave. The propagation of N-waves have been extensively studied, in the scalar case, by Liu and Pierre [46] and Dafermos [12], Chapter XI.

4. The L^1-stability of constants

This section is devoted to the analysis of Claim 1. It turns out that it holds true for the viscous, relaxation and Rosenau models. The proof for the relaxation case is cumbersome and will not be presented here; we refer to [57] for the details.

Remark that, up to the choice of a moving frame,[15] we may always assume that $f(u_0) = f'(u_0) = 0$. Also, a translation in the unknown makes $u_0 = 0$. Thus we shall only treat the case where

$$b \in L_0^1(\mathbb{R}), \qquad f(0) = f'(0) = 0.$$

We use a notation similar than above:

$$\ell_0(b) := \lim_{t \to +\infty} \|S_t b\|_1.$$

[15] In a moving frame, a constant remains a constant!

The contraction property implies that ℓ_0 is Lipschitz continuous on L^1, with

$$|\ell_0(b_1) - \ell_0(b_2)| \leqslant \|b_1 - b_2\|_1. \tag{80}$$

4.1. *The viscous case*

We recall first that the Claim holds true for the heat equation

$$\partial_t z = \partial_{xx} z. \tag{81}$$

As a matter of fact, if $b = \partial_x \beta$ with $\beta \in W^{1,1}(\mathbb{R})$, then

$$\|z(t)\|_1 = \|K^t * b\|_1 = \|(\partial_x K^t) * \beta\|_1 \leqslant \|\partial_x K^t\|_1 \|\beta\|_1 \leqslant \frac{c}{\sqrt{t}} \|\beta\|_1.$$

Hence, $\ell_0(b) = 0$. Since $\partial_x W^{1,1}(\mathbb{R})$ is dense in $L_0^1(\mathbb{R})$, (80) tells that $\ell_0 \equiv 0$ on $L_0^1(\mathbb{R})$. We notice however that the decay rate $t^{-1/2}$, which occurs for a data in $\partial_x W^{1,1}(\mathbb{R})$, does not extend to general data. Indeed, given any decreasing function Φ with $\Phi(+\infty) = 0$, we may find a data b in $L_0^1(\mathbb{R})$ such that

$$\overline{\lim_{t \to +\infty}} \frac{\|z(t)\|_1}{\Phi(t)} > 0.$$

We now pass to general viscous conservation laws

$$\partial_t z + \partial_x f(z) = \partial_{xx} z.$$

Given a data $b \in L_0^1 \cap L^\infty$, we have $\|z\|_\infty = \|b\|_\infty$ and therefore

$$|f \circ z| \leqslant N_f(\|b\|_\infty)|z|^2,$$

where $N_f(\|b\|_\infty)|$ is twice the Lipschitz constant of f' on the interval $[-\|b\|_\infty, \|b\|_\infty]$. Recalling the dispersion property (12) that

$$\|z(t)\|_2 \leqslant c\|b\|_1 t^{-1/4},$$

we deduce

$$\|f \circ z(t)\|_1 \leqslant c^2 N_f(\|b\|_\infty)\|b\|_1^2 t^{-1/2}.$$

Let us now write that z is a mild solution

$$z(t) = K^t * b - \int_0^t (\partial_x K^{t-s}) * (f \circ z(s))\, \mathrm{d}s.$$

Young's inequality gives

$$\begin{aligned}\|z(t)\|_1 &\leqslant \|K^t * b\|_1 + \int_0^t \|\partial_x K^{t-s}\|_1 \|f \circ z(s)\|_1 \, \mathrm{d}s \\ &\leqslant \|K^t * b\|_1 + c_1 \int_0^t \|f \circ z(s)\|_1 \frac{\mathrm{d}s}{\sqrt{t-s}} \\ &\leqslant \|K^t * b\|_1 + c_2 N_f(\|b\|_\infty)\|b\|_1^2 \int_0^t \frac{\mathrm{d}s}{\sqrt{s(t-s)}}.\end{aligned}$$

The integral is a constant, independent on time, while the norm of $K^t * b$ decays to zero (see above for the linear case). Passing to the limit, we thus obtain

$$\ell_0(b) \leqslant c_3 N_f(\|b\|_\infty)\|b\|_1^2. \tag{82}$$

However, since $\ell_0(b) = \ell_0(z(t))$ for every t, we may apply (82) to $z(t)$ instead, and get

$$\ell_0(b) \leqslant c_3 N_f(\|b\|_\infty)\|z(t)\|_1^2.$$

Passing to the limit, we improve (82) into

$$\ell_0(b) \leqslant c_3 N_f(\|b\|_\infty)\ell_0(b)^2. \tag{83}$$

Fix a real number $R > 0$ and consider the ball B_R defined by $\|b\|_\infty < R$ in $L_0^1 \cap L^\infty$. On the connected set B_R, (83) tells that either $\ell_0(b) = 0$, or $\ell_0 \geqslant 1/(c_3 N_f(R))$. Since ℓ_0 is continuous and takes the value zero (for $b \equiv 0$), this implies that $\ell_0 \equiv 0$ on B_R; hence on the union $L_0^1 \cap L^\infty$ of these balls. This ends the proof of the Claim in the viscous case.

THEOREM 6. *Given a constant $u_0 \in \mathbb{R}$ and a data $a \in u_0 + L_0^1(\mathbb{R}) \cap L^\infty(\mathbb{R})$, the solution u of the viscous model satisfies*

$$\lim_{t \to +\infty} \|u(t) - u_0\|_1 = 0.$$

REMARK. The above theorem holds true in higher space dimensions d. Its proof is actually easier when $d \geqslant 2$, because of the stronger dispersion effect. The integral with the nonlinearity actually decays to zero in L^1-norm (instead of being bounded by $c_3 N_f(\|b\|_\infty)\|b\|_1^2$), giving the result that $\|z(t) - K^t * b\|_1$ tends to zero, no matter the mass of the initial disturbance is: The solution of the viscous conservation law is asymptotic to mK^t, the solution of the heat equation with same initial mass m. The same phenomenon is true in one space dimension provided $f''(0) = 0$, since then $\|f \circ z(t)\|_1$ is bounded by $c\|z(t)\|_3^3$, which decays[16] as t^{-1}. However, it fails when $d = 1$ and $f''(0) \neq 0$, since then the solution is asymptotic to the nonlinear diffusion wave $D_m(xt^{-1/2})$, the (fundamental) solution of

$$\partial_t D_m + \frac{f''(0)}{2}\partial_x(D_m^2) = \partial_{xx} D_m, \qquad D_m(0) = m\delta.$$

[16] Compare with the bound $ct^{-1/2}$ in the general case.

See [15] for a far more complete analysis of the asymptotic behavior in the viscous model and related questions. See also [36] for a refined analysis in the case of the Burgers equation ($f(u) = u^2/2$), using diffusive N-waves.

4.2. *The Rosenau model*

The situation here is a bit more difficult, in the sense that a bounded solution of the Rosenau model

$$\partial_t z + \partial_x f(z) = K * z - z$$

is not a mild solution in general: The integral term in the Duhamel's formula does not make sense in L^1 because the semigroup does not convert the distribution $\partial_x f(z)$ into an ordinary function.

Our strategy is therefore different. The way to make use of the assumption of zero-mass is to introduce the potential p such that

$$\partial_t p = H * z - f(z), \qquad \partial_x p = z, \quad p(\pm\infty, t) = 0.$$

Since z is bounded, p is Lipschitz, and one may apply the chain rule to derivatives of nonlinear functions of p. In particular, using the entropy inequality

$$\partial_t \frac{z^2}{2} + \partial_x g(z) \leqslant zLz$$

with $g'(s) = sf'(s)$, one finds the remarkable inequality

$$\partial_t \frac{1}{2}\left(z^2 + p^2\right) + \partial_x\left(g(z)\right) \leqslant zLz + pLp - pf(z).$$

Assuming that $p(0)$ is square-integrable, we may show that $p(t)$ remains so. Thus, integrating, we obtain

$$\begin{aligned}\frac{\mathrm{d}}{\mathrm{d}t}\int_{\mathbb{R}} \frac{1}{2}\left(z^2 + p^2\right)\mathrm{d}x + N[z]^2 + N[p]^2 &\leqslant -\int_{\mathbb{R}} pf(z)\,\mathrm{d}x \\ &\leqslant \|p\|_\infty \|f \circ z\|_1 \leqslant \|z\|_1 \|f \circ z\|_1,\end{aligned}$$

with the notation of Section 1.4. We now appeal to the lower bound (40), whose consequence is the inequality

$$\|z\|_2^2 \leqslant c\left(N[z]^2 + N[p]^2\right).$$

Using also the decay of the L^1-norm of z, we thus obtain

$$\frac{\mathrm{d}}{\mathrm{d}t}\int_{\mathbb{R}} \frac{1}{2}\left(z^2 + p^2\right)\mathrm{d}x + \omega\|z\|_2^2 \leqslant \|b\|_1 \|f \circ z\|_1$$

for some positive constant ω.

With the same notation as in the viscous case, there comes

$$\frac{\mathrm{d}}{\mathrm{d}t}\int_{\mathbb{R}}\frac{1}{2}\left(z^2+p^2\right)\mathrm{d}x+\omega\|z\|_2^2\leqslant\|b\|_1 N_f\left(\|b\|_\infty\right)\|z\|_2^2. \tag{84}$$

Fix now a positive real number $R>0$; we work with data in the ball B_R. If $\|b\|_1$ is small enough, then (84) implies that

$$\frac{\mathrm{d}}{\mathrm{d}t}\int_{\mathbb{R}}\frac{1}{2}\left(z^2+p^2\right)\mathrm{d}x+\frac{\omega}{2}\|z\|_2^2\leqslant 0,$$

and therefore that $t\mapsto\|z(t)\|_2^2$ is integrable. Since it is a nonincreasing function of time, it must decay at least as t^{-1}. In other words, there holds

$$\left\|z(t)\right\|_2=\mathcal{O}\left(t^{-1/2}\right), \tag{85}$$

a result that improves the dispersion estimate (43) in the case of small data of zero-mass.

The next step is another kind of dispersion estimates. Loosely speaking, the entropy inequality

$$|z|_t+\left(f(z)\operatorname{sgn} z\right)_x\leqslant L|z|$$

implies

$$\frac{\mathrm{d}}{\mathrm{d}t}\int_{\mathbb{R}}|xz|\,\mathrm{d}x\leqslant\int_{\mathbb{R}}|z|\left(L^*|x|\right)\mathrm{d}x+\int_{\mathbb{R}}f(z)\operatorname{sgn}(xz)\,\mathrm{d}x.$$

On the one hand, the last term is bounded by $N_f(R)\|z\|_2^2$. On the other hand,

$$L^*|x|=-\partial_x\left(\check{H}*|x|\right)=-\check{H}*\operatorname{sgn},$$

with $\check{H}(x):=H(-x)$. When K decays fast enough at infinity (typically, a finite second moment is sufficient), then $L^*|x|$ makes sense as an integrable function. One therefore obtains

$$\frac{\mathrm{d}}{\mathrm{d}t}\int_{\mathbb{R}}|xz|\,\mathrm{d}x\leqslant c\left(\|z\|_2+\|z\|_2^2\right),$$

from which we infer

$$\int_{\mathbb{R}}|xz|\,\mathrm{d}x=\mathcal{O}\left(\sqrt{t}\right). \tag{86}$$

Let us apply the Cauchy–Schwarz inequality twice:

$$\|z\|_1^2\leqslant\left\|(1+|x|)z\right\|_1\int_{\mathbb{R}}\frac{|z|}{1+|x|}\,\mathrm{d}x=\mathcal{O}\left(\sqrt{t}\right)\int_{\mathbb{R}}\frac{|z|}{1+|x|}\,\mathrm{d}x=\mathcal{O}\left(\sqrt{t}\right)\|z\|_2.$$

Since $t \mapsto \|z(t)\|_2$ is square-integrable, we deduce that

$$\int_1^{+\infty} \|z(t)\|_1^4 \frac{\mathrm{d}t}{t} < \infty.$$

But since $\|z\|_1$ is a nonincreasing function of time, this property implies that $\ell_0(b) = 0$.

At this stage, we have proved the L^1-stability for those data which have some regularity and/or decay at infinity, and that are small in L^1. The smoothness and decay are non-essential assumptions, because we know that the attraction basin of the null function is closed in the L^1-topology. Hence, there remains to get rid of the smallness assumption, a goal that can be reached exactly in the same way as in the viscous case. This completes the proof of the Claim, that we rewrite as follows.

THEOREM 7 [67]. *Assume that the kernel K belongs to $L^1((1+x^2)\,\mathrm{d}x)$ and satisfies*

$$K \geqslant 0, \qquad \int_{\mathbb{R}} K(x)\,\mathrm{d}x = 1.$$

Then the constants are L^1-stable for the Rosenau model, under zero-mass initial disturbances: given a constant β and a bounded initial datum $b \in \beta + L_0^1(\mathbb{R})$, the solution of the Cauchy problem

$$\partial_t z + \partial_x f(z) = K * z - z, \qquad z(x,0) = b(x),$$

satisfies

$$\lim_{t \to +\infty} \|z(t) - \beta\|_1 = 0.$$

REMARK. Contrary to the viscous case, this proof does not extend to the multidimensional setting, because it makes use of the primitive p of z.

OPEN PROBLEM 3. Prove that constants are L^1-stable under zero-mass initial disturbances for the Rosenau model, when $d \geqslant 2$.

5. Multidimensional stability

Consider a viscous[17] conservation law in dimension $d \geqslant 2$. For the sake of simplicity, we limit ourselves to $d = 2$. We call x and y the space variables:

$$\partial_t u + \partial_x f(u) + \partial_y g(u) = \Delta u. \tag{87}$$

The fluxes are smooth functions. We give a planar shock wave ϕ in the x direction and assume as above that it is steady: $\phi = \phi(x)$ satisfies $\phi_{xx} = (f \circ \phi)_x$ and $\phi(\pm\infty) = u_\pm$.

[17] Much of the analysis of this section hold for various types of diffusion.

The Rankine–Hugoniot condition tells that $f(u_+) = f(u_-)$, and we may assume that $f(u_\pm) = 0$, $u_+ < u_-$ and $f(u) < 0$ in (u_+, u_-). Integrating once, we have

$$\phi_x = f \circ \phi. \tag{88}$$

At last, the possibility of choosing a frame moving uniformly in the y direction allows us to substract a linear function to g, thus permits to fix

$$g(u_\pm) = 0.$$

Given a bounded initial datum $a \in \phi + L^1(\mathbb{R}^2)$, the problem of L^1-stability does not receive the same answer as in the case of a single space dimension. For a shifted profile $\tau_h\phi$, in the x direction, does not belong to the same class $\phi + L^1$, while a shift in the y-direction does not modify ϕ. Hence there is no mean to absorb the mass

$$m := \int_{\mathbb{R}^2} (a - \phi)\,\mathrm{d}x\,\mathrm{d}y \tag{89}$$

of the initial disturbance through a translation, as it happened for $d = 1$.

Therefore, while we still may hope for an L^1-stability result when $m = 0$, we must turn towards a more complicated description when $m \neq 0$. We shall describe here, at a formal level, the asymptotics that was proved by Goodman and Miller [22], and improved by Hoff and Zumbrun [29]. We emphasize that the obstruction encountered here is strictly limited to the L^1 context; the asymptotics will involve correctors that tend to zero as $t \to +\infty$ in every space L^p, but L^1. However, taking in account these correctors allows for a better decay rate in every L^p.

5.1. *Formal asymptotics*

The fundamental idea of Goodman and Miller is that, under the additional assumption that the shock is not characteristic (there holds $f'(u_+) < 0 < f'(u_-)$), the disturbances are convected towards the shock front, and then remain in this zone where the front is stiff, where they are diffused with respect to the y variable. Hence, we expect that the disturbance will, as in the heat equation, be governed by some function

$$\delta(y,t) = \frac{1}{\sqrt{t}} D\left(\frac{y}{\sqrt{t}}\right),$$

where D is smooth and has a rapid decay. Remark that the y integral of such a function is constant in time.

Typically, δ acts as a translation: Our solution is asymptotic to $\phi(x + \delta(y,t))$, or to $\phi(x) + \delta(y,t)\phi'(x)$, which amounts the same in L^1. The following formula shows that such a shift is well suited for absorbing the mass m:

$$\int_{\mathbb{R}^2} \big(\phi(x + \delta(y)) - \phi(x)\big)\,\mathrm{d}x\,\mathrm{d}y = [u] \int_{\mathbb{R}} \delta(y)\,\mathrm{d}y.$$

We shall focus in this section on the derivation of δ, through an asymptotic expansion.

The expansion that we are looking for defines an approximate solution v of (87), in the sense that

$$N[v] := \partial_t v + \partial_x f(v) + \partial_y g(v) - \Delta v$$

is integrable in space *and* time on $\mathbb{R}^2 \times (0, +\infty)$. That goal needs to incorporate several terms beyond those which are meaningful in L^1. To be precise, we need taking into account all the differential monomials in δ that are integrable in space and time:

$$v(x, y, t) := \phi(x) + \delta(y, t)\phi_x(x) + \delta_y \theta(x) + \frac{1}{2}\delta^2 \sigma(x) \\ + \delta_{yy}\rho(x) + \delta_y \delta \psi(x) + \delta^3 \chi(x).$$

The coefficients $\theta, \dots$ are smooth functions with rapid decay at infinity, so that all terms but the first one are "localized" in the zone where the profile varies significantly. Hence, the integrability properties are inherited from those of the monomials in δ. Noting that $\partial_y^m \delta = t^{-(m+1)/2} D^{(m)}(y/\sqrt{t})$, we find that the first corrector $\delta\phi'$ is integrable but does not decay in L^1. The two next correctors decay like $t^{-1/2}$, while the last three ones decay like t^{-1}. Higher-order terms, being integrable in space and time, would be useless for our purpose.

When computing $N[v]$, only terms of order at least $t^{-1/2}$ are present. Since we wish that $N[v]$ be integrable in space and time, we have to determine $\theta, \dots$ in such a way that the terms A of order $t^{-1/2}$ and B of order t^{-1} vanish. Let us begin with A:

$$A = \delta_y \big(\theta f'(\phi) - \theta_x + g(\phi)\big)_x + \frac{1}{2}\delta^2 \big(\sigma f'(\phi) - \sigma_x + \phi_x^2 f''(\phi)\big)_x.$$

In order to kill these terms, we take

$$\sigma = \phi_{xx}$$

and solve

$$P\theta := \theta_x - f'(\phi)\theta = g(\phi). \tag{90}$$

Notice that the integration constant in (90) has been fixed at zero, because of $g(u_\pm) = 0$.

Solvability of (90). This is a first-order linear differential equation. Since ϕ_x is a nontrivial solution of the homogeneous equation, the nonhomogeneous one can be solved explicitly. We find

$$\theta = \phi_x z, \qquad z_x = \frac{g(\phi)}{\phi_x} = \frac{g(\phi)}{f(\phi)}.$$

Since $g(u_\pm) = 0$ and $f'(u_\pm) \neq 0$, the function g/f is bounded over $[u_+, u_-]$ and z grows at most linearly at infinity. Hence θ inherits from ϕ_x a rapid decay. Notice that θ is defined up to a multiple of ϕ_x. Since $\int_{\mathbb{R}} \phi_x \, dx = [u]$ is nonzero, we may normalize by $\int_{\mathbb{R}} \theta \, dx = 0$, but this is compulsory.

More generally, given a function $h(x)$ with exponential decay at infinity, the equation $PK = h$ amounts to $K = k\phi_x$ with $k_x = h/\phi_x$. Though k might grow exponentially at infinity, its growth rate is less than the decay rate of ϕ_x, and thus K has a rapid decay at infinity. This has been well known in the spectral analysis of wave fronts since the seminal work by Sattinger [55]: though the essential spectrum of P reaches the origin when the operator is considered within L^p-spaces, it is shifted to the left when working in weighted spaces with exponentially growing rates. In such spaces, P is onto.

The diffusion equation for δ. We now turn towards the t^{-1} terms. We have

$$B = (\delta_t - \delta_{yy})\phi_x + \delta_{yy}\big(\theta g'(\phi) - (P\rho)_x\big) + \delta\delta_y\big(g(\phi)_x + f''(\phi)\phi_x - P\psi\big)_x + \delta^3\left(\frac{1}{6}f'''(\phi)\phi_x^3 - P\chi\right)_x.$$

The way to cancel all terms in B is the following. One chooses solutions ψ and χ of

$$P\psi = g(\phi)_x + f''(\phi)\phi_x, \qquad P\chi = \frac{1}{6}f'''(\phi)\phi_x^3.$$

That makes sense since the right-hand sides are exponentially decaying at infinity.

It is not possible, however, to cancel the term $\theta g'(\phi) - (P\rho)_x$ by a choice of ρ; this would imply an exponential *growth* of ρ, because $\partial_x P$ is not onto. An equation $(PK)_x = h$, with a rapidly decaying data h, admits a decaying solution if and only if $\int_{\mathbb{R}} h(x)\, dx = 0$. This is not the case here, since

$$\begin{aligned}\int_{\mathbb{R}} g'(\phi)\theta \, dx &= \int_{\mathbb{R}} g'(\phi)\phi_x z \, dx = -\int_{\mathbb{R}} g(\phi) z_x \, dx \\ &= -\int_{\mathbb{R}} \frac{g(\phi)^2}{f(\phi)} \, dx = -\int_{u_-}^{u_+} \left(\frac{g(\phi)}{f(\phi)}\right)^2 du.\end{aligned} \tag{91}$$

What we can do is thus to cancel the zero-mass function[18]

$$g'(\phi)\theta - \frac{\phi_x}{[u]}\int_{\mathbb{R}} g'(\phi)\theta \, dx,$$

by solving

$$(P\rho)_x = g'(\phi)\theta - \frac{\phi_x}{[u]}\int_{\mathbb{R}} g'(\phi)\theta \, dx.$$

[18]This function has an anti-derivative that decays exponentially at infinity.

After having determined the correctors, there remains $B = (\delta_t - \beta\delta_{yy})\phi_x$, where, thanks to (91),

$$\beta := 1 + \frac{1}{[u]} \int_{u_-}^{u_+} \left(\frac{g(\phi)}{f(\phi)} \right)^2 \mathrm{d}x.$$

Hence, we reach our target by choosing for δ a solution of the diffusion equation

$$\delta_t = \beta \delta_{yy}. \tag{92}$$

Notice that the diffusion coefficient β is strictly larger than the unit. Hence, the transverse diffusion in the perturbation of the front is stronger than the diffusion of the conservation law itself.

Having in mind an L^1-asymptotics of u, the only freedom degree in the choice of δ is that of its mass $\mu := \int_{\mathbb{R}} \delta(y)\,\mathrm{d}y$. Because of

$$\lim_{t \to +\infty} \int_{\mathbb{R}^2} \big(v(x,y,t) - \phi(x)\big)\,\mathrm{d}x\,\mathrm{d}y = \mu[u],$$

we use a one-parameter family v^m of approximate solutions, where $m := \mu[u]$. Since most of the correctors in the definition of v are negligible in L^1, we may as well consider the family

$$w^m(x,y,t) = \phi(x) + \delta_m(y,t)\phi_x(x),$$

where δ_m is a solution of (92), whose mass equals $m/[u]$. We warn however that $\|N[w^m]\|_1$ does not tend to zero as $t \to +\infty$. Notice that we may choose $v^0 = w^0 \equiv \phi$.

Approximate Liapunov functions. Given a real number m and a function $u \in \phi + L^1(\mathbb{R}^2)$, we define

$$J_m[u;t] := \left\| u - v^m(t) \right\|_1.$$

When $t \mapsto u(t)$ is a solution of (87), we immediately have

$$\frac{\mathrm{d}}{\mathrm{d}t} J_m\big[u(t);t\big] \leqslant \int_{\mathbb{R}^2} N\big[v^m\big](t)\,\mathrm{sgn}\big(v^m(t) - u(t)\big)\,\mathrm{d}x\,\mathrm{d}y \leqslant \left\| N\big[v^m\big](t) \right\|_1. \tag{93}$$

Since $N[v^m]$ is integrable in both space and time, the right-hand side is integrable over $(0, +\infty)$ (it is an $\mathcal{O}(t^{-3/2})$). Therefore, $J_m[u(t);t]$ admits a limit j_m as $t \to +\infty$, and there is a tendency to decay:

$$J_m\big[u(t);t\big] \geqslant j_m - \frac{c}{\sqrt{t}}. \tag{94}$$

Of course, we expect that $j_m = 0$ when m is given by (89).

5.2. *Stability results*

The previous analysis, though giving a formal asymptotics in L^1 (and this asymptotics can be continued up to every order), does not provide a rigorous proof. There are several reasons for that flaw, one being the fact that $u(t) - \phi$ does not remain in a compact set of L^1, contrary to the one-dimensional case.

Therefore, the method followed in [22] and [29] uses the spectral analysis of the linearized operator around the profile:

$$\mathcal{L}v := \Delta v - \partial_x\big(f'(\phi)v\big) - g'(\phi)\partial_y v.$$

Since its coefficients depend only on the x variable, it is conjugated to a collection of one-dimensional operators, parametrized by the Fourier variable ξ, dual to y

$$\mathcal{L}_\xi v := \big(\partial_x^2 - \xi^2\big)v - \partial_x\big(f'(\phi)v\big) - \mathrm{i}g'(\phi)\xi v,$$

that is,

$$\mathcal{L}_\xi = \partial_x P - \xi^2 - \mathrm{i}g'(\phi)\xi.$$

Notice that $\mathcal{L}_0$ is precisely the linearized operator encountered in the stability analysis of the one-dimensional case. It has the simple eigenvalue $\lambda = 0$, as we have the equation

$$\mathcal{L}_0\phi_x = 0,$$

and ϕ_x decays rapidly at infinity.

A fundamental difficulty in the stability of viscous shock fronts is that this eigenvalue is embedded in the essential spectrum, when $\mathcal{L}_0$ is associated to a domain L^p. As mentioned above, Sattinger [55] found that the replacement of L^p by a weighted space shifts the essential spectrum to the left, so that $\lambda = 0$ becomes an isolated eigenvalue. In addition, Krein–Rutman's theorem and the fact that ϕ_x has a constant sign imply that zero is the largest eigenvalue and is simple. Therefore standard perturbation theory shows that, within weighted spaces, $\mathcal{L}_\xi$ has a unique small eigenvalue $\lambda(\xi)$ when ξ is small, which is simple. A classical calculus that we leave to the reader gives

$$\lambda(\xi) \sim -\beta\xi^2,$$

where β is the diffusion coefficient found above, confirming our formal asymptotics.

The spectral properties of $\mathcal{L}$ are then converted into estimates for the semigroup that it generates. The nonlinear remainder in (87) is treated with the use of a Duhamel's principle. In [22] the linear estimates were obtained by a direct application of Sattinger's method, so that the stability result was proved under the assumption that the initial disturbance has a rapid decay at infinity. Also, for some obscure reason, the result held only for shocks of moderate amplitude. These flaws were removed in [29], where the linear estimates were established by a careful analysis of the elliptic and parabolic Green's functions. There, a stability result is obtained for every noncharacteristic viscous shocks, and for initial disturbances that are small in $L^1 \cap L^\infty$ and have a finite moment in the x direction. Additionally, the stability result holds true for nonlinear and nonisotropic diffusions

$$\sum_{j,k} \partial_j \big(a_{jk}(u)\partial_k u\big)$$

instead of the Laplacian.

THEOREM 8 (Hoff and Zumbrun [29]). *Assume $d > 1$, that the functions f, g and a_{jk} are smooth, with $(a_{jk}(u))_{1 \leqslant j,k \leqslant n}$ being positive definite. Assume that $f'(u_+) < 0 < f'(u_-)$ and that $\phi = \phi(x)$ is a steady viscous shock profile ($\phi(\pm\infty) = u_\pm$). Let σ, $M > 0$ be given.*

Then there exist $\theta, C > 0$ such that, given an initial datum a with Hölder norm (see [29]), bounded by M and satisfying

$$\|a - \phi\|_1,\ \|a - \phi\|_\infty \big\|x(a - \phi)\big\|_1 \leqslant \theta,$$

the corresponding solution u satisfies

$$\big\|u(t) - \phi - \delta_m(t)\phi'\big\|_p \leqslant C\theta t^\omega, \quad \omega = -\frac{d-1}{2p'} - \frac{1}{2} + \sigma, \tag{95}$$

for all $p \in [1, \infty]$, where δ_m is as above.[19]

If $d > 2$, one may take $\sigma = 0$.

OPEN PROBLEM 4. Prove that a planar scalar multidimensional noncharacteristic shock front is L^1-asymptotic stable, up to the corrector $\delta_m(y,t)\phi'(x)$ described above, regardless of the size of the integrable initial disturbance.

REMARKS.

- The stability of constants holds true in every space dimension. See [58]; the proof is easier in dimension $d \geqslant 2$ than in one space dimension, because of stronger dispersion. Therefore, it would be sufficient to consider data a which satisfy comparisons

$$\tau_\alpha \phi \leqslant a \leqslant \tau_\beta \phi.$$

- When the shock is characteristic, the integral in (91) diverges. Hence the perturbation of the front obeys to a different asymptotic. Hence the problems:

OPEN PROBLEM 5. Prove that a planar scalar multidimensional characteristic shock front is L^1-asymptotic stable under zero-mass initial disturbances.

OPEN PROBLEM 6. Consider a planar scalar multidimensional characteristic shock front. To what kind of asymptotics does the perturbation δ of the front obey?

[19] With a convenient generalization if the diffusion is nonlinear.

- The viscous model does not at all exhaust the interesting situations. Each other model should be studied in details. The most basic question concerns of course the inviscid multidimensional conservation law.

OPEN PROBLEM 7. Analyze the L^1-stability of the multidimensional inviscid shock waves, under bounded integrable initial disturbances. Describe the asymptotics of the shock front when the excess mass $\int_{\mathbb{R}^d}(a-U)\,dx\,dy$ is nonzero.

6. Discrete shock profiles

We go back to a single space variable and consider a discretization by finite differences, instead of "continuous" models. The numerical scheme will be chosen in a conservative form. Writing u_j^n for the approximation of $u(j\Delta x, n\Delta t)$, one computes inductively:

$$u_j^{n+1} = G(u_{j-1}^n, u_j^n, u_{j+1}^n) := u_j^n + F(\lambda; u_{j-1}^n, u_j^n) - F(\lambda; u_j^n, u_{j+1}^n). \tag{96}$$

The numerical flux F depends on the mesh ratio $\lambda := \delta t/\delta x$. The consistency with

$$\partial_t u + \partial_x f(u) = 0 \tag{97}$$

is expressed by

$$F(\lambda; u, u) = \lambda f(u). \tag{98}$$

Though we have considered in (96) only three-points difference schemes, the present section applies to much more general schemes, where F involves an arbitrary number of points. The only important properties that we require are the translational invariance (clear in Formula (96)), and the *monotonicity* in a strict form. The latter means that G is monotone increasing with respect to each of its arguments. In particular, the derivatives of F satisfy[20]

$$F_1(\lambda; u, v) > 0 > F_2(\lambda; u, v), \qquad F_2(\lambda; u, v) + 1 > F_1(\lambda; v, w) \tag{99}$$

for every values u, v, w in some relevant interval I. The monotonicity of the scheme implies readily the comparison property: If $u_j^0 \leqslant v_j^0$ for every j, then $u_j^n \leqslant v_j^n$ for every j and n. Besides, the conservative form of the scheme implies the conservation of mass: If $v^0 - u^0 \in \ell^1$, then $v^n - u^n \in \ell^1$, and there holds

$$\sum_{j\in\mathbb{Z}} (v_j^n - u_j^n) = \sum_{j\in\mathbb{Z}} (v_j^0 - u_j^0). \tag{100}$$

[20]We choose here a strict form of monotonicity, which serves for several purposes. Its drawback is that it rules out such schemes as the upwind (Godunov) scheme. We shall analyze this borderline case in Section 6.4.

The comparison principle and (100) imply together the contraction property (easy exercise):

$$\sum_{j\in\mathbb{Z}}\left|v_j^n-u_j^n\right|\leqslant\sum_{j\in\mathbb{Z}}\left|v_j^0-u_j^0\right|, \tag{101}$$

with equality if, and only if, $v_j^0-u_j^0$ has a constant sign.

We notice that (98) and (99) imply the Courant–Friedrichs–Lewy condition

$$\lambda\left|f'(u)\right|<1, \quad u\in I, \tag{102}$$

because of $0<F_1(\lambda;u,u)<1$ and $-1<F_2(\lambda;u,u)<0$.

EXAMPLES.

Lax–Friedrichs scheme. The iteration is defined by

$$G(a,b,c)=\frac{a+c}{2}+\frac{\lambda}{2}\big(f(a)+f(c)\big).$$

It has the form (96), with the numerical flux

$$F_L(a,b)=\frac{a-b}{2}+\frac{\lambda}{2}\big(f(a)+f(b)\big)=:h_+(a)+h_-(b).$$

Since $G=G(a,c)$ does not depend on b, the scheme is not strictly monotone, and the results below do not apply directly. However, iterating twice the scheme, and relabeling $v_k^m:=u_{2k}^{2m}$, we obtain a new form of the same scheme:

$$v_k^{m+1}=H\big(v_{k-1}^m,v_k^m,v_{k+1}^m\big), \quad H(a,b,c):=G\big(G(a,b),G(b,c)\big).$$

In this new form, the scheme is strictly monotonous as long as the CFL is satisfied, since it implies $\partial_a G>0$ and $\partial_c G>0$. Besides, using the identity $h_+(u)+h_-(u)=u$, we obtain the conservative form

$$H(a,b,c)=b+F'(a,b)-F'(b,c),$$
$$F'(a,b):=h_+(a)-h_-\big(h_+(a)+h_-(b)\big).$$

Hence the results below are valid for the Lax–Friedrichs scheme.

Godunov scheme. The Godunov scheme, described in Section 6.4, is a generalization of the upwind scheme when the flux f is nonmonotone. It is monotone but not strictly, except in the case where f is strictly monotone. We shall provide in Section 6.4 an extension of the classical results to this scheme.

Lax–Wendroff scheme. The Lax–Wendroff scheme is second-order accurate. As such, it cannot be monotone, even in a nonstrict form, according to a theorem of Harten, Hyman and Lax [28].

6.1. *Shock profiles*

Given a shock $(u_-, u_+; s)$ of (97), a "discrete" shock profile is an exact solution of (96), having the form of a traveling wave:

$$u_j^n = u(j - s\lambda n).$$

Such a solution depends only on the variable $(j - \lambda sn)\Delta x = x - st$, hence it has the velocity s. We shall see that the dimensionless parameter

$$\eta := s\lambda$$

plays a fundamental role in the analysis. We notice for the moment that (102) and the Lax condition imply

$$|\eta| < 1. \tag{103}$$

Plugging our ansatz in (96), we obtain a functional difference equation for the profile:

$$u(x - \eta) = u(x) + F\bigl(\lambda; u(x-1), u(x)\bigr) - F\bigl(\lambda; u(x), u(x+1)\bigr). \tag{104}$$

This equation shows that the profile u must be defined *a priori* on a set of the form $X + \mathbb{Z} + \eta\mathbb{Z}$. When η is irrational, such a set is always dense in $\mathbb{R}$, and is it reasonable[21] to ask that u be defined on $\mathbb{R}$ and satisfy (104) for every real number x.

When $\eta = p/q$ is rational, one may restrict the domain of u to a discrete subset of $\mathbb{R}$, for instance $\mathbb{Z} + \eta\mathbb{Z} = \frac{1}{q}\mathbb{Z}$. However, there is a good reason why the profile would actually be considered as a function of a real variable instead of a discrete variable. Let us assume $\eta = 0$ as a typical example of the rational case. Then (104) reduces to $F(\lambda; u(x-1), u(x)) = F(\lambda; u(x), u(x+1))$, which can be viewed as a discrete dynamical system

$$F(\lambda; u_{j-1}, u_j) = F(\lambda; u_j, u_{j+1}), \tag{105}$$

which tells that $F(\lambda; u_j, u_{j+1})$ is independent of j. Letting $j \to \pm\infty$, we find that its constant value equals $F(\lambda; u_\pm, u_\pm) = \lambda(f(u_\pm) - su_\pm)$. Hence, we recover the Rankine–Hugoniot condition as a necessary condition for the existence of a profile. Similarly, the monotonicity assumption tells that (96) is compatible with some discrete form of entropy

[21] The name "discrete shock profile" is hardly satisfactory, since a profile could have to be defined on a dense subset!

conditions. Using them as above, we find that the Oleinik condition, in its strict form, is a necessary condition for the existence of a profile.

Because of (99), the identity $F(\lambda; u_j, u_{j+1}) = \lambda(f(u_-) - su_-)$ may be written in the explicit form

$$u_{j+1} = g(\lambda, u_-; u_j), \tag{106}$$

where $g(\lambda, u_-; \cdot)$ is increasing. Since it fixes the points $u_\pm$, it is a diffeomorphism of the interval (u_+, u_-). In addition, the Oleinik condition implies that $g(\lambda, u_-; \cdot)$ has no other fixed point in this interval. Actually, there holds $g(\lambda, u_-; u) < u$ on (u_+, u_-). Therefore, every point $u^* \in (u_+, u_-)$ gives rise to a unique trajectory (a discrete shock profile) such that $u_0 = u^*$. This trajectory is decreasing and tends to $u_\pm$ at $\pm\infty$. Given a decreasing parametrization

$$y \in [0, 1] \mapsto u(y) \in [u_0, u_1], \tag{107}$$

we may define a continuous and monotone profile $u : \mathbb{R} \to (u_+, u_-)$, where $u(k + y)$ is the kth iterate of (106) from the initial value $u(y)$ ($k \in \mathbb{Z}$, $y \in [0, 1)$).

The extension of the previous analysis to other rational values of η is nontrivial, and we skip the proof of Jennings' theorem (see also the work by Fan [26]):

THEOREM 9 [31]. *Given a conservative, strictly monotone difference scheme* (96) *in an interval I, and a strict shock $(u_-, u_+; s)$ of* (97), *with $u_\pm \in I$. Assume that $\eta := s\lambda$ is rational. Then there exists a continuous and monotone discrete shock profile for this shock.*

6.2. *The function V*

The previous construction does not provide a unique up-to-a-shift continuous discrete profile, since the parametrization (107) is arbitrary. We now show that a canonical profile can be fixed, up to a shift. To understand what is going on, we go back to the case of an arbitrary η.

Applying the contraction property to u and shifts $\tau_h u$, we find that (101) must be an equality, and therefore each restriction of u to a set $x_0 + \mathbb{Z} + \eta\mathbb{Z}$ is monotonous nonincreasing and two such restrictions are comparable. There follows that, up to a rearrangement ρ such that $\rho(x + h) = \rho(x) + h$ for every $h \in \mathbb{Z} + \eta\mathbb{Z}$, u may be chosen monotonous.

Given a shock profile $u : \mathbb{R} \to I$, we define as in [59]

$$V(h) := \sum_{j\in\mathbb{Z}} \big(u(j + h) - u(j)\big).$$

Since u has bounded variation, V is well defined. For integer values, we have easily

$$V(m) = m[u], \quad m \in \mathbb{Z}.$$

Remark also

$$V(-\eta) = \sum_{j\in\mathbb{Z}} \big(F(\lambda; u_{j-1}, u_j) - F(\lambda; u_j, u_{j+1})\big)$$
$$= \lambda\big(f(u_-) - f(u_+)\big) = -\eta[u],$$

because of Rankine–Hugoniot condition. More generally, we have

$$V(h+1) = V(h) + [u], \qquad V(h+\eta) = V(h) + \eta[u],$$

from which there comes

$$V(k) - V(h) = (k-h)[u] \quad \forall k-h \in \mathbb{Z} + \eta\mathbb{Z}.$$

Besides, V is nonincreasing since u is so. We conclude that, when η is *irrational*, V satisfies

$$V(h) = h[u] \quad \forall h \in \mathbb{R}. \tag{108}$$

When $\eta = p/q \in \mathbb{Q}$, a profile u does not necessary yield (108), as this only needs to hold for $h \in \mathbb{Z} + \eta\mathbb{Z}$. But every homeomorphism $\rho : [0, 1/q] \to [0, 1/q]$ can be extended to a homeomorphism $\tilde{\rho}$ of $\mathbb{R}$ by $\tilde{\rho}(y+1) = \tilde{\rho}(y) + 1$. When composing u by $\tilde{\rho}$, one keeps a monotone discrete shock profile $\tilde{u}$, which yields a function $\widetilde{V} = V \circ \tilde{\rho}$. From Theorem 9, V is continuous. Therefore, there exists a unique ρ such that $\widetilde{V}(h) \equiv h[u]$, giving a unique profile $\tilde{u}$ satisfying (108) and such that $\tilde{u}(0) = u(0)$.

An immediate consequence of (108) comes with the remark that if $x < y$, then

$$u(y) - u(x) \leqslant \sum_{j\in\mathbb{Z}} \big(u(j+y) - u(j+x)\big) = V(y) - V(x) = (y-x)[u].$$

We have used the monotonicity of u. This inequality shows that the profile is globally Lipschitz.

We summarize the results that we have obtained in the following statement.

THEOREM 10. *Given a conservative monotone difference scheme* (96) *in an interval I, and a strict shock $(u_-, u_+; s)$ of* (97), *with $u_\pm \in I$. Let us suppose that there exists a discrete shock profile $u : \mathbb{R} \to I$. Then, up to a rearrangement ρ such that $\rho(x+h) = \rho(x) + h$ for every $h \in \mathbb{Z} + \eta\mathbb{Z}$:*

(i) *the profile may be chosen monotonous, and therefore we have $u_+ < u(x) < u_-$;*

(ii) *the function V satisfies* (108);

(iii) *the profile is uniquely defined by* (104), *the two properties above, and the value $u^* \in (u_+, u_-)$ that it takes at $x = 0$;*

(iv) *the profile is Lipschitz: if $y \geqslant x$, then*

$$(y-x)[u] \leqslant u(y) - u(x) \leqslant 0. \tag{109}$$

6.3. *Existence of discrete shock profiles; irrational case*

The existence of discrete shock profiles has been treated by Jennings [31], who proved it in the rational case and claimed it in the irrational case. It is widely accepted that his argument for the passage from rational to irrational values of η is too light for being "a proof". We fill below this gap, giving a complete proof of existence in the irrational case, taking for granted Theorem 9. Our result actually extends that of Fan [25], which is proved for the Godunov scheme with a convex flux.

For definiteness, we may assume that $\eta > 0$. Recall that $\eta < 1$. Define[22]

$$\widehat{F}(\lambda; v, w) := F(\lambda; v, w) - \eta v - \lambda\big(f(u_-) - su_-\big).$$

Given a shock profile for $(u_-, u_+; s)$, (104) gives (dropping the unnecessary λ)

$$\begin{aligned}&\widehat{F}\big(u(x), u(x+1)\big) - \widehat{F}\big(u(x-1), u(x)\big)\\&\quad = u(x) - u(x-\eta) - \eta\big(u(x) - u(x-1)\big),\end{aligned}$$

which may be integrated once, in the following way

$$\int_{x-1}^{x} \widehat{F}\big(u(y), u(y+1)\big)\,\mathrm{d}y = \int_{x-\eta}^{x} u(y)\,\mathrm{d}y - \eta\int_{x-1}^{x} u(y)\,\mathrm{d}y. \tag{110}$$

Notice that the constant of integration has been determined by letting $x \to -\infty$. Hence, (110) carries not only the fact that u solves (104), but also the values at infinity $u(\pm\infty) = u_\pm$.

We now follow Jennings strategy. Let $\eta = \lambda s$ be an irrational number, associated to a strict shock $(u_-, u_+; s)$, $u_\pm \in I$. We may find a sequence of strict shocks $(u_-^n, u_+^n; s_n)$, converging towards $(u_-, u_+; s)$ and such that η_n is rational. Hence, there exists a shock profile u^n for each element of the sequence. Each u^n is Lipschitz and monotone decreasing, unique up to a shift. Given some value $u^* \in (u_+, u_-)$, we fix these profiles by $u^n(0) = u^*$. We notice that the Lipschitz constant of u^n, being $u_-^n - u_+^n$, remains bounded as $n \to +\infty$.

From Ascoli's theorem, we may extract a subsequence, still labeled u^n, that converges uniformly on every compact interval of the line. Denote its limit by u. Uniform convergence allows us to pass to the limit in (110). Hence, u is a solution of (104).

There remains to identify the limits $u(\pm\infty)$, which exist since u is monotonous and bounded. Passing to the limit as $x \to \pm\infty$ in (110), we obtain on the one hand

$$f\big(u(\pm\infty)\big) - su(\pm\infty) = f(u_-) - su_-. \tag{111}$$

On the other hand, we have $u(0) = u^*$ and therefore $u(+\infty) \leqslant u^* \leqslant u(-\infty)$. At last, $u_+^n < u^n < u_-^n$ gives $u_+ \leqslant u \leqslant u_-$ and thus $u(+\infty) \geqslant u_+$ and $u(-\infty) \leqslant u_-$. Hence, $u(-\infty)$ is a zero of $v \mapsto f(v) - f(u_-) - s(v - u_-)$ that belongs to $[u^*, u_-]$. From the

[22] If η were negative, we would define $\widehat{F}(v, w) := F(v, w) - \eta w - \lambda(f(u_-) - su_-)$.

Oleinik criterion, there is only one zero of that function in this interval, namely u_-. Hence, $u(-\infty) = u_-$ and similarly $u(+\infty) = u_+$.

Therefore we may state the following.

THEOREM 11. *Given a conservative monotone difference scheme* (96) *in an interval I, and a strict shock $(u_-, u_+; s)$ of* (97), *with $u_\pm \in I$, there exists a discrete shock profile $u : \mathbb{R} \to I$. Up to a rearrangement ρ such that $\rho(y+h) = \rho(y) + h$ for every $h \in \mathbb{Z} + \eta\mathbb{Z}$ and $y \in \mathbb{R}$, it may be taken monotonous and such that* (108) *holds. Under both these properties, the profile is unique up to a shift.*

Continuous dependence. From the uniqueness of a discrete shock profile, we see that the whole sequence u^n converges, locally uniformly, towards u. Likewise, we have seen that $u^n(\pm\infty)$ tend to $u(\pm\infty)$. Viewing $\mathbb{R} \cup \{\pm\infty\}$ as a compact interval on which u^n and u are monotonous and continuous, we infer from Dini's theorem that actually u^n converges uniformly towards u. Hence, the profile, considered as an element of $\mathcal{C}_b(\mathbb{R})$ and normalized by $u(0) = u^*$ with $u^* \in (u_+, u_-)$ (say $u^* = (u_- + u_+)/2$), depends continuously on the strict shock $(u_-, u_+; s)$.

The tail of discrete profiles. It is remarkable that the present context permits that the profiles tend exponentially to their limits $u_\pm$, even for irrational parameters, provided that the shock satisfies the Lax inequalities in the strict form. This fact is far from clear in the case of systems (see [47]).

Given a profile u, we rewrite the right-hand side of (110) in the form

$$\int_{x-\eta}^{x} \mathrm{d}y \int_{x-1}^{x} \big(u(y) - u(z)\big)\,\mathrm{d}z,$$

from which we infer

$$\left|\int_{x-1}^{x} \widehat{F}\big(u(y), u(y+1)\big)\,\mathrm{d}y\right| \leqslant \eta\big(u(x-1) - u(x)\big). \tag{112}$$

Assume that $f'(u_+) < s$ and denote

$$\alpha := \left|\big(\partial_v \widehat{F}\big)(u_+, u_+)\right| = -\big(\partial_v \widehat{F}\big)(u_+, u_+),$$
$$\beta := \left|\big(\partial_w \widehat{F}\big)(u_+, u_+)\right| = -\big(\partial_w \widehat{F}\big)(u_+, u_+).$$

One has $\alpha + \beta = \eta - \lambda f'(u_+) > 0$. Denote also $w := u - u_+ > 0$. As $x \to +\infty$, the integral in (112) satisfies

$$\begin{aligned}
&\left|\int_{x-1}^{x} \widehat{F}\big(u(y), u(y+1)\big)\,\mathrm{d}y\right| \\
&\quad = \alpha \int_{x-1}^{x} w(y)\,\mathrm{d}y + \beta \int_{x}^{x+1} w(y)\,\mathrm{d}y + \mathcal{O}\big(w(x-1)^2\big) \\
&\quad \geqslant \alpha w(x) + \beta w(x+1) + \mathcal{O}\big(w(x-1)^2\big).
\end{aligned}$$

With (112), we deduce

$$\alpha w(x) + \beta w(x+1) \leqslant \eta\big(w(x-1) - w(x)\big) + \mathcal{O}\big(w(x-1)^2\big). \tag{113}$$

Let γ be the positive root of the quadratic equation

$$P(X) := X^2 + X(\alpha + \eta) - \beta\eta = 0. \tag{114}$$

Then the function W, defined by $W(x) := \eta w(x) + \gamma w(x+1)$, satisfies

$$W(x) \leqslant \frac{\gamma}{\beta} W(x-1) + \mathcal{O}\big(W(x-1)^2\big).$$

Since W decays monotonically to zero, this implies that $W(x) \leqslant C(\gamma/\beta)^x$ for some constant C. In particular, $W(x) \leqslant c(\gamma/\beta)^x$. We emphasize that the rate γ/β is strictly less than unity, since $P(\beta) = \beta(\alpha + \beta)$ is positive and $P(0) = -\beta\eta$ is negative. In summary, there holds:

THEOREM 12. *Let the strict shock* $(u_-, u_+; s)$ *satisfy the strict Lax inequalities*

$$f'(u_+) < s < f'(u_-).$$

Given a monotonous and conservative finite difference scheme (96) *in an interval I that contains $u_\pm$. Then the discrete profile associated to this shock tends exponentially to its limits $u_\pm$ at infinity.*

6.4. *The case of upwind scheme*

When f is monotonous, say increasing, it may be advantageous to consider the scheme defined by the numerical flux $F(\lambda; u, v) := \lambda f(u)$. For nonmonotonous functions f, it was generalized by Godunov [21] in the following way:

$$F(\lambda; u, v) = \lambda F_G(u, v) := \lambda \begin{cases} \min_{z\in[u,v]} f(z), & u \leqslant v, \\ \max_{z\in[v,u]} f(z), & v \leqslant u. \end{cases}$$

This upwind scheme is monotonous, but not strictly. In particular, the construction done above for steady shocks does not work because F is not invertible with respect to each of its arguments u and v. A direct analysis of discrete profiles for steady shocks can be done, however, even for *systems* of conservation laws. Under reasonable assumptions, the profile does exist, but it has the strange property of being equal to u_- or to u_+, away from an interval of unit length (see [59]). This reflects the lack of dissipation of the Godunov scheme, precisely at rest points (namely at solutions of $f'(u) = 0$).

The situation must change drastically as soon as η is different from zero. The profile, if it exists, must not be "compactly supported", contrary to the steady case. The existence and uniqueness of a discrete shock profile in this case is due to Fan [25]. We give here an alternate, somehow simpler, proof that follows the strategy already used in Section 6.3.

In a first instance, we modify the Godunov scheme by adding a small but positive amount of dissipation ε. This means replacing the numerical flux by a new one

$$F^{\varepsilon}(\lambda; u, v) := F(\lambda; u, v) + \varepsilon(u - v).$$

In other words, the linear term v that is present in $G(\lambda; u, v, w)$ is replaced by $\varepsilon u + (1 - 2\varepsilon)v + \varepsilon w$. Because of the CFL condition, the new scheme is strictly monotone for $\varepsilon > 0$ small enough.

Theorem 11 therefore applies to the modified scheme. Given a strict shock $(u_-, u_+; s)$, there exists a unique shock profile u^{ε} with the normalization $u^{\varepsilon}(0) = u^*$ (u^* chosen in (u_+, u_-)) and $V^{\varepsilon}(h) = h[u]$. It is monotone and Lipschitz continuous, the Lipschitz constant being bounded by $|[u]|$. Again, we may extract a subsequence, still denoted u^{ε}, converging uniformly on every compact set. Denote by u the limit. Pass to the limit in the integrated form (110) of the profile equations, which contains the boundary conditions at infinity. We obtain that u satisfies the integrated form of the profile equation of the Godunov scheme. Arguing as in Section 6.3 gives that u is monotone with $u(\pm\infty) = u_{\pm}$.

REMARK. The same strategy can be employed in order to show the existence of semidiscrete profiles. Consider the upwind semidiscretization

$$\frac{\mathrm{d}u_j}{\mathrm{d}t} = F_G(u_{j-1}, u_j) - F_G(u_j, u_{j+1}),$$

where F_G is the flux defined in the Godunov scheme. We have fixed $\Delta x = 1$. The semidiscretization is the formal limit of the Godunov scheme as Δt tends to zero.

Given a strict shock, one may prove the existence of semidiscrete profiles $u_j(t) = u(j - st)$ by passing to the limit as Δt tends to zero in the discrete ones. We just have shown that the latter exist, and we know that they have Lipschitz constants less than $|[u]|$. Hence, the uniform convergence on every compact set, up to extraction of a subsequence. Passing to the limit in the integrated form of the discrete profile equation, we obtain that the limit u satisfies the identity

$$\int_{x-1}^{x} F_G\big(u(y), u(y+1)\big)\,\mathrm{d}y - su(x) = f(u_+) - su_+.$$

We deduce on the one hand that u is continuously differentiable and satisfies the profile equation

$$su'(x) = F_G\big(u(x), u(x+1)\big) - F_G\big(u(x-1), u(x)\big).$$

On the other hand, letting x going to $+\infty$, we also get

$$f\big(u(+\infty)\big) - f(u_+) = s\big(u(+\infty) - u_+\big),$$

and we conclude as usual.

6.5. *Stability of discrete shock profiles*

In the sequel, we consider solutions and profiles that depend on a real variable x. One could restrict to $x \in \mathbb{Z} + \eta\mathbb{Z}$, especially if η is rational. In that case, the word "integrable" must be replaced by "summable", and $\|\cdot\|_1$ becomes the ℓ^1-norm.

Let U be a discrete profile of some shock: $u_j^n := U(j - \eta n)$ solves (96) exactly. Hence, U is a stationary solution of

$$u^{n+1}(x) = G\big(u^n(x+\eta-1), u^n(x+\eta), u^n(x+\eta+1)\big). \tag{115}$$

To begin with, let the initial datum a lie between $\tau_\alpha U$ and $\tau_\beta U$. The arguments of Section 3.2 work, and we only have to consider those data b that lie between $\tau_\alpha U$ and $\tau_\beta U$, such that the distance $\|S^n b - \tau_h U\|_1 \equiv \|b - \tau_h U\|_1$, S being the iteration operator defined by (115). But this is easily seen to imply that $b - \tau_h U$ has a constant sign (see the details in [31]). This, being true for every real number h, implies that b equals some shift of U. Hence the L^1-stability holds true for such a's. The standard density argument allows to extend this result to data such that $a - U$ is integrable, and that take values in the interval $[u_+, u_-]$.

Going beyond requires proving the L^1-stability of constant states under zero-mass initial disturbances. Of course, this depends on the numerical scheme under consideration. For instance, a scheme such that the one of Lax–Friedrichs is likely to fit this property, because of his rather strong dissipation. However, it is not so clear for the Godunov scheme, whom dissipativeness vanishes at points $\bar{u}$ such that $f'(\bar{u}) = 0$.

OPEN PROBLEM 8. Prove the L^1-stability of constants under conservative monotone difference schemes.

7. Initial-boundary value problems

The previous sections dealt with the pure Cauchy problem, the space domain having been $\mathbb{R}^d$, with often $d = 1$. An other interesting situation is that of the initial boundary value problem (IBVP). Let us consider for instance the viscous model

$$\partial_t u + \operatorname{div} f(u) = \Delta u$$

in a domain Ω. We supplement it with a bounded initial datum a and a (nonnecessarily homogeneous) Dirichlet boundary condition $u(x,t) = u_b$ on $\partial\Omega$. Here, b may depend on $x \in \partial\Omega$, but not on the time variable since we are searching asymptotic stability results.

The proof that the IBVP generates a unique semigroup S_t that satisfies the comparison principle and the L^1-contraction property is straightforward. However, and it is important in the analysis, S_t does no longer preserve the mass: When integrating the conservation law on Ω, we find that the mass varies with time, proportionally to the flux of $\nabla_x u - f(u)$ across the boundary. There is no reason why the latter would vanish.

The asymptotic analysis is pretty easy when Ω is bounded, because of the Poincaré inequality. The solution tends exponentially to the unique steady solution. Hence, we shall concentrate on unbounded domains, and more specifically to the case of a half-space $H := (0, +\infty) \times \mathbb{R}^{d-1}$ as a paradigm.

When $\Omega = H$, we distinguish the variable x normal to the boundary, and the tangential variables $y \in \mathbb{R}^{d-1}$. We fix a constant boundary value u_b, and a value at infinity, say $u_\infty = 0$. Hence, we are interesting in bounded data that are integrable. Our problem is to compare a solution $S_t a =: u(t)$ to a steady solution called "boundary layer", using the L^1-distance d.

The layer, denoted by U, is invariant under translations parallel to the boundary: $U = U(x)$. It thus satisfies the differential equation

$$U'(x) = f(U) - f(0), \tag{116}$$

together with the initial condition $U(0) = u_b$. There does not always exist such a boundary layer. On the one hand, if $f_1'(0) > 0$, the origin is a repulsor for (116) and the only possible boundary layer is obtained for $u_b = 0$, when it is constant. On the other hand, if $f'(0) < 0$, then a boundary layer exists (and is unique) provided there is no zero of $f - f(0)$ between u_b and the origin. This dichotomy shows the important role played by the hyperbolic wave velocity associated to the state at infinity.

The borderline case $f'(0) = 0$ corresponds to a characteristic boundary for the underlying inviscid model. In general, it is expected to create complications. For instance, if we were interested in decay rates,[23] it would certainly weaken drastically the optimal results, comparing to the noncharacteristic case. It turns out that, in the one-dimensional analysis done by Freistühler and the author [17], the proofs of L^1-stability of constants (say of the solution $\bar{u} \equiv 0$ with the homogeneous Dirichlet boundary condition) were significantly different, according to the case under consideration: characteristic ($f'(0) = 0$) or outgoing ($f'(0) < 0$), while the result itself (instability) is different in the incoming case ($f'(0) > 0$). Surprisingly enough, the characteristic case was the easiest one, because it could be converted in a pure Cauchy problem (see below). We give hereafter a new proof of stability of the outgoing case, by a comparison with the characteristic one. It does not only simplify the analysis, but it extends the stability result to several space dimensions.

7.1. *Stability of constants*

The one-dimensional case has been treated by Freistühler and the author in [17]. There is a complete description, as far as the L^1-asymptotic stability is concerned. It turns out that, as in the shock-stability problem, the L^1-stability of constants plays a fundamental role. Since constant are not solutions in general, this question makes sense only when $u_b = 0$. We notice also that the total mass is no longer a conserved quantity and therefore one must not assume that the initial disturbance have a zero mass.

The dichotomy mentioned above reflects in the following statement.

[23] But we deliberately ignored this aspect in this article, focusing only on the stability itself.

THEOREM 13. *Consider the viscous IBVP, with a homogeneous Dirichlet boundary condition* ($u_b = 0$).

(i) *Assume* $f_1'(0) \leqslant 0$. *Then the* L^1*-stability of* $u_0 \equiv 0$ *holds true*: *If* $a \in L^1(H)$, *then*

$$\lim_{t\to+\infty} \|u(t)\|_1 = 0.$$

(ii) *Assume on the contrary that* $f_1'(0) > 0$. *Then there exist bounded initial data* $a \in L^1(H)$ *such that*

$$\lim_{t\to+\infty} \|u(t)\|_1 > 0.$$

PROOF. We shall treat the first part (stability) of the theorem, and refer to [17] for the second one (instability), since the corresponding proof has a straightforward extension to the multidimensional setting.

Given a bounded integrable data a, the comparison principle tells that $S_t a_- \leqslant u(t) \leqslant S_t a_+$, where $a_\pm$ are the positive and negative parts of a. Hence

$$\|u(t)\|_1 \leqslant \|S_t a_-\|_1 + \|S_t a_+\|_1.$$

Therefore it is enough to consider data of constant signs. For instance, we focus on nonnegative data. Corresponding solutions will be nonnegative.

We begin with the characteristic case: $f_1'(0) = 0$. We may fix $f(0) = 0$. Choosing a frame that moves with a constant speed, parallel to the boundary, we may also fix $f_\alpha'(0) = 0$ for $\alpha > 1$. Let b and $v(t)$ be the extensions of a and $u(t)$ to the whole space $\mathbb{R}^d$ that are odd with respect to x. Similarly, the restriction of f to $\mathbb{R}^+$ is extended to $\mathbb{R}$ as a flux g, in such a way that g_1 be even, while g_α be odd otherwise. It is clear that the new flux is of class $W^{2,\infty}_{\mathrm{loc}}$ (locally bounded second derivative). One verifies easily that v is the solution of the pure Cauchy problem

$$\partial_t v + \operatorname{div} g(v) = \Delta v$$

with initial datum b. Since b is bounded, integrable with zero total mass, and since $g(w) = \mathcal{O}(w^2)$, the proof of the multidimensional version of Theorem 6 works out, and we obtain

$$\lim_{t\to+\infty} \|v(t)\|_1 = 0.$$

This solves the characteristic case.

Consider now the outgoing case $f_1'(0) < 0$. Define a new flux function F by $F_1(w) := f_1(w) - f_1'(0)w$ and $F_\alpha = f_\alpha$ otherwise. Define also a function

$$v(x, y, t) := \begin{cases} u\bigl(x + f_1'(0)t, y, t\bigr) & \text{if } x > -f_1'(0)t, \\ 0 & \text{otherwise.} \end{cases}$$

The distribution $T := \partial_t v + \operatorname{div} F(v) - \Delta v$ is supported by the hyperplane $x = f_1'(0)t$. More precisely, it is given by

$$T = \frac{\partial u}{\partial \nu}(0, y, t)\delta\big(x = f_1'(0)t\big).$$

Since u is nonnegative (recall that $a \geqslant 0$) and vanishes at the boundary, its normal derivative must be nonpositive and therefore we have $T \leqslant 0$. Since $v(0) = a$, we obtain that v is a subsolution of the characteristic IBVP (remark that $F_1'(0) = 0$):

$$\partial_t w + \operatorname{div} F(w) = \Delta w, \qquad w(x, y, 0) = a, \quad w(0, y, t) = 0.$$

The maximum principle applies to that kind of evolution problem and gives $v \leqslant w$. Thus we obtain

$$\|u(t)\|_1 = \|v(t)\|_1 \leqslant \|w(t)\|_1.$$

Since we already treated the characteristic case, we know that $\|w(t)\|_1$ tends to zero at infinity and this implies the desired result. □

7.2. *Stability of boundary layers*

Using Theorem 13, plus arguments of dynamical systems theory, one obtains the same kind of one-dimensional stability result than for viscous shocks:

THEOREM 14 ($d = 1$ [17]). *Let $U(x)$ be a nontrivial (that is, $u_b \neq u(+\infty)$) boundary layer for the viscous model. Then it is L^1-stable: Given any bounded initial data $a \in U + L^1$, one has*

$$\lim_{t\to+\infty} \|u(t) - U\|_1 = 0.$$

REMARKS.

1. Since U is nontrivial, $f'(0)$ must be nonpositive. Hence, the first part of Theorem 13 can be used.
2. Theorem 14 could not be true for constant U in general, because of the second part of Theorem 13.
3. When $f'(0) < 0$, U is itself integrable (say that $u(+\infty) = 0$) because it decays exponentially. Hence the assumption on the data is really that a is integrable. However, in the characteristic case, the boundary layer is *not* integrable, because it decays at best as $1/x$. Hence, the data are not integrable in this case. It would be interesting of course to prove L^p-stability results in this case, with p such that $U \in L^p(0, +\infty)$. For instance, $p > 1$ is relevant when $f'(0) = 0 \neq f''(0)$.

In several space dimensions, the situation is curious. Remember that, when studying the L^1-stability of viscous shock waves, we encountered a difficulty caused by the conservation of mass: The convergence $\|u(t) - U(\cdot - st)\|_1 \to 0$ could not hold unless the initial disturbance had zero mass. This is no longer true for boundary layers since mass conservation does not hold. Hence, we may expect a general stability result, regardless the mass of initial disturbance. However, we still miss a compactness argument, due to the translational invariance along directions that are parallel to the boundary. Thus we left the following

OPEN PROBLEM 9. Prove the stability of multidimensional viscous boundary layers, under bounded integrable initial disturbances.

Acknowledgments

I am happy to express my gratitude to Constantine Dafermos. I should not have found the courage to write this review without his constant interest. Also, he made me aware of techniques in dynamical systems theory, which have proved their efficiency in this domain. Thanks to his friendly pressure, this topic is now presented in a unified way, after having been disseminated in several smaller papers. This writing gave me the occasion to solve a few open question and to fix a few points of which the difficulty was underestimated in the literature. Special thanks go to H. Freistühler, whose collaboration was precious at some stage of this research. Last but not least, this article was polished during my stay in Isaac Newton Institute for Mathematical Sciences, in a warm atmosphere[24] and perfect technical as well as scientific environment.

This research was done from one part in accomplishment of the European IHP project "HYKE", contract HPRN-CT-2002-00282.

References

[1] C. Abourjaily and P. Bénilan, *Symmetrization of quasi-linear parabolic problems*, Rev. Un. Mat. Argentina **41** (1998), 1–13.

[2] J.-B. Baillon and R.E. Bruck, *Ergodic theorems and the asymptotic behavior of contraction semigroups*, Fixed Point Theory and Applications (Halifax, NS, 1991), World Scientific, River Edge, NJ (1992), 12–26.

[3] J.-B. Baillon and R.E. Bruck, *The rate of asymptotic regularity is* $\mathcal{O}(1/n)[\mathcal{O}(1/\sqrt{n})]$, Theory and Applications of Nonlinear Operators of Accretive and Monotone Types, Lecture Notes in Pure and Appl. Math., Vol. 178, Dekker, New York (1996), 51–81.

[4] P. Bénilan and M. Crandall, *Regularizing effects of homogeneous evolution equations*, Contributions to Analysis and Geometry (Baltimore, MD, 1980), John Hopkins Univ. Press, Baltimore, MD (1981), 23–39.

[5] H. Brézis, *Analyse fonctionnelle. Théorie et applications*, Masson, Paris (1983).

[6] G.-Q. Chen, D. Levermore and T.-P. Liu, *Hyperbolic conservation laws with stiff relaxation terms and entropy*, Comm. Pure Appl. Math. **47** (1994), 787–830.

[7] G.-Q. Chen and T.-P. Liu, *Zero relaxation and dissipation limits for hyperbolic conservation laws*, Comm. Pure Appl. Math. **46** (1993), 755–781.

[8] G.-Q. Chen and M. Rascle, *Initial layers and uniqueness of weak entropy solutions to hyperbolic conservation laws*, Arch. Ration. Mech. Anal. **153** (2000), 205–220.

[24] Special thanks to Louise, Konstantina and Debora.

[9] K. Chuey, C. Conley and J. Smoller, *Positively invariant regions of nonlinear diffusion equations*, Indiana Univ. Math. J. **26** (1977), 373–392.

[10] C. Dafermos, *Asymptotic behaviour of solutions of hyperbolic balance laws*, Bifurcation Phenomena in Mathematical Physics and Related Topics, C. Bardos and D. Bessis, eds, NATO Adv. Sci. Inst. Ser. C Math. Phys. Sci., Vol. 54, Reidel, Dordrecht (1980), 521–533.

[11] C. Dafermos, *Regularity and large time behaviour of solutions of a conservation law without convexity*, Proc. Roy. Soc. Edinburgh Ser. A **99** (1985), 201–239.

[12] C. Dafermos, *Hyperbolic Conservation Laws in Continuum Physics*, Grundlehren Math. Wiss., Vol. 325, Springer-Verlag, Heidelberg (2000).

[13] M. Di Francesco and P. Markovitch, *Relative entropy decay of solutions of the viscous Burgers' equation to self-similarity*, Preprint (2003).

[14] R.J. DiPerna, *Measure-valued solutions to conservation laws*, Arch. Ration. Mech. Anal. **88** (1985), 223–270.

[15] M. Escobedo, J.-L. Vazquez and E. Zuazua, *Asymptotic behaviour and source-type solutions for a diffusion-convection equation*, Arch. Ration. Mech. Anal. **124** (1993), 43–65.

[16] H. Freistühler and D. Serre, *L^1-stability of shock waves in scalar viscous conservation laws*, Comm. Pure Appl. Math. **51** (1998), 291–301.

[17] H. Freistühler and D. Serre, *The L^1-stability of boundary layers for scalar viscous conservation laws*, J. Dynam. Differential Equations **13** (2001), 745–755.

[18] C. Fries, *Nonlinear asymptotic stability of general small-amplitude Laxian shock waves*, J. Differential Equations **146** (1998), 185–202.

[19] C. Fries, *Stability of viscous shock waves associated with non-convex modes*, Arch. Ration. Mech. Anal. **152** (2000), 141–186.

[20] D. Gilbarg, *The existence and limit behavior of the one-dimensional shock layer*, Amer. J. Math. **73** (1951), 256–274.

[21] S. Godunov, *A difference scheme for numerical calculations of discontinuous solutions of the equations of hydrodynamics*, Math. Sb. **47** (1959), 271–306.

[22] J. Goodman and J. Miller, *Long-time behaviour of scalar viscous shock fronts in two dimensions*, J. Dynam. Differential Equations **11** (1999), 255–277.

[23] J. Goodman and Z. Xin, *Viscous limits for piecewise smooth solutions to systems of conservation laws*, Arch. Ration. Mech. Anal. **121** (1992), 235–265.

[24] E. Grenier and O. Guès, *Boundary layers for viscous perturbations of noncharacteristic quasilinear hyperbolic problems*, J. Differential Equations **143** (1998), 110–146.

[25] H. Fan, *Existence and uniqueness of travelling waves and error estimates for Godunov schemes of conservation laws*, Math. Comp. **70** (1998), 87–109.

[26] H. Fan, *Existence of discrete shock profiles of a class of monotonicity preserving schemes for conservation laws*, Math. Comp. **67** (2000), 1043–1069.

[27] B. Hanouzet and R. Natalini, *Weakly coupled systems of quasilinear hyperbolic equations*, Differential Integral Equations **9** (1996), 1279–1292.

[28] A. Harten, J.M. Hyman and P.D. Lax, *On finite-difference approximations and entropy conditions for shocks*, with an appendix by B. Keyfitz, Comm. Pure Appl. Math. **29** (1976), 297–322.

[29] D. Hoff and K. Zumbrun, *Asymptotic behaviour of multidimensional viscous shock fronts*, Indiana Univ. Math. J. **49** (2000), 427–474.

[30] D. Howard and K. Zumbrun, *Pointwise semi-group methods and stability of viscous shock waves*, Indiana Univ. Math. J. **47** (1998), 741–871.

[31] G. Jennings, *Discrete shocks*, Comm. Pure Appl. Math. **27** (1974), 25–37.

[32] S. Jin and Z. Xin, *The relaxation schemes for systems of conservation laws in arbitrary space dimensions*, Comm. Pure Appl. Math. **48** (1995), 235–276.

[33] S. Kawashima and S. Nishibata, *Weak solutions with a shock to a model system of the radiating gas*, Sci. Bull. Josai Univ., Spec. Issue no. 5 (1998), 119–130.

[34] S. Kawashima and S. Nishibata, *Shock waves for a model system of a radiating gas*, SIAM J. Math. Anal. **30** (1999), 95–117.

[35] S. Kawashima, Y. Nikkuni and S. Nishibata, *The initial value problem for hyperbolic–elliptic coupled systems and applications to radiation hydrodynamics*, Analysis of Systems of Conservation Laws (Aachen, 1997), Chapman and Hall/CRC Monogr. Surv. Pure Appl. Math., Vol. 99, Chapman and Hall/CRC, Boca Raton, FL (1999), 87–127.

[36] Y.J. Kim and A. Tzavaras, *Diffusive N-waves and metastability in the Burgers equation*, SIAM J. Math. Anal. **33** (2001), 607–633.

[37] C. Kondo and P. LeFloch, *Zero diffusion–dispersion limits for scalar conservation laws*, SIAM J. Math. Anal. **33** (2002), 1320–1329.

[38] G. Kreiss and H.-O. Kreiss, *Stability of systems of viscous conservation laws*, Comm. Pure Appl. Math. **51** (1998), 1397–1424.

[39] S.N. Kruzkhov, *Generalized solutions of the Cauchy problem in the large for nonlinear equations of first order*, Dokl. Akad. Nauk SSSR **187** (1969), 29–32.

[40] J.-P. Lasalle, *The Stability of Dynamical Systems*, CBMS-NSF Regional Conf. Ser. in Appl. Math., SIAM, Philadelphia, PA (1976).

[41] C. Lattanzio and P. Marcati, *Global well-posedness and relaxation limits of a model for radiating gas*, J. Differential Equations **190** (2003), 439–465.

[42] P.D. Lax, *Hyperbolic systems of conservation laws (II)*, Comm. Pure Appl. Math. **10** (1957), 537–566.

[43] P.-L. Lions, B. Perthame and E. Tadmor, *Formulation cinétique des lois de conservation multidimensionnelles*, C. R. Math. Acad. Sci. Paris **312** (1991), 97–102.

[44] H. Liu and E. Tadmor, *Critical thresholds in a convolution model for nonlinear conservation laws*, SIAM J. Math. Anal. **33** (2001), 930–945.

[45] T.-P. Liu, *Nonlinear Stability of Shock Waves for Viscous Conservation Laws*, Mem. Amer. Math. Soc. **56**(328) (1985).

[46] T.-P. Liu and M. Pierre, *Source-solutions and asymptotic behaviour in conservation laws*, J. Differential Equations **51** (1984), 419–441.

[47] T.-P. Liu and H.-S. Yu, *Continuum shock profiles for discrete conservation laws, I. Construction*; *II. Stability*, Comm. Pure Appl. Math. **52** (1999), 85–127; 1047–1073.

[48] C. Mascia and R. Natalini, *L^1 nonlinear stability of travelling waves for a hyperbolic system with relaxation*, J. Differential Equations **132** (1996), 275–292.

[49] R. Natalini, *Recent mathematical results on hyperbolic relaxation models*, Analysis of Systems of Conservation Laws (Aachen, 1997), Chapman and Hall/CRC Monogr. Surv. Pure Appl. Math., Vol. 99, Boca Raton, FL (1999), 128–198.

[50] J. Neveu, *Semi-groupes généralisés et processus de Markoff*, C. R. Math. Acad. Sci. Paris **240** (1955), 1046–1047.

[51] S. Osher and J. Ralston, *L^1 stability of travelling waves with applications to convective porous media flow*, Comm. Pure Appl. Math. **35** (1982), 737–751.

[52] P. Rosenau, *Extending hydrodynamics via the regularization of the Chapman–Enskog expansion*, Phys. Rev. A **40** (1989), 7193–7196.

[53] F. Rousset, *Viscous approximation for strong shocks of systems of conservation laws*, SIAM J. Math. Anal. **35** (2003), 492–519.

[54] T. Ruggeri and D. Serre, *Stability of constant equilibrium state for dissipative balance laws systems with a convex entropy*, Quart. Appl. Math. **62** (2004), 163–179.

[55] D.H. Sattinger, *On the stability of waves of nonlinear parabolic systems*, Adv. Math. **22** (1976), 312–355.

[56] S. Schochet and E. Tadmor, *Regularized Chapman–Enskog expansion for scalar conservation laws*, Arch. Ration. Mech. Anal. **119** (1992), 95–107.

[57] D. Serre, *Stabilité L^1 des chocs et de deux types de profils, pour des lois de conservation scalaires*, Preprint, ENS de Lyon, available (http://www.umpa.ens-lyon/~serre/PS/L1.ps).

[58] D. Serre, *Stabilité L^1 pour les lois de conservation visqueuses*, C. R. Math. Acad. Sci. Paris **323** (1996), 359–363.

[59] D. Serre, *Remarks about the discrete profiles of shock waves*, Mat. Contemp. **11** (1996), 153–170.

[60] D. Serre, *Stabilité L^1 d'ondes progressives de lois de conservation scalaires*, Séminaire EDP de l'Ecole Polytechnique (1998–1999).

[61] D. Serre, *Systems of Conservation Law, I*, Cambridge Univ. Press, Cambridge (1999).

[62] D. Serre, *Systems of Conservation Law, II*, Cambridge Univ. Press, Cambridge (2000).

[63] D. Serre, *Relaxations semi-linéaire et cinétique des systèmes de lois de conservation*, Ann. Inst. H. Poincaré Anal. Non Linéaire **17** (2000), 169–192.

[64] D. Serre, *Sur la stabilité des couches limites de viscosité*, Ann. Inst. Fourier **51** (2001), 109–129.

[65] D. Serre, *Stabilité des ondes de choc de viscosité qui peuvent être caractéristiques*, Nonlinear PDEs and Their Applications, Collège de France Seminar XIV, D. Cioranescu and J.-L. Lions, eds, Stud. Math. Appl., Vol. 31, Elsevier, Amsterdam (2002), 647–654.

[66] D. Serre, *Asymptotics of homogeneous oscillations in a compressible viscous fluid*, Bol. Soc. Brasil. Mat. **32** (Spec. issue dedicated to C. Dafermos) (2002), 435–442.

[67] D. Serre, *L^1-stability of constants in a model for radiating gases*, Comm. Math. Sci. **1** (2003), 199–207.

[68] D. Serre and K. Zumbrun, *Boundary layer stability in real vanishing viscosity limit*, Comm. Math. Phys. **221** (2001), 267–292.

[69] A. Szepessy, *An existence result for scalar conservation laws using measure-valued solutions*, Comm. Partial Differential Equations **14** (1989), 1329–1350.

[70] A. Szepessy and Z. Xin, *Nonlinear stability of viscous shock waves*, Arch. Ration. Mech. Anal. **122** (1993), 53–103.

[71] L. Tartar, *Compensated compactness and applications to partial differential equations*, Nonlinear Analysis and Mechanics: Heriot–Watt symposium, Vol. IV, R.J. Knops, ed., Pitman, London (1979), 136–212.

[72] L. Tartar, *The compensated compactness applied to systems of conservation laws*, Systems of Nonlinear Partial Differential Equations, J. Ball, ed., Reidel, Dordrecht (1983), 263–285.

[73] A.I. Volpert, *The space BV and quasilinear equations*, Math. USSR-Sb. **2** (1967), 225–267.

[74] K. Zumbrun and D. Serre, *Viscous and inviscid stability of multidimensional planar shock fronts*, Indiana Univ. Math. J. **48** (1999), 937–992.

Author Index

Roman numbers refer to pages on which the author (or his/her work) is mentioned. Italic numbers refer to reference pages. Numbers between brackets are the reference numbers. No distinction is made between first and co-author(s).

Abourjaily, C. 482, *550* [1]
Acquistapace, P. 411, *435* [1]
Adams, R. *279* [1]
Alber, H.D. 131, *166* [Al]
Albrecht, D. *79* [ADM96]
Alfors, L.V. *279* [2]
Alt, H.W. *279* [3]
Amann, H. 33, 77, *79* [AHS94]; *79* [Ama01]; *79* [Ama97]; 428, *435* [2]; *435* [3]
Ambrosio, L. 448, 449, *469* [1]; *469* [2]
Ames, K.A. 374, *375* [1]
Amosov, A.A. 298, 299, 330, 331, 334, *375* [2]; *375* [3]; *375* [4]; *375* [5]; *375* [6]; *384* [244]; *384* [245]; *384* [246]
Ancona, F. 92, 131, 152, *166* [AM1]; *166* [AM2]
Andreucci, D. 242, 257, 258, 261, *279* [4]; *279* [5]; *279* [6]; 429, *435* [4]
Andrews, G. 298, 300, 302, 308–310, *375* [7]; *375* [8]
Angenent, S. 393, *435* [5]
Antman, S.S. 310, *375* [9]
Antoncev, S.N. 177, *279* [7]; *279* [8]; *279* [9]; 297, 311, 334, 354, *375* [10]
Anzellotti, G. 177, *279* [10]; *280* [11]
Arendt, W. 5, 6, 10, 12–16, 20, 33, 40, 47, 48, 50–53, 55, 57, 61, 63, 64, 71, 72, 74, 76, *78* [[ABHN01]]; *79* [AB94]; *79* [ABB03]; *79* [ABH01]; *79* [AMPR03]; *79* [ARS94]; *79* [ArBu02]; *79* [ArBu03a]; *79* [ArBu03b]; *79* [Are94]; *79* [Are97]; *79* [AtE97]
Aronson, D.G. 75, *79* [Aro67]; 175, 185, 225, 231, 253, 260, *280* [12]; *280* [13]; *280* [14]; *280* [15]; *280* [16]
Auscher, P. 28, 74–76, *78* [[AT98]]; *80* [ABBO00]; *80* [ACT96]; *80* [AER94]; *80* [AHLMT02]; *80* [AHMT01]; *80* [AMN97]; *80* [AMT98]; *80* [AT01a]; *80* [AT01b]; *80* [Aus96]

Baconneau, O. 387, 424, 429, 431, 432, *435* [6]
Baillon, J.-B. 31, *80* [BC91]; 508, *550* [2]; *550* [3]
Baiti, P. 92, 94, 131, *166* [BaJ]
Ball, J.M. 300, 302, 308–310, *375* [8]
Barenblatt, G.I. 172, 233, 239, *280* [17]
Barthélemy, L. 74, *80* [ABBO00]
Batty, C. 5, 6, 10, 12–16, 47, 48, 50, 57, 76, *78* [[ABHN01]]; *79* [ABB03]
Baumann, P. 77, *80* [Bau84]
Bear, J. 254, 257, 258, *280* [18]; *280* [19]
Bebernes, J. 300, *375* [11]; *375* [12]
Becker, E. 312, *375* [13]
Belhadj, M. 447, *469* [3]
Belov, S. 298, 300, *375* [14]
Belov, V. 298, 300, *375* [14]
Benabidallah, R. 354, *377* [69]; *377* [70]
Bénilan, P. 74, *80* [ABBO00]; 237, *280* [20]; 482, 484, *550* [1]; *550* [4]
Berkson, E. 52, *80* [BG94]
Bernstein, S.N. *280* [21]
Berryman, J.G. 249, *280* [22]; *280* [23]
Berthelin, F. 440, 458, *469* [4]
Bertsch, M. 253, *280* [24]; *280* [25]; 298, *375* [15]
Bianchini, S. 102, 152, 154, *166* [BiB]
Biegert, M. 11, *80* [Bie03]
Blake, M. 13, *80* [BBN01]; *80* [Bla01]
Boccardo, L. 256, *280* [26]
Bombieri, E. 232, *280* [27]; *280* [28]
Bossavit, A. 225, 254, 257, *280* [29]
Bouchut, F. 440, 450, 458, *469* [4]; *469* [5]; *469* [6]
Bouleau, N. 60, *78* [[BH91]]
Bourgain, J. *80* [Bou86]

Brauner, C.-M. 387, 411, 417, 419, 422–424, 429, 434, *435* [7]; *435* [8]; *435* [9]; *435* [10]
Brendle, S. 13, *80* [BBN01]
Brenier, Y. 439, 447, *469* [7]
Bressan, A. 92–94, 102, 120, 131, 152, 154, 156, 158–160, *166* [B1]; *166* [B2]; *166* [B3]; *166* [B4]; *166* [B5]; *166* [BiB]; *167* [BC1]; *167* [BC2]; *167* [BCP]; *167* [BG]; *167* [BLF1]; *167* [BLF2]; *167* [BLY]; *167* [BLw]; *167* [BM]; 299, 300, *375* [11]; *375* [16]
Brézis, H. 177, 249, *280* [30]; 513, *550* [5]
Bruck, R.E. 508, *550* [2]; *550* [3]
Bu, S. 13, 33, 40, 51–53, 55, 57, 71, *79* [ABB03]; *79* [ABH01]; *79* [ArBu02]; *79* [ArBu03a]; *79* [ArBu03b]; *80* [BGD97]
Bukhvalov, A.V. 64, *79* [AB94]
Burkholder, D. 51, *80* [Bur83]

Caffarelli, L.A. 178, 259, 260, *280* [13]; *280* [31]; *280* [32]; *280* [33]; 429, 432, 433, *435* [11]
Campiti, M. 11, *80* [CM98]
Carillo, J. 446, *469* [8]
Carlson, D.E. 289, 294, *375* [17]
Castro, A. 387, *435* [25]; *436* [26]
Cercignani, C. 439, *469* [9]; *469* [10]
Chang, T. 328, *375* [18]
Chavent, G. 254, *280* [34]
Chen, G.Q. 311, 327–330, 340, 354, 365, *375* [19]; *375* [20]; *375* [21]; *375* [22]; *376* [23]; *376* [24]; 440, 446, 454, *470* [11]; *470* [12]; *550* [6]; *550* [7]; *550* [8]
Chen, S. 374, *376* [25]
Chen, Y.Z. 176, 233, *280* [35]; *280* [36]; *280* [37]; *280* [38]; *281* [39]; *281* [40]
Chen, Z. 299, *376* [26]
Cheng, K.S. *281* [41]
Chill, R. 16, 64, *80* [CFMP03]; *80* [CT03]; *80* [CT04]
Choe, H. *281* [42]; *281* [43]; *376* [27]
Chrzęszczyk, A. 365, *376* [28]
Chuey, K. 488, 491, *551* [9]
Chzhou, Y.-L. 242, *283* [104]
Cioranescu, I. 26, *80* [CZ76]
Clément, Ph. 31, 49, 52, 53, 56, 58, 71, *80* [BC91]; *81* [CL94]; *81* [CP01]; *81* [CPSW00]
Collela, P. 330, *376* [29]
Collins, R.E. 254, 257, 258, *281* [44]
Colombo, R.M. 93, 152, *167* [BC1]; *167* [BC2]
Conley, C. 488, 491, *551* [9]
Coulhon, T. 65, 66, 68, 71, 75, *79* [[VSC93]; *80* [ACT96]]; *81* [CD00]; *81* [Cou90]
Cowling, M. 31, 34, 64, *81* [CDMY96]; *81* [Cow83]
Crandall, M. 237, 253, *280* [14]; *280* [20]; 484, *550* [4]
Crasta, G. 152, *167* [BCP]
Crouzeix, M. 49, *81* [Cro03]

Da Prato, G. 20, *81* [DPG75]; 387, 391, 393, *435* [12]; *435* [13]
Dafermos, C. 92, 131, *167* [D1]; *167* [D2]; 289, 293, 298, 299, 301, 310, 364, 365, *375* [20]; *376* [30]; *376* [31]; *376* [32]; *376* [33]; *376* [34]; *376* [35]; *376* [36]; 448, 449, 454, *470* [13]; 520, 525, *551* [10]; *551* [11]; *551* [12]
Damlamian, A. 225, 254, 257, *280* [29]
Dan, W. 365, *376* [37]
Daners, D. 74, *81* [Dan00]
Danilyuk, I.I. 254, *281* [45]
Daskalopoulos, P. 177, *281* [46]
Dautray, R. 8, 11, *78* [[DL88]]; 439, *470* [14]
Davies, E.B. 37, 61–63, 65–67, 69, 70, 75, *78* [[Dav80]]; *78* [[Dav90]]; *81* [Dav87]; *81* [Dav97]
Dawson, J.M. 249, *284* [139]
Day, W.A. 374, *376* [38]
De Giorgi, E. 174, 180, 184, 185, 256, *281* [47]
de Lacerda Oliveira, M. 374, *381* [167]
De Lellis, C. 448, 449, *469* [1]; *470* [15]; *470* [16]
de Pagter, B. 49, 53, *81* [CPSW00]
del Pino, M. 177, *281* [46]
Denk, R. 19, 57, 77, 78, *81* [DDHPV04]; *81* [DHP01]
Deshpande, S.M. 463, *470* [17]
Desimone, A. 448, *470* [18]
Desjardins, B. 340, *376* [39]
Di Francesco, M. *551* [13]
DiBenedetto, E. 174, 176–179, 182, 185, 209, 211, 218, 219, 224, 227, 233, 234, 238, 239, 242, 243, 245, 249, 254, 257–263, 265, 267, 269–271, 273–278, *279* [3]; *279* [5]; *279* [6]; *280* [37]; *280* [38]; *281* [39]; *281* [40]; *281* [48]; *281* [49]; *281* [50]; *281* [51]; *281* [52]; *281* [53]; *281* [54]; *281* [55]; *281* [56]; *281* [57]; *281* [58]; *281* [59]; *281* [60]; *281* [61]; *281* [62]; *281* [63]; *281* [64]; *282* [65]; *282* [66]; *282* [67]; *282* [68]; *282* [69]; *282* [70]; *282* [71]; *282* [72]
Diller, D.J. 177, *281* [59]; *281* [60]; *282* [73]
DiPerna, R.J. 131, 159, *167* [DP1]; *167* [DP2]; 450, 457, *470* [19]; *470* [20]; *470* [21]; 498, *551* [14]
Dore, G. 27, 28, 36, 57, 78, *81* [DDHPV04]; *81* [DoVe87]; *81* [Dor00]; *81* [Dor93]; *81* [Dor99a]; *81* [Dor99b]
Doust, I. 31, *81* [CDMY96]
Drake, J.R. 249, *282* [74]

Ducomet, B. 311, 322, 326, 340, 348, *376* [40]; *376* [41]; *376* [42]; *376* [43]; *376* [44]; *376* [45]; *376* [46]; *376* [47]; *376* [48]; *376* [49]
Dungey, N. *78* [[DER03]]
Duong, X. 34, 71, 72, 77, *79* [ADM96]; *81* [CD00]; *81* [DM99]; *81* [DR96]; *81* [DS97]; *81* [DY01]; *81* [Duo89]
Durek, U. 365, *377* [50]
Durmagambetov, A.A. 297, *382* [209]
Dziubański, J. 76, *81* [DHZ94]

Eberhardt, B. 6, *81* [EG92]
Eberly, D. 300, *375* [12]
Ebmeyer, C. *282* [75]; *282* [76]
Edwards, R.E. 53, *78* [[EG77]]
Elliott, C.M. 374, *377* [51]
Engel, K.-J. 12, 16, *78* [[EN00]]; *82* [Eng04]
Ericksen, J.L. 300, *377* [52]
Escher, J. 387, 424, 426–429, *435* [14]; *435* [15]; *435* [16]; *435* [17]; *435* [18]; *435* [19]; *435* [20]
Escobedo, M. 528, *551* [15]
Esposito, L. 270, *282* [77]
Esteban, J.E. 177, *282* [78]
Evans, L.C. 177, 178, 259, 260, *280* [31]; *282* [79]; *470* [22]
Eymard, R. 446, 448, 463, *470* [23]; *470* [24]

Fabes, E.B. 232, *282* [80]; *282* [81]
Fan, H. 540, 542, 545, *551* [25]; *551* [26]
Fašangová, E. 64, *80* [CFMP03]
Fasano, A. 225, 254, 257, *282* [82]; *286* [179]
Favini, A. 12, *82* [FGGR02]
Feireisl, E. 311, 340–344, 348, 351, 352, 365, *376* [47]; *377* [53]; *377* [54]; *377* [55]; *377* [56]; *377* [57]; *377* [58]; *377* [59]; *377* [60]; *377* [61]; *377* [62]; *377* [63]; *377* [64]; *377* [65]; *470* [25]
Fendler, G. 35, *82* [Fen97]
Foias, C. 352, *377* [66]
Fonf, V.P. 14, *82* [FLW01]
Freistühler, H. 448, *470* [26]; 523, 547–549, *551* [16]; *551* [17]
Fremond, M. 225, 254, 257, *280* [29]
Frid, H. 322, *377* [67]
Friedman, A. 76, 77, *78* [[Fri64]]; 244, 257, *280* [32]; *280* [33]; *281* [61]; *281* [62]; *282* [83]; *282* [84]; *282* [85]
Fries, C. 448, *470* [26]; 508, *551* [18]; *551* [19]
Friesecke, G. 310, *377* [68]
Fujita-Yashima, H. 330, 354, *377* [69]; *377* [70]; *377* [71]
Fukushima, M. 10, 60, *78* [[FOT94]]; *82* [FT95]

Galaktionov, V.A. 429, *435* [21]
Gallouët, T. 446, 448, 463, *470* [23]; *470* [24]
Gariepy, R. 178, *281* [63]
Garnett, J.B. 31, *78* [[Gar81]]
Garofalo, N. 232, *282* [80]
Gaudry, G.I. 53, *78* [[EG77]]
Gerbeau, J.-F. 447, *469* [3]
Gianazza, U. 178, 273–275, 278, *282* [86]; *282* [87]
Gianni, R. 429, *435* [4]
Giaquinta, M. 177, *280* [11]
Giga, Y. 439, 447, *470* [27]
Gilbarg, D. 276, *282* [88]; 505, *551* [20]
Gillespie, T.A. 52, *80* [BG94]
Giusti, E. 232, *280* [28]
Glassey, R.T. 439, *470* [28]
Glimm, J. 92, 93, 112, 130, *167* [G]; 328, 330, *375* [21]; *377* [72]
Goatin, P. 159, *167* [BG]
Godlewski, E. 448, 463, *470* [29]
Godunov, S. 544, *551* [21]
Goldstein, G.R. 12, *82* [FGGR02]
Goldstein, J.A. 12, *82* [FGGR02]
Golse, F. 439, 450, 452, *469* [6]; *470* [30]; *470* [31]
Goodman, J. 478, 508, 510, 531, 535, *551* [22]; *551* [23]
Grad, H. 334, *377* [73]
Greenberg, J.M. 300, 302, 310, *377* [74]; *377* [75]
Greenwood, J.R. 249, *282* [74]
Greiner, G. 6, *81* [EG92]
Grenier, E. 340, *376* [39]; 509, *551* [24]
Grisvard, P. 20, 68, *78* [[Gri92]]; *81* [DPG75]; *82* [Gri67]; 387, 391, *435* [12]
Guerre-Delabrière, S. 57, *80* [BGD97]; *82* [Gue95]
Guès, O. 509, *551* [24]
Guo, B.L. 299, *377* [76]; *378* [77]

Haase, M. 23, 25, 26, 33, 38, 40, 46, 49, *79* [ABH01]; *82* [Haa02], *82* [Haa03a], *82* [Haa03b]; *82* [Haa03c]
Hadamard, J. 173, 230, 236, *282* [89]
Hamilton, R.S. 177, *282* [90]
Hanouzet, B. 477, 488, *551* [27]
Hansen, I. 365, *378* [78]
Hardt, R. 177, *282* [91]; *283* [92]
Harten, A. 539, *551* [28]
Hebisch, W. 76, *81* [DHZ94]
Heck, H. 77, *82* [HH03]
Hedberg, L.I. 43, *82* [Hed93]
Hempel, P. 72, *82* [HV86]
Henry, D. 373, *378* [79]; 388, 420, *435* [22]
Herbin, R. 446, 448, 463, *470* [23]; *470* [24]

Herrero, M.A. 227, *281* [64]; *282* [65]
Hieber, M. 5, 6, 10, 12–16, 19, 27, 33, 34, 47, 48, 50, 57, 71, 76–78, *78* [[ABHN01]]; *79* [AHS94]; *81* [DDHPV04]; *81* [DHP01]; *82* [HH03]; *82* [HM00a]; *82* [HM00b]; *82* [HP97]; *82* [HP98]; *82* [Hie96]
Hirsch, F. 60, *78* [[BH91]]
Hoff, A. 254, *283* [93]
Hoff, D. 323, 325–330, 333, 334, 339, 340, 344, 348, 351, *375* [22]; *376* [23]; *378* [80]; *378* [81]; *378* [82]; *378* [83]; *378* [84]; *378* [85]; *378* [86]; *378* [87]; *378* [88]; *378* [89]; *378* [90]; *378* [91]; *378* [92]; *378* [93]; *378* [94]; *378* [95]; *378* [96]; 478, 510, 531, 535, 536, *551* [29]
Hoffmann, K.H. 225, 254, 257, *283* [94]; 299, *376* [26]
Hofmann, S. 76, *80* [AHLMT02]; *80* [AHMT01]; *83* [MHL02]
Holden, H. 131, *167* [HR]
Holland, C.J. 249, *280* [23]
Howard, D. 509, *551* [30]
Hrusa, W.J. 362, 364, 365, *376* [33]; *378* [97]; *378* [98]
Hsiao, L. 298–301, 310–313, 322, 328, 364, *375* [18]; *376* [34]; *376* [35]; *376* [36]; *378* [99]; *378* [100]; *378* [101]; *378* [102]
Hu, J. 152, *167* [HLF]
Huang, C. *283* [95]
Hudjaev, S.I. 365, *383* [225]; 387, *435* [23]
Hulshof, J. 387, 411, 417, 419, 423, 424, 429, 434, *435* [7]; *435* [8]; *435* [21]
Hwang, S. 448, *470* [32]
Hyman, J.M. 539, *551* [28]

Illner, R. 439, *469* [10]
Iskenderova, D.A. 297, *378* [103]; *382* [209]
Ivanov, A.V. 225, 236, *283* [96]; *283* [97]; *283* [98]
Iwaniec, T. *283* [99]

Jabin, P.-E. 448, 450, 452, *470* [33]; *470* [34]; *470* [35]
Jaffré, J. 254, *280* [34]
Jennings, G. 478, 540, 542, 546, *551* [31]
Jenssen, H.K. 92, 94, 131, 132, *166* [BaJ]; *167* [Je]; *378* [91]
Jian, H.Y. 310, *378* [100]
Jiang, S. 298, 299, 311, 322, 330, 331, 334–337, 340, 347, 348, 351, 353, 354, 363, 365, 373, 374, *378* [104]; *379* [105]; *379* [106]; *379* [107]; *379* [108]; *379* [109]; *379* [110]; *379* [111]; *379* [112]; *379* [113]; *379* [114]; *379* [115]; *379* [116]; *379* [117]; *379* [118]; *379* [119]; *379* [120]; *379* [121]; *379* [122]; *379* [123]; *379* [124]; *379* [125]; *379* [126]
Jin, S. 477, *551* [32]
John, F. 91, *167* [J]; 366, 371, 373, *379* [127]; *379* [128]
Joly, J.L. 348, *379* [129]
Junk, M. 463, *470* [36]

Kalashnikov, A.S. 236, 242, *283* [100]; *283* [101]; *283* [102]; *283* [103]; *283* [104]
Kalton, N.J. 26, 35, 56, 59, *82* [KL00]; *82* [KW01]; *82* [Kal03]
Kamenomostkaya, S. *283* [105]
Kanel, Y. 310, 326, *379* [130]
Kaniel, S. 466, *470* [37]
Kartsatos, A.G. 411, *435* [24]
Kato, T. 19, 39, 40, 42, 44, 48, *78* [[Kat66]]; *82* [Kat78]
Kawashima, S. 297, 362, 363, 365, *379* [131]; *380* [132]; *380* [133]; *380* [134]; *380* [135]; *380* [136]; 477, 478, 491, 504, *551* [33]; *551* [34]; *552* [35]
Kawohl, B. 298, *380* [140]
Kazhikhov, A.V. *279* [9]; 297, 311, 328, 330, 334, 340, 354, *375* [10]; *380* [137]; *380* [138]; *380* [139]; *381* [164]; *383* [221]; *383* [222]
Keldyš, M.V. 43, *82* [Kel66]
Kim, H. *376* [27]
Kim, Y.J. 528, *552* [36]
Kinderlehrer, D. 244, 257, *283* [106]; *283* [107]
Kinnunen, J. *283* [108]; *283* [109]
Kircheim, B. 449, *469* [2]
Koch, H. 373, *380* [141]
Kohn, R.W. 448, *470* [18]
Kondo, C. 484, *552* [37]
Kratka, M. 340, 354, *376* [24]
Kreiss, G. 508, *552* [38]
Kreiss, H.-O. 508, *552* [38]
Kröner, D. 463, *470* [38]
Kruzhkhov, S.N. 156, *167* [K]; 387, *435* [25]; *436* [26]; 446, 448, *471* [39]; 477, 483, 484, *552* [39]
Kruzkov, N.V. 175, 177, 185, 257, *283* [110]; *283* [111]; *283* [112]; *283* [113]; *283* [114]; *283* [115]
Krylov, N.V. 185, 237, 260, *284* [116]; *284* [117]; 387, 411, *436* [27]
Kunstmann, P.C. 57, 64, 72, *82* [Kun99]; *83* [KS03]; *83* [KuWe01]; *83* [Kun02]
Kwapien, S. *83* [Kwa72]
Kwong, Y.C. 176, 233, 242, 243, 245, 249, *282* [66]; *282* [67]; *284* [118]

Lacey, M. 76, *80* [AHLMT02]; *83* [MHL02]
Ladyzhenskaja, O.A. 175, 177–180, 185, 186, 256, 257, 260, 276, *284* [119]; *284* [120]; *284* [121]; 403, *436* [28]
Lamberton, D. 58, *83* [Lam87]
Lancien, G. 31, 36, 56, *82* [KL00]; *83* [Lan98]
Lasalle, J.-P. 513, *552* [40]
Lattanzio, C. 491, *552* [41]
Lax, P.D. 92, 101, *167* [Lx]; 539, *551* [28]; *552* [42]
Le Merdy, C. 22, 31, 35, 37, 38, *83* [LeM00]; *83* [LeM98a]; *83* [LeM98b]
Lebeau, G. 373, *380* [142]
Lecumberry, M. 449, *469* [2]; *471* [40]
Lederman, C. 429, *436* [29]
LeFloch, P.G. 152, 159, *167* [BLF1]; *167* [BLF2]; *167* [HLF]; *167* [LF]; 440, 448, 454, *470* [11]; *471* [41]; 484, *552* [37]
Leis, R. 369, *380* [143]
LeVeque, R.J. 439, 463, *471* [42]
Leverett, M.C. 257, 258, *284* [122]
Levermore, D. *550* [6]
Lewicka, M. 159, *167* [BLw]
Lewis, J.L. *283* [108]; *283* [109]
Li, D. 365, *380* [144]
Li, S. 58, *81* [CL94]
Lieberman, G.M. 411, *436* [30]
Lin, L. 131, *167* [Ln]
Lin, M. 14, *82* [FLW01]
Lindenstrauss, J. *78* [[LT96]]
Lions, J.-L. 8, 11, *78* [[DL88]]; 179, 236, 257, *284* [123]; 439, *470* [14]
Lions, P.-L. 295, 339–341, 343, 344, 346, 348, 351, *376* [39]; *380* [145]; *380* [146]; *380* [147]; 439, 440, 446, 449, 450, 452, 455, 457, *470* [21]; *470* [30]; *471* [43]; *471* [44]; *471* [45]; *471* [46]; 502, *552* [43]
Liskevich, V.A. 63, 71, *83* [LM97]; *83* [LP95]
Liu, H. 478, 491, *552* [44]
Liu, T.-P. 102, 131, 152, 155, 159, *167* [BLY]; *167* [L1]; *167* [L2]; *167* [L3]; *167* [L4]; *167* [LY1]; *167* [LY2]; *167* [LY3]; *168* [LY4]; 326, 330, 333, 334, 371, 373, *378* [92]; *380* [148]; *380* [149]; *380* [150]; *380* [151]; *380* [152]; 508, 525, 543, *550* [6]; *550* [7]; *552* [45]; *552* [46]; *552* [47]
Lopes, M. 387, *435* [25]; *436* [26]
Lorenzi, L. 387, 424, 434, *436* [31]
Lumer, G. 10, *83* [LP79]; *83* [LS99]
Lunardi, A. *78* [[Lun95]]; 387, 389, 390, 393, 396, 398–400, 411, 416, 417, 419, 422–424, 429, 431, 432, 434, *435* [6]; *435* [7]; *435* [8]; *435* [9]; *435* [10]; *435* [13]; *436* [32]; *436* [33]; *436* [34]
Luo, T. 298, 300, 301, 311–313, 317, 322, 331–334, *378* [101]; *378* [102]; *380* [153]; *380* [154]; *380* [155]

Ma, Z.M. 60, *78* [[MR86]]
MacCamy, R.C. 300, 302, 310, *377* [74]; *377* [75]
Magenes, E. 254, *286* [179]
Majda, A. 330, *376* [29]
Makino, T. 333–335, 338, *380* [156]; *381* [163]; *381* [179]; *381* [180]
Malek, J. 442, 449, *471* [47]
Malý, J. 277, *284* [124]
Manavi, A. 71, *83* [LM97]
Manfredi, J.J. 253, *282* [68]; *282* [69]; *284* [125]
Mantegazza, C. 448, 449, *469* [1]
Marcati, P. 491, *552* [41]
Markovitch, P. *551* [13]
Marson, A. 92, 131, 152, 156, *166* [AM1]; *166* [AM2]; *167* [BM]
Martinez, C. 24, 28, *78* [[MS01]]
Martinson, L.K. 177, *284* [126]; *284* [127]
Mascia, C. 516, *552* [48]
Masmoudi, N. 340, *376* [39]; *380* [147]; *380* [157]
Matolcsi, M. 41, 42, *83* [Mat03]
Matsumura, A. 326, 339, 365, *380* [133]; *380* [158]; *381* [159]; *381* [160]; *381* [161]; *381* [162]
Matušů-Nečasová, Š. 334, 335, 338, 340, 348, *377* [59]; *381* [163]; *381* [180]
Mayer, U.F. 424, 427, *435* [14]
Maz'ja, V.G. 75, *83* [MNP85]; *284* [128]
McIntosch, A. 33, *83* [MY90]
McIntosh, A. 22, 28, 31, 44, 48, 72, 76, *79* [ADM96]; *80* [AHLMT02]; *80* [AHMT01]; *80* [AMN97]; *80* [AMT98]; *81* [CDMY96]; *81* [DM99]; *83* [MHL02]; *83* [McI82]; *83* [McI85]; *83* [McI86]; *83* [McI90]
McLeod, J.B. 310, *377* [68]
Meirmanov, A. 254, *284* [129]
Messaoudi, S.A. 364, *378* [97]
Metafune, G. 11, 12, 27, 64, *79* [AMPR03]; *80* [CFMP03]; *80* [CM98]; *83* [MPP02]; *83* [MPRS00]
Métivier, G. 348, *379* [129]
Meyer, Y. 450, *470* [21]
Michel, A. 446, 463, *470* [24]
Mielke, A. *83* [Mie87]
Miller, J. 478, 510, 531, 535, *551* [22]
Min, L. 340, *381* [164]
Mingione, G. 270, *282* [77]
Minty, G.J. 257, *284* [130]
Miyawaka, T. 439, 447, *470* [27]
Mizel, V.J. 310, *377* [75]

Mkrtchen, P.Z. *283* [98]
Mokhtar-Kharroubi, M. 16, *78* [[Mok97]]
Monakhov, V.N. *279* [9]; 297, 311, 334, 354, *375* [10]
Monniaux, S. 26, 27, *82* [HM00a]; *82* [HM00b]; *83* [Mon99]; *84* [MP97]
Morrey, C.B. *284* [131]
Moser, J. 75, *84* [Mos64]; 174, 175, 180, 185, 227, 230–232, 236, 256, 260, *284* [132]; *284* [133]; *284* [134]; *284* [135]
Mugnolo, D. 14, *84* [Mug02]
Mukoyama, T. 365, *381* [165]
Müller, S. 448, *470* [18]
Muñoz Rivera, J.E. 298, 357, 373, 374, *379* [118]; *381* [166]; *381* [167]; *381* [168]; *382* [187]
Murat, F. 256, *280* [26]; 449, 457, *471* [48]

Nagasawa, T. 297, 300, 311, 313, 320, *381* [169]; *381* [170]; *381* [171]; *381* [172]
Nagel, R. 12, 13, 16–18, *78* [[EN00]]; *78* [[Nag86]]; *80* [BBN01]
Nahmrod, A. 28, *80* [AMN97]
Nambu, N. 253, *280* [24]
Nash, J. 175, *284* [136]; 290, 339, *381* [173]
Natalini, R. 477, 488, 491, 516, *551* [27]; *552* [48]; *552* [49]
Navratil, G.A. 249, *282* [74]
Nazarov, S.A. 75, *83* [MNP85]
Necas, J. 442, 449, *471* [47]
Neubrander, F. 5, 6, 10, 12–16, 47, 48, 50, 57, 76, *78* [[ABHN01]]
Neveu, J. 476, *552* [50]
Nguyen, Z.B. 326, *384* [247]
Ni, W.M. *281* [41]; *284* [137]; *284* [138]
Nikkuni, Y. *552* [35]
Nikolaev, V.B. 340, 354, *381* [174]
Nirenberg, L. 244, *283* [106]
Nishibata, S. 477, 478, 491, 504, *551* [33]; *551* [34]; *552* [35]
Nishida, T. 297, 326, 333, 334, 339, *380* [134]; *380* [158]; *381* [159]; *381* [175]; *381* [176]
Nishihara, K. 326, *381* [160]; *381* [161]
Nishiura, Y. 424, 427, *435* [15]
Nollau, V. 23, *84* [Nol69]
Novotný, A. 330, 340–344, 348, 351, *377* [60]; *377* [71]; *381* [177]

Okada, M. 333–335, 338, 363, *380* [135]; *381* [163]; *381* [178]; *381* [179]; *381* [180]
Okazawa, N. 23, *84* [Oka00a]; *84* [Oka00b]; *84* [Oka99]
Okuda, H. 249, *284* [139]
O'Leary, M. *282* [70]
Oleinik, O.A. 242, *283* [104]; *284* [140]
Orsina, L. 256, *284* [141]
Osher, S. 478, 507, 512, *552* [51]
Oshima, Y. 60, *78* [[FOT94]]
Ossermann, R. *284* [142]
Otto, F. 448, 449, *470* [15]; *470* [18]; *470* [33]
Ouhabaz, E.M. 60, 70, 71, 74, 75, *79* [[Ouh04]]; *80* [ABBO00]; *84* [Ouh03]; *84* [Ouh92]; *84* [Ouh95]; *84* [Ouh96]

Padula, M. 330, 340, *377* [71]; *381* [181]
Pallara, D. 12, 64, *79* [AMPR03]; *80* [CFMP03]; *83* [MPP02]
Pan, R.H. 298, 300, *381* [182]; *381* [183]
Paplov, K.B. 177, *284* [126]; *284* [127]
Paquet, L. 10, *83* [LP79]
Pattle, R.E. 239, *284* [143]
Payne, L.E. 374, *375* [1]
Pazy, A. 62, *79* [[Paz83]]
Pego, R.L. 310, *381* [184]
Peletier, L.A. 253, *280* [14]; *280* [15]; *280* [24]; *280* [25]; *284* [138]; 298, *375* [15]
Perelmuter, M.A. 63, *83* [LP95]
Perthame, B. 439, 440, 442, 446–450, 452, 454, 455, 457, 463, 465, 467, *469* [3]; *470* [12]; *470* [30]; *470* [33]; *470* [34]; *471* [44]; *471* [45]; *471* [46]; *471* [49]; *471* [50]; *471* [51]; *471* [52]; *471* [53]; 502, *552* [43]
Petzeltová, H. 311, 340–344, 348, 351, 352, *376* [47]; *377* [59]; *377* [60]; *377* [61]; *377* [62]; *377* [63]; *377* [64]; *377* [65]; *470* [25]
Piccoli, B. 152, *167* [BCP]
Pierre, M. 525, *552* [46]
Pini, B. 173, 230, 236, *284* [144]
Plamensevskii, B.A. 75, *83* [MNP85]
Pohozaev, S.I. 249, *284* [145]
Ponce, G. 368, *382* [185]
Porzio, M.M. 225, 236, 256, *284* [141]; *285* [146]; *285* [147]
Post, R.S. 249, *282* [74]
Primicerio, M. 225, 254, 257, *282* [82]
Priola, E. 64, *83* [MPP02]
Prüss, J. 13, 19, 20, 25, 27, 33, 34, 52, 56, 57, 71, 76–78, *79* [[Prü93]]; *81* [CP01]; *81* [DDHPV04]; *81* [DHP01]; *82* [HP97]; *82* [HP98]; *83* [MPRS00]; *84* [MP97]; *84* [PS90]; *84* [PS93]; *84* [Prü84]
Puel, J.P. 256, *280* [26]
Pullin, D.I. 463, *471* [54]
Pulvirenti, M. 439, *469* [6]; *469* [10]

Qin, Y.M. 298, 340, 354, *382* [186]; *382* [187]; *384* [241]

Räbiger, F. 20, *79* [ARS94]
Racke, R. 310, 356, 363, 365, 366, 368, 373, 374, *379* [118]; *379* [119]; *379* [120]; *381* [168]; *382* [185]; *382* [188]; *382* [189]; *382* [190]; *382* [191]; *382* [192]; *382* [193]; *382* [194]
Raizer, Y.P. 312, *383* [236]
Ralston, J. 478, 507, 512, *552* [51]
Rascle, M. *550* [8]
Rauch, J. 348, *379* [129]
Raviart, P.-A. 448, 463, *470* [29]
Reissig, M. 374, *382* [195]
Reitz, R.D. 463, *471* [55]
Renardy, M. 13, *84* [Ren94]
Rhandi, A. 27, *83* [MPRS00]
Rieger, M.O. 365, 374, *382* [196]; *382* [197]
Riemann, B. 92, *168* [Rm]
Risebro, N.H. 92, 131, *167* [HR]; *168* [Ri]
Rivière, T. 449, *469* [2]; *471* [40]
Robinson, D.W. 34, 72, 76, *78* [[DER03]]; *80* [AER94]; *81* [DR96]; *81* [tER96a]; *82* [tER96b]; *82* [tER98]
Röckner, M. 60, *78* [[MR86]]
Rodrigues, J.F. *285* [148]
Rodriguez, A. 177, *282* [78]
Rohde, C. 448, *470* [26]
Rokyta, M. 442, 449, *471* [47]
Romanelli, S. 12, *79* [AMPR03]; *82* [FGGR02]
Rosenau, P. 477, *552* [52]
Rousset, F. 509, *552* [53]
Roytbund, V. 330, *376* [29]
Ruggeri, T. *552* [54]
Ruzicka, M. 442, 449, *471* [47]

Sachs, P. 178, 259, 260, *285* [149]; *285* [150]
Safonov, M.V. 185, 237, 260, *284* [117]; *285* [151]
Saint-Raymond, L. 439, *470* [31]
Saloff-Coste, L. 65, 68, *79* [[Sal02]]; *79* [[VSC93]]
Sanz, M. 24, 28, *78* [[MS01]]
Sargenti, G. 270, *285* [152]; *285* [153]; *285* [154]
Sattinger, D.H. 508, 533, 535, *552* [55]
Savaré, G. 249, 253, *285* [155]
Schaefer, H.H. 34, *79* [[Sch74]]
Scheidegger, A.E. 254, 257, 258, *285* [156]
Schmidt-Lainé, Cl. 387, 424, 434, *435* [10]
Schnaubelt, R. 10, 27, *83* [LS99]; *83* [MPRS00]
Schochet, S. 131, 132, *168* [Sc1]; *168* [Sc2]; 477, 491, 494, 504, 505, *552* [56]
Schwartz, J. 53, *84* [Schw61]
Seely, R. 28, *84* [See67]
Seidman, T.I. 310, *375* [9]
Sentis, R. 450, 452, *470* [30]
Serre, D. *168* [Se]; 326, 330, 334, 344, *378* [93]; *382* [198]; *382* [199]; 448, 449, 454, *471* [56]; 491, 509, 513, 516, 523, 525, 530, 536, 540, 544, 547–549, *551* [16]; *551* [17]; *552* [54]; *552* [57]; *552* [58]; *552* [59]; *552* [60]; *552* [61]; *552* [62]; *553* [63]; *553* [64]; *553* [65]; *553* [66]; *553* [67]; *553* [68]; *553* [74]
Serrin, J. 175, 185, 231, 260, *280* [16]; *285* [157]
Shelukhin, V.V. 297, 322, 326, 328, 330, *377* [67]; *380* [139]; *382* [200]; *382* [201]; *382* [202]
Shen, W. 299, 310, 362, 363, *382* [203]; *382* [204]; *384* [242]
Shi, P. 374, *382* [205]; *382* [206]
Shibata, Y. 356, 363, 365, *380* [136]; *382* [192]; *382* [193]; *382* [207]
Shillor, M. 374, *382* [205]
Simeoni, C. 467, *471* [51]
Simonett, G. 33, 77, *79* [AHS94]; *81* [DS97]; 387, 424, 426–429, *435* [16]; *435* [17]; *435* [18]; *435* [19]; *435* [20]; *436* [35]
Sinestrari, E. 390, 411, 417, *436* [34]; *436* [36]
Singer, I. 29, *79* [[Sin70]]
Skrypnik, I.V. 411, *435* [24]
Slemrod, M. 363, *382* [208]
Smagulov, Sh.S. 297, *378* [103]; *382* [209]
Smoller, J. 131, *168* [Sm]; 326, *378* [94]; *382* [210]; 488, 491, *551* [9]
Sohr, H. 19, 27, *84* [PS90]; *84* [PS93]
Solonnikov, V.A. 175, 178–180, 185, 186, 256, 257, 260, 276, *284* [120]; 403, 411, *436* [28]; *436* [37]
Sopolov, N.N. 387, *436* [38]
Souganidis, P.E. 440, 450, 457, *471* [44]; *471* [52]
Sourour, A.R. 20, *79* [ARS94]
Spanne, S. 259, *285* [158]
Spekels, J. 225, 254, 257, *283* [94]
Spruck, J. 177, *282* [79]
Stampacchia, G. 256, 257, *283* [107]; *285* [159]
Stewart, H.B. 398, 416, *436* [39]; *436* [40]
Straškraba, I. 326, 340, 348, *376* [47]; *377* [59]; *381* [177]; *382* [211]; *382* [212]; *383* [213]; *384* [248]
Štrkalj, Z. 53, 64, *83* [KS03]; *84* [SW00]
Stroffolini, B. 178, 270, 274, 275, *282* [77]; *282* [86]; *285* [160]
Stroock, D.W. 232, *282* [81]
Sukochev, F.A. 49, 53, *81* [CPSW00]
Sukorjanski, S.M. 257, *283* [115]
Szepessy, A. 326, *383* [214]; 498, 508, *553* [69]; *553* [70]

Tadmor, E. 440, 446, 449, 455, 457, *471* [45]; *471* [46]; *471* [57]; 477, 478, 491, 494, 502, 504, 505, *552* [43]; *552* [44]; *552* [56]
Takeda, T. 60, *78* [[FOT94]]

Tanabe, H. 44, 67, *79* [[Tan79]]
Tang, Q. 374, *377* [51]
Tao, T. 446, *471* [57]
Tarabek, M.A. 362, *378* [98]
Tartar, L. 310, 330, *383* [215]; *383* [216]; 449, 457, *471* [58]; 502, *553* [71]; *553* [72]
Tchamitchian, Ph. 75, 76, *78* [[AT98]]; *80* [ACT96]; *80* [AHLMT02]; *80* [AHMT01]; *80* [AMT98]; *80* [AT01a]; *80* [AT01b]
Temam, R. 177, *285* [161]; 330, 352, *377* [66]; *383* [217]
ter Elst, A.F.M. 34, 61, 71, 72, 74, 76, *78* [[DER03]]; *79* [AtE97]; *80* [AER94]; *81* [tER96a]; *82* [tER96b]; *82* [tER98]
Terreni, B. 411, *435* [1]
Tolksdorff, P. *285* [162]
Tomilov, Y. 16, *80* [CT03]; *80* [CT04]
Tomisaki, M. 10, *82* [FT95]
Tonegawa, Y. 177, *282* [91]
Triebel, H. 28, *79* [[Tri95]]
Trivisa, K. 327–330, *375* [22]; *376* [23]
Trudinger, N.S. 175, 185, 231, 232, 260, 276, *282* [71]; *282* [88]; *285* [163]; *285* [164]
Tzafriri, L. *78* [[LT96]]
Tzavaras, A.E. 298, *383* [218]; *383* [219]; 448, *470* [32]; 528, *552* [36]
Tzavaras, Th. 440, 454, *471* [53]

Uhlenbeck, K. 174, *285* [165]
Ukai, S. 340, *381* [164]
Ural'ceva, N.N. 175, 178–180, 185, 186, 256, 257, 260, 276, *284* [120]; *284* [121]; *285* [166]; 403, *436* [28]
Urbano, J.M. 258, 261, *282* [75]; *282* [76]; *285* [167]; *285* [168]; *285* [169]; *285* [170]; *285* [171]; *285* [172]

Vaigant, V.A. 340, *383* [220]; *383* [221]; *383* [222]
Valli, A. 326, 339, *382* [212]; *383* [223]; *383* [224]
Van Duijn, J. 254, *286* [173]
van Neerven, J.M.A.M. 12, 14, *79* [[Nee96]]
Varopoulos, N. 65, 68, *79* [[VSC93]]
Vasseur, A. 440, 447, 458, *471* [59]; *471* [60]; *471* [61]; *471* [62]
Vazquez, J.L. 177, 253, *282* [78]; *284* [138]; *286* [174]; 429, 432, 433, *435* [11]; *435* [21]; *436* [29]; 528, *551* [15]
Vega, L. 450, *470* [35]
Venni, A. 27, 31, 36, 57, 78, *81* [DDHPV04]; *81* [DoVe87]; *84* [Ven93]
Verdi, C. 254, *286* [179]
Verduyn Lunels, S.M. 298, *375* [15]
Vespri, V. 178, 179, 225, 236, 243, 245, 249, 253, 256, 259, 261–263, 265, 267, 269–271, 273–278, *282* [67]; *282* [69]; *282* [72]; *282* [86]; *282* [87]; *284* [125]; *285* [147]; *285* [154]; *285* [155]; *285* [160]; *286* [175]; *286* [176]; *286* [177]
Villani, C. 439, *471* [63]
Visintin, A. 254, *286* [178]; *286* [179]
Voigt, J. 61, 64, 72, *82* [HV86]; *84* [Voi92]; *84* [Voi96]
Volpert, A.I. *168* [V]; 365, *383* [225]; 484, *553* [73]
Von Wahl, W. 411, 417, *436* [34]
Vũ, Q.P. 16, *84* [Vu97]

Wang, D. 322, 330, *383* [226]
Wang, Y.G. 374, *376* [25]; *382* [195]; *383* [227]
Warma, M. 11, 12, *84* [War03a]; *84* [War03b]
Watson, S.J. 299, *383* [228]
Webb, G.F. 17, *84* [Web87]
Weis, L. 14, 35, 49, 53, 56–59, 71, *82* [KW01]; *83* [KuWe01]; *84* [SW00]; *84* [Wei00a]; *84* [Wei00b]; *84* [Wei95]
Wendroff, B. 131, *168* [W]
Westdickenberg, M. 448, 449, *470* [15]; *470* [16]
Wiegner, M. *286* [180]
Witvliet, M. 49, 53, *81* [CPSW00]
Wojtaszcyk, P. 14, *82* [FLW01]
Wolanski, N. 429, *436* [29]
Wolff, M. 23, *85* [Wol81]
Wu, L. 177, *286* [181]; *286* [182]

Xin, Z. 326, 331–337, 340, *379* [121]; *380* [150]; *380* [151]; *380* [155]; *383* [214]; *383* [229]; *383* [230]; *383* [231]; *383* [232]; 477, 508, *551* [23]; *551* [32]; *553* [70]
Xu, K. 463, *471* [64]
Xu, Y. 374, *382* [206]

Yagi, A. 28, 31, 33, *81* [CDMY96]; *83* [MY90]; *85* [Yag84]
Yan, L.X. 77, *81* [DY01]
Yanagi, S. 326, *381* [162]
Yanagisawa, T. 340, *383* [232]
Yang, T. 152, *167* [BLY]; *167* [LY1]; *167* [LY2]; *167* [LY3]; *168* [LY4]; 331–335, 337–339, *380* [151]; *380* [155]; *383* [233]; *383* [234]; *383* [235]
Yao, Z.A. 335, *383* [233]
Yu, H.-S. 543, *552* [47]

Zajączkowski, W.M. 339, *383* [224]
Zarnowski, R. 326, *378* [95]
Zel'dovich, Y.B. 312, *383* [236]

Zeng, Y. 330, *380* [152]
Zhang, H. 254, *286* [173]
Zhang, P. 330, 335–337, 340, 347, 348, 351, *379* [121]; *379* [122]; *379* [123]; *379* [124]; *379* [125]
Zhao, H.J. 337, *383* [234]
Zheng, S. 299, 310, 340, 354, 356, 362, 363, 365, *382* [193]; *382* [194]; *382* [203]; *382* [204]; *383* [237]; *383* [238]; *383* [239]; *383* [240]; *384* [241]; *384* [242]
Zhou, X. 177, *283* [92]; *286* [183]
Zhu, C.J. 335, 338, 339, *383* [233]; *383* [235]
Zhu, P. 299, 300, 310, *377* [76]; *378* [77]; *381* [183]; *382* [204]
Ziane, M. 326, *378* [96]
Ziemer, W.P. 178, 259, 260, 277, *284* [124]; *286* [184]
Zimmer, J. 365, *382* [197]
Zimmermann, F. *85* [Zim89]
Zinkiewitz, J. 76, *81* [DHZ94]
Zlotnik, A.A. 322, 326, 330, 331, 334, *375* [4]; *375* [5]; *375* [6]; *376* [48]; *376* [49]; *379* [126]; *383* [213]; *384* [243]; *384* [244]; *384* [245]; *384* [246]; *384* [247]; *384* [248]
Zsidó, L. 26, *80* [CZ76]
Zuazua, E. 373, *380* [142]; 528, *551* [15]
Zumbrun, K. 478, 509, 510, 531, 535, 536, *551* [29]; *551* [30]; *553* [68]; *553* [74]

Subject Index

$\omega(T)$ 5
ω-limit set 513
$\rho(A)$ 5
θ-Hölder continuous functions 398

A
A-bounded 33
accommodation of the degeneracy 188
Accurate Riemann Solver 113, 121
admissibility condition 299
analytic
  generator 26
  semigroup 387, 389, 420
asymptotically
  almost periodic 15
  stable 340
attractors 340, 351, 354
averaging lemma 449, 450, 452
axisymmetric 340, 348, 349

B
balance laws 289, 292, 294, 296
balanced exponential growth 17
blow-up 290, 291, 354, 363–366, 371
Boltzmann equation 334, 439, 446
boundary
  group 7
  layer 547
bounded
  holomorphic semigroup 7
  imaginary powers BIP 25
  variation condition 157
boundedness 225
$BUC(\mathbb{R}^n)$ 8
Burgers equation 99

C
C_0-semigroup 4
$C_0(\mathbb{R}^n)$ 8
capacity 43
Cauchy problem 6, 387
center
  manifold theory 424, 429
  manifolds 393, 395
center-unstable manifolds 395
Chapman–Enskog expansion 334
characteristic
  boundary 547
  shock 502, 518
classical solution 5
Clausius–Duhem inequality 292–294, 296
co-properties 476
commuting operators 19
compact
  resolvent 15, 16
  semigroup 10
compressible
  fluids 290, 339
  heat-conducting fluids 290, 295, 339
  Navier–Stokes 289–291, 296, 322, 334, 339, 351
  reacting flow 330
  viscous fluids 289
  viscous heat-conducting 294, 295
concentrated compactness 348
conditional Schauder basis 29
conservation law 89, 297, 302
conservative scheme 537
conserved quantity 89
constitutive relations 289, 293, 295, 297, 300, 301, 311, 365
contact discontinuity 106
continuous interpolation space 391, 393, 396
continuum physics 289, 291, 296, 376
contraction mapping 356, 369
countable spectrum 15
Courant–Friedrichs–Lewy condition 538

D
decay rates 332, 362, 363, 370, 373
degenerate 171, 187, 226, 257
diffusion waves 508
dilation theorem 37

Dirichlet
  Laplacian 9, 39, 68
  regular 8
discrete shock profile 539
dispersion 476, 479, 482, 498
dissipation mechanism 289, 296
Dore–Venni theorem 27
Duhamel principle 480
dynamic boundary conditions 12

E
elementary functional calculus 20
energy estimates 181
entropy 99, 441, 481
entropy
  admissible solution 100
  criterion 483
  dissipation measure 444, 445, 456
  flux 99
  solution 483
equilibrium curve 488
ergodic 14
error estimate 154, 160
Eulerian coordinates 291, 294, 295, 330, 331, 338, 340, 341, 353
eventually norm continuous semigroups 13
expansion of positivity 219, 241, 242
exponential
  decay 329, 355, 361, 374
  stability 322, 354, 363
exponentially stable 12
extinction
  in finite time 234, 247
  profile 249
  time 242

F
Fendler's dilation theorem 35
field equations 292
finite
  energy weak solution 341, 342, 345, 349–352
  volumes 462, 463
flux 89
form 38
form sum 42
Fourier law 295, 361
fractional powers 23
frame indifference 293
free boundary 322, 331–335, 338, 354, 424
free boundary
  heat equation 388
  problems 387, 388, 427
front tracking 92, 113
fully nonlinear
  parabolic equations 387
  problems 419
fundamental solution 172, 243, 244

G
Gaussian semigroup 8
Gearhart–Prüss theorem 13
generator 4
genuinely nonlinear 101, 291, 366, 371, 372
Gibbs principle 460
global
  attractor 340, 352
  existence 290, 291, 297, 299, 302, 311, 322, 330, 333, 334, 337–340, 348, 352, 354, 355, 361–368, 373
Godunov scheme 538
group type 22
growth bound 5

H
H^∞-calculus 28
Harnack estimate 230, 233, 236, 240
heat kernel 479
Hele–Shaw
  flow 388
  model 427
Helmholtz free energy 292, 297, 314, 362, 363, 371, 372
Hölder continuous 173, 179, 186, 209
holomorphic semigroup 7
homogeneous medium 292, 366
hydrodynamic limit 439

I
incompressible flow 298
instability 387
interaction potential 116
interpolation space 28, 45
intrinsic rescaling 176, 187
invariant manifold 387, 388, 395, 423
inviscid
  conservation law 476
  Euler equations 328
  shocks 501
irreducible 17
isentropic
  fluid 290, 341
  Navier–Stokes equations 296, 340, 351
  viscous flow 290, 322
isothermal fluid 340, 348

K
kinetic
  defect measure 452
  scheme 458, 468
Kruzkhov entropy 448, 481

L
L^1-stability 507
Lagrangian coordinates 291, 293, 326, 333, 335, 338, 352
Laplace transform 6
large-time behavior 289, 290, 327, 328, 330, 339, 340, 351, 354, 374
LaSalle invariance principle 513
Lax
  admissibility condition 101
  shock 502
Lax–Friedrichs scheme 538
Le Merdy's theorem 37
Liapunov function 513
lifting operator 417, 420
linearly degenerate 101
little-Hölder continuous functions 398
local
  estimates 180
  stable manifold 423
  unstable manifold 423, 434
logarithm 23
logarithmic estimates 181

M
m-accretive 36
Matolcsi's theorem 41, 42
maximal attractors 340, 354
maximum principle 9
method of weak convergence 339, 342
mild solution 5
monotone scheme 537
Mullins–Sekerka models 427
multiple singularity 261

N
N-wave 520
Navier–Stokes equations 289–291, 295, 296, 322, 326, 328, 330, 331, 334, 339, 340, 351
Neumann Laplacian 10, 40, 69
Newtonian fluid 295
noncharacteristic shock 502
noncommutative Dore–Venni theorem 27
nondegenerate locally bounded semigroup 6
nonlinear boundary condition 387, 401

O
Oleinik criterion 502
one-point singularity 259
operators associated with a form 39
optimal regularity 389–391, 415
oscillations 290, 325, 326, 347

P
parabolic Hölder spaces 401, 411
parabolic–hyperbolic conservation laws 440
periodic group 18
Perron–Frobenius theory 16, 17
phase transition 300
Piola–Kirchhoff stress 292
planar shock 500
point spectrum 15
pointwise instability 422
polytropic 295, 323, 327, 331, 334
porous medium equation 171, 224, 235
positive
  contraction semigroup 34
  semigroup 13
principle of linearized stability 388, 393, 415, 420

R
r-contractive semigroup 34
$R(\lambda, A) = (\lambda - A)^{-1}$ 5
radially symmetric 291, 366, 373, 374
radiative
  gas 477
  shocks 504
Raleigh quotient 249
Rankine–Hugoniot condition 97, 98, 483
rarefaction
  curve 103
  wave 104, 326
real interpolation spaces 390, 396
reference configuration 290–292, 297, 354
relaxation 477
relaxation shocks 504
renormalized solution 341, 344, 347, 350
rescaled iterations 212
Riemann problem 92, 101, 106
Rosenau model 491

S
$s(A)$ 12
saddle point theorem 394, 423
Schauder basis 14, 29
sectorial 18
sectorial operator 389, 390
self-similar solutions 432

semigroup 6
shape memory alloys 299
shock 105
shock
  curve 104
  formation 483, 491
  profile 501
  wave 322, 326
Simplified Riemann Solver 121
singular 171, 209, 228, 257
small perturbation 33
spectral
  bound 12
  projection 394, 395, 416
Spectral theorem 36
spherically symmetric 334, 340, 348, 349, 352, 354
stability 387, 396, 411
stability of constant states 524
stable open set 43
standing shock 500
stationary solutions 339
Stefan-like problem 253
Stefan-type problem 388
Steklov average 181
strictly hyperbolic 90
strip-type operator 22
strong solutions 297, 310, 340
subcharacteristic condition 489

T
tail of discrete profiles 543
tame oscillation condition 157
thermoelasticity 289, 290, 294, 354, 363, 365, 374
thermoviscoelasticity 289–291, 293, 310
Trotter's formula 42

U
UMD-space 26
unconditional Schauder basis 29
uniqueness 156, 158

V
variation of constants formula 390
viscoelasticity 300, 302, 310
viscosity 289, 290, 295–298, 300, 311, 322, 323, 327, 331, 334, 336, 337, 343
viscosity solution 153
viscous
  conservation law 479
  reactive gas 299
  shock 326, 503

W
weak solution 90, 95, 290, 322, 324, 326–328, 331–342, 345, 348–352, 365
Weis' theorem 14
well-posedness 290, 291, 327, 330, 339, 340, 374
Wentzell boundary conditions 11

Y
Young measure 300, 365

Z
zero dissipation limit 340
zero-mass 510

Kolibri 360 GmbH
Ferdinand-Jühlke-Straße 7
99095 Erfurt, Deutschland
produktsicherheit@kolibri360.de